GOVERNORS STATE UNIVERSITY LIBRARY
P9-CAO-229
3 1611 00320 6528

# TREATISE ON ANALYTICAL CHEMISTRY

## *PART I*

## THEORY AND PRACTICE

SECOND EDITION

# TREATISE ON ANALYTICAL CHEMISTRY

## *PART I*

## THEORY AND PRACTICE

**SECOND EDITION**

**VOLUME 4**

Edited by PHILIP J. ELVING
*Department of Chemistry, University of Michigan*

Associate Editor: VICTOR G. MOSSOTTI
*U.S. Geological Survey, Menlo Park, California*

*Editor Emeritus:* I. M. KOLTHOFF
*School of Chemistry, University of Minnesota*

AN INTERSCIENCE® PUBLICATION
*JOHN WILEY & SONS*
*New York—Chichester—Brisbane—Toronto—Singapore*

An Interscience® Publication


***Library of Congress Catalog Number:*** 78–1707

ISBN 0-471-01836-8

Printed in the United States of America

10 9 8 7 6 5 4 3 2 1

# TREATISE ON ANALYTICAL CHEMISTRY

*PART I*

## THEORY AND PRACTICE

## VOLUME 4:

**SECTION E**

**Principles of Instrumentation for Analysis** ***Chapters 1–9***

**AUTHORS OF VOLUME 4**

H. R. BRAND
STEVEN P. BRIMMER
DONALD A. BURNS
GALEN W. EWING
KENNETH W. GARDINER
CARL F. IJAMES
THOMAS L. ISENHOUR
DAVID O. JONES
PETER C. JURS
MARVIN MARGOSHES
HARRY B. MARK, JR.
VICTOR G. MOSSOTTI
SAM P. PERONE
EDWARD H. PIEPMEIER
THOMAS H. RIDGWAY
CHARLES L. WILKINS

Professor Philip J. Elving died on March 16th, 1984. The Publisher gratefully acknowledges Professor Elving's many contributions to the successful publication of the TREATISE; he will be missed.

# AUTHORS OF VOLUME 4

**H. R. Brand**

*University of California, Lawrence Livermore Laboratories, Livermore, California 94550, Chapter 8.*

**Donald A. Burns**

*Chemical Instrumentation, Technicon Instruments Corporation, Tarrytown, New York 10591, Chapter 5.*

**Steven P. Brimmer**

*Department of Chemistry, University of Cincinnati, Cincinnati, Ohio 45221, Chapter 3.*

**Harry B. Mark, Jr.**

*Department of Chemistry, University of Cincinnati, Cincinnati, Ohio 45221, Chapter 3.*

**Thomas H. Ridgway**

*Department of Chemistry, University of Cincinnati, Cincinnati, Ohio 45221, Chapter 3.*

**Galen W. Ewing**

*Chemistry Department, Carleton College, Northfield, Minnesota 55057, Chapter 4.*

**Kenneth W. Gardiner**

*Applied Science Program, University of California, Riverside, California 92502, Chapter 6.*

**Carl F. Ijames**

*Department of Chemistry, University of California, Riverside, California 92521, Chapter 7.*

**Thomas L. Isenhour**

*Department of Chemistry, University of North Carolina, Chapel Hill, North Carolina 27514, Chapter 9.*

**Peter C. Jurs**

*Department of Chemistry, Pennsylvania State University, University Park, Pennsylvania 16802, Chapter 9.*

**Marvin Margoshes**

*Chemical Instrumentation, Technicon Instruments Corporation, Tarrytown, New York 10591, Chapter 5.*

**Victor G. Mossotti**

*U.S. Geological Survey, 345 Middlefield Road, MS 90-C, Menlo Park, California 94025, Chapter 1.*

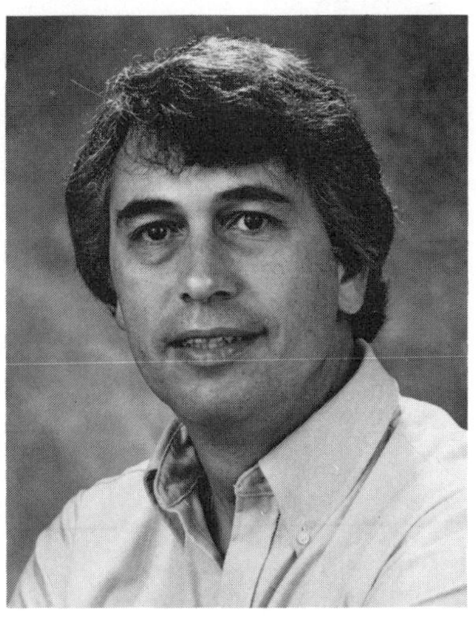

**Sam P. Perone**

*Lawrence Livermore Laboratories/Box 802, L-325, University of California, Livermore, California 94550, Chapter 8.*

**Edward H. Piepmeier**

*Department of Chemistry, Oregon State University, Corvallis, Oregon 97331, Chapter 2.*

**Charles L. Wilkins**

*Department of Chemistry, University of California, Riverside, California 92521, Chapter 7.*

## Preface to the Second Edition of the Treatise

In the mid-1950s, the plan ripened to edit a "Treatise on Analytical Chemistry" with the objective of presenting a comprehensive treatment of the theoretical fundamentals of analytical chemistry and their implementation (Part I) as well as of the practice of inorganic and organic analysis (Part II); an introduction to the utilization of analytical chemistry in industry (Part III) was also considered. Before starting this ambitious undertaking, the editors discussed it with many colleagues who were experts in the theory and/or practice of analytical chemistry. The uniform reaction was most skeptical; it was not thought possible to do justice to the many facets of analytical chemistry. Over several years, the editors spent days and weeks in discussion in order to define not only the aims and objectives of the Treatise but, more specifically, the order of presentation of the many topics in the form of a table of contents and the tentative scope of each chapter. In 1959, Volume 1 of Part I was published. The reviews of this volume and of the many other volumes of Part I as well as of those of Parts II and III have been uniformly favorable, and the first edition has become recognized as a contribution of classical value.

Even though analytical chemistry still has the same objectives as in the 1950s or even a century ago, the practice of analytical chemistry has been greatly expanded. Classically, qualitative and quantitative analysis have been practiced mainly as "solution chemistry." Since the 1950s, "solution analysis" has involved to an ever increasing extent physicochemical and physical methods of analysis, and automated analysis is finding more and more application, for example, its extensive utilization in clinical analysis and production control. The accomplishments resulting from automation are recognized even by laymen, who marvel at the knowledge gained by automated instruments in the analysis of the surfaces of the moon and of Mars. The computer is playing an ever increasing role in analysis and particularly in analytical research. This revolutionary development of analytical methodology is catalyzed by the demands made on analytical chemists, not only industrially and academically but also by society. Analytical chemistry has always played an important role in the development of inorganic, organic, and physical chemistry and biochemistry, as well as in that of other areas of the natural sciences such as mineralogy and geochemistry. In recent years, analytical chemistry—often of a rather sophisticated nature—has become increasingly important in the medical and biological sciences, as well as in the solving of such social problems as environmental pollution, the tracing of toxins, and the dating of art and archaeological objects, to mention only a few. In the area of atmospheric science, ozone reactivity and persistence in the stratosphere is presently a topic of great priority; extensive analysis is required both for monitoring atmospheric constituents and for investigating model systems.

One example of the increasing demands being made on analytical chemists is the growing need for speciation in characterizing chemical species. For example, in reporting that lake water contains dissolved mercury, it is necessary to report in which oxidation state it is present, whether as an inorganic salt or complex, or in an organic form and in which form.

As a result of the more or less revolutionary developments in analytical chemistry, portions of the first edition of the Treatise are becoming—and, to some extent, have become—out-of-date, and a revised, more up-to-date edition must take its place. In recognition of the extensive development and because of the increased specialization of analytical chemists, the editors have fortunately secured for the new edition the cooperation of experts as coeditors for various specific fields.

In essence, it is the objective of the second edition of the Treatise, as it was of the first edition (whose preface follows this one), to do justice to the theory and practice of contemporary analytical chemistry. It is a revision of Part I, which mirrors the development of analytical chemistry. Like the first edition, the second edition is not an extensive textbook; it attempts to present a thorough introduction to the methods of analytical chemistry and to provide the background for detailed evaluation of each topic.

*Minneapolis, Minnesota* I. M. KOLTHOFF
*Ann Arbor, Michigan* P. J. ELVING

## Preface to the First Edition of the Treatise

The aims and objectives of this Treatise are to present a concise, critical, comprehensive, and systematic, but not exhaustive, treatment of all aspects of classical and modern analytical chemistry. The Treatise is designed to be a valuable source of information to all analytical chemists, to stimulate fundamental research in pure and applied analytical chemistry, and to illustrate the close relationship between academic and industrial analytical chemistry.

The general level sought in the Treatise is such that, while it may be profitably read by the chemist with the background equivalent to a bachelor's degree, it will at the same time be a guide to the advanced and experienced chemist—be he in industry or university—in the solution of his problems in analytical chemistry, whether of a routine or of a research character.

The progress and development of analytical chemistry during most of the first half of this century has generally been satisfactorily covered in modern textbooks and monographs. However, during the last fifteen or twenty years, there has been a tremendous expansion of analytical chemistry. Many new nuclear, subatomic, atomic, and molecular properties have been discovered, several of which have already found analytical application. In the development of techniques for measuring these and also the more classical properties, the revolutionary progress in the field of instrumentation has played a tremendous role.

It has been difficult, if not impossible, for anyone to digest this expansion of analytical chemistry. One of the objectives of the present Treatise is not only to describe these new properties, their measurement, and their analytical applicability, but also to classify them within the framework of the older classifications of analytical chemistry.

Theory and practice of analytical chemistry are closely interwoven. In solving an analytical chemical problem, a thorough understanding of the theory of analytical chemistry and of the fundamentals of its techniques, combined with a knowledge of and practical experience with chemical and physical methods, is essential. The Treatise as a whole is intended to be a unified, critical, and stimulating treatment of the theory of analytical chemistry, of our knowledge of analytically useful properties, of the theoretical and practical fundamentals of the techniques for their measurement, and of the ways in which they are applied to solving specific analytical problems. To achieve this purpose, the Treatise is divided into three parts: I, analytical chemistry and its methods; II, analytical chemistry of the elements; and III, the analytical chemistry of industrial materials.

Each chapter in Part I of the Treatise illustrates how analytical chemistry draws on the fundamentals of chemistry as well as on those of other sciences; it stresses for its particular topic the fundamental theoretical basis insofar as it affects the analytical approach, the methodology and practical fundamentals used both for the development of analytical methods and for their implementation for analytical service, and the critical factors in their application to both organic and inorganic materials. In general, the practical discussion is confined to fundamentals and to the analytical interpretation of the results obtained. Obviously then, the Treatise does not intend to take the place of the great number of existing and exhaustive monographs on specific subjects, but its intent is to serve as an introduction and guide to the efficient utilization of these specialized monographs. The emphasis is on the analytical significance of properties and of their measurement. In order to accomplish the above aims, the editors have invited authors who are not only recognized experts for the particular topics, but who are also personally acquainted with and vitally interested in the analytical applications. Only in this way can the Treatise attain the analytical flavor which is one of its principal objectives.

Part II is intended to be very specific and to review critically the analytical chemistry of the elements. Each chapter, written by experts in the field, contains in addition to a critical and concise treatment of its subject, critically selected procedures for the determination of the element in its various forms. The same critical treatment is contemplated for Part III. Enough information is presented to enable the analyst both to analyze and to evaluate a product.

The response in connection with the preparation of the Treatise from all colleagues has been most enthusiastic and gratifying to the editors. It is obvious that it would have been been impossible to accomplish the aims and objectives cited in the preface without the wholehearted cooperation of the large number of distinguished authors whose work appears in this and future volumes of the Treatise. To them and to our many friends who have encouraged us we express our sincere appreciation and gratitude. In particular, considering that the Treatise aims to cover all of the aspects of analytical chemistry, the editors have found it desirable to solicit the advice of some colleagues in the preparation of certain sections of the various parts of the Treatise. They would like at this time to acknowledge their indebtedness to Professor Ernest B. Sandell of the University of Minnesota for his interest and active cooperation in the organizing and detailed planning of the Treatise.

*Minneapolis, Minnesota* I. M. KOLTHOFF
*Ann Arbor, Michigan* P. J. ELVING

## Acknowledgment

In view of the wide scope of the Treatise, it has been considered essential to have the advice and aid of experts in various areas of analytical chemistry. For the section on "Principles of Instrumentation for Analysis," the editor has been fortunate to have the cooperation of Dr. Victor G. Mossotti of the U.S. Geological Survey; his collaboration is acknowledged with gratitude.

P.J.E.

# PART I. THEORY AND PRACTICE

## CONTENTS—VOLUME 4

## SECTION E. Principles of Instrumentation for Analysis

# TREATISE ON ANALYTICAL CHEMISTRY

## *PART I*

## THEORY AND PRACTICE

SECOND EDITION

# SECTION E: Principles of Instrumentation for Analysis

Part I
Section E

Chapter 1

# THE INFORMATIONAL STRUCTURE OF ANALYTICAL CHEMISTRY

BY VICTOR G. MOSSOTTI, *U.S. Geological Survey, Menlo Park, California*

**Contents**

# I. INTRODUCTION

It has been said, with some degree of accuracy, that the chemistry has gone out of analytical chemistry. For years, the activities of the analytical chemist were labor and time intensive, and a large proportion of analytical work was based on solution chemical procedures. Today the analytical chemist finds himself in the midst of a revolution in electronic-signal and data processing, a revolution that has touched nearly every aspect of analytical chemistry and left many laboratory chemists in a mild state of shock.

Traditionally, the emphasis of the chemist's training has been on solution chemistry, and little attention is given to electronics. Accordingly, chemistry and electronics have been united in a shotgun marriage. Present-day commercial chemical instrumentation and the high degree of laboratory automation have opened a gap between the analytical chemist and the final chemical measurement. As recent as the early 1960s, the analytical chemist took pride in calibrating the weights he used with a double-pan balance; now it is not uncommon for a chemist to view his instrumentation as a set of black boxes and to place a large amount of trust in the commercially available software packages for data collection and manipulation. If we as chemists are to take full advantage of the technology available and are to maintain a balance between the application of such technology and the point of diminishing returns, we must learn to perceive our rapidly changing profession in terms of the immutable, fundamental concepts of information measurement and control. This chapter is intended to provide such a perspective.

## A. ABOUT THE VOLUME

This volume provides a profile of all major aspects of electronics operating to approximately 30 MHz, with special emphasis on applications to problems in analytical chemistry. The approach in each chapter is to focus on design principles and on practical technology. This present chapter provides a treatment of noise limitations from the most fundamental measurement principles. The two chapters following this introductory chapter give an overview of analog electronics, with special emphasis on operational systems. These chapters cover the basics of power supplies, amplification, passive and active filtering, and diode circuits. Included are essential concepts in the design and analysis of control systems, focusing on system stability with $s$-plane and Bode analysis. An exhaustive chapter on transducers is followed by an introductory chapter on digital measurements. The next four chapters review in considerable detail the application of computers in analytical chemistry. This review contrasts the applications of analog and digital computers and includes a comparison of various computer architectures, from large main-frame systems to minicomputers and microcomputers. Also, attention is given to such input/output techniques as programmed data transfer, vectored interrupt, and data break. In this review, an entire chapter is devoted to a perspective on software that covers assemblers, compilers, system-service routines, device handlers, and interpreter programs, including applications in real-time control of chemical instrumentation. A chapter stressing real-time data acquisition and interfacing is followed by one focusing on the design of laboratory-automation systems. The final two chapters cover clinical and industrial process control and automation; in these chapters, considerable emphasis is given to the electronic sample-control and management techniques needed in high-volume applications.

In spite of a complete neglect in this volume of electronic packaging, an art and a technology that represents a substantial component of the final cost of any laboratory instrument, the reader who fully understands the material presented here very likely will be able to design and build a prototype circuit for a given application. Although such design work may be possible, however, it is not our main goal. Instead, this volume is intended to prepare the reader to identify a design philosophy for a particular measurement or control problem, to specify the particular off-the-shelf modules suitable for implementation of a given measurement technique, to analyse and troubleshoot an existing system, and to recognize the limitations of his or her analytical-chemical measurements.

## B. ABOUT THE CHAPTER

Although the reader should have no particular difficulty in grasping the principles of a given signal-manipulation technique, there may exist a requirement of perspective that is not satisfied by simply recounting the table of contents for the volume. How do we identify the important signals and signal components associated with a given analytical measurement? What sort of signal manipulations are necessary along the path of information flow, and on which signal components do we carry out these manipulations? Is true integration ideal, or would an exponentially weighted integration be more suitable? Do we add, subtract, multiply, divide, average, convolute, deconvolute, amplify, attenuate, selectively filter, differentiate, or integrate signal components in isolation from one another? And on what basis can we select between frequency-domain and time-domain filtering? If frequency-domain filtering is selected, is band-pass filtering or phase-sensitive detection more appropriate? Could novel measurements, such as cross-correlation or cross-power detection, be designed to yield information signals that otherwise have not been available for analytical purpose? Finally, under what conditions can we expect selective modulation of various system parameters, derivative methods, interferometric techniques, resonance detection, Fourier methods, and so on, to result in analytical improvements?

In considering a general model for any analytical method or measurement technique, I choose to recognize two distinct parts. The first and, in general, the most elusive part of this model is that in which the material-transport phenomena and chemical reactions and mechanisms for each individual physical and chemical process are resolved and evaluated; most research efforts in chemistry are addressed to this part of the model. The second part of the model is that which characterizes the net relation between the parameters that describe the microscopic processes and the actual experimental observations and measurements. My general concern in this chapter is with this second part of the model which I refer to as the informational structure of the measurement system.

In the main body of the chapter to follow, I show that the informational structure of a large class of analytical measurements can be formulated in general terms and applied to the understanding of many critical measurement problems. Once we have examined certain measurement principles in terms of a general model, I give special attention to analytical spectrometric methods because of the extended practical application of such methods in the analytical-chemical laboratory.

## II. ELEMENTARY MODEL FOR INFORMATION TRANSFER

In this section, I examine whether chemical instrumentation, as currently practiced with a great diversity of hardware, has fully participated in the benefits of modern signal-recovery technology already applied so successfully in some of the more visible areas, such as communications engineering. We now enjoy virtually errorfree data transfer across continents and even between planets; astoundingly low signal levels are recoverable in high-noise environments. Are we making the best use of this highly developed technology, or are there fundamental differences between the measurement demands in the chemistry laboratory and in information transfers across solar systems? I introduce a basic model that I subsequently use to explore the relations between the quantilies of chemical interest and the factors that limit our ability to measure such quantities. This simple model, based on the concepts of information and communication theory, should provide a perspective on the generation and transfer of various analytically useful signals.

## A. CONCEPTS OF INFORMATION AND ZERO-ENTROPY ENERGY

The origin of modern information theory is usually ascribed to Shannon (19), although very important contributions were made by earlier workers. There have been hundreds of publications since Shannon's now-classic paper. Although much existing theory has been developed by mathematicians (2,6,12), where possible, I emphasize the phenomenological concept of information, with specific application to the measurement problem in quantitative analytical chemistry. My intent in this chapter is to be brief, and so I omit many theoretical details not important for the exposition.

Information flow is essential within any viable system, whether it be a communication system, an instrument for chemical analysis, an automated factory, or a biological system. Information, originating at a source, is always at least one input of such a system. Because the system is stimulated into action by information, it is essential to make clear what we mean by this term. *Information* is a representation of knowledge, and the amount of information can be measured by what it *does*, just as a force may be defined in terms of the acceleration it causes or could cause. If we consider ourselves in the position of a receiver of the information, we can explore the potential effect that receipt of the information could have on our future actions.

Let us suppose that we purchase a newspaper containing a fixed, predetermined weather report, known in advance to all potential readers. The report is printed on the front page in large boldface type with the words "Cloudy Today;" no other words are allowed. Only one message could be selected, and so the report would be completely predictable. Clearly, the amount of information conveyed by this weather report would be zero, and the newspaper would have value only to a recycling center. Now let us suppose that one of two messages may have been selected for publication: "Cloudy Today" or "Sunny Today." Because of the unexpectedness of this report, the newspaper now contains a finite amount of information. The point is that the informational source selects a particular message from a set of possible alternative messages and that the amount of information is related to the number of these possible alternative messages and to the probabilities of their selection.

Of the many possible approaches to the quantitative measurement of information, the most intuitively satisfying is based on the degree of unexpectedness in the information source. If only one message is permitted, then the probability of that particular message being generated is unity, and the informational content of the message is zero. If the number of different messages is infinite, however, the informational content of a given message event is infinite. Using arguments that follow a similar line of intuitive reasoning, Shannon showed that information can be measured in terms of its unexpectedness and that the form of the relation between informational content $I$ and the number of possible messages $M$ is logarithmic as follows:

$$I = \log M \tag{1}$$

If the base of the logarithm is 10, the informational measure is the decimal digit, sometimes called a decit; if the base is 2, the unit is the binary digit or bit.

Such messages are selected without attention to semantic content; only the number of possible alternative messages is significant. The interpretation of such a message as "The coast is clear" may be entirely different to a sailor in a sailboat or to an adventuresome adulterer answering the telephone. I have no concern in this discussion with the individual interpretation of a particular message.

Otherwise stated, a *message* is a selected amount of information. A single sentence in a book may be considered a message; one chapter in the book or even the entire book may also be considered a single message. The informational content of a given message is the number of bits it contains, and when more than one source supplies information to a system, the amounts of information from the various sources are summed.

Shannon also showed that when information is measured in terms of its unexpectedness, it has the mathematical form of entropy as defined in thermodynamics and statistical mechanics. *Entropy* is a measure of the degree of disorder of a system; it is maximum when a sequence of events is completely unpredictable. If the entropy of a system is zero, the system is completely predictable and furnishes no information whatever; such a source is called a zero-entropy energy source. Thus the direct-current (dc) flow from a battery, the sine-wave signal from an oscillator, or the predetermined printing of "Cloudy Today" on a newspaper page are all predictable and are all energy sources having zero entropy and thus conveying zero information.

If information is to be transmitted between two points, a zero-entropy energy quantity must be modulated to carry the message. Any operations on information can involve only operations on the physical quantities carrying a message, and the only way to represent information physically is in the form of a "signal." Information is carried by unexpected variations in energy flow over time, and so the only way a signal can carry information is for an informational source to control some property of a zero-entropy source, so that this property becomes unpredictable. In many cases, the message energy is used to control an electrical generator because of the highly developed technology available for transmitting and handling electrical signals. Electrical energy, guided by wire conductors, is used for telegraph, for telephone, and for a great assortment of signals in electronic circuits. In concept, the nature of the energy medium is unimportant, and, in general, message and signal energies may or may not have the same form. The mechanical energy of vibrating vocal cords modulates the acoustic waves that carry human speech. Other modulations in the mechanical domain include the position of a throttle, of a needle on a gauge, or of a pen on a strip-chart recorder. The energy in electromagnetic fields carries the information for radio, television, and radar signals. Light waves, which also are electromagnetic, carry many kinds of information; the scope of examples ranges from the complex images originating from a painting by an old master to an Indian's smoke signal. Finally, in the chemical domain, the variable concentration of chemical species is really a chemical modulation that can be used to carry an information-containing message. This somewhat novel view of the chemical species is an important concept in the model presented in this chapter.

## B. SHANNON'S MODEL

We talk with our mouths and listen with our ears, an arrangement generalized in the communication model proposed by Shannon and illustrated in Fig. 1.1. Information, originating at a source (e.g., the human brain), is converted into an audio signal by a *transmitter* (the mouth) for communication over a *channel* (air path). The *receiver* (the ear), on sensing the transmission, performs an operation inverse to that of the transmitter by converting the information signal into a form suitable for its destination, namely, the device for which the information is intended. In general, this *destination* may be a human brain, or it may be a pseudoinformational source that serves as the input for the next link in an information-processing chain. This simple five-element symmetrical model, consisting of an informational source, two transducers, a transmission channel, and a destination, constitutes the most elementary information-processing unit. An important point to recognize is that this elementary information-processing unit, in its canonical form, can be viewed as a building block for much more complex communication or measurement systems. In many practical situations, the output from one such five-element unit is the input for the next. All complex measurement systems can be analyzed and resolved in terms of a network of elementary information-processing units such as that proposed by Shannon.

The informational source provides the driving signal for the system. In an instrument designed for quantitative chemical analysis, the information of interest can be viewed as being accommodated in the chemical domain in the form of the concentration of a particular chemical species. Accordingly, the measurement system is driven by a chemical signal, and the task of the analytical system, much like all other measurement systems, is to transform the input

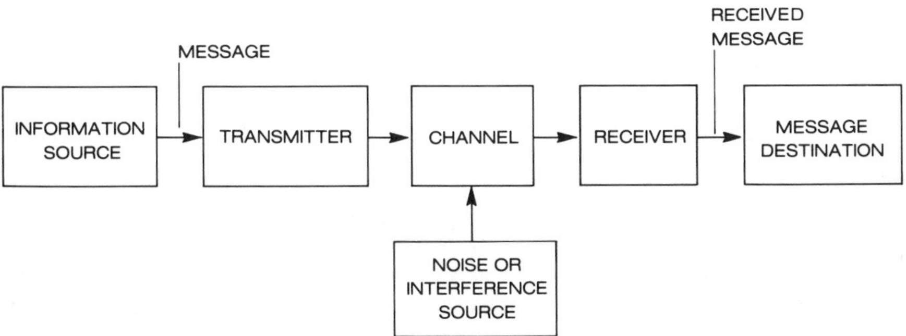


**Fig. 1.1.** Shannon's model for an elementary communication system.

informational signal for interpretation by a receiving device, commonly a human brain. Various methods can be imagined by which such a transformation might be achieved. For example, a direct biological system could be used for the detection of acid concentration; a person could taste a solution to determine whether it is acidic. Though generally highly sensitive, however, biological systems are commonly nonspecific, have extremely poor quantitative resolution and dynamic range, and are exceedingly difficult to engineer for quantitative purposes. Furthermore, recruiting a person to do the tasting could be difficult, especially if the person succumbs to poisoning after filing his report. Solution-chemical procedures have been used for years to convert chemical signals into mechanical or optical signals that can be quantitatively interpreted by human beings. Examples of this conversion are everywhere in the chemistry laboratory: the number of standard weights on the pan of a balance, the differential volume reading on a burette, or the color change of a solution at the endpoint of a titration all represent such chemical-to-mechanical or chemical-to-optical transformations.

From Shannon's model, we can see that the measurement problem in quantitative analytical chemistry is fundamentally identical with all other measurement problems, including those in interplanetary data transmission, satellite tracking, and long-range radar detection. What practical engineering limitations do we face in the chemistry laboratory, and how closely have we approached the point of diminishing returns in our development of chemical instrumentation? Such are the questions to be explored in the remainder of this chapter.

## III. MEASUREMENT LIMITATIONS

### A. SYMBOLIC REPRESENTATION OF INFORMATION

In human and animal communications, and, in fact, in all forms of measurement, information transfer cannot take place without a system of signs and symbols. Many lower animals rely on such signs for communication. Commonly, behavioral displays are used to convey disposition or emotional state; the repertoire of symbols in the lower animals seems to be somewhat limited, and therefore the amount of information also is limited. In digital information transfers, the modulation of a zero-entropy energy source, whether a voltage signal, a light beam, or a smoke signal, effectively creates a symbol that can exhibit two distinguishable states. In analog data transfers, a symbol is created that can exhibit a multiplicity of states, the number of which depends on the number of distinguishable energy levels, which in turn is determined by the resolution of the instrument and by the dynamic range over which energy excursions can be observed. In this treatment, I define a *symbol* as some observable quantity that the system's users regard by general consent as exhibiting a set of selectable states which can be understood by reference to an agreed-upon coding system—a dictionary, so to speak. A *sign* can be regarded as a set of symbols organized into a "word" of information. A *code* is a mapping whereby the transformation may or may not provide a one-to-one symbol-for-symbol representation. It follows that encoding is the application of a set of transformational rules to a message and that decoding is the inverse operation; encoding is built into the transmitter, and decoding into the receiver. The "state" dictionary—which, in effect, is a decoder—provides the understanding that we call *meaning*.

### B. SAMPLING THEOREM AND BANDWIDTH-TIME TRADE-OFF

In the 1920s, two fundamental concepts were developed–the Nyquist sampling theorem (20) and the Hartley bandwidth-time (5) relation. The sampling theorem states that a function of time $f(t)$, band limited with its highest frequency at $B$ Hz, can be completely specified for a time $T$ sec by $2BT$ numbers spaced at uniform intervals throughout the measurement period. Figure 1.2 illustrates such discrete sampling of a continuous record. Otherwise stated, the function can be completely specified without loss of information by a set of discrete values spaced at uniform intervals $\Delta t_N$ sec apart, where $\Delta t_N$,called the Nyquist interval, is given by

$$\Delta t_N = \frac{1}{2B} \tag{2}$$

(The corresponding frequency-domain sampling theorem is discussed below in Section V.A.)

It follows from the Nyquist sampling theorem that the maximum rate at which independent informational symbols can be generated by a source band limited to $B$ Hz is $2B$ symbols per second. There is a certain amount of intuitive satisfaction in this theorem when we consider that to obtain $2BT$ uniformly spaced numbers, the function must be sampled twice per cycle of the highest-frequency component in the function over an interval of $T$ secs. The signal is said to be *undersampled* if the time between samples is longer than one-half the fundamental period of the highest-frequency component in the signal. If a signal is undersampled, events occurring between samples go unobserved. If the rate at which statistically independent samples can be observed is determined by the bandwidth of the signal, an increase in sampling rate does not yield more information about the signal, and the signal is said to be oversampled. The usefulness of oversampling is limited by the classical uncertainty principle (further discussed below in Section V.A.).

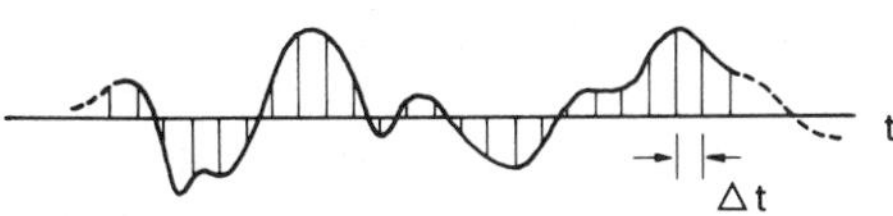


**Fig. 1.2.** Discrete sampling of time-domain signal.

The rate at which information is generated by a source or processed by a transmission channel is given by the product of the informational-symbol content and the rate at which statistically independent symbols are generated. The informational content of a given symbol, as we have seen, depends on the total number $M$ of possible states allowed for the symbol where $I = \log_2 M$ (in bits). Therefore, from the Nyquist sampling theorem, the maximum rate at which information can be carried by a band-limited signal, known as the informational capacity $C$, given by

$$C = 2B \log_2 M \text{ (in bits per second)} \tag{3}$$

The number of symbol transitions generated or processed per second is called the baud; for a binary signal, 1 bit/sec = 1 baud. From the above, Hartley recognized that the quantity of information $Q$ that can be transmitted over a band-limited channel within a time $T$ is given by

$$Q = 2\,BT \log_2 M \text{ (in bits)} \tag{4}$$

Thus to transmit a given amount of information, a minimum bandwidth-time product is needed for a given value of $M$. In general, any measurement problem can be resolved into three factors that can be engineered with some degree of independence, namely, the length of time allotted for the observation, the rate at which the information symbols can be processed, and the number of distinguishable states of the symbol. More will be said about practical limitations on the distinguishable states of an analog signal in the following paragraphs.

## C. CHANNEL NOISE

For various reasons in many practical systems, two separate signals may be present at the receiver when only one is wanted. The unwanted signal, usually referred to as noise, is regarded as a nuisance and an interference. In some cases, the noise signals cannot be distinguished in any intrinsic way from the wanted signals; only the relative importance to the person receiving the signal enables a distinction between wanted signal and noise.

It is important to distinguish between noise, delay, and distortion. Signals have to travel with finite velocity, and so some finite transmission time is always required for the signal to traverse the channel; this interval is called the *delay*. If a particular input signal always generates the same output signal, but the output signal differs in some definite way from the input signal, then this effect can be called *distortion*. If the distorting function has an inverse then, at least in principle, the distortion can be corrected by an inverse functional operation on the received signal. For example, attenuation of the signal can be made up for by amplification. If distortion of the signal occurs on the channel such that the original signal cannot be reconstructed at the receiver, the information is lost.

The disturbances in a signal limit the discernibility of one signal level from another and thus limit the number of distinguishable signal levels in the transmission. If the average signal power is given by $S$ and the average noise power by $N$, then the number of distinguishable signal levels $M$ is given by

$$\begin{aligned} M &= \frac{(S+N)^{1/2}}{N^{1/2}} \\ &= \left(1+\frac{S}{N}\right)^{1/2} \end{aligned} \tag{5}$$

The signal-to-noise ($S/N$) ratio requirements of a transmission channel vary widely for different applications. High-fidelity reproduction of music and quantitative representation of analytical-chemical information both require a high $S/N$ ratio, whereas telephone communication can be conducted at a somewhat lower $S/N$ ratio. With complex receivers, radar signals can be detected at $S/N$ ratios less than unity. To the extent that the $S/N$ ratio limits the number of distinguishable levels a signal can exhibit, it also limits the rate at which an information-processing unit can handle information. From equations 3 and 5, we obtain the following simple expression, which characterizes the informational capacity of a noisy channel:

$$C = B \log_2\left(1+\frac{S}{N}\right) \text{ (in bits per second)} \tag{6}$$

To increase the rate at which a particular information-processing unit can handle information, it is necessary to increase either the bandwidth of the device or the $S/N$ ratio at the receiver. It follows, also, that the total amount of information collected during the observation time $T$ through a channel disturbed by noise is given by

$$Q = BT \log_2\left(1+\frac{S}{N}\right) \text{ (in bits)} \tag{7}$$

For binary signals, the number of distinguishable signal levels is fixed at 2. From equation 4, we see that the total information in a binary message $T$ secs long is $2BT$. From equation 3, we see that the band limit determines the system's informational capacity. From this point of view, it is clear why efforts are continually made to increase the speed of digital computers.

In attempting to increase the rate of information flow between two points, we should not discount the possibility of engineering plural channels for a simultaneous transfer of information. For example, the picture and sound information that together furnish the information to a television transmitting station may go over different kinds of facilities and over quite different geographic routes; the video might go over a microwave relay system, and the audio over a wide-band channel in an underground cable. Because the principles controlling the transfer of information over parallel channels are identical to those controlling information flow over a single channel, we have no further interest here in the properties of parallel transmission.

## D. COMPATIBILITY OF SERIALLY LINKED ELEMENTS

The most important principle of this chapter and probably the most fundamental design principle provided by information theory is the need for *compatibility* with respect to the informational capacities of all the serially linked elements within a system. The problem is much like that of pouring water from a high-volume spigot into a funnel. If the funnel cannot handle the water as fast as it is supplied, the funnel overflows, and water is lost. The key consideration that determines whether two information-processing elements are compatible at their common interface has to do with the relative rates at which information is generated and processed. The informational capacity of the device receiving information should equal or exceed the capacity of the informational source. If this condition is not satisfied, then either no information is received, or what is received is either incomplete or inaccurate. When a source supplies information to an information-processing channel, the information can be recovered without error at the receiving end of the channel if the informational capacity of the channel equals or exceeds that of the source. The importance of this simple concept can hardly be overemphasized.

From the above, we see that an analog informational source having an infinite number of closely spaced analog values would require a measurement channel with an infinite informational capacity for exact information recovery at the receiving point. Nature offers a great many examples of such informational sources: the distance between two points, the mass of an object, the interval between two events, the pressure of a gas, the temperature of a flame, the wavelength of light, and, of course, the concentration of a particular chemical species in an analytical sample. The signals associated with these measurement problems all have two characteristics in common. All are continuous functions that can assume any of an infinite number of values, and all are analog signals in the space of the original information.

## E. SIGNAL CODING

At first glance, it may appear that any effort at errorfree recovery of an analog signal by a practical measuring device would be futile. Unless the $S/N$ ratio at the receiver and the bandwidth of the channel are infinite, information from an analog source would be lost, at least in part. Within the present perspective, we are prepared to explore how signals that are analog in the space of the informational source can be processed to within an arbitrarily small error by a transmission channel disturbed by noise. Again, the engineering concept is very simple. When the rate at which information is generated by the source exceeds the processing capacity of the channel, then the only option is to decrease the rate at which information is applied to the channel. In terms of the analog of pouring water into a funnel, if you can't make the funnel larger, then pour at a slower rate. Information is "poured" at a slower rate onto the channel by decreasing the amount of information per symbol.

When analog signals are applied with one-to-one symbolism onto the measurement channel, each symbol in the informational source space is represented by a symbol in the channel space; such a measurement is said to be *uncoded*. In this case, each and every possible amplitude value in the space of the source is represented on the measurement channel by a state of the measurement symbol. For example, suppose we are interested in measuring the temperature of a melt with a thermocouple. Each of the infinitely closely spaced analog temperature values is represented by a voltage at the output of the thermocouple. If the thermocouple transducer is linear, then there is an exact one-to-one correspondence between the temperature values and voltage values. Any noise in the measurement system limits our ability to distinguish between neighboring voltage values, and the uncoded temperature measurement suffers from error.

As an alternative measurement scheme we can, at least in principle, decrease the rate at which information is applied to the measurement channel by distributing the original information over a multiplicity of symbols in the source domain; such a measurement is said to be *coded*. In a coded measurement, a set of symbols, each having finite informational content, is transmitted for each original message symbol. The number of channel symbols per message is selected in accordance with the desired precision in the measurement.

As an example of how such coding can work, let us consider another measurement problem for which coding is well known and highly successful. Suppose our task is to transmit a voltage signal representing one of the 26 letters of the English alphabet; suppose also that the transmission is to be over a long distance and through a noisy environment. If the transmission is to be analog, the original signal must have at least 26 distinguishable voltage levels, each of which is assigned to represent a specific letter of the alphabet. If one letter per second is transmitted onto the channel, the informational capacity of the uncoded source signal is $\log_2(26)$=4.7 bits/sec. Now suppose the noise in the channel is such that only 16 voltage levels can be distinguished within the dynamic range of the measurements. The informational capacity of the channel is $\log_2$ = 4.0 bits/sec, a rate insufficient for the needs of the measurement. This difficulty can be overcome by reducing the rate of information generation by the source to that which can be processed without error by the channel. For example, instead of representing the 26 letters of the alphabet by a single informational symbol having 26 distinguishable levels, two separate eight-level symbols transmitted one at a time, could represent as many as 64 different alphanumeric characters. The requisite channel data rate needed to process each individual octal symbol is 3 bits/sec, a rate within the 4 bits/sec actual capacity of the channel. An even greater margin of protection would be achieved if a set of five binary symbols were used to represent the letters of the alphabet. Because the signal could appear in only one of two distinguishable states, the requisite channel capacity would be reduced to 1 bit/sec, a rate well exceeded by the capacity of the channel.

In summary, as long as signal-level thresholds are distributed across the signal dynamic range so as to be unambiguously identifiable in the presence of existing noise, the signal can be transmitted over an infinite distance without error. For example, if after transmission over an extended distance, the binary signals become distorted and noisy, an intermediate relay station (receiver-transmitter) can be used to intercept the signal and retransmit a fresh, clean signal, free of noise and distortion. Under such measurement conditions, the set of five binary signals representing a 32-element alphanumeric-character set could be transmitted over an infinitely long channel through a high-noise environment and be detected at the receiving end of the channel without error. And what price does nature extract for our errorfree measurement? The tariff is denominated in units of time; it takes five times longer to process five symbols individually than one; such is the nature of a coded measurement.

## F. IMPLEMENTATION OF CODED MEASUREMENTS

Communication systems may be classified as discrete, continuous, or mixed. In a *discrete* system, information is transmitted in the form of sequences of pulses, as in telegraphy. In a continuous system, transmission is by a continuous functions over an interval, as in voice communication. In a *mixed* system, both descrete and continuous information are transmitted. The vertical and horizontal synchronizing pulses of television furnish information in discrete form; in the audio channel, information is furnished in continuous form. Both are central parts of the program information.

Many messages are naturally discrete in their original form. One class of naturally discrete informational sources includes those that are binary in their natural form and cannot be reduced to a more primitive code. Typical examples include "pregnant?," "cloudy or sunny?," and "by land or by sea?" Some signals are naturally discrete but are not binary. The number of members of a social group, such as a family, the number of legs on a insect, the number of photons per unit time incident on a detector, or the number of species in a mixture of hydrocarbons are all examples of such messages. Although some signals can easily be rendered binary by a simple threshold selection in the domain of the original signal, not all messages that are originally discrete are necessarily easy to code. And if a transmission is to be truly coded, the coding process must be conducted in the domain of the orginal information or, at least, upstream from the noisy channel.

Analog-to-digital conversion (Chapter 2;18) can be conceptualized as a two-step process, quantization and coding. *Quantization* is the process of dividing a continuous analog signal into a set of discrete parts. If we quantize a population of inhabitants by using an approximate nonlinear quantization, we might use two discrimination thresholds, one at 1000 and one at 100,000 residents. We would thereby divide the population scale into three parts that we could name "village," "town," and "city," respectively. In certain cases, it might be useful to quantize a signal nonlinearly by dividing the signal dynamic range into segments of unequal size, to compensate for known nonlinearities in a particular signal. such *nonlinear* quantization could be useful in a digital liquid-level sensor in which the sensing device actually measures depth in a tank and in which the volume is related to the depth by a nonlinear function. With the use of nonlinear quantization, each signal quantum could be made to represent the same volume of fluid as that of any other quantum, and the nonlinearity of the signal could be removed.

If the quanta are all of equal size, the analog-to-digital conversion is said to be *linear*. The transfer function shown in Fig. 1.3, which itself is nonlinear, is that of an ideal quantizer with eight output states. The resolution of the quantizer is characterized by the informational content per symbolic word at the output of the analog-to-digital converter (ADC), expressed in bits; in this case, it is a 3-bit quantizer. The decision points must be precisely set in order to divide the analog signal into the correct quantized values. The analog decision points are halfway between the center points of each output code word. The quantizer staircase function is the best approximation that can be made to a straight line drawn through the origin and the full-scale point. Notice that the line passes through the center points of each quantum range. At any part of the input range of the quantizer, there is a small range of analogy values within which the same output code word is produced. This small range, designated by "Q" is the difference between any two adjacent decision points and is known as the analog

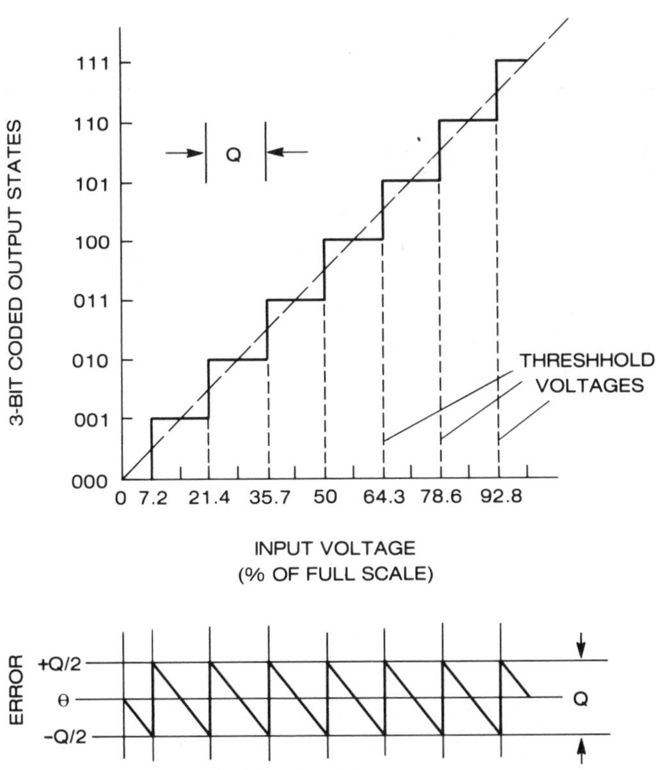


**Fig. 1.3.** Ideal analog-to-digital-converter transfer function.

quantization size. The quantization size is given by dividing the full-scale range by the number of output states.

If the analog input is moved through its entire range of analog values, the quantization uncertainty due to rounding off of the variable to conform to the nearest quantization level results in a sawtooth error function. This quantization error can be reduced only by increasing the number of quantizer output states. The error, which is zero only at the analog values corresponding to the centerpoints of the quanta, ranges from 0 to $\pm Q/2$. Messages that are naturally discrete can be coded free of such quantization noise if the number of quantization states equals or exceeds the number of states in the original signal.

In general, complete flexibility exists in the assignment of symbolic codewords to the output states of an ADC. A wide variety of coding systems have been designed, each having their own characteristic properties. A code is said to be analytic (4) if there exists an analytic relation between the value of the integers coded and the symbols used for their representation. Analytic codes are especially well suited for carrying out arithmetic operations. Binary code is popular because it lends itself well to direct arithmetic computations, whereas the Gray code (4) and binary-coded decimal, which are closely related to binary code, do not have analytic properties. Arithmetic with binary numbers is discussd in Chapter 8 on digital electronics.

## IV. SIGNAL STRUCTURE IN ANALYTICAL SPECTROMETRY

The emphasis in this section is on formulating and evaluating the informational structure of an analytical spectrometric system of general utility. The system selected for representation, namely, a flame-emission/atomic-absorption (FE-AA) spectrometer (21), was chosen because the noise problems that appear in various other types of spectrometric measurements also appear in FE or AA spectrometry. As such, a model representing the spectrometry of flames can provide a general picture of the informational structure of a large class of spectrometric measurements.

In the paragraphs that follow, the basic Shannon model is applied to FE-AA spectrometric measurements. The nature of the noise problems in optical measurements and in the spectrometry of time-dependent analytical cells is discussed in some detail. Finally, several signal-recovery techniques are explored with a view toward demonstrating how such a model can be used for the design of new techniques and for the refinement of existing analytical methods.

### A. SHANNON'S MODEL CONFIGURED FOR ANALYTICAL SPECTROMETRY

FE spectroscopy has been used for many years as a rich source of analytical information on ions in solution. A storehouse of relatively simple analytical FE spectrometric techniques, developed almost solely by empirical methods, is available for application in industrial, clinical, and research laboratories. Over the years, progress in analytical FE technology has been hindered by the lack of a valid and complete kinetic model of the flame plasma and measurement system. The extreme complexity of microscopic flame processes has precluded our understanding of all except the most primitive flame behavior. It is this fundamental complexity of practical flame systems that has made the empirical approach to analytical FE spectrometry the most attractive and rewarding to many investigators. Although it is true that the basic Shannon model is structured around net observable signals, the basic architecture of the model can be formulated to identify the functions to which the microscopic processes give form. Shannon's model also can provide a bridge between the microscopic processes responsible for signal generation and the directly observable signals; it identifies the important informational and noise signals in the measurement; and it can be used to portray schematically the relation between noise signals, informational signals, and the system responses.

In this interpretation of Shannon's model, the analytical sample is viewed as an information-generation and -storage device in which the physical concentration of the analyte species is regarded as the informational symbol, represented by a number. The execution of a chemical analysis is equivalent to converting the analytical information into a channel-compatible symbol and recovering that signal at the receiving end of the system. Thus the process of signal recovery can be viewed as an effort to determine those parameters that characterize the signal components carrying the information of interest. Normally, one of the main objectives in chemical analysis is the recovery of a faithful replica of the informational signal generated in the chemical domain; this objective is called accuracy.

If we define the chemical signal as the concentration of a particular species, then the action of the flame, in converting the chemical signal into an optical signal, can be thought of as a transducer that transforms the information of interest from the chemical domain into the optical domain. In general, the device assigned to perform this chemical-to-optical conversion is called the analytical cell. This conversion is the most critical of all signal manipulations and must be matched to the signal-recovery technique for optimization of a given analytical problem. The skeletal features of the flame model are schematically portrayed in Fig. 1.4.

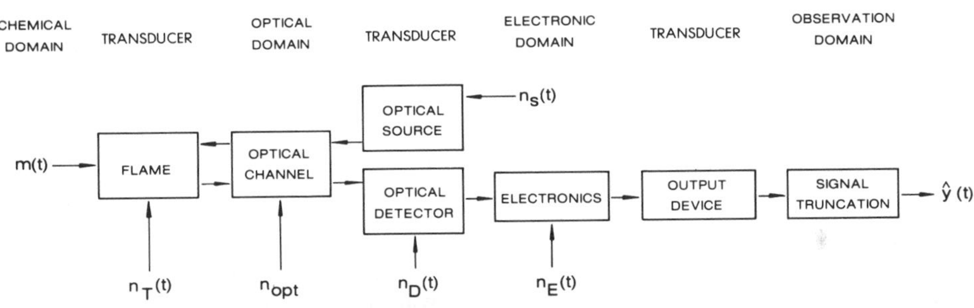


**Fig. 1.4.** Model for information transfer in FE-AA system.

For the purposes of this treatment, we consider the input chemical signals to be generally time dependent and deterministic. Although most solutions analyzed by FE spectrometry are stable, the general time dependence extends the applicability of the model to studies of kinetics, relaxation, and transients, as well as to refined signal-recovery techniques.

The conversion of the input chemical signal to the output optical signal by the analytical cell can be characterized by a set of linearly related impulse-response functions. The basic notion of linear-response theory is that if it is known how a system responds to an infinitely brief unit perturbation at its input (the impulse response), then the output can be determined for any form of driving function at the input. This concept is schematically represented in Fig. 1.5*a* for a noise free system, in which $m(t)$ represents the chemical driving signal to the flame, $h(\tau)$ the impulse-response function[10], and $y(t)$ the output. Although the notion of an infinitely brief unit pertubation (rigorously defined in Section VI.B.2, equation 14) (10, p. 49;11, p. 20) is something of an abstraction, the impulse-response function is a readily measured quantity (11, p. 239). For a linear system, the output $y(t)$ is given by the convolution of the input driving function with the impulse-response function, as given by equation 8 in which the convolution product is denoted by an asterisk between the functions under the convolution integral:

$$\begin{aligned} y(t) &= \int_{-\infty}^{t} m(t-\tau)\, h(\tau)\, d\tau \\ &= m(t)*h(\tau) \end{aligned} \tag{8}$$

The convolution integral, operating over the entire history of the input from minus infinity to time $t$, is a summation of the contribution of the input applied $\tau$ sec earlier to the output at time $t$, with the input weighted over its history by the impulse-response function. This formulation is illustrated for a hypothetical

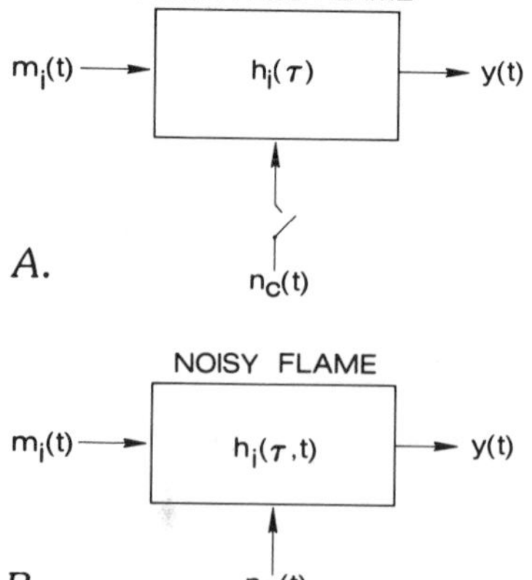


**Fig. 1.5.** (*a*) Response of time-invariant (noisefree) linear system with impulse-response function $h(\tau)$, input $m(t)$, and output $\hat{y}(t)$. (*b*) Response of time-varying (unstable) linear system.

case in Fig. 1.6, in which the input signal is resolved into two discrete unit impulses spaced in accordance with the sampling theorem and applied to a general analytical cell. Since the impulse applied to the system $\tau$ sec earlier, the output at time $t$ for the situation in Fig. 1.6 is the sum $m(t_1)h(t - t_1) + m(t_2)h(t - t_2)$.

An alternative way to describe the response of a system relies on the convolution theorem (10, Chapter 2), which states that functions that are related by convolution in the time domain are multiplicative in the frequency domain, and vice versa. Thus Fourier transformation of both sides of equation 8 yields the simplified expression

$$Y(\omega) = M(\omega)H(\omega) \tag{9}$$

where $Y(\omega)$ and $M(\omega)$ represent the direct Fourier transforms of the output and input functions, respectively, and $H(\omega)$ represent the frequency-transfer function (10) for the system (where $\omega$ is the radial frequency, in radians per second). Strictly speaking, the frequency-transfer function, a complex quantity, is defined by the ratio $Y(\omega)/M(\omega)$. From this point of view, the system can be considered to be an active filter operating on the amplitude and the phase of the signal components applied at its input.

In the case of a flame sėrving as an analytical cell, the response function for the generation of each flame informational signal is controlled by a sequence of processes acting to convert the input chemical signal into an output optical signal. The mechanisms controlling the response of the flame could hardly be more complex. First, the analytical solution is nebulized into an aerosol that passes into the flame. As each small droplet enters the flame environment, a sequence of complex processes occur, namely, desolvation, residue sublimation, salt dissociation, molecular reduction, and a host of competing chemical reactions, all depending on the nature of the reactions in the flame gases and on position in the flame. At any given point in the flame, a great variety of species may be present. Many reactions occur concurrently, forming a set of products. Any particular species may be formed as the result of more than one reaction, and any given reaction may proceed through several different products. Processes may be coupled by reactions taking place between reactants and intermediate products to form additional products; excitation and deactivation

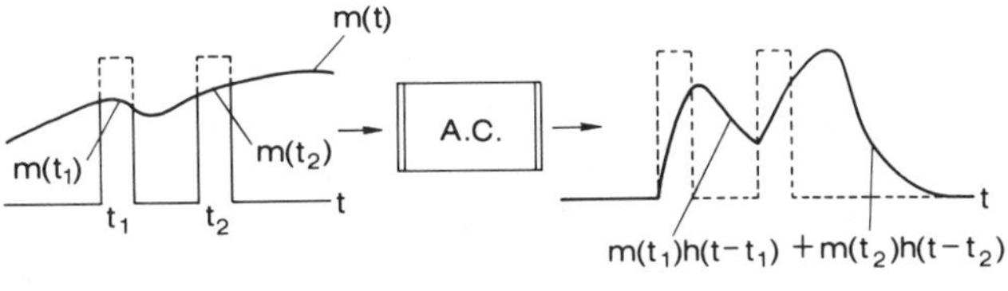

**Fig. 1.6.** Schematic representation of output of linear system for hypothetical input signal resolved into two Nyquist pulses spaced $1/2B$ sec apart.

phenomena may occur at any stage. Owing to the presence of finely divided salt particles, solid oxides, aggregations, and droplets, chemical reactions may take place in many phases or at phase interfaces. As the transient particles stream through the chemical gradients of the flame, many reactions may be quenched before they reach completion. The collection of steady-state processes at a fixed point in a given flame depend critically on radial and vertical position in the flame, and for each different type of flame, the detailed processes differ markedly. Moreover, for a given flame type, the net results depend on burner design and on the constituent-dependent physical properties of the liquid sample, such as surface tension and viscosity.

In an explicit mathematical statement of the model, the action of each separate controlling step in the development of the flame signal would be resolved and characterized by a response function. However, even in our ignorance of the detailed microscopic processes in the flame plasma, a composite impulse response can easily be measured (10) that characterizes the net analytical performance of the flame. Response characteristics of analytical interest include sensitivity, specificity, background, stability, response time, and relaxation time, all of which are determined by the microscopic processes that generate the output optical signal and all of which are measured by the composite impulse-response function.

## B. NOISE

The components of interest in the output signal are those that represent the characteristic optical spectra of analyte species in the analytical solution. The interfering signal components are those representing the spectra of species other than the analyte and those created by disturbances in the flame system. It is useful to distinguish between those interfering components that are deterministic, commonly referred to as background, and those that appear randomly in the signal. Strictly speaking, the presence of background per se does not represent a fundamental limitation on the measurement; it is fluctuations in the background that limit our ability to distinguish one signal level from another. Historically, most research efforts in FE spectroscopy and, possibly, most studies in other areas of instrumental analytical chemistry have focused on phenomena that directly relate to analyte-signal generation; the noise parts of the informational structure of the various analytical systems have been relatively neglected. The fluctuating part of the signal is discussed below.

### 1. Types of Noise

System disturbances can be divided into two broad types. In the first type, the *additive* channel noise, the temporal components of the disturbances are not only statistically independent of the informational signals but also can appear at any point in the system quite independently of the signals of interest. This type of noise (13, p. 9;22) includes shot noise associated with dark current in the detector, Johnson noise in the electronics, power-line hum, and impulse noise.

Although additive noise can cause serious problems in long-distance signal transfers, there is no fundamental limitation on the degree to which a signal can be freed of additive noise. If the source of the additive noise can be localized, in many cases such noise can be accommodated on the channel without loss of information by suitable signal coding upstream from the point of noise introduction (see Section V.B below).

Unlike additive channel noise, the second type of noise is *multiplicative* (2, Chapter 9;23) with respect to the informational and interfering signals and always leads to broadening of the informational-signal base band. A distinctive feature of multiplicative noise is that the disturbance is propagated through the system on the signal of interest; the presence of multiplicative noise requires finite informational-signal power in the channel.

We can identify two categories of multiplicative noise. In the first category, the noise-in-signal is regarded as inherent to quantum signals. Such noise manifests itself at the output of detectors that rely on the photoelectric effect. This Poisson (7, p. 153) noise is generally termed *fundamental noise* in the literature of chemical instrumentation. Fundamental noise is most accurately modeled as originating in the chemical domain as a quantum characteristic of the chemical signal.

The most vexing class of multiplicative noise and certainly the most difficult to characterize falls into the second category. In contrast to fundamental noise, this second type commonly originates from temporal perturbations on the parameters of the transducing elements in the system. There are many mechanisms by which *nonfundamental noise* enters into the system. For example, the pneumatic supply to a burner system controlling the rate of solution uptake into the flame may be unstable. This instability can cause variations in the analyte number density in the flame that will be reflected in the magnitude of the

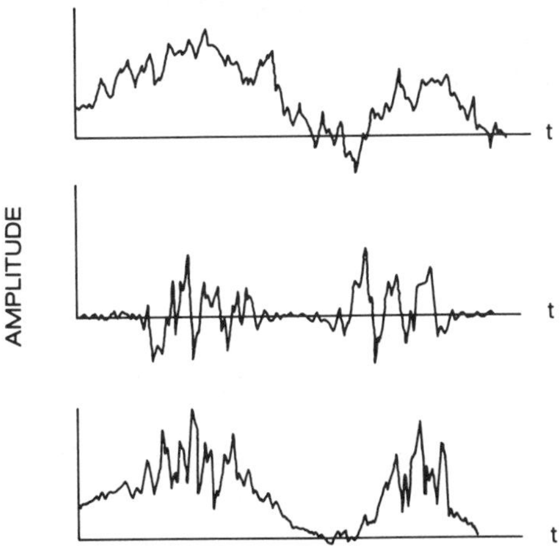


**Fig. 1.7.** Examples of nonstationary noise signals.

analytical signal in the optical channel. Other potential sources of nonfundamental multiplicative noise include flame turbulence, detector high-voltage instability, and, in the case of an absorption measurement, primary-optical-source instability.

In many cases, the nonstationary statistics of nonfundamental multiplicative noise, as well as the multiplicativity of the noise, makes such noise extremely difficult to characterize or to circumvent. Figure 1.7 shows typical examples of nonstationary noise signals.

When noise is modeled as multiplicative with the informational signal in the time domain, the analytical-cell response functions are regarded as time dependent with stochastic elements in their architecture. Such multiplicative noise is acknowledged in our schematic model (Fig. 1.4) by the noise driving function acting on the flame, denoted by $n_T$. Nonfundamental noise is discussed further in Section IV.B.4 below.

### 2. Noise Representation in the Frequency Domain

Inasmuch as temporal excursions of the noise signals in the system are random, these signals are most usefully characterized in terms of certain statistical properties. One of the most intuitively satisfying and generally applied techniques for characterization of the fluctuating part of a signal is the representation of the signal in the frequency domain in terms of its power-spectral-density function (3) (also referred to as the power-density spectrum). The power-spectral-density function is a measure of the signal power per unit bandwidth as a function of frequency. The signal power within any frequency band is given by the integral of the power-density spectrum over the frequency band of interest.

The power-density spectrum of a signal differs substantially from its Fourier transform. Unlike the direct Fourier transform, which is a complex quantity whose existence is not guaranteed, the power-density spectrum is a real quantity that is unambiguously defined in terms of the measurable signal autocorrelation function (2,10,11,3). An important property of the power-spectral-density function is its relation to the autocorrelation function; specifically, the two functions are related by Fourier transformation, as follows:

$$\begin{aligned} P_y(\omega) &= 2\int_{-\infty}^{\infty} R_y(\tau)e^{-i\omega t}d\tau \\ &= F[Ry(\tau)] \end{aligned} \tag{10}$$

where $F$ is the direct Fourier transformation operator, $\tau$ (in seconds) is the lag variable, $\omega$ is the radial frequency (defined by $\mathrm{f} = \omega/2\pi$ Hz), and $i$ is the complex operator. Also in Equation 10, the autocorrelation function $R_y(\tau)$ is given by

$$\begin{aligned} R_y(\tau) &= \lim_{T\to\infty} \frac{l}{T}\int_{-T/2}^{T/2} y(t)y(t+\tau)\,dt \\ &= < y(t)y(t+\tau) > \end{aligned} \tag{11}$$

where $T$ represents the data window and the angular brackets denote the expected value.

The physical significance of the autocorrelation function becomes apparent if we consider the details of the measurement procedure. After signal digitization, the apparent autocorrelation function is calculated as a sum of discrete lagged products, a procedure approximating the integral equation 11. Figure 1.8 illustrates this calculation for a hypothetical signal. In this calculation, a relative shift of $\tau$=r seconds is applied to the time-series data, and the lagged products, formed across the original series and the shifted series, are averaged to give the value of the autocorrelation function at a lag value of $r$. One important property of the autocorrelation function is its ability to reveal the extent of the statistical dependence between various sample values in the data field. If low-frequency components predominate in the original signal, the signal fluctuates slowly, and the data values retain a measure of similarity over even extended lag values. If high-frequency components predominate in the signal, the converse is true, and the autocorrelation function diminishes to zero at relatively small lag values. In effect, the autocorrelation function acts to reformat the phase of all frequency components in the original time-domain signal to a phase angle of zero. Thus with all frequency components converted to cosine components, the component amplitudes add constructively at a lag value of zero to give the autocorrelation function its maximum value at $R(0)$. From the definition of the autocorrelation function, we an recognize that its value at a lag value zero equals the mean square value of the signal:

$$R_y(0) = \langle y(t)^2 \rangle \tag{12}$$

It also follows from equation 11 that the autocorrelation function equals the square of this mean value as the lag variable becomes large:

$$R_y(\infty) = \mu_y^2 \tag{13}$$

From the above, it follows that the autocorrelation function is an even function of the lag variable $\tau$ and that all imaginary coefficients of the Fourier transform of the autocorrelation function are zero. Fourier transformation of the autocorrelation function merely reveals the frequency structure of the autocorrelation function in a more intuitively satisfying format, that of the power-density spectrum.

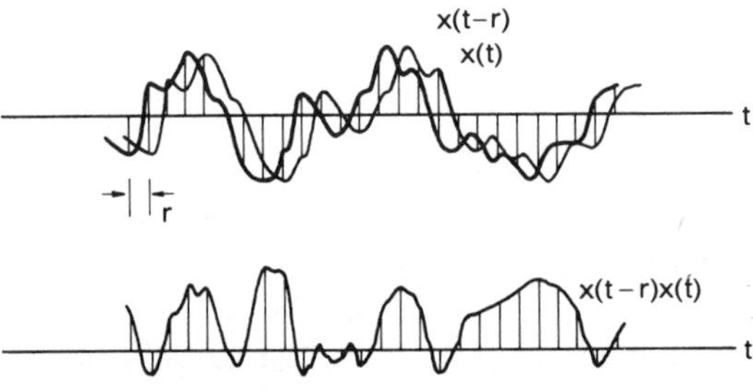


**Fig. 1.8.** Calculation of autocorrelation function for lag value $r$.

Because all phase information in the signal components is lost in the autocorrelation averaging process, the power-density spectrum is not unique for a particular time-domain signal and may, in fact, represent an infinite number of time-domain functions, both deterministic and stochastic. Clearly, the power-density spectrum is a statistical property of a signal, whereas the complex Fourier transform is the deterministic frequency-domain representation of a signal. Finally, it should be noted that most treatments use two-sided power-density spectra in which the frequency variable is allowed to range from $-\infty$ to $+\infty$. This expedient not only affords a satisfying symmetry with the time domain, but it also allows Fourier analysis in exponential form and leads to a more compact form of expression. The value of the physically realizable power-density spectrum $P(\omega)$, which exists only in the first quadrant, equals twice that of the power-density spectrum defined in equation 10.

There is a qualitative difference between the characteristics of the power-density spectra for random and for periodic signals. For random signals, the presence of finite power density at a given frequency does not necessarily guarantee the existence of a particular component in the signal because the magnitude of the power-density spectrum at a given frequency represents the average power contributions from frequency components within the frequency-resolution element centered at the selected frequency. In contrast, the spectrum of a truly periodic signal has infinitely narrow peaks of infinitely large power density at the fundamental and harmonic frequencies making up the signal.

The concentrated power density in the spectra associated with periodic signals (line spectra) can be represented mathematically by the Dirac delta function $\delta(\zeta)$, a singularity function most appropriately defined in terms of the sampling process, as follows(10, p. 49;11, p. 20):

$$\int_{-\infty}^{+\infty} f(\zeta)\delta(\zeta)\,d\zeta = f(0) \tag{14}$$

The Dirac delta function, also referred to as the unit impulse, the test function, the sampling function, or, simply, the delta function, is an abstraction that can be represented graphically by an upward arrow having unit area. We have reason to use the delta function in the discussion that follows.

### 3. Fundamental-Noise Model

Fundamental noise is inherent to quantum signals. The stochastic structure of such signals originates in the domain of the original informational source and is not a property introduced during signal recovery. However, the fundamental-noise properties of such signals are preserved by systems that respond to the quantum characteristics of the signal, including systems that rely on the photoelectric effect or on photochemically induced current pulses in any form (photomultiplier tubes, ionization detectors, silicon detectors, etc.).

Quantum signals can be modeled in the time domain as a series of Poisson distributed (*7, p. 153*) unidirectional pulses. For the purposes of this discussion, we assume that the pulses composing the signal are uniform in strength. The essence of the Poisson model for quantum signals is that the probability of occurrence of a pulse within a short interval $d\tau$ is independent of pulse events in the preceding interval. Also, the probability of occurrence of a pulse within the interval $d\tau$ equals $\alpha\, d\tau$, where $\alpha$ is the average number of pulse events per unit time. In general, the signal of interest is the quantity $\alpha$.

The autocorrelation function of a Poisson process is well known and is given by (*11, p. 193*)

$$R_m(\tau) = \alpha\delta(\tau) + \alpha^2 \tag{15}$$

The corresponding power-density spectrum is given by Fourier transformation of equation 15:

$$P_m(\omega) = \alpha + 2\pi\alpha^2\delta(\omega) \tag{16}$$

Because the stochastic process under consideration consists of unidirectional pulses, it is not surprising that the power-density spectrum has two terms, a pulse of strength $\alpha^2$ at the origin (dc) and a constant term $\alpha$. From equation 16 we see that the presence of finite noise power in the signal is as much a fundamental part of the total signal as is the signal of interest.

A significant property of fundamental noise is that the power spectal density is independent of frequency and equals the strength of the signal of interest. Because of its white spectral characteristic, there are no "clean" spectral regions in which signal recovery can be conducted free from noise for Poisson signals. Not only is the noise power-spectral density multiplicative with the signal of interest, but the ratio of the signal strength to the noise-in-signal spectral density also increase with the square root of the signal strength. (Recall that the voltage and the current signals are proportional to the square root of the power signal.) The only way to reduce the noise power-spectral density is to reduce the strength of the signal of interest, clearly an unacceptable alternative.

One interesting property of multiplicative noise is that the noise power is directly proportional to the strength of the signal. It follows, therefore, that an analytical measurement could be made on the basis of the statistical properties of the noise (16). But what can be said regarding the quality of such an unconventional measurement? The answer depends on the stability of the statistical measure of the noise power, a topic discussed below in Section V.A.4.a.

Fig. 1.9 shows the power-density spectrum of a photomultiplier-tube dark-current signal electronically summed with a 0.5280 mHz square-wave signal; the square-wave components were added to illustrate the spectral nature of a deterministic additive disturbance in the signal. The power spectra shown here were measured in my laboratories in accordance with the procedures outlined in the treatment by Blackman and Tukey(3). The spectrally white behavior of the fundamental noise generated by the photomultiplier tube, and all peaks in the spectrum of Fig. 1.9, closely agree with theoretical expectations. The bandwidth-time limitations on the recovery of signals limited by fundamental noise are discussed below in Section V.A.3.a.

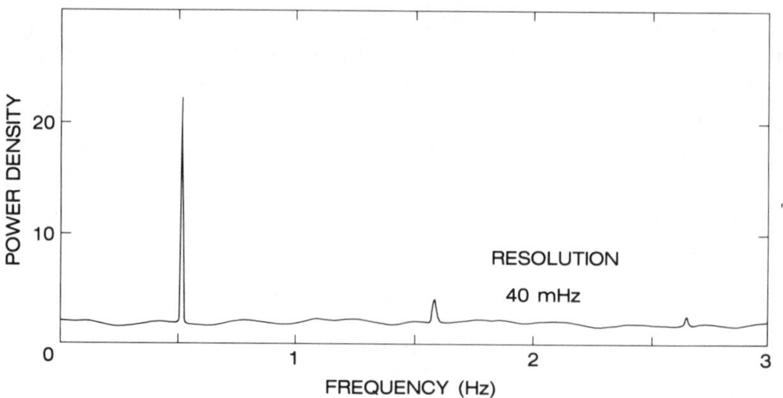


**Fig. 1.9.** Power-density spectra for 0.528 mHz square-wave electronically summed with photomultiplier-tube dark-current signal.

## 4. Nonfundamental-Noise Model

Analytical-cell instability often can be the dominant source of noise in an analytical system. The techniques most subject to multiplicative noise are those in which material transport and (or) sample atomization are required in the generation of the signal; such techniques include sample atomization with flames and electrothermal devices; arc-, spark-, and microwave-driven discharges; certain electrochemical cells; certain mass-spectrometry sources; and various detector cells in chromatography.

Before attempting to represent analytically the behaviour of nonfundamental noise, I review the experimentally observed statistical properties of such signal disturbances. The FE noise spectra in Figs. 1.10–1.13, observed over three spectral regions, were taken from premixed acetylene-nitrous oxide FE signals generated by atomic manganese and by the CN radical at 4031Å(14,9). These FE noise spectra were obtained under conditions such that the shot noise from the detector and the Johnson noise from the amplifier were constant in all the signals analyzed. For this technique, a background signal free of nonfundamental noise was optically summed with the atomic-manganese signal at a level suitable to maintain the optical-signal strength constant (photomultiplier-detector output at 20 nA).

Figure 1.10 shows the power-density spectra of signals for which the signal-to-background ratios range from 0 to 5 (9). (Modulation and signal-recovery frequencies are typically selected within the frequency range 10–500 Hz). As the absolute contribution of the Mn emission was increased, the relative power spectral density within the range 200–450 Hz increased strikingly, indicating a multiplicative relation between the magnitude of the manganese emission and the noise-in-signal.

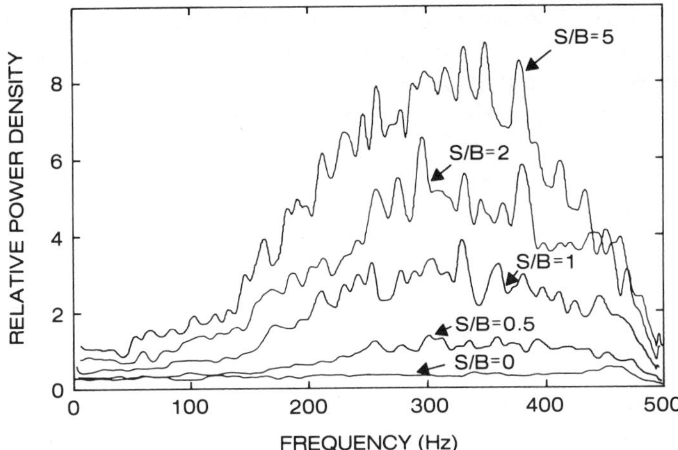


**Fig. 1.10.** FE noise power-density spectra for Mn 4031 Å with CN background for various signal-to-background ($S/B$) ratios.

Inasmuch as the noise power recovered in the final measurement is related to the integral of the noise power-density spectrum over the measurement band pass, the structure of the noise spectrum within the band pass of the measurement has a predominating influence on the quality of the analytical measurement. Dc signal-recovery techniques, with the information of interest located at 0 Hz, collect informational and noise components over a frequency band extending outward from the origin of the signal spectrum. Several frequency modulation techniques translate the informational and noise signals from the

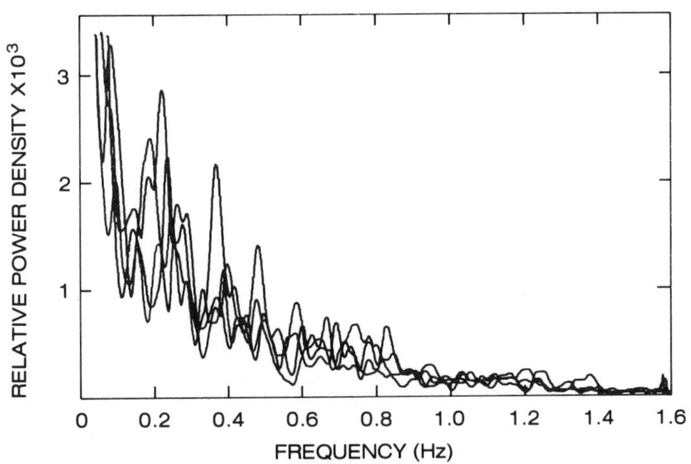


**Fig. 1.11.** Repetitive measurements of Mn 4031 Å FE noise power-density spectrum.

region near 0 Hz to a higher frequency, with no change in the spectral distribution around the modulation frequency. Thus the frequency components in the noise spectrum within the range 0.001–1 Hz fall within the measurement band pass of many spectrometric measurements.

Figure 1.11 shows repetitive spectral analyses of the fluctuating part of the manganese emission signal from 0 to 1.6 Hz (9). Nonstationarity of the signal is evident in the poor reproducibility of the spectral features. Such nonstationary statistics, typical of nonfundamental multiplicative noise, can be extremely vexing if we are trying to isolate and control noise-in-signal exhibiting such properties. Figure 1.7 shows examples of nonstationary behavior in the time domain. Several approaches have been developed for the treatment of nonstationary signals. Possibly the most realistic technique(*2, Chap. 9*) treats nonstationary power-density spectra as time dependent. Because we are interested in the general properties of multiplicative noise, we rely on the averaging of spectral data across an ensemble of signals. Figure 1.12 (9) shows examples of such ensemble-averaged power-density spectra.

We now turn our attention to isolation of a typical source of nonfundamental noise. The upper spectrum in Fig. 1.13 was obtained when deionized water was nebulized into the flame, and the lower spectrum when the flame was operated dry. Also, the spectra are undistorted by signal truncation; the signal-observation time was at least 10 times longer than the period corresponding to the lowest frequency measured. (Signal truncation is discussed below in Section V.A.2.) The most striking feature in the spectra in Figs. 1.12 and 1.13 is the rapid increase in the power-spectral density at frequencies below 1 Hz, a pattern exhibiting $1/f$-type frequency dependence. Comparison of the spectra in Fig. 1.13 gives a clear indictment of the flame as the main source of nonfunda-

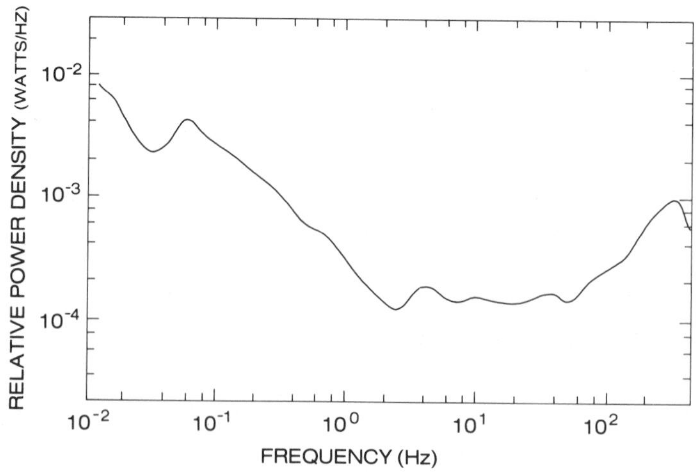


**Fig. 1.12.** Ensemble-averaged Mn 4031 Å FE noise power-density spectrum.

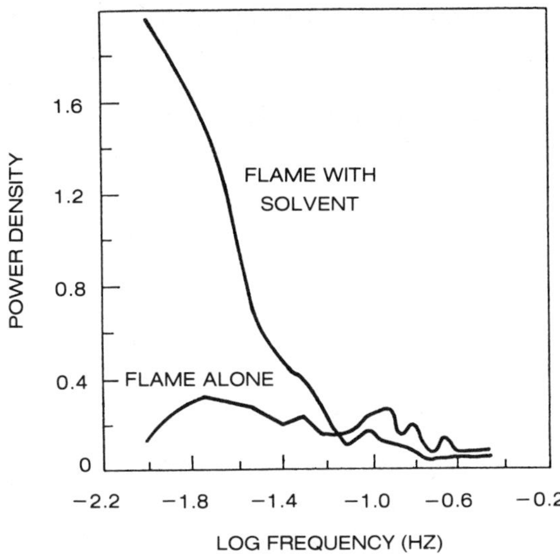


**Fig. 1.13.** Noise power-spectral-density functions for FE OH background at 3064 Å.

mental noise in the system studied and further suggests that processes related to sample nebulization are chiefly responsible for the $1/f$-type dependence of the noise spectra.

Although the spectra in Figs. 1.10–1.13 are unique to the particular systems studied, and although for a given system a special noise study would be necessary to assess the relative contributions from and the spectral features of fundamental, nonfundamental, and additive noise, the noise patterns illustrated in Figs. 1.10–1.13 are useful for identifying problems common to a large class of instrumental measurements. The simple model developed in the following paragraphs was designed to represent a typical example of how nonfundamental noise can be introduced and propagated through a system.

The viewpoint taken in our model is that the disturbances introduced into the system by the analytical cell are the result of temporal perturbations of the parameters of the analytical cell (17). Figure 1.5*b* illustrates this notion of a time-varying linear system.

In the particular case of a flame as an analytical cell, for a given chemical input a displacement of the optical power level occurs at the output. In our simple model, we assume that the flame exhibits no second-order characteristics; thus we can relate the output signal to the input signal by a first-order time-varying differential equation:

$$\frac{dy(t)}{dt} + k(t)y(t) = m(t) \tag{17}$$

In the situation represented by equation 17, $k(t)$ is a stochastic process having an intense $1/f$-type spectral density similar to that of the upper curve in

Fig. 1.13. For the purposes of our discussion, we further assume that the second term in equation 17 is much more significant than the first and that no correlation exists between the deterministic input signal and the time-varying coefficient $k(t)$. If the noise-in-signal generated by $k(t)$ dominates the stochastic behavior of the output signal, we can write the autocorrelation function $R(\tau)$ of the output signal as follows:

$$
\begin{aligned}
< y(t)y(t+\tau)> &= < m(t)m(t+\tau)>< \frac{1}{k(t)} \cdot \frac{1}{k(t+\tau)} > \\
R_y(\tau) &= R_m(\tau)R_{1/k}(\tau)
\end{aligned}
\tag{18}
$$

Fourier transformation of both sides of equation 18 yields the power-density spectrum of the output signal $P_y(\omega)$, in the following form:

$$P_y(\omega) = F[R_m(\tau)\, R_{1/k}(\tau)] \tag{19}$$

Applying the convolution theorem to the right side of equation 19, we obtain

$$P_y(\omega) = P_m(\omega) * P_{1/k}(\omega) \tag{20}$$

where the asterisk denotes frequency convolution of the noise and informational-signal power-density spectra. In effect, the cell introducing the multiplicative noise is regarded as a frequency mixer in which the power spectra of the various informational signals are convolved with the spectra of the noise signals, subject to the restrictions imposed by process kinetics (normally acknowledged by the inclusion of derivative terms, as in equation 17. Such frequency-domain convolutions, which increase the base band of the analyte and noise signals, cause the noise to be carried through the system as side bands on the informational signals, and some information to be carried through the system on the increased base band of the noise signal. This process accounts for the fact that for many techniques, the relative precision of an analysis does not necessarily improve as the concentration of the analyte is increased.

Nonfundamental noise-in-signal can have surprising consequences on the efficacy of signal recovery, especially if the design of the signal-recovery technique was predicated on the existence of fundamental-noise behavior. When the stochastic structure of the analytical signal is predominantly nonfundamental, however, the dependence of the analytical precision on signal-observation time and on bandwidth can deviate severely from that expected from a system limited by fundamental noise. The problem of signal recovery in the presence of both fundamental and nonfundamental noise is discussed below.

## V. SIGNAL RECOVERY

Over the years, progress in FE-AA technology, as well as in many other areas of chemical instrumentation, has been marked by improvements designed to increase directly the level of desired information in the signal relative to the level of undesired informational and noise components. The natural tendency has been to develop techniques that directly improve the ratios of signal to background and signal to noise in the optical signal. From our understanding of the informational structure of FE-AA systems, we can see that the ultimate design limitations on these systems are not strictly determined by the total amount of information and noise in the signal but by the distribution of the informational and noise signals in the informational spectrum and by the coding of the information onto the measurement channel. We also recognize that instrumental design is best optimized with respect to the total information in a measurement rather than around a particular signal characteristic or instrumental parameter. In this section, we examine those measurement conditions that have a critical influence on analytical detection limits and precision, and interpret such measurement conditions in terms of the Hartley bandwidth-time relation and in terms of equation 4. We also consider the use of selective signal-recovery techniques and, finally, review the feasibility of coding in the chemical domain.

## A. ANATOMY OF DC MEASUREMENTS

In considering the process of signal recovery, we must distinguish between operations that continuously modify the signal and those that are conducted discontinuously by the user. Here the term "user" refers either to a human being or to an automatic device substituted into the role traditionally served by human beings. Continuous operations include signal demodulation and filtering; discontinuous operations include signal truncation and mean-value estimation. We outline demodulation and filtering in only the most general way, without reference to any particular technique (see the following two chapters for specific techniques); however, the philosophy behind signal truncation and mean-value estimation is reviewed in more detail.

### 1. Signal Filtering

As a matter of convenience, the last stages of signal manipulation, before the signal is physically observed by the user, are usually demodulation and low-pass filtering. These operations convert the signal to dc and output the dc signal to a display device that can be read by the user (meter, recorder, etc.). In terms of the Fourier spectrum of the signal, the demodulation step (rectification) concentrates the information being sought into the lowest resolution band near 0 Hz. The dc informational signals can be represented by delta functions (cf. Section IV.B.2, equation 14) of finite strength located at the origin. Figure 1.14 shows a hypothetical power-density spectrum for typical noise and informational signals. The filtering step attenuates all frequency components above the band limit $B$ of the filter.

The output-power signal from the filter can be represented in the frequency domain as the product of the power-transfer function for the system with the power-density spectrum of the input signal, as follows:

$$P_y(\omega) = |H(\omega)|^2 \cdot P_m(\omega) \tag{21}$$

Ideally, the true mean value of the output signal is given by

$$\mu_y = H(0)\mu_m \tag{22}$$

where the frequency-transfer function $H(\omega)$ is defined in equation 9.

In Fig. 1.14, the idealized spectral band pass $\Delta F$ equals twice the bandwidth $B$ (i.e., $\Delta F = 2B$). If we assume for the moment that the noise power-density spectrum is white, a reduction in the measurement band pass (increase in the time constant) reduces the noise power appearing in the observed signal. Because the information of interest has infinite spectral density at the origin, the strength of the measured signal does not decrease as the measurement band pass is reduced. It seems, therefore, that the smaller the measurement band pass $B$ (frequency-domain resolution, $\Delta F = 2B$), the better would be the $S/N$ ratio. However, as the band pass is reduced, the opportunity for the signal to "wiggle" within a given time frame (time-domain resolution, $\Delta t = 1/2B$) is degraded, and the signal more nearly approaches a zero-entropy energy source. Thus, although the delta functions representing deterministic information are infinitely narrow in width, the signal base band must be finite if the signal is to be permitted to exhibit excursions. A necessary consequence of a finite signal base band is that noise power from the neighborhood near 0 Hz contributes to variance in the observed signal. This requirement for a finite signal base band is predicted by equation 4.

The conflicting requirement for simultaneous high resolution in the time domain and in the frequency domain are summarized by the classical uncertainty principle. In actual practice, the analyst must choose between an extended instrumental time constant (reduced bandwidth) and nose-in-signal. The consequences of his choice, in terms of the quality of the measurement, depend on the nature of the noise-in-signal.

In summary, the output filter places a limitation on the highest-frequency component in the signal and thus on the total amount of noise power contributing to variance in the observed signal. Our desire to reduce the noise power

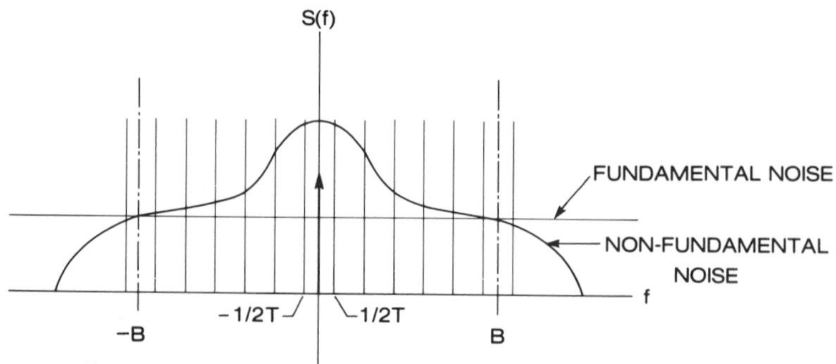


**Fig. 1.14.** Hypothetical power-density spectrum partitioned into $\eta = 2BT$ elementary resolution elements. Vertical arrow denotes delta function.

would dictate a low-filter band pass, whereas our desire to optimize the quality of the measurement in terms of the total information would make us want to comply with equation 4 and increase the band pass. This conflict is a ramification of the classical uncertainty principle, which requires that $\Delta F \Delta t = 1$ as a minimum if the measurement is to have any meaning. As we have seen earlier, nature extracts a price for the solution to this conflict, and that price is denominated in units of time.

## 2. Signal Truncation

The task of the chemist, in conducting his or her activities in the time domain, is to measure the contribution of the analyte to the strength of the dc signal. The actual physical estimation of the mean value of this signal is made by a comparison of the output-signal level to the signal level obtained with an analytical blank. This activity, equivalent to estimating the signal power at 0 Hz, is based on our knowledge of an important property of the true signal mean value: it does not fluctuate.

The final signal averaging is initiated after the system has adjusted to any changes caused by switching on the chemical input (either blank or analytical solution). As such, the final signal-averaging step, which is a form of filtering, is discontinuous and distinct from the hardware filtering of the signal upstream from the output display device. Note that there are two mean value estimations (blank and analytical signals) and that the errors in each of these estimations, at best, sum quadratically.

Because the analyst has a limited amount of either sample or patience, the strength of the signal must be estimated by viewing and averaging the signal over a finite record length. This time-domain signal truncation, depicted in Fig. 1.15, is the last refinement of the model of Fig. 1.4.

Two general conditions affect the time taken to achieve a desired accuracy in a measurement, namely, (1) whether the system has its own internal time constant governing integration time (e.g., microammeter or rate meter), or (2) whether the system is used with some form of accumulating storage in which no information is lost (integrator, pulse counting, etc.). For the purposes of our discussion, these two types of systems are referred to as *capacitive* and *digital*, respectively. If a capacitive system is used, information is lost because the output response is limited by the energy-storage capacity of the output device. Strictly speaking, information is lost if the user is not present. However, to some

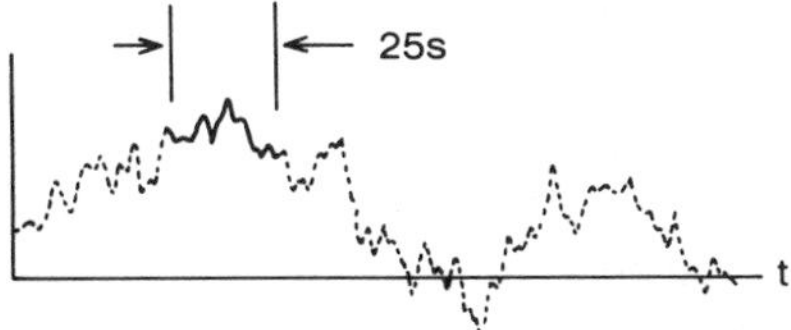


**Fig. 1.15.** Truncation of signal in the time domain.

extent, even a capacitive device can approach the performance of a digital device because it is not necessarily true that all past readout information is lost when the user is present; the alert user remembers the average location of the meter needle and averages the needle excursions over the total observation period. Indeed, in a capacitive system the results of the final signal-processing operation are more subjective, depending on how the meter excursions are interpreted by the user, whereas a digital system generally removes the element of subjectivity and performs a true integration as the final signal-processing step.

The strip-chart recorder is a readout device of special interest because it is commonly used in many analytical laboratories. Depending on the specific hardware implementation, a strip-chart recorder can be used as either a capacitive or a digital readout system, although it is not commonly thought of in such terms. If the instantaneous pen deflection is taken as the output reading, the device is used as a meter, and the measurement limitations are those of a capacitive system. If the strip-chart tracing is observed for an extended period and an estimation made of the average output reading, no information is lost, and the final-integration limitations approach those of a digital system.

When the signal is observed for a finite period, in effect, the signal is multiplied in the time domain by a gate function $g_T$, having a value of unity during the observation period $T$ and of zero at all other times. Corresponding to this multiplicative operation in the time domain is a convolution in the frequency domain of the signal spectrum with the Fourier transform of the gate function, represented as follows:

$$\begin{aligned} g_T \cdot y(t) &\leftrightarrow T\left[\operatorname{sinc}\left(\frac{\omega T}{2}\right)\right] * Y(\omega) \\ &= \hat{Y}(\omega) \end{aligned} \tag{23}$$

where the notation $\leftrightarrow$ indicates a Fourier transform pair. Sinc($\zeta$) is used to identify the function $(\sin \zeta)/\zeta$, commonly called the sampling function (11, p. 21). Depending on the contour of the function sinc($\zeta$), this convolution can considerably distort features in the spectra of the noise and informational signals. As the observation time is increased, the function sinc($\zeta$) approximates a delta function, and the actual noise power-density spectrum remains undistorted under the convolution integral. As the observation time is decreased, the function sinc($\zeta$) is broadened, and the spectral density at a given frequency is effectively distributed over an idealized equivalent frequency-resolution element of width $W_e = 1/T$ (3, p. 20).

We can represent graphically the relation between the variables controlling measurement quality by partitioning the spectral range from $-B$ to $B$ Hz into $\eta$ elementary resolution elements $1/T$ Hz wide (called the Nyquist cointerval), where $\eta = T/\Delta t$. (Since $\Delta t = 1/2B$, $\eta = 2BT$.) Figure 1.14 depicts a typical measurement with $\eta = 12$ (e.g., $B = 0.5$ Hz, corresponding to an output time constant of $1/\pi$ sec, $\Delta F = 1$ Hz, and $T = 12$ sec.

If we consider the elementary resolution element centered at the origin, the lowest frequency component that can be distinguished from the mean value of the signal appears at $\pi/T$ rps, corresponding to $1/2T$ Hz. This result, which is in keeping with the frequency-domain-sampling theorem (symmetrical with the time domain-sampling theorem; cf. Section III.B), is intuitive when we consider that frequency components having periods longer than the observation time cannot influence the strength of the signal and thus, cannot be distinguished from its mean value. This condition can be understood graphically in the time domain by time shifting the observation window in Fig. 1.15. Thus signal power present at frequencies less than $1/2T$ Hz is assigned to signal power at 0 Hz, and the frequency-domain interpretation of the sample mean-value estimate is given by

$$\hat{\mu}_y^2 = \int_{-\pi/T}^{\pi/T} \hat{P}_y(\omega)\, d\omega \tag{24}$$

Therefore, the final signal-truncation and signal-averaging step, in effect filters the output signal down to $1/2T$ Hz.

The questions of ultimate concern to the analytical chemist are how good is the mean-value measurement, and what are the effects of bandwidth and observation time on the quality of the measurement? In the following section, we explore how the properties of noise in the frequency range from the origin to $B$ Hz determine the effects of bandwidth and observation time on the stability of the measurement.

### 3. Fundamental-Noise Limitations

From equation 4, we determined that the signal bandwidth must be finite if the signal is to have other than zero-entropy energy. We also found in Section Iv.B.3 that a legitimate part of the signal generated by a quantum process is fundamental noise. Accordingly, the signal values collected during observation of a signal disturbed by fundamental noise can be regarded as a sample of a random variable. (In the context of this discussion, the term "sample" refers either to the truncated segment of signal of record length $T$ sec or to the set of statistically independent signal values collected over the observation time $T$; the collection of such samples is referred to as an ensemble of samples.) As such, any quantity computed from these sample values is also a random variable. For example, consider the mean value $\hat{\mu}_y$ and the variance $\hat{\sigma}_y^2$ for a sample consisting of $\eta$ statistically independent observations. If a series of different samples, all of size $\eta$, were selected from the same parent random variable, the values of $\hat{\mu}_y$ and $\hat{\sigma}_y^2$ computed from each sample generally would differ. Thus the sample mean $\hat{\mu}_y$ and the sample variance $\hat{\sigma}_y^2$ would also be random variables.

At this point, we are dealing with three random variables, which we designate $\{y(t)\}$, $\{\hat{\mu}_y(k)\}$, and $\{\hat{\sigma}_y^2(k)\}$, where $k$ represents the sample index. Each of these random variables can be characterized by a sampling distribution, a mean value, and a variance. From the point of view of the analytical chemist, who commonly must draw conclusions on the basis of one or two signal samples, our interest is in the statistical properties of estimates of the sample mean and the sample variance. Specifically, we seek answers to the following questions: what is the relation between the precision in the average value of the sample mean and the sample variance, and how is this analytical precision related to the signal bandwidth and the signal-observation time?

The following discussion is a frequency-domain interpretation of relations that rest on a rigorous mathematical foundation. However, the interpretation here, offered without proof, is based strictly on intuition and plausibility arguments. To facilitate our efforts, we review the following notation:

$\mu_y$ = the true mean value for $\{y(t)\}$

$\sigma_y^2$ = the true variance for $\{y(t)\}$

$\{\hat{\mu}_y(k)\}$ = the sample mean (estimator for $\mu_y$)

$\{\hat{\sigma}_y^2(k)\}$ = the sample variance (estimator for $\sigma_y^2$)

$<\hat{\mu}_y>$ = the average value of the set of sample means

$\text{Var}[\hat{\mu}_y]$ = the variance in the average value of the means

The true mean value $\mu_y$, a property of the parent random variable $\{y(t)\}$, is the quantity of analytical interest. The quantity $\hat{\mu}_y$, an estimator of $\mu_y$, is based on a limited number of observations. The variance in the average value of the means $\text{Var}[\hat{\mu}_y]$, a measure of the analytical precision, is here referred to as the analytical variance.

### a. Mean-Value Estimators

Our first concern in this section is with the relation between the true mean value and its estimators. We begin by focusing our attention on the lowest elementary resolution element shown centered at 0 Hz in Fig. 1.14. At the outset, we need to distinguish between deterministic and random power components that occupy the same spectral-resolution element containing the signal of interest. Because the deterministic interferences have infinite spectral density and reside at the same frequency as the signal of interest, a reduction of the instrumental band pass has very little value for signal resolution, and the interferences are recovered in the output along with the analyte signal. The influence of such spectral interferences are threefold: (1) the effective dynamic range of the output readout device is reduced; (2) interpretation of the optical spectrum is complicated, with resulting inconveniences in the qualitative identification and location of the spectral lines; and (3) the interfering signals act as a vehicle for the propagation of multiplicative noise through the system. Under ordinary conditions, such interfering signals are common to both the blank signal and the analytical signal and are accounted for in the final calculation.

For the purposes of this discussion, we expect to recover such interferences along with the signal of interest.

Because all deterministic-signal power within the lowest resolution element is considered a part of $\mu_y$, we need concern ourselves only with the noise power within this element. If the parent random variable $\{y(t)\}$ follows a normal distribution, signal deviations around the mean are equally likely to be positive or negative, and for a large number of samples, we can expect that the mean-value estimations will be randomly distributed about the true mean. Formally, the quantity $\hat{\mu}_y$ is said to be an unbiased (2, p. 124) estimate of $\mu_y$ such that

$$\langle\hat{\mu}_y\rangle = \hat{\mu}_y = \mu_y \quad \text{(for large } \eta) \tag{25}$$

It follows that the error in the estimation is determined solely by the variance across the ensemble of estimations.

### b. Analytical Resolution

There are two questions relating to the several measures of signal variance that are of interest to the analytical chemist. One concerns the variance in highly precise quantitative measurements in which the intrinsic $S/N$ ratio is high; the other question concerns the variance in measurements at the qualitative detection limit where the intrinsic $S/N$ ratio is low. Our interest is in the relationship between analytical precision and experimentally measurable quantities such as sample variance, signal bandwidth, and signal-observation time. We first consider the quantitative analytical variance.

#### (1) Quantitative Analysis

We can identify the noise contributions to the quantitative analytical variance (i.e., variance in the average value of the means) once we recognize that noise power within the lowest resolution band cannot be distinguished from the mean and that the filtering action of the final signal-averaging step removes all frequency components above $1/2T$ Hz. It follows, then, that the variations in the sample mean are due only to the noise power within the lowest resolution band.

It is useful to return to the notion of the frequency domain partitioned into elementary resolution bands, as depicted in Fig. 1.14. The noise power in the lowest resolution element can be evaluated by straightforward inspection of Fig. 1.14. If, as we have assumed, the noise-in-signal has a unilateral white power-density spectrum of strength $\alpha$, then the noise power is uniformly distributed between each of the elementary resolution bands. The noise power in the lowest resolution element is given by the integral of the power-spectral-density function over the range 0–$1/2T$ Hz. Thus

$$\text{Var}[\hat{\mu}] = \frac{\alpha}{2T} \tag{26}$$

We take this opportunity to define the quantitative analytical resolution $k_q$ as the mean value of the signal divided by the analytical standard deviation (square root of the analytical variance). The quantity $k^{2_q}$ is given by

$$k^{2_q} = \frac{\hat{\mu}_y^2}{\mathrm{Var}[\hat{\mu}_y]} \tag{27}$$

$$= 2T\,\frac{\hat{\mu}_q^2}{\alpha} \tag{28}$$

For quantitative analytical work with a relative precision of about 1%, the value of $k_q^2$ would have to be at least $10^4$. From equation 28, the quantitative analytical resolution depends on the partitioning of the spectrum as determined by the frequency -domain sampling theorem (measurement time $T$) and on an intrinsic property of the signal (the signal-to-noise spectral-density ratio $\hat{\mu}_y^2/\alpha$); $k_q^2$ is independent of the signal bandwidth.

If the quantitative analytical resolution is independent of the signal bandwidth, then why not operate at the lowest possible bandwidth, so that the signal at least appears to have less noise-in-signal? First, note that the finite-bandwidth requirement imposed by the classical uncertainity principle is satisfied so long as the bandwidth is at least equal to or greater than $1/2T$ Hz. If $B > 1/2T$ Hz, the filtering action controlling the highest-frequency component in the signal is the final signal-averaging process. As a matter of practical expedient, a bandwidth fixed at a value somewhat greater than $1/2T$ Hz is needed for the system to be able to respond quickly to the difference in signal strength between the analytical sample and the blank. Moreover, in a capacitive system, the final signal integration is commonly done by the operator viewing the signal excursions about the mean in real time. The judgment of the operator regarding when to begin and when to conclude the visual integration is influenced by observations of signal excursions. To reasonably estimate the mean value, the operator must observe more than a single informational symbol. Thus the bandwidth must be sufficiently high to permit a reasonable number of independent measurements (10–25) to be observed within the time span given to the measurement. In a digital system, once the final integration process has been initiated, the analytical variance depends only on the integration time and not on the bandwidth, provided the noise power-density spectrum is white. In some output amplifiers, a dynamic time constant is an option that allows the system to respond quickly to changes in analytical input while providing reasonable filtering of the noise once the signal has reached a steady-state level. We have more to say about the output filter bandwidth in Section V.A.3.d.

We now determine how the quantitative analytical variance $\mathrm{Var}[\hat{\mu}_y]$ relates to the directly measurable sample variance $\hat{\sigma}_y^2$. First, note that the noise in the lowest resolution element represents $(1/\eta)$th of the total band limited white noise carried by the signal. The remaining fraction of the signal noise $(\eta - 1)/\eta$ is contributed from the region $1/2T$ to $B$ Hz, as summarized below:

$$\mathrm{Var}[\hat{\mu}_y] = \frac{1}{\eta}\alpha B \quad \text{(noise power in the range 0–1/2}T\text{ Hz)} \tag{29}$$

and

$$\hat{\sigma}_y^2 = \frac{\eta - 1}{\eta}\alpha B \quad \text{(noise power in the range 1/2}T - B\text{ Hz)} \tag{30}$$

From equation 29 and 30, we obtain

$$\mathrm{Var}[\hat{\mu}_y] = \frac{\hat{\sigma}_y^2}{(\eta - 1)} \tag{31}$$

Thus if the sampling for the sample means is normally distributed, the variance in the average of the mean values across an ensemble of samples is simply the variance of the individual samples divided by the number of statistically independent measurements in each sample, less one. If conditions are such that $\hat{\sigma}_y^2$ is independent of $\eta$, an increase in $\eta$ results in improved analytical results. But as we will see, for fundamental noise, $\hat{\sigma}_y^2$ is directly proportional to the signal bandwidth and there is no net improvement in analytical results as the bandwidth is increased. The main value of equation 31 is that the relationship provides a means for estimating $\mathrm{Var}[\hat{\mu}_y]$ on the basic of a single data record.

We can use equation 30 to develop a direct measure of the white-noise spectral density $\alpha$, as follows:

$$\alpha = \frac{\eta - 1}{\eta}(\hat{\sigma}_y^2/\hat{B}) \tag{32}$$

where $\hat{B}$ is the specific band limit used for measurement of the sample variance $\hat{\sigma}_y^2$.

Equations 27 and 31 can also be used to cast the quantitative analytical resolution in terms of directly measurable quantities:

$$k_q^2 = (\eta - 1)\frac{\hat{\mu}_y^2}{\hat{\sigma}_y^2} \tag{33}$$

### (2) Qualitative Analysis: the Detection Limit

We now consider the noise limiting the detection of a signal under qualitative analytical conditions when the intrinsic $S/N$ spectral-density ratio is relatively low. For this discussion, we must identify the noise that obscures the quantitative detection of a signal. In practical detection-limit work, the analyst observes alternately the blank and the analytical signal, each for one sample period. Usually the instrumental time constant is adjusted such that the instrument responds quickly to the difference in signal levels between the blank and the analyte. In this mode, it is the task of the analyst to determine whether he or she can discern the analyte signal over and above the noise in the sample. Thus it is the ratio of the signal strength to the sample-standard deviation $\hat{\mu}_y/\hat{\sigma}_y$ that determines whether the signal can be detected.

At this point, we define the analytical resolution at the detection limit $k_d^2$, as follows:

$$k_d^2 = \frac{\hat{\mu}_y^2}{\hat{\sigma}_y^2} \tag{34}$$

Since $\sigma^2 = [(\eta - 1)/\eta]\alpha B$,

$$k_d^2 = \frac{\eta - 1}{\eta}\frac{1}{B}\frac{\hat{\mu}_y^2}{\alpha} \tag{35}$$

$$= \frac{1}{B}\frac{\hat{\mu}_y^2}{\alpha} \quad \text{(for large } \eta\text{)} \tag{36}$$

Observe that if $\eta$ is small, an erroneous value for $k_d^2$ is obtained. This is equivalent to making the signal a zero-entropy energy signal. Observe also that the quantity $k_d^2$ quickly becomes independent of $\eta$ (the number of degrees of freedom) as $\eta$ becomes reasonably larger than unity. The qualitative-analytical resolution $k_d^2$ depends only on the signal bandwidth and on the intrinsic $S/N$ spectral-density ratio. Contrary to conventional wisdom, the detection limit, when measured under the usual qualitative-analytical conditions, is independent of the signal-observation time and of the number of degrees of freedom in the measurement. This observation implies that for detection-limit work, the product $2BT$ should be fixed at a reasonable number, possibly 10 or 15, and that the bandwidth should be kept as low as possible within the constraint $\eta = 2BT$.

If a qualitative-analytical procedure is conducted on the basis of an ensemble of measurements for both the blank and the analyte, and if a truly digital system is used, the qualitative resolution may take a form closer to that of the quantitative resolution. If quantitative-measurement conditions are maintained, increased observation time will then favor the detection limit.

From the definition of $k_d^2$ and from equation 33, we can relate the quantitative resolution to the qualitative resolution by

$$k_q^2 = (\eta - 1)k_d^2 \tag{37}$$

The purposes of our statistical models are well served if the theoretical results provide guidelines for our laboratory efforts. Measurement conditions should be brought under close scrutiny in the selection of a relation for estimating the analytical precision or the qualitative-detection limit, and the sort of confidence usually reserved for calculations with deterministic functions should not be attached to precision or detection-limit values based on equations 31 and 34.

### c. Bandwidth Selection

What guideline can we establish regarding the selection of the signal bandwidth $B$? At first glance, it appears that equation 29 which indicates that the analytical variance is independent of $B$, conflicts with equation 4, which indicates that the total information in the measurement increases to infinity as the bandwidth is increased to infinity. If the $S/N$ ratio in equation 7 were independent of the signal bandwidth, the total information $Q$ would, indeed, increase in direct proportion with $B$. For white noise, however, the symbol noise power $N$ (nose disturbing the measurement of each statistically independent analog level) is the quantity $\alpha B$ because the measurement time frame $\Delta\tau$ for each symbol is limited by the sampling theorem ($\Delta t = 1/2B$); thus the frequency limit of the lowest resolution element coincides with the band limit $B$. In effect, there is only one resolution element in the spectrum between 0 and $B$ Hz when each symbol is measured in the shortest time possible. Thus the symbol $S/N$ ratio is degraded as the bandwidth is increased. We should not be surprised, therefore, to see the sample information approach a limit as $B \rightarrow \infty$. In exploring this limit, we first cast the symbol $S/N$ ratio (cf. equation 7) in terms of the symbol signal power $\hat{\mu}_y^2$. The total sample information is then given by

$$Q = BT \log_2\left(1 + \frac{\hat{\mu}_y^2}{\alpha B}\right) \tag{38}$$

We can now use equation 28 to rewrite the quantity $\hat{\mu}^{2y}/\alpha B$ in the above equation:

$$Q = BT \log_2\left(1 + \frac{k_q^2}{2B\hat{T}}\right) \quad \text{bits} \tag{39}$$

where $\hat{T}$ is the time used in the measurement of $k_q^2$. If we define the unitless variable B′ by

$$B' = \frac{B}{(\hat{\mu}_y^2/\alpha)} = \frac{2B\hat{T}}{k_q^2} \tag{40}$$

and if we normalize the informational capacity $Q/T$ by $(\hat{\mu}_y^2/\alpha)$, we obtain

$$\begin{aligned} C' &= \frac{Q}{\hat{T}(\hat{\mu}_y^2/\alpha)} \\ &= \frac{Q}{(T/\hat{T})(k_q^2/2)} \end{aligned} \tag{41}$$

then

$$C' = B' \log_2\left(1 + \frac{1}{B'}\right) \quad \text{bits/sec} \tag{42}$$

The limit of equation 42 as $B' \to \infty$, from the definition of $e$ ($\lim_{x\to 0}(1+x)^{1/x} = e$), is given by

$$\lim_{B'\to\infty} C' = \lim_{B'\to\infty} B'\log_2\left(1+\frac{1}{B'}\right) = \log_2 e = 1.44 \quad \text{bits/sec} \tag{43}$$

The normalized sample informational capacity $C'$ is plotted as a function of $B'$ in Fig. 1.16. For all practical purposes, $C'$ becomes relatively unresponsive to the signal bandwidth when the quantity $B'$ reaches a value of about 2. Thus we can select a value for $B'$ in accordance with our signal-transfer requirements, such that $B'_{\zeta\%} = \epsilon$, where $\zeta\%$ is the value of $C'$ as a percentage of its limit. From equation 40,

$$B'_{\zeta\%} = 2\,\frac{B_{\zeta\%}\hat{T}}{k_q^2} = \epsilon \tag{44}$$

Therefore,

$$B_{\zeta\%} = \epsilon\left\{\frac{k_q^2}{2\hat{T}}\right\} \tag{45}$$

$$= \epsilon\,\frac{\hat{\mu}_y^2}{\alpha} \tag{46}$$

Thus when signal conditions favor a quantitative measurement (e.g., $k^2 = 10^4$, $\mu^2/\alpha = k^2/2\hat{T} = 500$ for $\hat{T} = 10$ sec), the informational capacity can be extended to perhaps 81% of the theoretical limit $C_{\max} = 1.44(k^2/2\hat{T}) = 1.44(500) = 720$ bits/sec by fixing the bandwidth with $\epsilon = 2$, as follows. From equation 45,

$$B_{81\%} = 2\,\frac{k_q^2}{2\hat{T}} = 2(500)\cdot 10^{-3} = 1\ \text{kHz} \tag{47}$$

In the case described above, $C_{81\%} = 585$ bits/sec.

If there exists no particular requirement for high-speed signal transfer, we can freely select $\epsilon = 0.001$, for example. In this case, $B_{0.69\%} = 0.5$ Hz and $C = 5$ bits/sec. It should be emphasized that use of a bandwidth such that $\zeta = 1\%$ has no influence on the quantitative analytical resolution $k_q^2$, which is a function only of the measurement time; only the rate of information transfer is influenced by $B$. The matter of bandwidth selection on the basis of $B_{\zeta\%} = \epsilon(\mu^2/\alpha)$ is a one of analog information-transfer rate. The consequences of bandwidth choice are most important in high-speed signal transfers, interfacing of analog instrumentation, and automation.

If the signal is nearly obscured by noise, as is the case with analytical work near the detection limit (e.g., $k_d^2 = 3$ for $B = 0.5$), and if qualitative measurement procedures prevail, then from equations 36 and 46, $B_{\zeta\%}$ can be related to $k_d^2$ as follows:

$$B_{\zeta\%} = \epsilon(2\hat{B}\hat{T} - 1/2\hat{T})k_d^2 \tag{48}$$

$$= 2\hat{B}k_d^2 \quad \text{for } large\ 2\hat{B}\hat{T} \tag{49}$$

$$B_{81\%} = 2(0.5)3 = 3\ Hz$$

The information transfer rate corresponding to the qualitative measurement conditions used in the hypothetical case above is given by

$$C_{81\%} = \hat{B}k_d^2 \log\left(1 + \frac{\hat{B}k^2}{B}\right)$$

$$= 0.5 \log\left(1 + 0.5\,\frac{3}{3}\right)$$

$$= 0.88 \quad \text{bits/sec}$$

Once we recognize that for fundamental noise the rate of information transfer reaches a limit with increasing bandwidth, the only remaining alternative for increasing the total information in a measurement is the observation time. Optical spectrometers, which sequentially pass an isolated segment of the spectrum to the detector, waste much of the available spectral information because

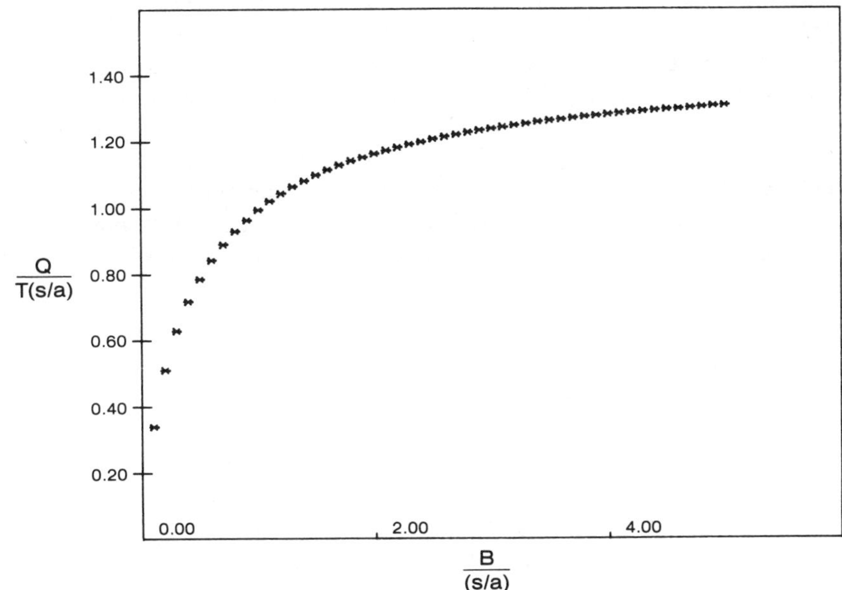


**Fig. 1.16.** Informational capacity (normalized) $C/(S/\alpha)$ as a function of normalized bandwidth $B/(S/\alpha)$ for fundamental noise.

only one particular resolution element is observed at a given time during a spectral scan. The well-known multiplex advantage makes use of the linear dependence of information on observation time by passing the entire available spectrum to a detector at a given time (8). The most notable success with spectral multiplexing, Fourier-transform infrared spectrometry, uses thermal detectors that do not generate quantum noise. When quantum detectors are used in spectral multiplex techniques, the multiplex advantage can be largely eroded by the increased fundamental noise generated from the entire optical signal falling on the detector.

### 4. Nonfundamental-Noise Limitations

In this section, we are interested in how the analytical resolution responds to signal-observation time and bandwidth when the measurement is disturbed by nonfundamental noise. Note that the relations in equations 29 and 30 are based on a uniform power-density spectrum and that the predictive value of these equations is not preserved if the noise power-density spectrum is not both white and Gaussian. We now have the somewhat more ambitious purpose of developing a measure of analytical stability that is generally applicable to problems in which the noise spectrum may assume any shape. Our first task is to explore the influence of nonfundamental noise on the analytical variance $\mathrm{Var}[\hat{\mu}]$.

Perhaps the most commonly used technique for dealing with $1/f$-type noise is to recalibrate the instrumental system at close intervals. This recalibration is generally done by normalizing the raw analytical values with the results from a single analytical reference or control sample measured in sequence with the unknown. This technique forces the integral of the $1/f$-type noise spectrum to converge at the origin. Although this procedure has the effect of whitening the noise spectrum, we are still left with the following question: how long should we observe the signal before recalibration? If the measurement time is increased, the instrument has a greater chance to drift out of calibration. If the observation time is reduced for the purposes of recalibration, fewer statistically independent measurements are collected. Also, if the noise spectrum is other than white, how many statistically independent observations are we collecting within a given measurement period?

#### a. Equivalent Number of Degrees of Freedom

Because the techniques devised to deal with $1/f$-type noise have the common feature of tending to whiten the noise spectrum, the natural tendency is to attempt to retain the form of equation 31 by defining a quantity equivalent to $\eta$. This quantity, which we denote $\eta_e$ in the following paragraphs, is here referred to as the equivalent number of degrees of freedom.

First, note that the spectral partitioning, which depends only on the observation time $T$, is independent of the shape of the power-density spectrum. We continue to use the quantity $\eta$ to partition the time domain in accordance with the sampling theorem [$T$ sec partitioned by $\eta$: $\Delta t = (T/\eta) = 1/2B$]. Secondly, observe that the sum of the variances $\hat{\sigma}_y^2 + \mathrm{Var}[\hat{\mu}_y]$ for noise represented by power-density spectra of any general shape is given by the integral of the power-density spectrum over the signal base band. This integral can be written in discrete form in terms of the partitioned power-density spectrum as follows:

$$\hat{\sigma}_y^2 + \mathrm{Var}[\hat{\mu}_y] = \sum_{i=1}^{\eta} \hat{P}_y(f_i) W_e \tag{50}$$

Our tactic is to model the behavior of the noise in terms of the equivalent behavior of white noise. If we normalize the quantity $\hat{P}_y(f_i)W_e$ in equation 50 by the amount of white noise power within one elementary frequency element (right side of equation 29, i.e., $\alpha B/\eta$), then the normalized quantity $\eta\,\hat{P}_y(f_i)/\alpha B$ approaches unity if the noise power-density spectrum approximates a white continuum of strength $\alpha$. With this simplification, the normalized sum of the variances $\{\hat{\sigma}_y^2 + \mathrm{Var}[\hat{\mu}_y]\}\eta/\alpha B$ can be closely modeled by the chi-square variable $\chi_\eta^2$, with $\eta$ degrees of freedom (2, p. 103).

If the chi-square variable represents the sum of the $\eta$ statistically independent quadratic components, each having a variance of unity, then the expected value of $\chi^2$ is given by

$$\mu_{\chi 2} = \eta \tag{51}$$

Also, from the definition of chi square, it is a simple exercise to show that the variance in $\chi^2$ is given by

$$\mathrm{Var}[\chi^2] = 2\eta$$

Thus the relative stability of the variable $\chi^2$ is

$$\begin{aligned} \frac{\mathrm{Var}[\chi^2]}{[\mu_{\chi 2}]^2} &= \frac{2\eta}{\eta} \\ &= \frac{2}{\eta} \end{aligned} \tag{53}$$

By rearrangement of equation 53, we can define the *equivalent number of degrees of freedom* (3, p. 24) as follows:

$$\eta_e = 2\,\frac{[\mu_{\chi^2}]^2}{\mathrm{Var}[\chi^2]} \tag{54a}$$

$$= 2\frac{\left\{\int_a^b \hat{P}(f)\,df\right\}^2}{\int_a^b \hat{P}^2(f)\,df} \tag{54b}$$

where $a$ and $b$, the limits of the frequency range of interest, are expressed in units of $W_e = 1/T$ Hz. The equivalent number of degrees of freedom also can be calculated from power-density spectra in discrete form:

$$\eta_e = 2\frac{\left\{\sum_{i=1}^{n}\hat{P}(f_i)W_e\right\}^2}{\sum_{i=1}^{n}\hat{P}^2(f_i)} \tag{54c}$$

where $n$ is the total number of resolution elements in the evaluation.

A few comments regarding the quantity $\eta_e$ are in order. If we eye equation 54 narrowly, it becomes evident that the equivalent number of degrees of freedom is denominated in units of the sample index. Much as the quantity $M$, defined in equation 5, resolves the informational symbol-ordinate into $M$ distinguishable levels, the quantity $\eta_e$ measures the resolution of the symbol-index coordinate.

We now have a useful tool for assessing the influence of bandwidth and time on measurement quality for signals disturbed by noise having power-density spectra of general shape.

### (1) Quantitative Analysis in the Presence of 1/*f*-Type Noise

Equations 31 and 33 can be used with $\eta_e$ to approximate the analytical resolution for a measurement in which the noise power density spectrum has been whitened. Thus

$$\mathrm{Var}[\mu_y] = \frac{\hat{\sigma}_y^2}{(\eta_e - 1)} \tag{55}$$

and

$$k_q^2 = (\eta_e - 1)\frac{\hat{\mu}_y^2}{\hat{\sigma}_y^2} \tag{56}$$

We now calculate the equivalent number of degrees of freedom in a measurement disturbed by noise having a prewhitened unilateral power-density spectrum of strength $\alpha$ from 0 to $B$ Hz and of strength zero for frequencies above $B$ Hz. (The physically realizable unilateral power-density spectrum equals twice the spectral density of the bilaterable power-density spectrum.) We measure $\eta_e$ in units of the elementary resolution bandwidth $W_e = 1/T$ Hz.

$$\eta_e = 2\frac{\left\{\left(\int_0^{BT}\alpha\, df\right)\right\}^2}{\int_0^{BT}\alpha^2\, df} \tag{57}$$

$$= 2BT \tag{58}$$

This is the expected result. It follows that if the frequency range 0–B Hz can accommodate $BT$ elementary resolution bands $1/T$ Hz wide, then there will be two equivalent degrees of freedom per elementary resolution band (for unilateral power-density spectra). Thus the resolution element accommodating the dc signal at the origin, covering the range 0–1/2$T$ Hz, has one degree of freedom, and the remaining region of the spectrum from 1/2$T$–$B$ Hz contains 2$BT$–1 degrees of freedom. (The same result is obtained with bilateral spectra except that only one degree of freedom is assigned to each of the 2$BT$ elementary resolution elements.) Only one degree of freedom is needed to specify the dc signal, whereas two degrees of freedom are needed for each elementary frequence—one for amplitude and one for phase.

Equation 55, for Gaussian white noise, can now be cast in terms of the bandwidth-time product:

$$\text{Var}[\hat{\mu}_y] = \frac{\hat{\sigma}_y^2}{(2BT - 1)} \text{ (white-noise limited)} \tag{59}$$

If the noise exhibits a power-density spectrum approximated by $P(f) = 1/f$, the situation is quite different. The equivalent number of degrees of freedom over the frequency range contributing to sample variance (1/2$T$ to $B$ Hz) can be obtained from equation 54b:

$$\eta_e = 2\frac{\left\{\int_{1/2}^{BT} (1/f)\,df\right\}^2}{\int_{1/2}^{BT} (1/f^2)\,df} \tag{60}$$

$$= 4(\ln 2BT)^2 \quad \text{for } \eta > 3 \tag{61}$$

and from equation 55,

$$\text{Var}[\hat{\mu}_y] \simeq \frac{\hat{\sigma}_y^2}{(\ln 2BT)^2} \quad \text{(prewhitened } 1/f\text{-type noise limited)} \tag{62}$$

Thus even with prewhitening, $1/f$-type noise presents us with a measurement situation in which, unlike white noise-limited measurements, a point of diminishing returns is quickly reached as the measurement time is increased. If the noise spectrum were strictly white, an increase in the instrumental time constant and observation time would result in an increase in the output ($S/N$) ratio and a corresponding improvement in the analytical precision. With prewhitened $1/f$-type noise, however, a decrease in bandwidth and a corresponding extension of the record length does not lead to an improvement in the ($S/N$) ratio or in statistical confidence because more low-frequency noise power is concentrated into the measurement band pass. Intuitively, we can see that the longer the observation time, the greater the opportunity the signal has to drift, an effect that tends to offset the gain in $S/N$ ratio or in statistical confidence.

### (2) Qualitative Analysis in the Presence of $1/f$-Type Noise

Equation 35, with $\eta$ replaced by $\eta_e$, shows the dependence of $k_d^2$ on the equivalent number of degrees of freedom. Interestingly, if $\eta_e$ is larger than unity, the increasingly high noise power-spectral density at low frequencies associated with $1/f$-type noise has relatively little influence on qualitative-analytical results, provided qualitative measurement conditions are maintained. For example, an operator viewing a meter gives the greatest weight to the signal activity immediately after switching the chemical input from the blank to the sample. Under such conditions, the signal is not viewed long enough to have any opportunity to drift, and signal detection is limited by the signal deviation around the sample mean.

#### b. Bandwidth Selection

Much like white noise, nonfundamental noise is a continuous function of frequency, a property that couples the symbol $S/N$ ratio to the bandwidth. Unlike white noise, however, the strength of the $1/f$-type-noise spectral density critically decreases with increasing frequency, and the observed $S/N$ ratio approaches a constant value as the bandwidth is extended. Thus for pure $1/f$-type noise, the rate of information transfer is more nearly optimal with the bandwidth set at the highest frequency consistent with the readout instrumentation. If, however, the nonfundamental noise contains intense frequency components in the range 250–450 Hz, as in Fig. 1.10 and 1.12, the bandwidth should be selected on the basis of a relationship similar in form to equation 45.

## B. SELECTIVE MODULATION

Selective signal recovery depends on imparting a certain uniqueness to the informational symbol of interest and using this uniqueness to recognize the special symbol in the presence of potential interferences. The uniqueness of the symbol has the effect of increasing the informational capacity per symbol, an increase that should be reflected in an improved $S/N$ ratio at the point where the signal is recovered. Selective modulation differs from signal coding by maintaining a one-to-one symbol representation throughout the information transfer, as opposed to coding, which uses a multiplicity of symbols for each original informational symbol.

Examples of selective signal recovery are found throughout the animal kingdom. We have no problem understanding a conversation in the presence of background music, when the voice and music signals are transmitted simultaneously over the same medium. Our experience allows us to "tune into" an audio signal having a particular frequency composition. Insects seeking a mate will home in on creatures of like persuasion by following a pheromone gradient in the presence of a complex background of chemical signals. Selective pattern recognition in the optical domain is so effective that many species invest considerable energy in achieving the opposite effect, that of signal deselection; this effect, known as camouflage, makes use of background noise to increase the error rate in signal detection. In such matters of signal processing, the animal brain seems to get a higher score than do digital computers or analog circuitry.

### 1. Selective Modulation in the Optical Domain

The task of a selective-modulation device is to provide a means for "spotting" a signal in the presence of signals of like structure. The simplest selective-modulation techniques used in analytical spectrometry rely on manipulation of the desired analytical informational signal into a region of the informational spectrum relatively free of unwanted information and noise power. In conventional ac spectrometry, dc to ac signal conversion is commonly carried out with a mechanical chopper positioned between the optical source and the detector. In AA spectrometers, the primary source is electronically modulated, or a mechanical chopper is used between the source and the analytical cell to allow the circuitry to distinguish the primary-source signal from the potentially interfering FE signals. Ideally, the signal representing the analyte, namely, the primary-source emission, is converted into an ac signal at a fixed frequency, whereas the interfering FE signal remains a dc signal. An even higher degree of

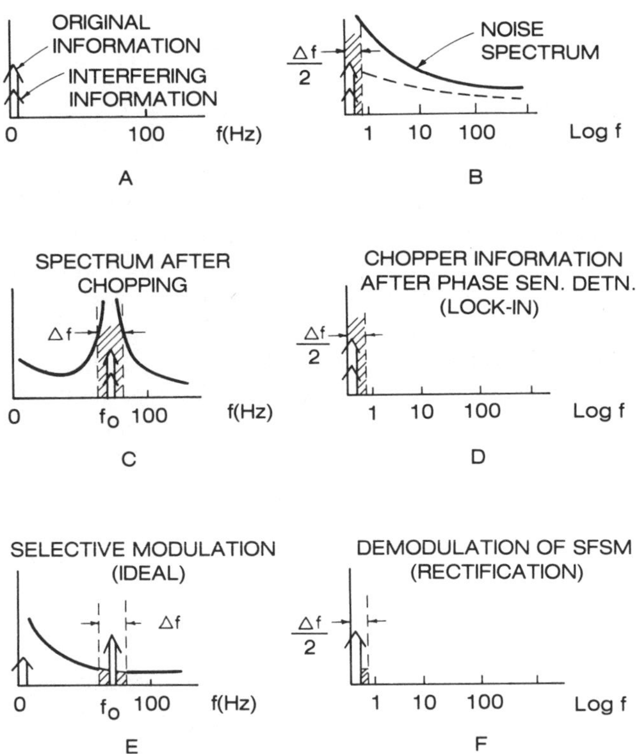


**Fig. 1.17.** Schematic representation of frequency-domain-signal structure with various signal-recovery techniques. (*a*) and (*b*) dc recovery; (*c*) and (*d*) mechanical optical chopping with ac amplification or phase-sensitive detection; (*e*) and (*f*) analyte-selective modulation.

spectral selectivity can be achieved by Zeeman modulation of the excited states of atoms in the primary source or in the flame (22). In any case, the product modulator generates frequency components that are the sum and difference of the modulation frequency $f_o$ with all frequency components in the original signals. This operation effectively shifts the Fourier spectrum up and down the frequency axis by an amount $f_o$, as depicted in Fig. 1.17*c*. Only additive signals entering the system downstream from the chopper are not frequency translated to the chopping frequency. Under ideal conditions, additive and multiplicative noise associated with the background originating downstream from the chopper, and additive noise and low-frequency drift originating in the detector and circuitry, are discriminated against in the output. When multiplicative noise is introduced into the signal by the analytical cell, the optical-chopper frequency shifts the multiplicative-noise spectrum to the recovery frequency along with all interference and noise signals originating upstream from the chopper. On demodulation, either with a simple envelope detector or with synchronous detection, the translated spectrum is retranslated to 0 Hz with no change in the frequency distribution around 0 Hz.

### 2. Selective Modulation in the Chemical Domain

The success of selective modulation depends on the purity of the modulation with respect to selectivity. If the dominating noise originates in the analytical cell, signal modulation, to be effective, is introduced in the chemical domain upstream of the noise source in such a way that the amplitude of the modulation is independent of the influence of gain variations in the analytical cell. Figures 1.17*e* and *f* illustrate this type of selective modulation.

Modulation techniques that induce selective alterations in the chemistry of the analyte independent of the sample matrix, or that cause periodic changes in the free-atom populations of the analyte species while maintaining constant the nominal steady-state population densities of all other species in the flame, are a challenge to engineer. Over the past 25 years, various modulation devices, all relying on mechanical or electromechanical interruption of solution flow through the burner, have been reported for selective flame-signal modulation (15). One of the simplest modulators uses periodic piezoelectric throttling of the solution flowing through the fluidic channel feeding the burner-nebulizer (15). The resulting ac optical signal reflects the chemical upset in the steady-state population densities of many flame species, including that of the analyte. The phase part of the signal generated by a particular flame species indicates the shift on modulation in dynamic balance of the population density and excitation of the particular flame specie relative to that of other species. With a suitably designed solvent system (15), the ac component of the background signal can be minimized. Alternatively, the phase of the ac signal relative to the piezoelectric driving signal can be used to distinguish background emission from atomic emission. Figure 1.18 illustrates the results of analyte modulation with phase-sensitive detection of the magnesium emission at 2852 A.

In the example (15) in Fig. 1.18, an ac component is generated on top of the regular dc flow of solution to a total-consumption nebulizer-burner for FE spectroscopy. On the positive half-cycle of the modulation, the increased solvent load on the flame cools the flame and results in a general decrease in background-emission intensity. At the same time, the increase in the analyte number-density due to the increased flow of analytical solution in the flame results in an increase in analyte-emission intensity. With this sort of response, which depends on the particular solvent system and the particular flame chemistry, the background emission is always out of phase with respect to the atomic emission. The simplest recovery system for such signals would use a phase-sensitive detector (PSD) controlled by a reference signal in phase with the atomic emission. Because the PSD generates a negative-going output for signal components 180° out of phase with the reference signal, a scan over a structured background-band system would appear inverted at the amplifier output; the spectrograms in Fig. 1.18*b* represent the 180° out-of-phase mode of signal

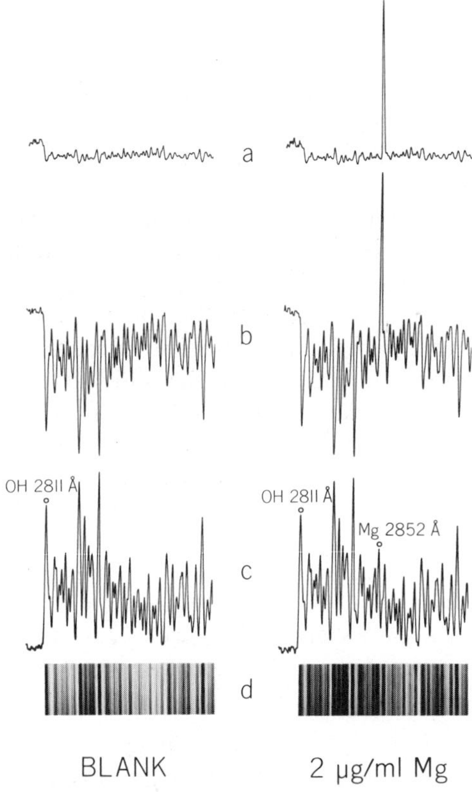


**Fig. 1.18.** Recovery of Mg 2852 Å flame emission in the spectral region from 2800 to 2900 Å. (*a*) Piezoelectric modulation (PM) of solution flow to flame with phase-sensitive detection (PSD) and out-of-phase background clipped at base of Mg peak; (*b*) PM-PSD with background 180° out of phase with Mg signal; (*c*) mechanical optical chopper with PSD; (*d*) photographic.

recovery for 2 μg Mg/ml in 95% ethanol; spectrograms obtained by the conventional chopper recovery method are shown for comparison in Fig. 1.18*c*.

The spectral scans in Fig. 1.18*a* illustrate how an even greater amount of background discrimination can be achieved by taking advantage of the knowledge that all the signal components that drive the output in a negative direction arise from background emission. If the output of the PSD is zeroed at the wavelength of interest by reference to a blank solution, all output more negative than the preset zero level can be confidently discarded. A simple zero-suppress and diode-clipping circuit on the amplifier output can be used for this purpose. Figure 1.18*a* shows the result of clipping the 180° out-of-phase background at 2855 Å. This clipped background technique simplifies spectral interpretation and increases the useful dynamic range of the output readout device. Photographically recorded spectrograms of the same region and at the same dispersion are included in Fig. 1.18*d* to emphasize that background emission from the flame is not actually suppressed in the optical signal. (The same spectral slit-width and flame conditions were used in the preparation of each part of Fig. 1.18.)

In addition to the possible benefits of selective modulation for FE spectroscopy, there are special applications for selective modulation in AA and atomic fluorescence spectrometry (15). In conventional ac AA spectroscopy, in which a line source is used, the total optical power transmitted by the cell is detected and the spectral band pass is determined by the width of the line emitted by the source. When a continuum source is used, the task of narrowing the band pass falls on the monochromator. In AA measurements with selective modulation in the chemical domain, the ac signal detected is a measure of the algebraic sum of changes in the optical power emitted and in the optical power absorbed per half-cycle of modulation. Because nonabsorbed radiation is undetected, the optical pass band observed in the spectrum of the primary source is determined by the width of the absorption line in the flame. The ability of modulation techniques to narrow effectively the optical band pass in AA measurements can be advantageous in the application of low-intensity sources or low-resolution monochromators with continuum-source atomic absorption. The format of a selective-modulation AA measurement resembles that of a conventional FE measurement; the base line is at the electronic zero level and remains flat even for a spectral scan over varying continuum intensity. Scale expansion is simplified, sensitivities are improved, and nonabsorbed radiation passed by the monochromator does not influence the shape of the working curves. The basic limitations on spectral discrimination by selective modulation are imposed by shot noise, photocathode fatigue, and anode saturation.

## C. CODING IN THE CHEMICAL DOMAIN

If we are to overcome the problems associated with noise introduced in the early data-domain conversions in an instrumental chemical measurement, we must reduce the rate of generation of information in the chemical domain to that which can be handled without error by the measuring device; that is, we must code the chemical signal in the chemical domain.

If the chemical information is to be represented digitally before it is applied to the measurement channel, we need an analog-to-digital converter that has a chemical signal as both its input and its output. Much like analog-to-digital conversion in electronics, chemical analog-to-digital conversion can be conceptualized in two steps. The first step, that of quantization, is a matter of threshold detection in the chemical domain. The second step, coding the detected threshold, must also be conducted in the chemical domain if the results are to be immune to noise hazards in the measurement channel.

Examples of threshold detection are not unknown in chemistry. Probably the most common example is the use of litmus or pH paper to quantize approximately the pH of a test solution. With litmus paper, the color is either blue or red, and once the threshold for a color change has been reached, the information, though limited, is highly immune to noise outside the chemical domain. In this example, a small range of acid concentration triggers a reaction on the surface of the paper, a reaction that is signaled by the production of a symbolic chromophore, either red or blue. The litmus paper also acts as a transducer by presenting the chromophore for observation by broad-band reflection spectroscopy; in short, the chemist looks at it. The room lights can flicker (much like primary-source power-supply instability in a reflection spectrometer), the lab bench can shake, and the test solution can slosh, but the chemist still sees the color as either red or blue. The single-threshold detection and one-bit coding are highly immune to noise.

End-point detection in volumetric analysis provides another example of single-threshold detection. In a titration, feedback is used in the chemical domain to adjust the chemical signal to the threshold point. In an automated titration, or in any automated solution-chemical operation in which the chemical signal is adjusted as part of a servomechanism to a decision point, the task of the electronic detection system is to flag the appearance of the one-bit symbol signaling the point of decision. It should be no surprise that the most effective marriage of electronics and chemistry as regards the quality of quantitative results is in the automation of solution-chemistry procedures.

In general, the engineering of a high-resolution multiple-threshold chemical quantization system is exceedingly difficult. The species of interest would have to catalyze one of many reactions to completion, each at a critical concentration. Each reaction would have to generate, as one of its products, a distinguishable species that would flag the particular concentration threshold. Finally, it is difficult to imagine any but the most inefficient codes for transmission of the concentrational information; each threshold would be signaled by a unique symbol, so that for 0.1% resolution, 1000 reactions and 1000 symbols would have to be used. Clearly, the chemist comes by his measurement problems legitimately.

## VI. SUMMARY

Quantitative demands on the analytical chemist are continually increasing in direct relation to the growing sophistication of all areas of the physical sciences and engineering. Concomitant with the demands for high accuracy in chemical analysis is the pressure brought about by the growing volume of samples for analysis and by the decreasing amount of time allotted for a given quantitative chemical measurement. As a recourse, the chemist has turned to the world of electronics. And as a natural hazard of unfamiliar territory, the chemist has found himself without a complete perspective on the limits of his tools. At times he feels intimidated by the items in his toolbox; and at times he has warmed himself before the electronically produced illusion of effortless chemical measurements.

One of the contentions of this chapter is that the business of analytical chemistry can be understood in terms of the same basic information-processing model applicable to all other engineering measurements, whether the problems originate in an industrial process-control environment, on a machinist's bench, or in outer space. With such a model, we have a basis for exploring the relations between the actual signals measured and the microscopic phenomena responsible for the generation of the signals, and have a tool for speculating on new signal-generation and recovery techniques.

A source of difficulty for the analytical chemist is that conceptually the basic analytical measurement is singularly easy to grasp; the simplicity of the question "How much of substance $x$ ?" is so disarmingly easy to recite that it distracts from any perspective on the true difficulty of the measurement. The challenge presented by analytical chemical measurements is unique. And although quantitative analytical measurements are fundamentally the same as any other measurement problem, in practice, chemical measurements are especially difficult to engineer because of a lack of rapid detailed temporal control over the chemical-domain signal.

## REFERENCES


1. Beauchamp, K. G., *Signal Processing*, Wiley, New York, 1973.
2. Bendat, J. S., and A. G. Piersol, *Random Data: Analysis and Measurement Procedures*, Wiley, New York, 1971.
3. Blackman, R. B., and J. W. Tukey, *The Measurement of Power Spectra*, Dover, New York, 1959.
4. Chu, Y., *Digital Computer Design Fundamentals*, McGraw-Hill, New York, 1962, Chaps. 1 and 2.
5. Daniels, H. E., *J. R. Stat. Soc.*, *B*, 27, 234 (1965).
6. Eckschlager, and V. Stepanek, *Information Theory as Applied to Chemical Analysis*, Wiley-Interscience, New York, 1979.
7. Feller, W., *An Introduction to Probability Theory and its Applications*, Vol. I, 3rd ed., Wiley, New York, 1968.
8. Fellgett, P., *J. Phys. Radium* 19, 187 (1958).
9. Hareland, W. A., "Analysis and Characterization of Noise in Flame Spectrometry," Thesis, University of Minnesota Library, Minneapolis, 1976.

10. Kaplan, W., *Operational Methods for Linear Systems*, Addison-Wesley, Palo Alto, Calif., 1962.
11. Lathi, B. P., *An Introduction to Random Signals and Communication Theory*, International Textbook, Scranton, Pa., 1968.
12. Liteany, C., and I. Rica, *Statistical Theory and Methodology of Trace Analysis*, Ellis Horwood, Chichester, U.K., 1980.
13. Mertz, L., *Transformations in Optics*, Wiley, New York, 1965.
14. Mossotti, V. G., and F. N. Abercrombie, *Preprints Proc. XVI Colloq. Spectrosc. Int., Heidelberg, 1971*, Paper No. 166, p. 441.
15. Mossotti, V. G., F. N. Abercrombie, and J. A. Eakin, *Appl. Spectrosc.*, 25, 331 (1971).
16. Mossotti, V. G., and J. Maurus, "Flame Spectrophotometric Analysis Based on Noise," *Nat. Soc. Appl. Spec. Meet., Dallas, Sept. 1972*.
17. Mossotti, V. G., and S. Prager, 1973, "Time-dependent Linear Model for Multiplicative Noise in Analytical Cells," *XVII Colloq. Spectrosc. Int., Florence, Italy, Sept. 16–22, 1973*.
18. Schmid, H., *Analog/Digital Conversions*, Van Nostrand Reinhold, New York, 1970.
19. Shannon, C. E., "A Mathematical Theory of Communication," *Bell Syst. Tech. J.*, 27, 204–237 (1948) (also available in *Scientific American*, July 1949 and in paperback through University of Illinois Press, Urbana, Ill., 1972).
20. Stearns, S. D., *Digital Signal Processing*, Hayden, New York, 1975.
21. *Treatise on Analytical Chemistry*, Vol. 7 (Sect. H), Part I, 2nd ed., P. J. Elving, E. J. Meehan, and I. M. Kolthoff, Eds., Wiley, New York, 1981, Chaps. 7 and 8.
22. Yu, F. T. S., *Optics and Information Theory*, Wiley-Interscience, New York, 1976.
23. Zadeh, L. A., "Frequency Analysis of Variable Networks," *Proc. IRE*, 38, 291 (1950).

Part I
Section E

Chapter 2

# ANALOG ELECTRONICS

BY EDWARD H. PIEPMEIER, *Oregon State University, Corvallis, Oregon*

## Contents

## I. INTRODUCTION

Analytical instruments usually display their data as digital numbers or as scale positions on a chart or meter. Several forms of permanent recording of the data are also available. Prior to the display and recording of data, the physical or chemical quantities to be measured are usually converted into electrical signals because such signals are easily processed and modified. In some instruments the signal may be processed and fed back to control some part of the experiment. Eventually the electrical signals are processed so that the final display of data is in a form that may be conveniently interpreted by the observer. The general design techniques to accomplish these tasks are covered in other chapters. This chapter presents the *hardware design techniques* that are used to implement the types of signal processing discussed in other chapters.

The versatility available in processing electrical signals is apparent in the frequency with which scientists and engineers have built their own electronic circuits and instruments when commercial instruments have not been available. The reliable use of electronic circuits and, in some cases, even the proper interconnection of separate electronic modules or instruments require an understanding of electronic and design principles. Although some of the basic principles may have been studied in physics courses, they are briefly reviewed with emphasis on those that are used in the circuits described.

## II. BASIC ELECTRICAL VARIABLES

An electrical signal may physically exist as the motion of **electrical charges** (**current**), as the separation of unlike charges (which causes a potential difference), or as **electromagnetic** waves. Electrical charges occur in matter, which is made up of atoms containing negatively charged electrons and equally but oppositely charged positive protons. The removal of an electron from an atom leaves a positively charged **ion**. The attachment of an electron to a neutral atom produces a negatively charged ion. In the crystal lattice structure of semiconductor materials, the presence of an impurity atom that has fewer bonding electrons than its neighbors produces a hole that has the characteristics of a positively charged particle.

The charge on singly charged particles is $1.602 \times 10^{-19}$ **coulombs** (C). The charge on one chemical equivalent (1 mole of singly charged particles) is the **Faraday**, equal to 96,487 C.

Like charges forcibly repel each other and oppositely charged particles attract each other. Therefore energy is required to separate unlike charges or to bring like charges close to each other. For discussion purposes, infinity is usually chosen as a convenient reference point in space where the potential energy of a charged particle and the force on it are negligible. Then the **electrical potential** that exists at a point near a group of charges is equal to the amount of potential energy gained (or lost) by a particle of unit charge that is moved at rest from infinity to rest at the point near the charges. The unit of potential is the **volt** (V).

The terms **potential difference** and **voltage difference** (e.g., between two points) are sometimes used to emphasize that the potential of a point (in a circuit) is a *relative* quantity and is always measured with respect to some other (reference) point (infinity in the above case). It is common to omit the word "difference" and speak simply of the **potential of a point**. In this case, some reference point is always implied. In a circuit the reference point is called the circuit **common**, or **ground** when the common is connected to earth ground.

When electrical charges pass from a point of high potential energy to a point of lower potential energy in a circuit, potential energy decreases and is converted to kinetic energy. The amount of energy $U$ that is converted when an amount of charge $Q$ moves from one point to another is proportional to the voltage $V$ or potential (difference) between the two points and is independent of the path taken by the charge.

$$U \text{ (joules, J)} = Q \text{ (coulombs, C)} \times V \text{ (volts, V)} \tag{1}$$

The **power** transferred to the charge carriers depends upon how rapidly energy is converted and is found by dividing both sides of equation 1 by time to obtain

$$P \text{ (watts, W)} = I \text{(amperes, A)} \times V \text{(volts, V)} \tag{2}$$

where the current $I$ is the rate of flow of charge per unit time between the two points. The kinetic energy is converted by collisions to heat. Equation 2 is important in practical work because it can be used to calculate the rate at which heat must be dissipated by a device if thermal destruction is to be avoided.

Electrical voltage and current signals are processed by circuits made up of components whose electrical behavior may be described by relatively simple mathematical relationships among the *currents* passing through them and the *voltage* across their terminals. A **resistor** resists the flow of current. For a **linear resistor**, a two-terminal device, the relationship is described by **Ohm's law**:

$$V = IR \text{ (volts, V = amperes, A} \times \text{ohms, } \Omega) \tag{3}$$

where $V$ is the voltage between the two terminals that is required to force the current $I$ through the resistor, and $R$ is a proportionality constant called the **resistance**. **Dynamic resistance** $R_d$ is the inverse slope of the current versus voltage ($i$ vs. $v$) characteristic curve of the resistor

$$R_d = \frac{dv}{di} \text{ (ohms = volts/ampere)} \tag{4}$$

For a linear resistor $R = R_d$. For a **nonlinear** resistor, such as a diode, the *dynamic resistance* may be used in circuit equations, such as those discussed below, to calculate *small changes* in current and voltage that occur in circuits containing nonlinear resistors.

A **capacitor** has the capacity to store electrical charge on its plates. The amount of charge that accumulates on the plates of a capacitor is the time integral of the current $i$ which flows through it:

$$q = \int i(t)\, dt \text{ (coulombs = amperes} \times \text{seconds)} \tag{5}$$

As charge accumulates, it produces a potential difference between the plates. The amount of charge necessary to produce a given voltage is a measure of the **capacitance** $C$ of the capacitor or

$$q = Cv \text{ (coulombs = farads} \times \text{volts)} \tag{6}$$

Equations 5 and 6 may be combined to show that the voltage is proportional to the time integral of the current and that the instantaneous current is proportional to the time rate of change of the voltage or

$$v = \frac{1}{C}\int i\, dt \quad \text{and} \quad i = C\frac{dv}{dt} \tag{7}$$

These relationships make the capacitor very useful in signal integration and differentiation, and in timing circuits, for example.

**Inductance** is a magnetic influence that resists a *change* in current. A linear **inductor** induces a voltage across its terminals that is directly proportional to the rate of change of current through it

$$v = -L \frac{di}{dt} \text{ (volts = henries} \times \text{amperes/second)} \tag{8}$$

where $L$ is the inductance in **henries**, and the minus sign indicates that the induced voltage is in a direction that opposes the change in current. Inductors are used in filters to help reduce current fluctuations and are used with capacitors at high frequencies to tune circuits to a desired frequency.

Ideally, components should exhibit resistance, capacitance, or inductance in order to simplify the design and behavior of electrical circuits. Most real components behave ideally at moderate voltages, currents, and frequencies. However, for applications requiring very low or very high values of voltage, current, or frequency, the fact that any real component exhibits more than one of these ideal electrical characteristics may have to be considered. In such cases it is usually possible to express the electrical behavior of the real component in terms of a simple circuit made up of two or more ideal components. Such a circuit is called an **equivalent circuit**.

Even if a circuit is built with ideal components, **stray** resistance, capacitance, and/or inductance may be present due to the wires and physical support materials. The extent of these and other imperfections are important when considering the characteristics of real components.

## III. CHARACTERISTICS OF REAL COMPONENTS

### A. RESISTORS

**Resistors** are made from various materials including pressed carbon granules, metal films, and wire. The resistance of a material increases with length and decreases with increasing cross-sectional area. **Carbon** resistors are available in a wide range of values from 1 to 22 MΩ (±10, ±5, ±2, ±1%) and from $\frac{1}{8}$ to 2 W. The capacitance between the two ends of the resistor is negligible compared to stray capacitance. The inductance of an ordinary carbon resistor is also negligible, little more than that of an equal length of conducting wire. Carbon resistors are electrically noiser than other types of resistors and they have a relatively high temperature coefficient. Sufficient heat from a soldering iron may permanently change the resistance of some brands.

**Metal film** resistors are available with values from 10 Ω to 2000 MΩ (±1, ±0.5, ±0.25, ±0.1%) with temperature coefficients of ±2 ppm/°C (0.0002%/°C) to +100 ppm/°C and with voltage coefficients of less than a fraction of 1 ppm. Power ratings of metal film resistors are similar to carbon resistors, although some very high precision units may have power ratings from only $\frac{1}{20}$ to 1 w.

Noise levels may be less than 0.1 $\mu$V/V per decade of frequency range. Metal films that are deposited as a helix or an insulating substrate have an inductance that may become significant at frequencies above 1 MHz. Nonhelix units are available for higher-frequency applications.

**Wire-wound** resistors are used for low resistance, high precision, or high power dissipation. They are available in values from $\frac{1}{10}\Omega$ to 100 k$\Omega$ ($\pm$0.1 to $\pm$10%) with power ratings up to and exceeding 200 W. Other characteristics are similar to those of metal film resistors although the inductance may be higher. "Noninductive" units are wound to minimize the net self-inductance of the windings. The temperature rise at high power levels must of course be considered since a 100°C temperature rise may cause a 0.1–1% change in resistance.

When a resistor of a few ohms or less is connected into a circuit, the resistance of the connections may significantly increase the total resistance of the circuit. For low resistance values or high currents (hundreds of milliamperes), the resistance of the hookup wire may also have to be considered. Number 22 American Wire Gauge (AWG) copper wire has a resistance of 5 m$\Omega$/cm; No. 12 AWG, 0.5 m$\Omega$/cm. Therefore, a current of 100 mA is accompanied by a 5 mV drop across 10 cm of No. 22 gauge wire.

For resistors with values above several megohms, leakage of current on the surface of the resistor may decrease its effective value, or leakage through or on the surfaces of the insulator mounting materials for the terminals may become significant. Surface currents are influenced by ambient humidity and can be minimized by the proper choice of insulating materials.

The **skin effect** is the increase in the effective resistance of a conductor with increasing frequency. At higher frequencies the current density is not uniform throughout the cross section of a conductor, but diminishes along the axis compared to regions near the surface. The skin effect increases with conductivity, frequency, and the cross-sectional area of the conductor. For No. 28 gauge copper wire the skin effect increases the effective resistance by 1% at 100 kHz. Wires made from several strands of smaller gauge wire show less skin effect than a solid wire of the same cross section, and are therefore desirable for high-frequency applications.

Ordinary resistors can withstand several hundred volts, provided that they can dissipate the power. Higher-voltage units are available with ratings to several tens of kilovolts.

## B. CAPACITORS

**Capacitors** are made by placing two conducting sheets called **plates** close to each other and separating them by a nonconducting medium, or insulator. When current flows, a change in the net charge on one plate induces an equal excess of opposite charge on the other plate. Because the charges attract each other but are separated by a nonconductor, it is theoretically possible for a charged condition to exist indefinitely when there is no external conductor connecting the two plates.

When the plates are parallel planes and are close enough together to neglect edge effects, then the capacitance is directly proportional to the area of the plates and the dielectric constant of the insulator (close to unity for air) and inversely proportional to the distance separating them. The proportionality constant is called the **permittivity** of free space. Variable plate capacitors which use air to separate the metal plates may be reproducibly adjusted by rotating a shaft which changes the area of overlap of the plates. Usually a decade of capacitance range is covered with maximum values from a few to a few hundred picofarads. Voltage breakdown ratings range from a few hundred volts to more than 2000 V.

The capacitance of a capacitor is increased when an insulator with polarizable molecules (a dielectric) is placed between the conductors. This is commonly done to minimize the physical size of a capacitor. **Mica** (dielectric constant 5–7) is often substituted for air in variable plate capacitors, somewhat reducing the reproducibility and linearity of adjusting the capacitance. **Trimmer** capacitors consist of metal sheets usually separated by mica. The spacing between the plates is adjusted by a screw and the adjustment may not be reproducible or linear. Adjustments usually cover less than a decade of range and maximum values range from a few picofarads to about a nanofarad (1000 pF). Voltage breakdown ratings may be less than 50 V to higher than a few hundred volts.

Other dielectric materials include paper, ceramics, plastics, oil, and metal oxides. When stability is important, mica or NPO ceramic capacitors are preferred for low capacitances, whereas polystyrene or polycarbonate capacitors are preferred for high capacitances. The variety of capacitors that are available results mainly from several size and performance considerations. Large capacitance values and high voltage ratings generally require large sizes. Capacitance tolerance values, temperature ranges, aging and lifetime considerations, and cost may also be important. For capacitance values above 1$\mu$F, **electrolytic capacitors** are usually used because of their smaller size. These capacitors use a very thin metal oxide on aluminum or tantalum foil plate as the dielectric, and immerse the plate in an electrolyte solution to make close electrical contact with the oxide. An electrolytic capacitor must usually be used so that its polarity is always in one direction. A reversed polarity reduces the metal oxide destroying the dielectric and a gas may be evolved which can cause the capacitor to explode unless safety vents are provided.

Electrolytic capacitors are available with capacitances up to tens of thousands of microfarads (−20 to +150%, ±20%, ±10%) at voltage ratings from 3 to 600 V, depending upon the capacitance value. The capacitance may change significantly with time and temperature (several percent per degree), and also may vary with frequency. Leakage currents are usually quite high because of the relatively low resistance of the metal oxide, and may be as high as several microamperes. The series resistance of the thin foil plates may be as high as 1 Ω. Inductance of the plates becomes significant at frequencies above 10 kHz. The effect of inductance can be minimized in some applications by connecting several capacitors in parallel, with each capacitor having one or two decades of capacitance less than the preceding one. The smallest capacitor, having the least inductance, would handle the fastest transients, and the larger ones would handle the slower signals.

**Polystyrene** capacitors, available from 5 pF to 0.5 $\mu$F, are especially stable and have relatively low temperature coefficients of a few hundred parts per million per degree centigrade. Oil-filled capacitors are usually used for high voltage applications.

Dielectric materials cause power losses and heating at high frequencies. The **power factor** of a capacitor is an efficiency factor that indicates the extent of power losses due to this cause as well as leakage and resistive losses in the plates. Typical power factors may be a few percent to a fraction of 1% at audio frequencies.

A dielectric material between the plates of a capacitor may continue to produce a small voltage across the capacitor just after it has been discharged. This effect is called **dielectric absorption** and may produce unwanted voltages, especially in critical applications such as signal integration. Mixed and inhomogeneous dielectrics such as waxed paper are particularly troublesome. Teflon (trademark) and polystyrene are dielectrics that are relatively free from this effect.

Capacitance between different components and conductors in a circuit, including shielding containers, is called **stray** capacitance. It may be particularly troublesome at frequencies above 1 MHz, and when resistors above 1 M$\Omega$ are used.

The capacitance $C$ between two parallel wires is given by

$$C = \frac{0.121\ L}{\log_{10}(D/r)} \tag{9}$$

which indicates that the capacitance in picofarads is proportional to their length $L$ in centimeters, and varies rather slowly with the ratio of the distance $D$ between the wires to their radii $r$. The magnitude of the stray capacitance between two wires is indicated by the 1 pF capacitance between two No. 22 gauge wires (diameter 0.071 cm) 10 cm long and 0.5 cm apart. The capacitance between the two conductors of a coaxial cable is proportional to the length of the cable and inversely proportional to the logarithm of the ratio of the radius of the outer conductor to the radius of the inner conductor. The capacitance of typical 50 $\Omega$ coaxial cable is about 1.2 pF/cm.

Feed-through capacitors are coaxially shaped devices that may be used where a wire passes through a metal shield. The outer terminal of the capacitor is bolted to the shield, and the inner terminal passes through the body. The feed-through capacitor provides capacitance between the wire and the metal shield through which it passes. The capacitance tends to short high-frequency signals to the shield, while allowing low-frequency or dc signals to pass through.

## C. INDUCTORS

**Inductance** is a magnetic influence that resists a *change* in the current through a device. The effect is caused by the induced voltage or electromotive force, emf, that occurs whenever a conductor is in a *changing* magnetic field. A *constant* current surrounds itself with a constant magnetic field and there is no induced emf. However, a *change* in current causes a change in the magnetic field which induces an emf in any conductor within the field, including the conductor carrying the changing current. In the latter conductor the emf is in a direction that opposes the change in the current and this influence is called **self-inductance**.

A practical inductor may be a coil of wire consisting of several, or often many, turns of wire. The inductance of a coil is approximately proportional to the square of the number of turns of wire, the cross-sectional area of the coil, and approximately inversely proportional to the length of the coil. A 10-turn coil of No. 22 wire 1 cm long wound on a 1 cm diameter air core has an inductance of approximately 2 $\mu$H.

Self-inductance of a single wire may become significant at frequencies above 1 MHz. Self-inductance is proportional to the length of the wire. A 10 cm length of wire has an inductance of approximately 20 nH. The inductance of a coil may be increased by winding it on a core of magnetic material such as iron or ferrite. The inductance can be varied by moving the core in and out of the coil. The changing magnetic field induces undesirable currents, called eddy currents, in an iron core because it is a conductor. The iron core is usually laminated to minimize these currents. Eddy currents cause power losses that increase with signal frequency. For iron the power losses become intolerable above kilohertz frequencies. Eddy currents in ferrite are much less than in iron, and ferrite is used for higher frequencies.

At high currents the resistive and magnetic power losses may cause thermal destruction of the coil wire insulation. Moreover, core material saturates at high magnetic fluxes, so that further current increases are relatively ineffective in producing the desired result. Therefore an inductor usually has a current rating above which it should not be operated. The capacitance between coil windings and the resistance of coil winding wire are other imperfections that must also be considered. The **quality factor** $Q$ of an inductor is a measure of the coil resistance at a given frequency and is approximately equal to

$$Q = \frac{L2\pi f}{R} \left( \text{unitless number} = \frac{\text{henries} \times \text{hertz}}{\text{ohms}} \right) \tag{10}$$

Its exact value depends upon the interwinding capacitance and stray capacitance between the coils and other parts of the circuit including the ground wires. Because of these imperfections and the relatively large size and weight associate with inductors used below kilohertz frequencies, the use of inductors in this frequency range is usually avoided in signal-processing circuits in favor of other circuits having equivalent properties. Inductors are still used in power circuits such as high-voltage or high-current power supplies, and in high-frequency circuits where their size is relatively small.

Inductors for radio-frequency operation with dimensions less than 3 cm are available with values from a fraction of a microhenry with current ratings up to several amperes, to several hundred millihenries with current ratings of a fraction of an ampere. Typical quality factors vary from 20 to 200. Larger inductors for operation at lower frequencies in power supplies, often call filter chokes, and weighing more than a pound are available with values up to tens of henries and with current ratings up to tens of amperes.

**Mutual inductance** exists between any two conductors, and is often the way in which power-line frequency noise is picked up by a circuit. The changing magnetic field around one conductor that carries a changing current cuts across another conductor, inducing an emf in that conductor. Mutual inductance can be minimized by placing two circuits a long distance away from each other. The mutual inductance between two adjacent circuits is related to and is always less than the square root of the product of the self-inductances of the two circuits. The mutual inductance can therefore be minimized by designing the circuits themselves to have minimum self-inductance. This may be done by keeping small the area enclosed by a circuit, and by using short wires. The use of twisted-wire leads produces a circuit having successive loops where the induced emfs are in opposite directions and therefore tend to cancel each other in the same wire.

## D. TRANSFORMERS

**Transformers** consist of two circuits wound in such a way as to maximize their mutual inductance. Signals and power can be transferred from the **primary** circuit to the **secondary** circuit via the changing magnetic fields shared by the circuits. Transformers are used to multiply or divide periodically varying (alternating) voltages. They are commonly used in power supplies to step up or down the ac power line voltage and to eliminate a direct current path between the primary and secondary circuits. They are used to amplify minute periodic voltage signals prior to the input stage of an amplifier so that the input voltage is above the noise level of the input stage of the amplifier.

The secondary voltage $v_2$ induced in the secondary winding and the primary voltage $v_1$ are in direct proportion to the ratio of the number of turns $N_2$ and $N_1$ in the respective windings:

$$\frac{v_2}{v_1} = \frac{N_2}{N_1} \tag{11}$$

The *power* delivered to a load connected across the secondary of a transformer is, in the ideal case, equal to the power supplied to the primary winding of the transformer. It is usually less because of power losses in the transformer. Therefore, the current-voltage product of the secondary is equal to or less than the current-voltage product of the primary. A transformer that produces a higher voltage at the secondary than provided at the primary also supplies proportionally less current to the secondary than that supplied by the primary. A transformer is sometimes used for **impedance matching** because it can change the apparent resistance or ac impedance of a device connected across one of its windings. For instance, a resistor $R_2$ connected across the terminals of the secondary winding appears as a resistance $R_1$ when viewed across the terminals of the primary, where

$$\frac{R_1}{R_2} = \left(\frac{N_1}{N_2}\right)^2 \tag{12}$$

To obtain maximum coupling efficiency, the two windings of a transformer are wound together on an iron or ferrite core. This common arrangement has the disadvantage of capacitive coupling between the two circuits. Therefore, a change in the average dc level of the primary circuit may be fed through by capacitive coupling to the secondary circuit or vice versa. When maximum isolation from this type of capacitive feed through is desired, for instance, in medical applications, the two circuits may be wound on opposite sides of a toroidal (or similar) shaped iron core.

**Autotransformers** use different sections of the same coil winding for both the primary and secondary. For instance, the primary voltage may be applied to the two ends of 100 windings of the coil whereas the secondary is taped off a smaller or larger number of windings to produce a smaller or larger voltage, respectively. Usually one side of the secondary is connected to one side of the primary to provide a common voltage reference point between the two circuits. It is important to recognize for *safety* purposes and in circuit design that autotransformers have a direct current path between the secondary and primary.

## IV. BASIC CIRCUITS

### A. BASIC RESISTOR CIRCUITS

Two simple dc circuits are particularly useful in instrumental work, the parallel resistance circuit and the series resistance circuit. Both circuits are helpful when solving loading problems that may occur when two instruments are connected together. A series resistance circuit may be used as a voltage divider, to produce a voltage that is a known fraction of a given voltage. A parallel resistance circuit may be used as a current divider, or current bypass circuit. Another use for these circuits is the synthesis of a single new resistance from several resistors that happen to be on hand.

These circuits may be mathematically treated using Ohm's law, $V = IR$, with the help of **Kirchhoff's** two laws. The first of Kirchhoff's laws states that *the algebraic sum of all voltage sources equals the sum of the IR drops (potential differences across the resistors) around a loop*:

$$\Sigma V_{\text{source}} = \Sigma\, IR \tag{13}$$

The second law states that *the algebraic sum of the currents flowing into any point in a circuit is zero*:

$$\Sigma I_{\text{in}} = 0 \tag{14}$$

For purposes of this summation the values of currents flowing into the point are given one algebraic sign and the values of currents flowing out of the point are given the other algebraic sign. If in a circuit analysis the numerical value of an unknown current turns out to be negative then the current is actually flowing in a direction opposite to the assumed direction.

The polarities of the voltage sources in a circuit usually suggest the direction of flow of the current carriers through the other components. By historical convention, current is assumed to flow away from the positive terminal of a voltage *source* into the external circuit, and through a resistor from its positive terminal to its negative terminal. These directions are the directions that positive charges would flow. Although the historical convention appears to contradict the physical picture of electrons flowing in a wire, it is nonetheless algebraically consistent and widely used.

Consider the **parallel** resistor circuit shown in Fig. 2.1. The polarity of the battery indicates that conventional current flows through the resistors in the directions indicated by the arrows. The voltage difference is the same across all resistors. The currents are different, however. Using Kirchhoff's law about currents we find for point A,

$$\Sigma I = I_T - I_1 - I_2 - I_3 = 0 \tag{15}$$

In this equation, a negative sign is placed in front of a current that is assumed to be flowing out of a point. Rearranging and using Ohm's law to substitute for the currents $I_1$, $I_2$, and $I_3$ gives

$$I_T = \frac{V_T}{R_1} + \frac{V_T}{R_2} + \frac{V_T}{R_3} = V_T\left[\frac{1}{R_1} + \frac{1}{R_2} + \frac{1}{R_3}\right] \tag{16}$$

If the total effective resistance of the parallel combination between points A and B is called $R_p$ then Ohm's law gives

$$\frac{I_T}{V_T} = \frac{1}{R_p} \tag{17}$$

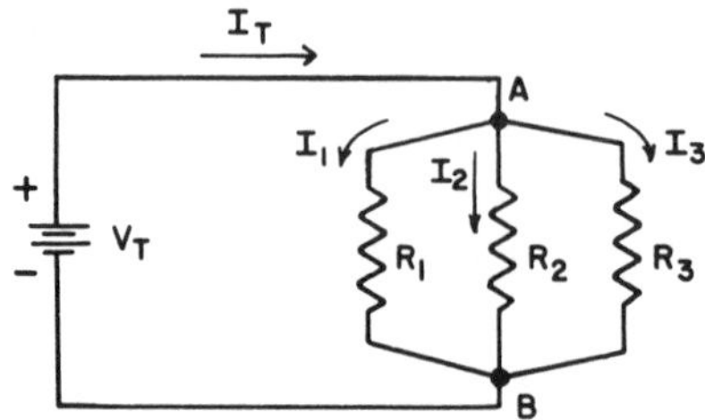


**Fig. 2.1.** A parallel resistor circuit.

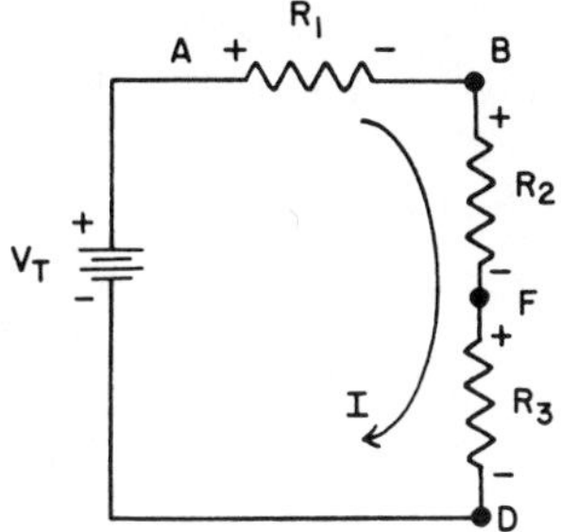


**Fig. 2.2.** A series resistor circuit.

which when combined with equation 16 gives

$$\frac{1}{R_p} = \frac{1}{R_1} + \frac{1}{R_2} + \frac{1}{R_3} + \dots \tag{18}$$

where the dots indicate that additional resistors could be added in parallel. When only two parallel resistors are involved equation 18 may be rearranged to give

$$R_p = \frac{R_1 R_2}{R_1 + R_2} \tag{19}$$

When the two resistors that are used are equal, then $R_p = R/2$.

A **series** resistor circuit is shown in Fig. 2.2. The polarity signs indicate the polarity of the voltage difference across each component. The positive sign at point B indicates that B is positive with respect to point F, and the negative sign at point B indicates that B is negative with respect to point A. Polarity signs, one positive and one negative, occur in pairs. When only one sign is shown, it always implies that the sign of opposite polarity occurs at a common reference point indicated by the symbols ⏚, ⏚, or ⏚. The latter two indicate a chassis connection and an earth ground, respectively.

The same current flows through all components in the series resistor curcuit. Kirchhoff's law about voltages around a loop gives

$$V_T = IR_1 + IR_2 + IR_3 \tag{20}$$

Upon rearrangement,

$$\frac{V_T}{I} = (R_1 + R_2 + R_3) \tag{21}$$

This has the same form as Ohm's law. The effective resistance $R_s$ between points A and D is the sum of the individual resistances. When only two resistors are used and they are equal, then $R_s = 2R$.

## B. VOLTAGE DIVIDER

The series resistor circuit is often used as a voltage divider when a smaller voltage is desired that is a constant fraction of a larger voltage. The output voltage is usually taken between one end of the series circuit, for example, point D in Fig. 2.2, and another point, F or B, in the circuit. When the resistor that is between the two output terminals is designated $R_{\text{out}}$ and the total series resistance of the voltage divider circuit is $R_{\text{total}}$, then the output voltage is

$$V_{\text{out}} = V_{\text{in}} \frac{R_{\text{out}}}{R_{\text{total}}} \tag{22}$$

using Ohm's law and equating the currents. This equation is called the **voltage divider equation.**

## C. VOLTAGE SUMMING CIRCUIT

The circuit in Fig. 2.3 produces an output voltage that is proportional to the sum of several input voltages. According to Kirchhoff's current summing law,

$$i_0 = i_1 + i_2 + i_3 \tag{23}$$

Applying Ohm's law to express each current as the ratio of the voltage drop across each resistor to its resistance gives

$$\frac{V_0}{R_0} = \frac{V_1 - V_0}{R_1} + \frac{V_2 - V_0}{R_2} + \frac{V_3 - V_0}{R_3} + \dots \tag{24}$$

where the dots indicate that other input resistors and voltages could be added in a similiar manner. Rearranging equation 24 gives

$$V_0\left(\frac{1}{R_0} + \frac{1}{R_1} + \frac{1}{R_2} + \frac{1}{R_3} + \dots\right) = \frac{V_1}{R_1} + \frac{V_2}{R_2} + \frac{V_3}{R_3} + \dots \tag{25}$$

which indicates that the output voltage is a weighted sum of the input voltages. When the resistors are all equal equation 25 reduces to

$$V_0 = \frac{1}{n+1}(V_1 + V_2 + V_3 + \dots + V_n) \tag{26}$$

where $n$ is the number of input resistors, equal to 3 in Fig. 2.3.

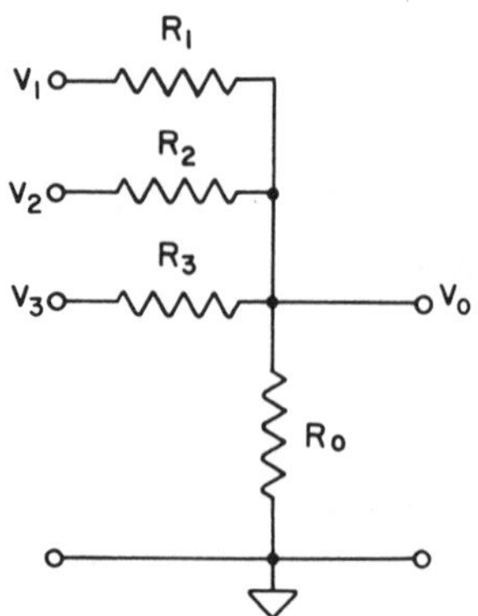


**Fig. 2.3.** A voltage summing circuit.

## D. LOADING AND EQUIVALENT CIRCUITS

When two instruments are connected together, the input of one may *load* the output of the other so that the instruments no longer behave as expected. For instance, a relatively simple voltage divider (consisting of two resistors in series) produces a voltage that is less than expected when too heavy a load is connected across its output terminals. If one of the instruments is a measuring device, then the measurement may be in gross error. Some relatively simple concepts of circuit analysis may be applied to identify and correct such problems. The discussion and equations below apply to ac signals when resistance is generalized to ac impedance (Section IV.H) and vector or complex number arithmetic is used.

**Thevenin's equivalent circuit** theorem states that when a load resistor is connected between any two terminals of any network of resistors and voltage sources, then the behavior or the network may be described by an equivalent circuit consisting of a single voltage source in series with a single resistor. **Norton's equivalent circuit** theorem states that the same network may also be described by a single current source in parallel with a single resistor. For a given circuit the resistors used in Thevenin's and Norton's equivalent circuits are equal. Norton's equivalent current source is equal to the voltage of Thevenin's equivalent voltage divided by the equivalent resistance. Although the two equivalent circuits may be arbitrarily interchanged, Thevenin's circuit is usually used if the output voltage of the network is of primary interest, and Norton's circuit is usually used if the current is of primary interest.

These theorems also apply if ac signals of a given frequency are used and the word *resistor* is replaced by the word *impedance*. The theorems may be applied as a first approximation to any circuit, even those containing nonlinear components, provided that the loading does not change the voltage between the terminals by more than, say, 10–20%.

In effect, then, the behavior of the output of any signal source, amplifier, power supply, voltage divider, and so forth may be described by the behavior of a simple equivalent circuit. The same theorems may also be applied in an identical manner to the input terminals of any electronic instrument. The problem of one instrument loading another then reduces to considering what happens when two simple equivalent circuits are connected together.

The equivalent circuit of the input to many measuring instruments is simply a resistor. For instance, the ohms-per-volt specification for a VOM (volt-ohm-meter) is multiplied by the full-scale voltage reading to obtain the value of the input resistance of the meter. When a VOM is connected to a voltage source to measure its voltage, a voltage divider is produced—the output resistance of the source is now in series with the input resistance of the meter, $R_m$. The output voltage of the loaded source is a fraction of the output voltage of the source with no load (Fig. 2.4). From the voltage divider equation,

$$\begin{aligned} V_{\text{out}} &= V_{\text{ideal}} \frac{R_m}{R_m + R_{\text{out}}} \\ &\simeq V_{\text{ideal}} \times 1, \quad \text{if } R_{\text{out}} \ll R_m \end{aligned} \tag{27}$$

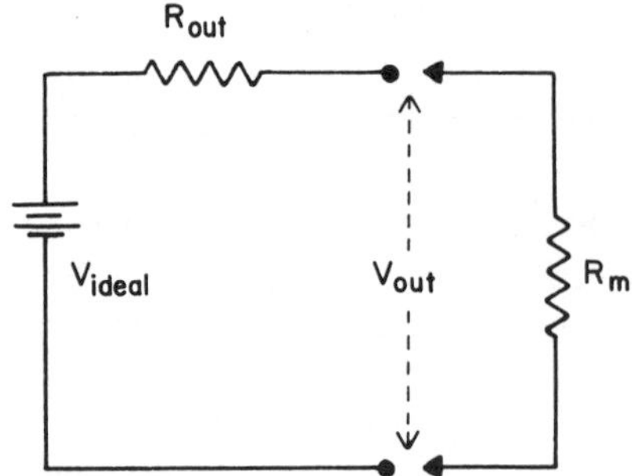


Fig. 2.4. Loading a voltage source with a resistor.

This equation shows that the load resistance must be much greater than the output resistance if the output voltage is to be an accurate representation of the ideal voltage.

Over the current range where the output resistance is linear, the output resistance and the value of the ideal voltage at no load may be determined by sequentially applying two or more values of load resistance and measuring each output voltage. (The input resistance of the voltmeter may have to be considered in parallel connection with the applied load resistance to obtain an accurate value of the total load resistance.) If one load resistance is sufficiently high, then $V_{out} = V_{ideal}$ and the voltage divider equation may be used directly with the other value of $R_m$ to calculate $R_{out}$. If two pairs of values of $R_m$ and $V_{out}$ are used, the values of $V_{ideal}$ and $R_{out}$ can be determined by substituting each set into the voltage divider equation and simultaneously solving the two equations. If several pairs of $R_m$ and $V_{out}$ are available, they may be used to plot a curve of $V_{out}$ versus $V_{out}/R_m$. Rearranging equation 27 gives

$$V_{out} = V_{ideal} - R_{out} \frac{V_{out}}{R_m} \tag{28}$$

which shows that $V_{ideal}$ is the intercept (for $R_{m=\infty}$) of the curve and $-R_{out}$ is its slope. Care must be taken not to overload a voltage source with too low a resistance load. Some sources tolerate $R_m = 0\Omega$; others blow a fuse, burn out, or turn off automatically.

**Current loading** may occur when the input terminal of the loading device is connected to an active component such as a transistor. The input equivalent circuit may then be considered a resistor in parallel with a constant (input leakage or bias) current source, or a Norton's equivalent circuit. Some digital voltmeters and operational amplifier follower circuits are examples. In some cases, the equivalent resistance is so large as to draw much less current from a voltage source than the current supplied (or required) by the input equivalent current source. Then the loading is essentially due to the current source alone. In other cases, the equivalent resistance may dominate the current source. Oscilloscopes are examples where a 1–MΩ resistor is usually placed between the input terminals to dominate the current source caused by the input current of the transistor amplifier circuit, because such current sources may be highly temperature dependent. On the other hand, determining the loading characteristics of a digital voltmeter that has a very high input resistance and a small input bias or leakage current may require consideration of both the equivalent input resistance and the equivalent input current source.

## E. LOADING OF A VOLTAGE DIVIDER

When a voltage divider is used, its total resistance must be large enough so as not to load the voltage source significantly. The competing requirement is that the two series resistors that make up the divider must be small enough so that the output of the divider is itself not significantly loaded. The Thevenin's equivalent voltage for a divider is its ideal output voltage $V_{out} = V_{in}R_{out}/R_{total}$ Thevenin's equivalent output resistance of a voltage divider is equal to the resistance of the two resistors calculated as if they were connected in *parallel*. Therefore, the equivalent output resistance is *always less than the smaller of the two resistors*. The equivalent output resistance may also be expressed as $R_{total}b(1 - b)$, where $b$ is equal to $R_{out}/R_{total}$, the output voltage expressed as a fraction $b$ of the input voltage. Since $b$ goes from zero to unity, the function $b(1 - b)$ has a maximum value of $\frac{1}{4}$ (when the two divider resistors are equal). Therefore, the output resistance of a voltage divider is always less than $\frac{1}{4}$ the total resistance.

## F. BASIC CAPACITOR CIRCUITS

Single parallel and series capacitor circuits are primarily used for the synthesis of a single new capacitance from several capacitors that happen to be available. Because precision capacitors are usually not available, it is sometimes desirable to *trim* a capacitor by adding smaller capacitors in parallel with it.

The charge that is stored on the plates of a capacitor is directly proportional to the voltage difference between the plates and is given by

$$Q \text{ (coulombs)} = C \text{ (farads)} \times V \text{ (volts)} \tag{29}$$

where $C$ is a proportionality constant, the capacitance.

When two capacitors are connected in parallel as shown in Fig. 2.5*a*, the

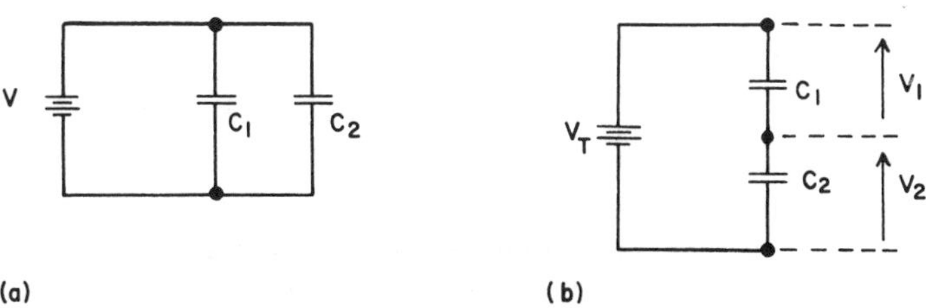


**Fig. 2.5.** (*a*) Parallel and (*b*) series capacitor circuits.

total charge $Q$ on both capacitors is equal to the sum of the charges on them, or

$$Q_{\text{total}} = C_1V + C_2V = (C_1 + C_2)V \tag{30}$$

Because the voltage across both capacitors is the same, the effective capacitance $C_p$ is the sum of the two capacitances. This additive nature of paralled capacitors may be thought of as being due to the area of their plates being effectively added together.

When two capacitors are connected in series, as shown in Fig. 2.5*b*, the current that passes through them as they become charged to their individual voltages causes the same amount of charge $Q$ to be redistributed in each of them. Because the total voltage across the pair is the sum of the individual voltages across each capacitor,

$$V_{\text{total}} = V_1 + V_2 = \frac{Q}{C_1} + \frac{Q}{C_2} = Q\left(\frac{1}{C_1} + \frac{1}{C_2}\right) \tag{31}$$

Comparing equations 29 and 31 we find that the effective capacitance $C_s$ of the series combination is related to $C_1$ and $C_2$ by

$$\frac{1}{C_s} = \frac{1}{C_1} + \frac{1}{C_2} \tag{32}$$

The capacitance of two capacitors in series is less than the smallest capacitance in the series.

The series circuit may not produce the desired results when the capacitors have significant leakage currents. For slowly varying signals or signals that are mostly of one polarity, when electrolytic capacitors are used, each capacitor charges through the leakage resistance of the other capacitor. The result may be a voltage division between the two components, which is closer to the ratio of their leakage resistances than to the ratio of their inverse capacitances. This is one reason that the voltage rating of one of the capacitors might be unexpectedly exceeded.

## G. THE RESPONSE OF *RC* VOLTAGE DIVIDERS AND FILTERS

A resistor in series with a capacitor is a circuit that is widely found in signal-processing circuits including amplifiers, sample-and-hold circuits, boxcar integrators, signal averagers, and peak detectors, as well as in timing circuits, signal conditioners, and power supplies. The circuit may also be used to explain capacitive loading effects including problems with stray capacitance. The circuit is called a **low-pass filter** if the *capacitor* is connected to the common point between the input and output (Fig. 2.6) and a **high-pass filter** if the *resistor* is connected to common.

Summing the instantaneous voltages around the input loop in Fig. 2.6 $v_r(t) + v_c(t) = v(t)$ and substituting the expressions shown in Fig. 2.6 gives the differential equation in $q$, the charge on the capacitor,

$$\frac{dq(t)}{dt}R + \frac{q(t)}{C} = v(t) \tag{33}$$

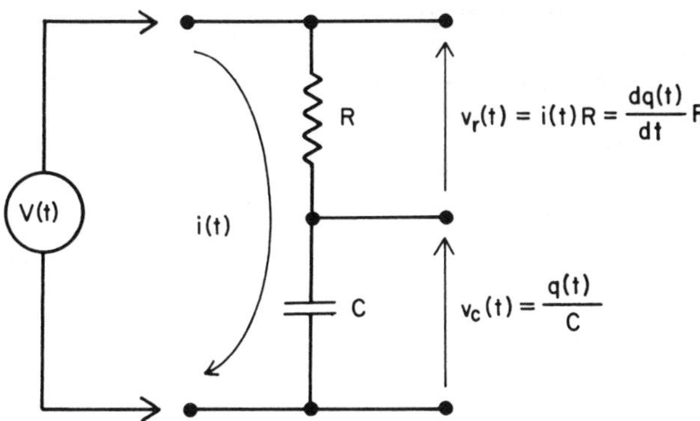


**Fig. 2.6.** An *RC* voltage divider circuit.

which describes the electrical behavior of the circuit. The general solution of the equation for any type of input voltage $v(t)$ is discussed below. When the input voltage is a step function $v(t)$ = constant $V_o$, then a particular solution to the differential equation is

$$q(t) = CV_o\left[1 - \exp\left(\frac{-t}{RC}\right)\right] + q(0) \tag{34}$$

when the initial charge on the capacitor is $q(0)$. The product $RC$ (ohms × farads = seconds) is called the **time constant**. After one time constant, the capacitor charge (and voltage) has changed by 63% of the total change it will make from its initial value of $q(0)$ to its final value $CV_o + q(0)$; after 3 time constants 95%, after 4.6 time constants 99%, and after 7 time constants 99.9%. From this equation expressions for the current and voltages may be obtained:

$$v_c(t) = \frac{q(t)}{C} = V_o\left[1 - \exp\left(\frac{-t}{RC}\right)\right] + v_c(0) \tag{35}$$

$$i(t) = \frac{dq(t)}{dt} = \frac{V_o}{R}\exp\left(\frac{-t}{RC}\right) \tag{36}$$

$$v_r(t) = V_o - v_c(t) = V_o\exp\left(\frac{-t}{RC}\right) \tag{37}$$

Equations 35 and 37 show that the capacitor voltage changes exponentially from its initial value of $v_c(0)$ toward the new value $V_o + v_c(0)$, whereas the voltage across the resistor suddenly changes by the amount $V_o$ and exponentially decays to zero. The sum of these two voltages always equals the applied voltage. For instance, when the input voltage steps up and down periodically the voltage wave forms in Fig. 2.7 can be observed.

When the output voltage is observed across the resistor (Fig. 2.7*c* and *d*), sudden changes of magnitude $V_o$ occur each time the input voltage suddenly

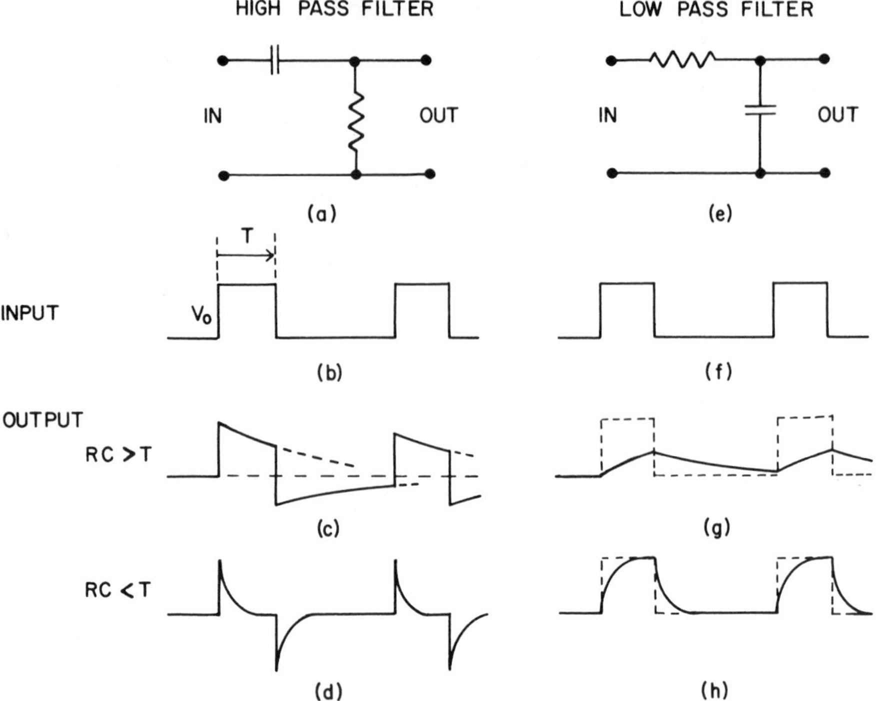


**Fig. 2.7.** Responses of a high-pass filter and a low-pass filter to step function inputs, for large and small *RC* time constants.

changes by $V_o$. Between the sudden changes, the output exponentially decays towards zero. When the *RC* time constant is much less than the pulse width and time between pulses (Fig. 2.7*d*), the output appears as a series of spikes of alternating polarity. This is a way to obtain negative and positive voltage spikes even though the input pulses are all of one polarity. The high-pass filter also removes the dc component of the signal after several time constants have passed from the start of a periodic signal.

The areas of the positive and negative parts of the output signal are then equal and independent of the polarity or magnitude of the dc component of the input signal.

When the output voltage is observed across the capacitor (Fig. 2.7*g* and *h*), the output voltage never changes suddenly because it takes time to charge the capacitor through the resistor. Note that the sum of the wave forms in Fig. 2.7*g* and *c* (and also *h* and *d*) equals the input wave form. When the *RC* time constant is long compared to the time between pulses (Fig. 2.7*g*), then the output voltage approximately equals the **time average** of the dc level of the input signal. Capacitance due to **stray capacitance**, cable capacitance, or amplifier input capacitance may form a low-pass filter with the output resistance of a voltage or current source. This may cause the leading and trailing edges of a square wave or pulse signal to appear rounded as shown in Fig. 2.7*h*.

The low-pass filter and circuits that behave in a similar manner are often used as **weighted integrators** in sample-and-hold circuits and boxcar "integrators." The purpose of the capacitor in this case is to integrate the signal during a short period of time and store the result for a longer period of time. The exact behavior of the circuit is expressed by the general solution of differential equation 33:

$$q(t_1) = \frac{1}{R}\int_0^{t_1} \exp\left(\frac{(t - t_1)}{RC}\right) v(t)\, dt + [q(0) - v(0)C]\exp\left(\frac{-t_1}{RC}\right) \qquad (38)$$

The duration of the integration is from $t = 0$ to $t = t_1$. The last term in this equation represents the initial conditions at (just before) $t = 0$. The integral term shows that the charge on the capacitor, and hence its voltage, is proportional to a weighted integral of the input voltage $v(t)$, where the input voltage at any time $t$ is weighted by the factor $\exp[(t - t_1)/RC]$. When the integration period $t_1$ is much less than the $RC$ time constant this factor is essentially unity, and the output approaches a nonweighted integral. For $t_1 \leq 0.1\ RC$ the approximation $\exp[-t_1 - t)/RC] \simeq 1 - [(t_1 - t)/RC]$ is valid to better than 0.5% relative error. Values of $v(t)$ nearest the end of the integration time $t_1$ are weighted most heavily whereas values of the input voltage that occurred before then are weighted less heavily.

**Other circuits** that respond in this manner are a resistor and capacitor connected in parallel being charged by a current signal rather than a voltage signal, an operational amplifier integration circuit where a resistor is connected in parallel with the integrating feedback capacitor (cf. Section VI.C.5), and a resistor-resistor voltage divider that is loaded with a capacitor. The resistance to be used to evaluate the time constant of the latter circuit is the Thevenin's equivalent circuit output resistance of the voltage divider. Some recorder damping circuits used to reduce the noise level of a signal respond as an $RC$ low-pass filter.

The response of the low-pass $RC$ filter is so commonly found in instruments that its **distortion** of three other types of signals is worth mentioning.

For a linearly changing signal (Fig. 2.8) the output never reaches the true value. However, the *slope* of the output, often the desired characteristic of this type of signal, reaches 63% of the true value in a time $t$ equal to $RC$, 95% in $3RC$, 99% in $4.6RC$, and 99.9% in $7RC$. In some experiments a sudden (step

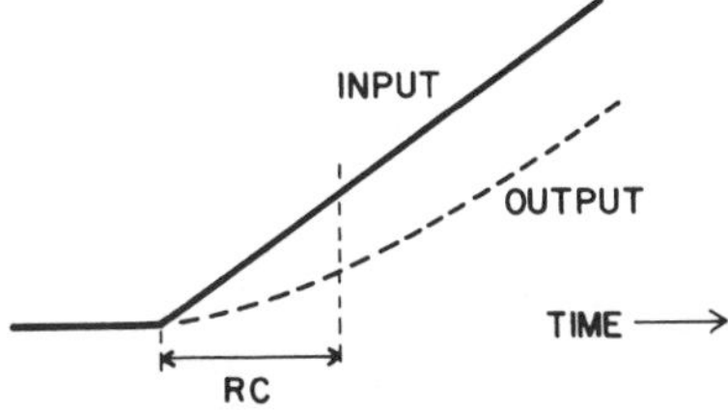


**Fig. 2.8.** Simple low-pass filter distortion of a signal that suddenly begins to increase linearly with time.

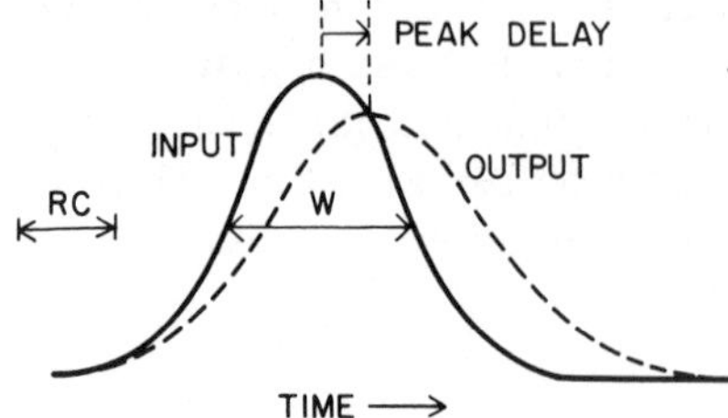


**Fig. 2.9** Distortion of a Gaussian-shaped peak by a relatively long time constant: peak is delayed, broadened, skewed, and deceased in height.

function) change may occur at the start of a sloping signal. The result depends upon the size of the step $a$, the slope $b$, as well as the $RC$ time constant:

$$\frac{\text{output slope}}{\text{input slope}} = 1 + \left(\frac{a}{bt} - 1\right)\exp\left(\frac{-t}{RC}\right) \tag{39}$$

When distorted by an $RC$ time constant, a Gaussian-shaped peak (Fig. 2.9) undergoes a decrease in peak height, an increase in width, a delay in the time that the peak occurs, and an increase in the trailing edge. Table 2.I indicates the extent of some of these distortions. In order to make the table values generally useful for a Gaussian peak, the time constant $RC$ is divided by the width $w$ of the peak measured between the one-half height points. Both $RC$ and $w$ must be measured in the same units of time. For instance, consider a peak whose true width is 5 s on a strip-chart recorder. Let the instrument that observes the peak have a 0.5 s $RC$ time constant. Then $RC/w = 0.1$. From the table the observed peak height is 98% of the true height, the peak is delayed by $5 \times 0.08 = 0.4$ s, and the width at half-height is 1.02 times the true width, a relative increase in width of 2%. Although not indicated by the table, for relatively large time constants the width of the peak near its base increases even more than at its half height, developing a noticeable tailing or skewing to one side. McWilliam and Bolton (26) have discussed $RC$ time-constant distortion of Gaussian peaks in considerable detail.

TABLE 2.I

Distortions of a Gaussian Peak of half-height width $w$ by a Simple Low-Pass Filter of time constant $RC$

| Normalized time constant, $RC/w$ | Ratio of observed to true peak height | Ratio of peak delay to true half-height width | Ratio of observed to true half-height width |
|---|---|---|---|
| 0.5 | 0.78 | 0.30 | 1.22 |
| 0.25 | 0.91 | 0.18 | 1.13 |
| 0.1 | 0.98 | 0.08 | 1.02 |
| 0.05 | 0.995 | 0.04 | 1.005 |
| 0.025 | 0.9987 | 0.02 | 1.003 |
| 0.020 | 0.9992 | 0.017 | 1.001 |

SOURCE: Values obtained from information given by McWilliam and Bolton (26).

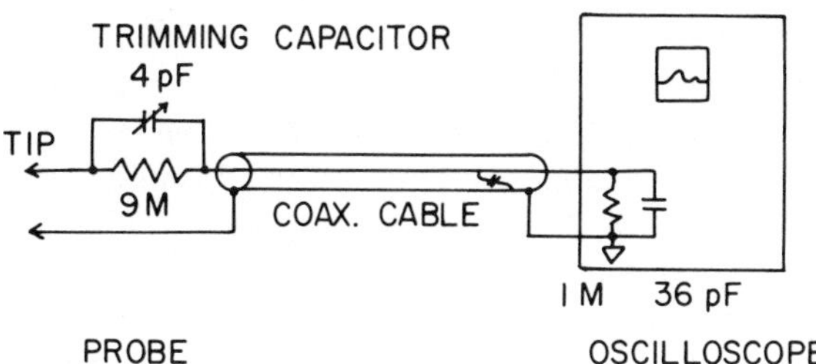


**Fig. 2.10.** Circuit diagram of an oscilloscope probe used to reduce loading of the signal voltage source. Trimming capacitor adjusts the ratio of the capacitor voltage divider to equal that of the resistance voltage divider.

Hand-held **probes** used to sample a voltage signal for an oscilloscope or other measuring instrument are important applications of *RC* voltage divider circuits. A "×10" (times ten) probe divides the signal voltage by 10 before it enters the input to an oscilloscope. It also reduces the resistive and capacitive loading of the source by a factor of 10 relative to the input resistance and the cable and input capacitance of the oscilloscope. The circuit for a probe is shown in Fig. 2.10.

A typical probe contains a 9 MΩ resistor that is connected in series with the 1 MΩ input resistance of the oscilloscope. This provides a ×10 voltage divider for slowly varying signals. For fast signals, the input capacitance of the oscilloscope ordinarily would form a low-pass filter with the resistor voltage divider and distort the signal. To prevent such distortions by a probe, a small trimming capacitor is also placed in series with the input capacitance of the oscilloscope and cable, thereby forming a capacitive ×10 voltage divider. For example, a 4 pF trimming capacitor would be required to compensate for a 36 pF oscilloscope input and cable capacitance. The signal source to which the probe is connected sees a probe with an input capacitance of only 3.6 pF and an input resistance of 10 MΩ.

The capacitor and resistor dividers are in parallel with each other. No distortion occurs when the resistor and capacitor dividers divide the signal voltage by exactly the same factor. To ensure that the factors are the same, the trimming capacitor should be adjusted each time that a probe is connected to a different oscilloscope input. The adjustment of the capacitor is made while observing a square wave. When the trimming capacitance is too small, the square wave is rounded as shown in Fig. 2.7*h*. When it is too large, the rising edge of the square wave overshoots as in Figs. 2.7*c* or *d*. Proper adjustment produces a square wave with minimal distortion of the rising and falling edges.

## H. IMPEDANCE, LAPLACE TRANSFORMS, AND BODE DIAGRAMS

Understanding the behavior of circuits that involve resistors, capacitors, and inductors is simplified by introducing the concept of an **impedance** or a **reactance**. Impedance is defined in the same form as resistance; that is, it is the ratio of the *voltage* across two terminals of a device or network to the *current* passing through those two terminals. The goal behind this particular definition is to provide parameters for capacitance and inductance which may be used in an Ohm's law type of relationship to simplify the analysis of circuits.

For a resistor, Ohm's law shows that the impedance is simply its resistance. For a capacitor or inductor the impedance is called **reactance**. In the time domain this ratio involves a time derivative or integral. The derivative or integral in the ratio may be removed by mathematical transformations. The transformed ratio of voltage to current is used in an Ohm's law expression where resistance is replaced by reactance or impedance. The analysis of the circuit that would otherwise involve differential equations reduces to *algebraic manipulations* like those used to analyze resistor networks.

The Laplace transformation is the most generally applicable transformation (Table 2.II) that may be used to convert the differential equation problem into an algebraic one. In this case the voltages and current wave forms are transformed from the time domain $v(t)$ and $i(t)$, into the Laplace domain $V(s)$ and $I(s)$ by the multiplication-integration transformation:

$$I(s) = \mathcal{L}\,[i(t)] \equiv \int_{t=0}^{\infty} i(t)\exp(-st)\,dt \tag{40}$$

Differentiation in the time domain becomes simple multiplication by $s$ in the Laplace domain when the value of the function being differentiated is initially zero (at $t = 0$). Scaling constants are transferred without change from one domain into the other. For a capacitor the time-domain relationship

$$i(t) = C\,\frac{dv(t)}{dt} \tag{41}$$

becomes

$$I(s) = CsV(s) \tag{42}$$

TABLE 2.II
Useful Properties of Laplace Transforms

| | |
|---|---|
| Differentation | $\frac{df(t)}{dt} \rightarrow sF(s) - f(0+)$ initial condition |
| Integration | $\int f(t)\,dt \rightarrow \frac{F(s)}{s}$ |
| Linearity | $Af(t) \rightarrow AF(s)$<br>$f_1(t) \pm f_2(t) \rightarrow F_1(s) \pm F_2(s)$ |
| Scale change | $f(at) \rightarrow \frac{1}{a}F\left(\frac{s}{a}\right)$ |
| Initial value | $\lim_{t\to 0} f(t) \rightarrow \lim_{s\to\infty} sF(s)$ |
| Final value | $\lim_{t\to\infty} f(t) \rightarrow \lim_{s\to 0} sF(s)$ |
| Steady-state substitution for sinusoidal wave forms | $s \rightarrow j\omega$ or $\sqrt{-1}\omega$ |

when $v(t = 0) = 0$. Rearranging this equation gives the capacitive reactance:

$$X_c(s) = \frac{V(s)}{I(s)} = \frac{1}{sC} \tag{43}$$

For an inductor

$$X_L(s) = \frac{V(s)}{I(s)} = sL \tag{44}$$

The ease with which the Laplace transformation technique may be used depends upon the availability of extensive tables of transformations, just as the use of the logarithm transformation to multiply and divide numbers by addition and subtraction depends upon the availability of logarithm tables. Fortunately such tables have been prepared for many of the voltage wave forms used in electronic circuits.

As an example consider the low-pass $RC$ filter, which can be treated simply as a voltage divider in the Laplace domain. The output voltage across the capacitor may be expressed in the form of the voltage divider equation:

$$V_{\text{out}}(s) = V_{\text{in}}(s)\frac{X_c(s)}{X_c(s) + R} = V_{\text{in}}(s)\frac{1}{1 + RCs} \tag{45}$$

When the input voltage is a step function of magnitude $V_o$, its Laplace transform is $V_o/s$ so that the output voltage becomes

$$V_{\text{out}}(s) = \frac{V_o}{s(1 + RCs)} \tag{46}$$

Tables of the Laplace transform show that the inverse of the expression on the right is $V_o[1 - \exp(-t/RC)]$, which gives

$$v_{\text{out}}(t) = V_o\,[1 - \exp(-t/RC)] \tag{47}$$

identical to equation 35.

When the input voltage is a sine wave $V_p \sin(\omega t)$, its Laplace transform is found in tables to be $\omega/(s^2 + \omega^2)$. Using equation 45 the output voltage becomes

$$V_{\text{out}}(s) = \frac{V_p\omega}{(s^2 + \omega^2)} \times \frac{1}{1 + RCs} = \frac{V_p}{\omega} \times \frac{1}{\left(1 + \dfrac{s^2}{\omega^2}\right)(1 + RCs)} \tag{48}$$

The inverse Laplace transform gives

$$v_{\text{out}}(t) = \frac{V_p RC\omega}{1 + (RC\omega)^2}\exp\left(\frac{-t}{RC}\right) + \frac{V_p \sin(\omega t + \phi)}{[1 + (RC\omega)^2]^{1/2}} \tag{49}$$

where $\phi \equiv -\tan^{-1} RC\omega$ and the scaling constant $1/\omega$, which transfers without change between domains, happen to cancel with an $\omega$ in each term of the numerator of the inverse transform. The first term in this solution is the transient response of the circuit when a sine wave is switched on at $v_{\text{in}} = 0$. It decays to zero for times longer than several $RC$. The second term is the readily observed steady-state output response, consisting of a sine wave of the same frequency as the input voltage but shifted in phase by $\phi$ radians and reduced in peak amplitude by the factor $1/[1 + (RC\omega)^2]^{1/2}$.

A general substitution that may be used when the voltages and currents in a circuit are **steady-state** sinusoidal wave forms is

$$s = j\omega \tag{50}$$

where $j \equiv \sqrt{-1}$. This relationship is readily derived by twice differentiating a sine wave and using the Laplace operator $s \equiv d(\ )/dt$

$$f(t) = \sin(\omega t) \tag{51a}$$

$$sf(t) = \omega \cos(\omega t) \tag{51b}$$

$$s^2 f(t) = -\omega^2 \sin(\omega t) \tag{51c}$$

$$s = \omega\sqrt{-1} \tag{51d}$$

For steady-state sinusoidal wave forms the capacitive reactance becomes

$$X_c(j\omega) = \frac{V(j\omega)}{I(j\omega)} = \frac{1}{j\omega C} = -\frac{j}{\omega C} = \frac{1}{\omega C}(-j) \tag{52}$$

The capacitive reactance is therefore a complex number that may be represented by a vector of length $1/\omega C$ and direction $-j$, or $-90°$ from the real axis on a complex number graph. The magnitude, $1/\omega C$, of the reactance for sinusoids is equal to the ratio of the output to input *peak voltages*. The phase angle indicates that the voltage lags the current in time by 90°; that is, corresponding points on the voltage wave form occur one-quarter cycle later in time than the points on the current wave form.

The analogous inductive reactance $X_L$ is given by

$$X_L(j\omega) = j\omega L \tag{53}$$

which is a complex number represented by a vector of length $\omega L$ and direction $+j$, or $+90°$ from the real axis. The voltage therefore leads the current by 90°, or one-quarter cycle.

The **transfer function** for a circuit is the ratio of output to input voltages, and may be expressed as a complex quantity. The transfer function for a low-pass filter when sinusoidal voltages are used is

$$\frac{v_{out}}{v_{in}} = \frac{X_c}{X_c + R} = \frac{1}{1 + (R/X_c)} = \frac{1}{1 + j\omega RC} = \frac{1}{1 + (\omega RC)^2}(1 - j\omega RC) \tag{54}$$

The last expression in parentheses is a complex number that represents a vector with a real component of unity length and a $j$ component of length $-\omega RC$. The magnitude of the vector is therefore $[1 + (\omega RC)^2]^{1/2}$ which gives the magnitude for the transfer function in equation 54 as

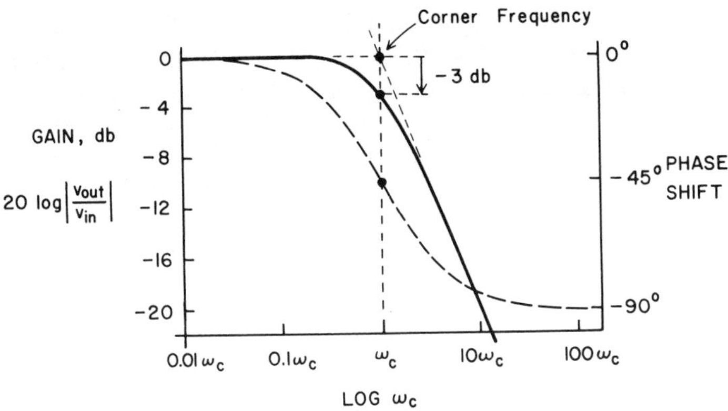


**Fig. 2.11.** Bode plot showing the frequency response of a simple low-pass filter.

$$\left|\frac{v_{\text{out}}}{v_{\text{in}}}\right| = \frac{1}{1 + (\omega RC)^2}\,[1 + (\omega RC)^2]^{1/2} = \frac{1}{[1 + (\omega RC)^2]^{1/2}} \tag{55}$$

The magnitude of the transfer function for sinusoids is the ratio of the peak values of the two voltages. The angle that the vector makes with the real axis is $\phi = \tan^{-1}(-\omega RC) = -\tan^{-1}(\omega RC)$, which indicates that if $v_{\text{in}}$ is assumed to be along the real axis (no phase shift) then $v_{\text{out}}$ is in the direction of the vector. Therefore a phase difference of $\phi$ radians exists between these two sinusoidal voltages, with the output voltage lagging behind the input voltage.

The magnitude and phase angle of the transfer function versus frequency is shown in Fig. 2.11. This representation where the magnitude and frequency axes are plotted using logarithmic scales is called a **Bode diagram**. The **corner frequency** $\omega_c$ is defined as the frequency where $\omega_c = 1/RC$ or $\omega_c RC = 1$. At this frquency the magnitude of the transfer function is the square root of 2, or 0.707, and the phase angle is −45°. At low frequencies the capacitive reactance is high, $\omega RC \ll 1$, the output voltage essentially equals the input voltage, and the phase angle approaches zero degrees. At high frequencies $\omega RC \gg 1$, the output voltage is inversely proportional to frequency, and the phase shift approaches −90°.

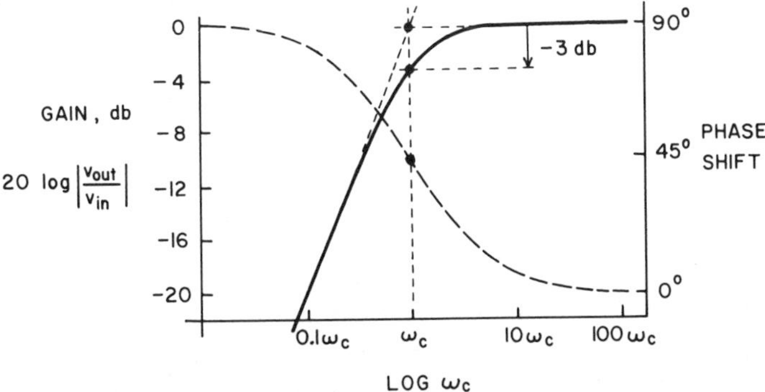


**Fig. 2.12.** Bode plot showing the frequency response of a high-pass filter.

The slope of the Bode diagram in this region is −20 db/decade where the **decibel**, db, is defined as db = 20 $\log_{10} |v_{out}/v_{in}|$. Similar relationships for the transfer function for a high-pass filter (Fig. 2.12) may be derived:

$$\frac{v_{out}}{v_{in}} = \frac{j\omega RC}{1 + j\omega RC} \tag{56a}$$

$$\left|\frac{v_{out}}{v_{in}}\right| = \frac{\omega RC}{[1 + (\omega RC)^2]^{1/2}} \tag{56b}$$

$$\phi = \tan^{-1}\left(\frac{1}{\omega RC}\right) \tag{56c}$$

## I. RESONANCE CIRCUITS

**Resonance circuits**, consisting of an inductor and capacitor connected in series or in parallel, are used in radio circuits, tuned amplifiers, and oscillators, particularly in the high-frequency range, say, above 100 kHz. The impedance of a **series resonance circuit** is

$$Z = R + j\left(\omega L - \frac{1}{\omega C}\right) \tag{57}$$

where the series resistance is due mainly to the resistance of the wire of the inductor. The impedance has a minimum value, $R$, when $\omega L = 1/\omega C$. The frequency at which this condition is satisfied is the resonant frequency of the circuit

$$\omega_o = \left(\frac{1}{LC}\right)^{1/2} \tag{58}$$

At this frequency the voltage across $R$ and the voltage applied to the circuit are equal. The voltages across the inductor and capacitor must therefore be equal and opposite, or exactly 180° out of phase with each other, so that their sum is zero. The voltage across the capacitor for an applied voltage of $v$ is

$$v_c = iX_e = \frac{V}{R} X_c = \frac{v}{j\omega_o CR} \tag{59}$$

since the current $i = v/R$ at resonance.
Then the magnitude

$$\left|\frac{v_c}{v}\right| = \frac{1}{\omega_o CR} = \frac{1}{R}\left(\frac{L}{C}\right)^{1/2} = \frac{\omega_o L}{R} = Q \tag{60}$$

where $Q$ is defined as the **quality factor** of the circuit. Quality factors vary from about 20 to 200 for typical inductors designed to resonate in circuits used in the audio-frequency range to those used in the megahertz-frequency range. Equation 60 shows that a capacitor used in a resonant circuit must have a voltage rating much higher than the peak value of the applied sinusoidal voltage. The quality factor is also a measure of how sharply the circuit is tuned, that is, of how fast the impedance rises as the frequency deviates from the resonant frequency. For instance, the current through the circuit for a constant applied voltage drops (due to an increase in impedance, $Z$) to 0.7 of its maximum value (−3 db) when the frequency deviates by $\pm 1/(2Q)$.

$Q$ is also equal to $\omega_o$ times the ratio of the energy stored in the circuit (e.g., in the charged capacitor) to the rate of energy dissipation by the circuit. This relationship is used to define $Q$ in more complicated resonant circuits, and in other resonance systems such as wave-guide cavities where discrete components of inductance and capacitance cannot be readily identified.

The **parallel resonance circuit** where a capacitor is in parallel with the series combination of inductance and (coil) resistance behaves in a manner similar to the series resonance circuit. The $Q$ is essentially the same. In this case, however, the *current* through the capacitor, rather than the voltage across it, is greater at resonance by a factor of $Q$ than the applied current coming from the driving source. The resonance frequency is not quite the same as in the series circuit but is essentially the same if $Q$ is large, and differs by a relative difference on the order of only $1/Q^2$. The current drawn from a voltage source is a *minimum* at resonance and increases to 1.4 times its minimum value (3 db) when the frequency deviates by $\pm 1/(2Q)$. The impedance of this circuit therefore decreases on either side of resonance.

The sharp variation in resonance circuit impedance with frequency may be used to provide a circuit whose output voltage is a sharp function of frequency. For instance, replacing one leg of a resistor-resistor voltage divider with a resonance circuit gives a voltage divider whose transfer function varies sharply with frequency. The transfer function has either a maximum or minimum value at the resonant frequency, depending upon which type of resonant circuit is used and which leg of the divider is used for the output voltage.

## J. TRANSMISSION LINES

**A transmission line** is any pair of conductors that is used to transmit signals from one point to another. Common examples are parallel wire lines, coaxial cables, a single wire near a ground strip plane, and parallel flat strip lines. Their impedance characteristics cause signal delays, and may cause reflections and amplitude and phase shift distortions unless proper precautions are taken. Distortions become significant when the transmission line becomes as long as a large fraction of the shortest wavelength in the frequency range of the signal being transmitted. For instance, the impedance of telephone lines becomes important even at audio frequencies, and the impedance of cables used in laboratories becomes important above the 10-MHz region or when pulse rise times become as short as the order of nanoseconds.

Uniform transmission lines may be characterized by an inductance $L$ per unit length, a series resistance $R$ per unit length (including both sides of the line), a capacitance $C$ per unit length, and a leakage conductance $G$ (the inverse of leakage resistance) between the conductors per unit length. A signal applied at one end of a transmission line is attenuated by the factor $\exp(-kx)$ where $x$ is the distance from the beginning of the line and $k$ is the attenuation constant per unit length of line. The attenuation constant is in general a *function of frequency*:

$$k = \left( \frac{(RG - LC\omega^2) + [(R^2 + L^2\omega^2)(G^2 + C^2\omega^2)]^{1/2}}{2} \right)^{1/2} \tag{61}$$

A signal composed of many frequencies would therefore suffer distortion unless the attenuation were negligible at all frequencies, for instance, due to a relatively short line. If however, $GL = RC$, then $k = (RG)^{1/2}$, $k$ becomes independent of frequency, and the line is distortion free. Or if $G \ll C\omega$ and $R \ll L\omega$, then $k = [R(C/L)^{1/2} + G(L/C)^{1/2}]/2$ and the line is also distortion free at sufficiently high frequencies. In both cases the propagation time delay per unit length is

$$t = (LC)^{1/2} \tag{62}$$

Sometimes this delay may be used to advantage, such as when the leading edge of a fast signal is used to trigger the sweep of an oscilloscope, and the signal is delayed by a transmission line in the vertical amplifier so that its leading edge is fully displayed on the screen. The delay time for a typical 50Ω coaxial cable is about 6 ns/m.

When the line is very long, a signal eventually is attenuated to a negligibly small value. At this point adding more line does not change the load impedance that the signal source sees at the input terminals of the line. This load impedance is called the **characteristic impedance** of the line and is equal to

$$Z_o = \left( \frac{R + j\omega L}{G + j\omega C} \right)^{1/2} \tag{63}$$

When $R$ and $G$ are small enough or the frequency large enough to make the line distortion free, then the characteristic impedance becomes a pure resistance

$$R_o = \left( \frac{L}{C} \right)^{1/2} \tag{64}$$

Typical values for coxial cable and twisted wire pairs vary from 50 to 125 Ω. For parallell-wire television antenna lines it is typically 400 Ω.

When a rapid change in a signal, such as a pulse, occurs at one end of a transmission line, there is a delay before it reaches the other end. After reaching the other end, the rapid change may be reflected back to the source end of the line. Upon reaching the source end, the rapid change may again be reflected. The reflections are repeated at each end, but become weaker as the energy of the rapid change becomes stored in the capacitance of the line and dissipated in the cable resistance and in the load and the source resistances. Reflections appear as stairsteps following a step function change in a signal, and as additional pulses following a short pulse signal. Experimental results showing the influence of reflections when pulses are sent down a coaxial cable are presented and discussed by Walters and Bruhns (43). Reflections may be prevented by making the transmission line appear to be infinitely long to the signal source. This may be done, as suggested in the preceding paragraph, by using a pure resistance equal to the characteristic impedance (equation 64) as the load at the end of the transmission line. A transmission line with such a load is said to be **properly terminated**.

A small reflection at the receiving end may occur owing to a small mismatch of the impedances of the line and the load. A reflection may also occur at discontinuities in a nonuniform transmission line, especially at a cable connection. A further reflection of this signal at the source end can be prevented by using a signal source whose output impedance is resistive and equal to the characteristic impedance of the line.

Transmission lines, for a given frequency, that are terminated by a short circuit or open circuit may be used at very high frequencies in place of capacitors or inductors. For instance, the impedance of a short-circuited line is $jZ_o \tan(2\pi\ell/\lambda)$, where $\ell$ is the length of the line and $\lambda$ the wavelength. A length between one-quarter and one-half wavelength of the *specified frequency* causes the tangent to be negative so that the shorted line behaves as a capacitor (equation 52). If it is less than a quarter wavelength, the tangent is positive and the shorted line behaves as an inductor. Inductors of this type have a higher $Q$ than a coil of wire at the very high frequencies where they are normally used.

## V. SEMICONDUCTOR DEVICES

### A. THE *pn* JUNCTION

The electrical characteristics of integrated circuits, amplifiers, transistors, diodes, and other semiconductor devices are due to the ways in which several unique properties of *pn* junctions are used in these devices. A *pn* junction occurs in the region where *p*-type and *n*-type semiconductor materials are adjacent to each other. Pure crystalline semiconductor materials that contain no impurities have very few electrons in their conduction band because of a forbidden energy gap between their conduction energy band and their valence electron energy band; essentially all their valence electrons are strongly tied up in maintaining the atomic bonds which hold their crystal lattice structure together, and very few have enough energy to cross the gap to become conducting electrons.

When impurity atoms having more than the required number of valence electrons are introduced into the crystal structure in place of the regular atoms, then there are more than enough electrons to satisfy the chemical or spatial bonding requirements of the crystal lattice structure. These extra electrons may easily enter the conduction band, reducing the electrical resistance of the semiconductor material. The **majority carriers** of electricity in this case are these negative charges (electrons) and the semiconductor material is called ***n*-type**.

When impurity atoms having fewer than the required number of valence electrons are introduced into the crystal structure, then there is a **hole** in the normal valence bond structure of the crystal lattice. A valence electron in an adjacent crystal bond may move in to fill the hole, leaving behind a hole in the location that it just left. Again another electron may move into the new hole, leaving behind a hole of its own. In this way the hole may effectively move about the crystal lattice as though it were a charged particle that conducts electric current. The absence of an electron to complete the crystal bond causes a real or induced positive charge at the location of the hole. The positive holes become the major carriers of electricity and the semiconductor material is aptly called ***p*-type**. The introduction of the desired impurities into *n*- or *p*-type material is called **doping**.

When *n*-type and *p*-type materials are adjacent to each other in the same crystal, the majority carriers from the *p*-type material are attracted across the junction to fill up holes in the *n*-type material. This attractive force caused by the crystal lattice as it attempts to fill its valence bonds is eventually balanced by the potential difference across the junction that is caused by the redistribution of charge when the extra electrons enter the *p*-type material. This potential difference is called the junction potential and amounts to a few tenths of a volt. Some of the carriers that cross the *pn* junction combine with each other to produce a region of graded concentration of carriers, the center of which is relatively void of majority charge carriers. This region approaches the very high resistance of the pure semiconductor material, and because of its reduced concentration of majority carriers it is called a **depletion region** or **depletion layer**.

A useful and important property of the graded depletion region of a *pn* junction is the way in which its effective width expands and contracts with changes in a voltage applied across the junction (Fig. 2.13). A negative voltage applied to the *p*-type material (**reverse bias**) supplies extra electrons to that side of the junction. The extra electrons flow toward the depletion region where they combine with the large number of holes in the *p*-type material, causing the depletion region to increase in thickness. The effective width of the depletion

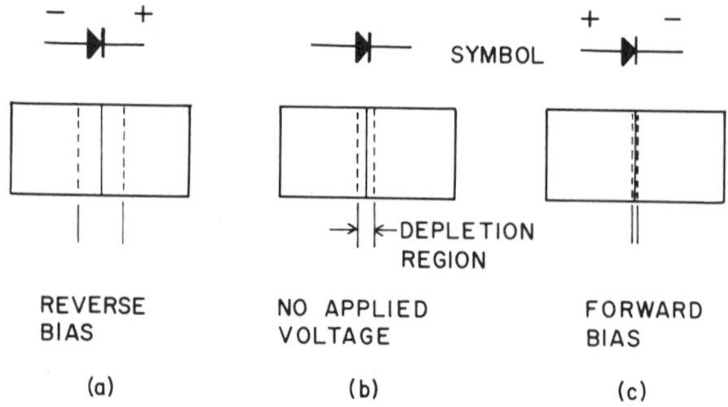


**Fig. 2.13.** Behavior of the depletion region with bias voltage.

region increases with the reverse bias voltage. The small current that flows is limited by the high resistance of the depletion layer. The applied voltage and the resistance of the depletion region both increase by about the same factor and the reverse bias leakage current therefore remains relatively constant over a wide range of applied voltages.

When a positive voltage is applied to the $p$-type material (**forward bias**), electrons are attracted away from the depletion region, regenerating more holes in that region and causing its width to decrease. The applied voltage (negative on the $n$ side, positive on the $p$ side) is also in a direction that reduces the height of the energy barrier of the junction, making it easier for the majority carriers from each side of the junction to cross the depletion region. The result is a relatively high flow of current which increases exponentially with applied voltage.

A quantitative approximation for the current-voltage relationship (Fig. 2.14) is given by

$$I = K \exp\left(\frac{-qV_j}{kT}\right)\left[\exp\left(\frac{qV}{kT}\right) - 1\right] \tag{65}$$

where $K$ depends upon carrier lifetimes, carrier diffusion coefficients, and physical dimensions; $q$ is the charge on an electron; $k$ is the Boltzmann constant; $T$ is the absolute temperature; $V_j$ is the junction potential; $V$ is the applied voltage; and $I$ is the resulting current. When the applied voltage to the $p$-type material is positive (forward bias), the exponential term containing $V$ quickly becomes greater than unity and rapidly increases with voltage, causing the current to rise exponentially when the applied voltage is greater than about a tenth of a volt. This exponential relationship is commonly used to make logarithmic and exponential amplifier circuits.

Eventually, as the voltage across a reverse biased $pn$ junction is increased, the electric field strength across the narrow depletion layer becomes so high that its resistance breaks down, allowing relatively high currents to flow. (See line $a$ in Fig. 2.15.) The breakdown is reversible unless the power generated (current-voltage product) is high enough to cause thermal destruction of the junction. Devices that are designed so that their $pn$ junctions operate in the breakdown region without being destroyed are called **zener diodes**.

Several mechanisms can cause breakdown, namely, thermal breakdown,

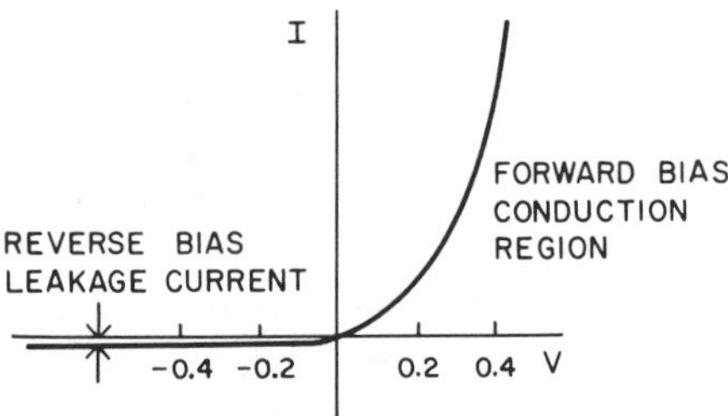


**Fig. 2.14.** Current-voltage characteristic curve for a $pn$ junction.

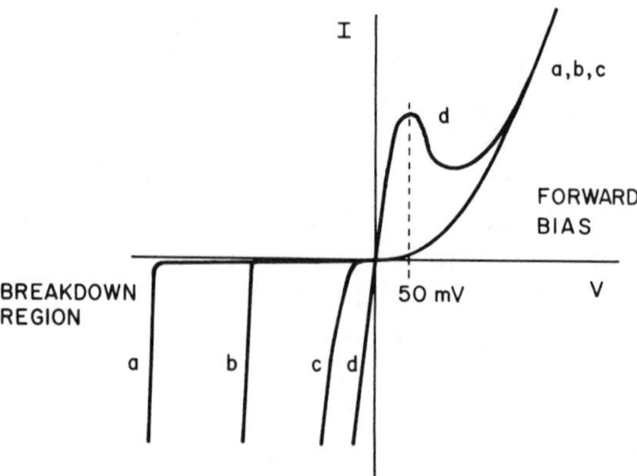


**Fig. 2.15.** Current-voltage characteristic curves including the breakdown region for a diode, curves *a* and *b*; a backward diode, curve *c*; and a tunnel diode, curve *d*.

avalanche multiplication, and tunneling. **Thermal breakdown** is encouraged by the exponential increase in reverse bias current with temperature. The current-voltage power generated by high reverse bias voltages causes an increase in temperature that in turn increases the reverse current, causing further increases in power and temperature. When the junction is unable to dissipate the heat to limit the temperature rise, thermal runaway occurs, rapidly leading to thermal destruction of the junction. Thermal breakdown may be recognized by a negative resistance region in current-voltage characteristic curves.

Even when a junction is designed to prevent thermal breakdown, breakdown due to either avalanche multiplication or tunneling can occur. **Avalanche breakdown** occurs when the *electric field* in the depletion region becomes so high that the charge carriers gain enough kinetic energy between collisions to break crystal structure bonds. The collision produces two more charge carriers (a hole and a conduction electron) which themselves gain enough energy to produce further carriers by collision. This avalanche-type increase in carriers rapidly increases the conductivity of the region and the current through it.

The avalanche breakdown voltage can be increased by increasing the effective width of the depletion region, since the electric *field* (V/cm) and not the applied voltage alone is the important variable. The physical width of the depletion region increases as the concentration of impurity atoms near the junction decreases. Therefore the avalanche breakdown can be controlled by adjusting the concentration of impurity atoms during the construction. Avalanche breakdown voltages can be decreased to about 8 V, below which the junction becomes so narrow that tunneling begins to join the avalanche mechanism in causing breakdown.

**Tunneling** can occur when the depletion region is extremely narrow (on the order of 10 nm) and the majority carriers on one side of the junction have an energy equal to unoccupied energy levels on the other side. High doping concentrations and a very sharp physical transition from $p$- to $n$-type materials produce a very narrow depletion region across which electrons may pass by tunneling *through* the energy barrier of the junction rather than by gaining sufficient energy to go over its top. In a reverse biased junction, the energy levels satisfy the energy level requirements for tunneling; valence bond electrons in the $p$-type material can tunnel through to the equal energy levels of the empty conduction band in the $n$-type. The application of a reverse bias distorts the forbidden energy gap in such a way as to decrease the physical distance between corresponding points on both sides of the junction which have identical energy levels. This makes it easier for electrons of a given energy to tunnel through to the other side into an empty level of the same energy. This effect is only partially offset by the increase in total width of the depletion region which accompanies the increased reverse voltage bias. Eventually a breakdown voltage is reached, usually between 0 and 5 V, where the tunneling current shows a marked increase with reverse voltage.

In the range of 4–8 V reverse bias, the avalanche and tunneling mechanisms occur simultaneously. The avalanche breakdown voltage *increases* with temperature, whereas the tunneling breakdown voltage *decreases* with temperature. When a *pn* junction has a breakdown voltage between 5 and 6 V, the temperature coefficients compensate for each other at a particular current, and the overall breakdown voltage is relatively constant with temperature. Zener diodes are available in this voltage range with voltage temperature coefficients less than 5 ppm/°C, when operated at a specified current.

The concentration of impurity atoms may be increased during construction until tunnel breakdown occurs for reverse bias voltages of only a few tenths of a volt. The junction then conducts more easily in the reverse biased direction due to tunneling than in the forward biased direction. Such a device is called a **backward diode** (line $c$ in Fig. 2.15).

When the doping is increased even more, a relatively large number of crystal valence bonds are empty in the $p$-type and a large number of conduction electrons are available in the $n$-type material. In this case the conduction electrons of the $n$-type material can tunnel through to the equal energy levels of the unfilled valence bonds in the $p$-type material. These energy levels are about the same energy for bias voltages up to 50–100 mV of forward bias. Above this voltage the energy levels no longer overlap as much and the tunneling current decreases with voltage, causing a negative resistance region (line $d$ in Fig. 2.15). The forward bias current then returns to the lower ordinary level of a forward biased *pn* junction. This type of junction is called a **tunnel diode**.

On either side of the relatively *nonconducting* depletion region are the higher conductivity $p$- and $n$-type regions. The *pn* junction therefore has the appearance of a **capacitor**. The capacitance of a *pn* junction in a typical discrete device such as a diode is on the order of a fraction to a few picofarads, and becomes significant in high-frequency and short delay time applications of semiconductor devices and circuits. Junctions in integrated circuits may be smaller and their capacitances are correspondingly smaller.

## B. SEMICONDUCTOR-METAL JUNCTIONS

The junction between a semiconductor and a metal may have the rectifying characteristics of a *pn* junction, or it may conduct current equally well in both directions as though it were a homogeneous conductor. Its behavior depends upon how it is made. If the contact potential causes a depletion of majority carriers on the semiconductor side, a depletion layer forms which behaves like a *pn* junction. An example is a **Schottky diode**, which is known for its low forward voltage drop (0.3 V) and its very short turn-off time. Other examples are **point-contact diodes**, and copper oxide and selenium rectifiers. Doping the semiconductor with high concentrations of impurity atoms near the junction to increase the majority carriers results in an ohmic contact.

## C. FIELD-EFFECT TRANSISTORS

The **field-effect transistor** (**FET**) is important because it may be used as an amplifier with a high input impedance and low input bias current, and also because it may act as a variable ohmic resistor. A **junction FET** (or **JFET**) consists of a bar of *n*- or *p*-type semiconductor material with ohmic connections at each end. Current is carried through the bar by the majority carriers. The end of the bar where the carriers originate is called the **source**; the other end is called the **drain**. A **gate** of opposite type semiconductor material is placed around or on opposite sides of the bar at some point between the source and drain. The gate forms a narrow channel whose effective width and length are controlled by varying the size of the depletion region that forms at this *pn* junction. As the reverse bias of the gate-channel *pn* junction is increased, the relatively nonconductive depletion region moves farther into the channel, squeezing off the conductivity of the channel. FETs are appropriately called *n*- or *p*-channel depending upon the types of semiconductor material used for the channel.

The drain-source characteristic current-voltage curves (Fig. 2.16) have three

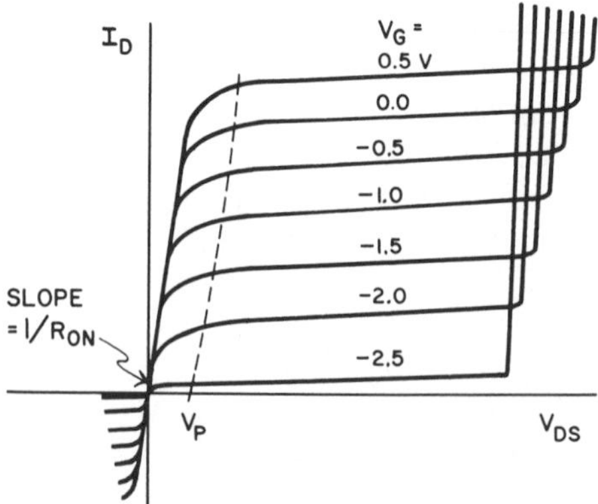


**Fig. 2.16.** Characteristic curves for an FET showing drain current $I_D$ versus drain-source voltage $V_{DS}$ for different values of gate-source voltage.

distinct regions, namely, a resistive region, a current saturation region, and a voltage breakdown region. Consider the case where the gate and source are at the same potential (zero gate voltage, $V_G = 0$). As the drain voltage $V_{DS}$ increases from zero the reverse bias of the gate-channel junction increases, especially near the drain end of the gate. The exact shape of the current-voltage curve in this region depends upon the physical shape and doping gradients of the channel, but it can be made to behave as either a nonlinear or linear resistor. Eventually the depletion region extends across a significant part of the drain end of the channel and the current-voltage curve rounds off as the channel conductivity decreases. At the pinch-off voltage $V_p$, the depletion region has extended over the entire channel at the drain end and the drain current $I_D$ is now limited mainly by the resistance of the depletion region. Further increases in drain voltage cause the pinch-off region to extend along the channel toward the source end, lengthening the depletion region and increasing its overall resistance. The drain voltage and depletion region resistance increase by about the same factor, causing the current to remain relatively constant (saturated) over a moderate voltage range, indicated by the horizontal parts of the curves in Fig. 2.16. At higher voltages, avalanche breakdown occurs. Because the breakdown occurs across the drain-gate junction and depends upon the voltage across that junction, the drain-source voltage where breakdown occurs in Fig. 2.16 varies with the gate-source voltage $V_G$.

The curves pass through the origin and extend into the third quadrant, indicating that currents may pass in *either* direction between the drain and source. This region helps to determine how an FET behaves in circuits where it is used as an on-off switch.

The reverse biased gate-channel *pn* junction accounts for the low input bias current and the high input resistance of the gate when operated in the first quadrant. At high frequencies, the relatively high capacitance of the *pn* junction may become a significant part of the input impedance. In the third quadrant, used in switching applications, the gate-channel junction becomes forward biased when the drain voltage exceeds the gate voltage, causing large gate currents to flow.

The region near the origin in Fig. 2.16 indicates that an FET may be used as a gate-voltage-controlled variable resistor, provided the drain-source voltage and current are not too high. Typical values of resistance are on the order of 100 Ω and greater. Application of a sufficiently large gate voltage can completely pinch off the entire channel, providing an extremely high OFF resistance between the drain and source for all voltage signals below the breakdown region.

Since an electric field, rather than a current, is used to cause a change in the conductivity of the channel of an FET, current is not required to flow across the gate-channel junction, except for the transient current needed to charge the gate capacitance. Therefore a thin insulating layer can be deposited between the gate electrode and the channel without destroying the effect of the electric field applied to the gate. An FET with an insulated gate is called an **IGFET**.

The most common IGFET is the **MOSFET** (metal oxide-semiconductor

FET), which is constructed by forming a narrow semiconductor channel on a substrate of opposite type semiconductor material. An insulating layer of silicon oxide is coated over the channel and a metal gate deposited over the oxide. The insulation provides an extremely high gate-channel resistance, on the order of $10^{15}\ \Omega$, for gate voltages of either polarity. The metal gate and the channel act as opposite plates of a capacitor; a voltage on the metal gate induces an equal and opposite voltage on the channel side. A positive charge on the metal gate induces a negative charge in the channel. If the channel is $n$-type material, an increase or enhancement in the concentration of majority charge carriers (electrons) occurs, resulting in an increase in conductivity of the channel. A negative gate voltage induces a positive charge in the channel, depleting the concentration of majority carrier electrons. These modes of operation are called the **enhancement mode** and **depletion mode**, respectively. A MOSFET of this type may be used in either mode but is historically called a depletion mode MOSFET.

A MOSFET that operates *only* in the enhancement mode is called an enhancement mode MOSFET. In this case the source and drain regions are heavily doped, whereas the channel and substrate are opposite-type semiconductor materials. With zero gate voltage, the drain channel is a reverse biased *pn* junction, giving the drain-source channel a very high resistance and a small leakage current. An enhancement type voltage applied to the gate induces an increase in the majority carriers in the channel, increasing its conductivity, just as with the depletion mode MOSFET. However, a gate voltage of opposite polarity simply encourages the already OFF condition of the channel.

Symbols for the various types of FETs are shown in Fig. 2.17. MOSFETs have a fourth terminal connected to the substrate base. This terminal is usually connected to the source or to common. It may also be used in special feedback circuits to minimize capacitive effects at high frequency.

Special care must be exercised when handling MOSFETs because of their extremely high gate resistance and very thin gate insulation. Ordinary electrostatic charges that may build up on a person or tools produce voltages that can

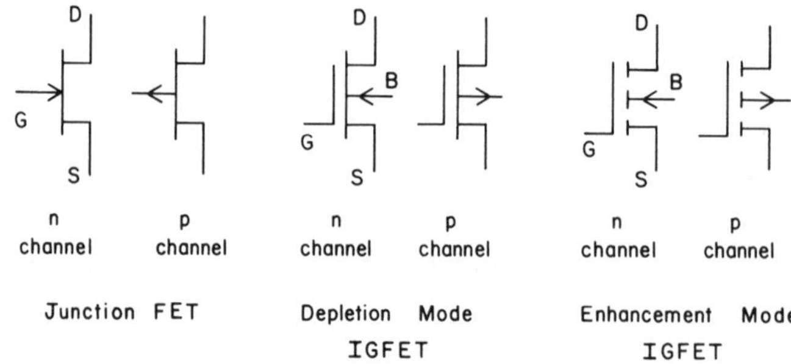


**Fig. 2.17.** Symbols for field-effect transistors. Junction FETs are used with a reverse biased gate junction. Depletion mode FETs respond to gate voltages of either polarity. Enhancement mode FETs respond to gate voltages that forward bias the gate (e.g., positive gate voltage for an $n$-channel FET.)

puncture the thin gate insulation. In dry air, drafts may cause the unconnected gate itself to accumulate a sufficient charge to destroy the insulation. Therefore MOSFETs may come with a wire wrapped around all leads or with the leads inserted into black electrically conducting styrofoam. This protection should not be removed until just before the MOSFET is ready to be connected into a circuit which supplies an electrical path to prevent electrostatic buildup. Even then the gate should not be touched unless the path is also able to prevent high voltages from being transferred to the gate. It is common practice to ground the circuit and also to ground the wrists of the person handling the MOSFETs. The soldering iron should also be grounded if soldering is necessary.

## D. BIPOLAR JUNCTION TRANSISTORS

The bipolar junction transistor, or just **bipolar transistor**, or **junction transistor**, has found use in amplifier, control, and switching circuits. Unlike the FET, a significant current is drawn by the input terminal of a junction transistor. The junction transistor consists of two *pn* junctions separated by a very thin common **base** (Fig. 2.18). The collector-emitter voltage $V_{CE}$ is chosen to encourage the flow of majority carriers from the emitter region toward the collector. Most of the voltage drop occurs across the depletion regions of the forward biased base-emitter *pn* junction and the reversed biased collector-base *pn* junction. There is relatively little voltage drop across the base region, and the current, which must pass through the base region, is therefore controlled by *diffusion* of the charge carriers across the base. The charge carriers are emitted into the base region by the **emitter**, and after reaching the **collector** are immediately attracted across the reverse biased junction by its large voltage drop (its polarity is designed to be in the proper direction for this attracton). The concentration of carriers in the base is essentially zero at the collector junction, and is a maximum value at the emitter junction. The base-emitter junction behaves like a forward biased *pn* junction (equation 65). Increasing the externally applied forward voltage bias of the emitter-base junction lowers the voltage barrier that the majority carriers in the emitter must exceed to cross that junction. This causes an exponential increase in concentration of carriers in the base region at

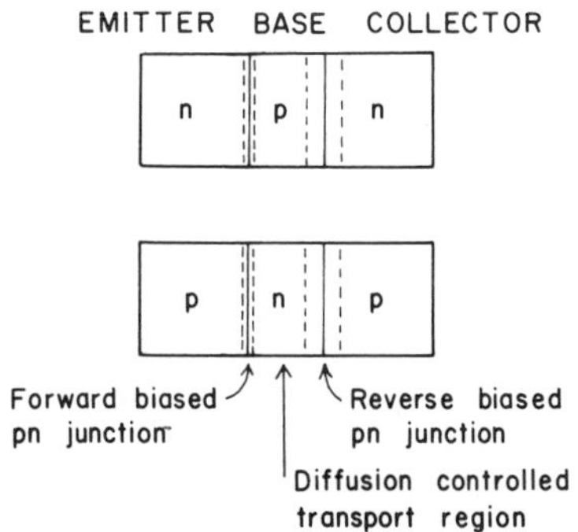


**Fig. 2.18.** Schematic showing the important regions present in a bipolar junction transistor.

the emitter junction, resulting in an increase in concentration gradient, and hence diffusion current, across the base from the emitter to collector. In this way the collector current may be controlled by the base-emitter voltage. Because the current is exponentially related to the base-emitter voltage, the voltage is usually limited to a working range of several tenths of a volt, as with a forward biased diode. Higher base voltages may be required, however, in power transistors where the resistance of the base material in series with the junction may become significant.

Some of the carriers in the base region diffuse to the base connection causing a small base current to flow in the external circuit. The emitter current is the sum of the base and collector currents. The base current is normally only 1–10% of the total current, and because of geometric (and other) considerations it is a constant (but temperature-dependent) fraction of the collector current. The junction transistor therefore *acts* as a current amplifier although the *cause* of its mode of operation is better explained by the influence of the base voltage upon the emitter current, and consequently the collector current. The ratio of collector current to base current is called the **dc current gain** and is given the symbol $h_{fe}$. An older term for the dc current gain is **dc beta**, $\beta$.

A condition known as **saturation** occurs when the collector-emitter voltage is insufficient to remove immediately all the charge carriers that diffuse to that end of the base. This occurs for collector voltages of less than a few tenths of a volt. When this occurs the concentration of charge carriers builds up throughout the entire base, saturating it with carriers. When the collector voltage drops below the base voltage, the collector-base junction becomes forward biased, causing even more carriers to be emitted into the base region, now from the collector as well as the emitter. Switching the base voltage to zero does not stop the current until the charge carriers have all cleared out of the base region. This time is called the **storage time**. A saturated transistor takes longer to turn off its collector current than if it were not allowed to saturate.

A transistor may be connected with any one of its three terminals being common to both the input circuit and the output circuit. The characteristic curves for the **common-emitter circuit** are shown in Fig. 2.19. Even at zero base

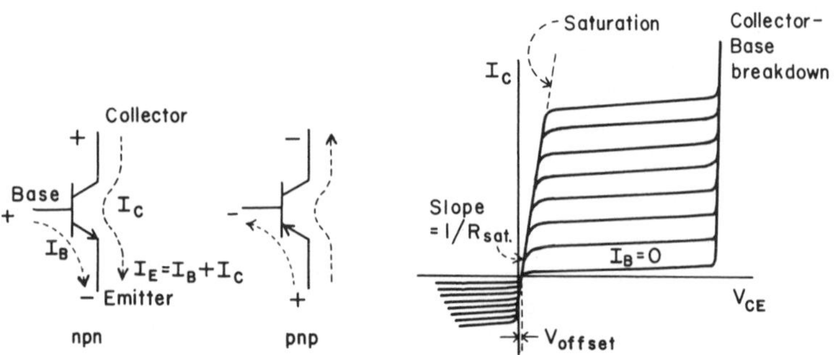


**Fig. 2.19.** Circuit symbols and characteristic curves for a bipolar junction transistor.

current, a small current flows between the emitter and collector, owing to the leakage current across the reverse biased collector-base *pn* junction. The characteristic curves extended into the third quadrant show that current may flow in either direction between the collector and emitter for a given base current. In effect, the collector and emitter reverse positions when operating in the third quadrant. This allows the transistor to be used as a bidirectional switch. Its ON saturation resistance is equal to the inverse slope of the steeply rising portion of the curves. The curves do not pass through the origin, but intersect the voltage axis at a small **offset voltage** $V_{\text{offset}}$ indicated in Fig. 2.19. The offset voltage must be considered in precision switching applications.

Most of the heat in a bipolar transistor is generated at the reverse biased collector-base junction, which has a high voltage drop and passes all of the collector current. Since this is also the junction where voltage breakdown occurs, it is not uncommon for the main cause of breakdown to be thermal breakdown. Therefore the collector of a transistor is usually connected to its case to improve heat dissipation.

## E. THYRISTORS

Diodes that can be switched on and/or off by a control signal connected to an additional terminal are used to control the average ac current of a load, such as a storage capacitor in a power supply, or a heater, and are also used to switch off a power supply to prevent damage if a preset voltage or current level are exceeded. They may also be used in circuits that suppress transient voltages on a power line. Such devices are part of a general class of four-layer *pnpn* semiconductor devices called **thyristors**. The most common thyristors are the **silicon-controlled rectifier, SCR**, and the **triac** (a bidirectional SCR). Symbols for these devices are shown in Fig. 2.20.

Devices with special characteristics may be made by controlling the exact geometry and level of doping of each layer. However, their general structure is shown in Fig. 2.21. Terminal connections to either one or both gates may be omitted, depending upon the particular type of device.

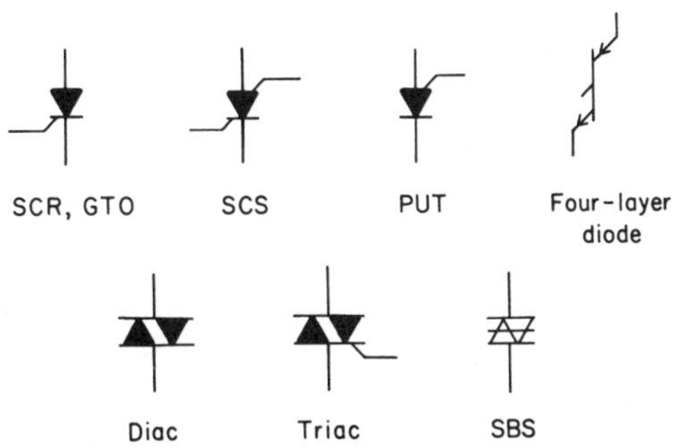


**Fig. 2.20.** Symbols for several types of thyristors.

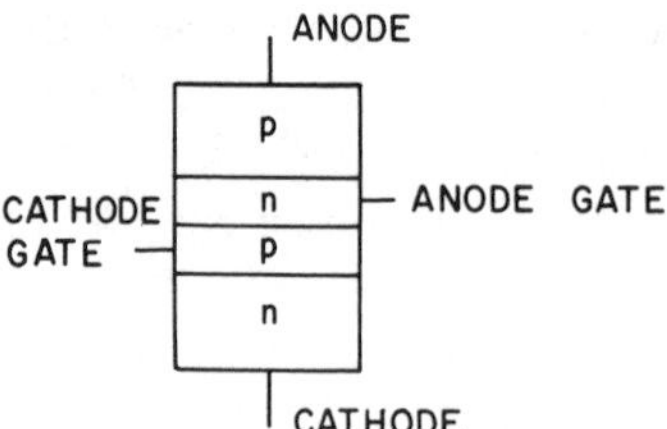


**Fig. 2.21.** Schematic diagram showing the important regions present in a thyristor.

To understand the operation of the device, consider the case where the anode is made positive with respect to the cathode. The *pn* junction in the center is therefore reverse biased, whereas the other two *pn* junctions are forward biased, as is the base-emitter junction of a transistor. The current that flows is limited to the leakage current through the depletion layer of the reverse biased center junction. When the current across any of the junctions is increased to a critical level, the rapid breakdown of the depletion layer of the reverse biased *pn* junction occurs. For instance, if the cathode gate voltage is made sufficiently positive with respect to the cathode a large increase occurs in the emission of majority carrier electrons from the *n*-type cathode into the *p*-type cathode gate (similar to an *npn* transistor base). The carriers diffuse across the narrow cathode gate and are immediately attracted across the reverse biased center junction by the large positive voltage of the *n*-type anode gate. These electrons neutralize the electron-poor side of the depletion layer on the *n* side of the reverse biased junction and make the anode gate more negative with respect to the *p*-type anode. This reduces the voltage barrier of this junction, allowing more holes to be emitted by the anode into the anode gate region (similar to a *pnp* transistor base). The holes diffuse across the *n*-type anode gate, are attracted across the reverse biased junction, neutralize the hole-poor side of the depletion layer, and make the cathode gate even more positive. This circuitous process rapidly increases the current until saturation is reached. Since the resistance of the depletion layer of the reverse biased junction is now effectively destroyed, the device behaves like a conducting diode, even after the external gate voltage is shut off.

Other ways to increase a junction current sufficiently to trigger the turn-on mechanism include applying a negative anode gate voltage with respect to the anode (or a positive anode voltage with respect to the anode gate), increasing the anode voltage until voltage breakdown of the reverse biased junction occurs, or injecting radiant energy into the junction through a transparent window. Also, if a fast rising voltage step function is applied to the anode or cathode, the junction capacitances may feed a fast rising voltage pulse through to a gate, which may be sufficient to cause the device to turn on. This is the so-called "$dV/dt$" effect. If the junction becomes too hot, the leakage current (which approximately doubles for every 10°C rise in temperature) may be sufficient to cause it to turn on.

Once the thyristor is turned on, it stays on until the depletion region of the central junction is restored, by clearing it of excess charge carriers. The quickest way to turn off the current is to reverse the anode-cathode voltage, which reverse biases the other two junctions, and allows the charge carriers to diffuse away from the center junction region or recombine with oppositely charged carriers. This process normally takes from 1 to 10 $\mu$sec before the forward voltage can be applied again without causing current flow to resume. The delay is called the **turn-off time**. The thyristor may also be turned off by reducing its anode voltage to zero and at the same time reducing its current to below a **holding current** threshold level. Some devices are made so that a relatively large current pulse at a gate removes sufficient carriers to turn the device off. The **gate turn-off switch, GTO**, is an SCR which may be turned off by applying a reverse gate current, provided the anode current is not too large.

A **four-layer diode** has only the anode and cathode terminals available. A **silicon-controlled switch, SCS**, has both gates available. A **programmable unijunction transistor, PUT,** is really a thyristor with only an anode gate. The PUT switches on when the anode voltage exceeds the anode gate voltage by a few tenths of a volt. A **SUS** (silicon unilateral switch) is a PUT with a built-in zener diode to fix the anode gate voltage with respect to the cathode.

Bidirectional thyristor switches may conduct current in both directions and therefore find use in ac power control circuits. The **diac** has only two terminals. The **triac** has one gate, and gate voltages of either polarity can turn it on in either direction. Usually positive voltages are used to turn on positive currents and negative voltages are used to turn on negative currents. The **silicon bilateral switch, SBS**, is often used to trigger triacs. The SBS is essentially a bidirectional SUS.

A simple circuit that uses an SCR to supply a variable amount of power to a heater is shown in Fig. 2.22. The output voltage of a low-pass $RC$ filter is used to trigger the SCR. The time at which the SCR is turned on during each positive half cycle is determined by the phase and amplitude of the output of the low-pass $RC$ filter, which in this case is controlled by $R$. Since the phase shift of the low-pass filter cannot exceed 90°, the trigger point for this circuit can be controlled only over the first half of a half cycle. The half-cycle power to the heater can therefore be varied from full to half power. The power is turned off when $R$ is so large that insufficient voltage or current is supplied to the gate to turn the SCR on. The diode prevents reverse biased gate voltages and currents, and $R_1$ limits the gate current to safe levels.

Large transients occur when an SCR or triac is turned on during the central part of a half cycle. To eliminate such transients, special control circuits are used to switch the device on only when the wave form crosses the horizontal axis, at zero voltage. The average amount of power supplied to a load is then varied by controlling the average number of full half cycles that the device is turned on during a given time period.

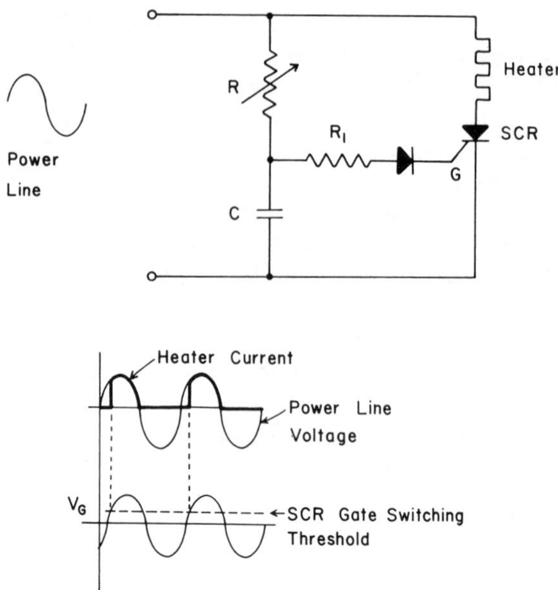


**Fig. 2.22.** A simple SCR circuit used to control the average current passed through a heater.

## F. UNIJUNCTION TRANSISTORS

The **unijunction transistor, UJT**, has been used in relatively simple timing circuits that have a jitter (precision) on the order of 0.1%, and as trigger level detectors. The low output impedance of a typical timing circuit in which they are used allows them to drive SCRs. The UJT consists of a bar of *n*-type semiconductor material with a resistance on the order of several kilohms (Fig. 2.23*a*). About midway on the bar is placed an emitter E, of *p*-type mate-

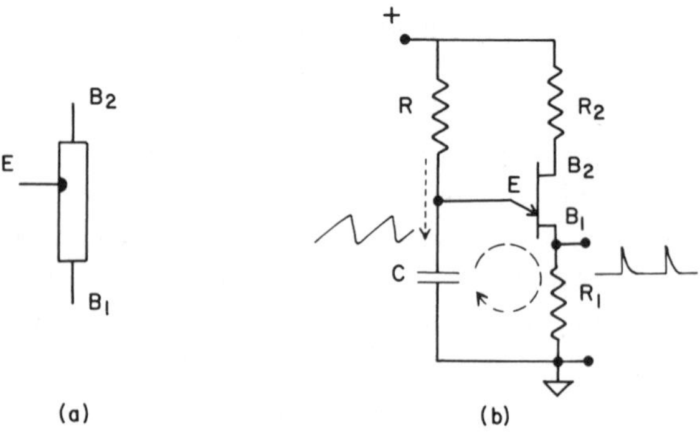


**Fig. 2.23.** (*a*) Schematic of important regions in a unijunction transistor and (*b*) a typical oscillator circuit using a unijunction transistor.

rial. When a voltage is applied across the bar, the bar acts as a voltage divider with the emitter located so that a fraction of about 0.6 of the applied voltage appears between the emitter and $B_1$. This fraction is called the **intrinsic standoff ratio**. When the emitter voltage is less than this fraction of the applied voltage, the junction is reverse biased and only a very small leakage current flows from the emitter. When the emitter junction is forward biased, holes are emitted into the $B_1$ region, decreasing the resistance in this region. The decrease in resistance of this leg of the voltage divider causes the junction to be forward biased even more, increasing its emission of holes and further decreasing the resistance.

Consequently the emitter-$B_1$ region resistance quickly drops to a few ohms as it becomes saturated with charge carriers, and its voltage drops to 1–2 V. The emitter-$B_1$ region can be switched off when its current decreases sufficiently to allow the region to be cleared of excess charge carriers.

The intrinsic standoff ratio is usually very stable and has a temperature coefficient of only about 0.06%/°C. The effect of this small temperature dependence can be offset by placing a carbon resistor of several hundred ohms in series with $B_2$, provided that the resistor and the UJT are kept at the same temperature.

A typical timing circuit is shown in Fig. 2.23*b*. When the emitter junction is reverse biased, the capacitor charges exponentially through $R$. When the voltage across the capacitor reaches a value that forward biases the emitter junction, the emitter-$B_1$ region switches on, quickly discharging the capacitor through $R_1$. Because $R_1$ is normally 47–100 Ω, a short (1–100 μsec) positive pulse is obtained across $R_1$. The equivalent output resistance of this pulse generator decreases during the rise time of the pulse from about $R_1$ to a value on the order of a few ohms. It therefore can drive heavy loads and is commonly used to trigger SCRs. A similar negative pulse can be obtained by placing $R_1$ between the capacitor and ground. A *negative-going* pulse is obtained at $B_2$ but its output impedance is somewhat higher owing to the resistance of the $B_2$ region.

After the capacitor has discharged, the emitter-$B_1$ current falls below its holding value and the emitter-$B_1$ region turns off again. The resistor $R$ must not be too small or it may provide enough current to the ON emitter-$B_1$ region to keep that region turned on. Moreover, resistor $R$ must not be too large or it will not be able to supply the peak point current $I_p$ necessary to cause the *pn* junction to switch on, or the resistor current will be a large fraction of the emitter leakage current and the charging time will become highly temperature dependent. For high precision the average current through $R$ during charging must be several decades higher than the leakage current through the capacitor. This usually precludes the use of electrolytic capacitors (which have high leakage currents) with megohm value resistors to produce long time constants.

When a timing circuit of this type is used to switch on an SCR at a reproducible point in time during a power line frequency half cycle, the dc power supply voltage for the circuit in Fig. 2.23*b* is replaced by a low-voltage (10–20 V. peak) rectified power-line frequency sine wave. The time constant may be adjusted to discharge the capacitor several times during a half cycle (only the first pulse influences the SCR) but the capacitor always ends a half cycle in the discharged condition. This occurs because the voltage at $B_2$ drops to zero at the end of the half cycle, forward biasing the emitter junction and discharging the capacitor. The circuit is then ready to restart at the beginning of its timing cycle when the next half cycle begins.

## VI. CIRCUITS USING ACTIVE DEVICES

### A. ANALOG SIGNAL GATES

**Analog signal gates** are used in sample and hold circuits, peak detectors, signal averagers, phase-sensitive detectors, lock-in amplifiers, and in other circuits to transmit or block the passage of analog signals during selected time intervals. Other terms for an analog signal gate are analog or linear transmission gate, analog or linear gate, analog or linear switch, and sampling gate. Equivalent circuits for an ideal analog gate are shown in Fig. 2.24. When the gate is *closed*, the analog signal at the input appears at the output; when it is *open* the signal does not appear at the output. The signal at the control terminal determines whether the gate is closed or open. The simplest analog gate is a manual switch where the control signal is mechanical force that causes the contacts to be opened or closed. **Electromechanical relays** such as the **reed relay** are analog gates that use an electrical control signal to activate a magnetic coil that opens or closes the contacts. Compared to semiconductor gates mechanical gates usually have a very low ON resistance between their contacts, a very high OFF resistance, a very high OFF leakage current, and can switch high voltages. The time required to activate a mechanical switch is relatively long, on the order of 1 ms for a relatively fast reed relay. For all except those with mercury wetted contacts, the **contacts bounce** a few times upon opening and closing. This adds an additional millisecond time period to the total transition time and causes several pulses to appear in the output signal during each transition. Electromechanical switches have limited useful lifetimes. Even a lifetime of $10^7$ operations would amount to a life of only 1 yr at three operations per second.

Semiconductor gates make use of diodes, FETs, or junction (bipolar) transistors as switches. Each gate may have more than one switch, to help compensate for nonideal behavior of a single switch. Several types of analog gate circuits that

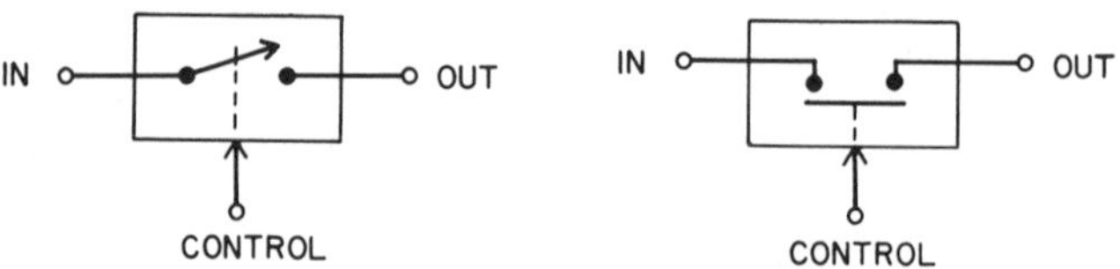


**Fig. 2.24.** Equivalent symbols for analog signal gates.

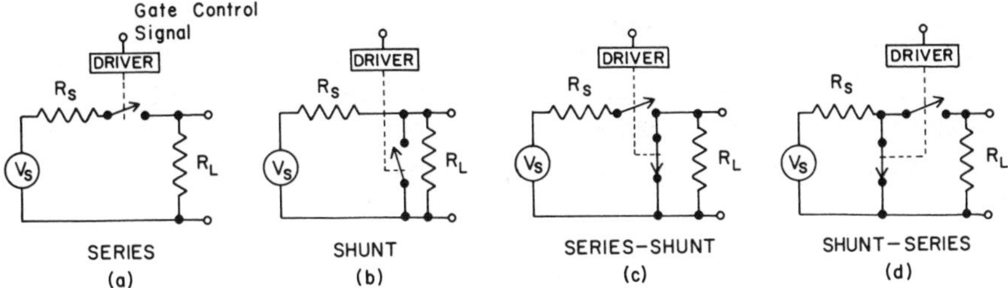


**Fig. 2.25.** Several types of analog signal gates.

use one or two switches are shown in Fig. 2.25. The characteristics of these circuits as gates may best be understood by considering their behavior as voltage dividers when the gate is OPEN and CLOSED. For the **series switch**, Fig. 2.25*a*, the gate-output voltage $v_o$ is equal to the voltage divider fraction $R_L/(R_s + R_c + R_L)$ of the input signal voltage $v_s$, where $R_c$ is the switch contact resistance, $R_s$ is the output resistance of the signal source, and $R_L$ is the load. When the ON resistance of the switch behaves as a linear resistor, the shape of the signal is undistorted but its amplitude is decreased. If the decrease in amplitude must be less than 0.1%, then the sum of $R_s + R_c$ must be 1000 times less than $R_L$ when the switch is ON. When the switch is OFF, the contact resistance must be at least 1000 times greater than the load resistance $R_L$ if the output signal is not to exceed 0.1% of the *maximum* input signal when the gate is OPEN. Therefore the OFF/ON ratio of the switch resistance must be greater than $10^6$. If small as well as large signals are to be rejected by the gate when it is OPEN then the open contact resistance may have to be $10^4$ or $10^5$ times greater than the load $R_L$ if a large input is not to be *fed through* the OPEN gate and appear as a small signal. An IGFET can be an excellent switch. However, even an $n$-channel FET usually meets the above requirements because its ON drain-source resistance is on the order of 10–100Ω and is usually linear when the gate-source voltage is constant, and its OFF drain-source leakage current, typically 0.1 nA, provides an effective OFF resistance as high as $10^8$–$10^{10}$Ω. A typical load resistor for the circuit would be about 100 kΩ. When the FET is off, the leakage current would cause only a 10$\mu$V drop across the load resistor.

The OFF resistance of the **shunt switch** in Fig. 2.25*b* must be 1000 times greater than the load resistance $R_L$ if the switch resistance is not to decrease the amplitude by more than 0.1% when the gate is CLOSED. The source resistance $R_s$ may have more of an influence than that! When the gate is CLOSED, the switch ON resistance must be more than 1000 times less than the source resistance if the output voltage is to be less than 0.1% of the *maximum* input signal. It must be considerably less than that if both small and large signals are of interest, just as with the CLOSED series-switch gate. It may be necessary to increase artificially the source resistance $R_s$ by placing a resistor in series with the voltage source to meet that requirement.

The **series-shunt** and **shunt-series** circuits (Fig. 2.25) provide a less demanding OFF/ON resistance for each of the two switches. Or, for the same ON/OFF resistance ratio for each of its switches, these gates can be designed to provide a gate with less amplitude distortion when closed and less capacitive feed-through of large signals from the input to output terminals when OPEN than either the series or shunt circuits alone. The series-shunt circuit has the advantage of having the load always connected to either the source or the common terminal, except perhaps during a switching transition. The shunt-series circuit has the advantage of allowing several switches to feed their signals to the same load alternately.

Series-shunt gate circuits implemented with FETs are shown in Fig. 2.26. To ensure minimum signal distortion during transmission through a CLOSED gate, the control circuit for the series FET may be designed to maintain the gate-source voltage $V_{GS}$ equal to zero when this FET is on; the gate voltage is allowed to track the source voltage. The ON resistance is then constant and not dependent upon the signal level. The $R_1D_1$ resistor-diode circuit provides one way of maintaining $V_{GS} = 0$, the resistor essentially shorting the gate to the source when the diode is reverse biased and the FET is on. When the series FET is off, the resistor $R_1$ loads the source because the diode is forward biased. Diode $D_2$ and $R_2$ keep the shunt FET gate-source voltage close to zero when the control voltage is negative and prevents the flow of forward biased gate current. The gate in Fig. 2.26 switches (block or transmit) voltages as positive as $+V_c$ and as negative as $-V_c + V_t$, where $V_t$ is the magnitude of the most negative gate-source or gate-drain voltage necessary to keep the FETs off.

Gates often come in integrated circuit form with more than one gate on a chip. To provide versatility they may or may not include gate driver circuits that keep the FET gate-source voltage constant. When the gates are arranged to connect any of several signals to a given output terminal or channel, the integrated circuit is called a **multiplexer**.

Depletion mode MOSFETs may be used as switches without concern for forward biasing the insulated gate. Their gate leakage currently is typically decades of magnitude less than that of a junction FET but their drain-source leakage is comparable to that of the FET. Just as with the junction FET, their gate-source voltage must be constant if their ON resistance is to be constant with signal voltage variations.

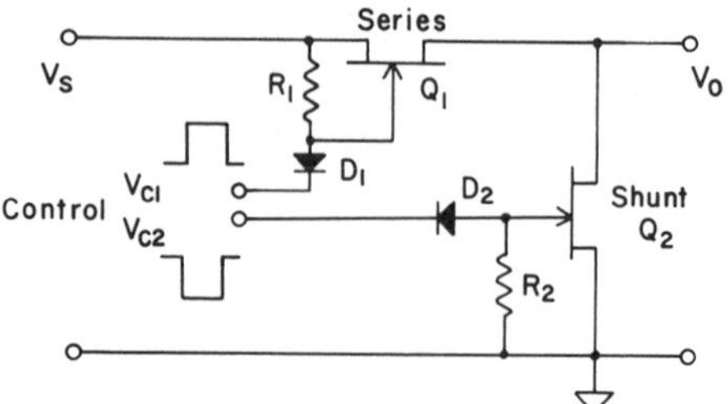


**Fig. 2.26.** A series-shunt gate implemented with JFET switches.

An enhancement mode $n$-channel MOSFET is OFF whenever its gate-source voltage and gate-drain voltage are zero or negative, and is ON when its gate-source voltage is sufficiently positive, say, $V_t$. A positive gate voltage $+V_c$ allows all signals less positive than $V_c - V_t$ to pass, and a negative gate voltage $-V_c$ blocks all signals more positive than $-V_c$. When operated with a fixed positive gate voltage, the ON drain-source resistance varies with the signal since $V_{GS}$ is not constant. To minimize this effect, a $p$-channel and an $n$-channel enhancement mode MOSFET may be operated in parallel with each other. The $n$-channel resistance increases and the $p$-channel resistance decreases for a positive-going change in signal. When properly matched, the ON resistance of the parallel combination remains relatively constant with signal level throughout the entire useful signal range of $-V_c$ to $+V_c$. Complementary gate control signals are required since $+V_c$ turns on the $n$-channel and $-V_c$ turns on the $p$-channel enhancement mode MOSFET. A **complementary MOSFET** of this type is called a **CMOS** or **COSMOS** and may have an inverter amplifier integrated into its structure to provide a complementary gate control signal so that only one external gate control signal is necessary.

A **switching transient** occurs at the output terminal of a gate when a rapid change in the control voltage is fed through the gate-channel capacitance. Several techniques may be used to reduce the magnitude of a switching transient. One way is to short to common via a low impendance path. For instance, the shunt switch $Q_2$ in Fig. 2.26 could be turned on slightly before $Q_1$ is turned off. Then the switching transient from $Q_2$ would be shorted to the signal source via $Q_1$ which is still on, and the transient that occurs when $Q_1$ turns off would be shorted to common by $Q_2$. When switching the gate to the other state $Q_1$ is turned on slightly before $Q_2$ is turned off. The signal source must have a low output impedance and be able to withstand the momentarily heavy load that occurs when both switches are momentarily on.

Another technique to reduce switching transients is to use two complementary ($n$-channel and $p$-channel) FETs in parallel for each switch. A positive-going control signal is applied to the gate of the $n$-channel FET and a negative-going control signal to the gate of the $p$-channel FET, when the FETs are turned on simultaneously. The resulting positive and negative transients cancel each other. A slight mismatch in the gate-channel capacitances of the two FETs can be reduced by connecting a small trimming capacitor between the gate and drain of one of the FETs. Unlike the previously discussed technique, this method requires exactly synchronized, complementary gate control pulses. Also, to reduce transients when MOSFETs are used, each substrate terminal is reverse biased and held at a fixed voltage by connecting it to common or to an appropriate power supply terminal, decoupled, for example, with a 100 kΩ, 0.01 $\mu$F low-pass filter.

Switching transients are short pulses which constitute a smaller fraction of the total signal at *lower* switching frequencies, and therefore are usually less troublesome at such frequencies. The amplitude of the transients in low switching frequency applications can be reduced by slowing down the rise time of the gate control signal. Turning a shunt switch on before opening a series switch and turning it off after the series switch is closed may also help because the transients of both switches are shunted to ground or through the low output impedance of the signal source. Synthetic transients of opposite polarity can be coupled into the output voltage to compensate partially for the switching transients. For some applications transients may be accepted as part of the total signal and then later subtracted to obtain the signal of interest.

When switching times of less than about 0.1 $\mu$s are needed, then **diode switches** are used. They require matched diodes and very careful control of the gate control signal to minimize offset voltages, but switching times of less than 1 ns are obtained for use with sampling oscilloscopes and other high-frequency sampling applications. A four-diode series switch analog gate is shown in Fig. 2.27. The control voltages $+V$ and $-V$ must be greater than the positive and negative excursions of the input signal, respectively. When the control voltages are equal in magnitude, the resistors are equal, and the forward and reverse current-voltage characteristics of the diodes are matched, then there is no voltage offset at the output whether the switch is OPEN or CLOSED. When the gate is CLOSED (to pass the signal) all four diodes are in a forward biased conducting state (of low dynamic resistance), and a signal of either polarity is conducted through the gate to the output. When the control voltages are reversed to OPEN the gate, all the diodes are reverse biased and have a high resistance for the OFF condition of the series switch. The rise times of the control voltages for both terminals should be equal to minimize switching transients.

**Bipolar junction transistors** may be used as switches to gate signals having both positive and negative polarities. Their ON and Off collector-emitter resistances are comparable to those of the FET. Their main disadvantage is that their collector-emitter current-voltage curve does not pass through the origin as it does for an FET. There is a temperature-dependent collector-emitter offset voltage on the order of 1–10 mV at zero collector-emitter current. When a

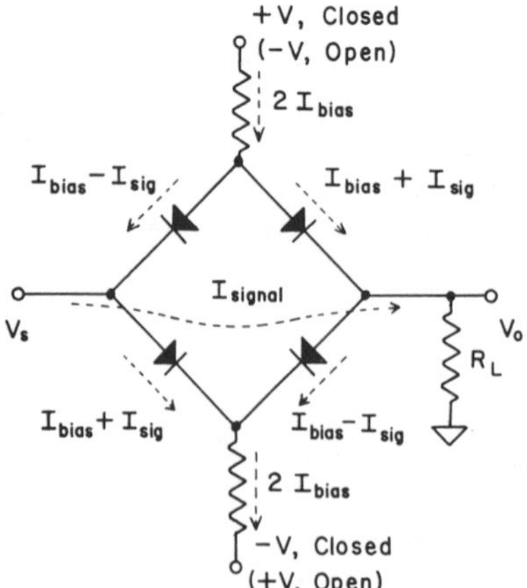


**Fig. 2.27.** A fast series-type analog signal gate made from diode switches.

bipolar junction transistor is operated in an inverted mode (e.g., an *npn* connected as a *pnp*) the offset voltage may be reduced by a factor of 10. The offset may cause negligible error for large signal voltages, but is particularly troublesome for small signals because its variation with time and temperature make compensation for it difficult. This difficulty has helped the FET to become the more popular of the two types of switches for many applications.

## B. AMPLIFIERS

Both current and voltage amplification are important in electronic signal processing and control systems. Amplifiers are used to increase the *power level* (current-voltage product) of a signal so that it is not severely attenuated or distorted while being processed or used to control a meter or other devices. The signal power level may be increased by increasing the magnitude of the *voltage* of the signal, its *current*, or *both*. The way in which the signal power is amplified depends both upon the value of the input impedance of the amplifier *relative* to the output impedance of the signal source, and upon the value of the output impedance of the amplifier relative to the impedance of the load that it is driving. An amplifier that might nominally be a voltage amplifier may in fact become a current amplifier if the external source and load impedances are changed, without making any changes in the characteristics of the amplifier itself. Impedance considerations are therefore important when dealing with amplifiers.

A **voltage amplifier** has a relatively *high input impedance* so that it does not significantly load down a voltage source. It also has a relatively *low output impedance* compared to the load so that its output voltage is relatively independent of changes in the load. A **current amplifier** has just the opposite characteristics. It has a relatively *low input impedance* so that essentially all of the current from a current source is pulled through the amplifier input (rather than being bypassed through the Norton's equivalent impedance of the current source), and has a relatively *high output impedance* so that changes in the load do not significantly influence the output current of the amplifier.

Other useful amplifiers include a **current-to-voltage converter** amplifier which has a relatively low input impedance compared to the source impedance, and a relatively low output impedance compared to the load. A **voltage-to-current converter** amplifier has a relatively high input impedance and a relatively high output impedance. A **power amplifier** usually takes advantage of the fact that maximum power is transferred to the load when the output impedance of the amplifier is equal to the load impedance. The term "power amplifier" is usually reserved for an amplifier which is able to supply more power, for example, to a control circuit of an external device, than the power ordinarily needed for voltage or current signal processing. Rarely does a power amplifier supply an amount of *power* that is independent of the load.

To summarize, the output of an amplifier supplies either a *current signal*, *power signal*, or *voltage signal* to a load depending upon whether its output impedance is, respectively, much higher, about equal to, or much lower than the load impedance. The amplifier responds to a signal source as though it were a *current source*, *power source*, or *voltage source* depending upon whether the input impedance of the amplifier is much less than, about equal to, or much greater than the output impedance of the source.

### 1. Basic Amplifier Circuits

Integrated circuit amplifiers and other commercially produced amplifiers are available and widely used for signal processing and control circuits. It is not uncommon, however, to find a situation where an inexpensive and relatively simple one- or two-transistor amplifier circuit such as those discussed below may solve a problem or improve the characteristics of an available commercial amplifier. An understanding of the basic transistor amplifier circuits discussed below will help to determine when such amplifiers may be used and will also provide a background for discussing the more complex commercial amplifiers which use these circuits.

A typical amplifier circuit consists of a **load**, usually a resistor, and an **active device** such as a transistor or vacuum tube (Fig. 2.28*a*). The characteristic curves of the active device (Fig. 2.28*b*) and the resistance of the load interact with each other to determine the amplifying and impedance characteristics of the amplifier circuit. The voltages and currents in the circuit that result from this interaction can be determined with the help of the equation that results when the voltages are summed around the output circuit loop (Fig. 2.28*a*):

$$V_{DD} = V_{DS} + V_L = V_{DS} + I_D R_L \tag{66}$$

where $V_{DD}$ is the power supply voltage, $I_D$ is the current through the load resistor $R_L$ and into the top terminal of the active device, and $V_{DS}$ is the voltage across the two output terminals of the active device expressed either as a function of the input signal voltage (or signal current as could be the case for a

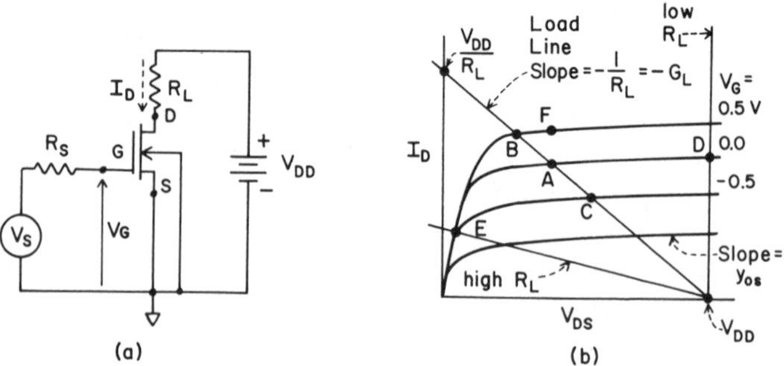


**Fig. 2.28.** (*a*) A common source FET amplifier circuit. (*b*) The operating points of the circuit are determined by the intersections of the load line and the characteristic curves of the FET.

bipolar junction transistor.) Equation 66 is now superimposed on the characteristic curves by rearranging the equation to give

$$I_D = \frac{V_{DD}}{R_L} - \frac{V_{DS}}{R_L} \tag{67}$$

On a graph with axes $I_D$ versus $V_{DS}$ (Fig. 28*b*), this equation represents a straight line with slope $-1/R_L$. This line is called the **load line** and it intersects the $I_D$ axis at maximum current $I_D = V_{CC}/R_L$ (where $V_{DS} = 0$, equivalent to the case where the active device is a short circuit), and the $V_{DS}$ axis at the power supply voltage $V_{DS} = V_{DD}$ (where $I_D = 0$, equivalent to the case where the active device is an open circuit). All possible operating points for this circuit, regardless of the active device that is used, lie on the load line. When the characteristic curves of a particular active device are plotted on the graph, the intersections of these curves with the load line determine the actual locus of operating points.

For example, the depletion mode MOSFET may be used in a **common-source configuration** as a simple voltage-to-current converter amplifier (Fig. 28*a*) having an extremely high input impedance, because of the insulated gate, and moderate output impedance. This circuit handles both positive and negative input voltages without the need for special bias circuits, but provides only a unidirectional output current of varying magnitude through the load. When the signal voltage $V_G$ at the gate is zero, the current and voltages are determined by operating point A (Fig. 2.28*b*). This particular operating point (zero signal) is called the **quiescent** point. A signal of $V_G = +0.5$ V would move the operating point to B and $V_G = -0.5$ V would move it to C. Changing the load resistance to a very small value near 0 Ω would make the load line vertical and the quiescent point would move to point D. Notice that because of the horizontal nature of the characteristic curves, the current $I_D$ did not change very much from point A to D even though the load resistance changed dramatically. This indicates that the output impedance of this amplifier circuit is very large. The slope of this almost horizontal characteristic curve of the active device is called the **output admittance,** $Y_{os}$. (Admittance is the inverse of impedance, just as conductance is the inverse of resistance.) Smaller values of this variable give higher output impedances and better constant current control when changes in the load resistance occur. When the load resistance becomes too high, the quiescent point moves into the region below the pinch-off voltage, point E, where the output admittance is high and the current changes significantly when the load resistance changes. The operating region is therefore usually chosen to be in the region where the characteristic curves are horizontal. The output admittance $Y_{os}$ in specification sheets is given in units of $\mu$mhos for a typical horizontal region of the characteristic curves. The proportionality factor relating the change in load current for a given change in signal (gate) voltage is proportional to the spacing between the horizontal characteristic curves in Fig. 2.28*b*. This vertical spacing is called the **mutual conductance, forward transadmittance**, or **transconductance** $g_m$ defined as $\Delta I_D/\Delta V_G$. The actual change in current in going from point A to point B in Fig. 2.28*b* is slightly less than the vertical spacing between points A and F because of the slight downward slope from point F to B on the constant gate-voltage curve. The actual change for a given load resistance therefore slightly depends upon the load resistance $R_L$ or its conductance $G_L$ and the output admittance, and can be found from Fig. 2.28*b* using geometric considerations and taking into account the slopes of the characteristic curve $Y_{os}$ and the load line, $G_L$:

$$\left(\frac{\Delta I_D}{\Delta V_G}\right)_{R_L} = Y_{fs}\frac{G_L}{G_L + Y_{os}} = Y_{fs}\frac{1}{1 + Y_{os}R_L} \cong Y_{fs} \text{ for } R_L Y_{os} \ll 1 \tag{68}$$

When $R_L$ is not zero a change in output voltage accompanies the change in current. The circuit may then be considered a voltage amplifier with a voltage gain found by multiplying equation 68 by $-R_L$:

$$\text{Voltage gain} = \frac{\Delta V_{\text{out}}}{\Delta V_G} = \frac{-R_L \Delta I_D}{\Delta V_G} = Y_{fs}\frac{-R_L}{1 + Y_{os}R_L} \cong -Y_{fs}R_L \tag{69}$$

where the minus sign accounts for the decrease in $V_{DS}$ with an increase in signal voltage. The transconductance and the load resistance are therefore the major influences on the voltage gain of an FET common-source amplifier. These approximate results are also applicable to the vacuum tube pentode or any other active device that has equally spaced *horizontal* characteristic curves for constant input *voltage* signals.

For a junction FET (Fig. 2.29) it may be desirable to bias the gate-source voltage to a nonzero value so that the quiescent point occurs at some other point along the load line. This may be done by adding a resistor between the source and ground. Since a current flows through this resistor, the source is at a positive voltage with respect to ground, causing the gate voltage to be negative with respect to the source voltage when the gate voltage is zero with respect to ground. A large capacitance may be placed in parallel with the source resistor to act as a short circuit to **bypass** ac signals around the resistor. Otherwise the resistor would reduce the gain of the circuit to ac signals.

An analysis of the operating characteristics of a **bipolar junction transister common emitter circuit** (Fig. 2.30*a*) could be done using the same type of characteristic curves and equation parameters as were used for the FET. A more

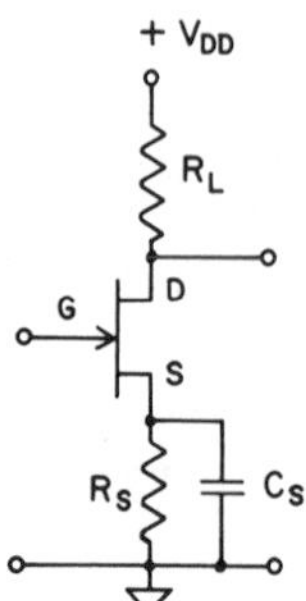


**Fig. 2.29.** A self-biasing circuit for a JFET common-source amplifier.

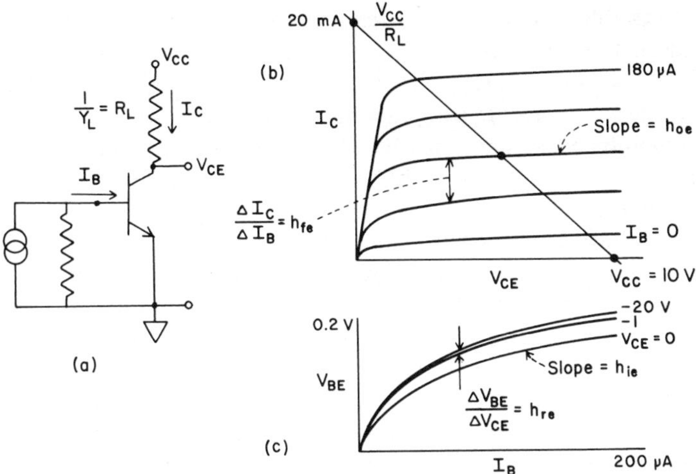


**Fig. 2.30.** Characteristic curves and a common emitter amplifier circuit for a bipolar junction transistor.

convenient set of curves and parameters exists, however. Instead of using collector current-voltage curves for constant base *voltages*, curves at constant base *currents* are used because they are more evenly spaced than the other curves. Typical characteristic curves are shown in Figs. 30*b* and *c*. This circuit is a current amplifier with a gain close to $h_{fe}$. It may also be considered a current-to-voltage converter, with a transfer characteristic that can be derived from the load line and characteristic curves in Figs. 2.30*b* and *c*. The use of this circuit as a *linear* voltage amplifier is limited to input voltage signal amplitudes of less than about 10 mV because of the nonlinear current-voltage characteristic curves (Fig. 2.30*c*) of the forward biased base-emitter *pn* junction.

The quiescent (no signal) operating point may be placed at a selected point on the load line by adding a resistor between $V_{cc}$ and the base to bias the base current to the desired magnitude (Fig. 2.31*a*). This configuration is relatively

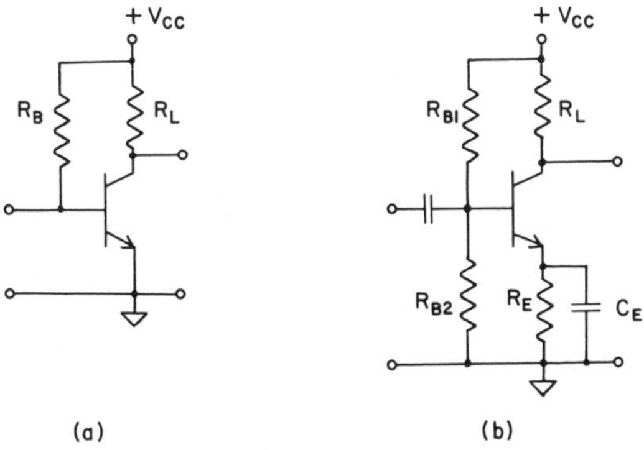


**Fig. 2.31.** Common emitter amplifier circuits with (*a*) a simple base current biasing resistor, and (*b*) a temperature-compensating biasing system, with capacitive coupling to the signal source.

thermally unstable and two other resistors may be added to the circuit to improve thermal stability as shown in Fig. 2.31*b*. Design considerations for the values of these resistors are given in most textbooks. The circuit is included here to aid the reader in recognizing the significance of transistor circuit components. **Coupling capacitors** must be added to the input and output terminals to isolate the dc levels of one amplifier stage from the other stages. At sufficiently high frequencies they act as short circuits to ac signals.

The output impedance of the bipolar junction transistor when driven by an ideal current source is the inverse of the **output admittance** $h_{oe}$ (the slope of the collector current-voltage curve) provided that the base current, in the common emitter circuit, is constant. However, changing the collector (output) current changes the collector-emitter voltage which in turn influences the base-emitter voltage. A measure of the influence is called the **reverse transfer voltage ratio** $h_{re} = \Delta v_{BE}/\Delta v_{CE}$. If the constant current source is not ideal (its output resistance is not infinite) this change in base voltage causes the base current to change, shifting the output operating point from one constant base current curve to another in Fig. 2.30*b*. The output impedance therefore differs from the inverse of $h_{oe}$ by an amount indicated in Table 2.III below.

Interdependencies of the input and output currents and voltages also cause the input impedance to vary from $h_{ie}$ and depend upon the load resistance unless the load resistance is zero. Similar interdependencies are present for the FET although their effects are sometimes not as apparent because the approximations needed to approach ideal behavior (input impedance $= 1/Y_{is}$, output impedance $= 1/Y_{os}$) are often met in FET circuits.

A **circuit model** for an active device is often used to display these interdependencies and to help derive equations for voltage and current gains and input and output impedances. Thevenin's or Norton's equivalent circuit models can be used for the input terminals and for the output terminals. The output resistance of a transistor is usually much greater than the load. Therefore Norton's model of a current generator in parallel with a resistance is convenient to use for the output terminals because the output resistance for this model can be essentially ignored when it is much greater than the load. For the input, Norton's model is usually used for an FET, whereas Thevenin's model is commonly used for a bipolar junction transistor. The admittance ($y$) parameters and hybrid ($h$) parameters are obtained in a straightforward way using four-terminal linear network analysis. The **admittance ($y$) parameters** are derived by considering the input voltage $V_{GS}$ and output voltage $V_{DS}$ as independent variables and the input and output currents as dependent variables, in a set of linear equations shown in Fig. 2.32*a*. The admittance parameters are the coefficients of the independent variables. For each equation, each admittance parameter is the partial derivative of the dependent variable with respect to the independent variable by which it is multiplied. For instance, $y_{fs} = \partial I_D/\partial V_{GS}$. They are called admittances because each is the ratio of a change in current to a change in voltage.

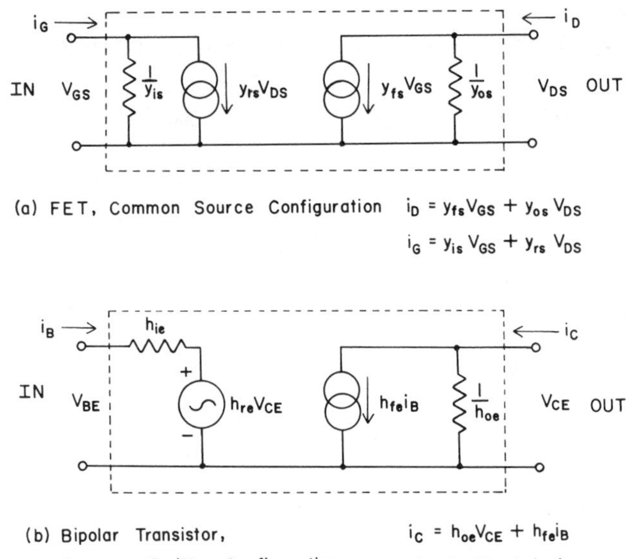


**Fig. 2.32.** Small signal equivalent circuits for (*a*) an FET connected in the common-source configuration, and (*b*) a bipolar junction transistor connected in the common-emitter configuration.

The **hybrid (*h*) parameters** are derived by considering the output voltage and input current as independent variables and the input voltage and output current as dependent variables (Fig. 2.32*b*). They are termed hybrid because they are a mixture of voltage change and current change ratios.

The circuit models and the parameters assume that the interrelationships among the four circuit variables are linear. They are therefore valid over an operating region where the parameters do not change significantly. This usually means that the output voltage signal amplitude must be less than a few tenths of a volt. Of course larger output signals may be used for these simple circuits if nonlinear results and deviations from calculated values are tolerable. Usually, however, more complex circuits with large negative feedback to minimize distortion (such as operational amplifier circuits) are used when large signals are processed.

These circuit models assume that the frequency is low enough (say, below 1 MHz) so that interelectrode capacitances may be neglected. At high frequencies these effects become important and the parameters must be treated as complex numbers. A detailed analysis of high-frequency models is found in advanced texts. In general, the gain of a transistor decreases at high frequencies, and phase shifts become important. The input and output impedances take on a capacitive nature. Specification sheets for FET's often include the capacitive components $C_{iss}$ and $C_{rss}$ of the admittance parameters $Y_{is}$ and $Y_{rs}$ because these components become more important than the resistive components at

TABLE 2.III

Definitions and Circuit Variable and Parameter Relationships for Common-Emitter and Common-Source Circuits

| Common-emitter relationships | Common-source relationships |
|---|---|
| $h_{ie} = \dfrac{\Delta v_{be}}{\Delta i_b}$ = input impedance | $y_{is} = \dfrac{\Delta I_G}{\Delta V_{GS}}$ = input admittance |
| $h_{oe} = \dfrac{\Delta i_c}{\Delta v_{ce}}$ = output admittance | $y_{os} = \dfrac{\Delta I_D}{\Delta V_{DS}}$ = output admittance |
| $h_{re} = \dfrac{\Delta v_{be}}{\Delta v_{ce}}$ = reverse transfer voltage ratio | $y_{rs} = \dfrac{\Delta I_G}{\Delta V_{DS}}$ = reverse transfer admittance |
| $h_{fe} = \dfrac{\Delta i_c}{\Delta i_b}$ = forward transfer current ratio (current gain) | $y_{fs} = \dfrac{\Delta I_D}{\Delta V_{GS}}$ = forward transadmittance or transconductance |
| $Z_{\text{out}} = \dfrac{h_{ie} + R_s}{\Delta h + h_{oe}R_s} = \dfrac{1}{h_{oe}}$ for $R_s = \infty$ | $Z_{\text{out}} = \dfrac{y_{is} + Y_s}{\Delta y + y_{os}Y_s} = \dfrac{1}{y_{os}}$ for $R_s = 0$ |
| $= \dfrac{h_{ie}}{\Delta h}$ for $R_s = 0$ | $\Delta y = y_{is}y_{os} - y_{rs}y_{fs}$ |
| $\Delta h = h_{oe}\,h_{ie} - h_{fe}\,h_{re}$ | $Z_{\text{in}} = \dfrac{y_{os} + Y_L}{\Delta y + y_{is}Y_L} = \dfrac{1}{y_{os}}$ for $R_L = 0$ |
| $Z_{\text{in}} = \dfrac{h_{ie} + \Delta h R_1}{1 + h_{oe}R_L} = h_{ie}$ for $R_L = 0$ | $Y_s$ = voltage source admittance $(1/R_s)$ |
| $R_s$ = voltage source resistance | $Y_L$ = load admittance $(1/R_L)$ |
| $R_L$ = load resistance | |
| Current gain $\dfrac{\Delta I_c}{\Delta I_b} = \dfrac{h_{fe}}{h_{oe}R_L + 1} \cong h_{fe}$ for $Y_L >> h_{oe}$ | |
| Voltage gain $\dfrac{\Delta v_{ce}}{\Delta v_{be}} = \dfrac{-h_{fe}R_L}{h_{ie} + R_L\,\Delta h} \simeq -h_{fe}\dfrac{R_L}{h_{ie}}$ | |

moderate frequencies. For instance, the input impedance of an FET may be so high as to have a negligible loading effect compared to even the small picofarad input capacitance. However, at low frequencies even the input capacitance is so small that it is not significant until signal sources with output resistances of many megohms are used. Table 2.III summarizes the important relationships for the common-emitter and common-source circuit configurations. Other circuit configurations such as the common base, common collector, and common drain may also be analyzed using four-terminal network analysis to obtain a new set of parameters for each configuration. The reader is referred to textbooks for such analyses. It is more convenient, however, to use the common-emitter or common-source parameters to calculate the characteristics of these other circuit configurations because these parameters are more easily measured or available in specification sheets. Parameters for one configuration may be converted to those for another configuration using simple relationships presented in textbooks. The results of an analysis of the very useful common-collector configuration are now presented in terms of common-emitter parameters.

The **emitter-follower (common-collector)** configuration (Fig. 2.33) is a widely used circuit for output stages because of its low output impedance. The collector is common to both input and output signals since it is at a constant voltage with

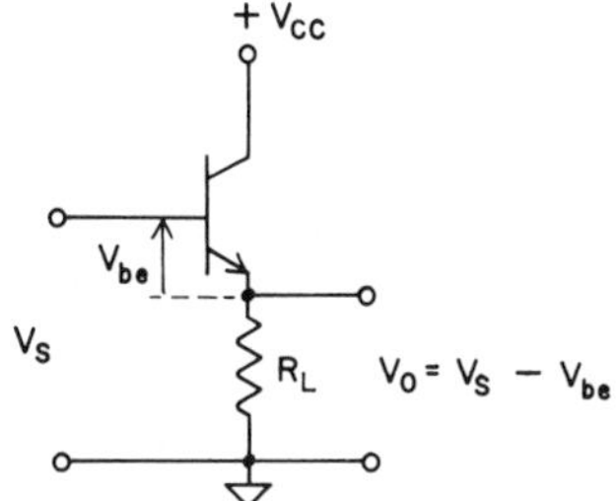


**Fig. 2.33.** An emitter-follower (common-collector) amplifier circuit.

respect to both. The emitter follower has other desirable characteristics including a high input impedance and good linearity over a wide voltage range about equal to the power supply voltage $V_{cc}$.

A positive signal voltage forward biases the base-emitter *pn* junction, turning the transistor on, causing current to flow in the load resistor. An increase or decrease in signal voltage causes corresponding changes in the current through the load and hence the output voltage. Since according to Fig. 2.33

$$v_o = v_s - v_{be} \tag{70}$$

the output voltage $v_o$ is always less than the input voltage $v_s$ by a small amount $v_{be}$, usually 0.1–1 V. The voltage gain is

$$A_v = \frac{dv_o}{dv_s} = \frac{d(v_s - v_{be})}{dv_s} = 1 - \frac{dv_{be}}{dv_s} \tag{71}$$

which is close to unity when the change in base-emitter voltage $dv_{be}$ is relatively small compared to the change in signal voltage $dv_s$, as it typically is. An expression that indicates how the voltage gain varies with load resistance due to the interaction of the input and output variables is

$$A_v = \frac{R_L}{R_L + [h_{ie}/(1 + h_{fe})]} \tag{72}$$

which shows that the gain is close to unity when the load resistance $R_L$ is much greater than the transistor input resistance $h_{ie}$ (typically a few thousand ohms) divided by 1 plus the current gain $h_{fe}$ (typically 10 for a power transistor and 100 for other transistors).

The current gain for the circuit is approximately $h_{fe}$, or more accurately,

$$A_i = \frac{(1 + h_{fe})}{R_L h_{oe} + 1} \cong 1 + h_{fe} \text{ for } R_L h_{oe} \ll 1 \tag{73}$$

The relatively high input impedance is approximately the load resistance times the current gain:

$$Z_{in} = h_{ie} + \frac{(1 + h_{fe})R_L}{R_L h_{oe} + 1} \cong (1 + h_{fe})R_L \tag{74}$$

The low output impedance depends on the voltage signal source output

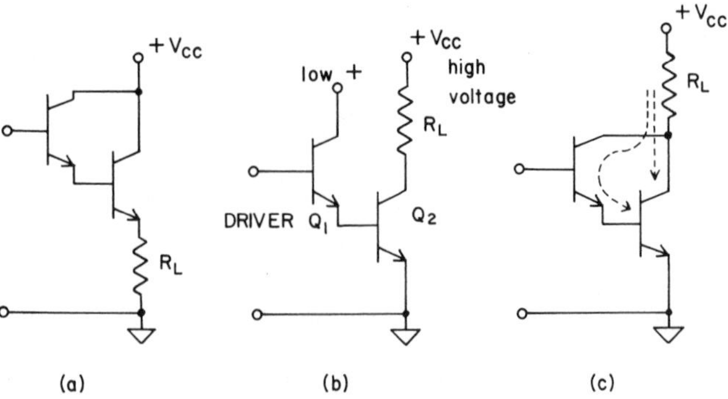


**Fig. 2.34.** Cascading two transistors to increase the current gain: (*a*) and (*c*) Darlington configurations, and (*b*) cascaded emitter-follower circuits.

resistance $R_s$ and is approximately equal to the total resistance in the base circuit divided by the current gain:

$$Z_{\text{out}} = \frac{h_{ie} + R_s}{h_{oe}(h_{ie} + R_s) + (1 + h_{fe})} \cong \frac{h_{ie} + R_s}{1 + h_{fe}} \tag{75}$$

These equations show that a transistor with high $h_{fe}$ produces an emitter-follower circuit having the best characteristics.

An emitter follower is sometimes used to drive a high-voltage power transistor, which often has a low current gain, $h_{fe}$. The cascaded emitter-follower configuration is shown in Fig. 2.34*a*. The equivalent current gain of this circuit is the product of the current gains of the two transistors. When this combination is used as an inverter amplifier, two configurations of the driver transistor, $Q_1$, are possible. One configuration, which connects the two collectors together, is called the Darlington (Figs. 2.34*a* and *c*). The other is the emitter-follower driver configuration in Fig. 2.34*b*. For the circuits in Figs. 2.34*a* and *c*, the two transistors may be packaged together in one housing to simulate a single power transistor with high gain. The Darlington configuration in Fig. 2.34*c* has a higher current gain than the circuit of Fig. 2.34*b*, since both collector currents pass through the load in the Darlington configuration. The true emitter-follower driver configuration (Fig. 2.34*b*), on the other hand, allows the driver transistor to use a lower power supply voltage than the power transistor, and thermally separates the driver from the power transistor.

A push-pull configuration of complementary *pnp* and *npn* transistors (Fig. 2.35) is often used when high power or large currents are required. Each transistor acts as an emitter follower for signals of one polarity or the other, and is essentially off for signals of opposite polarity. The transistor that is off acts as a load resistor that is very large, and therefore dissipates little power. When the top transistor is on, it pushes current into the load; when the bottom transistor is on it pulls current from the load.

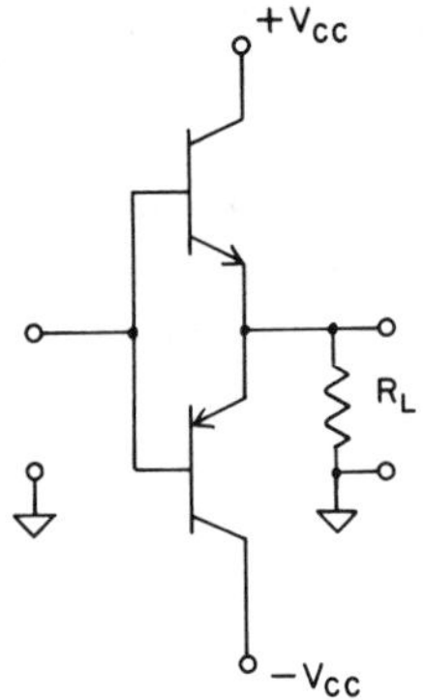


**Fig. 2.35.** A complementary *npn-pnp* push-pull driver circuit.

## 2. Difference Amplifiers

A **difference** or **differential amplifier** is widely used as the input stage to an operational amplifier, and as a voltage comparator whose output is high or low depending upon which of the two input voltages is more positive. Its complementary outputs can provide perfectly symmetrical signals for driving analog switches.

A basic differential amplifier is shown in Fig. 2.36*a*. The constant current source shown in Fig. 2.36*a* may be approximated by a simple resistor, but much better results are obtained with constant current sources made with active

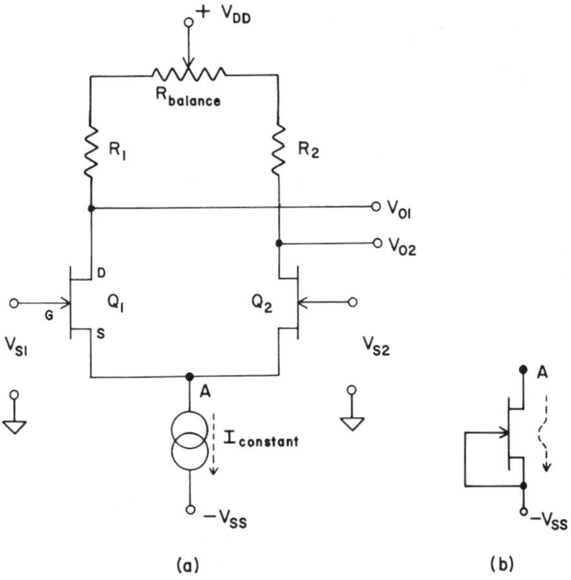


**Fig. 2.36.** An FET difference amplifier circuit (*a*), and a constant current circuit (*b*).

devices such as an FET (Fig. 2.36*b*) with a constant gate-source voltage (e.g., with the gate connected to the source so that $V_{GS} = 0$). The FET must be selected or designed to provide the desired current.

The output of the differential amplifier is proportional to the *difference* between its two input signals. The output may be taken between the two drains (**differential output mode**), for instance, to drive another differential amplifier stage, or between one drain and ground (**single ended output mode**). The gain of the single-ended output mode is one-half that of the differential output mode. The two outputs may be considered as separate signals, in which case the signals are complementary or mirror images of each other. The amplifier may be used with one input grounded and one active (**single-ended input**), or both inputs active. To understand the operation of the circuit, connect $v_{S2}$ to common. Transistor $Q_1$ acts as a voltage inverter amplifier with output $v_{O1}$ and input $v_{S1}$. An increase in $Q_1$, causing a corresponding decrease in the source current of $v_{S1}$ increases the source current of $Q_1$, causing a corresponding decrease in the source current of $Q_2$, since the sum of the two source currents is held constant by the constant current source. Because the drain currents through the resistors are equal to the corresponding source currents, the current through $R_2$ decreases by the same amount as the current through $R_1$ increases. The voltages $v_{O2}$ and $v_{O1}$ therefore change equally but in opposite directions to each other. By recognizing the symmetry of the circuit it is seen that a change in the other input voltage $v_{S2}$ causes the output voltages to change in just the opposite directions to those described. Thus either output voltage is proportional to the difference in the two input signals.

A well-balanced differential amplifier with adjacent transistors is relatively independent of power supply fluctuations and thermal changes because changes in the operation points of both transistors compensate each other. A finite output **voltage offset** that occurs when both inputs are 0 V (grounded) may be reduced by a **voltage balance** potentiometer in the drain circuit. Voltage rebalancing is occasionally necessary to compensate for minor changes in the components over a period of time. Input currents are usually not the same at both inputs and are more difficult to compensate because of their large thermal coefficients. Because the bias current causes a voltage drop when it passes through the output resistance of the voltage source that drives an input, a difference in bias currents between the two inputs causes an apparent voltage offset when the voltage source output resistances are relatively large (even though the resistances may be equal). *It is usually better to choose an amplifier with small intrinsic bias currents than to try to compensate large bias currents.*

When the transistors are matched, all the resistors identical, and the constant current source ideal, then it should be possible to apply the same signal commonly to both inputs and obtain no output signal. Because of imperfections, however, a small output signal is usually observed. The ratio of this output voltage to the common input voltage is called the **common mode gain**, and is ideally zero. The **differential mode gain** is the ratio of the output voltage to the input voltage measured as a difference in voltage between the two input terminals. This gain is of the order of magnitude of the gain for a simple voltage inverter amplifier. The **common mode rejection ratio (CMRR)** is a measure of how closely the differential amplifier behaves as an ideal differential amplifier:

$$CMRR = \frac{\text{differential mode gain}}{\text{common mode gain}} \tag{76}$$

The CMRR is ideally infinite. In practice, values of CMRR = $10^5$ (or 100 db = 20 log $10^5$) are obtained by adding a second differential stage to increase the differential mode gain, and using large negative feedback for the common mode signal (but not the differential signal) to reduce the common mode gain. Common mode feedback may be accomplished, for example, by connecting an inverter amplifier between the FET sources of the second stage and the FET sources of the first stage.

A differential amplifier has a limited voltage range beyond which its CMRR falls below its best (or specified) value. This range is called **common mode range**.

Even the best differential amplifier can have its effective CMRR defeated if the output impedances of the voltage signal sources are not identical or are not small enough to be neglected compared to the input impedances of the amplifier. For a differential amplifier with infinite CMRR and equal input impedances $R_{\text{in}}$, signal source resistances of $R_2$ and $R_1$ reduce the ideally infinite CMRR for the entire circuit to approximately

$$CMRR = \frac{R_{\text{in}}}{R_2 - R_1} \text{ for } R_2, R_1 \ll R_{\text{in}} \tag{77}$$

For $R_{\text{in}} = 1\ M\Omega$ a difference in signal source resistances of only 100 Ω would limit the effective CMRR to a maximum of 80 db ($10^4$).

When the *RC* time constants of the signal source output resistances in series with the differential amplifier input capacitances and cable capacitances are larger than a small fraction of the period of the highest signal frequency, then these time constants for both inputs must be matched if a high CMRR is to be maintained for the circuit. Otherwise the signal that is common to both channels at the voltage source end of the cables is attenuated and phase shifted differently, and therefore appears in a different shape at each of the two amplifier inputs. Subtraction by the amplifier of the two distorted "common mode" signals produces a finite difference at the amplifier output, when ideally the output should be zero when a common mode signal is present.

### 3. Amplifier Feedback

Feeding a part of an amplifier's output signal back to the input can improve the amplifying and impedance characteristics of the overall amplifier circuit. Unplanned feedback can cause undesirable effects such as amplifier oscillation and overshoot and ringing at the leading and trailing edges of a square pulse signal. **Feedback** is widely used not only in amplifiers, but in many other aspects

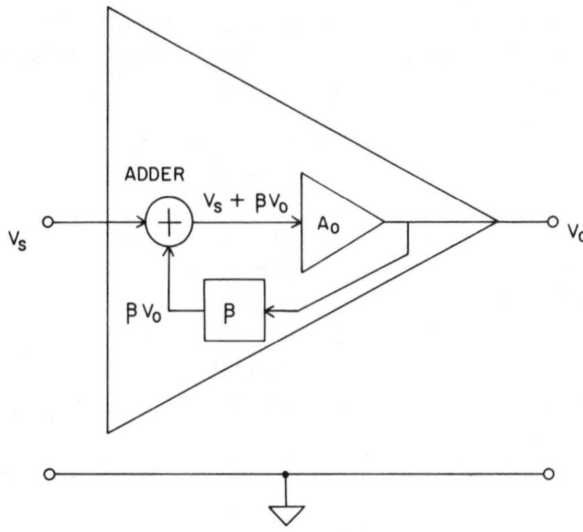


**Fig. 2.37.** An amplifier with feedback.

of instrumentation and control. A more thorough discussion of feedback is found in Chapter 3.

The general form of an amplifier with feedback is shown in Fig. 2.37. The voltage gain of the amplifier without feedback is $A_0$, and with feedback is $A_f = v_o/v_s$. A fraction $\beta$ of the output voltage is fed back and added to the input voltage signal and the result used as the input to the original amplifier without feedback. Then

$$v_o = (v_s + \beta v_o)\, A_o \tag{78}$$

and

$$v_o = v_s A_f \tag{79}$$

Substituting Equation 79 into 78 eliminates $v_s$ and then $v_o$ cancels to give the final relationship between the gain of the overall circuit with feedback and the gain of the amplifier without feedback

$$A_f = \frac{A_o}{1 - \beta A_o} \tag{80}$$

When $\beta A_o$ is greater than zero the feedback is called positive feedback. When $\beta A_o$ is equal to or greater than unity the circuit oscillates because the voltage fed back to the input of the amplifier is then sufficient to sustain the output voltage even when $v_s = 0$. When $\beta A_o$ is negative the feedback is called negative feedback and $A_f$ is less than $A_0$. When $\beta A_0$ is large and negative, equation 80 reduces to

$$A_f \cong \frac{1}{\beta} \tag{81}$$

thus essentially eliminating the effect of the gain $A_o$ of the amplifier upon the gain $A_f$ of the overall circuit. When $\beta$ is prepared from highly stable components, such as precision resistors, then an amplifier circuit with a very stable gain results. In practice $\beta A_o$ is made very large by choosing an amplifier with large gain $A_o$.

The effect of **noise** introduced by an amplifier inside the feedback loop is reduced by that part of the amplifier gain that occurs *prior* to the introduction of the noise. Consider the amplifier as having two stages with voltage noises $N_1$ and $N_2$ introduced at the input of each stage as shown in Fig. 2.38. Then

$$v_o = A_1 A_2(v_s + \beta v_o + N_1) + A_2 N_2 \tag{82}$$

which can be rearranged to give

$$v_o = \frac{A_1 A_2}{1 - \beta A_1 A_2}\left(v_s + N_1 + \frac{N_2}{A_1}\right) \tag{83}$$

Equation 83 shows that the noise introduced at the first stage is not attenuated relative to the signal $v_s$, although the noise introduced at the second stage is reduced by the factor $1/A_1$ relative to the signal voltage $v_s$. *This emphasizes the importance of having a relatively noisefree input stage of moderate gain even when feedback is used.*

It is worthwhile to look at several ways in which feedback can be implemented in order to understand especially how undesirable positive feedback can be prevented. An emitter follower (and source follower) is an example of a simple amplifier circuit with high negative feedback. The output signal that appears across the emitter resistor (Fig. 2.33) is subtracted from the input signal at the base to produce the base-emitter input voltage to the amplifier. The large negative feedback can be used to account for the circuit gain, which is slightly less than unity, for the improved linearity, and for the increased input impedance and lower output impedance compared to the voltage inverter (common emitter) amplifier. An emitter resistor added to the voltage inverter amplifier adds negative feedback which reduces the gain of the amplifier, but somewhat improves its linearity. Feedback paths from the collector to the base and from

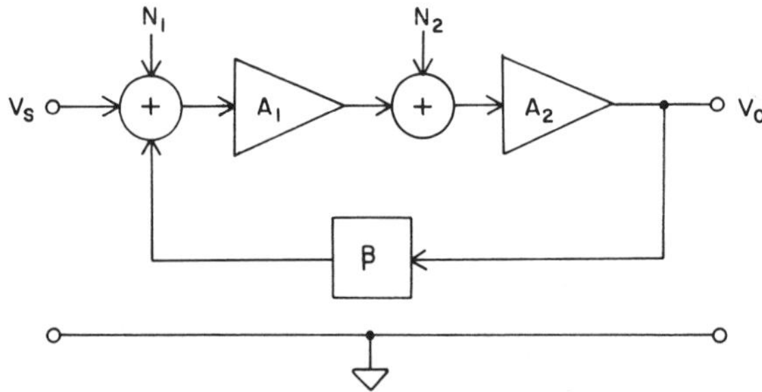


**Fig. 2.38.** A two-stage amplifier with feedback. Noise signals are shown introduced at the inputs of each stage.

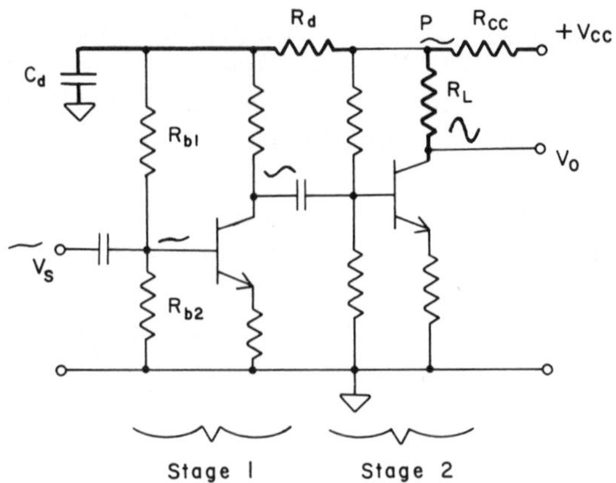


**Fig. 2.39.** A two-stage amplifier circuit showing the output resistance $R_{cc}$ of the power supply, and an *RC* decoupling network $R_d C_d$ that prevents oscillation.

the collector to the emitter introduce negative feedback when the path is through a resistor. However, when a path is via an *RC* network (stray capacitance included!) the network may produce a phase shift of 180° which changes a negative feedback path into a positive feedback path at a given frequency. Then if $\beta a_o$ exceeds unity, oscillation may occur.

An important positive feedback path, which can be readily eliminated, occurs when two or more amplifiers or amplifier stages are cascaded and connected to the same power supply as shown in Fig. 2.39. A small fraction of the output signal appears at the power supply terminal P because of the small but finite output resistance $R_{cc}$ of the power supply: $R_L$ and $R_{cc}$ act as a voltage divider. Then, assuming $R_d = 0$ and $C_d = 0$ for the moment, a fraction $R_{b2}/(R_{b1} + R_{b2})$ of this signal is fed back to the input terminal at the base of the first stage where it is added in phase with the input signal. This is a positive feedback path and may cause oscillations and certainly some distortion. Oscillation can be eliminated and the distortion reduced by **decoupling** the power supply from the first stage by a small resistor $R_d$ and large capacitor $C_d$ which act as a low-pass filter to attenuate the feedback signal. Values of 10 Ω and 1–10 $\mu$f are typical for ordinary signal processing amplifiers, but smaller resistances and larger capacitors may be needed for power amplifiers. Sometimes the resistance of the lead wires or conducting strips on a printed circuit board may be sufficient to act as $R_d$. *When several operational amplifiers are driven from the same power supply each one should be decoupled in this way from both the positive and negative power supplies.*

## 4. Commercial Amplifiers

Many types of amplifiers are commercially available to satisfy different needs. A **preamplifier** is designed to be placed very near a transducer which has a high output impedance so as to provide a signal with low output impedance that can be used to drive a cable. It should have low input noise and wide bandwidth, and might have moderate gain. It may have a differential output to drive shielded twisted-pair wire cables or double-shielded cables to minimize ground loop problems. **Fixed-gain** amplifiers may have a few selectable gains of usually less than 100. Some have selectable frequency bandwidth controls to reduce gain outside the desired frequency region in order to reduce noise.

**Power amplifiers** and **booster amplifiers** are designed to deliver large powers to a load. Maximum power transfer (and therefore maximum power efficiency) occurs when the output resistance of the amplifier is equal to the load resistance. Because they may dissipate considerable heat, power amplifiers should be isolated from sensitive circuits such as preamplifiers, and provided with adequate means for heat dissipation. Precise control can be achieved by including the power amplifier in a feedback loop with the power amplifier thermally isolated from the rest of the circuit.

An **instrumentation amplifier** is a dc amplifier that is capable of excellent performance in critical measurement systems. It has two inputs with very high CMRR (e.g., 100 db), high input resistance $> 10^{12}\ \Omega$), low input capacitance, very low bias currents, low offset voltages with very low drift, moderatly high selectable gains (up to 1000), that are accurate over a wide frequency range (0–10 kHz with a bandwidth of 1–10 MHz), and low output impedance. The inputs are sometimes designed to be **guarded** against leakage paths using triaxial cables.

An **operational amplifier** is an inverting or differential amplifier with a single mode output, very high gain ($>10^4$), and wide bandwidth. The high gain is constant up to 10–100 Hz and then drops off at a rate of one to two decades per decade increase in frequency. Unity gain is usually reached between 1 and 500 MHz. If it is a differential amplifier it should have a high CMRR (100 db) when it is used in circuits where the voltages of both inputs may change. An operational amplifier is designed to be used with a wide variety of negative feedback circuits where the overall circuit gain with feedback is usually less than 100. Other characteristics vary widely from one type of operational amplifier to another, influenced by cost, compromises, and other factors. Their best values are similar to those of the instrumentation amplifier.

## 5. Amplifier Specifications

This section introduces important amplifier specifications and summarizes them for convenience along with those presented above. It should be recognized that external feedback circuits may cause the characteristics of the amplifier *circuit* to differ from those of the amplifier itself.

**Open-loop gain** is the voltage gain of an amplifier without any external feedback circuits connected to it. It is often given in decibels, db = 20 log(voltage ratio or gain). A gain of 100 db equals a voltage ratio of $10^5$ (where $100/20 = 5$).

**Frequency response** may be specified as a bandwidth or as a gain-bandwidth product. The **bandwidth** is the difference between the upper and lower frequencies at which the gain has dropped to 3 db = 0.7 times its maximum value. These frequencies are called **cutoff, rolloff**, or **corner** frequencies. For a dc amplifier the upper cutoff frequency is equal to the bandwidth because the lower cutoff frequency is zero (dc). The **gain-bandwidth product** is usually given as the **unity-gain bandwidth** (or unity-gain crossover frequency), the frequency at which the gain has dropped to unity. The bandwidth and unity-gain bandwidth for a heavy load may be rated less than the bandwidths for a light load and are usually also specified.

Many amplifiers are designed to have a frequency response similar to that of a simple low-pass *RC* filter so that signal distortions can be readily predicted using the equations presented above for the *RC* filter. When they do not respond in this way it may be desirable to determine the shape of the curve because signal distortions are intimately related to the shape of the gain-frequency curve. For instance, the phase shift at a given frequency is a function of a weighted integral of the frequency response curve.

**Slew rate** is the fastest possible rate of change of the output signal in volts per second. It is limited by the maximum current available at some point in the amplifier to charge a capacitance at that point.

**Settling time** is the maximum time (on the order of microseconds) required for the output voltage to enter and remain within a specified relative error band (say, ±0.01%) about the final value in response to a step function input change. It includes propagation delay, slew rate delay, and the time it takes for any overshoot damped oscillations to become less than the error band.

The **input bias** voltage can be considered as a small voltage in series with one of the inputs that makes the output voltage nonzero when the input is zero. The **input offset voltage** is the difference between the two input bias voltages. Usually the offset voltage can be balanced on zero by an external or internal potentiometer. It drifts away from zero because it depends on temperature, age, and power supply voltage.

The **input bias current** is the current required by an input even to hold the input voltage at zero. It must therefore be considered in addition to the input current caused by an input resistance. It is due to transistor bias currents and MOSFET gate insulation leakage paths. Although it can be balanced by injecting a current into the input, it is subject to considerable drift because it is highly temperature dependent. The difference between the two input bias currents is called the **input offset current** and is often equal to a large fraction of the bias current. A constant input bias current causes a constant voltage drop across the output resistance of the signal source, resulting in an *apparent* voltage offset.

**Leakage currents** at the inputs are due to insulation leakage paths and depend on humidity, temperature, and age. They may be **guarded** against by minimizing leakage paths and also by keeping both sides of any leakage path at the same potential so that current does not flow across the path. To minimize leakage at the input, it is best to run a *wire* from the input terminal on the amplifier box to the actual amplifier input stage, rather than use a printed circuit board path. To guard against cable insulation leakage a "triaxial" cable can be used. This is a coaxial cable with two shields. A special output from the amplifier drives the inner shield at the same voltage as the central input signal wire. Therefore the voltage drop across the insulation between the signal wire and the inner shield is minimal and very little leakage current flows from the signal wire. The outer shield is grounded but any leakage between the two shields does not load the signal source, but simply loads (insignificantly) the special amplifier output. When this guarding and shielding technique runs all the way from the cable into the amplifier, the amplifier is said to have a **guarded input**. A guard terminal on an integrated circuit amplifier is connected to a conducting guard ring that encircles the input terminal, with mounting insulation between the input and the guard ring. The guard terminal is meant to be connected to an amplifier output that keeps it at the same potential as the input terminal.

The **input resistance** is another source of input current *in addition* to the input bias current. It should be very high compared to the signal source output resistance to minimize loading and to prevent a significant reduction in the effective CMRR of the amplifier circuit. The **common mode rejection ratio** (CMRR) should be high ($100\text{ db} = 10^5$) if both inputs are to be active (ungrounded).

**Output ratings** include the maximum voltage range for a light load, and the maximum current range for a heavy load. The maximum voltage range may decrease as the current load increases (load resistance decreases), and it is usually specified for some given heavy load. Some amplifiers oscillate if too large a capacitance (e.g., cable capacitance) is connected between the output and common or ground. The maximum acceptable capacitance is usually specified. A small resistor connected between the output and a load capacitor prevents oscillation but the resulting *RC* low-pass filter distortion of the signal must be evaluated to determine its significance. Amplifiers with a 50 Ω output impedance are designed to drive transmission lines (e.g., coaxial cables) that have a 50 Ω characteristic impedance and that are properly terminated (e.g., with a 50 Ω load resistance).

**Sense terminals** of a power amplifier are sometimes present to allow the voltage to be observed directly at the load, and fed back to the amplifier with separate wires *other* than the lead wires used to carry the high current to the load. These voltage sense wires have negligible voltage drop along them because they do not carry the load current. Feedback circuits connected to the sense terminals are then used to adjust the output voltage to compensate for the voltage drop across the lead wires that occurs due to large currents. Power amplifiers of this nature are sometimes referred to as a type of *programmable* power supply.

**Overload recovery** time is the time it takes the output of an amplifier to respond correctly to an input signal that places its output within its useful operating range after its output has been driven into one of its current or voltage limits by too large a signal. When the output does not recover, the amplifier is said to be *hung up*.

## C. OPERATIONAL AMPLIFIER CIRCUITS

**Operational amplifier circuits** are widely used in signal processing and control applications because they are capable of high precision and accuracy (in the range of 0.1–1% or better). Operational amplifier circuits use differential amplifiers (called **operational amplifiers**) with very high gain, and external components connected in such a way as to provide large negative feedback (Section VI.B.3). Consequently the performance of the overall circuit depends heavily on the characteristics of the external components (e.g., capacitors, precision resistors) and to a much smaller extent upon the characteristics of the operational amplifier itself. Because of this heavy dependence on external component characteristics, it is common to describe the performance of an operational amplifier circuit in ideal terms. This allows the quick and easy design of operational amplifier circuits that operate with high precision. It should be kept in mind, however, that how closely a given circuit approaches ideal behavior depends on the characteristics of its particular operational amplifier and external components. The real and ideal performances of several of the circuits presented in this section are compared to help to evaluate the influence of the real characteristics of the operational amplifier upon the circuits in which it is used.

### 1. Operational Amplifiers

An operational amplifier is a differential voltage amplifier with very high voltage gain, usually greater than $10^4$. All the other desirable and undesirable characteristics of a differential amplifier discussed in Section VI.B.2 are important to the performance of operational amplifier circuits. In general an *ideal operational amplifier* has infinite gain, infinite common mode rejection ratio (CMRR), infinite bandwidth, no bias currents, no voltage offset, infinite input impedances, zero output impedance, and voltage and current ranges adequate to drive any load. The amplifier has two input terminals called the **summing point** (or inverting input) and the **follower input** (or noninverting input) (Fig. 2.40). Positive and negative signs at the input terminals do not represent the input voltage polarity but rather the noninverting and inverting nature, respectively, of the inputs. A small bias current flows into or out of the input terminals, and the amplifier should be chosen so that this current is negligible compared to other currents in the circuit, or too small to load the signal source

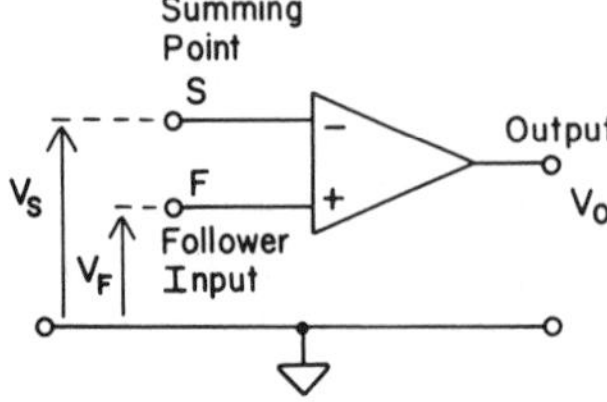


**Fig. 2.40.** An operational amplifier.

or significantly charge any circuit capacitors. The output terminal, on the other hand, can supply relatively large currents that may be required by the devices connected to it. The amplifier is usually indicated by a triangle with power supply and common (or ground) terminals omitted to simplify the diagram. Although their circuits need a common terminal, operational amplifiers themselves generally do not, because their terminal voltages are internally referenced to two power supply voltages of opposite polarity, each of which is itself referenced to the common terminal of the circuit. The common terminals are sometimes omitted from the entire diagram, showing only the signal terminals. Unless otherwise indicated, voltages shown are assumed to be voltage differences between a terminal and common. When only one input is shown it is usually understood to be the summing point, and the follower input is understood to be connected to common.

The amplifier produces an output voltage $v_o$ between its output terminal and common that is proportional to the voltages $v_f$ and $v_s$ at the follower and summing point input terminals, respectively. The relationship is usually close to linear and may be expressed by the equations

$$v_o = A_f(v_s + \Delta v_f) - A_s(v_s + \Delta v_s) \tag{84}$$

$$v_o = A_f v_f - A_s v_s + \text{(voltage offset term)} \tag{85}$$

where the proportionality constants $A_f$ and $A_s$ are called the gains and are essentially equal, $\Delta v_f$ and $\Delta v_s$ are the offset voltages of the individual input terminals, and the voltage offset term is equal to $A_f \Delta v_f - A_s \Delta v_s$. When the common mode gain is zero (infinite CMRR), $A_f = A_s = A$ and equation 85 reduces to

$$v_o = A(v_f - v_s) + A\ \Delta v \tag{86}$$

where $A$ is called the **open-loop gain** of the amplifier and $\Delta v = \Delta v_f - \Delta v_s$ is called the **voltage offset** of the amplifier. The voltage offset produces a relative error on the order of $\Delta v / v$, where $v$ is the *smaller* of the two voltages, either the output voltage or the input voltage of the overall circuit. A balance control can usually be used to reduce $\Delta v$.

An important characteristic of all operational amplifier circuits is the extremely small voltage difference that always exists between the two input terminals whenever the output voltage is not zero. The magnitude of the gain $A$ of an operational amplifier is usually greater than $10^5$ and may go higher than $10^8$. Therefore an output voltage of 0.1–10 V requires only a small voltage difference (say, $10^{-3}$ V) between the two input terminals. This voltage difference is usually *negligibly small* compared to the output voltage, and also small compared to the input voltage when the voltage gain of the overall circuit is not too large. The small voltage difference between the two input terminals is often assumed to be essentially zero for purposes of calculating the characteristics of a circuit having external components. It must be remembered that a minute voltage difference does exist electronically and that the assumption that it is zero is for computation convenience only. An operational amplifier circuit does not operate properly if the input terminals are electrically shorted to each other.

External electrical components may be connected between any of the input or output terminals, the common terminal, and signal source terminals. The particular terminal connections and types of components that are used depend upon the type of operation to be performed. Components that are connected between the output terminal and one of the input terminals form a **feedback loop** with the operational amplifier. Three types of operational amplifier circuits can be distinguished by the types of feedback that they use. Most circuits use large *negative* feedback. *Oscillator* circuits usually use a combination of both positive and negative feedback. Voltage and current *comparator* circuits usually use no feedback, and operate with the amplifier in one of its voltage or current limits except during transitions. The voltage overload or current overload recovery time limits the response time of such circuits and for such applications it may be desirable to use special **voltage comparator** amplifiers that are designed with faster recovery times, though they usually have lower gains.

Negative feedback is accomplished by connecting external components from the output terminal to the summing point. Two types of inputs to the circuit may then be used; a voltage input at the follower terminal and a current input at the summing point (follower input grounded).

### 2. Operational Amplifier Circuits Using Resistors

The **voltage follower circuit** (Fig. 2.41) has a high input impedance and is commonly used when a voltage source cannot supply sufficient current to drive a voltage measuring device or other load. The bias current of the follower input may load the voltage source, but amplifiers with very low bias currents may be used to minimize this problem. The voltage follower has one of the simplest feedback loops; a wire directly connecting the output terminal with the summing point input terminal, $S$. The fraction $\beta$ of the output voltage that is fed back and subtracted from the input voltage is unity (except for the small voltage difference between the input terminals required to produce a finite output voltage). Therefore, $\beta$ in equation 80 is essentially unity. The follower input is connected to the voltage source and the load is connected between the output terminal and some other point in the circuit such as common. The follower output voltage essentially equals the input voltage and the output terminal supplies the larger currents demanded by the load. The solid wire connecting

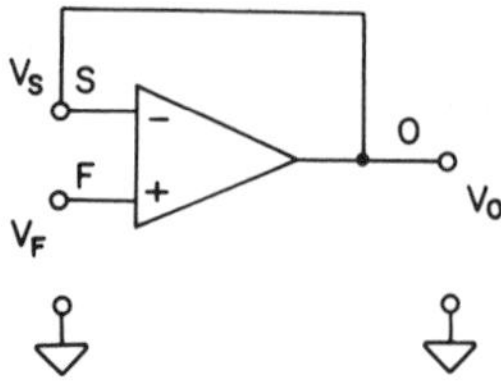


**Fig. 2.41.** A voltage follower operational amplifier circuit.

the output and summing point terminals forces these two terminals to be at the same potential. Since the voltage difference between the two input terminals is negligible, all three amplifier terminals are at essentially the same voltage.

The circuit accomplishes this result in the following way. Assume that initially the three terminals are at 0 V with respect to ground so that $v_f = v_o = v_s = 0$. Connecting the voltage source to the follower input terminal, F, changes the voltage $v_f$ to make it equal to the voltage of the source. Even a small voltage difference between $v_f$ and $v_s$ is greatly amplified by the operational amplifier causing a large change in the output voltage. The change in output voltage follows in the same direction as the change in input voltage (hence the term *follower input*). The new output voltage is fed back to terminal S by the solid wire connected to the output terminal. Since the input voltage is connected to the follower input, F, both voltages $v_s$ and $v_o$ quickly follow behind $v_f$ until an equilibrium condition is reached where a small voltage difference $(v_f - v_s)$ between the amplifier input terminals is just sufficient to sustain the larger output voltage. The equilibrium condition is given by equation 85. Since $v_s = v_o$, $v_o$ can be substituted for $v_s$ in equation 85 to give

$$v_o = A_f v_f - A_s v_o + (A_f \Delta v_f - A_s \Delta v_s) \tag{87}$$

Rearranging and solving for $v_o$ gives

$$v_o = \frac{A_f}{A_s + 1} v_f + \frac{A_f \Delta v_f - A_s \Delta v_s}{A_s + 1} \tag{88}$$

When $A_f \approx A_s$ the last term may be approximated by $\Delta v\, A_f/(A_s + 1)$ so that

$$v_o = \frac{A_f}{A_s + 1}(v_f + \Delta v) \tag{89a}$$

$$= \frac{A_f}{A_s}(v_f + \Delta v) \qquad \text{when } A_f \text{ and } A_s \gg \text{unity} \tag{89b}$$

$$= \frac{A}{A + 1}(v_f + \Delta v) \qquad \text{when } A_f = A_s \text{ (CMRR} \rightarrow \infty) \tag{89c}$$

$$= v_f + \Delta v \qquad \text{when } A \rightarrow \infty \tag{89d}$$

$$= v_f \qquad \text{when } \Delta v \ll v_f \tag{89e}$$

Equation 89b shows that the voltage gain of the follower circuit is essentially constant if $A_f$ and $A_s$ are large and their ratio $A_f/A_s$ remains constant, even if their magnitudes change. The importance of a very high CMRR and very high amplifier gain is indicated by equations 89c and 89d. The equations also show that the voltage offset $\Delta v$ is directly additive to the input voltage signal of the follower circuit. When the amplifier gain $A = 10^5$

$$v_o = 0.99999 v_f \tag{90}$$

which differs from unity gain by only 0.001%.

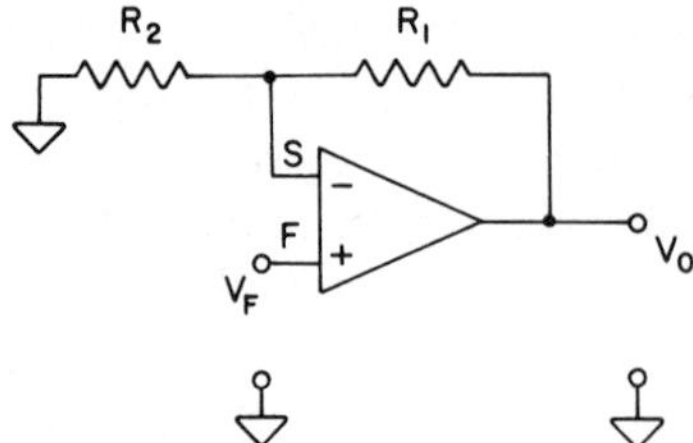


Fig. 2.42. A voltage follower-with-gain circuit.

When a greater output voltage is needed from a follower than is put into the follower, the **follower with gain circuit** in Fig. 2.42 can be used. The two resistors are a voltage divider whose input is the output of the amplifier. The output of the voltage divider is at point S. The operation of this circuit may be most easily understood by assuming the accurate approximation that the voltage at point S essentially equals the follower input voltage at point $F$. To make the voltages at points S and F essentially equal, the output voltage of the amplifier must be greater than the voltage at point S by an amount determined by the voltage divider equation

$$v_s = v_o \frac{R_2}{R_1 + R_2} \tag{91}$$

which indicates the fraction $\beta$ of the output voltage fed back and subtracted from the input voltage. Upon rearranging equation 91 and using the approximate substitution $v_f = v_s$, the output voltage of the follower with gain is related to the input voltage by

$$v_o = v_f \frac{R_1 + R_2}{R_2} = v_f \left( \frac{R_1}{R_2} + 1 \right) \tag{92}$$

A **voltage summing amplifier circuit** (Fig. 2.43) has practical application as a voltage zero suppression circuit, and may be obtained by connecting the input of a follower with gain circuit to the output of the voltage summing circuit of Fig. 2.3. A weighted sum may be obtained as indicated by equation 25. An unweighted sum may be obtained as indicated by equation 26. An unweighted

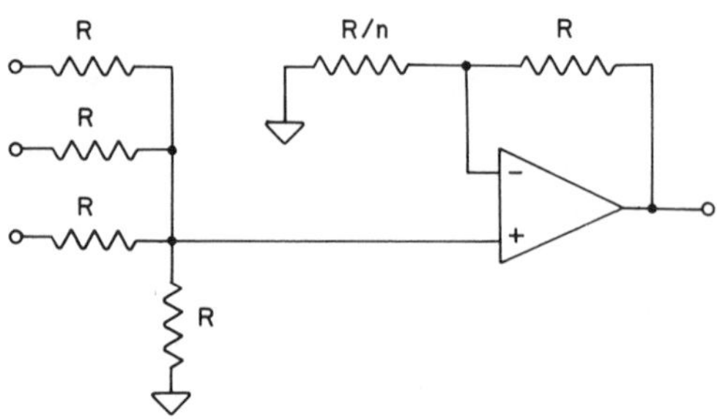


Fig. 2.43. A noninverting voltage summing operational amplifier circuit.

(or equally weighted) sum is obtained and the gain of the voltage summing amplifier circuit is unity for each input voltage when equal resistors are used as indicated in Fig. 2.43, where $n$ is the number of input resistors, three in Fig.2.43.

It is worthwhile to emphasize the need to use an amplifier with a high CMRR when *noninverting* or voltage follower circuits of this type are used. *Inverting* circuits, to be discussed below, have the follower input connected to a fixed voltage reference point such as common, and therefore do not require an amplifier with a high CMRR. Early work with operational amplifier circuits used, when possible, inverting circuits and avoided noninverting circuits, because operational amplifiers with a high CMRR were not as readily available as they are today. The advantages of noninverting circuits should not be overlooked; it may be possible in some cases to replace two inverting circuits with one noninverting circuit at a savings in cost and improvement in performance.

**The current-to-voltage converter circuit** (Fig. 2.44) finds use in spectrometer circuits, electrochemistry, chromatography, and other areas. A current-to-voltage converter is commonly used with a transducer that produces an electrical *current* (rather that a voltage) that is directly proportional to the signal of interest. Conversion of a current to a voltage that is proportional to it is needed when the input to the readout device must be a voltage. An example of such a transducer is a phototube, which produces a current that is directly proportional to the photon rate falling upon its photosensitive cathode.

A current-to-voltage circuit has a resistor $R_f$ connecting its output terminal to the summing point S. The follower terminal is connected to common as shown in Fig. 2.44a. The follower terminal is often omitted from the diagram as shown in Fig. 2.44b. On some amplifiers the follower terminal may not even be accessible. Equation 85 now becomes

$$v_o = A_f 0 - A_s v_s + (A_f \Delta v_f - A_s \Delta v_s) = -A_s v_s + (\text{voltage offset term}) \quad (93)$$

which indicates that the CMRR (equation 76) is not important for circuits where the follower input is connected to common, because $v_f = 0$, making $A_f$ unimportant (except as it influences the small voltage offset term). The output voltage $v_o$ has a polarity opposite to the polarity of the summing point voltage $v_s$. This difference in polarity of the two terminals requires that one end of the

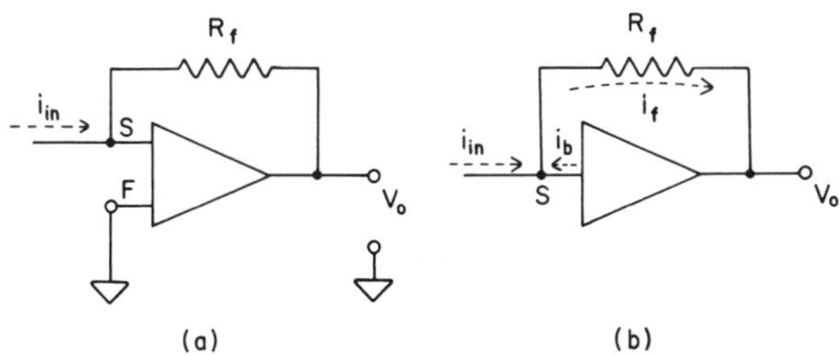


**Fig. 2.44.** Two representations of a current-to-voltage converter circuit.

feedback resistor be positive and the other end negative (when current flows through $R_f$), or that both ends be 0 V (when no current flows through $R_f$).

By Ohm's law the voltage drop across $R_f$ and the current $i_f$ through it are related by

$$v_s - v_o = i_f R_f \tag{94}$$

Combining equations 93 and 94 by substitution and rearranging gives

$$v_o = \frac{-A_s}{1 + A_s}(i_f R_f + \Delta v) \tag{95}$$

where $\Delta v$ is the voltage offset of the amplifier and the approximation $A_f = A_s$ has been made for the voltage offset term only.

Each of the three wires connected to the summing point S may carry a current into or away from the point. For convenience of discussion, we arbitrarily choose the direction of flow of current in each wire as indicated by the arrows in Fig. 2.44*b*.

Then these currents are related by

$$i_f = i_{\text{in}} + i_b \tag{96}$$

where $i_b$ is the bias current of the operational amplifier. Negative values for a current occur when the flow of current is in a direction opposite to the arbitrarily chosen direction. When this relationship is not satisfied excess charge is flowing into or away from the summing point, causing its potential to change. This potential change causes the output voltage to change to increase or decrease $i_f$ to bring the circuit back to its steady condition where equation 96 is valid.

Substitution of equation 96 into equation 95 gives

$$v_o = \frac{-A_s}{1 + A_s}[(i_{\text{in}} + i_b) R_f + \Delta v] \tag{97}$$

which shows how the output of the current-to-voltage converter circuit depends upon the signal input, amplifier bias current, and voltage offset. When the bias current and offset voltage are negligible and the amplifier gain is very large,

$$i_f = i_{\text{in}} \tag{98}$$

and

$$v_o = -i_{\text{in}} R_f \tag{99}$$

which represents the ideal behaviour for the current-to-voltage converter circuit. Equation 99 indicates that the output voltage is negative when the input signal current (flow of positive charge) is flowing into the summing point.

More than one signal current $i_j$ can flow into the summing point, in which case $i_{\text{in}}$ in the above equations can be replaced by

$$i_{\text{in}} = \Sigma_j i_j \qquad (100$$

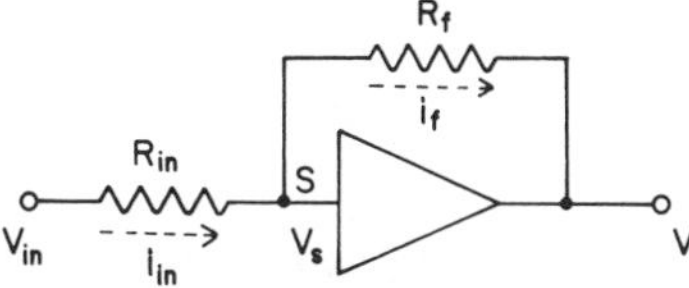


**Fig. 2.45.** A voltage inverter circuit (follower input has been grounded and is not shown).

The output voltage is then negative when there is a *net* flow of current into the summing point from the input current sources. A **current summing circuit** of this type has practical application as a zero suppression circuit. The output voltage is made zero for a given input current by withdrawing an equal current from the summing point by using a current source consisting of a resistor in series with a voltage source, either one of which may be varied to make adjustments in the zero point (i.e., to make the output voltage zero when the signal voltage is nonzero).

A **voltage inverter amplifier circuit** can be made by adding an input resistor $R_{\text{in}}$ to a current-to-voltage converter (Fig. 2.45).

From Ohm's law the input current is

$$i_{\text{in}} = \frac{v_{\text{in}} - v_s}{R_{\text{in}}} \tag{101}$$

which reduces to

$$i_{\text{in}} = \frac{v_{\text{in}}}{R_{\text{in}}} \tag{102}$$

if it is assumed that $v_s$ is essentially 0 V. Substituting this into equation 99 gives

$$v_{\text{o}} = -v_{\text{in}} \frac{R_f}{R_{\text{in}}} \tag{103}$$

which shows that the ideal behavior of a voltage inverter circuit is that of an amplifier with gain $-R_f/R_{\text{in}}$. The relationship that includes amplifier gain, bias current, and offset voltage can be obtained by combining equations 93, 94, 96, and 101 to eliminate $v_s$, $i_f$, and $i_{\text{in}}$ and solving for $v_{\text{o}}$ to obtain

$$v_{\text{o}} = \left[ -v_{\text{in}} \frac{R_f}{R_{\text{in}}} - i_b R_f + \Delta v \left( 1 + \frac{R_f}{R_{\text{in}}} \right) \right] \left[ \frac{1}{A} \left( 1 + \frac{R_f}{R_{\text{in}}} \right) + 1 \right]^{-1} \tag{104}$$

The right-hand factor is close to unity for large amplifier gain, provided the ratio $R_f/R_{\text{in}}$ is not too large. This equation shows that the bias current and voltage offset distort the ideal output voltage by adding an offset to it; however, even then a *change* in output voltage is still an accurately known factor $-R_f/R_{\text{in}}$ times the *change* in input voltage.

A **voltage summing amplifier circuit** may be obtained by connecting a resistor to the summing point for each input voltage as shown in Fig. 2.46. Each input voltage-resistor combination produces a current that is summed at the summing point. Again Ohm's law may be used to calculate each input current separately,

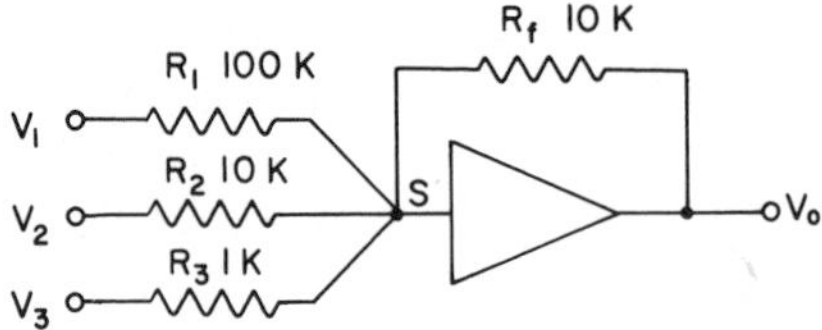


**Fig. 2.46.** An inverting voltage summing operational amplifier circuit.

since the voltage of the summing point is essentially 0 V. The algebraic sum of the input currents is set equal to the feedback current, giving the ideal relationship

$$v_o = -R_f \Sigma i_{\text{in}} = -R_f \sum_j R_j = -R_f \sum_j \left(\frac{v_{\text{in}}}{R_{\text{in}}}\right)_j \tag{105}$$

which shows that the output voltage is the negative of the weighted sum of the input voltages. This voltage inverting circuit does not require an amplifier with a high CMRR, as does the noninverting follower with gain voltage summing circuit.

### 3. Multiplexer and Programmable Amplifier Circuits

**Programmable amplifiers** whose gain can be precisely selected by electronic switches are shown in Fig. 2.47. The desired gain is selected by the control signal levels which open and close the switches $SW_1$ and $SW_2$. Important characteristics of electronic switches are discussed in Section VI.A. The series ON resistance of the switches (typically 50–100 Ω for an FET) should be included when calculating the total resistance of each circuit leg. For the inverter amplifier circuit one switch should be closed before the other is opened in order to maintain a closed feedback loop to prevent delays caused by amplifier limiting. For the inverter amplifier circuit, $SW_1$ could be omitted and the 10 kΩ resistor increased to 11.11 kΩ to give a total feedback resistance of 10 kΩ when both resistors are in parallel. For the programmable follower, opening both switches produces a gain of unity.

Placing one terminal of a switch (e.g, the FET source) at the summing point

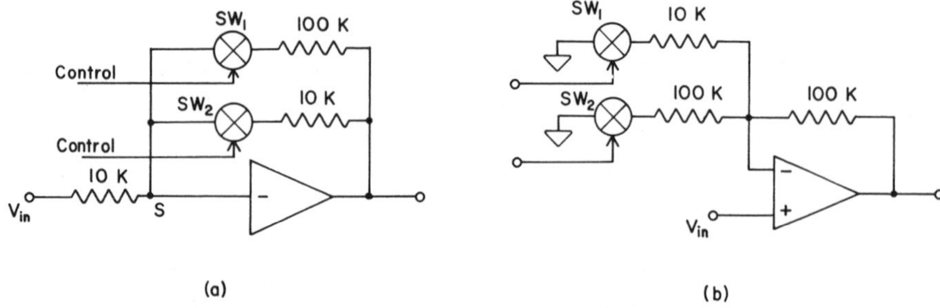


**Fig. 2.47.** Programmable amplifier circuits, which use switches to control the gain of (*a*) a voltage inverter amplifier circuit and (*b*) a follower with gain amplifier circuit.

or at common rather than at the amplifier's output provides a stable reference point for the control signal regardless of the amplifier output polarity. However, the interelectrode capacitance of a switch at the summing point may cause oscillation unless a feedback capacitor is used. The voltage drop across the switch is always small when the switch is conducting. Therefore the current-voltage characteristics of the switch near the origin apply when the switch is on, and the switch can conduct current in either direction. When the switch is off, however, the output voltage of the amplifier is effectively placed across the switch terminals and the control signal must be capable of maintaining the switch in its OFF condition for all values.

**Multiplexer** circuits use switches to select any of several possible input signals for processing. Typical inverter and follower circuits are shown in Fig. 2.48. The desired signal is selected by the control signal levels which turn the switches on and off. For the follower circuit none of the switch terminals is near common, and therefore all switches must be able to withstand the full voltage range of all input signals. The input impedance of this circuit is relatively high and includes the input impedance of the follower circuit in parallel with OFF impedances of all switches that are off. The input impedance of this inverting amplifier multiplexing circuit is constant for each input since one switch shorts the input resistor to common when the other has not shorted it to the summing point S. The inverting symbol, a small circle at the tip of the control signal arrow, indicates that the switches are in opposite states. The source and drain, or terminal, voltages of the switches in the inverter circuit are close to zero. Therefore standard digital logic voltage levels of about 0–5 V could be used to control $p$-channel switches since only a few volts is usually needed to turn off an FET. Moreover, the input voltages could be very large (100 V or more). If very large voltages are used, however, the switch control circuit must be designed to short terminals $S_1$ and $S_2$ in Fig. 2.48$b$ to ground during a control circuit power failure. Otherwise the full input voltage could appear across the terminals of the OFF switches, destroying them. Another way to prevent such destruction would be to place back-to-back zener diodes from

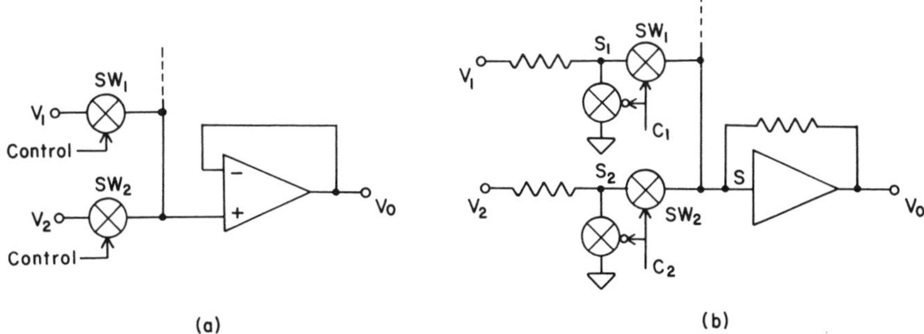


**Fig. 2.48.** Multiplexer circuits used to connect one of several signals to the input of ($a$) a follower circuit, and ($b$) a voltage inverter (or summing) circuit.

terminals $S_1$ and $S_2$ to common, or to replace each FET that is connected to common with two ordinary diodes placed in parallel with each other, but pointing in opposite directions to each other. The forward biased dynamic impedance of the diodes at the small voltages appearing at $S_1$ and $S_2$ would have to be several decades larger than the ON resistance of the switches to prevent a significant fraction of a signal from being bypassed to common when it is meant to got to the summing point. Input current signals could be multiplexed by removing the input resistors and connecting the currents to points $S_1$ and $S_2$.

**Low-level signals** (less than about 1 V) coming from sources located at some distance from the multiplexer usually require multiplexing both the signal and the common wires in order to remove interfering signals that may be picked up by the long transmission line or that may occur as a difference in voltage between the common of the amplifier and the common of the voltage source. The two wires for each channel may be connected to a difference amplifier or a flying capacitor multiplexer in order to eliminate the effect of interfering signals that are *common* to both the signal wire and the common wire for that channel. In the **flying capacitor multiplexer** (Fig. 2.49) a capacitor is first connected to the signal channel by switches $SW_1$ and $SW_1'$ to obtain a sample of the signal voltage $v_1$. Any interfering signal that is common to both wires of a channel is present at both ends of the capacitor, but the capacitor stores only the voltage *difference* between these two wires. The difference is the signal. These switches are then opened to disconnect the capacitor from all channels. Any interfering voltage that was common to both wires of the channel is no longer present at the capacitor. While the capacitor retains the signal voltage, it is connected by switches $SW_A$ and $SW_B$ to the follower amplifier circuit to produce an output voltage proportional to the signal voltage and free from the common mode interfering voltage present on the channel wires.

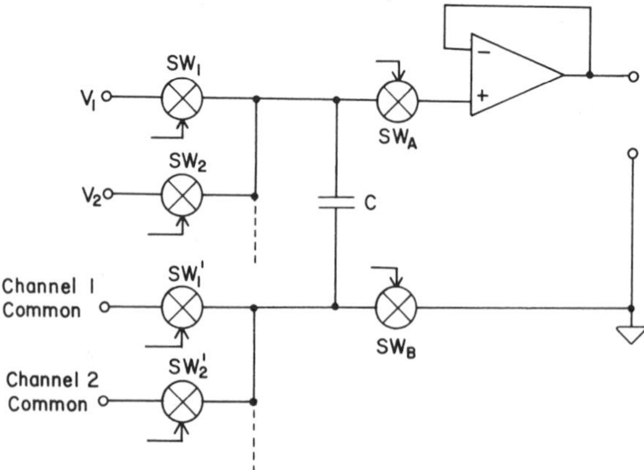


**Fig. 2.49.** A *flying capacitor* multiplexer circuit, used for low-voltage signals to eliminate a noise signal that is common to both wires of a transmission line.

## 4. Instrumentation and Difference Amplifier Circuits

A **voltage difference amplifier circuit** is shown in Fig. 2.50. Equating currents at S gives

$$\frac{v_1 - v_s}{R_1} = \frac{v_s - v_o}{R_2} \tag{106}$$

The voltage at F is found from the voltage divider equation to be

$$v_F = \frac{v_2 R_4}{R_3 + R_4} \tag{107}$$

Since the voltage difference between the amplifier inputs S and F is essentially zero, $v_S = v_F$. Using this identity to eliminate $v_S$ and $v_F$ from equations 106 and 107 by substitution gives, upon rearranging, the result

$$v_o = -v_1 \frac{R_2}{R_1} + v_2 \frac{R_4}{R_3 + R_4}\left(1 + \frac{R_2}{R_1}\right) \tag{108}$$

When $R_1 = R_3$ and $R_2 = R_4$ then

$$v_o = (v_2 - v_1)\frac{R_2}{R_1} \tag{109}$$

which is the equation for a voltage difference amplifier with gain $R_2/R_1$. An amplifier with high CMRR is needed because although $v_S = v_F$, both may vary considerably from 0 V.

The degree to which a signal that is common to both voltage sources is rejected by a difference amplifier depends on how well matched the voltage source output impedances are, since they are in series with $R_1$ and $R_3$. To minimize this potential problem, the input terminals may be bufferred with followers as shown if Fig. 2.51. The result is called an **instrumentation amplifier**. The addition of the variable resistor provides a means of adjusting the gain of a difference amplifier by changing only one resistor.

The improved circuit in Fig. 2.52 uses follower-with-gain amplifier circuits at the inputs. When $V_1$ and $V_2$ have the same voltage applied to them, this *common* voltage also appears at points A and B. Since A and B are at the same voltage, no current flows through $R_3$. Therefore $R_3$ effectively has no influence on the circuit for this *common* signal and each input follower has a gain of unity

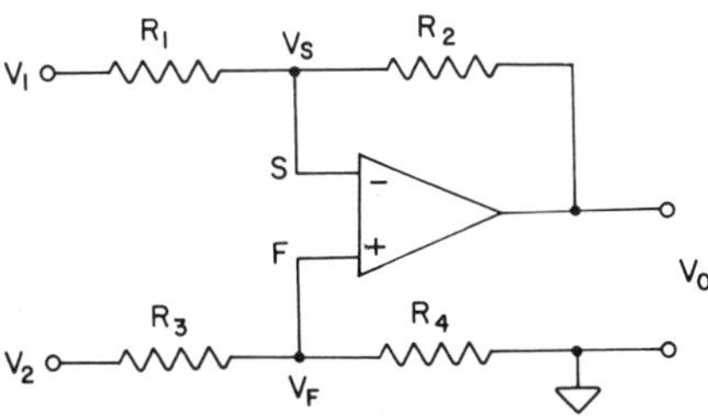


**Fig. 2.50.** A voltage difference operational amplifier circuit.

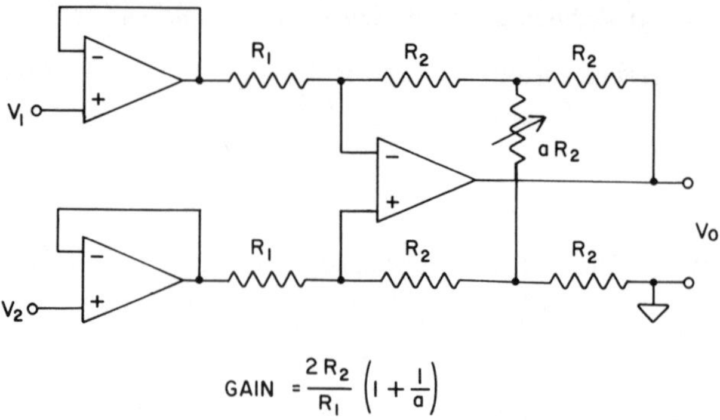


**Fig. 2.51.** A voltage difference amplifier circuit with voltage followers at each input, and a means of adjusting the gain by changing just one resistor.

for the *common* signal. Consequently, the cross-coupled input follower circuits do not influence the *common mode gain* of the overall amplifier. On the other hand, the cross-coupled feedback loops for the follower-with-gain circuits produce a gain of $(1 + a + b)$ for any *difference* signal $(v_1 - v_2)$. The CMRR for the overall amplifier is thereby increased, ideally, by the factor $(1 + a + b)$. This improvement is realized only if the CMRRs for the two input amplifiers are well matched. Typically a total CMRR of $10^5$–$10^6$ (100–120 db) can be achieved. Matched resistors do not have to be used for the cross-coupled input stage as they do for the difference amplifier stage. The overall gain can be adjusted by the single resistor $R_3$ without significantly influencing the CMRR.

Even the purpose of an amplifier circuit with a high CMRR can be defeated if the *RC* time constants of the transmission circuits are not well matched or negligible for ac or rapidly varying signals. Therefore, when a high CMRR is

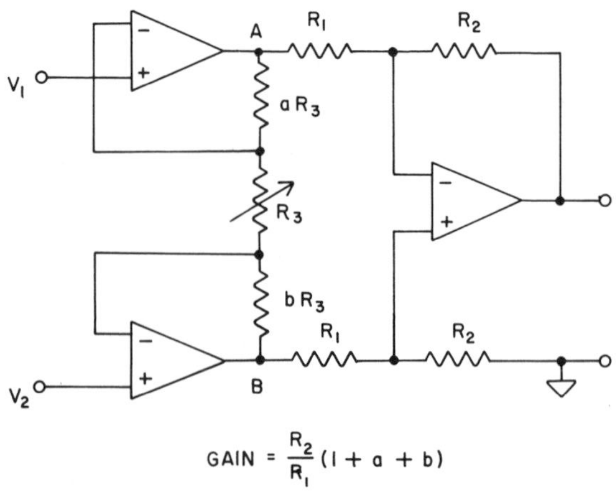


**Fig. 2.52.** A voltage difference amplifier with follower inputs and an improved CMRR.

necessary it is desirable to drive transmission lines with low output impedance amplifiers, so that their source resistances contribute insignificantly to the *RC* time constant caused in part by transmission line capacitance.

A difference amplifier can be used to eliminate most of the noise that may occur between the common terminal of the voltage source and the common terminal of the amplifier circuit. To do this the two wires (signal and common) from the voltage source are connected to the two input terminals with the common wire at $v_2$. The signal voltage $v_s$ is now equal to $v_1 - v_2$ so that equations such as equation 109 become

$$v_o = -v_s \times \text{amplifier circuit gain} \tag{110}$$

The noise voltage between the common terminals of the amplifier circuit and the voltage source is now $v_2$ and does not appear in the output voltage. The degree to which the noise is eliminated depends on how high a CMRR is maintained, as discussed above.

## 5. High-Frequency Response and Circuit Stability

The frequency response of an operational amplifier plays a major role in determining whether or not a circuit behaves as expected, produces large errors, or even oscillates out of control. A log-log plot of amplifier gain versus frequency (often called a **Bode plot**) for a compensated and uncompensated amplifier is shown in Fig. 2.53. The actual shape of the curve near a corner frequency is often approximated by extrapolating the two straight portions of the curve

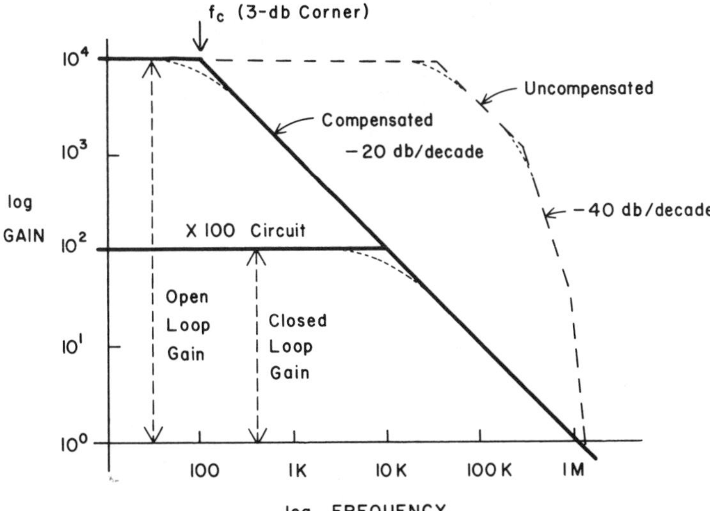


**Fig. 2.53.** A Bode diagram (log-log gain vs. frequency) for compensated and uncompensated operational amplifiers, and for an operational amplifier circuit with a gain of 100.

and causing them to intersect at a point at the corner frequency. Above its corner frequency of 100 Hz the gain for this **compensated amplifier** drops off inversely with frequency, −20 db/decade, losing one decade of gain for each decade increase in frequency. The *shape* of the curve for a compensated amplifier is similar to that for a low-pass *RC* filter and can be described by the magnitude of the complex number

$$|A(jf)| = \left| A \frac{1}{1 + j(f/f_c)} \right| = A[1 + (f/f_c)^2]^{-1/2} \qquad (111)$$

where $f$ is the frequency of interest, $f_c$ is the corner frequency, and $j = \sqrt{-1}$. The corner frequency for a compensated amplifier can be found by dividing the unity gain crossover frequency by the open-loop gain $A$.

An **uncompensated amplifier** consists of several amplifier stages each having its own corner frequency due to internal capacitances. Together they combine to give a Bode plot similar in shape to the one for the uncompensated amplifier in Fig. 2.53. Each stage has its own −90° phase shift above its corner frequency, and together they produce a −360° total phase shift at some frequency for almost any feedback network. To prevent this, an additional *RC* network is added to **compensate** the overall amplifier so that it has only one dominant −90° shift up to the frequency where the amplifier gain has dropped below unity. This reduces the gain of the amplifier at high frequencies, but allows it to be used with most feedback networks without taking special care in designing to prevent oscillations.

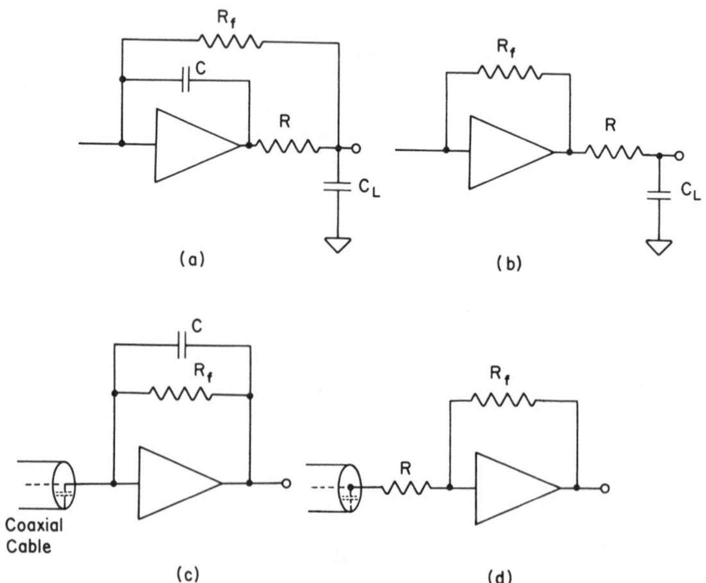


**Fig. 2.54.** Techniques to stabilize an operational amplifier circuit against oscillation.

Even with a compensated amplifier there are two major sources of instability: capacitance between the inverting input (summing point) and ground, and capacitance between the output and ground. Capacitance at the input (e.g., due to a coaxial cable) forms an *RC* low-pass filter with a feedback resistor. Load capacitance forms an *RC* low-pass filter with the output resistance of the amplifier. Either *RC* network causes an additional −90° phase shift at high frequencies and may produce oscillation when the capacitance is too large. The techniques shown in Fig. 2.54 introduce positive phase shifts that help to stabilize the circuit against oscillation.

The frequency response of an operational amplifier circuit with negative feedback is shown in Fib. 2.53. The voltage gain of an operational amplifier, without feedback, is called its **open-loop gain**, and the voltage gain of an operational amplifier *circuit* is called the **closed-loop gain**. The corner frequency of the circuit is higher than that of the compensated amplifier by a factor called the **loop gain**, which is the ratio of the open-loop gain to the closed-loop gain. The frequency response for a circuit (closed-loop gain) is flat over a frequency range which increases as the closed-loop gain of the circuit decreases. Although the frequency response would be flat to higher frequencies if an uncompensated amplifier were used, the circuit would still be subject to oscillation near 1 MHz because of the phase shift accompaning the high rolloff rate. When higher frequency response is needed for a circuit it is better to use a compensated amplifier with a higher unity-gain crossover frequency than an uncompensated amplifier with a lower unit-gain crossover frequency.

## 6. Input and Output Impedances

The input and output impedances of operational amplifier circuits with feedback are different from those of the operational amplifiers themselves. For instance, the input impedance of the voltage inverter amplifier circuit in Fig. 2.45 is usually equal to $R_{in}$ to a very good approximation, since the summing point is virtually 0 V. For a current-to-voltage converter the input impedance is very low, as would be desired for a circuit of this type. The circuit in Fig. 2.55 shows the effective differential input impedance $R_d$ between the two

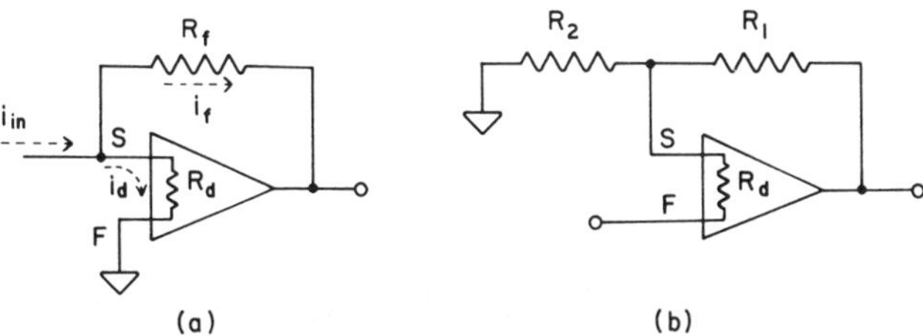


**Fig. 2.55.** Diagrams showing the relationship of the effective input resistance of an operational amplifier to the external circuit, for (*a*) a current-to-voltage converter, and (*b*) a follower with gain circuit.

input terminals of the amplifier. Summing the currents $i_{\text{in}} = i_d + i_f$, using Ohm's law $i_d = v_s/R_d$ and equations 94 and 95, gives

$$i_{\text{in}} = \frac{v_s R_d(1 + A) + v_s R_f}{R_f R_d} \tag{112}$$

The input impedance $Z_{\text{in}}$ is then found, using Ohm's law, to be

$$Z_{\text{in}} = \frac{v_s}{i_{\text{in}}} = \frac{R_f}{1 + A + (R_f/R_d)} \cong \frac{R_f}{1 + A} \text{ for } R_f < R_d \tag{113}$$

which shows that the input impedance is decades less than the feedback resistance $R_f$, since $A$ is usually large. For the voltage inverter amplifier circuit

$$Z_{\text{in}} = R_{\text{in}} + \frac{R_f}{1 + A + (R_f/R_d)} \cong R_{\text{in}} \tag{114}$$

A similar derivation for the follower-with-gain circuit (Fig. 2.55*b*) gives

$$Z_{\text{in}} = R_d\left(\frac{A}{1 + (R_1/R_2)}\right) \tag{115}$$

which shows that the effective input impedance is many decades greater than the differential input impedance of the amplifier.

The output impedance $Z_{\text{out}}$ of the follower-with-gain circuit is given by

$$Z_{\text{out}} = R_o\left(\frac{1 + (R_1/R_2)}{A}\right) \tag{116}$$

and for the inverter amplifier circuit

$$Z_{\text{out}} = R_o\left(\frac{R_f/R_{\text{in}}}{A}\right) \tag{117}$$

where $R_o$ is the output impedance of the operational amplifier itself. The output impedances of the overall circuits are decades less than the amplifier output impedance because of $A$.

The amplifier open-loop gain $A$ is reduced to an effective value $A_e$ by the finite output resistance $R_o$ of an amplifier when the amplifier is required to drive current through a feedback impedance or a load. For the voltage inverter circuit the output impedance forms a voltage divider with $R_f$, or $R_f$ in parallel with $R_L$. This reduces the amplifier gain by the voltage divider fraction (to an approximation with an accuracy on the order of $1/A$)

$$A_e = \frac{AR - R_o}{R_o + R} \cong \frac{AR}{R_o + R} \tag{118}$$

where $R$ is $R_f$ or the parallel equivalent of $R_L$ and $R_f$ when both are present. Since the open-loop gain is reduced at all frequencies, the unity-gain crossover frequency and the bandwidth of the amplifier circuit with feedback are also reduced by about the same factor.

### 7. Integrators, Differentiators, and Low-Pass Filter Circuits

**Integrators** are commonly used to measure areas under chromatographic and other peaks. They are also used to obtain the average value of a signal over a long period to minimize the influence of random fluctuations in the signal.

A typical operational amplifier **current integrator** is shown in Fig. 2.56*a*. The circuit is similar to the current-to-voltage converter except a capacitor rather than a resistor is used in the feedback loop. The voltage across a capacitor is equal to $1/C$ times the time integral of the current passing through it. Since as with the current-to-voltage converter the summing point is essentially 0 V and the input and feedback currents must be equal, this circuit produces an output voltage that is proportional to the time integral of the input current. A switch SW is used to discharge momentarily the capacitor to reset it (and therefore the output voltage) to 0 V before each run. In Fig. 2.56*b* two resistors have been added to provide a means of setting the output voltage to an initial value before the integration. When the switch is in the *Initialize* position the circuit is a voltage inverter amplifier whose output is $-v_{ic}R_2/R_1$. When the switch is in the *Hold* position the only current flowing through the capacitor is the bias current of the amplifier.

Integrator circuits require very low bias currents to prevent drift when there is no signal input current. The best way to determine very low values of summing point bias current is to measure this drift in output voltage.

A **voltage integrator** is obtained by placing a resistor in series with the summing point of a current integrator (Fig. 2.57). In addition to a very low bias current, a voltage integrator should have a very low voltage offset to prevent drift. As with all operational amplifier circuits the signal voltage to be integrated should be many decades greater than the voltage offset (and its drift).

When it is desirable to reduce the noise in a signal, a capacitor can be placed in parallel with the feedback resistor of a circuit (Fig. 2.58). The corner fre-

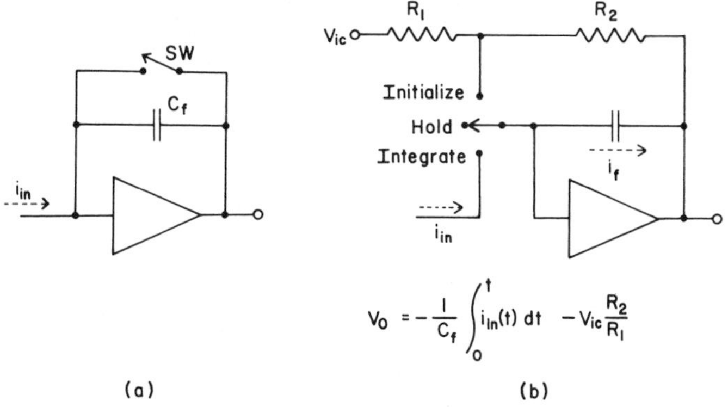


**Fig. 2.56.** A current integrator operational amplifier circuit (*a*) without and (*b*) with a voltage initialization circuit.

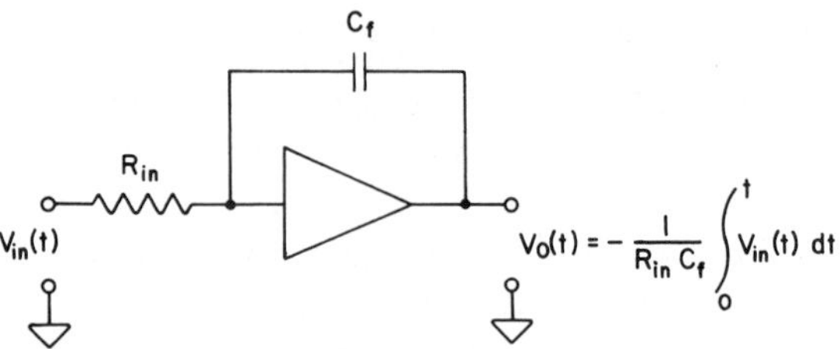


**Fig. 2.57.** A voltage integrator operational amplifier circuit.

quency of the circuit is reduced to $1/2\pi R_f C_f$. The circuit responds temporally as a low-pass filter with a time constant $R_f C_f$ (Section IV.G). In this case, however, a heavy load does not significantly influence the circuit as it would a simple *RC* filter without the amplifier. Changing the circuit from a current-to-voltage converter to a voltage inverter by adding an input resistor in series with the summing point does not change the temporal response. The low frequency gain of the circuit is the same as though the capacitor were not present. The corner frequency of the circuit is reduced to $1/2\pi R_f C_f$.

When $R_f C_f$ is relatively large this circuit becomes an approximate integrator. Its output does not have to be reset with a switch as with a true integrator, because the feedback resistor allows the circuit to respond slowly to changes in the input signal. The output is really a time-weighted integral whose response is quantitatively given by equation 38.

In kinetic experiments it is often desirable to obtain a voltage that is directly proportional to the slope (or derivative) of a reaction rate curve. A simple **voltage differentiator circuit** is shown in Fig. 2.59. Because the current through a capacitor is equal to the capacitance times the time derivative of the voltage across its terminals, and the feedback and input currents for the circuit must be equal, the output voltage is proportional to the time derivative of the input voltage.

The ideal differentiator circuit in Fig. 2.59 has several problems in actual practice. First, for some values of $R_f$ and $C_{\text{in}}$ the circuit oscillates. Although the oscillation may be at too high a frequency to be observed on a recorder or voltmeter, it significantly degrades the operation of the differentiator, causing incorrect results. Second, any sudden change in input voltage (such as is commonly found in noisy signals) causes an extremely large $de_{\text{in}}/dt$ value that produces a very large output voltage; the derivative of a noisy signal is an even

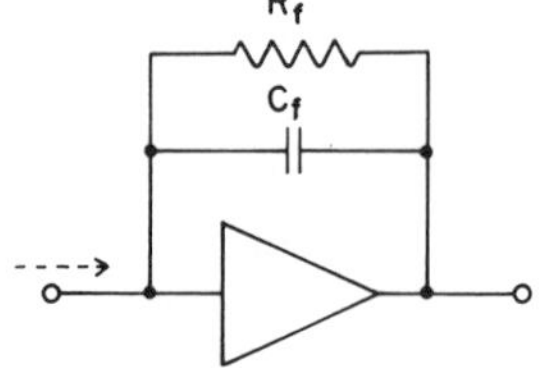


**Fig. 2.58.** A low-pass filter operational amplifier circuit.

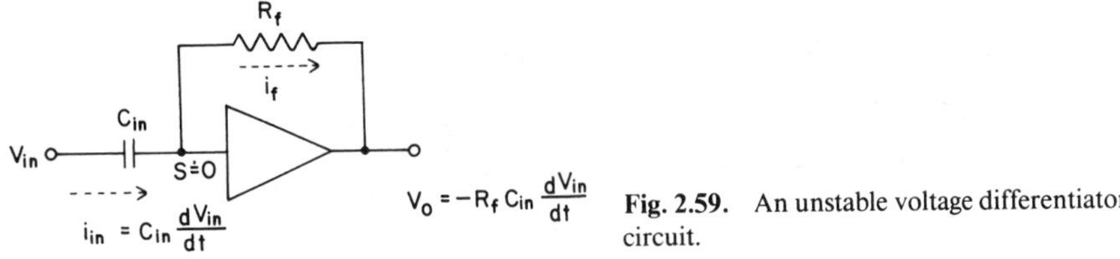


**Fig. 2.59.** An unstable voltage differentiator circuit.

noiser signal. The oscillation can be eliminated and the noise reduced by the circuit in Fig. 2.60. When $R_{in}C_{in} = R_f C_f$ the circuit is a first-order band-pass filter with a peak frequency of $1/2\pi R_f C_f$. A good compromise between accurate differentiation and low noise is obtained when the peak frequency is about 10 times the highest frequency that must be differentiated.

The feedback capacitor helps to reduce noise but is undesirable when observing discontinuities in a signal that sometimes occur when chemicals are mixed to start a reaction. When discontinuities are prone to occur at certain times, the circuit in Fig. 2.61 may be helpful. The feedback capacitor $C_f$ is chosen to minimize noise. A second feedback capacitor $C_f'$ is chosen to be 100 to 1000 times less than $C_f$, but large enough to prevent oscillation (>100 pF). Some time before the discontinuity occurs in the signal, capacitor $C_f$ is disconnected from the summing point and then connected to ground to improve the response time of the circuit and yet keep the voltages on both capacitors equal to each other. Just after the discontinuity, $C_f$ is reconnected to the summing point to continue its damping action.

## 8. Active Filters and Tuned Amplifier Circuits

Because of the ease with which they may be designed and their relative freedom from distortion due to loading, operational amplifier circuits with resistors and capacitors are widely used as **active frequency filters**, or simply **active filters**, which pass a predetermined range of frequencies while rejecting other ranges. The response of any circuit with an inductor can be duplicated by using only capacitors and resistors with operational amplifiers. Therefore inductors are usually not used in these circuits because of their large size and nonideal characteristics.

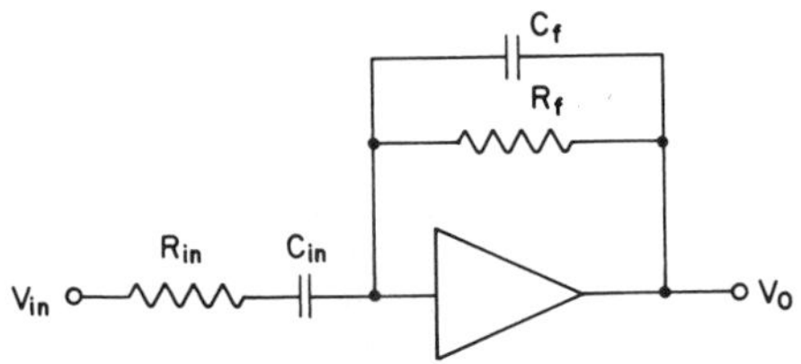


**Fig. 2.60.** A practical, but approximate, voltage differentiator circuit.

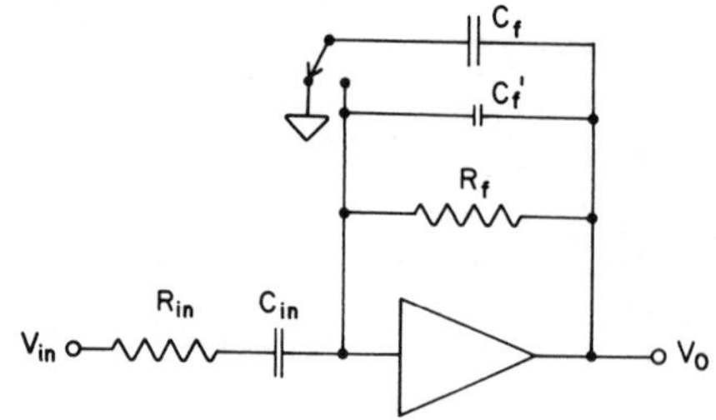


**Fig. 2.61.** A differentiator circuit for discontinuous signals.

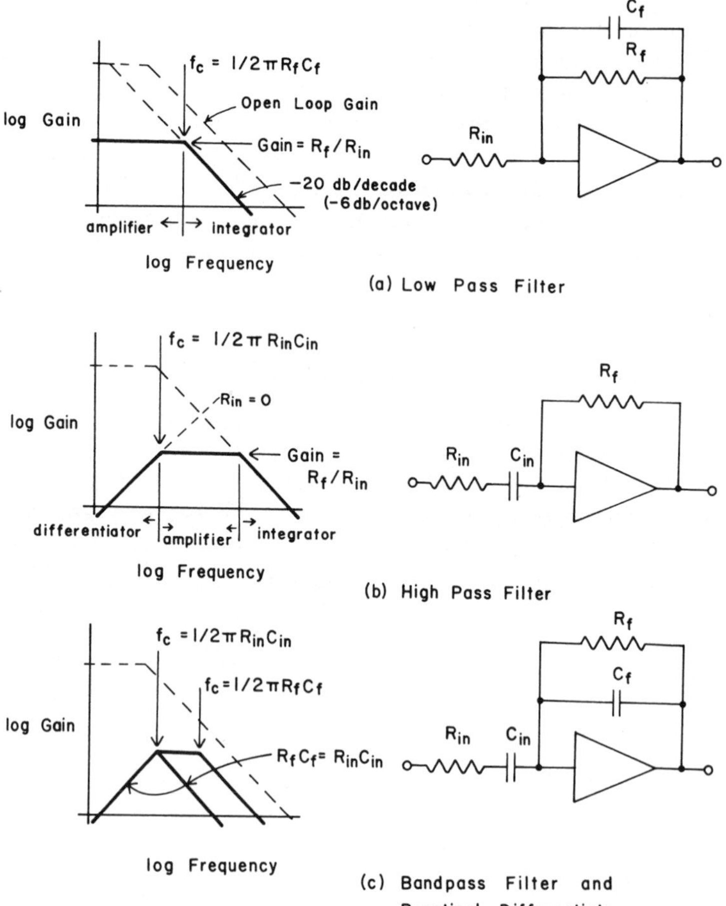


**Fig. 2.62.** First-order filters: (*a*) low pass, (*b*) high pass, and (*c*) band pass or practical differentiator.

**First-order filters** (Fig. 2.62) behave as a simple *RC* filter, with a gain that rolls off at −20 db/decade above or below a corner or cutoff frequency. **Second-order filters** (Fig. 2.63) are similar to first-order filters but roll off at −40 db/decade (Fig. 2.63*a*), giving a sharper boundary to the frequency pass band. The many variables in a second-order filter allow some variation in the shape of the frequency response curve near the corner frequency. The curves for frequencies greater than 10 times the corner frequency, $f_c$, essentially coincide. The differences are indicated by an equation parameter called the **damping factor**, which is a number that has practical values near unity and is a complicated function of the values of the passive components that make up the filter. When the damping factor is unity, the filter behaves as though it were two first-order filters in series (assuming negligible loading on both filters). The gain of this filter is down −6 db at the corner frequency and its response to a step function input signal is

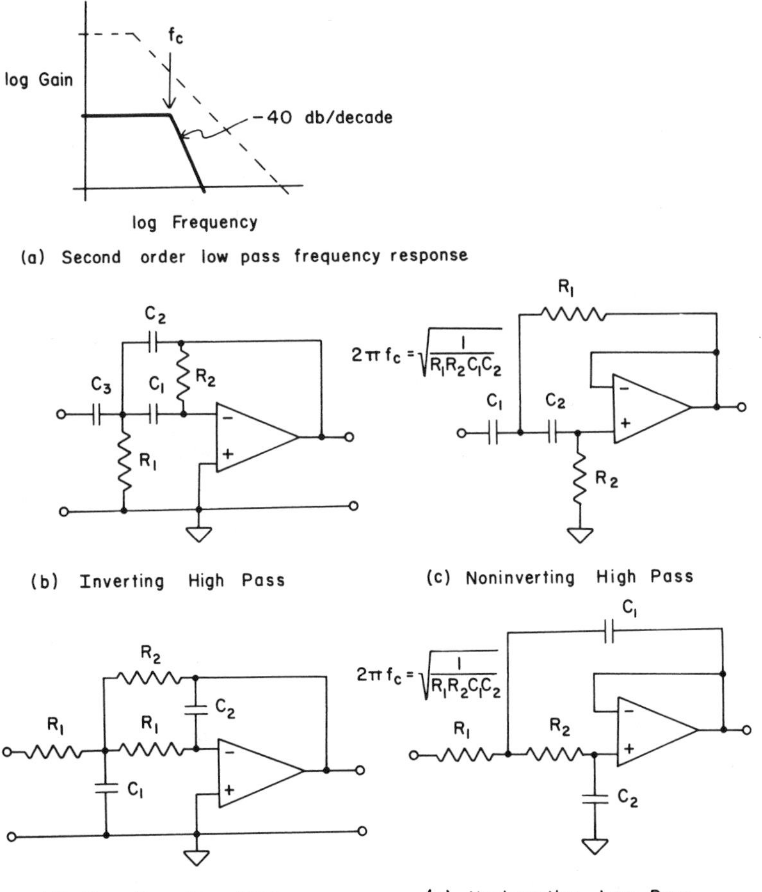


**Fig. 2.63.** Second-order filters using one operational amplifier and multiple-path feedback circuits.

to rise as rapidly as it can without any overshoot of the final value. A filter with a **Butterworth** response has a damping factor of 0.707 and extends the flat part of its frequency response curve as far toward the cutoff frequency as is possible with a second-order filter without causing the curve to peak or rise above its low frequency value in the region of the corner frequency. The curve is "maximally flat." The gain of this filter is down −3 db at the corner frequency. In response to a step function input the Butterworth filter rises to 90% of the final value in about two-thirds the time it takes a critically damped filter to do the same, but it overshoots the final value by 4% before settling down to the final value. Adjusting the component values to give a **Bessel** response (damping factor 0.87) produces a filter with characteristics that are a good compromise between the critically damped and Butterworth filters; overshoot is less than 1% and gain is down −5 db at the corner frequency. In addition, the phase shift of the Bessel filter is reasonably linear with frequency out to the cutoff frequency.

Higher-order filters that produce sharper rolloffs can be obtained by cascading several first- or second-order filters. The **Tchebyscheff** filter has a very sharp rolloff but the gain in the frequency band-pass region has considerable ripple. First-order filters are easy to design. High-order filters with preselected characteristics are usually available commercially.

**Narrow band-pass** filters are used in lock-in amplifiers to reduce the amplitude of noise in a periodic signal. This is particularly helpful when amplifying a

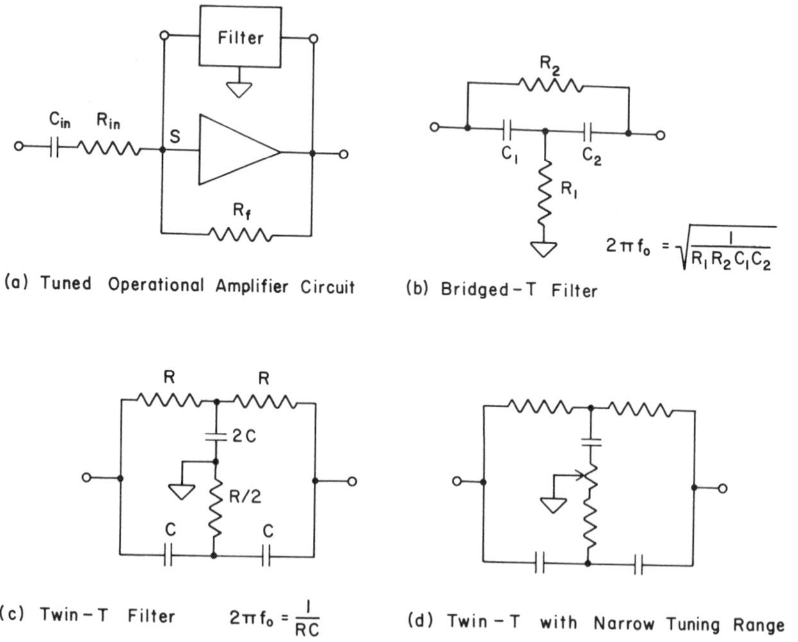


**Fig. 2.64.** A tuned operational amplifier circuit capable of using a bridged-T or twin-T band rejection filter.

small signal that is buried in noise, since the otherwise high noise might cause an amplifier to go into limit even though the signal itself would not. They are also used to convert a periodic signal (such as the reference signal to a lock-in amplifier) to a sinusoidal shape so that its phase can be shifted by a simple phase shift network.

Two common narrow band rejection filters are shown in Fig. 2.64. Another is shown in the feedback loop in Fig. 2.93. The filters do not pass signals at their central frequencies. The frequency rejection notch in the frequency response curve for the bridged-T filter is not as sharp as the twin-T filter. When placed in the feedback loop of a voltage inverter amplifier, the filters do not feed back the signal at their central rejection frequency, but reduce the overall circuit gain for all other frequencies. The amplifier acts as a narrow band-pass filter or **tuned amplifier** having highest gain (approximately $R_f/R_{\text{in}}$ at the central frequency of the filter.

The central frequency of a tuned amplifier and its gain at that frequency depend highly on the values of the *RC* components. A twin-T amplifier circuit shows greater dependence on component values than a bridged-T circuit. Because component values tend to drift with time, a means for occassionally calibrating or retuning a tuned circuit should be available.

## 9. The Generalized Voltage Inverter Network

The voltage inverter circuit in Fig. 2.45 can be generalized by replacing the input and feedback resistors with any impedance network as indicated in Fig. 2.65*a*. The voltage integrator and differentiator are examples. When the impedance networks become more complex, it may be advantageous to use Laplace transforms and the appropriate transfer impedances $Z_{\text{in}}$ and $Z_f$ of the networks. The transfer impedance of a network is the ratio of the input voltage to the output current when the output is effectively grounded (Fig. 2.65*b*) as it would be when connected to a summing point. For a resistor, the transfer impedance is simply its resistance, from Ohm's law. The transfer impedances are appropriate because each network has one terminal connected to the sum-

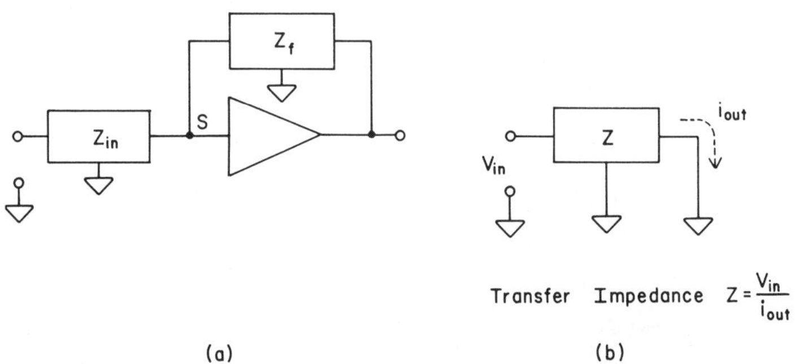


**Fig. 2.65.** A generalized voltage inverter operational amplifier circuit.

ming point, a virtual ground. Because the output currents of both networks are essentially equal (though in opposite directions), we obtain for the circuit in Fig. 2.65*a*

$$V_o(s) = -\frac{Z_f(s)}{Z_{in}(s)} V_{in}(s) \tag{119}$$

where the variables are expressed as their Laplace transform functions. Equation 119 is a generalized form of equation 103 and can be used to evaluate the circuit response characteristics for complex networks. It finds application in the analysis of active filters. However, the design of active filters and tuned amplifiers goes beyond the use of circuits with current summing points. Details are presented in references found in the bibliography.

## D. CONTROL CIRCUITS

Two of the most important uses of operational amplifiers are (1) control of voltage across a device (allowing the current through the device to vary in order to maintain voltage control) and (2) control of current through a device (allowing the voltage to vary as necessary to maintain current control). An important chemical example is the electrochemical potentiostat for controlling the voltage of a working electrode with respect to a reference electrode. Other variables (mechanical, optical, chemical, etc.) may also be controlled by using transducers to convert between the domains of the variables and the electrical domains.

The basis of operation of any control system is to sense a variable and compare its value against a reference value. Any small difference between the two values is called the error signal, and is greatly amplified and used in some way to maintain the controlled variable very near to the desired value (i.e., to within the corresponding magnitude of the error signal).

The voltage follower circuit (Fig. 2.41) controls the voltage across a load by comparing this voltage with a reference (signal) voltage connected between the follower input and common. The error signal is the very small voltage difference between the two input terminals of the amplifier. This circuit requires an amplifier with a high CMRR. A voltage control circuit that does not require an amplifier with high CMRR is shown in Fig. 2.66. The reference voltage is

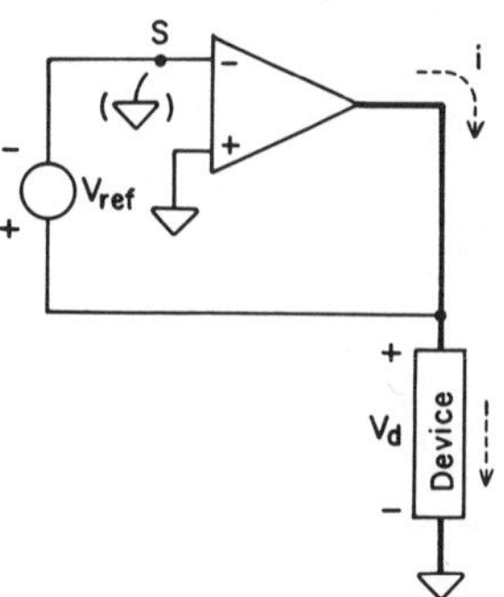


**Fig. 2.66.** A voltage control operational amplifier circuit.

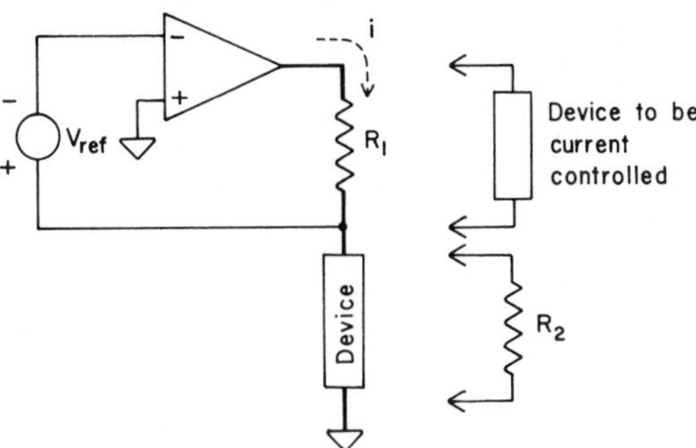


**Fig. 2.67.** A voltage control circuit with a current sampling resistor, and indictions of how to convert the circuit to a current control circuit.

connected in series with the voltage across the device in such a way that the polarities of the two voltages oppose each other. Therefore the side of the reference source that is connected to the inverting input of the amplifier is at 0 V when the two voltages are identical. Any error in this condition causes an error voltage signal that is amplified and used to maintain the voltage across the device nearly equal to the reference voltage [to within a relative error of $1/(1 + A)$].

The current through the device could be monitored by placing a resistor $R_1$ in the circuit as show in Fig. 2.67 and observing the voltage across the resistor. Resistor $R_1$ causes the output voltage of the amplifier to be larger (by the $iR$ drop) but the current through the device is the same because $v_d$ must equal $v_{ref}$.

Now notice that if the *device* in Fig. 2.66 is chosen to be a stable resistor $R_2$, the current that is required to make $v_d = v_{ref}$ is equal to $v_{ref}/R_2$, and does not depend on $R_1$. Therefore any device placed in the position of $R_1$ has a controlled current flowing through it that is determined by $v_{ref}$ and $R_2$, and not by the current-voltage characteristics of the device.

The circuits in Figs. 2.66 and 2.67 have the disadvantage that the reference voltage source must be floating (neither side can be connected to common). In order to develop a more convenient circuit that is commonly used as an electrochemical potentiostat, notice that a voltage follower could be placed in the $v_{ref}$ part of the feedback circuit without changing the operation of the circuit, as shown in Fig. 2.68. The voltage at point A follows the voltage at B. No current (except amplifier bias current) flows from B into the follower. The follower now allows the voltage source $v_{ref}$ to be replaced with a reference voltage source consisting of a resistor $R_3$ with a controlled current $i_3$ flowing through it as shown in Fig. 2.69. The current $i_3$ is controlled by and equal to the current $i_4$, which in turn is controlled by $R_4$ and $v'_{ref}$ through $i_4$ and $i_3$. The follower completes the current path for $i_3$ through its internal connections. (Only a negligible amplifier input bias current flows from point B into the follower.)

Other voltage signals could be converted to currents $i_5$, $i_6$, and so forth, shown in Fig. 2.70, and passed through S, which can still be considered a summing point for all the currents; $i_3 = i_4 + i_5 + i_6 + \ldots$. The net effect would be

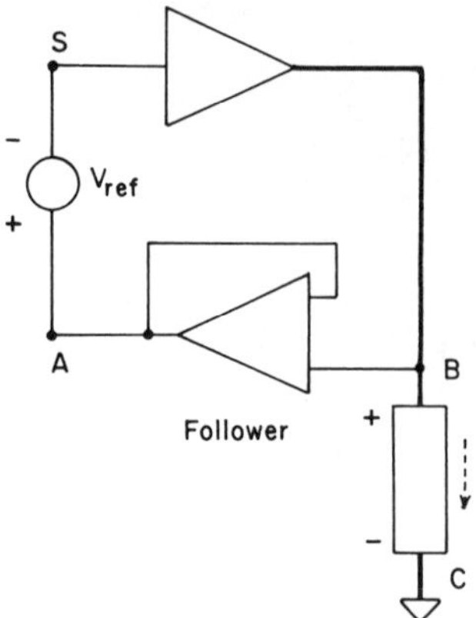


Fig. 2.68. A voltage follower added to a voltage control circuit.

that the negative value of the reference voltage across $R_3$, $v_{ref}$, would be the (weighted) sum of the $v'_{ref}$ voltage signals that produce these currents.

$$-v_{ref} = \frac{R_3 v'_{ref}}{R_4} + \frac{R_3 v''_{ref}}{R_5} + \frac{R_3 v'''_{ref}}{R_6} + \ldots \tag{120}$$

If the device were an electrochemical cell, the reference electrode could be placed at B and the working electrode at D in Fig. 2.70. The current through the device can be injected into the summing point of a current-to-voltage converter as shown in Fig. 2.70 in order to measure the current. In order to obtain accurate electrochemical information about the processes at the working electrode using the circuit in Fig. 2.70, the reference electrode must be designed so that its electrochemical potential (with respect to the solution) does not change significantly even though current passes through it. A typical experiment would control the voltage difference between the two electrodes, and observe the

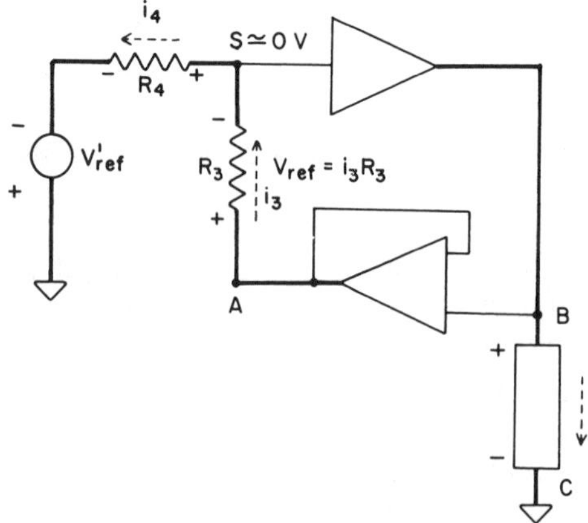


Fig. 2.69. A voltage control circuit that allows both the load and the reference voltage to be grounded.

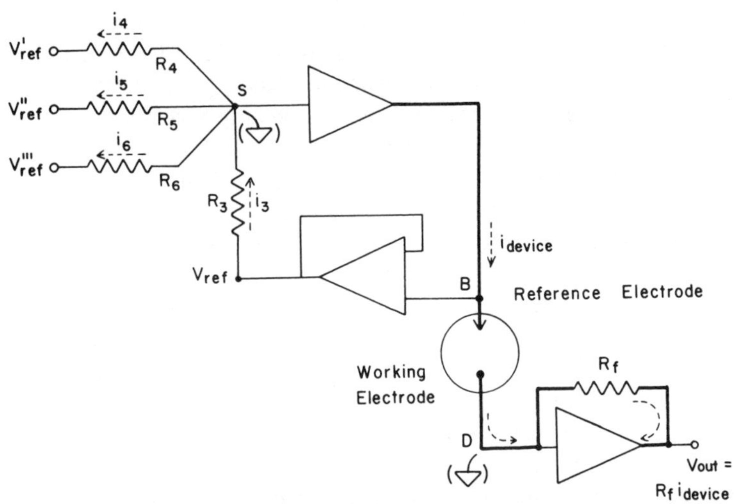


**Fig. 2.70.** A voltage control circuit where the voltage (potential) across the device, an electrochemical cell, is equal to the sum of several references voltages, and the current through the device is sensed with a current-to-voltage converter circuit.

behavior of the resulting current. Changes in the controlled voltage would cause equal changes in the potential of the working electrode because the reference electrode potential is constant. The current-voltage curves obtained in this way would be due to the processes occurring at the working electrode.

It would be better to prevent current from flowing through the reference electrode. This can be done by using third, counter electrode to supply the current to the working electrode (Fig. 2.71). The voltage that is controlled is still

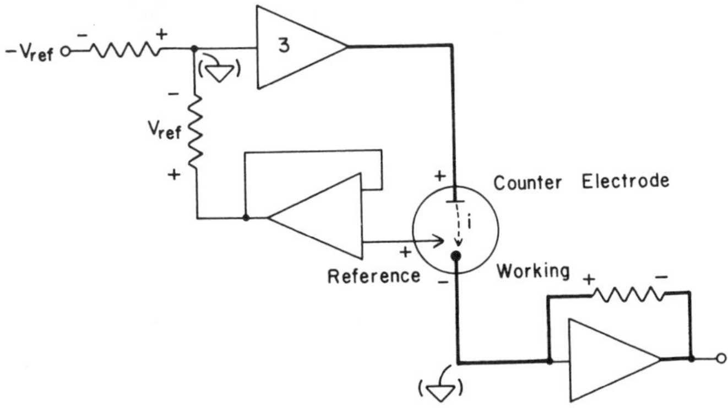


**Fig. 2.71.** A three-electrode potentiostat that controls the voltage (potential) between the working and reference electrodes by controlling the current between the counter and working electrodes. The corresponding voltage-current relationship depends upon the electrochemical behavior of the working electrode.

the voltage (potential difference) between the reference and working electrodes. Now, however, no current is required to pass between the reference and working electrodes, because the current to the working electrode flows from operational amplifier 3 via the counter electrode. Operational amplifier 3 indirectly controls the voltage difference between the reference and working electrodes by automatically adjusting the current so that the current density demanded by the electrochemical reactions at the surface of the working electrode is sufficient to maintain the desired potential difference between the reference and working electrodes.

The circuit in Fig. 2.71 has a complex feedback loop and is worthy of special attention because other useful circuits also have complex feedback loops. In this control circuit, amplifier 3 acts as both an error sensing and control amplifier. The negative feedback loop contains more than just a simple resistor; it also contains part of an electrochemical cell and a voltage follower. Eventually the feedback signal is converted to a current $i_3$ which must be held equal to the input currents at the summing point. The output current and voltage of operational amplifier 3 change in the proper direction to make the input and feedback currents at the summing point equal to each other. If the feedback loop were so complex that the output of amplifier 3 changed in the wrong direction to maintain control (positive feedback), then its direction could be reversed by using the follower input of amplifier 3 as the current summing point.

For instance, if a positive high-voltage signal were needed at the output of a voltage inverter operational amplifier circuit, a common emitter high-voltage transistor circuit could be used as shown in Fig. 2.72. The voltage inversion of the transistor circuit would require the *follower input* to be used as the current summing point. The amplifier would have to supply the base current for the transistor but its output voltage would remain close to about −10 V, as determined by the zener diode and the relatively small forward biased *pn* junction voltages of the base-emitter junction and diode D. The output voltage could swing close to −11 V, and close to +200 V (depending upon the voltage divider effect caused by $R$, and $R_f$ in parallel with any load resistance). Diode D is used

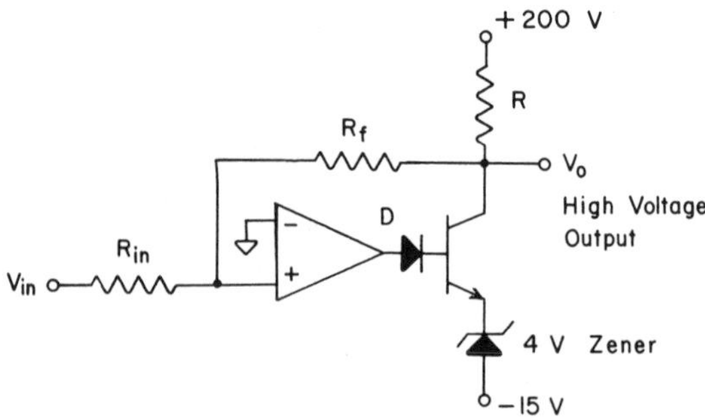


**Fig. 2.72.** A unipolar high-voltage inverter operational amplifier circuit. The summing point is at the noninverting (follower) input of the amplifier because of the additional voltage inversion of the high voltage transistor circuit.

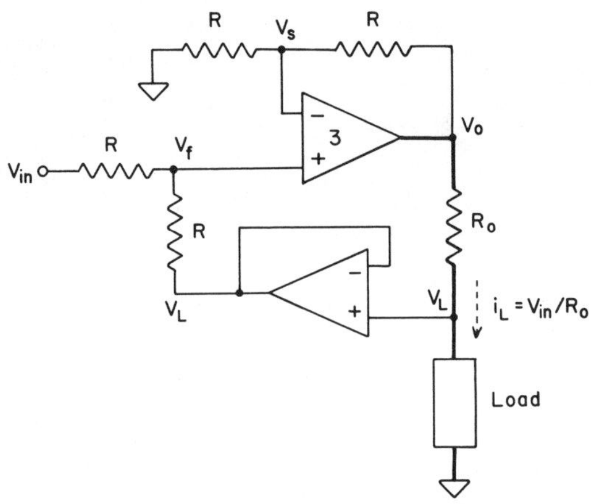


**Fig. 2.73.** A current control circuit that allows the load and the control signal to be grounded.

to protect the transistor base from large negative voltages that would arise when the transistor turned off while trying to produce a high-voltage output higher than the positive voltage limit.

**Control of current** to a load that is connected to common can be achieved by observing the voltage across a current-sampling resistor placed in series with the load, and adjusting the current until the voltage across the resistor equals the desired value. Since both ends of the current-sampling resistor float above ground, a difference amplifier with high input impedances and low input bias currents, such as an instrumentation amplifier, could be used to observe the voltage across the resistor. The output of the instrumentation amplifier could be used in a manner similar to that of the follower output in Fig. 2.69 to provide a voltage feedback signal to be used by the current control circuit. A simplier circuit requiring only two operational amplifiers is shown in Fig. 2.73. Because the resistors, except $R_o$, are all equal, $V_f = V_s = V_o/2$. Also, the voltage $V_f$ is the average of $V_{in}$ and $V_L$, or $V_f = (V_L + V_{in})/2$. Combining these relationships shows that the voltage drop across $R_o$ is $V_o - V_L = V_{in}$, so that $i_L = V_{in}/R_o$. Therefore the current through the load is proportional to the input voltage. Amplifier 3 provides both the current for the load and the current to drive the voltage divider, whose output is $V_s$.

## E. CIRCUITS WITH NONLINEAR TRANSFER FUNCTIONS

The mathematical relationship between the input and output of a system or circuit is called its **transfer function**. Its shape may be found experimentally by plotting a calibration curve for the system. In a measurement system the electrical output voltage or signal may not be linearly related to the number of units of the measured quantity. A nonlinear calibration curve and transfer function for this measurement system would result. A linear calibration curve could be obtained by passing the nonlinear electrical signal into a circuit that would compensate for the nonlinear relationship. Such a circuit would have an output

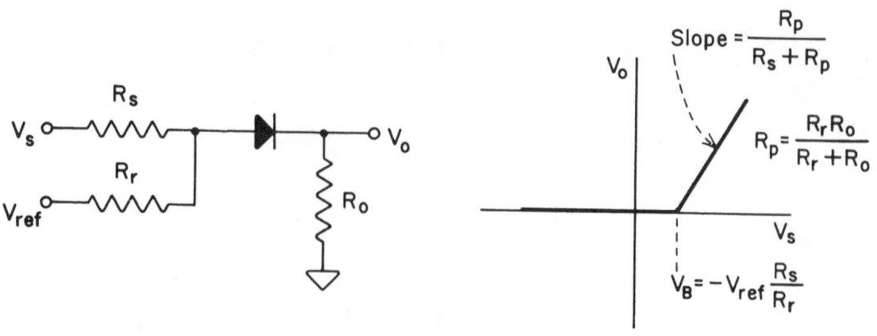


**Fig. 2.74.** A diode circuit with a nonlinear transfer function.

that was nonlinearly related to its input; its transfer function would also be nonlinear. Circuits with nonlinear transfer functions are also used for other purposes such as limiting currents and voltages to safe values, rectifying ac signals, performing analog multiplication and division of two or more voltages or currents, and wave from shaping.

Most **nonlinear circuits** are based on the nonlinear characteristics of the *pn* junction. For voltage signals less than a few tenths of a volt, the nonlinear exponential characteristics of a *pn* junction are important. For voltage signals much greater than the forward biased voltage drop of a *pn* junction, a diode may be considered as an ideal switch that has zero resistance when forward biased and zero conductance when reversed biased. A useful circuit with a nonlinear transfer function is shown in Fig. 2.74. The transfer function shown in the figure assumes that the signal is so large that the diode is behaving as ideal. The real transfer function has some curvature in the region of the change in slope. The reference voltage $v_{ref}$ is a negative voltage that reverse biases the diode, causing the output voltage to be zero until the signal voltage $v_s$ is more positive than $V_B$. When $v_s$ is more positive than the breakpoint voltage $V_B$, the diode conducts and the output voltage increases in direct proportion to the signal voltage. Such a circuit can be used at the input to a voltage summing operational amplifier circuit with $R_o$ used as the input resistor, since the summing point is essentially 0 V. Several of these circuits could be used with different breakpoints and slopes with the result as shown in Fig. 2.75. As each

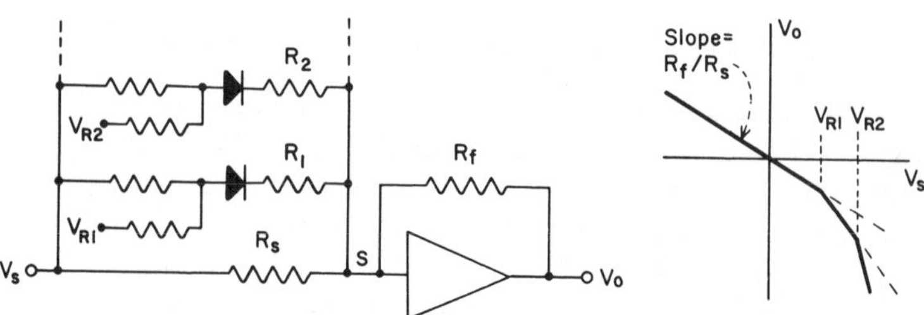


**Fig. 2.75.** An operational amplifier circuit that uses several diode circuits of the kind in Fig. 2.74 to cause a piecewise approximation to any desired shape for a nonlinear transfer function.

breakpoint voltage is exceeded, the slope of that branch of the circuit is added to the cumulative slope. If the reference voltages that determine the breakpoints are less than a few tenths of a volt apart, then the exponential shape of the current-voltage curve of the *pn* junction tends to round off the corners causing a relatively smooth looking transfer function. Placing the diodes in series with the reference voltages causes the curve to bend toward the $v_s$ axis instead of away from it as shown in Fig. 2.75.

### 1. Precision Rectifier Circuits

A **precision rectifier circuit** which uses the high gain of an operational amplifier essentially to eliminate the nonideal switch behavior of a *pn* junction even for millivolt signals is shown in Fig. 2.76. When the signal voltage is more positive than the breakpoint voltage, the output of the amplifier goes negative, forward biasing diode $D_1$ and limiting at about −0.5 V. The output voltage $v_o$ is essentially 0 V because it is due only to the leakage current of diode $D_2$ which flows through $R_L$ in parallel with $R_f$, and both are connected essentially to 0 V (common and the summing point). When the signal voltage is more negative than the breakpoint voltage, the amplifier becomes positive, reverse biasing $D_1$ and forward biasing $D_2$. The feedback current now flows through $R_f$ and the output voltage $v_o$ becomes the value necessary to make the feedback current through $R_f$ equal to the algebraic sum of the input currents through $R_r$ and $R_{in}$ (plus the leakage through $D_1$ and the amplifier bias current). The effective output resistance of the amplifier is increased by the dynamic resistance of $D_2$. Equation 118 shows that the effective open-loop gain $A_e$ is decreased as the output resistance is increased and as $R_L$ and $R_f$ are decreased. But as $R_L$ and $R_f$ decrease, the current through $D_2$ increases, causing its dynamic resistance to decrease. Because this tends to increase the effective open loop gain, an optimum value of $R_L$ and $R_f$ in parallel occurs. The optimum value is not critical and in practice satisfactory performance is usually obtained if the forward bias current through $D_2$ is above 1 mA or so. Any loads connected to $v_o$ must go to common and not to a power supply voltage if $v_o$ is to remain at zero when $D_2$ is reverse biased, because then the amplifier is not in control of $v_o$.

Precision rectifier circuits are used to convert an ac signal into a unipolar

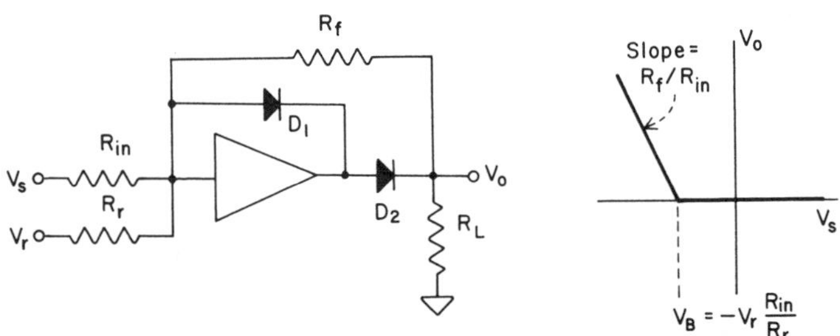


**Fig. 2.76.** A precision half-wave rectifier circuit with an adjustable breakpoint.

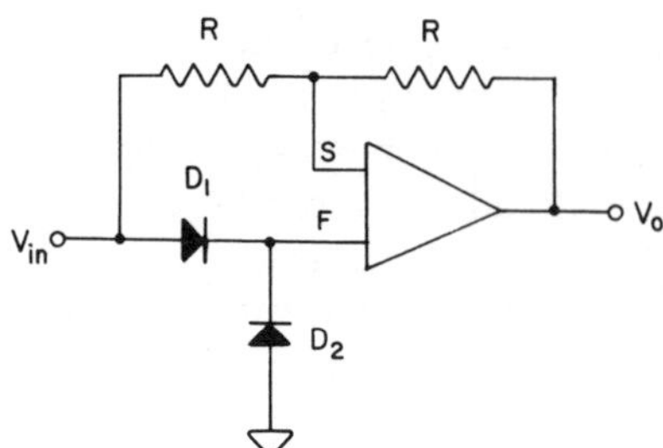


Fig. 2.77. A precision full-wave rectifier circuit using only one operational amplifier.

signal that can be sent through a low-pass filter to convert it to a dc signal. When the reference voltage $V_R$ is omitted from the circuit, the breakpoint occurs at the origin and a precision half-wave rectifier results. A precision full-wave rectifier, also called an **absolute value circuit**, results when the input signal is fed to a voltage summing amplifier circuit through a gain-of-one input and the output of the inverting half-wave rectifier is fed to the same voltage summing amplifier circuit through a gain-of-two input. A precision full-wave rectifier that uses one amplifier is shown in Fig. 2.77. For positive signals diode $D_1$ conducts and the circuit behaves as a voltage follower circuit. For negative signals $D_2$ holds the follower input at 0 V and the circuit acts as a voltage inverter circuit.

## 2. Logarithmic Functions

Circuits with **logarithmic transfer functions** are used to condense several decades of variation of input signal into a single range of a readout device. They are also used to produce a log-ratio amplifier circuit when the logarithm of the ratio of two signals is desired. A voltage proportional to the photon absorbance $A$ of a sample is obtained by applying two voltage signals $V_R$ and $V_S$ to a log-ratio circuit to obtain

$$A = \log\left(\frac{V_R}{V_S}\right) \tag{121}$$

The voltages $V_S$ and $V_R$ are signals proportional to the light flux passing through the sample and a reference, respectively.

Circuits with logarithmic transfer functions are shown in Fig. 2.78. They are

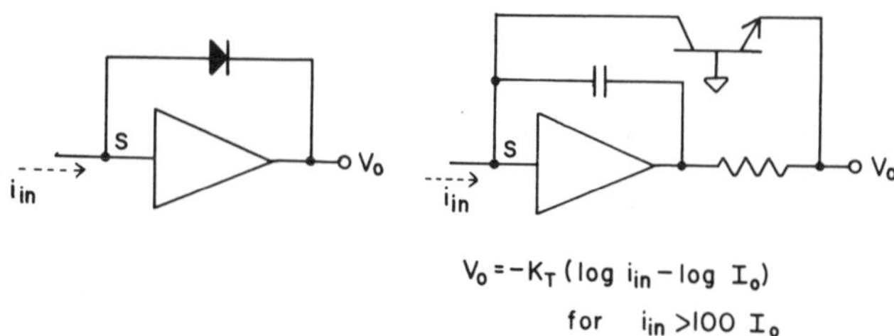


Fig. 2.78. Operational amplifier circuits that use ($a$) a diode and ($b$) a transistor to obtain a logrithmic transfer function.

both current-to-voltage converters with a logarithmic device in the feedback loop. A voltage input signal can be used instead of a current signal by putting an input resistor between the input voltage signal and the summing point.

The voltage $v$ across a *pn* junction is logarithmically related to the current $i$ flowing through it by

$$v = K_T[\log(i + I_T) - \log I_T] \tag{122}$$

where $I_T$ is a temperature-dependent term related to the reverse bias leakage current and $K_T$ is directly proportional to temperature and equal to 120 mV/decade of current at 300°K for silicon, and 59 mV/decade of current for germanium and most transistors. For an output voltage greater than about 0.2 V, $i \gg I_T$ and equation 122 becomes

$$v = K_T(\log i - \log I_T) \tag{123}$$

which has a slope $K_T$ that varies about 0.2%/°C at room temperature, and a voltage offset or intercept term which varies about −2 mV/°C. The useful dynamic range for the signal current is limited at the low end by the voltage offset and bias current drifts of the amplifier and the approximation $i \gg I_T$, and at the high end by the assumption that the resistive voltage drop across the semiconductor material, contacts, and leads is negligible. The useful dynamic range for most diodes is about two decades of current, although special diodes may provide four or five decades. The base-emitter voltage of a transistor is logarithmically related to collector current when the current gain $h_{fe}$ is sufficiently high. In practice the transistor gives a wider dynamic range than a diode. The interelectrode capacitances of the transistor load the summing point and the amplifier output, since the base is grounded, and might cause oscillation if the resistor and capacitor were not also included in the circuit.

Temperature compensation for both the intercept and slope terms in equation 123 should be used to obtain reasonable results. The effect of the intercept term can be minimized by using a log-ratio amplifier circuit (Fig. 2.79). A difference amplifier is used to subtract the output voltages of two *identical* log

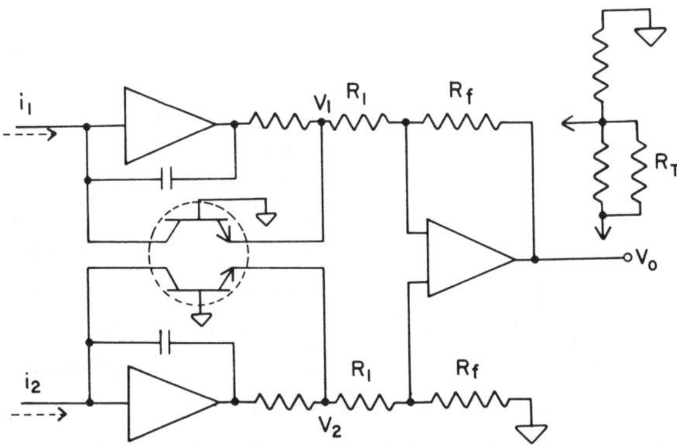


**Fig. 2.79.** A log-ratio circuit with temperature compensation.

amplifier circuits. When $h_{fe}$ is high and the transistors are perfectly matched to have the same $h_{fe}$ and $v_{be}$ characteristics, then the output voltage $v_o$ becomes

$$v_o = v_2 - v_1 \tag{124}$$

$$v_o = \frac{R_f}{R_1} K_T (\log i_1 - \log I_T - \log i_2 + \log I_T) \tag{125a}$$

$$= \frac{R_f}{R_1} K_T \log \frac{i_1}{i_2} \tag{125b}$$

When only one signal is of interest, the other is held constant. For instance, a constant $i_2$ is easily provided by a constant voltage and an input resistor connected to the summing point. When both transistors are fabricated on the *same* silicon chip matching of their characteristics is improved and their temperatures are held more nearly equal to each other. The temperature dependence of $K_T$ can be reasonably compensated over a given temperature range by using a thermistor $R_T$ in the gain control circuit of a voltage follower or inverter circuit following the difference amplifier, or by using matched thermistors in corresponding arms of the difference amplifier itself. An approximate compensation can also be achieved by using a moderate gain, say, $R_f/R_1 = 10$, and a thermistor voltage divider between $v_o$ and the upper $R_f$ instead of a direct connection, as indicated in Fig. 2.79. The thermistor $R_T$ must be placed in thermal contact with the transistors.

An antilog or exponential voltage transfer function can be obtained by placing the grounded base transistor in the input circuit with its collector connected to the summing point, and using a resistor in the feedback loop. Temperature compensation circuits should also be used.

The above circuits are useful for currents and voltages of one direction or polarity. Signals of opposite direction or polarity can be used by reversing the diode or by changing to a *pnp* transistor.

### 3. Multiplier, Ratioing, and Square Root Circuits

A **multiplier circuit** that produces a voltage proportional to the product of two voltages is used in modulation and demodulation circuits where the signal of interest is impressed upon a carrier signal for transmitting or $S/N$ enhancement purposes. A simple multiplier can be obtained by summing the output of two log circuits and following the result with an antilog circuit. Multiplier circuits are available commercially, however, and are usually easier to use. The operation of an integrated circuit multiplier is often based on the proportional relationship between the current through a transistor and its transconductance. The outdated quarter-square multiplier uses a multiple segment nonlinear function circuit such as indicated in Fig. 2.75. The approximation of this circuit is its major source of error. One of the most accurate analog multipliers is the triangle-averaging multiplier. This circuit adds a triangular wave to both input voltages. The final dc value must be filtered to remove the triangular wave. The slow response due to the low-pass filter is the main disadvantage of this multiplier.

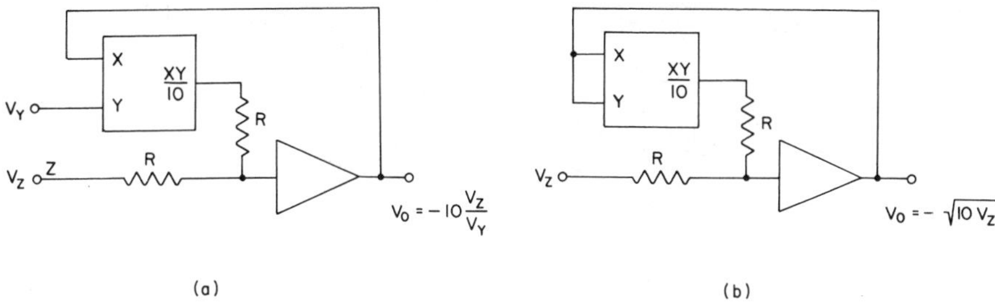


**Fig. 2.80.** Circuits that use a voltage multiplier unit in the feedback loop of an operational amplifier to obtain a voltage ratio (*a*) and the square root function (*b*).

When two voltages are multiplied together there are four possible combinations of polarities: ++, +−, −+, and −−. These combinations correspond to the four quadrants on an ordinary Euclidean graph of the two voltages. The log-ratio antilog multiplier mentioned above accepts voltage pairs in only one quadrant. A **four-quadrant multiplier** accepts any combination of polarities.

The accuracy specifications for commercial multipliers are usually given as a percentage (say, 1 or 0.5%) of full scale range, usually 10 V. For best results it is necessary to amplify smaller signals so that they are a large fraction of this range. Signals that are only a fraction of a volt might well be distorted in the nonideal response range of a multiplier.

A **voltage ratioing circuit** and a **square root circuit** can be obtained by placing a multiplier in the feedback loop of a voltage inverter operational amplifier circuit (Fig. 2.80).

## F. PEAK FOLLOWERS

A **peak follower** is used to observe and hold the maximum value of a signal. A simple peak follower is shown in Fig. 2.81. A voltage follower is used to buffer the capacitor from a load. The diode allows the voltage on the capacitor to follow a positive input voltage signal until the signal begins to decrease. When the input voltage becomes less than the capacitor voltage, the diode becomes reverse biased and stops conducting. The capacitor holds the most positive voltage of the signal. Leakage through the reverse biased diode and the bias current of the voltage follower circuit cause the capacitor voltage to decrease slowly from its maximum value. This decrease is called the drift rate. The response of a simple peak follower suffers from the forward biased voltage drop

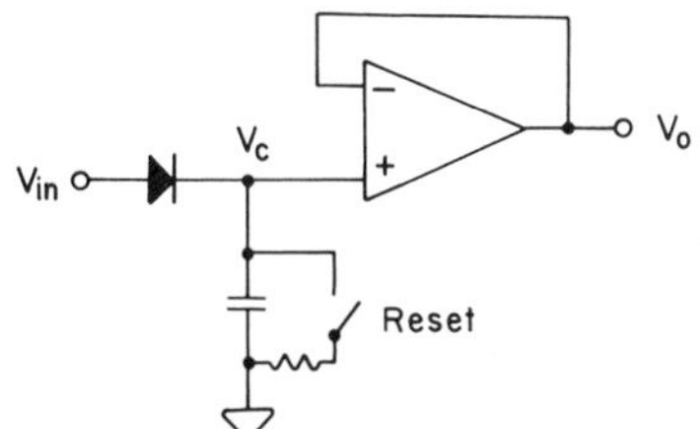


**Fig. 2.81.** A simple peak follower circuit.

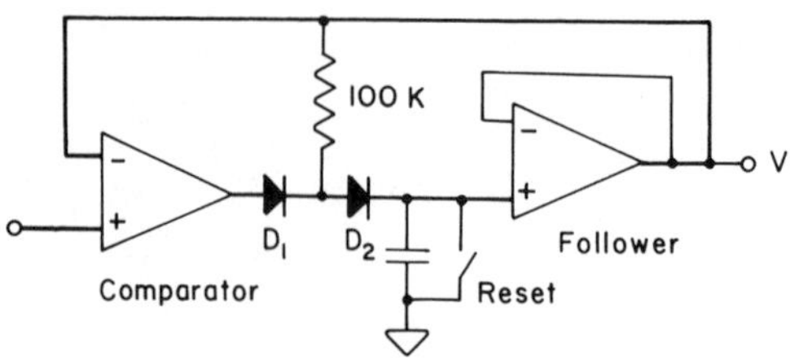


**Fig. 2.82.** A peak follower circuit that minimizes the nonideal behavior of switching diode $D_2$.

across the diode, which depends upon the required charging current. When sufficient time is allowed to charge the capacitor, this voltage drop eventually becomes negligible. Reducing the capacitance improves the response time but increases the drift rate. In order to reduce the effect of the voltage drop across the diode, the capacitor and diode may be enclosed in a feedback loop as shown in Fig. 2.82. The comparator is a high gain difference amplifier that charges the capacitor through the two diodes until the capacitor voltage equals the input voltage. The voltage drop across the diodes is effectively reduced by the gain of the comparator amplifier. When the input voltage drops below the capacitor voltage the output of the comparator drops to zero or below, reverse biasing $D_1$. Drift of the circuit is reduced by the resistor, which holds the anode of $D_2$ close to the capacitor voltage, so that the leakage current through $D_2$ is much smaller than it would be otherwise. The comparator should be able to respond quickly to a fast peak or be able to recover quickly from an output current or voltage limit that it will reach if the capacitor voltage does not rapidly follow close to the input voltage. Decreasing the capacitor may increase the response time at the expense of increasing the drift rate. If a relatively long holding time is required, a fast peak detector with a fast drift rate may be followed by a slow peak detector having a slower drift rate.

## G. SAMPLE-AND-HOLD CIRCUITS

**Sample-and-hold circuits** are used in boxcar integrators, sampling plug-in units for oscilloscopes, high-speed analog-to-digital converters for digital recorders and computers, fast analog storage registers and simulated delay lines, and "de-glitching" circuits to remove extraneous pulses from digital-to-analog converters. They have two operating modes that are determined by a control signal. In the *sample* mode the output terminal follows the input signal. When switched to the *hold* mode the output terminal retains the last value of the input signal that it had before the mode change was called for by the control signal. A basic sample-and-hold circuit is shown in Fig. 2.83 along with a voltage source. When the switch SW is closed the capacitor voltage follows the input voltage. When the switch is open the capacitor holds the last value of the input voltage (subject to drift caused by bias current of the follower circuit and leakage of the switch). The time it takes the capacitor to acquire a specified fraction (say, 0.99) of a step function signal voltage is called the **acquisition time**. It depends upon

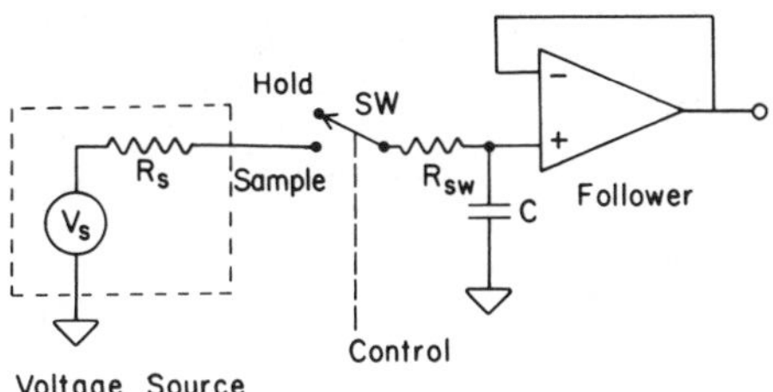


**Fig. 2.83.** A sample-and-hold circuit showing the switch resistance and the output resistance of a voltage source whose signal is being sampled.

the *RC* time constant of the capacitor and the series resistance, which includes the ON resistance of the switch and the output resistance of the voltage source. The **settling time** is the time for the output voltage to enter and stay within a specified fraction of its final value when a specified input step is applied. The settling time includes the acquisition time and the amplifier settling time. The **aperture time** is the time between the command to *hold* and the actual opening of the control switch. The aperture time determines how precisely the sampling time is controlled.

When the voltage source resistance is large, or capacitive loading of the voltage source causes it to oscillate, then a follower amplifier circuit could be placed between the input terminal and the switch. A feedback circuit that reduces the influence of *switch resistance* is shown in Fig. 2.84. The high gain difference amplifier, or comparator, amplifies the difference between the output voltage (equal to the capacitor voltage) and the input voltage, and uses the much larger voltage that results to charge the capacitor more quickly through the series resistance. Offset and CMRR errors in the output follower are automatically compensated. The performance of the circuit now depends upon the CMRR, offset errors, gain, and current driving characteristics of the comparator. The comparator should be able to recover quickly from current and voltage limits which may occur when the switch is closed and the output and input voltages differ widely, or when the feedback loop is opened by opening the mode control switch. An open feedback loop can be prevented by a second switch $SW_2$ placed between the output of the difference amplifier and its inverting input, as shown in Fig. 2.84*b*. The switches are in opposite states, as indicated by the inverting symbol, a small circle, at the tip of the control signal

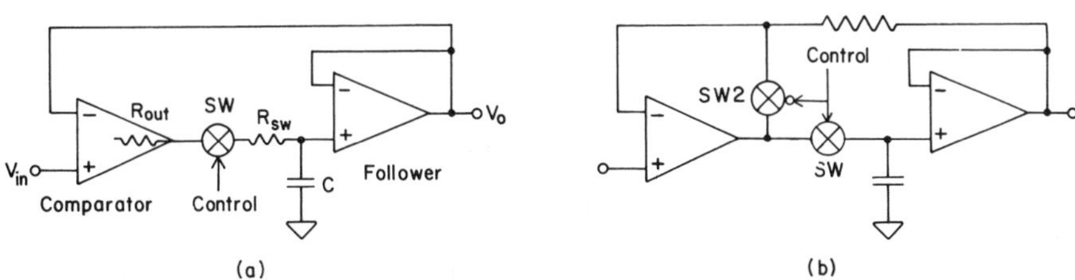


**Fig. 2.84.** Sample-and-hold circuits that (*a*) minimize the influence of switch resistance and in addition (*b*) prevent the comparator from saturating.

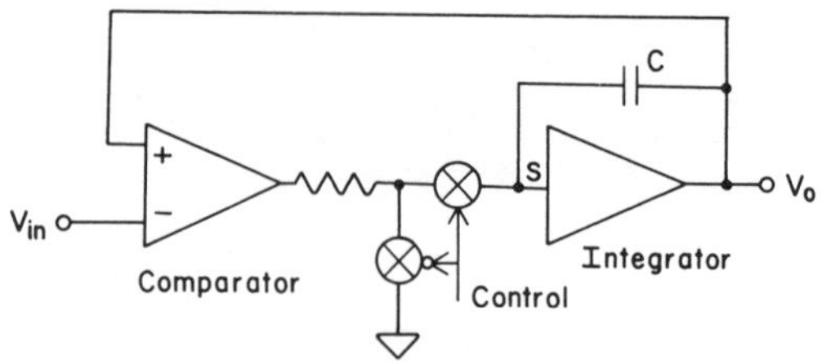


**Fig. 2.85.** A sample-and-hold circuit that allows one side of each switch to remain at 0 V.

arrow. When the mode control switch is opened, the second switch is closed to place the comparator in a follower configuration. A resistor $R$ is used to decouple the two amplifier outputs.

The capacitor can be placed in the feedback loop of the output operational amplifier as shown in Fig. 2.85. This forms an integrator configuration for the *output* amplifier circuit that stores a voltage equal to the input signal, but the circuit does not integrate the input signal. Just as with the previous circuits, current is supplied to the integrator by the input comparator amplifier to charge the capacitor whenever the output and input voltages are unequal. This circuit has an advantage of maintaining the switch terminals always near 0 V to minimize leakage across an open switch. Leakage to the control terminal could also be minimized by using a MOSFET.

The lower limit for the acquisition time for a simple sample-and-hold circuit is on the order of a few nanoseconds, considering a switch resistance of about 50 Ω and using the stray input capacitance of an amplifier of about 10 pF. Shorter sampling times would charge the capacitor up to only a fraction of its final output because of the exponential charging rate of an $RC$ circuit. The circuit in Fig. 2.86 is called an **error-sampled** sample-and-hold circuit and automatically corrects for the resulting error, making possible sampling times in the subnanosecond range.

In Fig. 2.86 assume that the output voltage is equal to the input voltage from the preceding sampling interval, $v_o = v_c$. The input voltage difference *change*

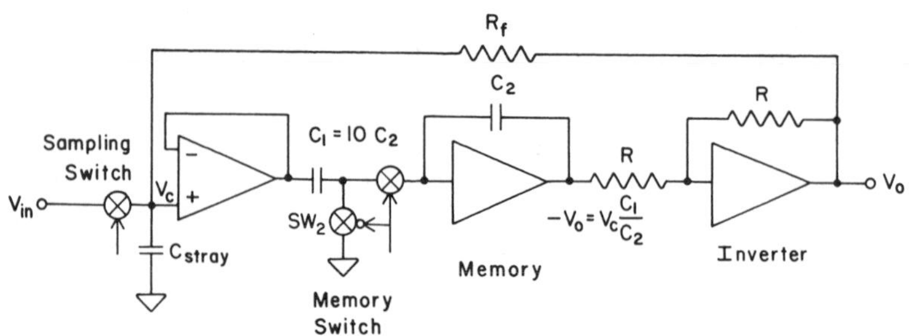


**Fig. 2.86.** An *error-sampled* sample and hold circuit for high speed applications in the nanosecond range.

between two successive sampling intervals is called the *error voltage*. When the sampling switch closes, the capacitor voltage $v_c$ begins to change by the amount of the error voltage. The sampling switch suddenly opens after a fixed time interval. During this time interval the capacitor voltage has only had time to charge to a *fixed* fraction, say, 0.1, of the final voltage. The memory switch also was turned on with the sampling switch but is now allowed to stay on until the output of the gain-of-ten voltage inverter amplifier formed by $C_1$, $C_2$, and the memory operational amplifier has had time to settle to its final value. Capacitor $C_2$ was grounded prior to this sampling interval by $SW_2$ so that connecting it to the summing point of the memory amplifier causes no current to flow unless the output of the input amplifier has changed. Since the output of this amplifier has changed by one-tenth the error voltage, this change is amplified by the ratio $C_1/C_2 = 10$, causing the voltage across $C_2$ to change by an amount *equal* to the error voltage. The voltage across $C_2$ therefore becomes equal to the new input voltage $v_{in}$. The memory switch turns off and $SW_2$ connects $C_1$ to ground. The output of the memory amplifier is inverted to the proper polarity and fed back to the input capacitor $C_{stray}$ via $R_f$. The time constant of $R_f$ and $C_{stray}$ is too long to charge $C_{stray}$ significantly during the time the memory switch is turned on, but it is short enough to allow $C_{stray}$ to charge to the true value of the last sampled input voltage before the next sampling interval. The sample-and-hold circuit is now ready for the next sampling interval. To speed up the charging of $C_{stray}$ by $v_o$ through $R_f$ it would be possible to allow $C_{stray}$ to be partially charged by $R_f$ during the time the memory switch were on, providing the gain $C_1/C_2$ were reduced to compensate for this additional change.

An **analog storage register** or **analog shift registor** consists of many capacitors, each with its own sample-and-hold circuit, that are multiplexed to observe sequentially the same input signal, and to drive sequentially the same readout amplifier. The circuits and capacitors are usually put on an integrated circuit chip that also contains the digital sequencing shift registers that drive the switches that determine when a capacitor is connected to the input signal and when its stored voltage is used to drive the output amplifier. External clocks (pulse generator) are connected to the integrated circuit to set the sequencing rates. Sampling rates can be in the megahertz range, so that the temporal history of a fast transient voltage signal can be stored or recorded in the register. During the recording of the transient input signal, the first capacitor is charged up to the voltage that the input signal has at the beginning of the recording. A fraction of a microsecond later, the first capacitor is disconnected and the next capacitor is charged to the voltage of the input signal during the next short time interval. This process continues until each of the capacitors has been charged to the voltage that the input signal had during a particular point (or submicrosecond window) in time. The voltages on the capacitors are then read out sequentially by momentarily connecting first one capacitor and then the next, and so forth, to the input of the readout amplifier. If the readout sequencing is done at the same rate as the sample and hold recording sequencing, then the output wave form of the readout amplifier has the same general shape as the input signal, though it may have a stairstep appearance. The affect is that of an **analog delay line**, which delays a wave form for a predetermined length of time.

The readout sequencing may also be done at a slower rate to stretch out the wave form on the time axis. Stretching the wave form in this manner may be done, for instance, to allow a relatively slow analog-to-digital converter to convert the wave form to digital numbers for storage in a digital memory.

How slowly the readout sequencing may be done depends on the droop rate of the stored voltages. Small capacitors are needed for fast sampling rates, but they also have higher droop rates than larger capacitors subject to the same switch leakage currents. Typically the readout sequencing may be done at a rate from 100 to 100,000 times more slowly than the recording rate, depending upon the design of the register.

## H. LOCK-IN AMPLIFIERS

A **lock-in amplifier** can be used to improve greatly the $S/N$ ratio of a slowly varying or dc signal if the signal can be modulated (chopped) at a stable frequency. The frequency at which the signal is modulated should be chosen to be well removed from $1/f$ noise, power-line frequency, and other frequencies that have relatively high noise levels in the system being used. The lock-in amplifier discriminates against any additional signal such as noise or a dc level that is added to the signal *after* the signal modulation step, and *before* the demodulation step. A signal that appears to be hopelessly buried in noise can often be recovered. The frequency should also be high enough to allow the lock-in amplifier to follow a changing signal without distorting the changes. The choice of frequency and the types of distortion that may occur are discussed in detail in reference 31a.

A lock-in amplifier is commonly used in atomic-absorption and fluorescence measurements to remove from the radiation source emission signal any dc emission signal from the sample cell (flame or plasma). To do this the light from the radiation source is modulated at a fixed frequency prior to passing the light into the sample cell. In infrared instruments the radiation is commonly modulated prior to its falling upon the radiation detector so that noise in the detector can be rejected by a lock-in amplifier.

The basic components of a measurement system that uses a lock-in amplifier are a signal modulator, a synchronous demodulator, and a low-pass filter (Fig. 2.87). The last two are usually contained in the lock-in amplifier and are the heart of its operation. The signal modulator chops the signal with an approximate square-wave pattern and sends a reference signal with the same frequency and phase to the lock-in amplifier to synchronize the demodulator. The modulated signal is amplified if necessary and sent to the demodulator which multiplies the modulated signal by +1 during one half cycle of the reference signal ($SW_1$ closed, $SW_2$ open) and −1 during the other half cycle ($SW_1$ open, $SW_2$ closed). One of many types of synchronous demodulator circuits is shown in Fig. 2.87. The reference and modulated signal are phase-locked (hence

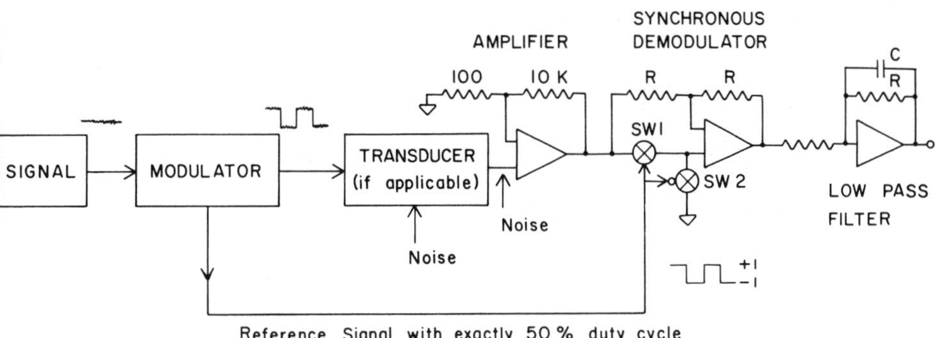


**Fig. 2.87.** A lock-in amplifier system.

the term "lock-in") to each other so that multiplication by +1 occurs during the ON part of the modulated signal and multiplication by −1 occurs during the OFF part of the signal. The result is then sent to a low-pass filter that time-averages the results. In this way what occurs during one half cycle is subtracted from the other half cycle. Noise that occurs *after* the modulation step (e.g., at a transducer or amplifier) is added to *both* half cycles and the subtraction process causes most of it to be eliminated from the final result. A small amount of such noise that does not average to zero over the time constant of the low-pass filter remains. The original signal (and its original noise) which appears during alternate half cycles is unaltered by the subtraction process and appears as the main component of the final filtered output.

Additional components of a lock-in amplifier may include selective band-pass amplifiers, and a phase-shift adjustment and wave-shaping device for the reference signal (Fig. 2.88). When the signal is buried in noise, a narrow band-pass tuned amplifier is often used in conjunction with the amplification step to filter out much of the noise so that relatively large noise spikes do not cause the amplifier to go into limit. The use of tuned amplifiers requires that the modulation frequency be highly stable. A tuned amplifier in the reference channel produces a sine wave from a reference signal of almost any shape, provided it has the proper frequency. A phase adjustment amplifier circuit provides the desired phasing of the reference and signal channels. The sine wave is then converted to a square wave with exactly equally spaced half cycles (50% duty cycle) to drive the synchronous demodulator. Four-quadrant analog multipliers may be used for the synchronous demodulation. However, these multipliers may be subject to inherent nonlinearities for signals on the order of a fraction of a percent of full scale.

In some cases when a signal source is modulated, the information contained in the signal pulse may change during the pulse. An example of this is the change in spectral profile that may occur during a pulse when a hollow cathode lamp is pulsed. The lock-in amplifier system with its associated filters and phase shifters produces a response that is a weighted average of the information. Reference 31b provides an example and discusses this effect in more detail.

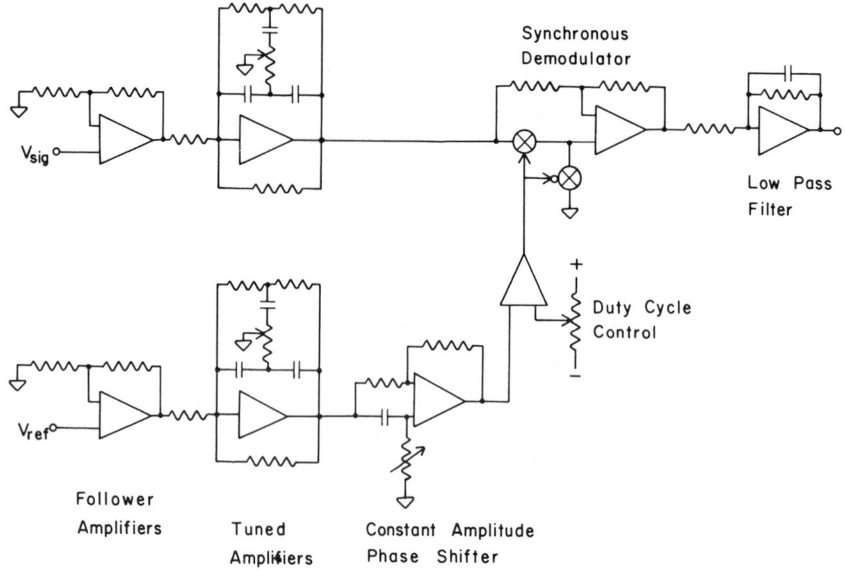


**Fig. 2.88.** Components of a lock-in amplifier.

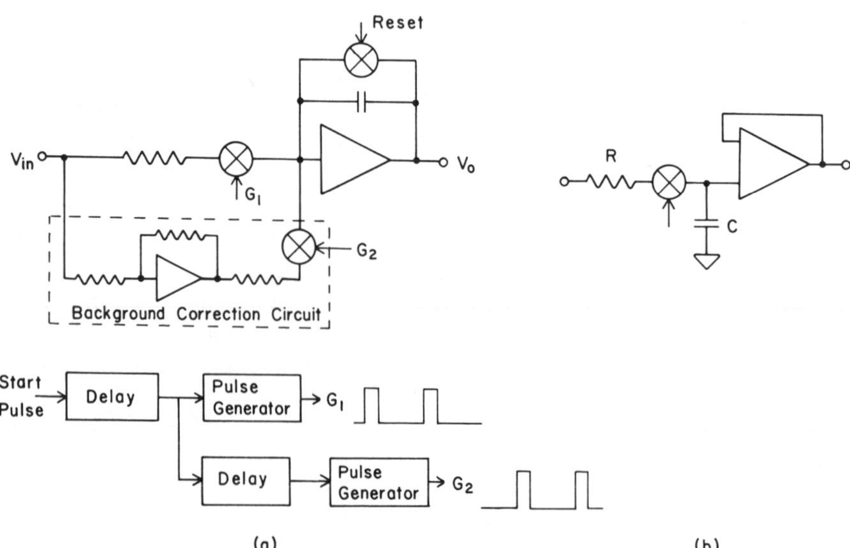


**Fig. 2.89.** Boxcar integrators that use an operational amplifier integrator (*a*) and an *RC* integrator (*b*).

## I. BOXCAR INTEGRATORS

When a repetitive signal with a short duty cycle is embedded in noise, a **boxcar integrator** may be used to improve the $S/N$ ratio. A gated integrator (Fig. 2.89*a*) is used to integrate the signal only during its short duration, each time the signal occurs. The output of the integrator after a predetermined number of pulses represents the final result. Using a low-pass filter instead of a true integrator eliminates the need for an integrator reset switch and allows the output to follow slow changes in amplitude of the input signal. An automatic background correcting pulse can be added by following the main signal integrating pulse with a second pulse of equal duration. The second pulse causes an inverted background signal to be integrated for a short time, resulting in the subtraction of the background signal from the main signal plus background. When a short and fast gate is required, an $RC$ integrator of the type shown in Fig. 2.89*b* may be used so that the amplifier response time does not limit the performance. This circuit is essentially a sample-and-hold unit to which a resistor has been added to provide an $RC$ integration response. Reference 31c discusses the weighting effects of an $RC$ integration relative to a true integrator.

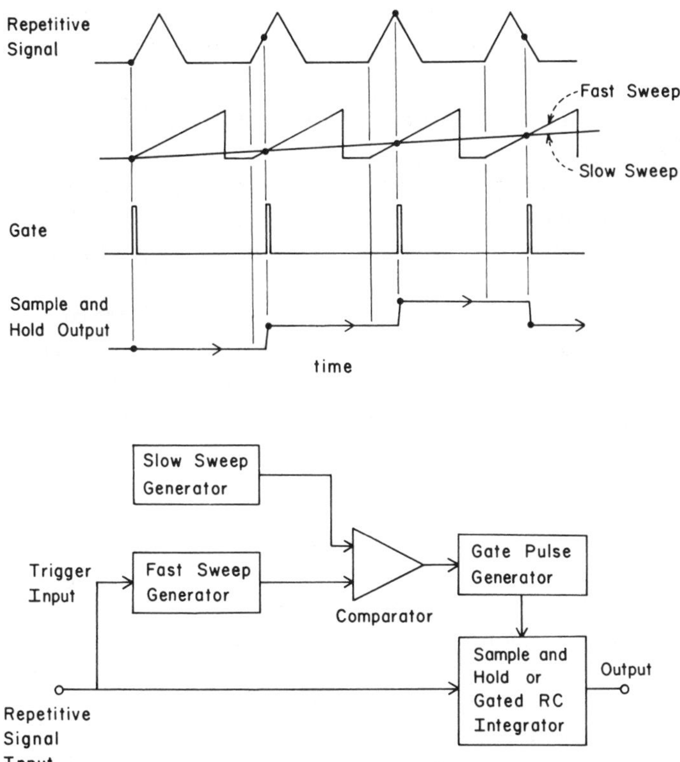


**Fig. 2.90.** A sampling technique used to record fast repetitive signals on a slow time base.

### J. RECORDING FAST, REPETITIVE SIGNALS ON A SLOW TIME BASE

The shape of a rapidly varying repetitive signal can be displayed on a much slower time scale by slowly scanning from one end of the repetitive signal to the other end with a sample-and-hold or boxcar *RC* integrator whose output is recorded on a relatively slow readout device. Sampling oscilloscopes use this technique to display the shape of a repetitive signal that is too fast to be recorded directly on the screen. Boxcar integrators used in their *RC* integration mode often have the ability to display the shape of a fast repetitive signal on a strip-chart recorder.

A common way to *scan* a repetitive signal is shown is Fig. 2.90. At the beginning of the experiment a slow sweep is started which lasts for the duration of the experiment. A fast scan is triggered each time the signal wave form to be recorded reoccurs. A comparator responds each time the fast scan reaches the voltage of the slow scan, and causes the control gate of the sample-and-hold or boxcar *RC* integrator to open for a short time. The gate opens later and later during each signal wave form, causing the output signal to slowly sweep out the shape of the wave form on the readout device. Sample-and-hold circuits may be used when the repetitive signal is relatively free from noise. Boxcar *RC* integrators can help to extract a signal wave form that is buried in noise. When used to do this, the slow scan is slow enough so that successive gate pulses overlap regions of the signal wave form to provide a time-averaged reduction in noise.

## VII. NOISE, SHIELDING, AND GROUNDING

Noise in electronic instruments may arise from any of several sources. **Johnson** or **thermal noise** is caused by the random thermal motion of electrons in resistive components. Differences in electron densities caused by the random motions produce observable voltage fluctuations across the resistance. The average of the voltage fluctuations is zero. Johnson noise is equally distributed across all frequencies. This type of distribution is called **white**, analogous to the continuum of frequencies in white light. A convenient measure of the magnitude of the Johnson noise voltage fluctuations is given by the Nyquist relationship:

$$\bar{v}_{\mathrm{rms}} = (4kTR\Delta f)^{1/2} = (5.5 \times 10^{-23}\ TR\Delta f)^{1/2}\ \mathrm{V} \tag{126}$$

which relates the root mean square (rms) voltage $\bar{v}_{\mathrm{rms}}$ (standard deviation of the noise voltages) obtained with a bandwidth $\Delta f$ Hz to the Boltzmann constant $k$, the absolute temperature $T$ of the resistor, and the value of the resistance $R$ in ohms. If an integration time is used to make the measurement then $\Delta f = 1/(2\tau)$. If a first-order low-pass filter is used to determine the bandwidth, then $\Delta f = 1/(4RC)$, somewhat greater than the upper cutoff frequency, $f_o = 1/(2\pi RC) = 0.16/RC$.

For a 1 MΩ resistor near room temperature and a bandwidth of 1 MHz, the predicted rms noise voltage is 0.13 mV, equivalent to a current noise of 0.13 nA through the resistor. Reducing the bandwidth to 1 Hz, typical of measurement systems using strip-chart recorders, reduces these values by three decades.

**Excess noise** is considered to arise from any source that does not represent a fundamental limitation such as Johnson noise. Excess noise usually varies with frequency. Semiconductor and other electronic components are subject to excess noise owing to imperfections and some causes that are not well defined. **One-over-$f$ ($1/f$)** noise, sometimes called **flicker noise** because of its large magnitude at low frequencies, is inversely proportional to frequency and is usually significant at frequencies below 100 Hz. Instrument drift and aging effects are commonly classified in this category. **Pink noise** does not have such a strong dependence upon frequency and is inversely proportional to the square root of frequency. It tends to be important in amplifiers above 100 Hz. **Popcorn noise** is sometimes present in transistors and is due to the transistor gain suddenly and randomly switching between two values. Noises of these types can be reduced by proper choice of components, and are particularly important at the front end of the measurement system where the signal is smallest. Further reduction in frequency-dependent noise can be obtained by modulating the signal (if possible) at a moderate and stable frequency where the noise is minimal, and using a narrow band-pass filter to reduce the noise contributions on either side of that frequency. The synchronous demodulator or lock-in amplifier uses this technique. These techniques do not reduce the noise already present in the signal prior to the modulation step, but do reduce the noise introduced by the amplification and electronic processing steps.

Another type of noise is due to interference noise picked up by a measurement system from outside sources such as power lines, flash lamp discharges, relay arcing, electrical motor brushes, and lightning. Interference noise can usually be reduced greatly by proper circuit layout and shielding. Sensitive input circuits should not be located near power-line transformers, near circuits that produce fast transients, such as digital circuits, or near circuits that carry changing high currents. The conductor that is used as the signal common (return path for the signal) should not carry other currents because these currents result in a voltage drop across the nonzero resistance of the conductor. This voltage drop appears as part of the signal.

**Shielding** is accomplished by completely surrounding a device with a conductor (called a shield.) The potential of the space completely enclosed by a *perfectly* conducting shield rises and falls with the potential of the shield. Because all points in this space rise and fall by the same potential (caused by the external noise source), any noise potential change appears on both the signal wire and its common wire. Since the *signal* is really the potential *difference* between these two wires, the noise that has been added equally to both wires does not appear in the signal.

In practice most measurement systems consist of several devices that are connected together by external wires so that no one device is completely enclosed. Shielded wires may be used for these connections to help to complete the shield around the entire measurement system. **Ground loop** difficulties arise when more than one wire connects the shields of two or more devices together. For instance, the shielding conductors on two shielded wires that connect two devices and the shields on the devices themselves form a complete electrical loop that acts like a transformer winding. Currents are readily induced in this loop by external sources of noise, especially power-line sources, causing voltage drops around the loop owing to the finite conductivity of the shielding materials. The induced loop currents and voltages may add to the noise in the signal by inductive and capacitive coupling between the shield and the signal path. The best solution to ground loop problems is to eliminate ground loops. When they cannot be eliminated, the area that they enclose should be reduced as much as possible to minimize the inductive coupling from the noise source to the ground loop. Moreover, material having the lowest practical resistance should be used throughout the loop to minimize the magnitude of the voltage differences caused by the currents.

Common components of a ground loop are the third-wire ground wires used for personnel safety purposes on the power line connecting cables of instruments, which are the source of the term *ground loop*. These wires run to an earth ground and are connected to the metal chassis of the instrument, which is usually used as part of the system shield. Although these wires prevent *dangerously* large voltage differences at a metal chassis, they can easily help to cause millivolt noise levels. Their long length helps to make a loop of large area, and their plug connections may have a small but significant resistance, so that any noise currents that they may carry cause relatively large voltage drops at these points. An especially troublesome situation occurs when two instruments are plugged into wall sockets that are supplied from different power transformers. Large noise voltages can develop because the ground wires for the two wall sockets may not be connected to each other except at a long distance from the wall sockets, causing a ground loop of very large area. Such situations should be avoided.

When the entire measurement system is shielded, one third-wire ground connection to the measurement system shield might be sufficient for safety purposes, since all metal chassis are electrically connected to each other. *An important exception* occurs when the shielded wires that connect the chassis are not able to carry high currents that might be caused by an electrical fault that might connect a hot power line to the shield. *When a noise shield is used as a safety shield, all its components must be able to carry all faulty currents greater than the line fuse rating, until the power line fuse blows.* Otherwise more than one ground wire may be necessary for safety, and the resulting ground loops tolerated and their area held to minimal size.

When any component of the measurement system lies outside the system's shield, that part is coupled to earth ground by stray capacitance, which tends to hold the potential of that component at ground potential. It then becomes desirable to connect the system shield to ground so that the entire system (except the signal) remains at a fixed potential with respect to ground. Any external noise now finds the shield shorted to ground but can still change the potential of the component that is outside the measurement system, because its stray capacitance to ground is only weakly effective at holding that component at ground potential. A small amount of noise may have a negligible effect at an outside component such as a readout device, because the signal at this point may be much larger than the induced external noise. However, the external noise might be transmitted along the wires back to the more sensitive circuits that lie within the system shield. To minimize this effect, the signal's common wire is connected to the grounded system shield to hold this wire at ground potential. To minimize the effects of possible ground loops that might now involve the signal common wire inside the shield, this connection should be made at only one point on the system shield. To minimize the effects due to the nonzero resistance of the shield material, this point should be the same point at which the shield is connected externally to the earth ground wire. Both points should also be near the signal source since this is the region whose potential must be most tightly held fixed.

Noise picked up by the power-line leads can be capacitively coupled into the system through the power supply transformer. *LC* filters inserted into the power line can reduce noise transients. Capacitors and *RC* and *LC* filters with their common at ground and placed in both the common and "hot" dc power supply lines can also help.

The capacitive and inductive coupling of noise to a signal path from either a direct source of noise or via a shield that is subject to ground loop effects can often be minimized by using a **twisted-wire pair** for the signal wire and its common wire. The twisted configuration provides a minimum loop area for inductive coupling and helps the signal wire and its common wire to be equally influenced by capacitive coupling of the noise source. Further reduction in noise can be achieved by shielding the twisted-wire pair. The shield should be connected to the signal common only at the source end to prevent a ground loop. For short lengths, the shield could be used as the signal common as is often done when using ordinary coaxial cable. However, for longer runs a shielded twisted-wire pair provides somewhat better performance because the shield is not a perfect conductor and is subject to variations in potential along its length when subjected to noise sources. These variations do not become a direct part of the signal as they would be if the shield were the signal common, but are only indirectly coupled to both wires of the twisted-wire pair.

A noise signal that has been picked up equally by both the signal wire and its common wire so that it is common to both wires can be eliminated from the signal by connecting the two wires to the two inputs of a difference amplifier that has a high common mode rejection ratio (CMRR). Usually the amplifier has a gain of 100–1000 to increase the signal to help reduce the influence on the $S/N$ ratio of further noise pickup. Care must be taken to prevent deterioration of the overall CMRR as discussed above in the section on difference amplifiers. For instance, the amplifier must have a high input impedance relative to the output impedance of the signal source. The capacitive loading caused by any shield around the twisted-wire pair also must be considered. The effect of capacitive loading caused by the time constant of the output resistance of the source and the signal-wire-to-shield capacitance may be negligible for a relatively slowly varying signal. However, relatively fast noise pulses that are present on the signal common, and therefore fed through the signal source to the signal wire, are attenuated differently by the *signal common*, which has essentially no output resistance in series with the cable capacitance, and the *signal wire*, which has some output resistance in series with the cable capacitance. In this case the loading due to the cable capacitance would cause the fast noise pulse to become distorted on one wire but not the other. The noise would no longer have a common shape on both wires, and the common mode rejection of the difference amplifier would not help to remove the noise. To minimize this capacitive loading of the common mode noise signal, the shield is attached to the common *at the signal source*. The shield is thereby driven by the same noise pulse that appears on the twisted-wire pair at the source, and the effect of capacitive loading upon the common mode noise signal is minimized. The shield should be connected only at its signal source end, and at no other place, to eliminate ground loops.

A way to equalize the effects of capacitive loading of the two wires for a common mode signal is to drive both wires of the twisted-wire pair with amplifiers that have equal output impedances. Capacitive loading can be equalized for a fast signal by driving one wire of the twisted pair with the signal and the other wire with an equal signal that is 180° out of phase, using amplifier outputs that have equal output impedances. Temporal *RC* distortions of the signal are common to both wires and are eliminated by the CMRR of the difference amplifier at the receiving end.

When the noise frequency is high enough (rf, radio frequency) that it causes reflections in the shield, then both ends of the shield where reflections can occur should be terminated, as would be a transmission line, to minimize reflections. Since such terminations cause ground loops it may be necessary to use two shields, one for rf noise, properly terminated, and another for low-frequency noise, grounded at only one point. Since high frequencies tend to travel near the surface of a conductor, an rf shield should have a large surface area wire braid or mesh to reduce its effective resistance.

## VIII. WAVE FORM GENERATORS

Repetitive wave forms such as sine, square, and triangular waves are often required to scan an experimental variable over a range of values and to time or synchronize events. The type 555 timer is a popular integrated circuit that uses a comparator in the manner discussed below with external *RC* circuits to generate timing intervals that are relatively independent of power supply voltages. (Caution is advised when using 555 timer integrated circuits, because those from different manufacturers may have somewhat different characteristics.) Operational amplifier circuits are particularly useful for generating precise

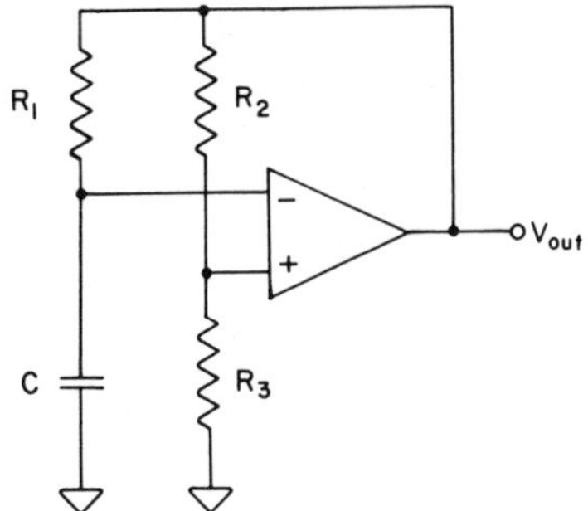


**Fig. 2.91.** A square-wave generator using a comparator or operational amplifier.

wave forms from subhertz to kilohertz frequencies. Special integrated circuits or circuits that use individual transistors may be needed for megahertz frequencies.

The comparator or operational amplifier circuit in Fig. 2.91 is a free-running multivibrator that generates a square wave with exactly a 50% duty cycle at its output. The capacitor charges toward $V_{out}$ until the capacitor voltage crosses over the threshold voltage at the noninverting input, determined by $V_{out}$ and the voltage divider $R_2$ and $R_3$. The output of the comparator then switches states, changing the polarity of the threshold voltage at the noninverting input. The capacitor now begins to charge toward the new voltage level of $V_{out}$, and continues to do so until its voltage crosses over the new reference voltage level. The comparator then changes states again. Polarized capacitors may be used by connecting one end to a power supply voltage instead of common. The circuit allows low frequencies to be obtained with relatively small capacitors because it can allow the capacitor to charge for a period equal to several $RC$ time constants before switching occurs. Other circuits have ON and OFF times equal to about one time constant. The frequency of this circuit is relatively independent of power supply voltages because the threshold voltage and the capacitor charging voltage are directly proportional to each other.

The commonly used circuit in Fig. 2.92 generates both a square wave and

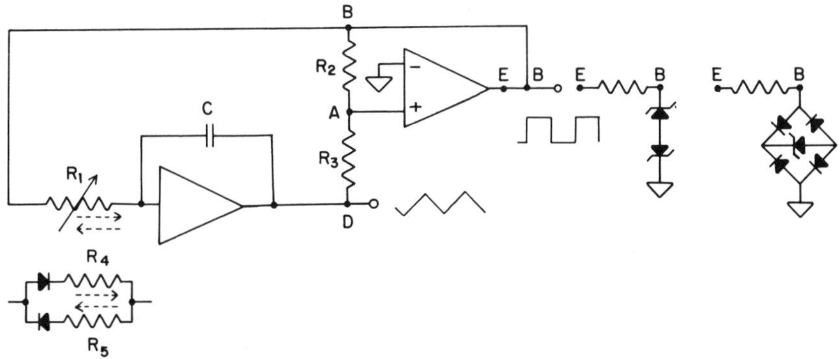


**Fig. 2.92.** A triangular-wave and square-wave generator circuit.

triangular wave whose amplitudes are indepenent of frequency when $R_1$ is used to adjust the frequency. The wave form generator uses an integrator to generate a triangular wave by alternately integrating the two constant voltage levels that occur at the output of a voltage comparator. The comparator switches states when the output of the integrator causes the voltage at point A to cross over the voltage at the inverting input of the comparator, 0 V in Fig. 2.92. The integrator output voltage at which this crossover point is reached is determined by the voltage divider $R_2$ and $R_3$ and the saturated output voltage of the comparator. Consider the case where $R_2 = 2R_3$ and the saturated comparator output voltage at point B is initially +12 V. The integrator produces a negative-going output voltage until it reaches a voltage slightly more negative than −6 V. At this point the voltage at point A becomes slightly negative with respect to the inverting input, causing the comparator output to change states to −12 V. The direction of the integration is now reversed and the integrator produces a positive-going ramp voltage until it reaches a voltage slightly more positive than +6 V, causing the comparator to switch states again. Slight rounding of the triangular wave form at its peaks occurs during the switching time of the comparator. For high frequencies, the comparator should have a fast slewing rate, and a short saturation recovery time.

The wave forms of the generator in Fig. 2.92 are symmetrical when the two output voltages of the comparator are of the same magnitude. When this is not the case, either of the zener diode voltage regulator circuits shown in Fig. 2.92 may be added to the output of the comparator. The zener diode in the diode bridge circuit allows the zener to regulate both positive and negative voltages. A matched pair of zener diodes placed back-to-back could also be used in place of the bridge circuit, but they might not be as well matched as desired. A dc offset may be added to the triangular wave by connecting the inverting input of the comparator to a dc voltage equal to the desired voltage offset. Unequal positive- and negative-going ramps, and a square wave with correspondingly unequal on and off periods, may be obtained by replacing $R_1$ with the $R_4R_5$ resistor/diode network shown in Fig. 2.92. When the voltage at point B is positive, $R_4$ controls the integration current and therefore the rate of the negative-going ramp. In a similar manner $R_5$ controls the rate of the positive-going ramp, which occurs when point B is negative. When one resistor of the $R_4R_5$ pair is much smaller than the other, the triangular wave becomes a sawtooth wave.

Sine-wave generators, or **oscillators**, use a frequency selective network in a positive feedback loop that has a gain slightly greater than unity. For instance, the sinusoidal output of a noninverting tuned amplifier (Section VI.C.8) may be connected to its input. If the gain of the amplifier is unity at the central frequency where no phase shift occurs, the signal is sustained. In practice it is impossible to maintain a gain of exactly unity with zero phase shift. A gain less than unity does not sustain oscillation, and a gain greater than unity causes the output voltage to increase until the amplifier limits. Therefore, to build a workable circuit, a nonlinearity is introduced into the feedback loop so that the overall circuit gain is greater than unity at low voltages and less than unity at higher voltages. A somewhat distorted sine wave results, with the quality of the sine wave dependent partly upon the degree of nonlinearity and mostly upon the quality of the tuned filter. A single-transistor oscillator takes advantage of slight variations in the gain of a simple transistor amplifier with the magnitude of the input signal.

A useful and extreme example of a nonlinearity is the case where a square wave of the desired frequency is passed through a tuned amplifier to produce a sine wave. The sine wave is fed back to the input of a voltage comparator to produce the square wave. An oscillator of this type, but with lower distortion, that uses a distorted sine wave instead of a square wave is shown in Fig. 2.93. In this case, the final sine-wave output of the tuned amplifier circuit is sent back to a voltage inverter circuit with clipping diodes to generate a symmetrically clipped sine wave, rather than a square wave. The clipped sine wave has less distortion than a square wave and when it is filtered it produces a lower-distortion sine wave at the output. The gain of the circuit in Fig. 2.93 is adjusted by $R_3$, $R_4$, or $R_5$ so that about 20% of the sine wave is clipped off. The distortion for this amount of clipping is indicated by the amplitudes of the harmonic frequency components, which amount to less than a few tenths percent of the fundamental frequency. The frequency of oscillation may be tuned by a factor of 4 or 5 by changing only $R_1$ by 2%. The amplitude of the sine wave is essentially constant over this tuning range.

Other common sine-wave oscillators include the phase-shift oscillator, Wien bridge oscillator, *LC* tank circuit oscillator (useful at high frequencies), and the twin-T oscillator. These circuits are more difficult to tune and usually require special feedback circuits to maintain a constant amplitude over the tuning range. They are discussed in several of the references.

The oscillator with by far the most stable frequency is a crystal-controlled oscillator. Stabilities of one part in $10^6$ are readily obtained and greater stabilities are obtained with special care. A quartz crystal sandwiched between two conducting plates deforms mechanically when a voltage is applied between the plates. When the voltage is removed, the energy that was stored by the mechani-

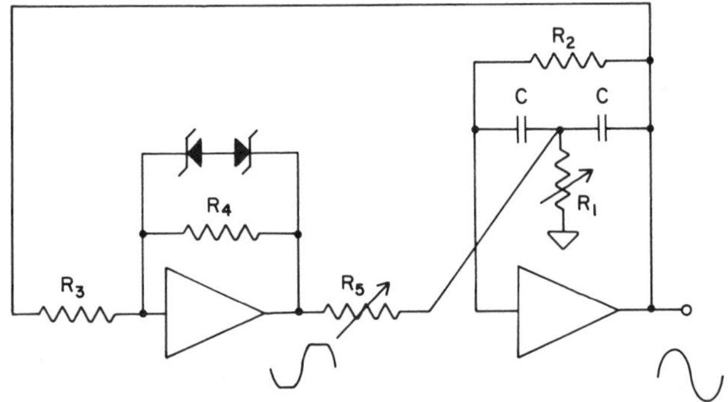


**Fig. 2.93.** An easily tuned sine-wave generator.

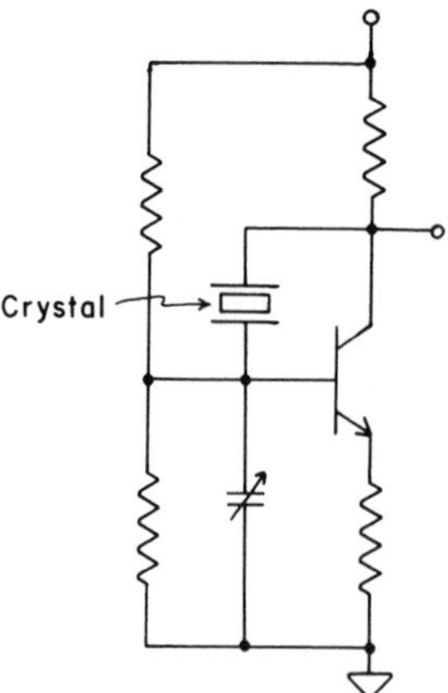


**Fig. 2.94.** A crystal-controlled oscillator.

cal strain tends to return the crystal to its original dimension. However, mechanical inertia temporarily causes the crystal to deform in the opposite direction, inducing a voltage in the circuit.

The mechanical vibration of the crystal can be sustained by placing it in the feedback loop of an oscillator. Because the interchange of mechanical and electrical energy between the circuit and the crystal occurs most easily at the mechanical vibration resonance frequency of the crystal, the circuit oscillates at that frequency. The oscillator frequency therefore essentially depends on the cut and physical dimensions of the crystal. A basic crystal oscillator is shown in Fig. 2.94. Crystal oscillator circuits may be designed so that minor adjustments in frequency can be made. Oscillations at frequencies that are harmonics of the fundamental resonance frequency may also be obtained.

## IX. CURRENT AND VOLTAGE READOUT SYSTEMS

A digital number or a physical position on a scale is usually used to represent and display an electrical signal in a form that is readily interpreted by an observer. Several of the most commonly used instruments that convert an electrical voltage or current to a digital number or scale position are moving coil meters and recorders, digital circuit meters and printers, oscilloscopes, and servo recording potentiometers such as an *XY* recorder or a strip chart recorder. This section considers instruments that produce an analog scale position proportional to a voltage or current. Section III of Chapter 8 considers instruments that produce a digital number or digital scale position by combining digital and analog circuits.

The best precision and resolution of an instrument with an analog scale position readout is about 0.1% of the full scale span, whereas an instrument with a digital number readout may have much higher resolution. Zero suppression and scale expansion techniques can be used to increase the readout resolution,

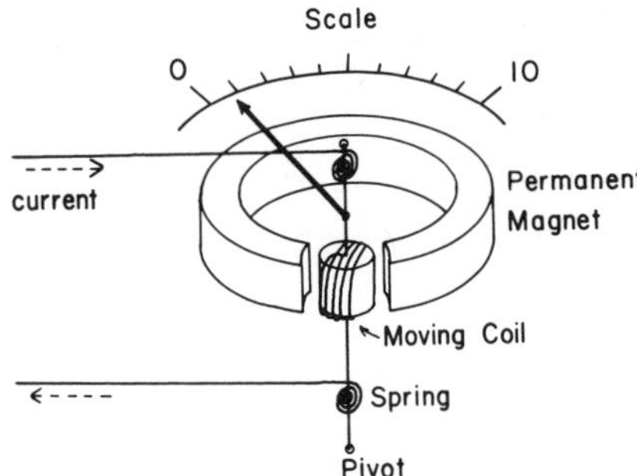


**Fig. 2.95.** Schematic of a D'Arsonval meter movement.

with respect to the measured quantity, by displaying over the full span of the readout scale only a fraction of the full range of values of the quantity to be measured.

## A. MOVING COIL METERS

A widely used system of converting a current and a voltage to a mechanical scale position is a moving coil meter, such as the **D'Arsonval** meter movement shown in Fig. 2.95. The current to be measured is passed through a coil of wire that has been wound many times around a loop. The coil is mounted on pivots so that it may rotate. A needle is attached to the coil to indicate its position of rotation on a scale. A fine spring holds the coil in a position at the low end of the scale, or at the center of the scale. When a current is passed through the coil, it induces a magnetic field around the coil. The induced field opposes the field of the permanent magnet, forcing the coil to rotate against the opposing force of the spring.

The linearity of the relationship between scale position and the amount of current depends upon the design of the permanent magnet, the coil, the spring, and the pivots. Good meters may be expected to deviate from a linear relationship by no more than 0.5% of the full scale reading. It is not uncommon to find meters with deviations of 2–3%. Mechanical wear of the pivots with much use and driving the meter off scale frequently may cause the linearity to deteriorate with time. A similar meter movement called a **taut-band** movement replaces both pivots with fine coiled springs designed to hold the coil in position as well as provide the restoring force against the magnetic force.

Magnetically shielded meters should be used near magnetic fields or near wires carrying large currents, because significant magnetic fields are generated around such wires.

## B. VOLTMETERS

A moving coil meter *responds* to current. The coil is wound with fine wire and has some resistance. When a current passes through the coil, a small voltage drop also occurs across the coil. The scale position may be calibrated to indicate either the current passing through the coil or the voltage drop across it. For example, a meter movement that has a coil resistance of 100 Ω and a full scale deflection of 200 μA has a voltage drop of 0.02 Volts across the coil when the meter reads full scale. A voltage of 20 mV could be accurately measured if its source was not significantly decreased by the loading caused by the 100-Ω coil resistance.

When a meter is to be used to measure larger voltages, a resistor is placed in series with the coil of the meter. The total resistance, including the coil resistance, is directly proportional to the full scale voltage reading. If the above meter were to have a full scale deflection of 2 V, a resistance of 9900 Ω would be placed in series with the coil to give a total resistance of 10,000 Ω.

Because loading is an important consideration and the input resistance of a voltmeter is proportional to the full scale reading, the resistance for a 1 V full scale reading is usually indicated on the face of the meter. The voltmeter in the example has an **ohms/volt rating** of 5000 Ω/V. The total resistance of the voltmeter is obtained by multiplying the ohms/volt rating by the full scale voltage. If this meter were designed to measure a full scale voltage of 1000 V, it would have a high input resistance of 5 MΩ.

## C. CURRENT METERS OR AMMETERS

A given meter movement may be used to measure currents larger than the current required to give a full scale deflection by placing a low value of resistance in parallel with the meter coil to bypass the extra current. To convert a 100 Ω, 200 μA meter into a meter that measures currents up to a full scale value of 2000 μA (2mA), a resistor of 11.1 Ω is placed in parallel with the coil. The voltage across both the coil and the resistor is the same. Therefore, a full scale voltage of 20 mV forces 1800 μA through the external resistor and 200 μA through the coil.

The total resistance of an ammeter is inversely proportional to the full scale current value to be displayed. When the parallel resistance is much lower than the coil resistance, then essentially all of the current passes through the parallel resistor and the coil resistance may be ignored. Then the meter may be considered as a voltmeter, and the parallel resistor as a **current sampling resistor** that converts the current to a voltage.

## D. BAR GRAPH DISPLAYS

Analog type displays are available that use closely spaced photodiodes to produce a single bar graph. Each photodiode corresponds to an increment of signal. Adjacent photodiodes are energized so that the length of the lighted region of the bar graph is proportional to the signal. Internally an analog-to-digital converter converts the signal to digital signals that are decoded by digital circuits which then light up the proper photodiodes. An advantage of such analog type displays is their much faster response time compared to D'Arsonval meters.

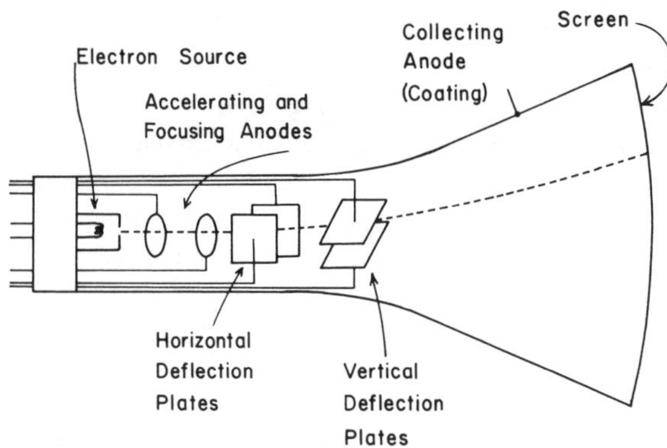


**Fig. 2.96.** Schematic of a cathode ray tube used in an oscilloscope.

## E. OSCILLOSCOPES

An oscilloscope produces simultaneous scale deflections of a visible point in a two dimensional plane. A *cathode ray tube* (CRT) is used to produce a beam of electrons that is focused to a small spot on a screen (Fig. 2.96). The screen temporarily phosphoresces visible light when the electrons strike the screen. A voltage signal to be displayed on the screen is amplified by a high-voltage amplifier within the oscilloscope. The resulting proportional high-voltage signal is applied to deflection plates within the CRT. The magnitude and direction of the electric field that is produced within the CRT causes the electron beam to be deflected proportionally in the proper direction. Separate high-voltage signals applied to separate sets of plates are used for horizontal deflection and for vertical deflection.

When the fluorescing spot is moved rapidly over a path, the result appears to the eye to be a line on the screen. The line appears stationary if the spot is swept repeatedly and rapidly over the same path. The result of a single sweep can be stored as an image on a screen by a **memory or storage oscilloscope**. One type of memory oscilloscope has a special CRT that uses the electron beam spot to sensitize the phosphor. Then the second source of electrons, called a **flood gun**, uniformly floods the entire screen, causing the sensitized phosphor to glow to illuminate the path of the original electron beam spot. The trace can be visible for minutes to hours, or erased at any time, to prepare to record another event.

Oscilloscopes are commonly used to display a selected time window of the temporal history of a voltage. The voltage to be displayed is connected to the vertical high voltage amplifier. A linearly increasing voltage is connected to the horizontal high voltage amplifier to sweep the spot across the screen at a predetermined rate. The horizontal position of the spot is therefore proportional to time, and the unit that produces the linearly increasing voltage is called a **time base generator**. The time base generator contains an integrator circuit that integrates a constant voltage or current to produce the linearly increasing voltage. At the end of the horizontal sweep, the integrator is reset to its initial condition (Fig. 2.56*b*) by transistor switches. The switches are driven by a **trigger circuit** that causes the integrator to begin the horizontal sweep when an internal or external event occurs. The event, for instance, might be the crossing of a preset voltage level by the signal voltage to be displayed. Such an event can be detected with a voltage comparator. **Delayed sweep** functions allow the time base sweep to start at a predetermined time after a trigger signal has been received, or, optionally, at the occurrence of a second trigger signal that occurs only after the main trigger signal has occurred.

**Dual beam** oscilloscopes have CRTs with two separate electron beams so that two fast signals can be simultaneously displayed on the vertical axis. **Dual trace** oscilloscopes have one electron beam in the CRT, but display two separate signals on the screen by either of two methods. In the **alternate mode**, first one signal and then the other is swept across the screen. In the **chopped mode**, the two signals are rapidly alternated by a 100 KHz or 1 MHz multiplexer into the vertical high-voltage amplifier *during* each sweep of the spot across the screen. In the alternate mode, two sweeps are required to display both signals; in the chopped mode, only one sweep is needed since the spot is rapidly jumping back and forth between the two signals as it moves horizontally across the screen.

When a repetitive signal is available, two time base generators may be combined with the alternate sweep mode to display the signal first on a slow time base, and then on a fast time base to expand some details of the signal. The start of the fast sweep may be delayed compared to the start of the slow sweep so that any narrow time window of the slow sweep may be displayed across the entire screen by the fast sweep.

Some oscilloscopes have **delay lines** that delay the signal so that a fast rising leading edge of the signal which triggers the time base is delayed enough that the leading edge does not reach the screen until after the horizontal sweep has started. This allows the entire leading edge to be displayed.

## F. POTENTIOMETERS

One of the most accurate voltage scale position measuring instruments is a **potentiometer**. A potentiometer has a stable voltage source and voltage divider that can be adjusted with high precision (Fig. 2.97). By adjusting the movable center tap of the voltage divider, any voltage may be set at the output terminals.

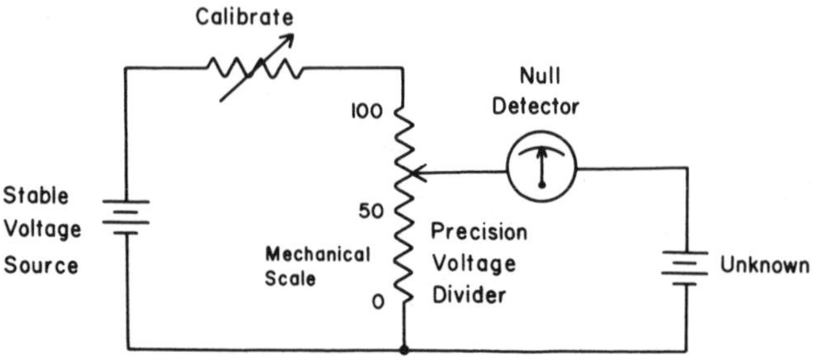


**Fig. 2.97.** A potentiometer for measuring voltages without loading the voltage source.

A sensitive meter or difference amplifier is used to compare the potentiometer voltage with the voltage to be measured. The voltage divider is adjusted until the two voltages are precisely equal, within the ability of the meter to detect a difference between them. Differences of about 0.01 mV can ordinarily be detected with a very sensitive meter. Smaller voltage differences require special amplifier circuits.

An important characteristic of this method is that no current is drawn from either voltage source when the two voltages are equal. This eliminates loading effects when a meter is used, except of course when the two voltages are not equal.

A relatively simple and inexpensive potentiometer with a linearity of ±0.1% of full scale may be made using an adjustable 10-turn helical voltage divider. A fine wire made of a high-resistance material is wound into a long coil. The long coil is wound into a 10-turn helix. A wiper arm connected to a central shaft moves along touching the coil as the shaft rotates. The movable wiper arm becomes the central electrical connection between the two resistor arms of the voltage divider. The rotational position of the wiper arm is linearly related to the value of resistance between the point where it contacts the coil, and either end of the coil. Therefore, the reading of a 10-turn dial that is connected to the shaft is linearly related to the output voltage of the voltage divider. Potentiometers of this type are not only used to measure voltage, but are often used as precisely variable voltage sources when loading is negligible.

The word *potentiometer* by common usage has come to mean both a voltage measuring instrument as well as an adjustable voltage divider that can be used to make the instrument. An adjustable voltage divider may also be used simply as a variable resistor and is often referred to as a "pot."

## G. SERVO RECORDING POTENTIOMETERS

The scale adjustment of a potentiometer can be automated with either a moving coil meter mechanism or a servo motor. If the scale position is recorded with a pen, then a continuous recording of a varying voltage may be readily made. With a servo motor, a relative precision on the order of ±0.1% of full scale can be achieved, with typical response times of one second full scale for a 25-cm scale. With a moving coil meter, a relative precision on the order of 1% can be achieved with typical response times of a few milliseconds full scale for a 5-cm scale.

In a recording potentiometer, Figure 2.98, the wiper of the voltage divider (or **slide wire** as it is commonly called) is mechanically linked to the shaft of a servo motor or a moving coil meter. When the voltage to be measured and the voltage of the potentiometer are not equal, the voltage difference between them is electronically amplified to drive the motor or moving coil. The motor always turns in the direction necessary to make the two voltages equal because the direction of the motor is designed to depend upon the polarity of the difference between the two voltages. The position of the wiper is mechanically coupled to a pen that records its position on a chart paper.

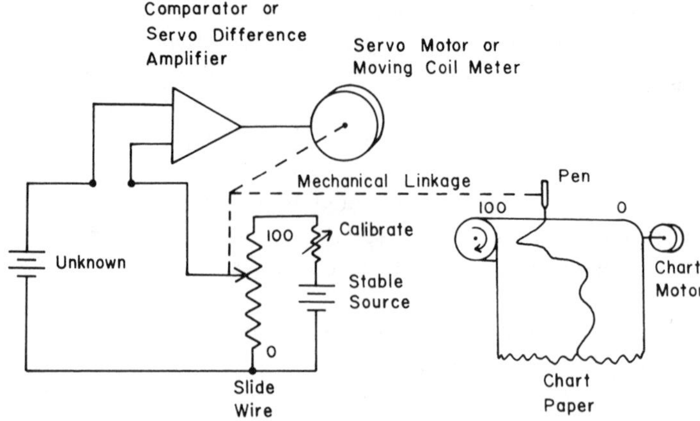


**Fig. 2.98.** A servo recording potentiometer.

When a moving coil is used to drive the wiper of the slide wire, a current is needed to maintain the deflection of the coil at the equilibrium position where the input voltage and slide wire output voltage are equal. A very small (and negligible) difference always exists between these two voltages at the equilibrium position. This small voltage difference is amplified by a high gain comparator or operational amplifier to supply the voltage and current necessary to maintain the proper rotational position of the moving coil. Fast response times require that the mechanical inertia and friction of the moving coil system be small. One way to accomplish this is to connect a small mirror to the coil and use it to deflect a light beam that projects a small spot on light-sensitive chart paper. Ultraviolet light is usually used to minimize interference from room light.

### 1. Zero Control

One end of the chart paper of a recorder is ordinarily adjusted to correspond to 0 V, and the other end is equivalent to the full scale positive voltage indicated by a range switch (10 mV, 100 mV, etc.). It is often convenient to adjust the 0 V position of the pen to the center or other end of the chart scale. Negative voltages may then be recorded. An adjustment of this sort may be made by placing a variable voltage source in series with the potentiometer output voltage and the input voltage as shown in Fig. 2.99. As it happens, the same voltage source that is used for the potentiometer may conveniently be used for the zero adjustment control.

Consider the example where the recorder range switch has been set to 100 mV full scale. When the wiper (point C) of the zero control is set at point B, 0 V is recorded at one end of the chart scale. When the wiper (point C) is located in the middle of the zero control plot, 50 mV appears between points B and C. The voltage at point A, with respect to C, is zero when the wiper of the potentiometer slide wire (point A) is in its middle position (also indicating a midway

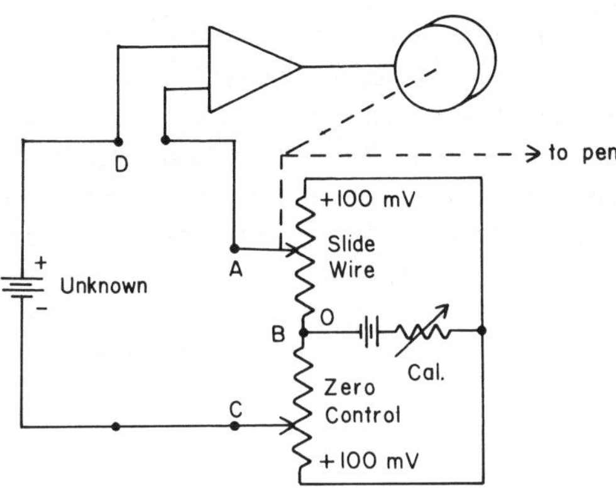


**Fig. 2.99.** A servo recording potentiometer with a zero control for the scale position of the pen.

position on the chart scale). The voltage between A and B then exactly equals the voltage between B and C, and the polarities cause them to cancel. A positive voltage at point D, with respect to C, causes the potentiometer slide wire to move in the direction away from point B until the voltages at points D and A are equal. A negative voltage at point D causes the potentiometer slide wire to move toward point B until the resulting negative voltage between points A and C equals the negative voltage between points D and C.

### 2. Damping and Dead Band

Mechanical inertia may cause the pen to **overshoot** the intended position on the chart paper. Having overshot the position moving in one direction, it may return and overshoot again going in the other direction. Oscillation of this kind may continue indefinitely or may eventually dampen and stop.

On the other hand, if the signal that the servo motor receives becomes too small as the equilibrium chart position is being approached, mechanical friction may cause it to stop short of the intended position, resulting in a **dead band**. For instance, the pen may stop 1/2 division short of its correct position on one side of the chart scale if the pen is coming from one direction, and may stop 1/2 division short on the other side if it is coming from the other direction. A dead band of one division would result. Oscillation can be prevented and dead band and overshoot minimized by properly adjusting the temporal response characteristics and gain of the amplifier that drives the servo motor.

## X. POWER SUPPLIES

The common type of power supply for measurement systems is one that converts to dc the ac electrical power that is fed into buildings by power companies. Portable instruments often use batteries. A battery voltage may be converted into a higher dc voltage by means of a dc-to-dc converter.

All these power sources provide a voltage that varies to some extent with time. Usually the unregulated power is regulated by feedback circuits so that the final output voltage remains constant even though the input voltage source or the load changes. For some applications the output current is regulated instead of the voltage. Regulator circuits are available in integrated circuit packages that can be simply connected between the unregulated output terminals and the load to be regulated. The power that must be dissipated by the regulator circuit is (at least) equal to the current supplied to the load multiplied by the voltage difference between the unregulated output and the voltage across the load.

The basic components used to convert an ac line voltage to a dc voltage are a transformer, a rectifier circuit, and a filter. The transformer is used to increase or decrease the amplitude of the ac voltage to the required level. A transformer with separate windings for the primary and secondary also gives dc isolation between the power supply and the power line. Insulation leakage and some capacitive coupling are still present, however, to an extent that depends on the design of the transformer. They may cause noise feed-through and, in some medical applications, safety hazards.

The rectifier circuit switches the current paths from the transformer so that current passes in only one direction through the load. The simplest rectifier circuits consist of diodes, which require no external control circuits. For high current applications, SCRs or triacs, which require control pulses to switch them on, or occasionally power transistors, which also require control circuits, are used. Their control capability allows the rectifier circuit to become part of the regulation circuit.

A **half-wave diode rectifier** circuit is shown in Fig. 2.100*a*. The diode conducts current on one half cycle but is reverse biased and conducts only leakage current on the opposite half cycles. The peak voltage across the load is equal to 1.414 times the rms voltage of the transformer secondary minus the forward biased voltage drop across the diode, which varies from 0 to about 1 V depending upon the load current.

A **full-wave rectifier circuit** is shown in Fig. 2.100*b*. This circuit is equivalent to two half-wave rectifier circuits 180° out of phase with each other, feeding the load. It requires a center tapped transformer that produces twice the voltage required across the load, or two transformers in series, but 180° out of phase with each other with respect to common. A **bridge full-wave rectifier** is shown in Fig. 2.100*c*. Although this circuit does not require a center tapped transformer it has two diodes conducting in series with each other during each half cycle. This doubles the voltage drop and the power loss due to the diodes, making this an electrically inefficient circuit for high current applications compared to the circuit in Fig. 2.100*b*.

A **voltage doubler rectifier circuit** is shown in Fig. 2.100*d*. It is essentially two half-wave rectifiers connected to a single ac source. The diodes conduct on opposite half cycles, one providing a positive voltage with respect to point A and the other providing a negative voltage with respect to point A. The output voltage of each rectifier circuit is filtered by a capacitor to provide a dc level as

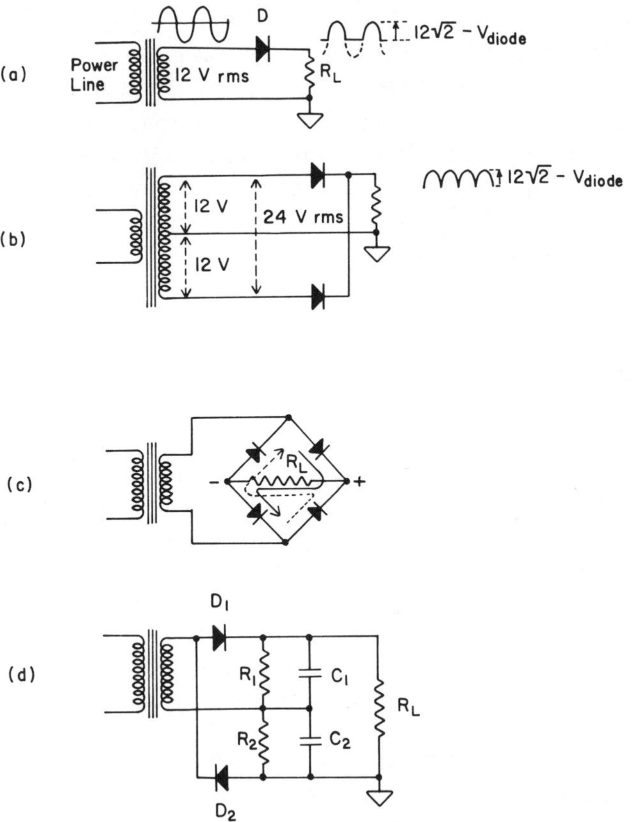


**Fig. 2.100.** Power supply rectifier circuits. (*a*) Half wave, (*b*) full wave, (*c*) bridge full wave, and (*d*) a voltage doubler.

discussed in the next section. The filtered dc voltage outputs of the two rectifier circuits are connected in series to provide twice the voltage to the load that a filtered single half-wave or bridge rectifier circuit would provide from the same ac source. The capacitors are essential to the operation of the circuit because they maintain the dc levels developed during opposite half cycles so that the dc levels may be added together. Resistors $R_1$ and $R_2$ help to equalize the voltages across $C_1$ and $C_2$ since their leakage currents and capacitances are usually not well matched.

A **filter circuit** is used to smooth the periodic wave form of the output of a rectifier circuit into a dc voltage level whose value approaches the peak value of the periodic wave form. The simplest filter consists of a capacitor $C$ in parallel with a resistive load $R_L$. The capacitor charges during part of each half cycle of the periodic wave form, and acts as a storage device to help maintain a dc voltage between half cycles. When a load is present the capacitor partially discharges through the load between half cycles. This superimposes upon the dc level a sawtooth type of ripple voltage. The rms value of the ripple voltage divided by the average dc level is called the **ripple factor**. For a simple filter with capacitance $C$ and a resistive load $R_L$ the ripple factor is approximately equal to $1/(\sqrt{12}\,fR_LC)$. Here $f$ is the frequency of the periodic wave form. Since $f$ is equal to the line frequency for half-wave rectifier and twice the line frequency for a full-wave rectifier, the latter circuit produces less ripple. A larger capacitance also improves (decreases) the ripple. As the ripple increases, the average dc level decreases and is approximately given by

$$V_{\mathrm{dc}} = 1.4V_{\mathrm{rms}} - V_{\mathrm{diode}} - \frac{I_{\mathrm{dc}}}{2fC} \tag{127}$$

where $V_{\mathrm{rms}}$ is the root mean square value of the rectifier supply voltage, $V_{\mathrm{diode}}$ represents the voltage loss across the rectifier diodes, and $I_{\mathrm{dc}}$ is the average dc current drawn by the load.

When a power supply is switched on, a sudden surge of current flows to charge the filter capacitor, which has usually been discharged while the supply was off. The diodes and fuse must be able to handle this surge without destruction. A resistor may be placed between the rectifier circuit and the capacitor to limit the initial surge, but the power loss and voltage drop across the added resistor during operation must be tolerable.

Combinations of inductors and capacitors or resistors and capacitors give improved filtering. However, inductors for 60 Hz applications are expensive and heavy, and tend not to be used except in high-current applications where power losses of resistors become important. When series or shunt type voltage regulator circuits are used, it is only necessary to provide enough filtering so that the minimum dc voltage on the ripple wave form is always about 1 V greater than the output voltage of the regulator circuit. The ripple on the dc level of the input to the regulator is reduced by the regulator circuit as would be any other variation in the input wave form.

**Voltage** and **current regulation** may be done by using feedback control circuits similar to those discussed in the operational amplifier Section VI.D. The output voltage is compared with a stable reference voltage and any error is greatly amplified and used to adjust the effective resistance of a power transistor that is in series or in parallel with the load, thereby controlling the voltage across the load. In high-current applications the voltage drop across the leads connecting the power supply and load may be significant. To eliminate this effect the voltage *at the load* is sensed by two additional wires, which carry no current, connected between the load and the **voltage sense** terminals of the power supply. The power supply is designed to respond to keep this voltage constant instead of the voltage at its output terminals.

For **current control**, the current through the load is passed through a small value resistor to develop a voltage to be compared with the reference voltage. Any error is greatly amplified and the rest of the control circuit is the same as for the output voltage control case.

Inexpensive voltage regulator control circuits are available as integrated circuits containing a stable voltage reference, a high gain difference amplifier, and a series regulator transistor capable of handling fractions of an ampere. For higher currents the regulator may be used to drive an external series regulator power transistor. When several sensitive circuits are to be powered by one power supply, it may be helpful to place a voltage regulator of this type near each circuit to decouple it from power supply load variations caused by other circuits.

When high (ampere) currents must be supplied over a wide range of voltages, the power loss across the series or shunt power transistor may often be severe. In such cases it is desirable to regulate the voltage partially by controlling the rectifying section. For this type of regulation the diodes in the above circuits may be replaced by SCRs or power switching transistors. The switching times of these devices are controlled so as to control the amount of charge that is allowed to pass into the filter capacitor, thereby controlling the average dc voltage across the capacitor. Such power supplies are called switching power supplies and are rated according to their efficiency of supplying power to the load relative to the power drain from the power line.

Many power supplies offer protection against a short circuit load, or an output voltage going too high. Thermal protection circuits may also be present that shut down the power supply if it becomes too hot.

A **zener diode** is sometimes used as a shunt type of voltage regulator to supply a relatively constant voltage to a load. A typical circuit is shown in Fig. 2.101. The voltage $V_z$ at point Z is relatively constant. The current through $R_s$ equals the current to the load plus the current through the zener diode. When such a circuit is designed, the value of $R_s$ must be small enough so that some bias current always flows through the zener even when the load current is highest and the supply voltage $V_s$ lowest. When the load is removed from this circuit all of the current that it carried must flow through the zener. The zener must then be able to dissipate the resulting power ($P_{\max} = I_{z(\max)} V_z$).

An important specification for a zener diode is its dynamic resistance $R_z$, which is the inverse slope of its current-voltage curve in its zener region. The output impedance of the simple zener circuit is equal to the value of $R_s$ and $R_z$ in parallel, when the zener is biased in its zener region. Variations in the voltage across the load are approximately equal to variations in the voltage source multiplied by the factor $R_z/(R_s + R_z)$, which is always less than unity. Therefore to reduce ripple or to regulate the output voltage against changes in input voltage, $R_s$ should be as large as allowed.

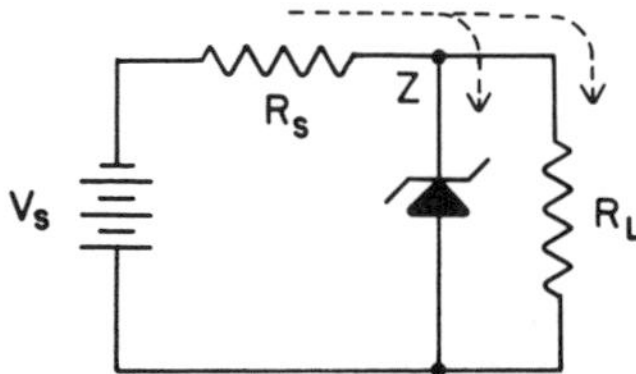


**Fig. 2.101.** A zener diode voltage regulator circuit.

TABLE 2.IV

Characteristics of Battery Cells

| Type of cell | Carbon-zinc dry cell | Alkaline-manganese cell | Mercury cell | Lead-acid (car) storage cell | Nickel-cadmium storage cell | Silver oxide-zinc storage cell | Lithium-iodine cell |
|---|---|---|---|---|---|---|---|
| Voltage of basic cell (V) | 1.5–1.6 | 1.5–1.6 | 1.35 | 2.1 fully charged | 1.2 | 1.45–1.50 | 2.8 |
| Voltage Performance | Decreases up to several tenths of a volt with increasing current drain, and low operating temperatures. Decreases continuously during discharge | Lower output resistance than carbon-zinc cell. Voltage decreases like a dry cell during initial discharge but tends to level off later | Very stable until last 5% of life, then drops rapidly. Low output resistance. Useful as a secondary voltage reference standard | Decreases slowly during discharge, and rapidly during final 1/3 of life. Very low output resistance. | Relatively constant | Decreases from 1.5 to 1.45 V in first 10% of discharge life, then slowly decreases during discharge | Relatively stable during discharge |
| Capacity of D cell | 1000–10,000 mA-hr depending upon design, current drain, and lowest allowable voltage. Rating may allow 0.8 V minimum | About twice the capacity of a carbon-zinc cell | About twice that of an alkaline cell | Rechargeable up to 500 times, 2500 mA-hr/charge. Units designed for car use are rated for many tens of A-hr | Rechargeable 500–2000 times, 4000 mA-hr/charge | Rechargeable 30–400 times | |

TABLE 2.IV

Characteristics of Battery Cells (*Continued*)

| | | | | | | | |
|---|---|---|---|---|---|---|---|
| Capacity behavior with current drain and temperature changes | Milliampere-hour capacity decreases with increased current drain and increased discharge temperature | Capacity does not decrease under heavy drain. Low temperature operating limit near 40°F (5°C) | Capacity decreases under heavy drain and at cool operating temperatures. 2/3 capacity near 40°F (5°C) | Useful at very low temperatures, but output resistance increases | | | Useful to −40°C |
| Other comments | Shelf-life is lengthened by cool storage temperatures | Longer service and shelf-life at cooler temperatures than carbon dry cell. Shelf-life over 2 yr | 1.4 V cells are available for applications where a constant voltage is not needed | Older models require considerable service care. Should be stored dry or fully charged. Sealed cells with gelled electrolytes are available | Sealed and unsealed cells available. May be stored charged or uncharged. Relatively fast self-discharge rate when not in use (1 month to 80% capacity) | High energy density. Used where light weight is desired. Three times higher energy density than NiCd | Self-life over 5 yr. Shelf-discharge only 10% in 10 yr for units used in pacemakers |
| Relative Price | 1–2 | 4 | 35 | — | 20 | — | — |

Zener diodes are often used as constant reference voltage sources. Since the temperature coefficient and voltage across the zener diode vary with current, a constant current should be supplied to the zener, and the zener followed by a circuit that draws little current from it. Zener diode voltages between 5.5 and 8.5 V are often used because such devices can be designed with a low temperature coefficient. An operational amplifier circuit (e.g., a follower with gain) with low drift can then be used to provide a constant voltage of some other value.

**Batteries** (one or a group of parallel and/or series electrochemical cells) are used as power sources for instruments located in remote places such as outer space, undersea, and where mobility and portability are required. Moreover, a battery is free from line frequency noise, and a measurement system powered by batteries may be completely enclosed in a shield to minimize external noise pickup. Their primary disadvantage is that they must be replaced or recharged periodically. In remote areas they may be recharged by solar cells, or wind and hydroelectric generators.

Several types of battery cells are available. **Primary cells** require replacement when their voltage drops below the required level. The carbon-zinc dry cell, zinc chloride heavy-duty cell, and alkaline manganese cell are examples of primary cells. Others, called **secondary cells** or **storage cells**, may be recharged. The lead-acid, silver-zinc, and nickel-cadmium cells are examples of storage cells. Table 2.IV compares the characteristics of some of the commonly available cells. Batteries with voltages that are integral multiples of the values in the table are available. Hybrid batteries containing more than one type of cell are also available. An article by Lyman (18) gives a good review of the types of batteries available in 1975.

Voltage regulators may be placed between a battery and its load, just as with any other type of power supply. However, power loss across the regulator control elements must be considered when determining the power available to the load and the service life of the battery.

The dc voltage of a battery can be converted into a much higher (or lower) value by using a **dc-to-dc converter** circuit. A typical converter uses a power transistor driven by an oscillator to produce an ac voltage in the kilohertz range. The ac voltage is then converted into the desired value by a transformer. The resulting ac voltage at the secondary of the transformer is rectified and filtered to produce the desired dc voltage just as occurs in a power supply driven by a power line. A moderately high frequency is desired to reduce the ripple factor with transformers and filter components of smaller size and weight than would be necessary at 50 or 60 Hz. Too high a frequency results in high power losses that occur across the devices during slow switching times that are typical of power transistors and diodes.

## REFERENCES

1. Bair, E. J., *Introduction to Chemical Instrumentation*, McGraw-Hill, New York, 1962.
Includes a section on circuit layout, construction, and trouble shooting.

2. Bleaney, B. I., and B. Bleaney, *Electricity and Magnetism*, Clarendon Press, Oxford, 1963.
An excellent college physics text which includes discussions of basic electrical quantities, ac theory, and a good introduction of the theory of transmission lines.

3. Bode, N. H., *Network Analysis and Feedback Amplifier Design*, Van Nostrand, Princeton, N.J., 1951.
A classic on amplifier feedback and network analysis.

4. *Burgess Engineering Manual*, Burgess Battery Company, Freeport, Ill., 1958.
A reference guide for the selection of standard carbon-zinc batteries. Includes curves of service hours versus current drain for final voltages from 0.8 to 1.2 V/cell.

5. Buus, R. G., "Electrical Interference," in *Design Technology*, Vol. 1, Prentice-Hall, Englewood Cliffs, N.J., 1970.
An excellent discussion of shielding, grounding, and ways to reduce interference in components, cables and connections.

6. Cleary, J. F., Ed., *Transistor Manual*, General Electric Co., 1964.
Theory and application of bipolar junction transistors in amplifier and switching circuits, with practical circuits. Includes a good section on unijunction transistors and circuits. Transistor specifications.

7. Cordos, E., and H. V. Malmstadt, "Dual Channel Synchronous Integration Measurement System for Atomic Fluorescence Spectrometry," *Anal. Chem.*, **44**, 2277 (1972).
A sequential two-channel boxcar integration system with automatic background correction.

7a. *D.A.T.A. Book*, Electronic Information Series, D.A.T.A., Inc., San Diego, Calif., 1982.
A series of books that give specifications on all semiconductor devices.

8. Diefenderfer, A. J., *Principles of Electronic Instrumentation*, 2nd ed., Saunders, Philadelphia, 1979.
Contains many diagrams illustrating analog devices and measuring techniques.

9. Duckworth, H. E., *Electricity and Magnetism*, Holt, Rinehart Winston, New York, 1960.
A college physics text with a good section on basic ac theory.

10. Eimbinder, J., Ed., *Application Considerations for Linear Integrated Circuits*, Wiley-Interscience, New York, 1970.
A collection of papers about the use of linear integrated circuits, including operational amplifiers, analog switches, voltage regulators, and transmission line drivers and receivers.

11. *Electronics Circuit Designer's Casebook*, McGraw-Hill, New York, no date; *Electronics Handbook of Circuit Design*, McGraw-Hill, New York, no date.
Collections of many types of practical circuits, originally published in *Electronics*.

12. Gillie, A. C., *Pulse and Logic Circuits*, McGraw-Hill, New York, 1968.
A clear book about *RC* wave-shaping circuits and coaxial cable transmission and delay lines.

13. *Handbook of Operational Amplifier Active RC Networks*, Burr-Brown Research Corp., Tucson, Ariz., 1966.
An introduction to the theory of operational amplifier low-, high-, and band-pass filter circuits, with cookbook equations for the design of practical filter circuits.

14. *Handbook of Operational Amplifier Applications*, Burr-Brown Research Corporation, Tucson, Ariz., 1963.
Theory of operational amplifiers with practical circuits.

15. Hnatek, E. R., *Applications of Linear Integrated Circuits*, Wiley, New York, 1975.
Includes many practical circuits using modern monolithic operational amplifiers, comparators, voltage regulators, and the integrated circuit type 555 timer.

16. Hunter, L. P., Ed., *Handbook of Semiconductor Electronics*, 2nd ed., McGraw-Hill, New York, 1962.
A thorough presentation of the physics of semiconductor devices, amplifier equivalent circuits, high-frequency amplifier design theory, and transistor oscillators.

17. Landee, R. W., D. C. Davis, and A. P. Albrecht, *Electronic Designer's Handbook*, McGraw-Hill, New York, 1957.
Includes sections on transformers, passive filter circuits, feedback, transmission lines, microwave wave guides, and complex network analysis.

18. Lyman, J., "Battery Technology," *Electronics*, **48**, 75 (April 3, 1975).
A good review of the types and characteristics of batteries available in 1975.

18a. Malmstadt, H. V., C. G. Enke, and S. R. Crouch, *Electronics and Instrumentation for Scientists*, Benjamin/Cummings, Reading, Mass., 1981.
An excellent book for students and practitioners, organized around system principles and electronic functions of integrated circuits and individual components.

19. Malmstadt, H. V., and C. G. Enke, *Digital Electronics for Scientists*, Benjamin, New York, 1969.
An analog section includes an introduction to null comparison measurements, servo systems, and operational amplifiers.

20. Malmstadt, H. V., C. G. Enke, and E. D. Toren, Jr., *Electronics for Scientists*, Benjamin, New York, 1963.
Contains sections on basic analog measuring techniques and devices.

21. Malmstadt, H. V., C. G. Enke, and S. R. Crouch, *Analog Measurements and Transducers*, Benjamin, Menlo Park, Calif., 1973.
Includes sections on basic electrical quantities, and important null comparison measuring techniques, with helpful introductions to servo systems and operational amplifier circuits.

22. Malmstadt, H. V., C. G. Enke, and S. R. Crouch, *Control of Electrical Quantities*, Benjamin, Menlo Park, Calif., 1973.
Includes good summaries of conduction mechanisms and the electrical characteristics of junctions, applications-oriented discussions of resistance and capacitance, the principles of analog switches and gates, and servo system and operational amplifier feedback control systems.

23. Malmstadt, H. V., C. G. Enke, and S. R. Crouch, *Digital and Analog Conversions*, Benjamin, Menlo Park, Calif., 1973.
Includes linear and nonlinear operational amplifier circuits and their applications in analog and digital measurement systems.

24. Malmstadt, H. V., C. G. Enke, S. R. Crouch, and G. Horlick, *Optimization of Electronic Measurements*, Benjamin, Menlo Park, Calif., 1974.
An excellent introduction and summary of $S/N$ enhancement measuring systems and principles. Analog components of these systems are discussed, including active $RC$ filters.

25. Malmstadt, H. V., C. J. Delaney, and E. A. Cordos, "Instruments for Rate Determinations," *Anal. Chem.*, **44**, 79A (1972).
Principles and circuits for measuring the rate of change of electrical signals, including integrating types of differentiators.

26. McWilliam, I. G., and H. C. Bolton, "Instrumental Peak Distortion. Part I. Relaxation Time Effects," *Anal. Chem*, **41**, 1755 (1969); McWilliam, I. G., and H. C. Bolton, "Instrumental Peak Distortion, Part II. Effect of Recorder Response Time," *Anal. Chem*, **44**, 1762 (1969).
Classic papers that consider transient signal distortions caused by single and double low pass filters, and servo recorders.

27. Millman, J. and H. Taub, *Pulse, Digital and Switching Waveforms*, McGraw-Hill, New York, 1965.
A well-known source book containing detailed discussions of wave shaping and diode and transistor switches and gates. Includes a good section on the response of transmission lines to several pulse shapes.

28. Morrison, R., *Grounding and Shielding Techniques in Instrumentation*, Wiley, New York, 1967.
A thorough treatment of grounding and shielding problems and their solutions.

29. *MP-System 1000, Operation and Applications*, 3rd ed., McKee-Pedersen Instruments, Danville, Calif., 1968.
A discussion of the theory of operational amplifiers, with many practical circuits.

30. Nikon, F. E., *Handbook of Laplace Transformation: Tables and Examples*, Prentice-Hall, Englewood Cliffs, N.J., 1961.

31. Page, L., and N. I. Adams, Jr., *Principles of Electricity*, 3rd ed., Van Nostrand, Princeton, N.J., 1963.
A physics text containing sections on the transient and ac response of circuits having resistance, capacitance, and inductance.

31a. Piepmeier, E. H., and L. de Galan, "Influence of the Temporal Response of the Detection System upon Atomic Absorption Analytical Curves for Steady State and Transient Absorption Cells and Sources," *Spectrochim. Acta*, **31B**, 163 (1976).
Consideration of the distortions caused by simple *RC* filters for signals containing transient information, and a discussion of the choice of frequency used for chopping transient signals.

31b. Piepmeier, E. H., and L. de Galan, "Profiles of the Calcium Resonance Emitted by a Modulated Hollow Cathode Lamp," *Spectrochim. Acta*, **30B**, 211 (1975).
Consideration of how a lock-in amplifier produces a weighted average response when the information contained within the signal changes during a cycle.

31c. Piepmeier, E. H., and L. de Galan, "Line Profiles Emitted by Cu and Ca Hollow Cathode Lamps Pulsed to One Ampere," *Spectrochim. Acta*, **30B**, 263 (1975).
A discussion of *RC* filter distortions when using boxcar integrators.

31d. Rich, A., "Shielding and Guarding," *Analog Dialogue*, **17**(1), 8 (1983). and "Understanding Interference-Type Noise," *Analog Dialogue*, **16**(3), 6 (1982).
Excellent summaries of noise and shielding problems, with diagrams of what to do and what to avoid.

32. Schmid, H., *Electronic Analog/Digital Conversions*, Van Nostrand Reinhold, New York, 1970.
Contains an excellent section on electronic switches.

33. Scott, W. T., *The Physics of Electricity and Magnetism*, Wiley, New York, 1962.
Includes thorough discussions of capacitance and electrostatic fields in the presence of conductors and dielectric materials.

34. *SCR Manual*, 5th ed., General Electric Co., Syracuse, N.Y., 1972.
Includes principles of *pnpn* devices (thyristors), and contains many practical circuits.

35. Sheingold, D. H., Ed., *Nonlinear Circuits Handbook*, Analog Devices, Inc., Norwood, Mass., 1974.
A thorough discussion of the practical application and theory of logarithmic, multiplier, ratioing, and power and root circuits. Also includes sections on the generation of nonlinear transfer functions, and low-distortion operational amplifier oscillator and function generator circuits.

36. Sheingold, D. H., Ed., *Analog-Digital Conversion Handbook*, Analog Devices, Inc., Norwood, Mass., 1974.
An excellent discussion of practical considerations for the application of analog multiplexers, sample-and-hold circuits, and analog switches. Includes an excellent section on the use of instrumentation amplifiers and proper grounding techniques to reduce interferences picked up by analog transmission lines.

37. Smith, J. I., *Modern Operational Circuit Design*, Wiley-Interscience, New York, 1971.
Includes detailed practical considerations for designing operational amplifier circuits. Contains a good section on sample-and-hold circuits.

38. Thomason, J. G., *Linear Feedback Analysis*, McGraw-Hill, New York, 1955.
A thorough treatment of the principles of feedback.

39. Tobey, G. E., J. G. Graeme, and L. P. Huelsman, Eds., *Operational Amplifiers*, McGraw-Hill, New York, 1971.
A thorough discussion of operational amplifier circuits, including active filters.

40. Turner, R. P., *Semiconductor Devices*, Holt, Rinehart, Winston, New York, 1966.
Includes the theory of operation of semiconductor devices.

41. Vassos, B. H., and G. W. Ewing, *Analog and Digital Electronics for Scientists*, Wiley-Interscience, New York, 1972.
Includes good review sections on ac circuit analysis and appropriate mathematics.

42. Wallmark, J. T., and H. Johnson, *Field-Effect Transistor*, Prentice-Hall, Englewood Cliffs, N.J., 1966.
Includes thorough treatment of the principles of operation of field-effect transistors.

43. Walters, J. P., and T. V. Bruhns, "Electronically-Controlled, Current-Injection Spark Source for Basic and Applied Emission Spectrometry," *Anal. Chem.*, **41**, 1990 (1969).
Contains experimental photographs of pulsed wave forms from transmission lines.

44. *Zener Diode Handbook*, International Rectifier Corporation, El Segundo, Calif., 1965.
Includes circuit design consideration and an excellent section on temperature coefficients.

Part I
Section E

Chapter 3

# OPERATIONAL AMPLIFIERS AND ANALOG CIRCUIT ANALYSIS IN INSTRUMENTATION

BY THOMAS H. RIDGWAY, STEVEN P. BRIMMER, AND HARRY B. MARK, JR., *Department of Chemistry, University of Cincinnati, Cincinnati, Ohio*

Contents

## I. INTRODUCTION

The operational amplifier is in great part responsible for the essentially revolutionary changes in scientific instrumentation that have occurred since the early 1960s. Since the operational amplifier (OA or op amp for short) is the subject of this chapter, perhaps a brief biography is in order. The principle that a high gain dc amplifier could be used to solve differential equations appears to have been first employed in 1938 by George A. Philbrick, who was then a young engineer working on servomechanism design problems for the Foxboro Corporation (1). Lovell of Bell Telephone Laboratory independently discovered the same properties, apparently while involved in the Western Electric M-IX anti-aircraft gun director project during World War II (1). The first open literature reference which identified the salient features, and which also utilized the term "operational amplifier," was a paper by Ragazzini, Randall, and Russel, which was submitted in early 1946 and appeared in 1947 (2). In 1946, Philbrick started his own company, George A. Philbrick Research, Inc., or GAP/R for short. Soon analog computers were being manufactured by GAP/R, Boeing

Airplane Company, Goodyear Aircraft Company, Berkeley Scientific Company, and the Reaves Instrument Company. The first devices were enormous by today's standards, being constructed of two or more vacuum tubes and being powered by ±300 V dual supplies. The initial uses were strictly as computing devices by the military and aircraft companies, although one might argue that the M-IX director was an instrumental control application. Applications in computing soon broadened to include hydrology, stability of railroad cars, and chemical process control (1). DeFord in his widely circulated but unpublished notes was the first chemist to fully appreciate the use of operational amplification in chemical instrumentation (3). Instrumental applications began to appear (4–6) and articles in the *Journal of Chemical Education* by Reilley (7) began to create widespread interest, but the November, 1963 *Analytical Chemistry* was the true watershed with separate articles on the applications of OAs to chemical instrumentation. At the time these articles were published, the typical system was composed of a \$1400 OA manifold containing 10 amplifiers, which along with its associated \$400 power supply effectively filled a 19 by 56 in. rack and required two strong men to move it. Even as these articles were published, the first solid-state devices were reaching production. These were discrete transistor devices housed in metal cans about the size and shape of a pack of cigarettes and powered by ±15 V and costing about \$85 each. Perhaps the real breakthrough, technologically speaking, was the development of the fabled 709 type OA by Robert Widlar, then with Fairchild Camera. This was a true integrated circuit OA and, although the first models were designed for military applications and cost over \$100 each, the way was open to the inexpensive OA. Today the same 709 in an eight-pin plastic DIP package can be purchased for a little as 8¢.

Because of the high cost and great bulk of the original operational amplifiers, the early designs tended to conserve amplifiers at all costs. Today even a field-effect transistor (FET) OA often costs less than two or three precision resistors or a good integrator capacitor, so that the ground rules for OA circuitry have changed somewhat and they are being utilized in applications, such as active filter circuits, which would have been unthinkable a decade before. Although the cost, shape, and size of the OAs have changed over the years, the basic properties have remained invariant.

This chapter deals with the "classical" OA; a discussion of the Norton or "current mirror" amplifier is not included. Although the reader has been introduced to the basic principles in Chapter 2, it is useful to review some of the fundamentals. Understanding and applying OAs is actually not terribly difficult, but problems arise if the reader decides to skip over sections that appear to be too elementary.

All authors have their favorite way to conceptualize their topic for the reader and we are no different. Today's OA is physically nothing more than a slice of "dirty silicon," although admittedly the dirt is special dirt, applied with great care in special patterns. Anyone who has ever looked at the specification sheet for an OA has found what is usually euphemistically termed an "equivalent circuit" diagram, which is a formidable assemblage of resistors and transistors with an occasional capacitor. What is not normally stated is that, in general, even an electrical engineer looking at a photomicrograph of the device would be hard put to identify what transistor is located where, and in fact, even to locate the clearly drawn transistors is often impossible. In applying OAs, it is not truly necessary to know what is actually inside the package, but it is necessary to understand what parameters influence the device's behavior and how to cope with these parameters.

## II. GENERAL CHARACTERISTICS

As our starting point we use a representation of the OA that is extremely simplistic, but that has been proved to be a useful representation. The OA in its simplest representation is a five-terminal device as shown in Fig. 3.1*a*, with two power supplies, an output, and two input terminals. As our first model for the behavior of the OA, we can visualize it as shown in Fig. 3.1*b*, complete with a resident Maxwell demon. The demon reads the potential difference between the two inputs with the voltmeter shown. The demon's job is to adjust the tap of the potentiometer between the power supplies. If the potential of the negative input is more positive than the positive input, the tap is moved toward the negative supply, whereas if the positive input potential is more positive than that of the negative input, the tap is moved toward the positive supply. Only if the two potentials are identical, that is, the voltmeter reads 0 V, is the tap left alone. Whenever there is a voltage imbalance between the inputs, the demon pulls on the tap trying to achieve a zero voltage difference. Let us now make a connection between the negative input and the output as shown in Fig. 3.2*a*. When the potential at the positive input changes, the demon moves the potentiometer tap until the voltage difference between the two inputs is 0 V, and as a result the output of the amplifier is identical to the voltage at the positive input. As previously noted in Chapter 2, this is known as the voltage follower configuration.

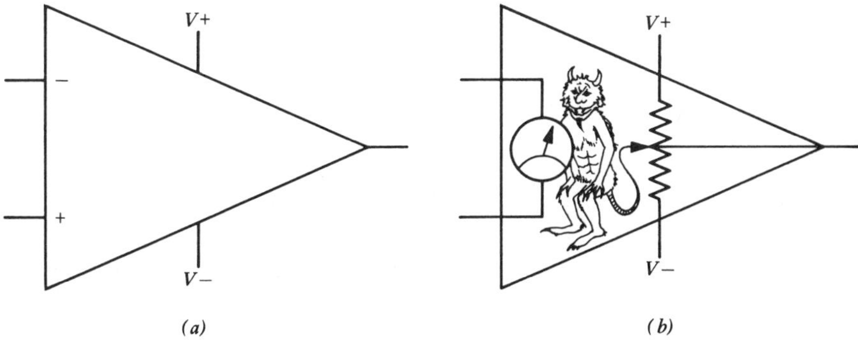


**Fig. 3.1.** (*a*) An outside view of an OA, simply a five-terminal device. (*b*) The Maxwell demon model of an OA. (The demon was drawn by Eileen Birch.)

There are three basic rules that govern the behavior of an *ideal* OA.

1. The output swings to maintain a zero potential difference between the inputs, if possible.
2. No current flows into or out of the inputs of the amplifier.
3. The sum of the currents at any circuit node is zero.

From these laws one can easily deduce the behavior of the amplifier configuration shown in Fig. 3.2*b*. The positive input is at ground; thus the potential of the negative input should also be at "ground" or what is called *virtual ground* (note that it is not actually grounded itself, however) so that the sum of the currents at the node connected to the negative input is given by

$$\sum i = \frac{E_1 - 0}{R_1} + \frac{E_0 - 0}{R_0} = 0 \tag{1}$$

which upon rearrangement leads to the relationship $E_0 = -E_1(R_0/R_1)$. If the positive input is at some potential $E_a$ instead of 0 V, then the voltage drops across resistor $R_1$ and $R_0$ are $(E_1 - E_a)$ and $(E_0 - E_a)$, respectively, and the sum of the currents expression yields

$$E_0 = -E_1\left(\frac{R_0}{R_1}\right) + E_a\left(1 + \frac{R_0}{R_1}\right) \tag{2}$$

This approach is successful for circuits containing only resistors, irrespective of the number of input signals.

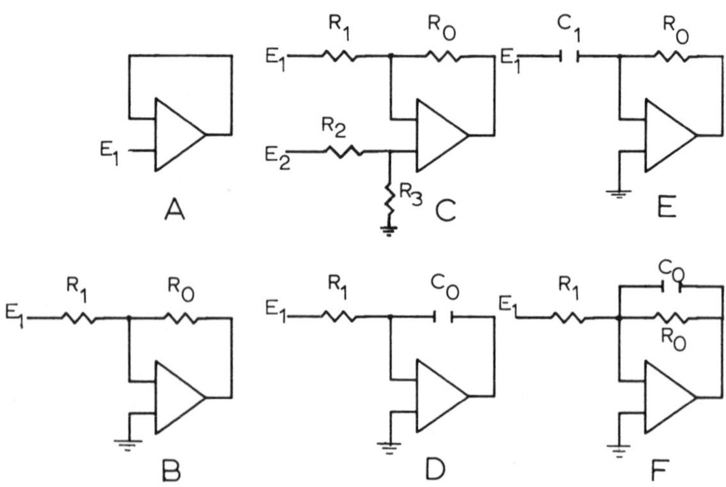


**Fig. 3.2.** A few simple OA circuits. (*a*) Voltage follower, (*b*) inverter with the gain of $R_0/R_1$, (*c*) a weighed subtractor, (*d*) an ideal integrator, (*e*) an ideal differentiator, (*f*) a low-pass filter.

The circuit in Fig. 3.2*c* is in a sense a combination of the preceding two examples. The potential $E_a$ is developed by the resistive divider $R_2$, $R_3$ from the input voltage $E_2$. Because no current flows into the positive input, we know that

$$E_a = \frac{E_2 R_3}{R_2 + R_3} \tag{3}$$

Substituting this result into the expression for the circuit in Fig. 3.2*c*, we have

$$E_0 = -E_1\left(\frac{R_0}{R_1}\right) + E_2\frac{R_3(R_0 + R_1)}{2R_1(R_2 + R_3)} \tag{4}$$

A common implementation of this circuit is to choose the resistance such that $R_0 + R_1 = R_2 + R_3$, in which case the expression simplifies to

$$E_o = -E_1\left(\frac{R_0}{R_1}\right) + E_2\left(\frac{R_3}{R_1}\right) \tag{5}$$

which is clearly a scaling *subtractor*.

When capacitors or inductors are considered, the situation becomes somewhat more complex because the relationships between current through the component and the voltage drop across them are differential equations:

$$I_{\text{capacitor}} = C\left(\frac{dE}{dt}\right) \tag{6}$$

$$E_{\text{inductor}} = L\left(\frac{dI}{dt}\right) \tag{7}$$

where $I$ is in amperes, $E$ in V, $C$ in farads, and $L$ in henries. For simple circuits, it is still possible to treat circuits containing these components in a fairly elementary manner. For example,the circuit in Fig. 3.2*d* is trivial if one remembers to add currents at the negative input node

$$\sum i = \frac{E_1}{R_1} + C_0\frac{d(E_0 - E_a)}{dt} \tag{8}$$

Because the potential at the negative input ($E_a$) is 0 V, one can rearrange the expression to give

$$dE_0 = \frac{-E_1}{R_1 C_0}\,dt \tag{9}$$

The answer we desire is the output voltage itself, not its derivative, so we must integrate the expression

$$E_0(t) = \frac{-1}{R_1 C_0}\int_0^t E_1\,dt \tag{10}$$

The circuit acts as an *integrator* for the input voltage signal.

Switching the position of the resistor and capacitor in the circuit does not unduly complicate matters and the resulting expression becomes

$$E_0 = R_0 C_1 \frac{dE_1}{dt} \tag{11}$$

which indicates that the circuit in Fig. 3.2*e* is a *differentiator* for the input voltage applied to it. Simple circuits of this type employing an inductor in place of the resistor or capacitor are also treated satisfactorily by this simplistic method. We run into problems when we adds a resistor in series with, or in parallel with, the capacitor or inductor as in Fig. 3.2*f*. We can still sum the currents at the negative input of the amplifier to give

$$\sum i = \frac{E_1}{R_1} + \frac{E_0}{R_0} + C_0\frac{dE_0}{dt} = 0 \tag{12}$$

$$\frac{E_0}{R_1C_0} + \frac{dE_0}{dt} = \frac{-E}{R_1C_0} \tag{13}$$

Depending upon how long it has been since the reader worked with differential equations, it may or may not be obvious that this is a fairly complex problem. We are now faced with three choices. We can attempt to solve the differential equations for this and any other circuits of this type (it is suggested that you attempt to develop a solution for this particular circuit for voltage step and ramp inputs to convince yourself of the utility of the alternatives), or we can use Laplace or Fourier transform methods. The Fourier transform approach is treated as a special case of a more general Laplace transformation.

## III. LAPLACE TRANSFORM CIRCUIT ANALYSIS

In the late 1800s Oliver Heavyside (1850–1925) began to develop a set of semiempirical methods, which he successfully applied to almost every phase of physics and applied mathematics. His method was essentially a collection of rules that he had found would generally work in any given case. Because he deemed it more important to spend his time solving problems than developing his rules within a rigorous mathematical framework, he was subjected to a fair amount of criticism. Once when a critic pointed out the insecure foundation of his methods, he replied, "Shall I refuse my dinner because I do not understand the process of digestion?" The work of Carson, Van der Pol, and others linked the "heavyside operator," as it was the known, to the outcome of the contour integration of functions of complex variables and formulated operational calculus on the basis of the finite integral transformation, which Laplace introduced in 1779. Because Laplace had shown the correspondence between the two functional domains and, we suspect, because his attitude was far more traditionally academic, the nomenclature has changed from the "heavyside operator" to the "Laplace transformation."

Irrespective of what name should be given to the technique, the basic concept involved is to convert an equation that is relatively intractable in its natural or initial form into another spatial representation where the equivalent equation is simpler. In contrast to the classical differential equation methods, which require that the general solution be fitted to the initial or boundary conditions, these conditions are automatically incorporated into the solution for any excitation. Assuming that the independent variable is time, or at least a function of time as is the case here, the Laplace transformation eliminates it from the differential equation and replaces it with an operator $s = a + jb$, which although it is complex can still be treated as an algebraic quantity. As a result of the transformation, the differential equation becomes an algebraic relationship that can be manipulated by the laws governing ordinary algebra. Once the transformation is carried out, certain manipulations can be performed that allow us to return to the time plane representation with the equation explicitly solved. Further advantages of the Laplace transformation are that discontinuous functions are readily treated and that in some instances one need not even return to the time plane, the Laplace representation being sufficiently informative by itself.

The Laplace transformation of the object or original function into the image or result function is given by

$$F(s) = \overline{f(t)} = \int_0^\infty f(t)e^{-s}dt = \lim_{\substack{a\to 0\\ b\to\infty}} \int_a^b f(te)e^{-st}\, dt \; (0<a<b) \tag{14}$$

The primary utility of the Laplace transformation for our purposes lies in the following observation. If a system, such as an OA circuit or assemblage of amplifier circuits, responds linearly to an excitation signal and if the excitation function in the time plane $X(t)$ and the system function in the time plane $F(t)$ both have Laplace transformations $X(s)$ and $F(s)$, then the response, $R(s)$, of the system in the Laplace plane is given by

$$R(s) = X(s)*F(s) \tag{15}$$

This is an exceedingly important statement since it means that we can predict the behavior of an electronic circuit stimulated by any signal, if we can represent both the excitation and the system as equations in Laplace space and are capable of returning to the time domain. Buried within the statement above are several assumptions that we examine briefly.

The most obvious assumption is that one can reverse the transformation process and return to the time plane, (i.e., convert the response image function into a response object function). It is important from a mathematical viewpoint to recognize that not every function $F(s)$ has an inverse Laplace transformation, but in electronic circuitry it is extremely unlikely that one will encounter a case of this type. The reversion to the time plane formally involves a complex integration:

$$f(t) = \mathcal{L}^{-1}\, F(s) = \frac{1}{2\pi j}\left(\lim_{R\to\alpha} \int_{a_1-jR}^{a_1+jR} F(s)e^{st}\, ds\right) \tag{16}$$

In practice one rarely resorts to the defining equation for either transformation or reversion since a vast body of transform pairs exist and one can simply look up the image function given the object function or vice versa; herein lies one reason for the popularity and utility of the method. It is important to recognize that the transform definition, equation 14, includes only positive values of the object variable (normally time) and actually is undefined for negative values. Technically, the initial point of validity is written as $t = 0_+$, meaning a time infinitesimally greater than zero, but for practical purposes one can consider the region of validity beginning at zero itself. Transform pairs are unique; that is, in the region of validity of the transform pairs, if two functions have the same object or image function, then they are identical. In other words, one may find several ways to return to the time plane from a single image function but these representations are, and must be, equivalent for all times greater than zero.

There is the restriction that the systems must be linear in our inherent definition. Actually this is a restatement of the requirement that the object function must have an image function, because the first requirement is a subset of the second. In practice one may often circumvent this requirement, particularly for system functions, by restricting the excitation function to small excursions, so that the system function can be linearized. This is a common approach when dealing with electrochemical systems as circuit elements for they are inherently nonlinear elements.

## A. TRANSFORMS OF COMMON FUNCTIONS

We endeavor to keep the number of direct applications of the defining equations to a minimum, but some examples are required in order to show the procedure and to develop the minimal sets of functions that we use later.

### 1. Transform of a Constant

$$F(s) = \int_0^\infty ke^{-st}\,dt = \frac{-k}{s}\,[e^{-st}]_0^\infty = \frac{k}{s} \tag{17}$$

### 2. Transform of a Step Function

The unit step function $U(t)$ has a value of zero for all argument values negative of zero and rises to a value of unity for argument values more positive than zero; the Laplace transform of the function is

$$F(s) = \int_0^\infty 1e^{-st}\,dt = \frac{-1}{s}\,[e^{-st}]_0^\infty = \frac{1}{s} \tag{18}$$

### 3. Transform of an Exponential Function

$$F(s) = \int_0^\infty e^{+at}e^{-st}\,dt = \left[\frac{1 - e^{-(s-a)t}}{sa}\right]_0^\infty = \frac{1}{s - a} \tag{19}$$

### 4. Transform of a Trigonometric Sine

The sine function has the exponential definition

$$f(t) = \sin(\omega t) = \frac{e^{j\omega t} - e^{-j\omega t}}{2j\omega} \tag{20}$$

To treat this case, we can use the superposition theorem which holds that

$$c_1 f_1(t) + c_2 f_2(t) = c_1 F_1(s) + c_2 F_2(s) \tag{21}$$

which when applied to the case of the sine wave yields

$$F(s) = \mathcal{L}\ \sin(\omega t) = \frac{1}{2j\omega}\left[\frac{1}{s - j\omega} - \frac{1}{s + j\omega}\right] = \frac{\omega}{s^2 + \omega^2} \tag{22}$$

## 5. Transforms of Delay and Shifting Functions

One of the more valuable properties of the Laplace transform is the ease with which one can treat delayed and repetitive functions. A wave form that is becoming common in instrumentation, owing to the proliferation of computers and digital to analog converters, is the staircase wave form. This is nothing more than a series of steps of height $A$ spaced $b$ sec apart and represents a discrete digital approximation to the continuous ramp function. In the time plane this signal can be represented by

$$F(t) = \sum_0^c AU(t - b) \tag{23}$$

where $U(x)$ is the unit step function which is zero for $x \leq 0$ and unity for $x > 0$. The problem of representing the staircase wave form is thus one of handling the delayed unit step function. If we use the convention ${}_aF(s)$ to represent the image function of a time function $F(t)$ whose transform is $F(s)$ but that has been delayed in time by an amount $a$, then

$${}_aF(s) = \int_0^\infty f(t - a)e^{-st}\, dt = \int_0^a e^{-st}\, dt + \int_a^\infty f(t - a)e^{-st}\, dt \tag{24}$$

A change of variables $z = t - a$ leads to

$${}_aF(s) = \int_0^\infty e^{-as}e^{-ts}f(z)\, d(z) = e^{-as}F(s) \tag{25}$$

Thus delay in time is equivalent to a multiplication by $e^{-as}$ in the image plane. The staircase function can now be represented in Laplace space by

$$F(s) = \sum_0^c \frac{A}{s}e^{-bcs} \tag{26}$$

Another common function is the square wave. This function can be constructed from step functions of alternating direction. If the wave form starts with amplitude $A$ and returns to zero at time $b$, then the time plane representation is

$$F(t) = A + \sum_{n=0}^{\infty} (-1)^n AU(t - bn) \tag{27}$$

and the Laplace representation is

$$F(s) = \sum_{n=0}^{\infty} \frac{A}{s}(-1)^n e^{-nbs} \tag{28}$$

Those readers who are familiar with series representations will recognize that for a summation to infinity the above equation can be rewritten as

$$F(s) = \frac{A(1 - e^{-sb})}{s(1 - e^{-2sb})} \tag{29}$$

Although this representation is compact and elegant, it is generally of little practical utility since the tables of transforms commonly found do not contain many entries of this form and the open series solution is preferable for practical applications.

The extension of our result to a centrosymmetric square wave, that is, one whose average value is zero with excursions to $^{\pm}A/2$ so that one can immediately write

$$F(s) = \frac{A}{s}\left[\frac{-1}{2} + \sum_{n=0} (-1)^n e^{-nbs}\right] \tag{30}$$

From this discussion the reader should be able to see that the extension to other repetitive wave forms such as triangular and sawtooth wave forms is quite simple. A number of examples are given in Fig. 3.3.

We make one further excursion into theory and consider the generalized transform for an arbitrary repetitive function with period $\phi$. The generating function $f(t)$ is assumed to have the transform $F(s)$ and the periodic function

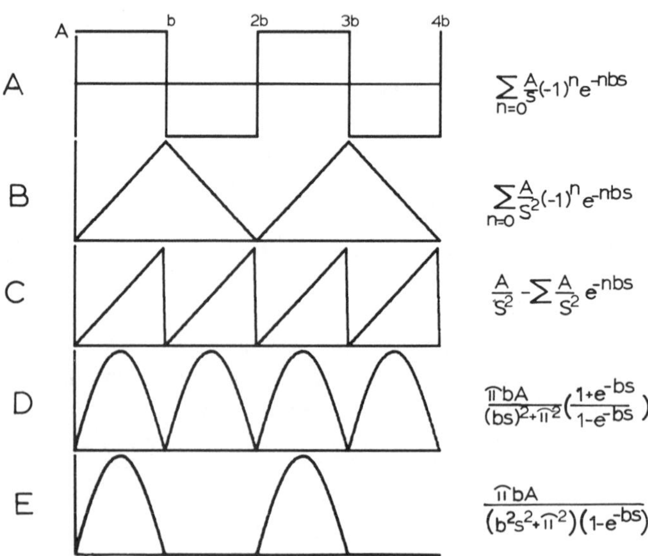


**Fig. 3.3.** The wave form and its corresponding Laplace plane representation. (*a*) Square wave, (*b*) triangular wave, (*c*) sawtooth wave, (*d*) rectified sine wave, (*e*) half-wave rectified sine wave.

$\phi f(t)$ has the property that $\phi f(t) = \phi(t + n\phi)$. The transform of the periodic form may then be written

$$_{\phi}F(s) = \int_0^{\infty} e^{-st}\phi f(t)\,dt = \int_0^{\phi} e^{-st}f(t)\,d(t) + \int_{\phi}^{2\phi} e^{-st}f(t)\,dt + \cdots \tag{31}$$

The general limits of the integral on the right-hand side of the equation are $n\phi$ and $(n + 1)\phi$. Changing variables to $x = t - n\phi$ and remembering that $f(x + n\phi) = f(x)$ leads to

$$_{\phi}F(s) = \sum_{n=0} e^{-ns\phi} \int_0^{\phi} e^{-sx}f(x)\,dx = \frac{1}{(1 - e^{-\phi s})} \int_0^{\phi} e^{-st}f(t)\,dt \tag{32}$$

An important example of a repetitive wave form is the half-wave rectified sine wave. This wave form is commonly found in power supplies and occasionally appears as a noise contribution in operational amplifier circuits. The time plane functionality is given by

$$f(t) = \begin{cases} \sin(\omega) & 0 < t < \pi \\ 0 & \pi < 2\pi \end{cases} \tag{33}$$

By application of our equation for periodic wave forms we arrive at

$$\phi F(s) = \frac{1}{1 - e^{-2\omega s}} \int_0 e^{-st}\sin(\omega t)\,dt \tag{34}$$

$$= \frac{-\omega}{1 - e^{-2\pi s}} \left\{ \frac{e^{-st}[s\,\sin(\omega t) + \cos(\omega t)]}{s^2 + \omega^2} \right\} \Bigg|_0^{\pi} \tag{35}$$

$$\phi F(s) = \frac{-\omega}{(1 - e^{-2\pi s})} \frac{1 + e^{-st}}{s^2 + \omega^2} = \frac{\omega}{(s^2 + \omega^2)(1 - e^{-\pi s})}$$

$$= \frac{\omega}{1 - e^{-2\pi s}} \frac{1 + e^{-\pi s}}{s^2 + \omega^2}$$

$$\phi F(s) = \frac{\omega}{(s^2 + \omega^2)(1 - e^{-\pi s})} \tag{36}$$

The transform for a full-wave rectified sine wave may be obtained from this result by noting that the full-wave rectified form is simply two half-wave rectified wave forms with one of them zero-shifted by a half period.

## 6. Transform of Damped and Modulated Wave Forms

Two other common wave forms that are often encountered in electronics are the exponentially damped function and the sine or cosine modulated function. Both these functions are treated simply by application of one theorem:

$$e^{+at}f(t) = F(s - a) \tag{37}$$

which can be readily proved by making the variable change $x = s \pm a$ and application of the definition of the Laplace transform. From this it can be seen that the formation of the image function of an exponentially damped wave form $x$, where the original function $f(t)$ has a transform $F(s)$, is simply a matter of replacing *all* $s$ terms in the transform with $s$ $a$ terms. For example, an exponentially damped ramp

$$f(t) = te^{-at} \tag{38}$$

can be transformed by noting that $t = 1/s^2$ and thus $\mathcal{L}te^{-at} = 1/(s + a)^2$. Extension to sine or cosine modulated wave forms is not much more complicated because

$$\cos(\omega t) = \frac{e^{j\omega t} + e^{-j\omega t}}{2} \tag{39}$$

and one can then write

$$f(t)\cos(Ut) = \tfrac{1}{2}f(t)e^{j\omega t} + \tfrac{1}{2}f(t)e^{-j\omega t} \tag{40}$$

$$f(t)\cos(\omega t) = \tfrac{1}{2}f(s - j\omega) + f(s + j\omega) \tag{41}$$

These and other useful transform pairs are to be found in Table 3.I below.

### 7. Transform of a Differential

If the differential of a function of time $f(t)$ is represented by $f'(t)$, then the transform is given by

$$\mathcal{L}f'(t) = \int_0^\infty e^{-st}\frac{df(t)}{dt}\,dt \tag{42}$$

Integration by parts leads to

$$\mathcal{L}f'(t) = e^{-st}f(t)\,{}_0^\infty + s\int_0^\infty e^{-st}f(t)\,dt = -f(0) + sF(s) \tag{43}$$

Actually, because the integral runs from 0+ and not zero, the function is evaluated at 0+. This is referred to as the differentiation theorem and can be expanded indefinitely to include higher derivatives, as for example,

$$f'(t) = s^2F(s) - sf(0) - f'(0) \tag{44}$$

## B. APPLICATION TO CIRCUIT ELEMENTS

As long as the ground rules for Laplace transformations are not violated, we can take the generalized form of Ohm's law;

$$E(t) = Z(t)I(t) \tag{45}$$

where $E$ and $I$ are potential and current as a function of time, and $Z$ is the circuit impedance as a function of time, and represent it in Laplace space as

$$E(s) = I(s)Z(s) \tag{46}$$

### 1. The Resistor, *R*

The special form of Ohm's law for resistances is

$$E(t) = I(t)R \tag{47}$$

Because resistance is not normally a function of time, the linearity theorem yields

$$E(s) = I(s)R \tag{48}$$

and it can be seen that the transform of a resistance is the resistance itself; that is, the object and image functions are identical: $\overline{Z}_R = R$.

### 2. The Capacitor, *C*

The normal expression for the current flow through a capacitor is

$$\frac{dE}{dt} = \frac{I(t)}{C} \tag{49}$$

If the differential expression just derived is utilized,

$$sE(s) - E(0+) = \frac{I(s)}{C} \tag{50}$$

Normally in treating electronic circuits one assumes that the circuit was initially at rest, that is, in this case that the initial voltage drop across the element was zero, so that one can write

$$E(s) = \frac{I(s)}{sC} \tag{51}$$

or equivalently, the transform of the image function of a capacitative impedance is

$$\overline{Z}_c = \frac{1}{sC} \tag{52}$$

### 3. The Inductor, *L*

The equation expressing current and voltage relationships in an inductor is usually written as

$$E(t) = \frac{L\,dI(t)}{dt} \tag{53}$$

Again applying the differential properties, one obtains

$$E(s) = Ls[I(s) - I(0+)] \tag{54}$$

As in the case of the capacitor, we normally neglect the initial conditions which allows us to write the impedance of an inductor in Laplace as

$$Z_L(s) = sL \tag{55}$$

TABLE 3.I
Some Useful Laplace Transform Pairs for Analytical Chemists

| $f(s)$ | $F(t)$ |
|---|---|
| 1. $\frac{1}{s}$ | 1 |
| 2. $\frac{1}{S^n}$ | $\frac{t^{n-1}}{(n-1)!}$ |
| 3. $\frac{1}{s+a}$ | $e^{-at}$ |
| 4. $\frac{1}{(s+a)(s+b)}$ | $\frac{-e^{at}-e^{-bt}}{(b-a)}$ |
| 5. $\frac{s+a}{s(s+a)}$ | $\frac{a}{b}+\left(1-\frac{a}{b}\right)e^{-bt}$ |
| 6. $\frac{1}{(s+a)(s+b)(s+c)}$ | $\frac{e^{-at}}{(b-a)(c-a)}+\frac{e^{-bt}}{(d-b)(c-b)}+\frac{e^{-ct}}{(a-c)(b-c)}$ |
| 7. $\frac{s}{s^2+a^2}$ | $\cos(at)$ |
| 8. $\frac{1}{s^2+a^2}$ | $\frac{1}{a}\sin(at)$ |
| 9. $\frac{s^2+b^2}{s(s^2+2b^2)}$ | $\cos^2(bt/2)$ |
| 10. $\frac{1}{s^2-a^2}$ | $\cosh(at)$ |
| 11. $\frac{s}{s^2-a^2}$ | $\cosh(at)$ |
| 12. $\frac{1}{(s^2+a^2(s^2-b^2)}$ | $\frac{a\sinh(bt)-b\sin(at)}{ab(a^2+b^2)}$ |
| 13. $\sqrt{s}$ | $-\frac{1}{2t^{3/2}\sqrt{\pi}}$ |
| 14. $\frac{1}{\sqrt{s}}$ | $\frac{1}{\sqrt{\pi t}}$ |
| 15. $\frac{1}{s^{3/2}}$ | $2\sqrt{t/\pi}$ |

| | |
|---|---|
| 16. $s^{-(N+1/2)}$ | $\dfrac{2^N t^{N-1/2}}{1\cdot 3\cdot 5\ldots(2N-1)}$ |
| 17. $\dfrac{1}{s^{\upsilon}}$ | $\dfrac{t^{\upsilon-1}}{\Gamma(\upsilon)}$ |
| 18. $\dfrac{s}{\sqrt{s}+a}$ | $\dfrac{1}{t^{3/2}\sqrt{\pi}}\left[a^2t-\dfrac{1}{2}\right]-a^3e^{a^2t}erfc(a\sqrt{t})$ |
| 19. $\dfrac{\sqrt{s}}{\sqrt{s}+a}$ | $\delta(t)-\dfrac{a}{\sqrt{\pi t}}+a^2e^{a^2t}\text{erfc}(a\sqrt{t})$ |
| 20. $\dfrac{1}{\sqrt{s}+a}$ | $\dfrac{1}{\sqrt{\pi t}}-ae^{a^2}t\,\text{erfc}(a\sqrt{t})$ |
| 21. $\dfrac{1}{\sqrt{s}(\sqrt{s}+a)}$ | $e^{a^2t}\text{erfc}(a\sqrt{t})$ |
| 22. $\dfrac{1}{s(\sqrt{s}+a})$ | $\dfrac{1}{a}(1-e^{a^2}t\,\text{erfc}(a\sqrt{t}))$ |
| 23. $\dfrac{1}{s^{3/2}(\sqrt{s}+a)}$ | $\dfrac{1}{a^2}[2a\sqrt{t/\pi}-1+e^{a^2t}\text{erfc}(a\sqrt{t})]$ |
| 24. $\dfrac{1}{s^2(\sqrt{s}+a)}$ | $\dfrac{1}{A^3}[a^2t-2a\sqrt{t/\pi}+1-e^{a^2t}\text{erfc}(a\sqrt{t})]$ |
| 25. $\dfrac{1}{s^{5/2}(\sqrt{s}+a)}$ | $\dfrac{1}{a^4}\left[\dfrac{4a^3t^{3/2}}{3\sqrt{\pi}}-a^2t+2a\sqrt{t/\pi}-1+e^{a^2t}\text{erfc}(a\sqrt{t}\,t)\right]$ |
| 26. $\dfrac{1}{s^3(\sqrt{s}+a)}$ | $\dfrac{1}{a^5}\left[\dfrac{a^4t^2}{2}-\dfrac{4a^3t^{3/2}}{3\sqrt{\pi}}+a^2t-2a\sqrt{t/\pi}+1-e^{a^2t}\text{erfc}(a\sqrt{t})\right]$ |
| 27. $\dfrac{1}{s^{7/2}(\sqrt{s}+a)}$ | $\dfrac{1}{a^6}\left[\dfrac{8a^5t^{5/2}}{15\sqrt{\pi}}+\dfrac{a^4t^2}{2}+\dfrac{4a^3t^{3/2}}{3\sqrt{\pi}}-a^2t+2a\sqrt{t/\pi}-1+e^{a^2t}\text{erfc}(a\sqrt{t})\right]$ |
| 28. $\dfrac{1}{s^4(\sqrt{s}+a)}$ | $\dfrac{1}{a^7}\left[\dfrac{a^6t^3}{6}-\dfrac{8a^5t^{5/2}}{15\sqrt{\pi}}+\dfrac{a^4t^2}{2}-\dfrac{4a^3t^{3/2}}{3\sqrt{\pi}}+a^2t-2a\sqrt{t/\pi}+1-e^{a^2t}\text{erfc}(a\sqrt{t})\right]$ |
| 29. $\sqrt{s+a}$ | $\dfrac{e^{-at}}{2t^{3/2}\sqrt{\pi}}$ |
| 30. $\dfrac{\sqrt{s+a}}{s}$ | $\dfrac{-e^{-at}}{\sqrt{\pi}}+\sqrt{a}\,\text{erf}\sqrt{at}$ |
| 31. $\dfrac{\sqrt{s+a}}{s+b}\ (b<a)$ | $\dfrac{-e^{at}}{\sqrt{\pi t}}+\sqrt{a-b}\,e^{-bt}\text{erf}\sqrt{(a-b)t}$ |

TABLE 3.I (*Continued*)
Some Useful Laplace Transform Pairs for Analytical Chemists

| | $f(s)$ | $F(t)$ |
|---|---|---|
| 32. | $\frac{\sqrt{s+a}}{s^{3/2}}$ | $e^{-at/2}\left[(1+at)I_o\left(\frac{at}{2}\right)+atI_1\left(\frac{at}{2}\right)\right]$ |
| 33. | $\frac{\sqrt{s+a}}{s(s-b^2)}$ | $\frac{1}{b^2}(a+b^2)^{1/2}e^{b^2t}\text{erf}(\sqrt{(a+b^2t)}-\sqrt{a}\text{erf}(\sqrt{at})$ |
| 34. | $\frac{\sqrt{s}}{s(s-a^2)}$ | $\frac{1}{\sqrt{\pi t}}+ae^{a^2t}\text{erf}(a\sqrt{t})$ |
| 35. | $\frac{1}{\sqrt{s}(s-a^2)}$ | $\frac{1}{a}e^{a^2t}\text{erf}(a\sqrt{t})$ |
| 36. | $s^{3/2}\frac{1}{(s-a^2)}$ | $\frac{1}{a^3}(e^{a^2t}\text{erf}(a\sqrt{t}-2a\sqrt{t/\pi})$ |
| 37. | $\frac{\sqrt{s+b}}{s\sqrt{s+a}}$ | $e^{-[(a+b)/2]t}I_o\left\{\left[\frac{a-b}{2}\right]t\right\}+b\int_0^{t}e^{-[(a+b)/2]u}I_o\left\{\left[\frac{a-b}{2}\right]u\right\}du$ |
| 38. | $\frac{1}{\sqrt{s+a}}$ | $e^{-at}/\sqrt{\pi t}$ |
| 39. | $\frac{1}{(s+a)\sqrt{s}}$ | $\frac{2}{a\sqrt{\pi}}e^{-at2}\int_o^a\sqrt{t}e^{x^2}\,dx$ |
| 40. | $\frac{1}{\sqrt{s+b}\sqrt{s+a}}$ | $e^{-[(a+b)/2]t}I_o\left[\left(\frac{a-b}{2}\right)t\right]$ |
| 41. | $\frac{1}{(s+b)\sqrt{s+a}}\ (a>b)$ | $\frac{e^{-bt}\text{erf}\sqrt{(a-b)t}}{\sqrt{a-b}}$ |
| 42. | $\frac{1}{(s+a)^{N+1/2}}$ | $\frac{2^{N-1/2}t^{N-1/2}e^{-at}}{1\cdot3\cdot5\cdots(2N-1)}$ |
| 43. | $\frac{\sqrt{s}}{(\sqrt{s}+a)^2}$ | $(2a^4t^2+5a^2t+1)e^{a^2t}\text{erfc}(a\sqrt{t})-2a\sqrt{t/\pi}(a^2t+2)$ |
| 44. | $\frac{1}{(\sqrt{s}+a)^2}$ | $(2a^2t+1)e^{a^2t}\text{erfc}(a\sqrt{t})-2a\sqrt{t/\pi}$ |
| 45. | $\frac{1}{\sqrt{s}(\sqrt{s}+a)^2}$ | $2\sqrt{t/\pi}-2ate^{a^2t}\text{erfc}(a\sqrt{t})$ |

| | |
|---|---|
| 46. $\frac{1}{s(\sqrt{s}+a)^2}$ | $\frac{1}{a^2}[1+(2at-1)e^{a^2t}\mathrm{erfc}(a\sqrt{t})-2a\sqrt{t/\pi}]$ |
| 47. $\frac{\sqrt{s}}{(\sqrt{s}+a)^3}$ | $(2a^4t+5a^2+1)e^{a^2t}\mathrm{erfc}(a\sqrt{t})-2a(a^2t+2)\sqrt{t/\pi}$ |
| 48. $\frac{1}{(\sqrt{s}+a)^3}$ | $2(a^2t+1)\sqrt{t/\pi}-ate^{a^2t}\mathrm{erfc}(a\sqrt{t})(2a^2t+3)$ |
| 49. $\frac{1}{\sqrt{s}(\sqrt{s}+a)^3}$ | $t(2a^2t+1)e^{a^2t}\mathrm{erfc}(a\sqrt{t})-2a\sqrt{t^3/\pi}$ |
| 50. $\frac{1}{s(\sqrt{s}+a)^3}$ | $\frac{1}{a^3}-\left(2a^2t-\frac{t}{a}+\frac{1}{a^3}\right)e^{a^2t}\mathrm{erfc}(a\sqrt{t})+2\sqrt{t/\pi}\left(t-\frac{1}{a^2}\right)$ |
| 51. $\frac{\sqrt{s}}{(s-b^2)(\sqrt{s}+a)}$ | $\frac{1}{(a^2-b^2)}[a^2e^{a^2t}\mathrm{erfc}(a\sqrt{t})+abe^{b^2t}\mathrm{erfc}(b\sqrt{t})-b^2e^{b^2t}]$ |
| 52. $\frac{1}{(s-b^2)(\sqrt{s}+a)}$ | $\left(\frac{1}{a^2-b^2}\right)[e^{b^2t}(a-\mathrm{b\ erf}(b\sqrt{t}))-ae^{a^2t}\mathrm{erfc}(b\sqrt{t})]$ |
| 53. $\frac{1}{\sqrt{s}(s-b^2)(\sqrt{s}+a)}$ | $\frac{1}{(a^2-b^2)}\left[e^{a^2t}\mathrm{erfc}(a\sqrt{t})+\frac{a}{b}e^{b^2t}\mathrm{erfc}(b\sqrt{t})-e^{b^2t}\right]$ |
| 54. $\frac{\sqrt{s+b}}{s+a}$ | $\frac{e^{-b^2t}}{\sqrt{\pi t}}-\frac{2}{\sqrt{\pi}}\sqrt{a^2-b}\;e^{-bt}e^{-(a^2-b)t}\int_0^{\sqrt{(a^2-b)t}}e^{\lambda^2}\,d\lambda$ |
| 55. $\frac{\sqrt{s+b}}{s(s+a)}$ | $\frac{1}{a^2}\sqrt{b}\mathrm{erf}(\sqrt{bt})+2\sqrt{a^2-b/\pi}e^{-bt}e^{-(a^2-b)t}\int_0^{(a^2-b)t}e^{\lambda^2}\,d\lambda$ |
| 56. $\frac{\sqrt{s}}{s+a^2}$ | $\frac{1}{\sqrt{\pi t}}-\frac{2a}{\sqrt{\pi}}e^{-a^2t}\int_0^{a\sqrt{t}}e^{\lambda^2}\,d\lambda$ |
| 57. $\frac{1}{\sqrt{s}(s+a^2)}$ | $\frac{2}{a\sqrt{\pi}}e^{-a^2t}\int_0^{a\sqrt{t}}e^{\lambda^2}\,d\lambda$ |
| 58. $\frac{1}{s^{3/2}(s+a^2)}$ | $\frac{1}{a^3}2a\sqrt{t/\pi}-\frac{2}{\sqrt{\pi}}e^{-a^2t}\int_0^{a\sqrt{t}}e^{\lambda^2}\,d\lambda$ |
| 59. $\frac{s^c}{(s-b)^a}$ | $\frac{t^{a-c-1}}{\Gamma(a-c)}\,{}_1F_1(a;a-c;bt)$ |
| 60. $\frac{1}{(s-b)(s+a)^{\upsilon}}$ | $\frac{e^{bt}\gamma[\upsilon,(a+b)t]}{\Gamma(\upsilon)(a+b)^{\upsilon}}$ |

## C. NETWORK ADMITTANCE, $Y$, AND IMPEDENCE, $Z$

It is often more convenient to talk about the reciprocal of the impedance in electronic circuits, particularly when dealing with operational amplifiers, where we are primarily concerned with the sum of currents at nodes in the circuits. The reciprocal, termed the admittance, is defined in Laplace space by

$$I(s) = E(s)Y(s) = \frac{E(s)}{Z(s)} \tag{56}$$

### 1. Application to Simple Networks (Kirchhoff's Laws)

We can treat any network, no matter how complex, by following three simple rules, generally referred to as Kirchhoff's laws.

1. The sum of the currents at any node in a circuit are zero.
2. For a collection of elements connected in series the total *impedance* is the sum of the individual *impedances*.
3. For a collection of elements connected in parallel the total *admittance* is the sum of the *admittances*.

Some of the most common electronic circuit elements are shown in Fig. 3.4, and, as we see later, most complex circuits can be decomposed into blocks containing these elements. The series resistor-capacitor combination in Fig. 3.4$a$ can be represented in transform space by

$$\overline{Z} = \overline{Z}_1 + \overline{Z}_2 = R + \frac{1}{sC} = \frac{RCs + 1}{sC} = \frac{R(s + 1/RC)}{s} \tag{57}$$

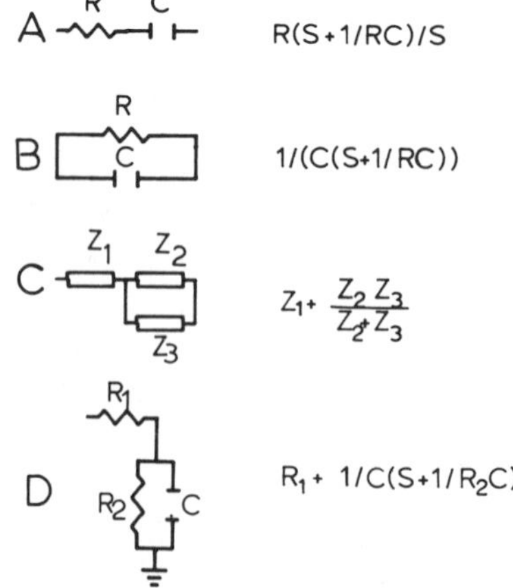


**Fig. 3.4.** Four commonly encountered network circuits and their Laplace plane representation. ($a$) Simple series $RC$, ($b$) $RC$ parallel, ($c$) a parallel impedance in series with another impedance, ($d$) an $RC$ parallel network in series with the resistance [a specific case of ($c$)].

The parallel $RC$ combination in Fig. 3.4$b$ can be treated similarly by

$$\overline{Y} = \overline{Y}_1 + \overline{Y}_2 = sC + \frac{1}{R} = \frac{RCs + 1}{R} = C\left(s + \frac{1}{RC}\right) \tag{58}$$

$$\overline{Z} = \frac{1}{\overline{Y}} = \frac{1}{C(s + 1/RC)} \tag{59}$$

It is useful to consider a generalized series parallel combination of the type shown in Fig. 3.4$c$ because, once we know how to handle the general case, we can treat any special $RC$ combination subcase by substituting the proper impedances. The total impedance $Z_t$ is composed of the sum of the series impedance and the impedance of the parallel network $Z_p$:

$$\frac{1}{\overline{Z}p} = \overline{Y}p = \overline{Y}_2 + \overline{Y}_3 = \frac{1}{\overline{Z}_2} = \frac{\overline{Z}_2 + \overline{Z}_3}{\overline{Z}_2\overline{Z}_3} \tag{60}$$

$$\overline{Z}t = \overline{Z}1 + \frac{\overline{Z}_2\overline{Z}_3}{\overline{Z}_2\overline{Z}_3} \tag{61}$$

From this development it immediately follows that the transform of the resistance $R_1$ in series with a parallel combination of $R_2$ and $C$ shown in Fig. 3.4$d$ is

$$\overline{Z}_t = R_1 + \frac{R_2(1/sC)}{R_2 + 1/sC} = R_1 + \frac{1}{C(s + 1/R_2C)} \tag{62}$$

We could have arrived at this result even more directly by remembering that we had just solved the parallel $RC$ case and the result could have been written by simple inspection.

## 2. Transfer Functions

Often we know the input voltage to a network and wish to know the current flow through one particular element in that network. This is called the transfer function of the network and it is related to the admittance. In the discussion to follow, the circuit element of interest is always given the subscript zero. A generalized three-element network is shown in Fig. 3.5$a$, where we wish to know the current flow through $Z_o$ as a result of applying $E_1$ to $Z_1$. The total current in the network is given by

$$\overline{I}_t = \frac{\overline{E}_1}{\overline{Z}_t} \tag{63}$$

and from Kirchhoff's first law we know that

$$\overline{I}_t = \overline{I}_0 + \overline{I}_2 = \frac{\overline{E}_o}{\overline{Z}_{o2}} \tag{64}$$

where $\overline{Z}_{o2}$ is the parallel combination of $\overline{Z}_o$ and $\overline{Z}_2$. Combining these two equations gives

$$\frac{\overline{E}_1}{\overline{Z}_t} = \frac{\overline{E}_o}{\overline{Z}_{o2}} \tag{65}$$

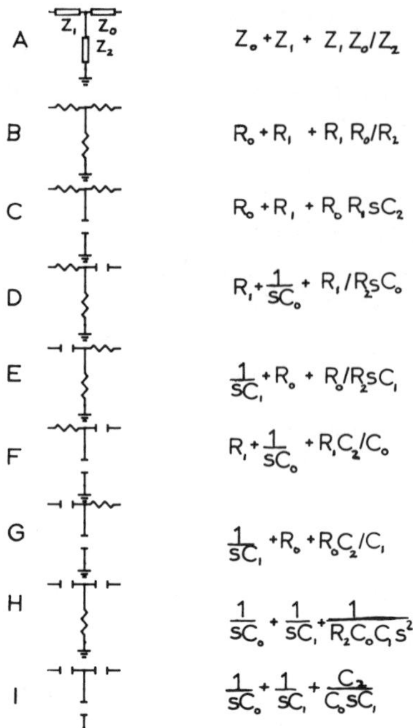


Fig. 3.5. Various cases of $t$ circuits, ($a$) being the general case and the rest showing how well they conform to the Laplace plane transformation.

from which we can write

$$\overline{I}_o = \frac{\overline{E}_o}{\overline{Z}_o} = \frac{\overline{E}_1\overline{Z}_{o2}}{\overline{Z}_t\overline{Z}_o} \tag{66}$$

The transfer impedance $Z_e$

$$\overline{Z}_e = \frac{\overline{E}_1}{\overline{I}_o} = \frac{\overline{Z}_t\overline{Z}_o}{\overline{Z}_{o2}} = \frac{(\overline{Z}_1 + \overline{Z}_{o2})\overline{Z}_o}{\overline{Z}_{o2}} = \overline{Z}_o\left(\frac{1 + \overline{Z}_1}{\overline{Z}_{o2}}\right) \tag{67}$$

$$= \overline{Z}_o + \overline{Z}_1 + \frac{\overline{Z}_1\overline{Z}_o}{\overline{Z}_2} \tag{68}$$

The transfer impedances for eight common networks are shown in Fig. 3.5.

## 3. System Responses for Simple Networks

The simple series $RC$ network in Fig. 3.5$a$ can be excited by a potential step of amplitude $\Delta E$ volts and the resulting current flow monitored. The image plane form of the response is

$$\overline{I} = \overline{E}\,\overline{Y} = \frac{\Delta E}{s}\left[\frac{s}{R(s + 1/RC)}\right] = \frac{\Delta E}{R(s + 1/RC)} \tag{69}$$

Upon return to the time plane the current response can be seen from Fig. 3.5*b* to be

$$I = \left(\frac{\Delta E}{R}\right) e^{-t/RC} \tag{70}$$

We can similarly investigate the response to a double potential step excitation, that is, one that stays at $\Delta E$ volts for $b$ sec and then returns to 0 V. In image space this is

$$\bar{I} = \bar{E}\bar{Y} = \frac{\Delta E}{s}(1 - e^{-sb}) \frac{s}{R(s + 1/RC)} \tag{71}$$

Upon return to the time plane this leads to

$$I = \frac{\Delta E}{R} e^{-t/RC} - U(t - b)e^{-(t-b)/RC} \tag{72}$$

It is common to see the $U(t - b)$ term omitted in the representation, the unit step function being assumed in the $e^{-(t-b)/RC}$ representation. Both responses are shown in Fig. 3.6 where the dashed continuation line shows the continued decay of the single step response.

As a second example let us consider the response to the application of a potential ramp of $\Delta E$ volts/sec to the circuit in Fig. 3.7, a resistor in series with

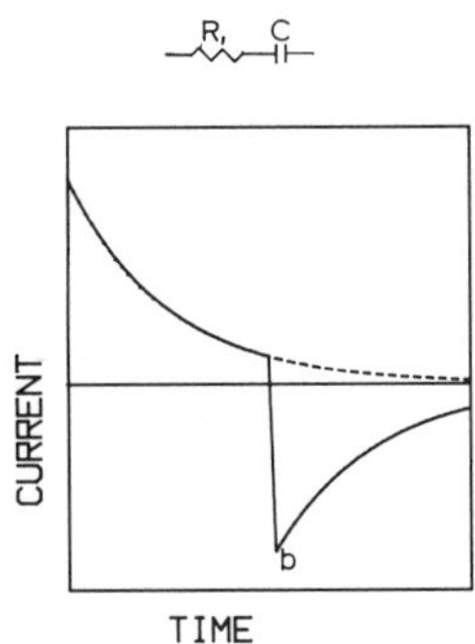


**Fig. 3.6.** Current response of a simple series *RC* network to a potential step excitation and double potential excitation, reversing at time *b*.

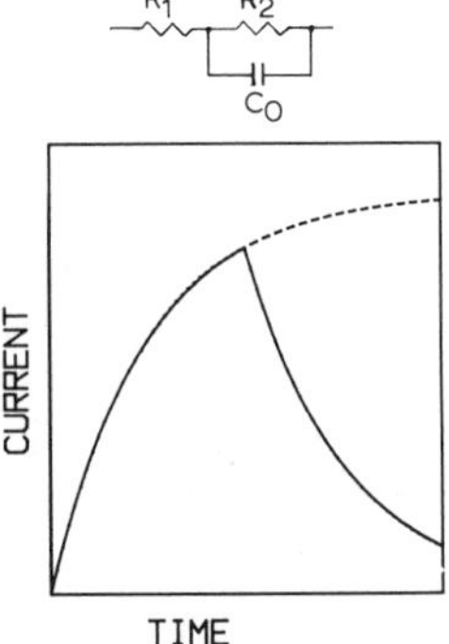


**Fig. 3.7.** Current response to a three-element network to a ramp excitation. The dashed line represents a continuous ramp; the solid line is the response to a ramp that holds at a constant potential at time *b*.

a parallel $R_2 - C_0$ network and determine the current flow through the capacitor.

$$\bar{I} = \overline{EY} = \frac{\Delta E}{s_2}\left[\frac{s}{R_1(s+a)}\right] = \frac{\Delta E}{R_1 s(s+a)} \quad a = \frac{R_1 + R_2}{R_1 R_2 C_0} \tag{73}$$

All that remains is to return to the time plane and there are numerous choices open to us. In order to demonstrate some of the flexibility of this method we examine several of the routes. In all these methods we deal only with the section of the image function that contains the Laplace variables, that is, the $1/[s(s+a)]$ portion of the image function.

### 4. Integration Theorem

The relationship

$$\frac{1}{s}F(s) = \int_0^t f(\theta)\, d\theta \tag{74}$$

has been invoked previously and in this case leads to

$$\frac{1}{s(s+a)} = \int_0^t e^{-e\theta}\, d\theta = \frac{1}{a}(1 - e^{-at}) \tag{75}$$

Factoring

$$\frac{1}{s(s+a)} = \frac{1}{a}\left[\frac{1}{s} - \frac{1}{s+a}\right] = \frac{1}{a}(1 - e^{-at}) \tag{76}$$

In both these cases we have used our knowledge of the transform pair

$$\frac{1}{s+a} = e^{-at} \tag{77}$$

which we previously derived. Even this relationship is derivable from one of the basic theorems of Laplace transforms, the translation of result function theorem, which asserts that for any object-image function pair, there exists the auxiliary relationship

$$e^{-at}f(t) = F(s+a) \tag{78}$$

### 5. Transformation Tables

One can also take the easy way out and search through the available transform pairs given in Table 3.I to find the object function. In this case the relationship is found in entry 3. It is advisable for the reader to try to reduce the transforms encountered by one of these methods and then to check the table to gain facility with the techniques, rather than just turning to the table for the solution. One eventually encounters systems that are not directly represented in the table but that can be reduced to available representations by these simple techniques.

All the approaches lead, naturally enough to the same result, which, when one substitutes the value for $a$, yields

$$I(t) = \frac{\Delta E \ R_1 R_2 C}{R_1(R_1 + R_2)}(1 - e^{-at}) = \frac{\Delta E R_2 C}{(R_1 + R_2)}\left[1 - \exp\left(\frac{-t(R_1 + R_2)}{R_1 R_2 C}\right)\right] \quad (79)$$

## 6. Extensions to Include Ideal Operational Amplifiers

We can use the methodology developed to this point to treat OA circuits. The only change in the formalism adopted at the beginning of the chapter is to sum the currents at the circuit nodes in transform space rather than in time space. This is perfectly valid, since we have already seen that the transform of the sum of a series of functions is the sum of the transforms of the individual functions. We are now in a position to return our attention to the circuit that proved troublesome previously. If the method developed so far is correct, we should also be able to deduce the same answers previously obtained for the circuits which we could treat by the simplified approach.

Returning to the circuit in Fig. 3.2$d$, we can write the image functions as

$$I(s) = \frac{E(s)}{R_0} + \frac{E(s)}{1/sC_1} = 0 \quad (80)$$

$$E_0(s) = -sC_1 R_0 E_1(s) \quad (81)$$

The differential relationship immediately allows us to write the time plane or object function as

$$E_0(t) = \frac{-R_0 C_1 dE_1}{dt} \quad (82)$$

which is the result previously derived. Similarly for the circuit in Fig. 3.2$c$ one has

$$I(s) = \frac{E_1(s)}{R_1} + \frac{E_0(s)}{1/sC_0} = 0 \quad (83)$$

$$E_0(s) = \left(\frac{1}{sC_0 R_1}\right)E_1(s) \quad (84)$$

Examining Table 3.I leads to

$$E_0(t) = \frac{-1}{R_1 C_0}\int_0^t E_1(t)\, dt \quad (85)$$

which is consistent with our previous result. Now let us consider the intractable case that was temporarily abandoned before. The circuit in Fig. 3.2$f$ can now be treated if we remember that admittances in parallel are linearly additive so that

$$I(s) = \frac{E_1(s)}{R_1} + \frac{E_0(s)}{R_0} + \frac{E_0(s)}{1/sC_0} = 0 \quad (86)$$

$$E_0(s) = \frac{-E_0(s)}{R_1 C_0(s + 1/R_0 C_0)} \quad (87)$$

We can now determine the response to any excitation we desire. If the form of $E_1(t)$ is a step of height $E$, then the response is

$$E_0(s) = \frac{-E/R_1C_0}{s(s+a)}, \qquad a = \frac{1}{R_0C_0} \tag{88}$$

We have already seen that the time plane form of $1/s(s+a)$ is $(1 - e^{-at})/a$ so that we may immediately write

$$E_0(t) = E\left(\frac{R_0}{R_1}\right)(1 - e^{-t/R_0C_0}) \tag{89}$$

With this background we are now prepared to attack any circuit and excitation signal combination that we may encounter, but there are still several features that we should investigate because they can make our lives simpler.

## 7. Special Input Functions

In treating analog signal processing there are three functions that have special properties. They are the unit impulse function $\delta(t)$, the unit step function $U(t)$, and the sine wave.

### a. The Unit Impulse Function

The unit impulse function $\delta(t)$ is defined as a function that has the value of zero for all values of time prior to $t = 0_+$ and zero for all times greater than $t = 0_{+}$, but it has the value of infinity at $t = 0_{+}$. Further, its integral is defined as unity:

$$\int_0^\infty \delta(t) = 1 \tag{90}$$

The Laplace transform of this peculiar function is

$$\mathcal{L}f(t) = \int_0^\infty e^{-st}\,\delta(t) = 1 \tag{91}$$

If one excites a system $H(t)$ with an excitation $X(t)$ which happens to be a unit impulse function, then the response function $R(t)$ is given in Laplace space by

$$R(s) = H(s)*X(s) = H(s)*1 = H(s) \tag{92}$$

or the response to an impulse excitation is the system itself. It is not possible to generate an ideal unit impulse function in the laboratory (although with high rise time pulse generators one can generate a respectable approximation to one), but it is a convenient fiction that is sometimes employed in discussing circuits and the application of an approximation sometime gives valuable insight into the properties of a circuit.

### b. The Unit Step Function

We have already employed the unit step function as an excitation in a number of examples, and we have employed the concept of the delay function to generate double steps, staircase, and square wave functions from the simple unit step function. Actually these applications are a special case of a more general property. We can generate *any* function by taking the proper combination of delayed unit step functions. For example, the ramp function is the limiting case of the staircase function when the steps become infinitely small and infinitely closely spaced together. Normally one can construct a signal more expeditiously using a combination of functions, but in principle the unit step function can always be employed to build the desired signal. This is true whether the desired signal is periodic or aperiodic, but for the periodic signal class the sinusoid is preferable as a generating function.

### c. Sinusoids and the Relationship between the Fourier and the Laplace Transforms

Any periodic function $f(t)$ can be expressed as the sum of a series of sine and cosine waves in the region $-T/2 < t < T/2$ by the series

$$f(t) = \tfrac{1}{2}a_o + \sum_{k=1} [a_k \cos(kt) + b_k \sin(kt)] = \tfrac{1}{2}a_o + \sum_{k=1} c_k \sin(kt + \phi_k) = \sum_{k=o} d_k e^{ikt} \quad (93)$$

which is known as the *Fourier series* representation of a periodic function (8). We can recast this expression into frequency notation using $\omega_0 = 2\pi/T$, where $T$ is the period of the basic function

$$f(t) = \frac{1}{2}a_o + \sum_{n=1} c_k \sin(n\omega_o + \phi_k) = \sum_{k=0} d_k e^{j\omega_o t} \quad (94)$$

From this expression it is apparent that we can predict the response of a circuit to any periodic excitation if we know two things: (1) the response of the circuit to a sinusoidal excitation and (2) the values of the constants $a_o$, $c_k$, and $\phi_k$. The constants for a number of periodic functions have been tabulated (9,10) and the procedures for treating untabulated cases are not particularly difficult (11), so that we could in principle use this procedure to treat periodic functions. There is a more fundamental relationship contained in these expressions which is the real reason for introducing the subject. We can go from the series representation of the exponential form of the Fourier series to an integral representation which is known as the Fourier transform of the function:

$$F(t) = \int_{-\infty}^{\infty} e^{-jtu} F(u)\, du \quad (95)$$

If we now restrict ourselves to functions that have the value of zero at times less than $t = 0_+$, as is the case for Laplace transforms,

$$F(t) = \begin{cases} e^{-xt}\Phi(t) & t > 0 \\ 0 & t < 0 \end{cases} \quad (96)$$

and then take the Fourier transform as before, we find that

$$F(t) = \int_0^{\infty} e^{-(x+jy)t}\Phi(t)\, dt = \int_0^{\infty} e^{-st}\Phi(t)\, dt = \int_0^{\infty} e^{xt} e^{-st} F(t)\, dt \quad (97)$$

From these relationships we can make the following observations: (1) it is possible to interconvert between Fourier and Laplace transformations; (2) because the Laplace transform can be expressed in terms of frequencies and phase angles, then any function that can be Laplace transformed can be thought of as being composed of a manifold of sine waves of varying frequencies and phase angles; and (3) by expressing the Laplace variable in terms of the complex variable $(a + j\omega_o)$, we can decompose the response of any circuit to a sinusoidal excitation into two components, a transient or start-up response (corresponding to the $e^{xt}$ term) and a steady-state response.

We can effectively demonstrate these features by applying a sinusoidal excitation to the circuit in Fig. 3.2*f* and examining the response. The transform of the circuit is given by

$$E_0(s) = \frac{E_1(s)}{R_1C(s + 1/R_0C)} \tag{98}$$

If we apply an input signal $E_1(t) = \Delta E \sin(\omega t)$ to the amplifier, then in image space

$$E_0(s) = \frac{\Delta E}{R_1C(s + a)(s^2 + \omega^2)}, \qquad a = \frac{1}{R_oC} \tag{99}$$

Resorting to the table of transform pairs we find that

$$E_o(t) = \frac{-\Delta E}{R_1C}\left(\frac{e^{-at}}{(a^2 + \omega^2)} + \frac{\sin(\omega t + \theta)}{a^2 + \omega^2}\right), \; \theta = -\tan^{-1}\left(\frac{\omega}{a}\right) \tag{100}$$

where the time plane solution is composed of a transient response (the portion containing the $e^{-at}$ term) and a steady-state response (the portion containing the sinusoidal component). In view of our previous discussion, it can be shown that one can obtain the steady-state response, that is, the response after any start-up transients have died away, from the transform by the following procedure:

1. Obtain the Laplace transform of the system.
2. Replace $s$ with $j\omega$.
3. Multiply the new form of the system transform by $E \sin(\omega t)$.
4. Treat the result as a rotating vector representation where terms multiplied by $j$ are considered to represent components 90° ahead of the excitation signal and real terms are considered in phase with the excitation signal.

Let us consider our system using this approach:

$$E_0(\omega t) = \frac{\Delta E \sin(\omega t)}{R_1C}\left(\frac{1}{j\omega + a}\right) \tag{101}$$

We now multiply numerator and denominator by $j\omega - a$, the complex conjugate of the denominator, in order to bring all $j$ terms into the numerator. For the time being we only carry through the "business end" of the equation.

$$\frac{1}{j\omega + a} = \frac{1}{j\omega + a}\frac{j\omega - a}{j\omega - a} = \frac{j\omega - a}{(j\omega)^2 - a^2} = \frac{a - j\omega}{a^2 + \omega^2} \tag{102}$$

We next plot the $a - j$ vectors and determine the phase difference $\theta$ and the vector length:

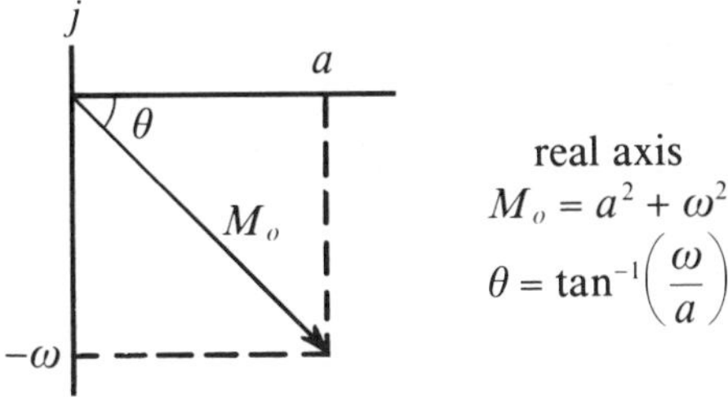


Combining the information, we have

$$E_0(\omega t) = \frac{\Delta E \sin(\omega t + \theta)}{R_1 C(\omega^2 + a^2)} \tag{103}$$

which is precisely the steady-state result obtained from the Laplace transform method. What we have just done is to revert to a simpler notation scheme where the impedance of a capacitor to a sinusoidal excitation is $1/j\omega C$, and that of a resistor is again simply $R$ and for an inductor $j\omega L$. This method is the one normally found in undergraduate physics texts.

## IV. BODE PLOTS OF OPERATIONAL AMPLIFIERS

Now that we know the frequency response of the circuit, we can gain insight into its behavior under all possible conditions. The simplest way to evaluate such a circuit is to construct a plot of log (gain) versus log (frequency), as shown in Fig. 3.8, which is known as a Bode gain plot or simply a Bode plot (12). As we use it here, gain is defined as

$$G = \frac{1}{R_1 C_1(\omega^2 + a^2)} \tag{104}$$

That is, we consider only the magnitude and not the sign of the amplification. Normally one also displays the phase shift versus log(frequency), as has been done in Fig. 3.8. Inspection of this plot shows that the gain at low frequencies is a constant ($R_0/R_1$) and in this region the circuit acts as a simple inverter. At high frequencies the gain decreases with a slope of $-1$ in log-log space. The breakover point occurs at a frequency where the capacitor $C_0$ and the feedback resistor $R_0$ have the same impedance, that is, $f = 1/2\pi R_0 C_0$. This circuit attenuates high-frequency signals and allows low-frequency signals to pass through at constant gain. The frequency where the breakover from one mode of operation to other occurs is called the cutoff frequency and the circuit is known as a low-pass filter. The region of constant gain is called the pass band and the region of attenuation, the stop band. Examining the phase angle plot, we see that the phase varies smoothly from 180° at low frequencies to 90° at high frequencies. At first glance it may seem that the limits should be 0° and −90°, but we must

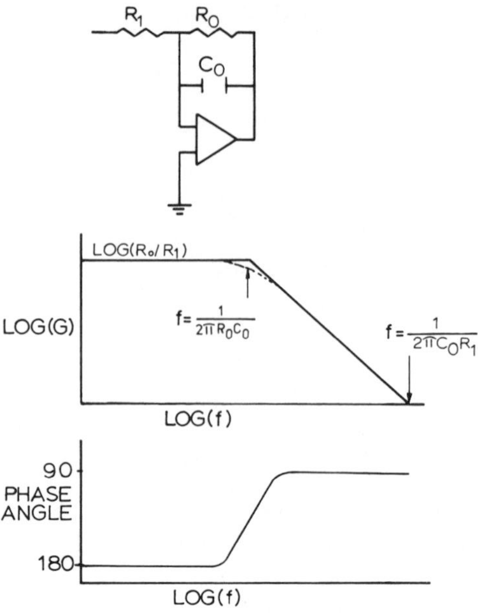


**Fig. 3.8.** Bode Plots of a low-pass filter. Top, log (gain) versus log (frequency). The dashed line represents the true response and the straight lines represent the approximations. Bottom, phase angle versus log (frequency).

remember the sign inversion of the circuit which provides an initial 180° phase shift.

Let us consider the integrator circuit of Fig. 3.2*d* whose log($G$) plot is shown in Fig. 3.9. The log($G$) plot decreases linearly at higher frequencies, having a potentially infinite gain at 0 Hz and zero gain and an infinite frequency. By comparing this plot with that of the previous circuit, we can see that at high frequencies, it was acting as an integrator. For this reason that circuit is sometimes spoken of as an AC integrator. The differentiator circuit of Fig. 3.2*e* is also treated similarly in Fig. 3.9. Here the gain increases linearly in log-log space as one goes to higher frequencies, and in fact the analysis indicated infinite gain at an infinite frequency. Such a circuit would be unstable in practice because any small high-frequency component would be amplified to such an extent that the circuit would oscillate, that is, swing from one bound to the other. A more practical circuit for differentiation is that in Fig. 3.10, which we analyze next.

The Laplace representation of the system function, that is, the transform of the circuit itself, is given by

$$X(s) = \frac{-sR_0}{R_1(s + 1/R_1C_1)} \tag{105}$$

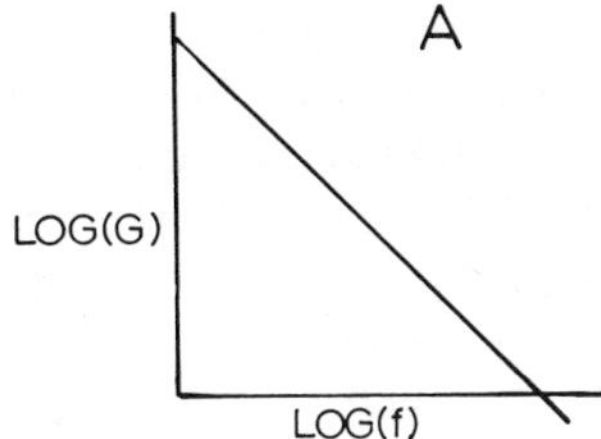


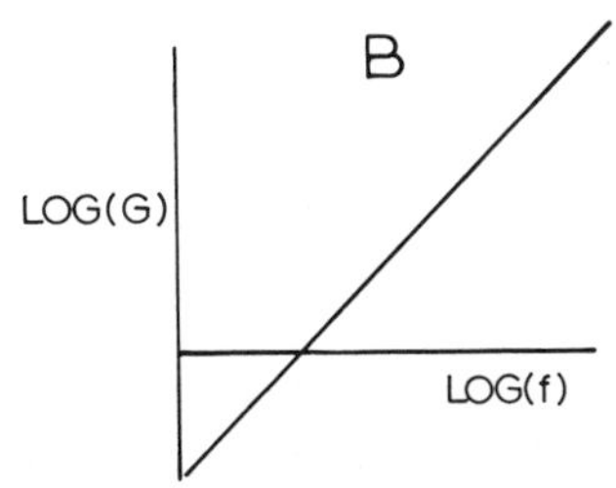


**Fig. 3.9.** (*a*) Bode plot of an ideal integrator. (*b*) Bode plot of an ideal differentiator.

Proceeding as before,

$$X(\omega t) = \left(\frac{-R_0}{R_1}\right)\frac{j\omega}{j\omega + a}, \; a = \frac{1}{R_1 C_1} \tag{106}$$

$$\frac{-R_1 X(\omega t)}{R_0} = \frac{j\omega}{j\omega + a}\frac{j\omega - a}{j\omega - a} = \frac{\omega + ja}{a^2 + \omega^2} \tag{107}$$

$$M_o = \frac{1}{\sqrt{a^2 + \omega^2}} \tag{108}$$

and the output becomes

$$E_0(t) = \frac{R_o}{R_1}\frac{\Delta E \sin(\omega t + \Theta)}{\sqrt{a^2 + \omega^2}} \; \Theta = \tan^{-1}\left(\frac{a}{\omega}\right) \tag{109}$$

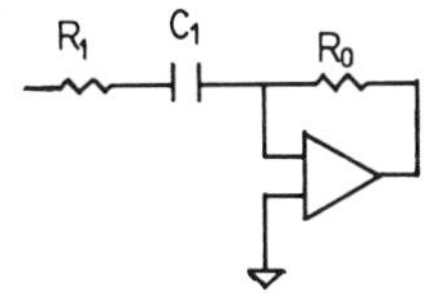


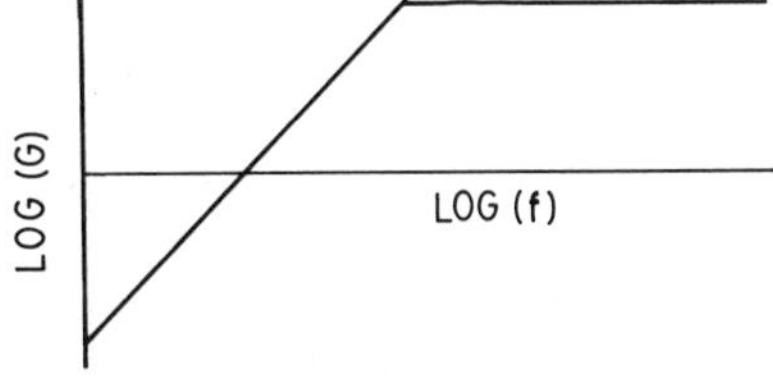


**Fig. 3.10.** Bode plot behavior for a practical differentiator (i.e., high-pass filter).

The log($G$) versus log($f$) plot of this circuit shown in Fig. 3.10 differs from that of the ideal differentiator in that at high frequencies the gain limits at a value of $R_0/R_1$, whereas at frequencies below the cutoff frequency the circuit acts as an ideal differentiator with a slope of +1 in log-log space. The cutoff frequency is again the frequency where the resistor and capacitor have the same impedance; in this case $C_1$ and $R_1$ are the dominant terms and $f = 1/2\pi R_1C_1$. The frequency where the gain of the circuit is unity is called the unity gain crossover frequency. For the differentiator, this is the frequency where the feedback resistor $R_0$ has the same impedance as the input capacitor, that is, $f_1 = 1/2\pi R_0C_1$.

## V. VERIFICATION OF CIRCUIT ANALYSIS

One can always determine the behavior of a circuit by carrying out these mathematical manipulations and evaluating the resulting mathematical expression. As the complexity of the circuit increases, there is increasing chance for errors to occur. Although there is no perfect way to verify that no errors have occurred in a derivation, there are two internal checks that can be performed which will detect most errors. If either fails, certainly an error has occurred.

### 1. Dimensionality Check

As in chemical equations, a dimension check is always advisable. The basic ground rules are as follows:

1. Each term in a summation must have the same dimensionality.
2. The dimensionality of both sides of an equation must be the same.
3. The arguments of any of these functions (the logarithm, exponential, trigonometric functions, gamma functions, Bessel functions, etc.) must be dimensionless.
4. The quantities $1/\omega$, $1/f$, $1/s$, $RC$, $L/R$, and $LC$ all have the dimensions of time. In conjunction with point 3 this implies that $s/RC$ or $t/RC$ is a valid argument for functions but $RC/f$ is erroneous.

A failure of this test normally indicates a mathematical error in the derivation.

### 2. LIMITING BEHAVIOR

Examination of a solution under certain special conditions not only serves as a test of the plausibility of a solution but also leads to a better intuitive feeling for the properties of the network or circuit. The solution generally simplifies to some predictable form when these special conditions are introduced. The most common special conditions are as follows:

1. *Special time*. Choose time equal to zero or infinity or some other case suggested by the form of the equations

2. *Specific frequency*. Choose $f$ equal to zero or infinity or some other value suggested by the form of the equations. For example, a capacitor acts as a short circuit at infinite frequency and as an open circuit at zero frequency, whereas an inductor acts as a short circuit at zero frequency and as an open circuit at high frequency.

3. *Specific component values*. High resistance, low capacitance, or high inductance acts as as an open circuit. Low resistance, high capacitance, or low inductance acts as a short circuit.

Failure of these tests can indicate either a manipulation error as in the case of the dimensionality test or, more seriously, an error in the initial model. As an example of the utility of the limiting behavior test, let us consider the behavior of the low-pass filter circuit (Fig. 3.8). As we let the value of $C_0$ get smaller and smaller, the circuit should approach a simple inverter. Similarly, as the frequency becomes lower and lower, the circuit should also approach the inverter as a limiting case. As the feedback resistor becomes larger and larger or the frequency higher and higher, the circuit should act as a simple integrator. Of course, this is just the limiting behavior that our model predicted.

The high-pass filter (Fig. 3.10) can also be treated in the same way. In this case at very high frequencies or large capacitance values the capacitor, $C_1$, can be treated as a short circuit and the system becomes a simple inverter while at low frequencies. With small capacitance values or low frequencies, the circuit becomes an ideal differentiator. Once again, this is the behavior predicted by the rigorous model.

The limiting behavior can be used not only to verify a rigorous solution, but also it often allows one to construct a log(gain)-log (frequency) plot without actually carrying out the detailed derivation. As an example of this approach, consider the circuit in Fig. 3.11. This circuit has some of the features of both the high-pass and low-pass filter and is known as a *band-pass* filter. If the frequency is high enough, both $C_1$ and $C_0$ act as very low impedances. This means that the input capacitance $C_1$ is negligible in comparison to the frequency independent input resistance $R_1$. The feedback capacitance $C_0$ is also small in comparison to the frequency-independent resistance $R_0$ so that $C_0$ dominates the high-frequency response.

For parallel $RC$ assemblages, the capacitance dominates the effective impedance at high frequencies whereas the resistance dominates at low frequencies. For series $RC$ assemblages, the capacitor dominates the low-frequency effective impedance, whereas the resistor dominates at high frequencies. If one chooses extreme enough frequencies only the dominant term need be considered. This poses the obvious question, what constitutes an extreme enough frequency? We can answer this by noting that for either parallel or series $RC$ pairs the equivalence frequency $f_e = 1/2\pi RC$, that is, the frequency at which the two components have the same impedance. The impedance of a capacitor decreases as the reciprocal of the frequency, so that at a frequency 10 times $f_e$, the capacitor represents an impedance that is either 10 times larger (lower frequency) or 10 times less (higher frequency) that the paired resistor. For graphic evaluation of a circuits performance, a factor of 10 is generally adequate.

As applied to our present example, there are four frequencies of interest:

1. The frequency where the series input components have the same impedance; $f_1 = 1/2\pi R_1 C_1$.

2. The frequency where the feedback parallel components are equivalent, $f_2 = 1/2\pi R_0 C_0$.

3. The frequency where the feedback resistor $R_0$ and the input capacitor $C_1$ have equal impedance, $f_3 = 1/2\pi R_0 C_1$.

4. The frequency where the feedback capacitor $C_0$ and input resistor $R_1$ are equivalent, $f_4 = 1/2\pi R_1 C_0$.

One can construct a sketch of the $\log(G) - \log(f)$ plot from these four frequencies for this circuit by noting that a high frequencies it should act as an integrator that has unity gain, that is, crosses the $\log(G) = 0$ axis, at $f_4$ and the log-log slope is $-1$. At low frequencies it acts as a differentiator with slope $+1$ and crosses the unity gain axis at $f_3$. Between the frequencies $f_1$ and $f_2$ the circuit has a gain $R_0/R_1$. Fig. 3.11 shows five possible shapes for this plot, in case the gain never even reaches the unity gain axis, let alone achieving the predicted gain $R_0/R_1$, but by using these special frequencies and using slopes of $+1$ or $-1$ the plot can still be constructed. The rigorous solution would be essentially

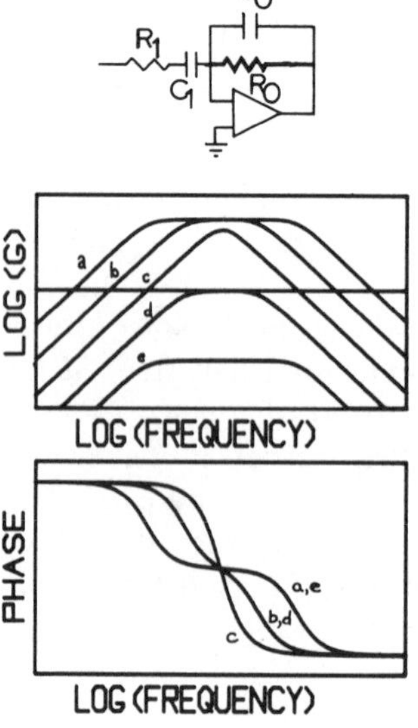


**Fig. 3.11.** Effect of relative component values on the Bode responses of a band-pass filter.

*a*. $R_0 = 10R_1$ where the *RC* pairs are chosen to give a wide band

*b*. $R_0 = 10R_1$ where the *RC* pairs are chosen to give a narrow band pass

*c*. $R_0 = 10R_1$ where $R_0C_0 = R_1C_1$, giving no pass band

*d*. $R_0 = R_1$ where the *RC* pairs are the same values as in *b*

*e*. $R_0 = 1/10R_1$ where the *RC* pairs are the same values as in *a*

indistinguishable from the approximations on the scale of the figures reproduced here. The points of maximum deviation occur at the intersections of the straight line segments and the rigorous solution is one-half of the approximation, that is, log(0.5) = −0.3 units low. In any case, when one is plotting curves of this sort, one must remember that the resistances are expressed in ohms and the capacitances in *farads* (*not microfarads*).

It is important to realize that although not all circuits are amenable to this approximate treatment, all can yield important information, when limiting behavior is examined. As an example, consider the circuit in Fig. 3.12. In this case our plotting procedure is inappropriate, but if we examine the high-frequency limit, it is apparent that the capacitance can be considered a short circuit. In this case the circuit acts as a unity gain follower. At very low frequencies the capacitor acts as an open circuit and the system reduces to an inverting amplifier with unity gain. At both frequency extremes the absolute value of the gain is unity but the sign is inverted at low frequencies. This means that the phase shift is 0° at high frequencies and 180° at very low frequencies.

We can examine how the gain and phase change in the intermediate region by applying the rigorous approach. The total current through the *RC* leg is, in image space,

$$\overline{I}_2 = \overline{E}_1/\left(R_2 + \frac{1}{sC}\right) = \frac{\overline{E}_1 sC}{R_2Cs + 1} = \frac{s\overline{E}_1}{R_2(s + a)}, \quad a = \frac{1}{R_2C} \tag{110}$$

and the potential drop across the resistance is

$$\overline{E}_a = \overline{I}R_2 = s\overline{E}_1\frac{R_2/R_2}{s + a} = \frac{\overline{E}_1 s}{s + a} \tag{111}$$

Summing the currents at the negative input gives

$$\frac{\overline{E}_1 - \overline{E}_a}{R_1} + \frac{(\overline{E}_0 - \overline{E}_a)}{R_1} = 0 \tag{112}$$

$$\overline{E}_0 = 2\overline{E}_a - \overline{E}_1 = \frac{\overline{E}_1(s - a)}{s + a} \tag{113}$$

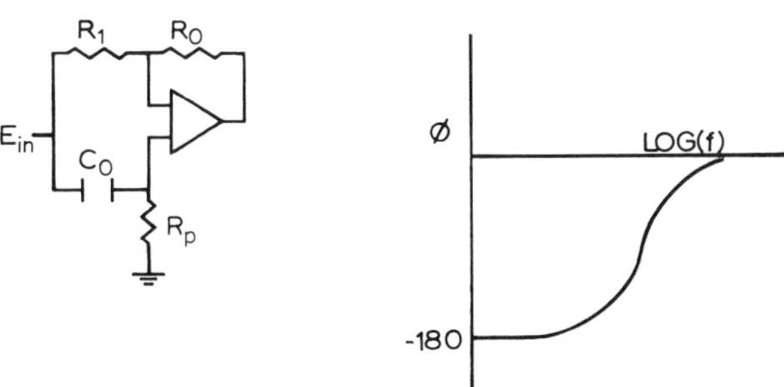


**Fig. 3.12** A simple phase shifter: when the resistance is varied at the positive input ($R_p$) the phase changes at the output.

Limiting our consideration to sinusoidal excitations and making the substitutions $s = j\omega$ leads to the following result:

$$E_0 = E_1 \sin(\omega t) + \tan^{-1}\left(\frac{2a}{\omega^2 - a^2}\right) = E_1 \sin\left(\omega t + \tan^{-1}\left(\frac{2RC}{(\omega RC)^2 - 1}\right)\right) \quad (114)$$

From this derivation, it can be seen that the absolute value of the circuit is unity irrespective of the frequency or the values of $R_2$ and $C_2$ but the phase angle varies continuously from 0° to almost −180°. One practical application of this type of circuit is to provide variable phase shift to a signal but to keep the gain constant by making $R_2$ a variable resistance. This is a very useful circuit for changing the phase relationship of a sinusoidal signal which has a fixed frequency and it finds application in a number of instruments where a fixed frequency signal is present, with NMR being a classic example. Note that the circuit also changes phase as the frequency changes and although zero phase shift is always possible with this type of circuit, that is, $R_2 = 0\ \Omega$, the degree to which one can approach −180° depends on the frequency and the actual values of $R_2$ and $C$. For example, if $2fR_2C = 100$, then $\Theta = -178.9°$.

The circuit shown in Fig. 3.13 is another one that does not immediately yield to inspection, but must be considered in detail before all its properties are fully appreciated (with experience, however, one finds that one can approximate even its properties). We can see that the input network is that found in Fig. 3.5*c*, whereas the feedback network $\overline{Z}_o$ is found in Fig. 3.5*h*. Using the transforms previously developed, we can write the image space response of the circuit to an object plane excitation signal $\overline{E}_1$:

$$\overline{Z}_1 = R_1 R_2 C_1\left(s + \frac{R_1 + R_2}{R_1 R_2 C_1}\right) \quad (115)$$

$$\overline{Z}_0 = \frac{C_2 + C_3}{s^2 C_2 C_3}\left(s + \frac{1}{R_3(C_2 + C_3)}\right) \quad (116)$$

$$\overline{E}_0 = -\overline{E}_1 \frac{(C_2 + C_3)(s + a)}{C_1 C_2 C_3 R_2 s^2 (s + b)} \quad (117)$$

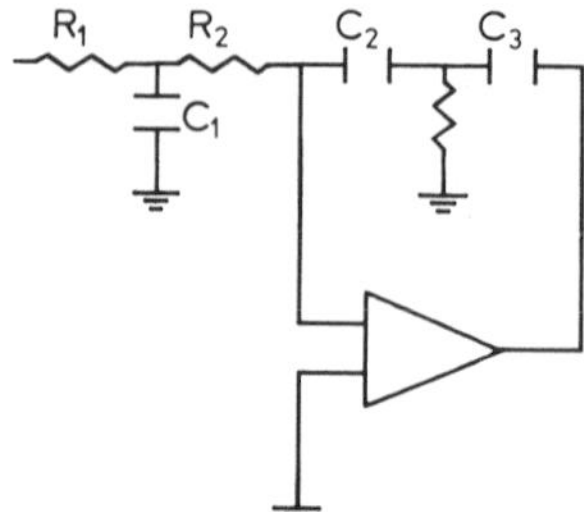


**Fig. 3.13.** A double integrator: a simple analog circuit that allows one to solve very complicated double integrations by simply entering the complicated function and monitoring the output for the numerical solution.

A very special case for this circuit exists when the component values are chosen such that $a = b$; then the response simplifies to

$$\overline{E}_0 = \frac{\overline{E}_1}{s^2} \frac{(C_2 + C_3)}{C_1 C_2 C_3 R_1 R_2} \tag{118}$$

which is the transform for a double integrator circuit with gain $(C_1 + C_2)/C_1 C_2 C_3 R_1 R_2$.

## VI. NONIDEAL OPERATIONAL AMPLIFIERS

Up to this point we have treated only ideal operational amplifiers; now we must begin to deal with real-world devices. Some of the problems are minor and easy to treat (although they may place undesirable restrictions on the application of operational amplifiers to real systems) whereas others are fairly complex.

To this point our Maxwell demon has been a very cooperative beast indeed, with the only limitation placed upon the circuit being that the output voltage could not exceed the power supply voltages. Typical real OAs are only a little less cooperative than this, with the true limits being restricted to roughly 1.5 V less than the power supply voltages. Unfortunately, there *are* restrictions upon the power supply voltages. Most typical OAs are designed to operate from ±12 to ±18 V supplies, with optimal performance being obtained at ±15 v. Some newer amplifiers are designed to operate successfully on ±5 v supplies to ease their use in digital systems; and a new class designed for battery operation in field instruments works with ±1.5 v supplies. These allow the output voltage swings to within 0.3 V of the supplies. The higher voltage end of the spectra is less well represented. A few amplifiers are designed to operate from ±36 V and some can be found that will handle ±150 V power supplies, but the vast majority of amplifiers are designed for ±15 V operation.

We have also assumed that the amplifier can supply an effectively infinite amount of current from its output, but unfortunately this is not the case. The vast majority of OAs are limited to between ±5 and ±10 mA output current. In terms of current handling ability, there are two terms that are sometimes encountered which the reader should be aware of. These are *sink* and *source*. *Sink* refers to an amplifier's ability to take current from an external network, and *source* refers to its ability to supply current to the external circuitry. There are some amplifiers that can handle significantly greater currents. One can find a number of amplifiers that can handle 100 mA currents, including some that operate from ±150 V supplies. A few amplifiers can also handle up to 1 A of output current but the vase majority of these are designed ±15 V operation, although there is at least one that can handle this current from ±36 V supplies. Unfortunately, most of the very high current (above 100 mA) amplifiers are distinctly nonideal in many other important features. The overall capability of an amplifier to drive loads is spoken of as the *compliance* of the amplifier, although often this term is used only with regard to its current capabilities.

The practical implications of finite compliance are that one must practice restraint in the choice of components used in a circuit containing an amplifier and in the kind of load that its is expected to drive. One can not, for example, expect an OA that has a maximum compliance of ±13 V at 10 mA to drive a 12 V dc motor that has an impedance of 10 Ω. Analogously one should not use 10 Ω resistors in a circuit when all one is trying to do is construct a unity gain inverter amplifier. In general, one should not go below $2K\Omega$ for most applications since a 10 V drop across this impedance represents 5 mA, which is from 50 to 90% of the current capacity of a typical amplifier, leaving very little capability for driving other components. One should not go overboard on the other extreme. It is quite difficult and expensive to find resistors with greater than 22 MΩ impedance and even when they are found, one encounters other difficulties. A real resistor actually can best be considered as an ideal resistor in series with an inductor (representing the lead inductance primarily) with a capacitor in parallel with these two components (the capacitance results from the practical problems of resistor construction). If one is working with wire-wound resistors the inductance can be quite high and the ratio of inductance to resistance is fairly constant. For non-wire-wound resistances the inductance is essentially independent of the resistance. These features are considered in some detail later, but the important feature here is that as the resistance increases the inductance can become more and more important. The prohibition against the use of excessively large resistances also arises from some of the other nonideal properties of OAs.

## A. INPUT PROPERTIES

We have been assuming that our friendly Maxwell demon has been working with an ideal voltmeter, that is, that no current was required to monitor the potential difference between the two inputs and that the meter was correctly calibrated. Neither supposition is true for real amplifiers.

The simplest problem to treat conceptually is the problem of offset voltage. Simply stated, the voltmeter is not correctly zeroed so that zero potential difference is registered when some difference, in fact, still exists. This is spoken of as the offset voltage of the amplifier and can range from several tens of millivolts to fractional microvolts depending upon the type of amplifier under discussion. Essentially, all operational amplifiers have provisions for nulling out this offset voltage (i.e., zeroing the demon's meter) but there are several points to bear in mind. First, it is going to be very difficult to null out an offset to better than 0.1% of its initial value with real components. Second, voltage offset drifts with both temperature and time. The temperature drift results from changing temperature properties of the elements that are used to construct the amplifier itself and from the temperature properties of the nulling components. The specification sheet of an amplifier specifies the temperature drift of the offset voltage, which may range from 0.1 to $50\mu$V per degree centigrade, but the temperature drift of the nulling components is not mentioned there. Let us see what effect offset voltage has on some simple circuits.

If a voltage follower has an offset voltage $e_{os}$, then the output voltage $E_0$ is $E_1 + e_{os}$. Changing to an inverting amplifier circuit, we can see that the sums of the currents at the negative input are given by

$$\frac{E_1 - e_{os}}{R_1} + \frac{E_0 - e_{os}}{R_0} = 0 \tag{119}$$

$$E_0 = -E_1\left(\frac{R_0}{R_1}\right) + e_{os}\left(1 + \frac{R_0}{R_1}\right) \tag{120}$$

and the offset voltage has been amplified by an amount $1 + R_0/R_1$; that is, the offset has been amplified even more than the signal $E_1$. One can determine what offset voltage can be tolerated by deciding what is the worst case error that can be tolerated in the output and then work backwards. For example, the circuit gain $R_0/R_1$ is 9 and a 1 mV output error is acceptable; then the largest permissible value for $e_o$ is 0.1 mV. It is important to remember that the offset voltage changes, with temperature and time and that just because this error is initially negligible does not mean that it remains so after the circuit has been in operation for a period of time. One common initial mistake is to assume that, because the instrument operates in an air-conditioned laboratory, one can ignore the temperature drift specifications. The temperature inside an instrument may be 30°C higher than the ambient temperature owing to heating from the components. Even if the instrument is cooled by a fan, the amplifier itself produces heat and the internal temperature may still be 30°C higher than the chassis air temperature.

The integrator is a common laboratory circuit that can be particularly sensitive to voltage offset, as shown below. If the input resistance is $R_1$ and the feedback capacitance is $C_0$, then the output for an input $E_1$ and offset voltage $e_{os}$ is

$$\frac{(E_1 - e_{os})}{R_1} + \frac{d(E_0 - e_{os})/C_0}{dt} = 0 \tag{121}$$

$$E_0 = e_{os} - \frac{1}{R_1 C_o}\int_0^t (E_1 - e_{os})\,dt \tag{122}$$

Normally, $E_0$ is large compared to $e_{os}$ so that only the integral portion is important and the relative error $E_r = (E_{\text{true}} - E_{\text{ideal}})/E_{\text{ideal}}$ is given by

$$E_r = \int_0^t \left(\frac{e_o}{E_1}\right) dt \tag{123}$$

This error can be relatively innocuous as long as $E_1$ is large compared to $e_{os}$ *at all times*, but consider the case of chronocoulometry where the current flowing through a cell is integrated to determine the total charge passed. If the potential is applied to an unstirred system, the current dies away as the reciprocal of the square root of time if no chemical kinetic complications occur. During the first part of the experiment, the offset contribution may be negligible; as the current approaches zero, the offset voltage contribution must begin to become significant at some point.

### 1. Input Currents

One of the cardinal rules for an ideal OA is that no current flows into or out of an amplifier; unfortunately this is not strictly true of a real amplifier, although the current flow may be negligibly small in some cases. The current flow that exists in a real amplifier is essentially a dc current and independent of the signal applied to the amplifier so it is fairly easy to treat. The current input for an amplifier is called the *bias current*, since it results from the current flow into the transistors in the input stage of the amplifier (base current for bipolar transistors or gate current for field effect transistors ranges from 1 $\mu$A to 1 fA). We can examine the effects of these currents for a simple inverting amplifier by representing them as simple current sources associated with an otherwise ideal amplifier as shown in Fig. 3.14*a*.

The current flow $i_{b2}$ through $R_2$ results in a voltage drop $e_a = i_{b2}R_2$ across that resistor, so that the potential at the positive input is $i_{b2}R_2$. Summing the currents at the negative node and including the current source $i_{b1}$ leads to

$$i_{b1} + \frac{E_1 - e_a}{R_1} + \frac{E_0 - e_a}{R_0} = 0 \tag{124}$$

$$E_0 = -E_1\frac{R_0}{R_1} - i_{b1}R_0 + e_a\left(1 + \frac{R_0}{R_1}\right) = -E_1\frac{R_0}{R_1} - i_{b1}R_0 + i_{b2}R_2\left(1 + \frac{R_0}{R_1}\right) \tag{125}$$

If the postive input is short circuited to ground, then $R_2 = 0$ and the error voltage becomes $i_{b1}R_0$, the larger the feedback resistor $R_0$ is, the greater the error becomes. If we think back to the current compliance of the amplifier, we can see that the proper choice of component values is a compromise between several parameters. The resistance $R_2$ was not added to the circuit gratuitously to make it more complicated, but because it allows us to minimize the effects of the bias currents. Because of the manufacturing process, the two bias current

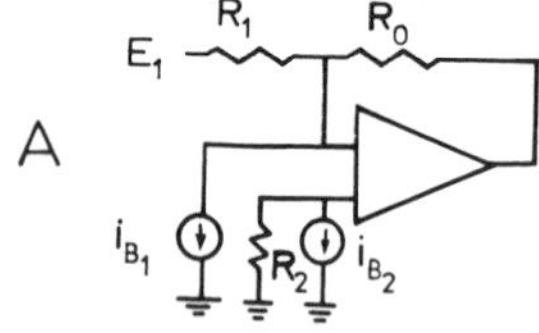


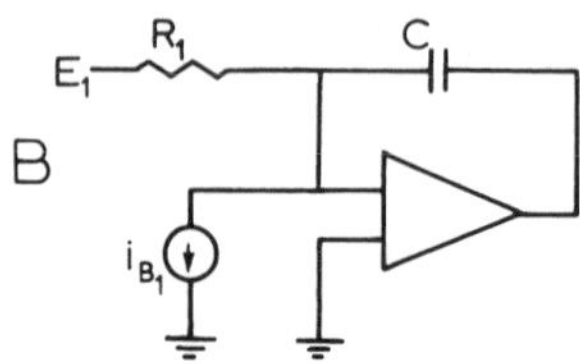


**Fig. 3.14.** Models for the effect of bias currents on inverting amplifier configuration. (*a*) Simple inverting amplifier; (*b*) integrator.

have the same sign and we can use the fact to minimize the bias current error. From inspection we can see that if $i_{b1}R_0 = i_{b2}R_2(1 + R_0/R_1)$ then the two error sources cancel out. In practice the two currents are not only of the same sign, but they also quite well matched in magnitude. The difference between the input bias currents, termed the input *offset current* $i_{os}$, is usually less than 10% of the bias currents. Choosing the value of $R_2$ such that it is the equal to the parallel combination of the input and feed back resistance, that is, $R_2 = R_1R_0/(R_1 + R_0)$, changes the results:

$$E_0 = -E_1\frac{R_0}{R_1} - i_{b1}R_0 + i_{b2}R_0 = -E_1\frac{R_0}{R_1} + i_{os}R_0 \tag{126}$$

The integrator is particularly subject to bias current errors. Let us treat an integrator, where we consider both the offset voltage and the bias current error contributions as shown in Fig. 3.14*a*.

The current at the negative input is given by

$$\frac{E_1 - e_a}{R_1} + i_{b1} + \frac{d(E_0 - e_a)}{C_0\,dt} = 0 \tag{127}$$

Rearranging and solving for $E_0$ leads to

$$E_0 = e_a - \frac{1}{R_1C_0}\int_0^t (E_1 - e_a)dt + \frac{1}{C_0}\int_0^t i_{b1}\,dt + \int_0^t de_a \tag{128}$$

We can assume that in general the offset voltage and bias currents are time invariant during the integration (usually valid if the internal temperature of the amplifier stays relatively constant), and under these conditions the response is

$$E_0 = e_a + \frac{t}{C}\left(i_b + \frac{e_a}{R_1}\right) + \frac{1}{R_1C}\int_0^t E_1\,dt \tag{129}$$

There are two time-varying error terms, both arising from changing the integrator capacitor $C_0$: one by the bias current $i_{b1}$ and the second due to charging the capacitor with the current equivalent to that producing a voltage drop $e_a$ across $R_1$. For any given amplifier we can minimize the errors by making both $R_1$ and $C_0$ as large as possible.

### 2. Effects of Finite DC Gain

We have assumed that the resident Maxwell demon could detect any infinitesimal voltage difference between the potentials of the two inputs and would adjust the output to achieve a precise null between the two voltages. Once again this has been a convenient fiction. What the demon really does is to amplify the voltage difference between the two inputs by an amount $A_0$, which is termed the dc open-loop gain of the circuit. The name arises in the following way. If the positive input is grounded and no connection is made between the output and the negative input, that is, the feedback loop is open or disconnected, then any signal $e_1$ at the negative input is amplified by an amount $-A_0$ at the output. The term dc refers to the fact that this is the gain that the amplifier has for steady-state signals. More about this feature later. Let us now investigate what this means and how it relates to our previous picture using an amplifier that is classically ideal in all its properties except for the gain parameter. In the inverting amplifier case with feedback impedance $Z_0$ and input impedance $Z_1$ responding to an input signal $E_1$ we have

$$\frac{(E_1 - e_1)}{Z_1} + \frac{(E_0 - e_i)}{Z_0} = 0 \tag{130}$$

$$E_0 = -E_1\frac{Z_0}{Z_1} + e_i\left(1 + \frac{Z_0}{Z_1}\right) \tag{131}$$

where $e_i = -E_0/A_0$, so that

$$E_0 = \frac{-E_0}{A_0}(1 + Z_0Z_1) - E_1\frac{Z_0}{Z_1} \tag{132}$$

$$E_0 = \frac{-E_1(Z_0/Z_1)}{1 + (1 + Z_0Z_1)/A_0} \tag{133}$$

As long as the dc gain of the amplifier $A_0$ is very large compared to $1 + Z_0/Z_1$, the result is identical with our simpler model. Even the cheapest OA has an $A_0$ of 20,000 or better, so the simple model is accurate except for large values of the ratio $Z_0/Z_1$.

For a noninverting amplifier with $Z_1$ connected from the negative input to ground, $Z_0$ from negative input to the output, and a signal $E_1$ applied to the positive input, the output voltage $E_0$ can readily be shown by analogous arguments to be

$$E_0 = E_1\frac{1 + Z_0/Z_1}{1 + (1 + Z_0/Z_1)/A_0} \tag{134}$$

In reading the literature on OAs one frequently finds the symbol $\beta$, the feedback factor, defined as

$$\beta = \frac{1}{1 + Z_0/Z_1} \tag{135}$$

used for convenience in discussion. The quantity $A\beta$ is the loop gain, which may be though of as the gain around the "loop" formed by the amplifier and its feedback network. Using this notation one can write the expression for either the inverting or noninverting amplifiers as

$$E_0 = E_1\frac{\text{Ideal Gain}}{1 + 1/A_0\beta} \tag{136}$$

where Ideal Gain is the transfer function of the circuit with the OA treated as an ideal device. The loop gain is the amount of gain left over to the amplifier after the ideal gain to make the amplifier look ideal. As the value $A_0\beta$ becomes smaller, the amplifier looks less and less ideal. The potential difference between the inputs due to finite gain in the amplifier given by $E_1$ (Ideal Gain)/$(A_0 + 1/\beta)$.

### 3. Use of the dB and Octave Notation for Gain

To this point, we have been representing our Bode plots as log(gain) versus log(frequency). This a fairly clumsy format and one that can lead to confusion when several variables are represented in a single figure. The electrical engineering literature uses alternative notation for this purpose which is worth adopting. The decibel or dB is defined as dB = $20\log_{10}$(variable) so a gain of $10^4$ is expressed as 80 dB and the Bode slope of an integrator is −20 dB/decade. If the value of a variable is one-half of its limiting value, then it is spoken of as being 3 dB down from its limiting value, and the frequency where this occurs is called the 3 dB point.

The equivalent notation for frequency is the octave and arises from musical terminology. The formal definition of the octave is $\log_2$(frequency) but it is rarely encountered by itself. Instead the common usage is to express the slope of a $\log(G)$ versus $\log(f)$ plot and the simple relationship is 20 dB/decade = 6 dB/octave.

## B. STABILITY AND THE DYNAMIC PROPERTIES OF OPERATIONAL AMPLIFIERS

Unfortunately the gain of an OA is not frequency independent. A well-behaved OA (we discuss the more ill-tempered variety later) has an open-loop gain that can be adequately represented in Laplace space by

$$A(s) = \frac{a_0 A_0}{s + a_0} \tag{137}$$

where $A_0$ is the dc open-loop gain. By now the function $1/(s + a)$ should be familiar enough that the reader recognizes it as the expression for a low-pass filter. At frequencies below $1/a_0$ the amplifier has an open-loop gain of $A_0$. At the frequency $1/a_0$, the gain is −3 dB down from $A_0$, that is, $A_0/2$, and this frequency is known as the 3 dB point of the amplifier. At frequencies above the 3 dB point, the gain decreases smoothly and crosses the unity gain axis at a frequency known as the unity gain crossover frequency.

Our ideal OA introduced a phase shift of either 0° (follower mode) or −180° (inverter mode), but this was because we could assume that the gain was constant. Thinking back to the low-pass filter, we remember that the phase angle of the circuit itself changed from 0° to −90° as we went from low to high frequency (taking into account the amplifiers assumed 180° phase shift converted this into a −180° to −270° phase shift). Since a real amplifier acts as a low-pass filter, we should not be surprised to discover that the phase shift of the amplifier itself exhibits a similar change at frequencies above the 3 dB point.

Now that we realize that the open-loop gain of an OA has a frequency dependency, we should examine the implications of this discovery. The first circuit to investigate is the inverting amplifier with gain. Because we are concerned with the frequency response properties of the circuit, we can ignore the

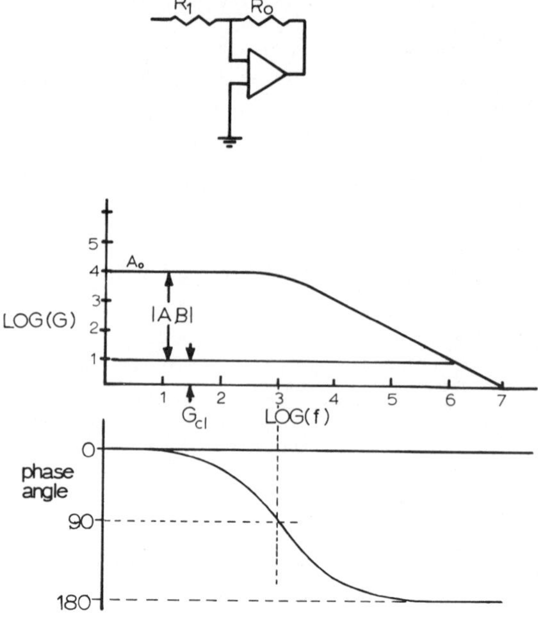


**Fig. 3.15.** The Bode plots showing the effect of a frequency-dependent amplifier open-loop gain with an inverting amplifier circuit.

DC nonideality terms such as offset voltages and bias currents. The Laplace plane representation for Fig. 3.15 is

$$\overline{E}_0 = -\overline{E}_1 \frac{R_0}{R_1} / 1 + \left(1 + \frac{R_0}{R_1}\right)\left(\frac{s + a_0}{A_0 a_0}\right) \tag{138}$$

More generally, we can write the response for any circuit using the $\beta$, Ideal Gain notation as

$$\overline{E}_0 = \frac{\overline{E}_1 \overline{G}}{1 + \dfrac{1}{\overline{A\beta}}} = \frac{\overline{E}_1 \overline{G}}{1 + \dfrac{(s + a_0)}{a_0 A_0 \overline{\beta}}} = \frac{a_0 A_0 \overline{E}_1 \overline{G\beta}}{a_0 A_0 \overline{\beta} + a_0 + s} \tag{139}$$

where $\overline{G}$ is now used for Ideal Gain. Remember that both $\overline{G}$ and $\overline{\beta}$ contain $s$ terms if either inductors or capacitors are involved in the circuit.

For the case of the inverter with gain neither $\overline{G}$ nor $\overline{\beta}$ contains $s$ so that the mathematical rearrangement is simple. We drop the bar symbol (–) over both $G$ and $\beta$ to indicate that neither term contains $s$. Note that the manipulation that follows is strictly valid for circuits containing only resistors.

$$\overline{E}_0 = \overline{E}_1 \frac{a_0 A_0 G\beta}{s + a_0(A_0\beta + 1)} \tag{140}$$

$$E_0(\omega t) = \frac{E_1(\omega t) a_0 A_0 G\beta \sin(\omega t + \Theta)}{\omega^2 + [a_0(A_0\beta + 1)]^2}, \quad \Theta = \tan^{-1}\left(\frac{\omega}{a_0(A_0\beta + a)}\right) \tag{141}$$

At very low frequencies, the $\omega$ term is negligible and the gain becomes $G/(a + a/A_0\beta)$, which is the original expression for a frequency-independent gain. As the frequency becomes so high that the $a_0(A_0\beta + 1)$ term is negligible in comparison with $\omega$, then the curve must roll off at 20 dB/decade (−6 dB/octave) as before. The 3 dB point is at the frequency where the gain is one-half the dc value, that is, where $\omega = a_0(A_0\beta + 1)$. At frequencies above the 3 dB point, we must take into account the rolloff in determining how good a job the amplifier is doing. If the dc gain of the circuit is 10 and the open-loop gain is 100,000 with a 3 dB point of 1 kHz, then from these results and our discussion of the effects of loop gain on accuracy, we know that prior to the 3 dB point, we have a highly accurate circuit. Above the 3 dB point, however, our loop gain begins to decrease; at 10 kHz we have an open-loop gain of 10,000 and this decreases to 1000 at 100 kHz, which is the minimum premissible value for 1% accuracy. In the meantime, the phase angle of the signal has shifted from −180° to −270°. Thus from a gain accuracy standpoint, the maximum frequency that can be treated correctly is 100 kHz, whereas from a phase angle standpoint, the error began at much lower frequency since the phase angle changed by 45° at the 3 dB point. Whether the phase change is important depends upon the application, but the gain error is an absolute criterion.

This phase change is the source of most of the stability problems encountered in operational amplifiers. Before we examine these problems and their cure, we must introduce the concept of poles and zeros.

## C. POLES AND ZEROS

The image plane representation of an operational amplifier circuit can always be cast into the form

$$G(s) = \frac{G\sum_{n=0}^{k}(a_n + s)}{\sum_{m=0}^{p}(b_m + s)} \tag{142}$$

where either $k$ or $p$ or both may be zero and the constants $a_n$ and $b_m$ may be either real, complex, or zero. Let us examine a slightly less general form of this equation

$$G(s) = \frac{G(a_0 + s)(a_1 + s)}{(b_0 + s)(b_1 + s)} \tag{143}$$

where both numerator and denominator have been expanded to contain two terms. If $s$ takes on the value $-b_0$ or $-b_1$, the value of $G(s)$ become infinite and a plot of $G(s)$ versus $s$ shows abrupt excursions to infinity at these values. Because the resulting plot resembles a series of posts or poles stuck in a plane, the values $b_0$ and $b_1$ have been termed *poles*. Similarly, when $s$ has the values of either $-a_0$ or $-a_1$, then $G(s)$ is zero and as a result $a_0$ and $a_1$ are termed the *zeros* of the equation. An ideal integrator (Fig. 3.2*d*) has a gain equation

$$G(s) = \frac{-1/RC}{s + 0} \tag{144}$$

whereas an ideal low-pass filter (Fig. 3.2*f*) has the equation

$$G(s) = \frac{-R_0/R_1}{(s + 1/R_0C)} \tag{145}$$

The inverter with gain employing a real OA has a gain equation

$$G(s) = \frac{R_0A_0a_0/(R_0 + R_1)}{s + a_0[1 + A_0/(1 + R_0/R_1)]} \tag{146}$$

All three equations are single pole equations with no zeros. The pole in the first example happens to occur at 0 Hz, but it still exists in the mathematical sense.

The ideal differentiator has the gain equation

$$G(s) = -RC(s + 0) \tag{147}$$

whereas a high-pass filter constructed with an ideal OA has the gain

$$G(s) = \frac{-R_0R_1(s + 0)}{(s + 1/R_1C)} \tag{148}$$

and the circuit in Fig. 3.8 using ideal amplifiers has the gain

$$G(s) = \frac{-R_0C_0(s + 1/R_0C_0)}{s} \tag{149}$$

The first two circuits have zeros at 0 Hz whereas the third has a zero at $1/R_0C$ Hz.

## D. APPROXIMATE GAIN AND PHASE ANGLE PLOTTING

### 1. Circuits with Poles Only

For a simple case containing only one pole and no zeros, the gain can be written as

$$G(s) = \frac{a_1}{s + b_1} \tag{150}$$

or converting to the frequency plane representation by making the substitution $s = j\omega$, we can rewrite the form as

$$G(j\omega) = \frac{a_1/b_1}{\sqrt{1 + (\omega/b_1)^2}} \tag{151a}$$

$$\theta = -\arctan\left(\frac{\omega}{b_1}\right) \tag{151b}$$

At low frequencies where $\omega/b_1$ is small compared to unity, we can write $\log[G(j\omega)]$ as

$$\log[G(j\omega)] = \log\left(\frac{a_1}{b_1}\right),\ \theta = 0° \tag{152}$$

At the frequency where $\omega = b_1$,

$$\log[G(j\omega)] = \log\left(\frac{a_1}{b_1}\right) - \frac{1}{2}\log(2),\ \theta = \arctan(1) = 45^\circ \tag{153}$$

Using the dB representation,

$$A(\text{dB}) = 20\log[G(j\omega)] = 20\log\left(\frac{a_1}{b_1}\right) - 3\text{ dB} \tag{154}$$

At frequencies where $\omega > b_1$,

$$A(\text{dB}) = 20\log\left(\frac{a_1}{b_1}\right) + 20\log\left(\frac{b_1}{\omega}\right) \tag{155}$$

$$A(\text{dB}) = 20\log(a_1) - 20\log(\omega),\ \theta = -90^\circ \tag{156}$$

A straight line approximation to the phase angle versus log(frequency) can be obtained by assuming constant 0° phase angle up to a frequency of $0.1b_1$, followed by a straight line segment connecting this point with −90° at the frequency $10b_1$ and having the value −45° at the frequency $b_1$, that is, a slope of −45°/decade. For frequencies higher than $10b_1$, the phase angle remains fixed at −90°. These plots are shown in Fig. 3.16*a*. The maximum estimate gain error in the approximation occurs at $b_1$, where the actual gain is 3 dB less than predicted. The frequency approximation has a maximum estimation error of 5.7% high at $0.1b_1$ and −5.7% low at $10b_1$.

For multiple pole systems the logarithmic representation is extremely convenient because one can simply add together the approximate results for each pole as shown in Fig. 3.16*b* for the case

$$G(s) = \frac{a_1}{(s + b_1)(s + b_2)} \tag{157}$$

## 2. Circuits with Zeros Only

For a simple single zero case written as

$$G(s) = s + a_1 \tag{158}$$

the frequency space representation is given by

$$A(dB) = 20\log[a_1\sqrt{1 + (w/a_1)^2}] \tag{159a}$$

$$\theta = \arctan\left(\frac{\omega}{a_1}\right) \tag{159b}$$

By analogy with the pole only case, it is simple to see how the straight line gain and phase angle approximations arise (shown in Fig. 3.16*c*). Here the slope for gain is +20 dB/decade and +45°/decade for the phase. The errors in this approximation are identical to those for the pole case, except that the signs are reversed. Once again, the presence of multiple zeros is treated by simple addition of the isolated zero response in log(frequency) space.

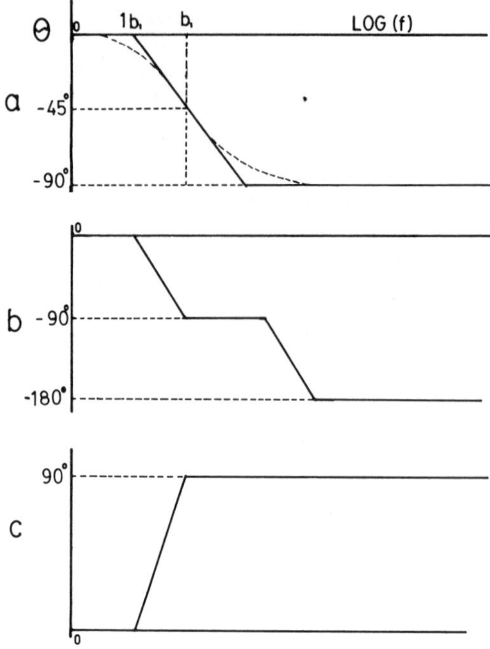


**Fig. 3.16.** Bode phase plots, using the approximation method. (*a*) Single pole; (*b*) double pole; (*c*) single zero.

### 3. Circuits with Combined Poles and Zeros

Circuits such as the ideal amplifier high-pass filter (Fig. 3.10) and the phase shifter circuit in Fig. 3.12 have both poles and zeros, and treatment of this case is simply an extension of the method we have just developed. The responses for the isolated pole and zero are simply added together to give the overall log(frequency) space response. We have now justified the previously intuitive treatment for gain plotting which was used earlier and extended it to include phase plotting as well. We now have all the tools necessary to address one of the major problems in OA circuitry, that is, stability or the lack of it.

## E. STABILITY AND OSCILLATION

Let us go back to the Ideal Gain representation of the amplifier, which stated in effect

$$E_0 = E_1\left[\frac{G_i(s)}{1 + 1/A(s)\beta(s)}\right] \tag{160}$$

where $G_i(s)$ is the ideal gain of the circuit in transform space.

The suspicious reader may have wondered just what the amplifier does in the paranoid circumstance where $A(s)\beta(s) = -1$, since the equation predicts an infinite or, perhaps better, an undefined output value. In point of fact the amplifier does its best to reflect the equation; it oscillates between its two potential bounds with a frequency such that $A(j\omega)\beta(j\omega) = -1$. A necessary and sufficient condition for instability, regardless of whether one is using an inverting, noninverting, or mixed configuration is that $A(j\omega)\beta(j\omega) = -1$ at some frequency [where $A(s)$ is the representation of the amplifier gain in transform space and $\beta(s)$ is the transform space representation of the *feedback factor*]. From this point of view we can restate our criterion for instability as

$$A(j\omega) = \frac{1}{\beta(j\omega)} \angle\, 180° = -1 \tag{161}$$

or verbally, the critical condition is when the open loop gain is equal to the reciprocal of the feedback factor *and* the phase difference *between them* is 180°. For either an inverting or a noninverting amplifier circuit, we can rewrite this condition using the value of $A(j\omega)$ as

$$A(j\omega) = \frac{1 + Z_o(j\omega)}{Z_1(j\omega)} \angle\, 180° = -1 \tag{162}$$

Reflecting back on our previous discussion of plotting the log(gain) and $\theta$ versus log(frequency) for poles and zeros; we can see that one way for this condition to occur is for a single zero circuit such as an ideal differentiator to be combined with a single pole amplifier response with both responses intersecting in the limiting slope regions as shown in Fig. 3.17a. A second way for this condition to exist, shown in Fig. 3.17*b*, is to have a simple frequency-independent gain circuit such as a follower or inverter with an amplifier whose high-frequency open-loop gain decreases at –40th dB/decade (or –12 dB/octave). In both cases, the rate of closure between the two curves is –12 dB/octave and the amplifier oscillates at the intersection frequency. Most, but not all, OAs are

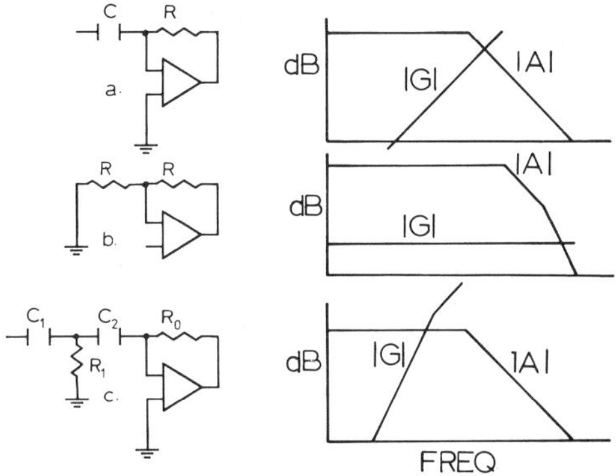


**Fig. 3.17.** Bode plots, help determine whether a circuit will oscillate. (*a*) An ideal differentiator oscillates. (*b*) Voltage follower with gain. The OA is uncompensated and oscillation occurs. (*c*) Oscillation due to 12 dB/octave slope.

internally compensated to guarantee a −6 dB/octave rolloff for just this reason. Some are not and require external compensation by the user; we discuss these types later.

For OAs with a −6 dB/octave rolloff, the main conditions for oscillation are for a closed-loop gain with slope of +6 dB/octave to intersect the −6 dB/octave rolloff of the amplifier open-loop gain. An alternative way, which is rarely encountered, is for the closed-loop gain to be increasing at +12 dB per octave and to cross the zero slope open-loop gain region. This latter functionality can be produced by the circuit in Fig. 3.17*c*, which may be recognized as the input section of the double integrator (Fig. 3.13) with a simple resistive feedback element. At frequencies where $1/R_0(C_1 + C_2)$ is greater than $\omega$, the Bode plot has a slope of +12 dB/octave; above this frequency, the slope drops off to +6 dB/octave.

### 3. "Parasitic" Inductance and Capacitance

In practice one frequently encounters circuits that on paper offer no problems but when constructed and powered up, immediately break into spectacular oscillation. This behavior may be due to "parasitic" inductances and capacitances; these are capacitances and inductances that were not purposefully added to the circuit and, as a result, are not shown in the circuit diagram. A common problem is a parasitic capacitance to ground which is often added inadvertently by the individual who lays out a printed circuit. One problem in OA circuits is that long runs leading to either the positive or negative inputs can act as antennae picking up broadcast noise signals, particularly line frequency components derived from lights and the power supplies. In order to avoid this, designers try to keep components physically as close as possible to the inputs. When this is impossible, one normally runs a "guard ring" around the long run

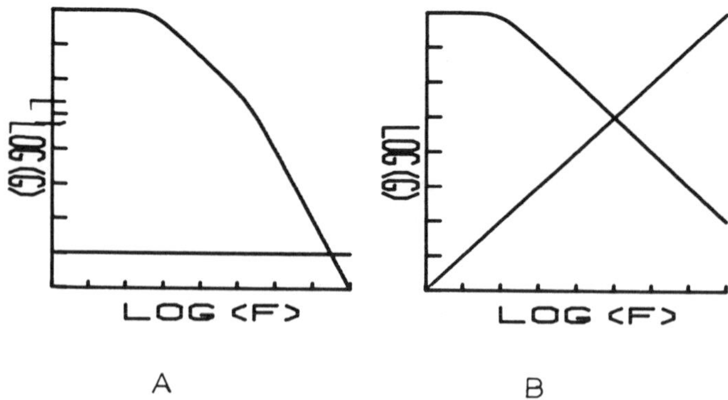


**Fig. 3.18.** (*a*) Bode plot of a circuit with unity gain showing no oscillation. (*b*) An ideal differentiator's Bode diagram, showing that the gain of +12dB/octave causes oscillation.

to the input. This guard ring is normally signal ground and can act as a capacitance between the input and signal ground.

In many cases, such as a simple inverter with gain, this causes no real trouble, but in some instances, inadvertent zeros or poles are generated. In Fig. 3.19, the capacitance shown is a parasitic capacitance of the type just discussed. The Bode plot shows breaks that occur at frequencies below the unity gain crossover frequency of the OA and interior to the amplifier's open-loop gain envelope. whether the break in a real circuit occurs within or without, the envelope obviously depends upon the resistive components and the physical layout of the circuit board. Normally the capacitances generated in this way are on the order of a few tens of picofarads so that one usually encounters trouble only when fairly large resistors are employed or the OA has a high (greater than a few megahertz) unity gain crossover frequency.

Even if the circuit designer were able to eliminate parasitic capacitances, often referred to in the application literature simply as "strays," there would still be capacitances in the circuit that are not represented by the normal circuit diagram. These capacitances are internal to the amplifier itself and cannot be avoided. Once again, we must expand our model for the OA to include these new properties. Figure 3.20 shows an expanded model that includes the input and output impedances of the amplifier. The common mode impedance, composed of a resistor paralleled by a capacitor, appears at both inputs to ground

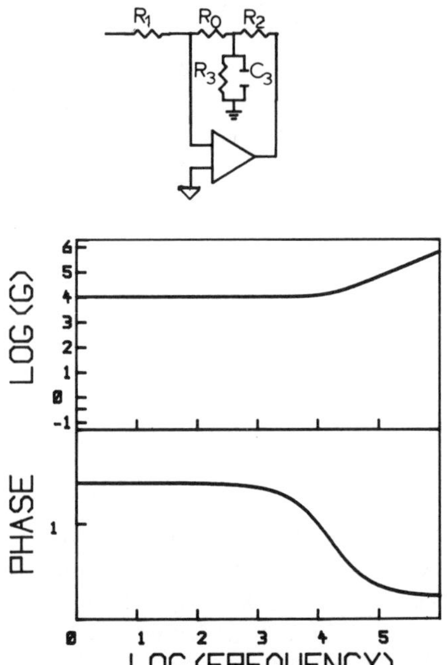


**Fig. 3.19.** Effect of parasitic capacitance on a high gain inverting amplifier.

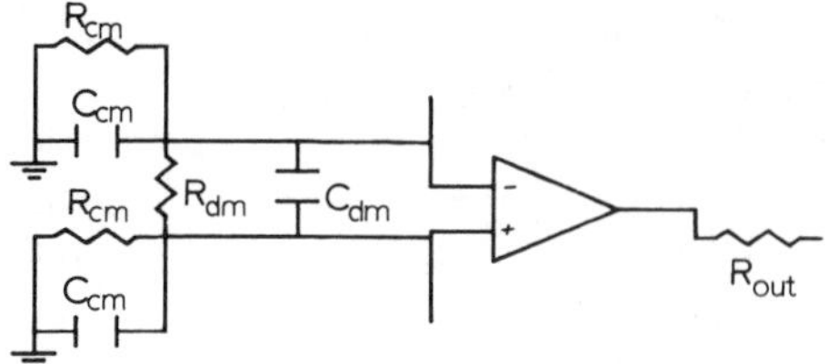


**Fig. 3.20.** Common mode, differential mode, and output impedance model for a real OA.

and the differential mode impedance, also a resistance in parallel with a capacitance, appears between both inputs. The specification sheets for most OAs describe the effective values for these impedances. For example, an AD517 lists the differential input impedance as 15 MΩ//2.5 pF, or 15 MΩ in parallel with 2.5 pF. There are a number of cases, particularly the industry standard 741 type amplifier, in which only the resistive component is specified; in this case, one is left in the dark about the shunt capacitance. Normally, one can estimate the shunt capacitances as being from 1 to 10 pF in magnitude when they are not specified.

There are two ways that one can take these impedances into account in determining circuit stability. The simplest is to include them as part of the input and feedback impedances. When using this approach one or more of the elements can often be ignored. For example, if the circuit is a simple inverter with gain where the positive input is physically connected to ground, the usual wisdom would be to ignore all these terms, because the impedances are either shorted out or represent inactive signal paths. In this approach, about the only time one considers the effects of these impedances is in noninverting configurations. This method, although simpler, is not particularly accurate and can lead to a false sense of security. A more accurate treatment involves incorporating these elements into the feedback factor in the following manner. The old $\beta$ is replaced by $\beta'$; defined as

$$\beta' = \frac{Z_0}{Z_1} + \frac{Z_0}{Z_{id}} + \frac{Z_0}{Z_{icm}} \tag{163}$$

and plotting the Bode response of this new factor. The effect can be more clearly seen if one evaluates the $Z_{id}$ and $Z_{icm}$ components in terms of their $R - C$ model, leading to

$$\beta' = 1 + \frac{Z_0}{Z_1} + Z_0 C_{id}\left[s + \left(\frac{1}{RC}\right)_{id}\right] + Z_o C_{icm}\left[s + \left(\frac{1}{RC}\right)_{icm}\right] \tag{164}$$

where it is now obvious that two new zeros have been added. What effect, is any, this will have on the stability obviously depends upon the other components in the circuit, but since the resistances are at least 1 MΩ (for the low impedance industry standard 741 type) and may be as high as several thousand megohms for field-effect transistor (FET) types, it can be seen that the break frequencies will lie above 10 KHz in any event and often far higher.

A second source of trouble arises from parasitic inductances. To this point we have not treated inductances except in the introduction because they have become fairly rare in recent OA electronic circuitry. The reason for this is that most inductors have a fairly high series resistance; that is, they are not ideal. As we see later, it is possible to use amplifiers, resistors, and capacitors in such a way as to emulate an ideal inductor. Inductances can enter a circuit inadvertently in at least two ways, through the inductance of the leads of resistors and capacitors (normally only important for high megohm resistors) and through the inductance of long wire leads connecting components mounted on switches. A frequent example of the latter is when one wishes to make a variable gain amplifier, which may have several orders of magnitude of dynamic range using a multiposition switch to change the feedback resistor. Another circumstance that can produce an inductance in series with a resistance is the use of a wire-wound potentiometer for a continuously variable resistance. In either case, the effective circuit is shown in Fig. 3.21 and the Bode plot of this circuit has the same form as the earlier examples because the gain is given by

$$G(s) = -\frac{L}{R_1}\left(s + \frac{R_0}{L}\right) \tag{165}$$

where the break frequency is now given by $R_0/L$

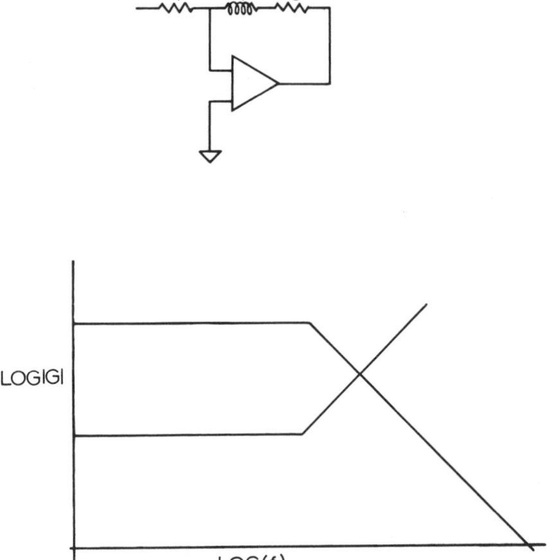


**Fig. 3.21.** Parasitic inductance in the feedback loop of an inverting amplifier and how it affects the Bode gain response.

## 2. Interactions Between the Output Impedance and the Load

We have not considered what might be connected to the output of the OA, except to state that the load must not place a current demand upon the amplifier that is excessive. In the revised model shown in Fig. 3.22, we have incorporated a resistance in series with the amplifier output to represent the finite output impedance of a real device. Although this parameter is often omitted in the "spec" sheets for an OA, it is often a source of instability when inductive or capacitive loads are involved. Typical values for small integrated circuit amplifiers are on the order of 1 K Ω; however, for discrete or modular amplifiers optimized for high current and/or broad bandwidth, the values are as low as 100 Ω. Looking at the circuit in Fig. 3.22, one can see that the voltage that is fed back to the amplifier is actually $E_0$, whereas the voltage generated is $E_0'$. The two are related by

$$E_0 = E_0' \frac{1}{1 + Z_0/Z_L + Z_0/(Z_1 + Z_f)} \tag{166}$$

which arises from a consideration of the total current flow at the two points. The $Z_1$ term is omitted if one is dealing with an inverting amplifier circuit and retained for noninverting applications. The net effect of this is to reduce the effective open-loop gain of the amplifier $A'$ to the form

$$A' = \frac{A}{1 + Z_0/Z_L + Z_0/(Z_1 + Z_f)} \tag{167}$$

where once again the $Z_1$ term is omitted for the inverting amplifier configuration. Usually this feature causes no trouble except in cases where fairly heavy inductive or capacitive loads are being driven, but occasionally one can get in trouble through the $Z_1 + Z_f$ term. The cause of the trouble is fairly obvious; the open-loop rolloff can be steeper than the assumed −6 dB/octave.

## 3. Stabilizing the Operational Amplifier Circuit

Irrespective of how the oscillation arises, there is always a way of eliminating the undesired behavior, although admittedly some solutions may be more obscure than others. The simplest to deal with are those arising from the obvious design elements of the circuit. Consider the case of the differentiator,

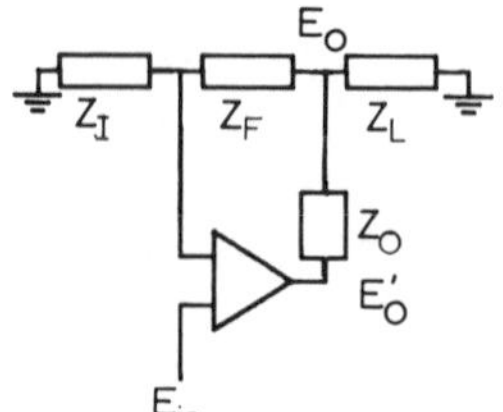


**Fig. 3.22.** A model for the effect of output impedance $Z_0$ or the behavior of $A$ inverting amplifiers under load $Z_1$.

which oscillates because of the intersection of the loop gain with the open-loop gain. This oscillation is most commonly eliminated by inserting a resistance in series with the input capacitance to yield what in effect is a high-pass filter as seen in Fig. 3.10. As long as the break frequency, $1/2\pi R_1C_1$, occurs before the loop gain intersects the open-loop rolloff, the circuit is stable (one normally tries for a frequency decade difference between the break frequency and the intersection frequency). A second way to achieve the same result is to place a capacitor $C_0$ in parallel with the feedback resistor. The resulting system transform is now given by

$$G(s) = \frac{s\; C_1/C_0}{s + 1/R_0C_0} \tag{168}$$

or

$$A\;(dB) = \log\left(\frac{C_1}{C_0}\right) + \log(\omega) - \tfrac{1}{2}\log\left[\omega^2 + \left(\frac{1}{R_0C_0}\right)^2\right] \tag{169}$$

The Bode plot of this circuit is identical to that of the high-pass filter except that now the break frequency is given by $1/2\pi R_0C_0$, and the limiting high frequency gain is $C_1/C_0$. This form of stabilized differentiator is not encountered as often as the high-pass filter for two reasons. First, capacitors are generally more expensive than resistors, and second, the bare input capacitance of the latter circuit can induce oscillations in the circuitry driving the differentiator. In many instances both approaches are taken producing a band-pass amplifier. This is done whenever possible to reduce the contribution from high-frequency noise components.

For inadvertent complications, or complications that arise due to capacitance strays that can not be readily eliminated, one can usually stabilize the amplifier by addition of capacitance to the feedback loop to form an overall circuit pole. Often there are several places where one might add the capacitance. Let us examine the circuit in Fig. 3.19. The intended circuit is a high gain inverting amplifier where a Thevenin T substitutes for a high megohm resistor. The attempt to shield the resistive components from noise sources has produced a parasitic capacitance between the resistive node and ground. The gain of the circuit is given by

$$G(s) = \frac{R_2R_0C_3}{R_1}\left[s + \left(\frac{1}{R_0} + \frac{1}{R_2} + \frac{1}{R_3}\right)/C_3\right] \tag{170}$$

$$A\;(dB) = \log\left(\frac{R_2R_0C_3}{R_1}\right) + \tfrac{1}{2}\log\,(\omega^2 + a^2) : a = \left(\frac{1}{R_0} + \frac{1}{R_2} + \frac{1}{R_3}\right)C_3 \tag{171}$$

$$\lim_{\omega\to\infty} A(dB) = \log\left(\frac{R_2 + R_0 + R_0R_2/R_3}{R_1}\right) \tag{172}$$

Because the purpose of this circuit is to emulate a large resistance, the usual form is to make $R_3$ much smaller than either $R_2$ or $R_0$ so that the break frequency is effectively given by $1/R_0C_3 + 1/R_2C_3$. At low frequencies compared to the break, the network acts as a high-impedance resistance, but as the frequency passes the break frequency,. the circuit begins to act as a differentiator with a +6 dB/octave slope because the higher frequencies begin to pass through the capacitance to ground. In order to eliminate this effect, we must either find an alternate path for these frequency components in the feedback loop or prevent them from reaching the loop itself.

We can attenuate the high-frequency components in the loop by bypassing the feedback loop with a capacitance. This is the most obvious solution and usually the easiest to implement, but not necessarily the best approach, as we shall see. If we incorporate this capacitance $C_c$, then the transform space gain becomes

$$G(s) = \frac{1}{R_1 R_0 R_2 C_3 C_c}\left[\frac{s+a}{s(s+a)+g}\right] : g = \frac{1}{R_0 R_2 C_3 C_c} \tag{173}$$

where $a$ retains its previous definition. Figure 3.23 shows several possible results, depending upon the choice of component values. In order for stabilization to be obtained, the integrator formed by the input resistance and $C_c$ must be chosen so that $1/R_1 C_c$ is less than the unity gain crossover frequency. By increasing the value of $C_c$, one can eliminate the rising portion of the gain

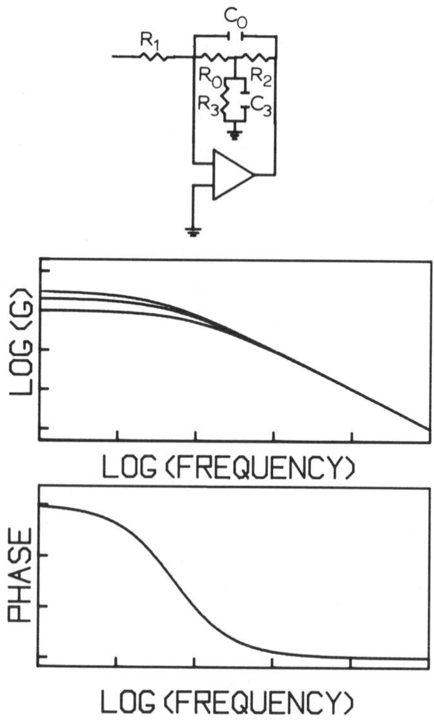


**Fig. 3.23.** Compensation for parasitic capacitance in a Thevinen T high gain circuit.

curve, but in doing this, one can disturb the high-frequency response of the amplifier yielding either higher or lower gain than the low-frequency response. Whether this is acceptable or not depends upon the use of the circuit.

A better place to insert the compensating capacitor is across $R_0$ alone. When this is done, as shown in Fig. 3.24$a$, the resulting gain is given by

$$G(s) = \frac{R_2C_0 + R_2C_3}{R_1C_0}\left(\frac{s+b}{s+a}\right): \qquad a = \frac{1}{R_0C_0}$$
$$b = \frac{R_2R_3 + R_0R_2 + R_0R_3}{R_0R_3(R_2C_0 + R_2C_3)} \tag{174}$$

From this form it is fairly obvious that if we make $a = b$, there is no frequency dependency to worry about; that is, the circuit has a flat gain out to the open-loop rolloff. If one sets up this equality, solves for $C_0$, and inserts the value into the gain equation the resulting frequency-independent gain is the same as the low-frequency gain for the uncompensated case; that is, perfect compensation is obtained. This is a somewhat superior approach to the initial method in that there is one less pole in the amplifier response.

Instead of attenuating the higher-frequency components by inserting capacitances in the feedback loop, one can introduce attenuation into the input stage as shown in Fig. 3.24$b$.

$$G(s) = \frac{R_0R_2C_3}{R_aR_bC_1}\left(\frac{s+b}{s+a}\right): a = \frac{R_a + R_b}{R_aR_bC_1}$$
$$b = \frac{1/R_0 + 1/R_2 + 1/R_3}{C_3} \tag{175}$$

The sum $R_a + R_b = R_1$ as in our other examples. If we again choose $a = b$, then we eliminate the frequency dependency of the response in this circuit, as in the preceding case. By making the proper substitutions into the gain equation, we find that the frequency-independent gain in this case is also equal to the low-frequency gain in the uncompensated case.

In a mathematical sense the latter two compensation schemes are perfectly identical, but in practical application the former is preferable for two reasons. In the final method one adds an extra component in series with the input. Moreo-

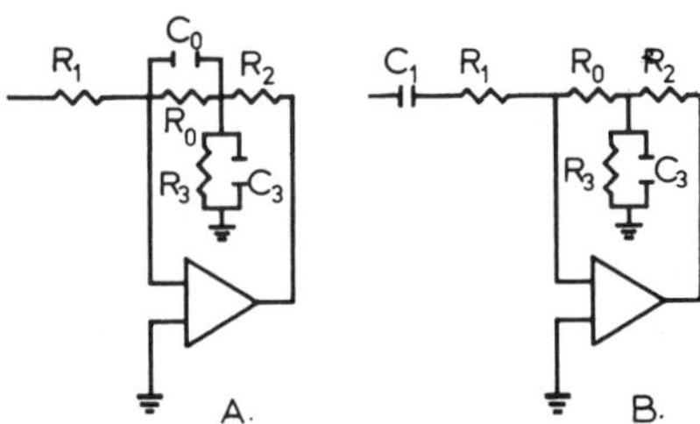


**Fig. 3.24.** Two methods for compensation for parasitic capacitance.

ver, one frequently has not planned ahead for this eventuality and the compensating components must be "space wired" or "kludged" into the circuit, but it is quite simple to place a small capacitance in parallel with a resistance already in place. A more fundamental objection to the latter method is that it attenuates only high-frequency components arriving at the input resistance of the amplifier and not any noise components generated by the amplifier itself.

### 4. Amplifier Input Noise: Voltage and Currents

The OA itself generates a certain amount of current and voltage noise which may be treated as if small noise generators were located at the amplifier inputs. The noise components result from three distinctly different processes and are inherent in all amplifying devices. The sources are Schottky or shot noise, Johnson or thermal noise, and flicker or $1/f$ noise. Shot noise results from the discrete particle nature of the charge carriers in semiconductors. Individual particles arrive at random times so the resultant current is an average current with a superimposed nondeterministic time-varying noise component. The spectral density of shot noise is constant from zero frequency to frequencies of the order of the inverse charge transit time of the semiconductor. The mean-square shot noise current is

$$I_n^2 = 2qI_{dc}\Delta f \tag{176}$$

where $q$ is the charge on an electron ($1.6 \times 10^{-19}$ $C$, $I_{dc}$ is the average current, and $\Delta f$ is the system bandwidth.

Johnson, thermal, or resistance noise is caused by the random motion of charges which is *independent* of their mean or average motion. The thermal noise in a resistor is given by

$$E_n^2 = 4KTR\Delta f \tag{177}$$

where K is the Boltzmann constant ($1.38 \times 10^{-23} J/°K$), $T$ the absolute temperature, and $R$ the resistance, and again $\Delta f$ is the system bandwidth.

There are two other noise sources that are not as well understood. Flicker noise has a $1/f$ spectral distribution and occurs in all semiconductors and a noise component that is peculiar to some integrated circuit OAs and is called "popcorn" noise. Popcorn noise is characterized by sudden noise *pulses* that may be several millivolts in magnitude and that appear randomly at the inputs of some operational amplifiers. This component is thought to be due to a manufacturing defect, but whatever the source, it can be highly distressing. The reason for the particular obnoxiousness of popcorn noise is that an impulse excitation contains all possible frequency components with equal amplitude and random phase relationships. If an amplifier and its associated circuit have an oscillation frequency, popcorn noise is sure to excite it. Fortunately, not all integrated circuit OAs are subject to popcorn noise. These noise components can also activate the complex differential and common mode input impedances of the OA.

## 5. Common Mode Signals

In many circuits the voltage at both inputs is at some potential other than ground or ground plus an offset voltage. The common mode voltage is defined as the average value of these voltages

$$E_{cm} = \frac{E^+ + E^-}{2} \tag{178}$$

An ideal OA produces an output voltage that corresponds only to the difference between these voltages $E_0 = A\ (E^- - E^+)$, but a real amplifier produces an output that depends on the common mode voltage. The common mode gain $A_{cm}$ is defined as

$$A_{cm} = \frac{\text{common mode voltage out}}{\text{common mode voltage in}} \tag{179}$$

The term common mode rejection ratio (CMRR) refers to the ratio of differential open-loop gain to common mode gain:

$$\text{CMRR} = \frac{A}{A_{cm}} \tag{180}$$

This definition obscures the fact that for most real amplifiers the CMRR is a fairly complex nonlinear function of the common mode voltages involved, and what one finds quoted in the manufacturers' literature is normally an average value over the expected signal range. The CMRR for any particular situation may be better or worse than this average value. Given the fact that the CMRR is a more nebulous quantity than we have encountered before, we can still roughly evaluate the effect of this parameter on the behavior of a real amplifier. The simplest way to evaluate its effect is to assume that the error is due to a small voltage source whose value is $e_{cm}$/CMRR, which is in series with the inverting input of the OA. We analyze the effects on a circuit that has a voltage $E_2$ applied directly to the noninverting input (obviously this could be derived from a voltage source and an impedance network to ground) and an impedance $Z_I$ between the input voltage $E_1$ and the inverting input, with a feedback impedance $Z_0$.

We can sum the currents at the inverting input node and write

$$E_o = -E_1\frac{Z_0}{Z_1} + E_a\left(1 + \frac{Z_f}{Z_1}\right) \tag{181}$$

where $e_a$ is the potential at the node given by

$$e_a = E^- - E^+ + E_{cm} = E^- - E_2 + \frac{E^- + E_2}{2\text{CMRR}} \tag{182}$$

Combining these expressions for $E_0$ leads to

$$E_o = \frac{-E_1\dfrac{Z_0}{Z_1} + E_2(1 - 1/\text{CMRR})}{1 + (1 + Z_0/Z_1)/A} \tag{183}$$

When $E_2$ is essentially ground as in the normal inverting configuration the effect of a finite CMRR is eliminated, but for other cases the errors associated with this term can be of major importance. The major cause of nonlinear behavior in some follower amplifiers is due to low CMRR effects, and the suitability of an amplifier for differential applications is primarily determined by the CMRR of the units under consideration.

### 6. Slew Rate, Settling Time, and Full Power Bandwidth

The response of real-world OAs is governed by a further set of additional parameters that can be easily overlooked by the user. The slew rate is maximum rate of change of the output of the amplifier while the rated output voltage and current are supplied. Any OA is specified as to its behavior while driving a certain specified load, its "rated load," which is typically between $10K$ and 500 Ω, depending upon the amplifier, although certain high-power amplifiers are rated under heavier load conditions. If one applied a potential step to either a voltage follower or simple inverter circuit and observed the output voltage with an oscilloscope, the output would be observed to have a linear slewing region as seen in Fig. 3.25. The slope of this essentially linear region is the slew rate normally expressed in volts per microsecond (warning: for some slow amplifiers the units are volts per second). This rate is related to another parameter, the full power band width $f_p$, by the expression

$$f_p = \frac{S_r}{2\pi E_{or}} \tag{184}$$

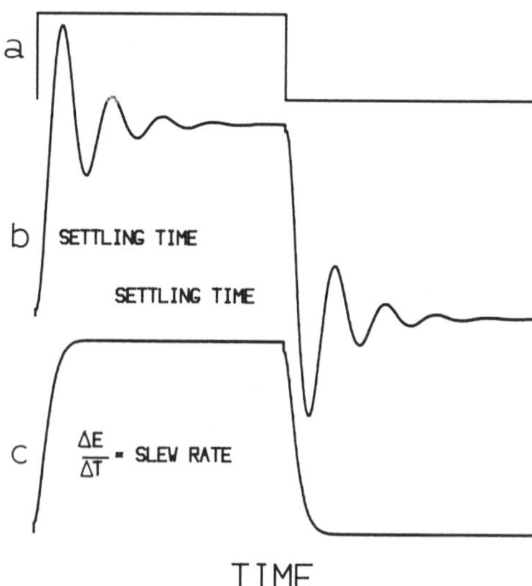


**Fig. 3.25.** Settling time and slew rate in OAs. ($a$) Input signal; ($b$) output with ringing; ($c$) damped response.

where $E_{or}$ is the rated output voltage and $S_r$ is the slew rate. These two parameters govern the suitability of an amplifier for a large amplitude excursion in a given frequency region. The unity gain crossover frequency $f_c$ is a *small* amplitude signal parameter and the $f_p$ of an amplifier is typically one to two orders of magnitude smaller than the $f_c$, although some amplifiers have $f_p$ values very near their $f_c$ and others have $f_p$ values as much as 77 dB down from $f_c$ (i.e., $f_c = 50{,}000 f_p$) so that some attention must be paid to these parameters.

An apparently similar but actually totally unrelated and very complex parameter is the settling time of the amplifier. This refers to the time required for an amplifier to reach its final output value, and we reiterate that there is no direct relationship between slew rate and settling time. One problem in discussing this parameter is defining what constitutes reaching the final value. One typically sees the times specified in terms of 1%, 0.1%, and occasionally 0.001% of final value, but different vendors use different levels and the same manufacturer may use different levels in specifying different amplifiers. If amplifier A reaches 1% faster than amplifier B, it may or may not reach 0.1% before amplifier B, depending upon the degree of damping in the amplifier itself (i.e., the amount of ringing or overshoot). To further complicate matters, the settling time normally depends strongly on both the magnitude of the excursion and the nature of the load being driven. There is also the phenomenon known as a "long tail," which refers to a very slow settling to the final value after perceptible oscillation has ceased. An amplifier may appear to settle out very rapidly to 0.1%, but may, in fact, require several *seconds* to reach 0.01%.

## 7. Amplifiers Requiring Open-Loop Gain Shaping

As we mentioned above, most OAs have a −20 dB/decade rolloff in their open-loop gain above the −3 dB point, which simplifies the user's design problems. A number of integrated circuit OA types do not have this friendly property and require the user to add additional components to ensure stable operation even for simple inverters and followers. A real OA actually consists of a number of gain stages, usually with a −20dB/decade rolloff for each stage so that the high-frequency rolloff may be −60 dB or higher. Internally compensated amplifiers have a built-in dominant low-frequency pole which guarantees the familiar −6 dB/octave rolloff at least up to the unity gain crossover frequency. Above this frequency the gain often drops off much more sharply; because the gain is less than unity, this normally causes no problem. Amplifiers without the built-in pole allow the user to tailor the response to their needs at the penalty of extra components. This is both a blessing and a curse. Let us consider a situation where an inverting amplifier with a gain of 100 is required. In order to maintain 1% accuracy, we require that the open-loop gain be 100 times the loop gain in the frequency range of interest or a minimum dC open-loop gain of 80 dB. With an amplifier that has an internal compensation network producing a uniform −20dB/decade rolloff down to the unity gain frequency, the maximum frequency would be either the −3 dB or corner frequency $F_c$. We can

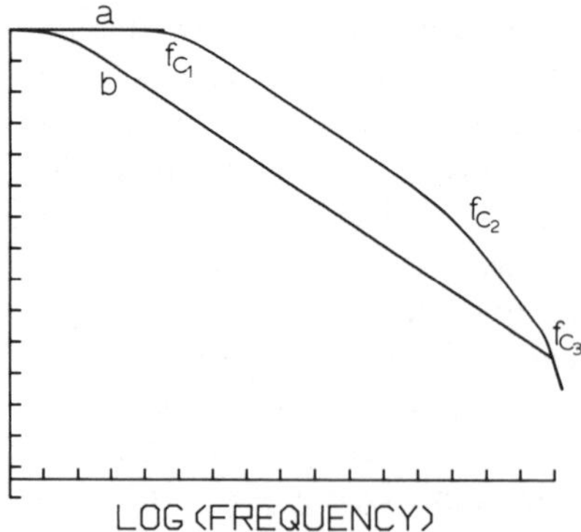


**Fig. 3.26.** Uncompensated OA open-loop gain. (*a*) Uncompensated response; (*b*) same amplifier with a compensation pole introduced.

extend this only by using an amplifier with either a higher dC gain or higher $f_c$. For uncompensated amplifiers the open-loop gain looks something like Fig. 3.26*a*, where there are several corner frequencies $f_{c1}, f_{c2}, f_{c3}$, and so forth. Depending upon the internal structure of the amplifier, the gain may drop off at an additional −6 dB/octave at each corner (i.e., −6, −12, −18 dB, etc.) or may actually reverse itself as −6 dB, −12 dB, and back to −6 dB again. As long as the loop gain is such that the intersection between loop gain and open-loop gain is in a −6 dB/octave region, the circuit is stable. For high loop gain situations this leads to a higher operating bandwidth than would be available for the same amplifier if it were compensated for a uniform −6 dB/octave rolloff over the whole range. This also explains why more compensation is typically required for low gain circuits than for high gain circuits, with unity gain being frequently the worst case situation. Amplifiers that require external compensation components are more difficult to work with, but they can provide optimal performance if the designer is careful.

A second type of compensation is sometimes required for amplifiers that are designed to drive heavy capacitive loads. Because of the finite output impedance of an amplifier, the presence of a capacitive load has the effect of radically decreasing the open-loop gain at high frequencies and to overcome this, some amplifiers designed for these applications have much flatter high-frequency gain than is usual and a much steeper high-frequency rolloff. Many of these amplifiers require an output loading capacitor to prevent oscillation; that is, there must be a certain minimum load capacitance.

## VII. ANALOG COMPUTATION

The OA can be employed to simulate systems and to determine coefficients and roots of equations. Any algebraic or linear differential equation is simply a collection of mathematical operations that individually can be implemented or emulated by an OA circuit. Because the individual elements of these equations can be implemented by OAs, in principle, and usually in practice, one can connect together a series of OA circuits into a super circuit that solves the equation of interest. What one is in fact doing is constructing a model whose Laplace transform is identical to the transform of the model or equation to be studied.

## A. LINEAR EQUATIONS

The mathematical task of solving two simultaneous linear equations with two unknowns is the simplest possible example of a problem that can be solved by means of an analog computer. Consider the following two equations:

$$\begin{aligned} x - 2y &= 3 \\ \frac{x}{3} + y &= 3 \end{aligned} \tag{185}$$

The first step in setting up an analog computer solution is to solve one equation for $x$ and the other for $y$:

$$\begin{aligned} x &= 3 + 2y \\ y &= 3 - \frac{x}{3} \end{aligned} \tag{186}$$

Each of these individual equations can be simply implemented with OAs as shown in Fig. 3.27. Note that the sign inversion of an OA has been taken care of by applying negated inputs to the first amplifier. We have somewhat arbitrarily assigned a scale factor of unity to the constant 3, making its representation 3 V. We could just as well have used 0.3 or 6 or any other convenient voltage, but the use of unit voltage as a scale factor makes it somewhat easier to follow what happens. We return to the question of scaling below. We now have two output voltages:

$$\begin{aligned} x &= -(2y - 3) \\ y &= -\left(\frac{x}{3} - 3\right) \end{aligned} \tag{187}$$

where we have taken care of the multiplications by constants by scaling the input resistors. All that is required to complete the solution is to invert the sign

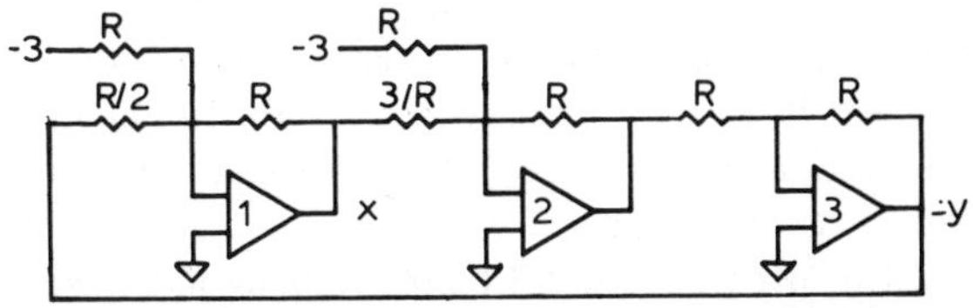


**Fig. 3.27.** Analog computer that can solve two simultaneous linear equations.

of the $y$ output, to connect this signal to the $y$ voltage input of OA1, and to connect the $x$ output (from OA1) to the $x$ input of OA2. The solutions to the equation, that is, $x$ and $y$, can be read from the outputs of OA1 and OA2, respectively. We can alter the constants and solve the resulting new equations essentially instantaneously. We can also change the nature of the equation by altering the scale factors for the $x$ and $y$ input resistors from their present values of $R/2$ and $3R$ respectively. The changing of these scale factors is not without some risk. Consider the set of equations

$$\begin{aligned} x - 2y &= 3 \\ 3x + y &= 3 \end{aligned} \tag{188}$$

which are a perfectly rational set of equations that should yield $y = -6/7$ and $x = 9/7$ as solutions. The most obvious method to set up this set of equations is to leave the circuitry around OA1 and OA3 unchanged and to replace the $3R$ resistor at the $x$ input of OA2 with a $R/3$ resistor. The setup of this equation simulator is perfectly reasonable, but if one implements it in practice and monitors the outputs of any of the amplifiers with an oscilloscope, one discovers that the amplifiers are oscillating. We have just encountered a situation where all the individual subcircuits are perfectly stable, but the resulting super circuit is unstable.

Because we are dealing solely with adder circuits, we know that the problem can not be related to the rate of closure of the open-loop gain with the loop gain. Obviously there must be a second criterion for stability that we have not identified. Let us compare the two implementations. In the first, stable case, the total gain of the circuit is $(-2)(-1/3)(-1)$ or $-2/3$ whereas for the second, unstable, circuit the gain is $(-2)(-3)(-1)$ or $-6$. If the total gain was $+6$, the answer would be obvious; any noise component would be amplified by six and fed back to undergo further amplification, and so forth. This is termed positive feedback. We have an odd number of inverting amplifiers so that the feedback should be negative. If we did view the oscillations on an oscilloscope, we would gain an important clue; the oscillations are at very high frequency and what we have failed to take into account is the phase shift of the amplifiers at frequencies above their $-3$ dB point. Figure 3.28 shows the closed-loop gain and the phase relationships for the two circuits. For both circuits, as the closed-loop gain of each stage begins to roll off, there is an additional 90° phase shift for the frequency components above the break frequency so that the total phase shifts become 270°, 360°, and eventually 450°. The important region is the 360° region, because this is the region where the gain becomes positive again. The two circuits differ by the magnitude of the gain in the region where the phase shift reaches 360°, that is, 0° again. The total gain for the stable case is less than 1 so no sustained oscillation can take place, where the unstable circuit has a region that the gain is greater than +1.

The solution to this problem should be obvious by now. We simply insert a pole in the circuit, that is, a feedback capacitor, preferably in the loop of the

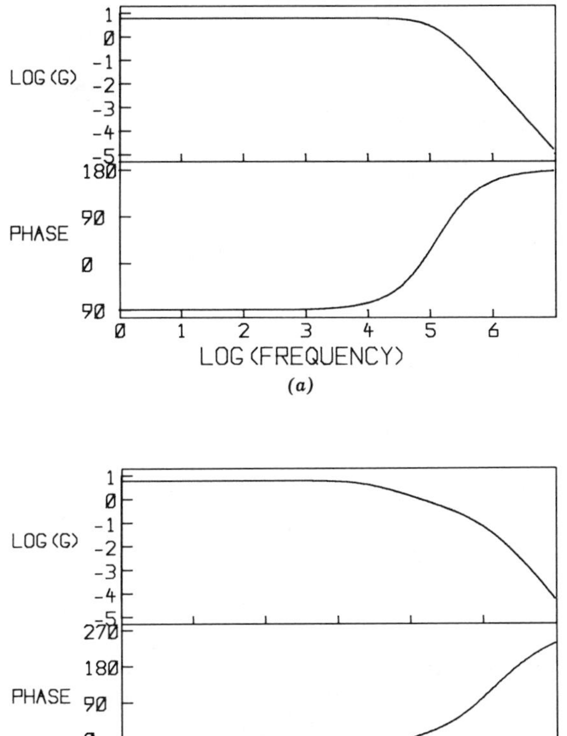


**Fig. 3.28.** Gain and phase shift for the computer loop in Fig. 3.27 as a function of frequency. (*a*) Without capacitance; (*b*) with $C = 2.2$ nF, $R = 10K$.

amplifier with a DC gain of 3, so the total gain of the circuit is less than unity in the region of positive feedback. Because we are interested only in a steady-state solution, almost any capacitor will do as long as it ensures stability. For dynamic circuits of the type we study next, the frequency response problem before oscillation begins. This problem is not peculiar to analog computer circuits; in fact, it is inherent in any feedback or closed-loop control system. A classic study was conducted by H. Nyquist (13) of Bell Laboratories in conjunction with the problem of oscillation in telephone circuits. Nyquist discovered that there is a simple way to predict such oscillations. For our purpose, we can state Nyquist's criterion in the following way. Express the voltage transfer function of the system $H(s)$ in the complex plane as $H(j\omega)$. Plot the real part of the gain on the horizontal axis and the complex part on the vertical axis, allowing the frequency to range from 0 to infinity. If the resulting curve encir-

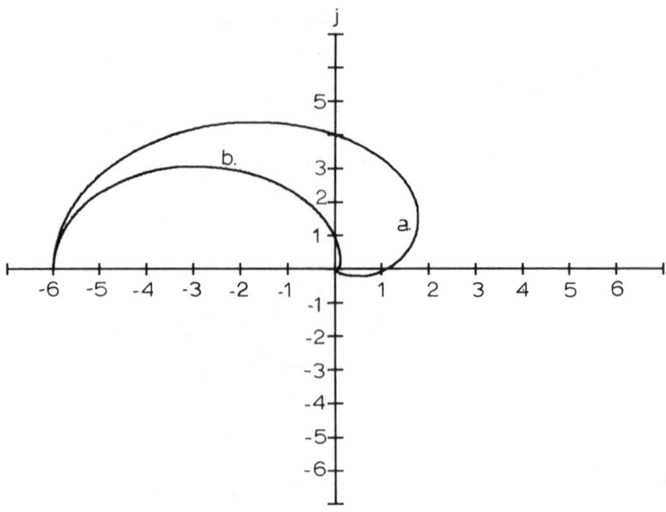


**Fig. 3.29.** Nyquist plot of computer loop in Fig. 3.27. (*a*) Unstable without stabilization capacitor; (*b*) after addition of capacitance to feedback loop of amplifier 3.

cles the point 1 + j0 in a clockwise manner as $\omega$ is increased, then the circuit is unstable. There are several exceptions to this simplification and the interest reader is referred to a fairly detailed discussion by Goldman (14) for a more complete treatment, but the simple criterion is adequate for our purposes. The Nyquist, or complex plane, plot for the two circuits is shown in Fig. 3.29. One can see that the unstable case indeed encircles the magic point. It is important to recognize that the Nyquist plot and the combination of the Bode gain and phase plots both contain the same information about the circuit stability; only the presentation differs. The Nyquist plot has the advantage that the stability indicator is obvious and unambiguous, but a little practice and experience with the Bode representation will make its stability criterion just as lucid to the reader. The choice of one over the other is strictly a matter of personal preference, although admittedly we prefer the Bode representation.

## B. DIFFERENTIAL EQUATIONS

To some extent chemistry is the study of applied differential equations, although often this fact is hidden by the experimental details, and sometimes the link is rather tenuous. Many simple chemical reactions can be expressed in terms of linear differential equations of the form

$$a_n \frac{d^n x}{dt^n} + \ldots a_2 \frac{d^2 x}{dt^2} + a_1 \frac{dx}{dt} + a_0 = f(t) \tag{189}$$

For simplicity's sake, we limit ourselves to first- or second-order cases, in which, using the definition of the relationship between time plane differentials and the Laplace transform representation, we can write

$$\mathcal{L}a_2\frac{d^2x}{dt^2}+a_1\frac{dx}{dt}+a_0-f(t)= \\ a_2 s^2X(s)-sX(0)-x'(0)+a_1 sX(s)-x(0)+a_0s-F(s) \tag{190}$$

One of the useful properties of the Laplace transformation which we mentioned earlier is that if two systems have the same Laplace plane representation, they have the same time plane representation. This means that if we construct a circuit whose transform is identical to the above equation, then this circuit behaves in exactly the same manner as the system we are interested in (or any other system which has the same transform for that matter). As a start, let us consider a simple irreversible first-order chemical reaction whose equation is given by

$$A \xrightarrow{k_f} B$$

$$\frac{dx}{dt}=k_f(A^\circ-x) \tag{191}$$

where $A^\circ$ is the initial concentration of $A$ and $x$ is the amount of $A$ that has reacted to form $B$ at any time $t$.

There are a number of analog computer solutions to this equation, one of which is shown in Fig. 3.30. We start with a summing amplifier OA1 which produces an output $(A^\circ-x)$ from the inputs $A^\circ$ (a constant voltage source) and $+x$ (which is not yet available). Next we invert and scale the signal by means of OA2 whose gain is $-R_0/R_1=-k_f$ to give $-k_f(A^\circ-x)$.

This signal is equal to the time derivative of the desired result $x$, so we next integrate the voltage with OA3 whose $RC$ (ohms and farads) time constant is unity to produce the solution $x$. We connect the output of OA3 back to the $x$ input of OA1 and the circuit is complete. The solution is dynamic, that is, time varying, and there are two ways to initiate the simulation. We can short-circuit the feedback loop of OA3, set the voltage at the $-A^\circ$ input to OA1 to the correct value, and open up the short circuit starting the simulation. Alternatively, we can apply 0 V to the initial concentration input of OA1 and then switch in the voltage representing $-A^\circ$, again initiating the simulation.

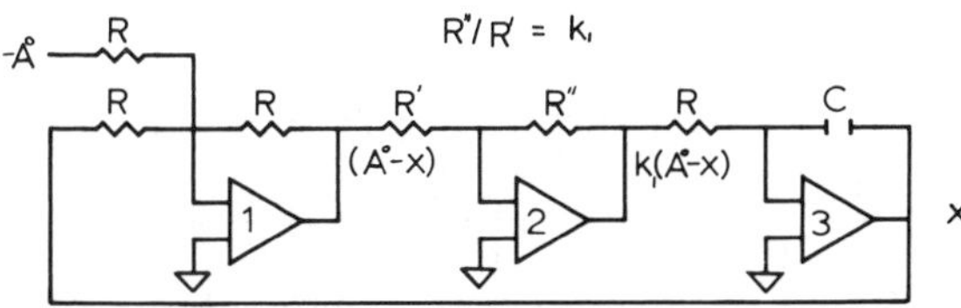


**Fig. 3.30.** An analog computer to simulate a simple first-order kinetics reaction solution to equation 191.

The literature on analog computation usually employs a somewhat different notation scheme than we have employed for our circuit diagrams. A multiple input inverter is symbolized by a triangle with one or more inputs. A circle with a constant within it just prior to the symbol represents the scaled gain of the summer. Placing a rectangular box on the flat side of the triangle converts the symbol into either an integrator or differentiator. One sometimes finds the symbol $dx/dt$ or $x$ within the triangle and usually the output leg of the symbol also defines what the operation is. One often wishes the integrator to start at some value other than zero as an initial condition. This is normally done by placing a charge on the feedback capacitor and is schematically represented by a signal line entering the top of the rectangular addition to the basic symbol.

Expansion to include a reversible case is simple. The basic relationship is

$$A \underset{k_b}{\overset{k_f}{\rightleftarrows}} B \tag{192}$$

where the differential rate equation is

$$\frac{dx}{dt} = k_f(A^\circ - x) - k_b(B^\circ + x) = k_fA^\circ - k_bB^\circ - k_fx - k_bx \tag{193}$$

and where $B^\circ$ is now the initial concentration of $B$. The implementation of this system is shown in Fig. 3.31, where the presence of OA4 allows us to measure the amount of $A$ remaining in solution.

The extension to two consecutive irreversible first-order reactions is also simple:

$$A \xrightarrow{k_1} B \xrightarrow{k_2} C \tag{194}$$

where the differential rate equations are

$$\frac{dx}{dt} = k_1(A^\circ - x) = -k_1(x - A^\circ) \tag{195}$$

and

$$\frac{dy}{dt} = k_2(x - y) = -k_2(x - y) \tag{196}$$

and where $x$ is the amount of $B$ formed from $A$ and $y$ is the amount of $C$ formed from $B$. As the complexity of the system increases, the number of

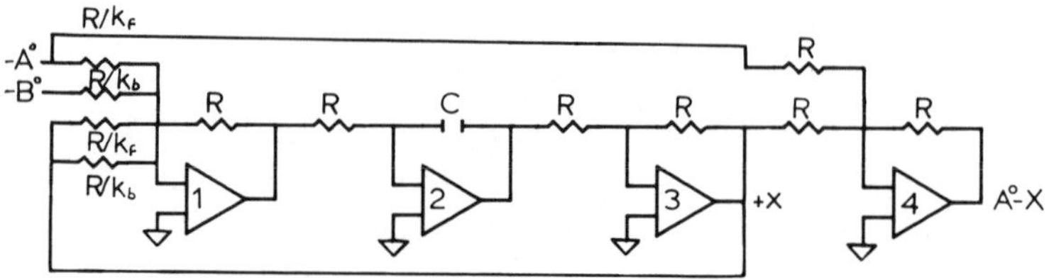


**Fig. 3.31.** An analog computer that simulates a reversible first-order kinetics reaction (equation 193).

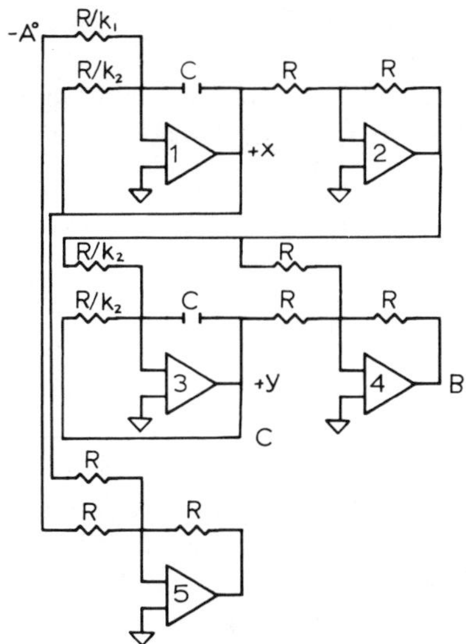


**Fig. 3.32.** An analog computer that simulates a consecutive first-order kinetics reaction (equations 195 and 196).

ways that it can be implemented increases as well. The circuit shown in Fig. 3.32 utilizes two inputs to each of the integrators to minimize the number of amplifiers involved. The amplifier is actually integrating the input currents at the negative input so that any number of inputs are permissible. The relative scaling between the inputs is accomplished by the input resistances.

## 1. Scaling Factors

The units of the system that we may be trying to simulate could be moles, meters, liters, or some other unit, whereas the response generated by our simulator is in volts. It is necessary to establish a conversion between the two systems. Not only must the absolute units be scaled, but often the time scale must be compressed or expanded as well. For example, a reaction with a half-life of either a nanosecond or a week would not be conveniently simulated in real time, that is, on the true time scale.

### a. Amplitude Scaling

When we convert from the true $x$ units of the system to be simulated to the voltage units $X$ of the simulator, we are making the transformation

$$X = a_x x \tag{197}$$

where $a_x$ is the scale factor to be employed. We do not have totally free choice in selecting the value for $a_x$. There are three factors that govern our choice: (1) the voltages that result must be within the range of voltages that the amplifiers can produce or tolerate, nominally ±10 V for most solid-state systems; (2) the scale factor should be as large as possible so that percentage errors due to noise and drift, for example, are minimized; and (3) the stability and loop-gain requirements of the amplifiers must not be ignored. Consider the two simultaneous equations

$$
\begin{aligned}
x - 2y &= 300 \\
\frac{x}{3} + y &= 300
\end{aligned}
\tag{198}
$$

These equations are identical to our first example except that the constants are a factor of 100 larger. If we choose to, we could proceed as before (assuming the availability of a −300 V signal), but the ideal output voltages would be $x = +540$ and $y = +120$ V which, if we were using ±15 V amplifiers, are clearly impossible so that a simulation error would result. All amplifiers would go to positive voltage saturation, and in fact amplifiers OA1 and OA2 would probably be destroyed by the overload voltage at their inputs. The most obvious solution is to use a scale factor of 100, that is, $a_x = a_y = \frac{1}{100}$, in which case the rescaled equations become

$$
\begin{aligned}
100X - 2(100Y) = 300 \text{ or } X = 3 + 2Y \\
\frac{100}{3}X + 100Y = 300 \text{ or } Y = 3 - \frac{X}{3}
\end{aligned}
\tag{199}
$$

which is precisely the condition we had in the first example. In order to convert the scaled results to the system results, we multiply both the $X$ and $Y$ output voltages by 100. Sometimes, however, it is not convenient to scale both equations by the same amount. We could just as well choose $a_x = \frac{1}{100}$ and $a_y = \frac{1}{30}$, in which case we have

$$
\begin{aligned}
100X - 2(30Y) = 300 \text{ or } X = 3 - 0.60Y \\
100\frac{X}{3} + 30Y = 300 \text{ or } Y = 20 - \frac{10}{9}X
\end{aligned}
\tag{200}
$$

We implement this by replacing the $-Y$ input of OA1 in Fig. 3.27 and with a $R/0.6$ resistor and the $+X$ input resistor of OA2 with a $0.9R$ resistor. Now the output voltages are $Y = +4$ V and $X = +5.4$ V.

b. Time Scaling

It is often necessary to scale the time response of a system in order to bring the simulation time scale to some convenient value, usually determined by the frequency response of the amplifiers and the recording media on the short time end and by the stability of the integrators and amplifiers and patience of the scientist on the long time end. We do this by making a second transformation from real time $t$ to simulation time $T$ by means of the relationship

$$
T = a_t t \tag{201}
$$

where $a_t$ is the time scale factor. For convenience of discussion we define the operator $p$ and its scaled equivalent $P$ such that

$$p = \frac{d}{dt} = \frac{a_t d}{dT} = a_t P \tag{202}$$

where $p$ represents differentiation with respect to real time and $P$ is differentiation with respect to simulation time. If we desire that the simulation time be 100 times longer than real time, then $a_t = 100$ or $T = 100t$. Consider the expression

$$\frac{dy}{dt} = 50 \text{ or } py = 50 \tag{203}$$

Using $a_t = 100$ this expression becomes $100Py = 50$ or $Py = 0.5$. If we integrate this expression and the time required for the output to reach 1 V is 1 sec, then the real time that would have been required to reach 1 V is 0.01 sec.

For complex simulations one sometimes needs operations that are not readily available with an OA, such as multiplication or division and occasionally log or exponential functions. One can purchase modules that perform these functions and the general computing notation is a tetrahedron with the symbol for the operation inside it. The log and exponential operations are actually generated by means of an OA and one or more transistors. These circuits take advantage of the exponential current-voltage relationship between current and voltage in diodes or *NPN* or *PNP* transistors.

Figure 3.33 shows some very elementary log and exponential circuits. More precise circuits are available that utilize several amplifiers and transistors and that are much better approximations to the ideal functions (12,15). Very accurate commercial circuits are also available. One thing that the reader should be aware of is that most of these circuits are monopolar; that is, they function properly only for one polarity of the input voltage or current.

An example of a system where multiplication is required is the irreversible first-order system with a competitive side reaction, which is represented by

$$\begin{aligned} A + R &\xrightarrow{k_1} C \\ R &\xrightarrow{k_2} B \end{aligned} \tag{204}$$

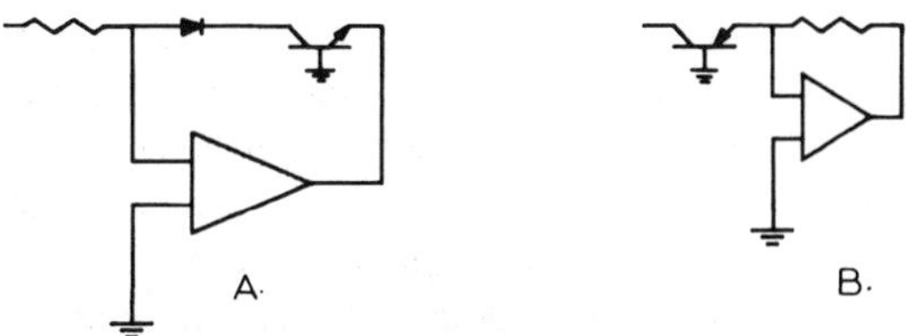


**Fig. 3.33.** Elementary transcendental function. (*a*) Logarithmic amplifier circuit; (*b*) antilogarithmic amplifier circuit.

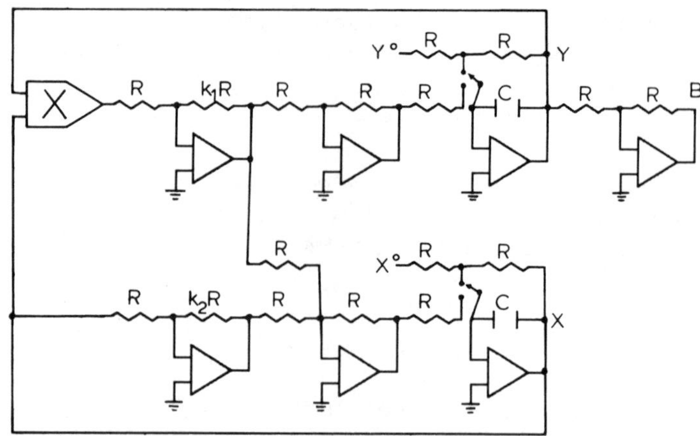


**Fig. 3.34.** Analog computer simulation of an irreversible first-order reaction with a competitive side reaction (equations 205 and 206).

The differential rate equations are

$$\frac{dR}{dt} = -k_1AR - k_2R$$
$$\frac{dA}{dt} = -k_1AR \tag{205}$$

which when transformed into simulation variables lead to

$$\frac{dx}{dt} = -k_1axy - k_2x$$
$$\frac{dv}{dt} = -k_1rxy \tag{206}$$

The resulting circuit is given in Fig. 3.34.

## VIII. FILTER CIRCUITS

Filter circuits are one of the most common applications of OAs in scientific equipment. Filters are employed to remove undesired frequency components from a signal. The undesired components may or may not be random components. For example, in spectroscopy, where the exciting light beam is chopped, the use of a band-pass filter allows one to reject the dark current and slowly changing signal components due to instabilities in the light source or building vibrations as well as high-frequency "true noise" components. In second harmonic alternating current polarography one excites an electrochemical cell with a sinusoidal potential of amplitude $\Delta E$ at a frequency $f_o$ and measures the strong response at $2f_o$. An electrochemical cell produces the

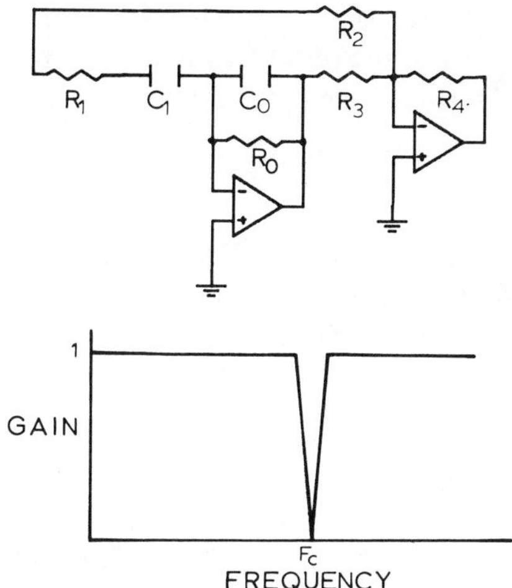


**Fig. 3.35.** A band rejection or notch filter.

strongest response component at the exciting frequency and the component of interest is less than 14Δ $En$ times the $f_o$ response or typically 0.14 times the $f_o$ component for a 10 MV excitation. One must extract the desired component from both the noise and the much stronger evoked response at $f_o$. If the unwanted component is at lower frequency one could use a high-pass filter. Alternately one could use a band-pass filter (Fig. 3.11) tuned at the frequency of interest, $2f_o$. One further possibility is to use a band reject filter, that is, one that strongly attenuates the center frequency, to remove the stronger $f_o$ component followed by further signal processing. A band reject filter can be constructed from a band-pass filter by the method shown in Fig. 3.35. A little thought shows that the two pairs of resistors, $R_1$, $R_0$ and $R_2$, $R_3$, must be perfectly matched to take into account the gain of the band-pass filter gain at the center frequency.

Because we must deal with real components in this type of circuitry, it is important to be able to determine how the performance of the filter changes if the values of the resistors and capacitors making up the filter change, owing either to initial mismatching or to aging or thermal effects. The above sentence carries within it the implication that a filter that is initially perfectly tuned to perform a given function may drift out of tuning and at some latter date fail to perform its design function. Unfortunately this implication is only too true.

## A. DEFINITION OF TERMS IN ACTIVE FILTERS

Before we proceed to look at active filters, we must define a few new terms and review some familiar ones. The transfer function of a filter is symbolized here as $H(s)$ in Laplace space. The gain or magnitude of the transfer function in response to a steady-state sinusoidal excitation is symbolized by *either* $H(j\omega)$ or, equivalently, by $G(s)$. The phase angle between excitation and response as a function of the exciting frequency is represented by $\phi(\omega)$. In some applications it is important to know how $\phi(\omega)$ changes as a function of frequency and this parameter is given the symbol $\tau(\omega)$ and is known as the *group delay*:

$$\tau(\omega) = \frac{d\phi(\omega)}{d\omega} \tag{207}$$

The pass-band gain $H_0$ is the gain of the circuit in the region of maximum amplification. In low-pass or high-pass filters, it is the frequency-independent gain value, whereas in band-pass filters it is the peak gain. The frequency region where the signal is frequency independent is termed the pass band, and the region or regions where the response rolls off is called the stop band.

### 1. Sensitivity Factors

The way that one property $U$ of a filter changes with alterations in a second property or parameter $v$ is termed its sensitivity, symbolized by $S_v^U$ and defined by

$$S_v^U = \frac{dU(\omega)/U(\omega)}{dv/v} = \frac{v}{U(\omega)}\frac{dU(\omega)}{dv} \tag{208}$$

where the parameters have been normalized relative to their nominal value so that $S_v^U$ actually represents relative changes from the nominal values of the circuit parameters. The sensitivity functions so defined are theoretically valid for infinitesimal changes from their nominal values. In practice the expressions derived using this approach are sufficiently accurate even for changes of 5% from nominal value and at least indicative for variations of up to 10%.

In dealing with a filter, one of the most important parameters is $S_v^G$, the sensitivity of the circuit gain with respect to changes in the parameter $v$. The gain is often expressed in the dB notation:

$$g(\omega) = 20 \log_{10}[G(\omega)] \tag{209}$$

One often wishes to know how many decibels the magnitude of the filter response changes for a given *normalized* change in the parameter $v$; that is, the gain change is absolute in terms of dB but the variation in the parameter $v$ is relative. Mathematically the generating equation is

$$S_v^g = \frac{dg(\omega)}{dv/v} = \frac{d[20 \log(G(\omega)]}{dv/v} \tag{210}$$

$$= 20 \log_e \frac{d[20\log_e(G(\omega)]}{dv/v} \tag{211}$$

$$= 8.7\frac{dG(\omega)/G(\omega)}{dv/v} = 8.7S_v^G \tag{212}$$

### 2. Basic Filter Formats

There are several basic ways of generating an active filter circuit. We have already become familiar with the simple single pole low-pass and high-pass filters and the simple form of the band-pass filter. We will also investigate the infinite gain multiple feedback, controlled voltage source, state variable, and negative immitance converter or gyrator types.

### a. Simple Pole-Zero Types of Filters

By now the reader is well aware that the circuit in Fig. 3.8 is a low-pass filter as well as an integrator for high frequencies. It is also known as a stabilized integrator, because it prevents the integrator from drifting to saturation by the integration of error currents arising from offset or bias currents or offset voltages. We reexamine it primarily to clarify the use of some of the new notation and concepts.

### b. Low-Pass Filters

The transfer function of the circuit in Fig. 3.8 is

$$H(s) = \frac{H_o \omega_o}{s + \omega_o} \tag{213}$$

where we now use the notation $\omega_o$ to represent the break or critical frequency in radians per second rather than the more familiar notation $a$, $b$, $c$, and so forth. The value of $\omega_o$ is still given by $1/R_0C_0$ and $H_o$ is still equal to $R_0/R_1$. The gain is

$$G(\omega) = \frac{H_o \omega_o}{\omega^2 + \omega_o^2} \tag{214}$$

and the phase angle is

$$\phi(\omega) = -\arctan\left(\frac{\omega}{\omega_o}\right) \tag{215}$$

The group delay, that is, the change in phase angle with respect to frequency is

$$\tau(\omega) = \frac{\cos^2(\phi)}{\omega_o} \tag{216}$$

The sensitivity of the gain with respect to the critical or break frequency $\omega_o$ and hence either $R_0$ or $C_0$ is given by

$$\begin{aligned} S^G_{\omega_o} &= \frac{H_o}{G(\omega)} \frac{dG(\omega)}{d(\omega_o)} = \frac{H_o}{H_o\omega_o/(\omega^2 + \omega_o^2)^{1/2}} \left[\frac{d(H_o\omega_o/(\omega^2 + \omega_o^2)^{1/2})}{dH_o}\right] \\ &= 1 - \frac{\omega_o}{\omega^2 + \omega_o^2} = 1 - \frac{G^2}{H_o^2} \end{aligned} \tag{217}$$

Similarly, the gain sensitivity with respect to $H_o$ or $R_1$ is given by

$$\begin{aligned} S^G_{H_o} &= \frac{H_o}{G(\omega)} \frac{dG(\omega)}{dH_o} = \frac{H_o}{H_o\omega_o/(\omega^2 + \omega_o^2)^{1/2}} \left[\frac{d(H_o\omega_o/(\omega^2 + \omega_o^2)^{1/2}}{dH_o}\right] \\ S^G_{H_o} &= 1 \end{aligned} \tag{218}$$

and the sensitivity of gain with respect to $R_o$ is

$$S_{R_0}^{G} = \frac{R_1}{R_0} S_{\mathrm{Ho}}^{G} = \frac{R_1}{R_0} \tag{220}$$

The other important sensitivity parameters are

$$\begin{gathered} S_{\omega_o}^{\phi} = -\frac{\sin 2\phi}{2\phi} \\ S_{\omega_o}^{\tau} = 2\sin^2\phi - 1 = -\cos 2\phi \end{gathered} \tag{221}$$

This circuit has a rolloff of –20 dB per decade at frequencies higher than $\omega_o$. There are several ways to increase the attenuation in the stop band. The simplest is to cascade two or more filters together, each one producing a –20 dB/decade attenuation, so that one can obtain –40 and –60 dB for two or three stage filters, for example. Although the cost of amplifiers has decreased dramatically, the cascaded system is not particularly common. There is a way to generate a 40 dB/decade rolloff in the stop band with one amplifier using networks that we have studied.

#### c. Complex Conjugate Pole Pair

The complex conjugate pole pair has the same functionality as two cascaded single pole filters. The result for the two cascaded single poles can be derived by considering the output of the first section as the input to a second section.

$$E_{o1} = \frac{-E_{i1}H_{o1}\omega_{o1}}{s + \omega_{o1}} \tag{222}$$

$$E_{o2} = \frac{-E_{i2}H_{o2}\omega_{o2}}{s + \omega_{o2}} = \frac{E_{i1}H_{o1}H_{o2}\omega_{o1}\omega_{o2}}{(s + \omega_{o1})(S + \omega_{o2})} \tag{223}$$

#### d. High-Pass Filters

We have already studied the single pole high-pass filter in some detail, first as a stabilized differentiator and then in its filter guise. The circuit is shown in Fig. 3.10. The relevant parameters are as follows:

$$S_{\omega_o}^{G} = -\frac{\omega_o^2}{\omega^2}\frac{G^2}{H_o^2} \tag{224}$$

$$S_{H_o}^{G} = 1 \tag{225}$$

$$S_{\omega_o}^{\phi} = \frac{+\sin 2\phi}{2\phi} \tag{226}$$

$$S_{\omega_o}^{\tau} = +2\cos^2\phi - 1 = \cos 2\phi \tag{227}$$

$$H(s) = \frac{H_o s}{s + \omega_o} \tag{228}$$

$$G(\omega) = \left(\frac{H_o^2\omega^2}{\omega^2 + \omega_o^2}\right)^{1/2} \tag{229}$$

$$\phi(\omega) = \frac{\pi}{2} - arctan\,\frac{\omega}{\omega_o} \tag{230}$$

$$\tau(\omega) = \frac{\sin^2 \phi}{\omega_o} \tag{231}$$

The circuit has already been discussed in sufficient detail and is not pursued any further.

The complex pole pair formulation can once again be formed either by cascading two simple single pole high-pass filters or by a single amplifier implementation such as that shown in Fig. 3.36. The results for this filter are as follows:

$$H(s) = \frac{H_o s^2}{s^2 + \alpha\omega_o s + \omega_o^2} \tag{232}$$

$$G(\omega) = \left(\frac{(H_o^2)\omega^4}{\omega^4 + \omega^2\omega_o^2(\alpha^2 - 2) + \omega_o^4}\right)^{1/2} \tag{233}$$

$$\phi(\omega) = \pi - \arctan\left[\frac{1}{\alpha}\left(2\frac{\omega}{\omega_o} + \sqrt{4 - \alpha^2}\right)\right] - \arctan\left[\frac{1}{\alpha}\left(\frac{2\omega}{\omega_o} - \sqrt{4 - \alpha^2}\right)\right] \tag{234}$$

$$\tau(\omega) = \frac{2\sin^2\phi}{\alpha\omega_o} - \frac{\sin 2\phi}{2\omega} \tag{235}$$

$$S_{\omega_o}^{G} = -\frac{\omega_o^2}{\omega^2}\frac{G^2}{H_o^2}\left(2\frac{\omega_o^2}{2} + \alpha^2 - 2\right) \tag{236}$$

$$S_{\alpha}^{G} = -\alpha^2\frac{\omega_o^2}{\omega^2}\frac{G^2}{H_o^2} \tag{237}$$

$$S_{\omega_o}^{\phi} = \frac{2\omega_o}{\alpha\omega\phi}\sin^2\phi + \frac{\sin 2\phi}{2\phi} \tag{238}$$

$$S_{\alpha}^{\phi} = \frac{\sin 2\phi}{2\phi} \tag{239}$$

$$S_{\omega_o}^{\tau} = \left(\frac{2\sin 2\phi}{\alpha\tau} - \frac{\omega_o}{\omega\tau}\cos 2\phi\right)\frac{\phi}{\omega_o}S_{\omega_o}^{\phi} - \frac{2\sin^2\phi}{\alpha\omega_o\tau} \tag{240}$$

$$S_{H_o}^{G} = 1 \tag{241}$$

### e. Band-Pass Filters

We have also briefly investigated the band-pass filter formed by incorporating the feedback loop of the single pole low-pass filter and the input network of the single pole high-pass filter in one circuit. Briefly, the resulting parameters and the sensitivity factors for this circuit are given by

$$H(s) = \frac{H_o\alpha\omega_o s}{s^2 + \alpha\omega_o s + \omega_o^2} \tag{242}$$

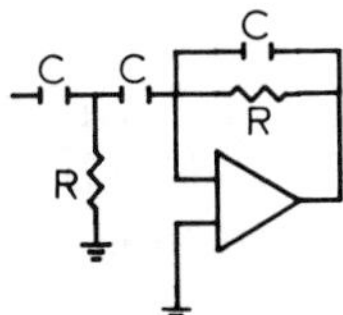


**Fig. 3.36.** Single amplifier implementation of a complex pole pair high-pass filter.

where $\omega_o$ is the center frequency in radians. The circuit in Fig. 3.11 is the case where $\omega_o = 1/R_0C_0 = 1/R_1C_1$ (curve c).

In this case the term $Q$, the figure of merit of the filter, is the reciprocal of $\omega$. A more meaningful definition is

$$Q = \frac{\omega_o}{\omega_2 - \omega_1} = \frac{f_o}{f_2 - f_1} = 1/\alpha \tag{243}$$

where $f_2$ and $f_1$ are the frequencies where the response is −3 dB from $H_o$, the pass-band gain which occurs at $F_o$.
The steady-state transfer function for a sinusoidal excitation is

$$H(j\omega) = \frac{H_o}{1 + jQ(\omega/\omega_o - \omega_o/\omega)} \tag{244}$$

and the remaining paramaters are

$$G(\omega) = \left(\frac{H_o^2}{1 + Q^2(\omega/\omega_o - \omega_o/\omega)^2}\right)^{1/2} \tag{245}$$

$$= \left(\frac{H_o^2\,\alpha^2\omega^2\omega_o^2}{\omega^4 + \omega^2\omega_o^2(\alpha^2 - 2) + \omega_o^4}\right)^{1/2} \tag{246}$$

$$\phi(\omega) =$$

$$\frac{\pi}{2} - \arctan\left(2Q\frac{\omega}{\omega_o} + \sqrt{4Q^2 - 1}\right) - \arctan\left(2Q\,\frac{\omega}{\omega_o} - \sqrt{4Q^2 - 1}\right) \tag{247}$$

$$\tau(\omega) = \frac{2Q}{\omega_o}\cos^2\phi + \frac{\sin 2\phi}{2\omega} \tag{248}$$

Therefore the transfer function for the two cascaded circuits has the form

$$H(s) = \frac{H_o\alpha\omega_o s}{s^2 + \alpha\omega_o s + \omega_o^2} \tag{249}$$

A complex conjugate circuit with the same functionality is shown in Fig. 3.40 below, where

$$H_o = \frac{R_0}{R_1 + R_2} \tag{250}$$

$$\omega_o = \frac{R_1 + R_2}{R_1R_2R_0C_1C_0} \tag{251}$$

The other properties of the circuit are identical to the preceding.

#### f. The Infinite Gain Multiple Feedback Configuration

There is one generalized configuration that is encountered commonly in −40 dB/decade roll of circuits in high-pass, low-pass, and band-pass filters. The generalized circuit diagram for this type of filter is shown in Fig. 3.37. Each element $Y_i$ is a single resistor or capacitor. The generalized transfer function for

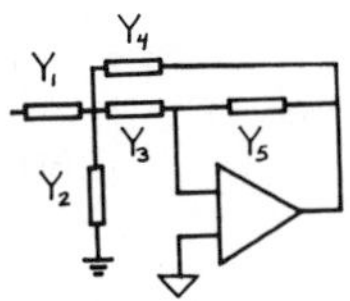


**Fig. 3.37.** General form of the infinite gain multiple feedback filter.

this circuit including the effect of finite amplifier gain is given by

$$E_o\frac{(s)}{E_1}(s) = \frac{-Y_1Y_3}{Y_5(Y_1+Y_2+Y_3+Y_4) + Y_3Y_4 + [(Y_3+Y_5)(Y_1+Y_2+Y_4) + Y_3Y_5](s + a_o)/a_oA_o} \tag{252}$$

where the tildes over the $Y_i s$ have been eliminated for simplicity and $a_o$ and $A_o$ have the same meanings as in the discussion of stability. The response circuit is complex enough that it warrants employing the simplifying assumption that the open-loop gain is effectively infinite. In fact, as long as the difference between the open-loop gain and the transfer function gain is 80 dB at all frequencies this assumption is valid. With this assumption made, and it is important enough that we occasionally recall it, the transfer function becomes

$$H(s) = \frac{-Y_1Y_3}{Y_5(Y_1 + Y_2 + Y_3 + Y_4) + Y_3Y_4} \tag{253}$$

This form is used to generate the three basic filter circuits. In each case the component retains the subscripts employed in Fig. 3.37.

## B. LOW-PASS FILTERS

The low-pass filter is constructed by employing $R_1$, $R_3$, $R_4$, $C_2$, and $C_5$. Note that all circuits employing this configuration produces a sign inversion. The transfer function is

$$H(s) = \frac{1/R_1R_3C_2C_5}{s^2+(s/C_2)(1/R_1 + 1/R_3 + 1/R_4) + 1/R_3R_4C_2C_5} \tag{254}$$

For this circuit and the following ones we follow the conventions that we have previously established. The relevant parameters for this circuit are as follows:

$$H_o = \frac{R_4}{R_1} \tag{255}$$

$$\omega_o = \left(\frac{1}{R_3R_4C_2C_5}\right)^{1/2} \tag{256}$$

$$\alpha = \sqrt{\frac{C_5}{C_2}}\left(\sqrt{\frac{R_3}{R_4}} + \sqrt{\frac{R_4}{R_3}} + \frac{\sqrt{R_3R_4}}{R_1}\right) \tag{257}$$

$$\phi = \pi + \phi_{LP} \tag{258}$$

$$\tau = \tau_{LP} \tag{259}$$

$$S_{R_3}^{\omega_o} = S_{R_4}^{\omega_o} = S_{C_2}^{\omega_o} = S_{C_5}^{\omega_o} = -\frac{1}{2} \tag{260}$$

$$S_{C_5}^{\alpha} = -S_{C_2}^{\alpha} = \frac{1}{2} \tag{261}$$

$$S_{R_1}^{\alpha} = \frac{1}{\alpha\omega_o R_1 C_2} \tag{262}$$

$$S_{R_3}^{\alpha} = \frac{1}{2} - \frac{1}{\alpha\omega_o R_3 C_2} \tag{263}$$

$$S_{R_4} = \frac{1}{2} - \frac{1}{\alpha\omega_o R_4 C_2} \tag{264}$$

$$S_{R_4}^{H_o} = -S_{R_1}^{H_o} = 1 \tag{265}$$

The other sensitivity parameters are identical to those of the complex conjugate pole pair. To tune a circuit of this type is not as complex as one might imagine and is in fact easier than the previous implementation. Since one cannot count on finding resistors and capacitors of the precise values desired, one must establish a rational tuning procedure. The available values for capacitors are far more restricted than the values for resistors, and it is more costly and complex to trim capacitors, so normally one chooses fixed capacitors of a convenient value and then selects the resistors and employs resistive trimmers whenever required. One tries to select the capacitors to have the same value (at least to the degree of effort and or cost that one is willing to expend). A plot of the circuit gain in decibels versus $\omega/\omega_o$ is shown in Fig. 3.38. From this it can be seen that

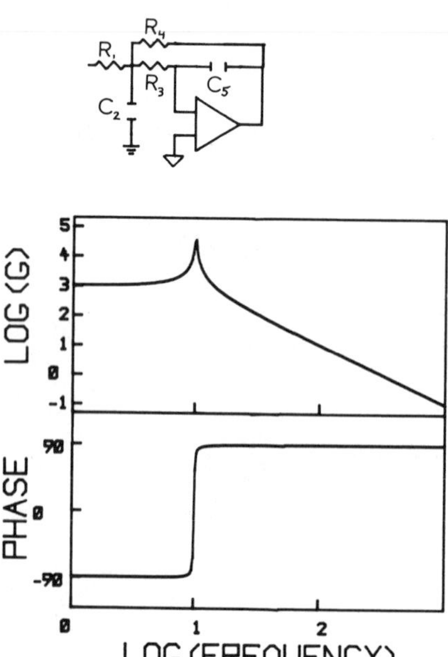


**Fig. 3.38** Infinite gain multiple feedback implementation of a low-pass filter, shown below.

at $\omega = 10\omega_o$ and at $\omega = 0.1\omega_o$ the gain is essentially independent of the parameter. Further, for $\alpha$ less than $\sqrt{2}$ there is a relatively distinct peak in the gain at a frequency $\omega_\alpha = \omega_o\sqrt{1-\alpha^2/2}$, which is called the *alpha peak frequency*. For alpha greater than $\sqrt{2}$ there is no peak and the −3 dB frequency can be used instead. The recommended tuning procedure is to set $\omega = 10\omega_o$ and adjust $R_3$ to give the desired gain. Next, one excites the circuit at $\omega_\alpha$ and adjusts $R_1$ to give the desired gain at the alpha peaking frequency. One can estimate the amount of trim required in $R_1$ and $R_3$ from the sensitivity equations. For a circuit where one wishes to have a known $H_o$ and the design procedure is to choose $C_2 = C$, where $C$ is some convenient value (i.e., one that is readily available commercially), normally a member of the series 1, 1.5, 2.2, 3.3, 4.7, 6.8, 10 for capacitances above 0.01 MF. For smaller values one can find capacitances in a more dense series with 20 entries from 1 to 10. Next one selects the value for $C_5 = KC_2$ where $K$ must be such that $C_5$ is also a readily available type, preferably polystyrene, silver mica, or NPO type ceramic for both members. One next calculates $R_4$, $R_4$, and $R_1$ from

$$R_4 = \frac{\alpha}{2\omega_o C}\left(1 \pm \sqrt{1 - \frac{4(H_o + 1)}{K\alpha^2}}\right) \tag{266}$$

$$R_1 = \frac{R_4}{H_o} \tag{267}$$

$$R_3 = \frac{1}{\omega_o^2 C^2 R_4 K} \tag{268}$$

One should restrain the tendency to employ $\alpha$ less than about 0.1, unless $H_o$ is less than 10 or one has an amplifier with phenomenal characteristics, because one must maintain the 80 dB loop gain. Normally one employs this circuit for simple stop-band filtering so that the use of an $\alpha$ greater than $\sqrt{2}$ (a maximally flat filter) is not particularly useful, and in such cases one may employ $H_o$ values of about 100, always keeping in mind the 80 dB separation requirements. The tuning of this circuit is simplified by the fact that $\omega_o$ is independent of $R_1$; hence one can tune for $\omega_o$ using $R_3$ and then adjust for using $R_1$ without affecting the value for $\omega_o$

## C. HIGH-PASS FILTERS

The high-pass implementation using the generalized circuit is formed by using $C_1$, $C_3$, $C_4$, $R_2$, $R_5$. As shown in Fig. 3.39 the transfer function is again of the same form as for the complex pole pair form and in this implementation becomes

$$H(s) = \frac{-(C_1/C_4)s^2}{s^2 + s(1/R_5)(C_1/C_3C_4 + 1/C_4 + 1/C_3) + 1/R_2R_5C_3C_4} \tag{269}$$

The relevant parameters for this implementation are

$$H_o = \frac{C_1}{C_4} \tag{270}$$

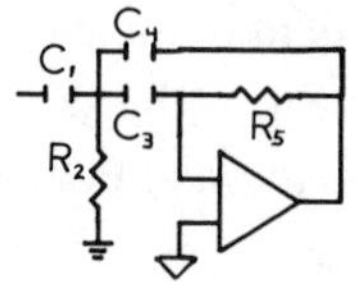


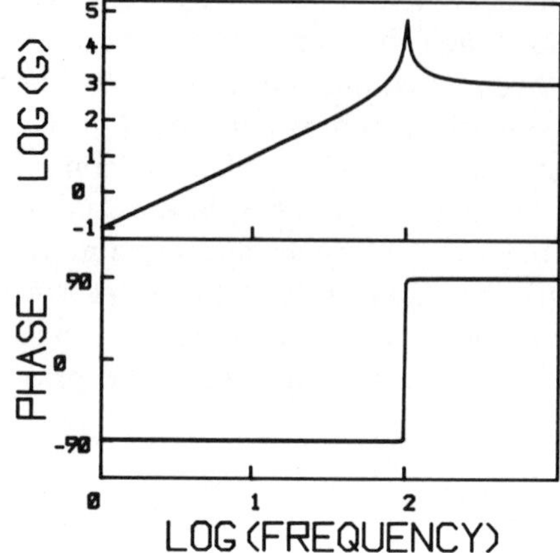


**Fig. 3.39.** Infinite gain multiple feedback implementation of a high-pass filter.

$$\omega_o = \left(\frac{1}{R_2R_5C_3C_4}\right)^{1/2} \tag{271}$$

$$\alpha = \sqrt{\frac{R_2}{R_5}}\left[\frac{C_1}{\sqrt{C_3C_4}} + \sqrt{\frac{C_3}{C_4}} + \sqrt{\frac{C_4}{C_3}}\right] \tag{272}$$

$$\phi = \pi + \phi_{HP} \tag{273}$$

$$\tau = \tau_{HP} \tag{274}$$

$$S_{R_2}^{\omega_o} = S_{R_5}^{\omega_o} = S_{C_3}^{\omega_o} = S_{C_4}^{\omega_o} = \frac{-1}{2} \tag{275}$$

$$S_{R_2}^{\alpha} = -S_{R_6}^{\alpha} = \frac{1}{2} \tag{276}$$

$$S_{C_3}^{\alpha} = \frac{1}{2} - \frac{1}{\alpha\omega_o R_5 C_3}\left[\frac{C_1}{C_3} + 1\right] \tag{277}$$

$$S_{C_4}^{\alpha} = \frac{1}{2} - \frac{1}{\alpha\omega_o R_5 C_4}\left[\frac{C_1}{C_3} + 1\right] \tag{278}$$

$$S_{C_1}^{\alpha} = \frac{1}{\alpha\omega_o R_5}\left[\frac{C_1}{C_3C_4}\right] \tag{279}$$

$$S_{C_1}^{H_o} = -S_{C_4}^{H_o} = 1 \tag{280}$$

Selection of the component values for this filter proceed much as before. Once one knows the required $H_o$, $\omega_o$, and $\alpha$ values one selects a convenient (i.e. available) value $C$ for $C_1$, and normally chooses $C_3$ to be identical to $C_1$ as well. The value for the other capacitance components $C_4$ is simply $C_1/H_o$. One can then compute the nominal values for $R_2$ and $R_5$ from

$$R_5 = \frac{1}{\alpha\omega_o C}(2H_o + 1) \tag{281}$$

$$R_2 = \frac{\alpha}{\omega_o C(2H_o + 1)} \tag{282}$$

Tuning this circuit is conceptually not much different from the low-pass case. The log (gain) verses $\log(\omega/\omega_o)$ plot for the high-pass filter is essentially the mirror image of the low-pass filter. The nominal tuning procedure is set $\alpha$ with *either* $R_2$ or $R_5$ at the frequency where the alpha peak occurs. This is not the alpha peaking frequency, because $\omega_o$ has not been set at this time. One then sets the exciting frequency at $10\omega_o$ and adjusts $\omega_o$ by adjusting $R_2$ and $R_5$ *simultaneously* by the *same* percentage so that $\alpha$ remains constant. From this discussion it is obvious that trimming the high-pass filter is far less fun than trimming the low-pass. The problem arises from our effort to avoid trimming the capacitors. This is normally justified since trimmer capacitors, also known as padders, are bulky, more expensive, and generally more obnoxious than resistive trimmers. A superior tuning procedure, at least in terms of experimental simplicity, would be to first tune $\omega_o$ at either $10\omega_o$ or $0.1\omega_o$ using $R_2$ and then set the value for alpha at the alpha peaking frequency using a trim on $C_1$ that does not alter $\omega_o$. As in the low-pass case, there must be 80 dB loop gain.

## D. BAND-PASS FILTERS

There are a number of ways to implement a band-pass filter using the general form of the five-element circuit in Fig. 3.40. Some are more practical than others in terms of sensitivities and tuning. A fairly practical form employs $R_1$, $R_2$, $R_5$, $C_3$, and $C_4$. Once again the format of the transfer function is identical to that of the complex conjugate pole representation, which in the present case reduces to

$$H(s) = \frac{-s(1/R_1C_4)}{s^2 + s(1/R_5)(1/C_3 + 1/C_4 + (1/R_5C_3C_4)(1/R_1 + 1/R_2)} \tag{283}$$

For this implementation the pass-band gain is given by

$$H_o = \frac{1}{(R_1/R_5)(1 + C_4/C_3)} \tag{284}$$

and the other parameters are

$$\omega_o = \left[\frac{1}{R_5C_3C_4}\left(\frac{1}{R_1} + \frac{1}{R_2}\right)\right]^{1/2} \tag{285}$$

$$\frac{1}{Q} = \alpha = \sqrt{\frac{1}{R_5(1/R_1 + 1/R_2)}}\left[\sqrt{\frac{C_3}{C_4}} + \sqrt{\frac{C_4}{C_3}}\right] \tag{286}$$

$$\phi = \pi + \phi_{BP} \tag{287}$$

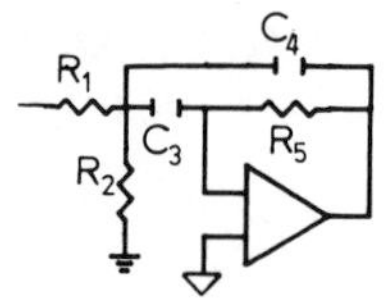


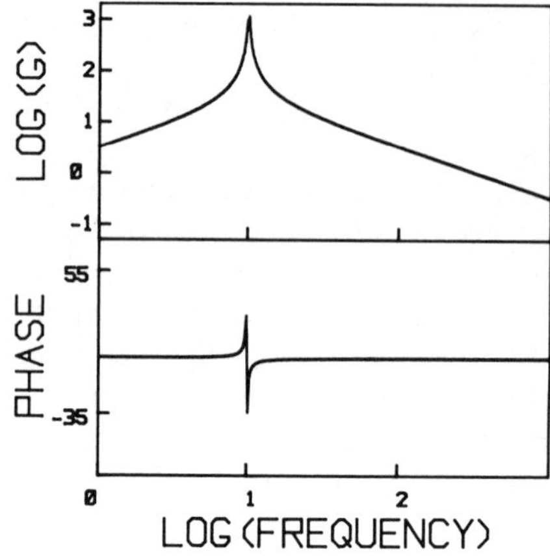


**Fig. 3.40.** Infinite gain multiple feedback implementation of a band-pass filter.

$$\tau = \tau_{BP} \tag{288}$$

$$S_{R_5}^{\omega_o} = S_{C_3}^{\omega_o} = S_{C_4}^{\omega_o} = -\frac{1}{2} \tag{289}$$

$$S_{R_1}^{\omega_o} = \frac{-1}{2\omega_o^2 R_1 R_5 C_3 C_4} \tag{290}$$

$$S_{R_2}^{\omega_o} = \frac{-1}{2\omega_o^2 R_2 R_5 C_3 C_4} \tag{291}$$

$$S_{R_1}^{Q} = \frac{R_1}{2(R_1 + R_2)} - \frac{1}{2} \tag{292}$$

$$S_{R_2}^{Q} = \frac{R_2}{2(R_1 + R_2)} - \frac{1}{2} \tag{293}$$

$$S_{R_5}^{Q} = \frac{1}{2} \tag{294}$$

$$S_{C_3}^{Q} = \frac{Q}{\omega_o R_5 C_4} - \frac{1}{2} \tag{295}$$

$$S_{C_4}^{Q} = \frac{Q}{\omega_o R_5 C_4} - \frac{1}{2} \tag{296}$$

Starting from a known, or desired, $H_o$, $Q$, and $\omega_o$ one chooses a convenient value for the capacitors, $C = C_3 = C_4$ and remembering that $Q = 1/\alpha$ one can then compute the nominal values for $R_1$, $R_2$, and $R_5$ from

$$R_1 = \frac{Q}{H_o\omega_o C} \tag{297}$$

$$R_2 = \frac{Q}{(2Q^2 - H_o)\omega_o C} \tag{298}$$

$$R_5 = \frac{2Q}{\omega_o C} \tag{299}$$

Because $R_1$ is normally much greater than $R_2$, one can use $R_2$ to trim the desired $Q$; then the value for $\omega_o$, the central frequency, is adjusted by a simultaneous equal percentage trim of $R_5$ and $R_2$ without affecting $Q$. Once again we are faced with simultaneous trims of two components, which sounds good on paper but is often rather frustrating in practice. In this case, there is no simple alternative that still retains the same basic configuration. Once again, one must be certain that the closed-loop gain at $\omega_o$ is 80 dB less than the open-loop gain at that frequency in order for our simplifying assumption to be valid. This circuit can make a lovely oscillator if one allows the open-loop and closed-loop gains to intersect.

## E. CONTROLLED SOURCE FILTER CIRCUITS

One difficulty with the circuits we have just been studying is that the input impedance may be quite low for the desired filter function with realizable components. In addition, the trimming proceedures for the high-pass and band-pass implementations leaves something to be desired. The noninverting voltage controlled voltage source (VCVS) circuit overcomes many of the limitations of the previous method. The VCVS is simply a fancy name for the follower with gain. The voltage transfer function for this circuit, reproduced in Fig. 3.41, is

$$H(s) = 1 + \frac{R_b}{R_a} = K \tag{300}$$

and the sensitivities are

$$S^K_Rb = 1 \tag{301}$$

$$S^K_Ra = -1 \tag{302}$$

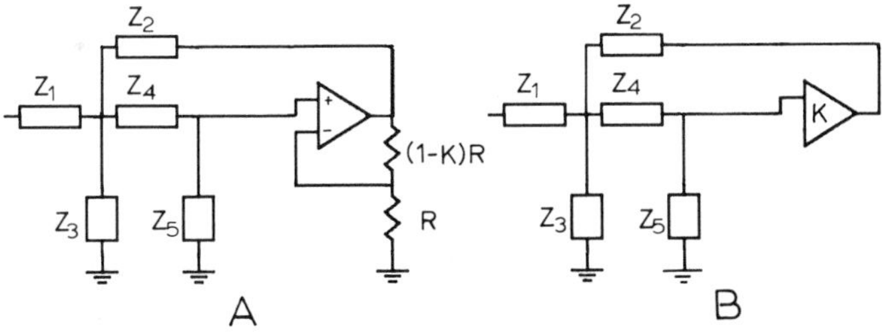


**Fig. 3.41.** The noninverting voltage-controlled voltage source filter (both with the gain of $K$). ($a$) Complete representation; ($b$) abbreviated representation.

Figure 3. 41 shows a generalized circuit based upon the VCVS employing five generalized admittances $Y_i$, which can be used to construct second-order low-pass, and high-pass, and band-pass filters. As in the previously discussed infinite gain multiple feedback circuitry each $Y_i$ is a single resistor or capacitor and the position numbering remains unchanged as we investigate the various circuit possibilities. The VCVS with gain $K$ is represented by a single input triangle with inscribed $K$. For the follower with gain implementation $K \geqslant 1$. The voltage transfer function for this circuit is

$$H(s) = \frac{KY_1Y_4}{Y_5(Y_1 + Y_2 + Y_3 + Y_4) + (Y_1 + Y_2(1 - K) + Y_3)} \tag{303}$$

where we have neglected the effect of finite amplifier open-loop gain. As in the other implementation this is valid as long as the difference between closed-loop gain and open-loop gain (i.e., the loop gain) exceeds 80 dB at all frequencies.

## 1. Low-Pass Filters

The VCVS implementation of the low-pass filter is obtained by employing $R_1$,$R_3$,$R_4$,$C_2$, and $C_5$. In this implementation $R_3$ is chosen to be infinitely large; that is, its admittance is zero. The transfer function for this circuit is

$$H(s) = \frac{K/R_1R_4C_2C_5}{s^2 + s[1/R_1C_2 + 1/R_4C_5 + (1 - K)/R_4C_5] + 1/R_1R_4C_2C_5} \tag{304}$$

which has the by now familiar form of the complex conjugate pole pair equation. The network parameters are

$$H_o = K \tag{305}$$

$$\omega_o = \left(\frac{1}{R_1R_4C_2C_5}\right)^{1/2} \tag{306}$$

$$\alpha = \left(\frac{R_4C_5}{R_1C_2}\right)^{1/2} + \left(\frac{R_1C_5}{R_4C_2}\right)^{1/2} + \left(\frac{R_1C_2}{R_4C_5}\right)^{1/2} - K\left(\frac{R_1C_2}{R_4C_5}\right)^{1/2} \tag{307}$$

$$\phi = \phi_{LP} \tag{308}$$

$$\tau = \tau_{LP} \tag{309}$$

$$S_{R_1}^{\omega_o} = S_{R_4}^{\omega_o} = S_{C_2}^{\omega_o} = S_{C_r}^{\omega_o} = -\frac{1}{2} \tag{310}$$

$$S_K^{H_o} = 1 \tag{311}$$

$$S_{R_1}^{\alpha} = \frac{1}{2} - \frac{1}{\alpha\omega_o R_1C_2} \tag{312}$$

$$S_{R_2}^{\alpha} = \frac{1}{2} - \frac{1}{\alpha\omega_o R_4}\left[\frac{1}{C_2} + \frac{1 - K}{C_5}\right] \tag{313}$$

$$S_{C_2}^{\alpha} = \frac{1}{2} - \frac{1}{\alpha\omega_o C_2}\left[\frac{1}{R_1} + \frac{1}{R_4}\right] \tag{314}$$

$$S_{C_5}^{\alpha} = \frac{1}{2} - \frac{1-K}{\alpha\omega_o R_4 C_5} \tag{315}$$

$$S_K^{\alpha} = \frac{-K}{\alpha\omega_o R_4 C_5} \tag{316}$$

One one has decided upon a value for $H_o$,$\omega_o$, and $\alpha$, one picks a convenient, commercially available value for $C_2$ amd $C_5$ (both equal to $C$). For this circuit we must choose $K > 2$ because

$$R_4 = \frac{\omega}{2\omega_o C}\left[1 + \sqrt{1 + \frac{4(H_o - 2)}{2}}\right] \tag{317}$$

$$R_1 = \frac{1}{\omega_o^2 C^2 R_4} \tag{318}$$

The trimming procedure is to set $\omega_o$, trimming $R_1$ and $R_4$ simultaneously (one could in principle trim $C_2$ and $C_5$ just as well) and then set $\alpha$ by adjusting $K$. A better procedure is to select $R_1 = R_2 = R$, in which case

$$S_{C_2}^{\alpha} = -S_{C_4}^{\alpha} = \frac{1}{2} - \frac{2}{\alpha\omega_o RC} \tag{319}$$

$$K = 3 - \alpha \tag{320}$$

This has the effect of minimizing the pass-band gain. It also produces the lowest drift circuit because the capacitors are normally the most temperature-sensitive components.

## 2. High-Pass Filters

The VCVS implementation of this circuit requires the use of finite values for $R_2$,$R_5$,$C_1$,$C_4$, with $R_3 = \infty$. The transfer function for this implementation of the high-pass filter is

$$H(s) = \frac{K s^2}{s^2 + s[1/R_5C_1 + 1/R_5C_4 + (1-K)/R_2C_1] + 1/R_2R_5C_1C_4} \tag{321}$$

The network parameters for this circuit are

$$H_o = K \tag{322}$$

$$\omega_o = \left(\frac{1}{R_2R_5C_1C_4}\right)^{1/2} \tag{323}$$

$$\alpha = \left(\frac{R_2C_1}{R_5C_4}\right)^{1/2} + \left(\frac{R_2C_4}{R_5C_1}\right)^{1/2} + \left(\frac{R_5C_4}{R_2C_1}\right)^{1/2} - K\left(\frac{R_5C_4}{R_2C_1}\right)^{1/2} \tag{324}$$

The sensitivity parameters for the circuit are given by

$$S_{R_2}^{\omega_o} = S_{R_2}^{\omega_o} = S_{C_1}^{\omega_o} = S_{C_4}^{\omega_o} = -1/2 \tag{325}$$

$$S_{R_2}^{\alpha} = \frac{1}{2} - \frac{1-K}{R_2C_1\alpha\omega_o} \tag{326}$$

$$S^{\alpha}_{R_5} = \frac{1}{2} - \frac{1}{R_5}\left(\frac{1}{C_1} + \frac{1}{C_4}\right)\frac{1}{\alpha\omega_o} \tag{327}$$

$$S^{\alpha}_{C_1} = \frac{1}{2} - \frac{1}{\alpha\omega_o C_1}\left[\frac{1-K}{R_2} + \frac{1}{R_5}\right] \tag{328}$$

$$S^{\alpha}_{C_4} = \frac{1}{2} - \frac{1}{\alpha\omega_o C_4 R_5} \tag{329}$$

$$S^{\alpha}_{K} = \frac{-K}{\alpha\omega_o R_2 C_1} \tag{330}$$

$$S^{H_o}_{K} = 1 \tag{331}$$

To design a filter of this type one first selects $H_o$, $\omega_o$, and $\alpha$. Normally one chooses $C_1 = C_4 = C$ and then calculates

$$R_2 = \frac{\alpha + \sqrt{\alpha^2 + 8(H_o - 1)}}{4\omega_o C} \tag{332}$$

$$R_5 = \frac{4/\omega_o C}{\sqrt{\alpha^2 + 8(H_o - 1)}} \tag{333}$$

Unless one has access to negative valued resistors, one must choose $K$ such that $R_2$ and $R_5$ are positive valued. The same trick employed in the low-pass case to decrease the temperature sensitivity of the circuit can be used here. If we keep the capacitors identical, and make the two resistors each equal to $R$, then the sensitivities become

$$S^{\alpha}_{C_1} = -S^{\alpha}_{C_2} = \frac{1}{2} - \frac{1}{\alpha\omega_o RC} \tag{334}$$

and $K = 3 - \alpha$. In their case, tuning the circuit is identical to the procedure for the low-pass filter.

### 3. Band-Pass Filters

There are several forms of band-pass filter that can be constructed from this basic format including $R_1$,$R_2$,$R_5$,$C_2$,$C_4$ and $R_1$,$R_2$,$R_5$,$C_4$ and connecting a second capacitor in parallel with $R_5$. Two additional forms can be generated from these by replacing resistors with capacitors and vice versa. The difficulty with these circuits is that they are quite complicated to tune and design and for these reasons are not commonly encountered. There is one VCVS format that does not follow the basic format. We investigate it because of its formal similarity to the first-order band-pass filter. This circuit is shown in Fig. 3.42, where the

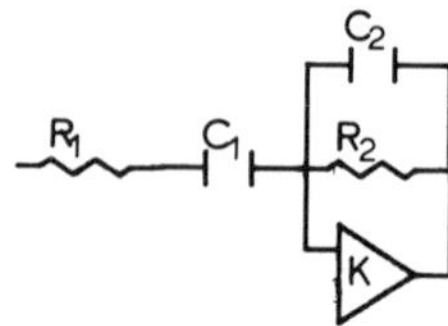


**Fig. 3.42.** Voltage-controlled voltage source implementation of a band-pass filter, to $K$.

triangle inscribed with a $K$ still indicates the voltage follower with a gain of $K$. The voltage transfer function is generated by noting that the voltage at the positive input is $E_o/K$, then summing the currents at the positive input. Assuming an ideal amplifier and using the notation $Z_o$ and $Z_1$ for the two generalized impedance, one can then write:

$$\frac{E_1 - E_0/K}{Z_1} + \frac{E_0 - E_O/K}{Z} = 0 \tag{335}$$

$$E_o(s) = \frac{K\,E_1(s)}{1 + (K - 1)Z_1/Z_0} \tag{336}$$

Further rearrangement and substitution of the $R$, $C$ values into the generalized impedances leads to the voltage transfer function:

$$H(s) = \frac{s\left(\dfrac{K}{1-K}\right)\dfrac{1}{R_2C_2}}{s^2 + s\left[\dfrac{1}{R_2C_2} + \dfrac{1}{R_1C_1} + \dfrac{1}{R_1C_2(1-K)}\right] + \dfrac{1}{R_1R_2C_1C_2}} \tag{337}$$

We can still cast this equation into the form of the complex conjugate pole pair representation, and using this format the relevant parameters become

$$H_o = \frac{K}{(1-K)(R_1/R_2 + C_2/C_1) + 1} \tag{338}$$

$$\omega_o = \left(\frac{1}{R_1R_2C_1C_2}\right)^{1/2} \tag{339}$$

$$\frac{1}{Q} = \alpha = \sqrt{\frac{R_1C_1}{R_2C_2}} + \sqrt{\frac{R_2C_2}{R_1C_1}} - \frac{1}{1-K}\sqrt{\frac{R_2C_1}{R_1C_2}} \tag{340}$$

$$\phi = \tau + \phi BP \tag{341}$$

$$\tau = \tau BP \tag{342}$$

$$S_{R_1}^{\omega_o} = S_{R_3}^{\omega_o} = S_{C_1}^{\omega_o} = S_{C_2}^{\omega_o} = -\frac{1}{2} \tag{343}$$

$$S_K^{H_o} = 1 + H_o\left(\frac{R_1}{R_2} + \frac{C_2}{C_1}\right) \tag{344}$$

$$S_{R_2}^{H_o} = -\,S_{R_1}^{H_o} = H_o\left(\frac{1-K}{K}\right)\frac{R_1}{R_2} \tag{345}$$

$$S_{C_1}^{H_o} = -\,S_{C_2}^{H_o} = H_o\left(\frac{1-K}{K}\right)\frac{C_2}{C_1} \tag{346}$$

$$S_K^Q = \frac{-K}{(1-K)^2}\frac{Q}{\omega_oR_1C_2} \tag{347}$$

$$S_{R_1}^Q = \frac{1}{2} - \frac{Q}{\omega_oR_1}\left[\frac{1}{C_1} + \frac{1}{C_2(1-K)}\right] \tag{348}$$

$$S_{R_2}^Q = \frac{1}{2} - \frac{Q}{\omega_oR_2C_2} \tag{349}$$

$$S_{C_1}^Q = \frac{-1}{2} + \frac{1}{\alpha\omega_oR_1C_1} \tag{350}$$

$$S_{C_3}^{Q} = \frac{-1}{2} + \frac{1}{\alpha\omega_o C_2}\left[\frac{1}{R_2} + \frac{1}{R_1(1-K)}\right] \tag{351}$$

Designing a filter of this type is fairly straightforward, but one must surrender the freedom to choose an arbitrary pass-band $H_o$. Assuming that one is designing for a given $Q$ and $\omega_o$ then one chooses $C_1 = C_2 = C$ (again this should be a value which is readily available commercially).

$$R_1 = R_2 = \frac{1}{\omega_o C} \tag{352}$$

$$K = \frac{3Q-1}{2Q-1} \tag{353}$$

and the absolute value of the pass-band gain becomes

$$|H_o| = 3Q - 1 \tag{354}$$

The central frequency is trimmed by varying $R_1$ and $R_2$ and $Q$ is trimmed by altering the VCVS gain $K$. Note that $K$ must be greater than unity or the system oscillates. If the resistors are trimmed simultaneously then $Q$ is not affected. In this circuit, the maximum value of Q should be 10 or less (but not less than about 1.5).

### 4. Final Thoughts on VCVS Filter Implementations

These circuits have the advantage of very high input impedances and output impedances on the order of 1 Ω. The high-pass and low-pass filters allow one to set $\omega_o$ without affecting $\alpha$, and one may alter $Q$, that is $1/\alpha$, without affecting $\omega_o$ by changing $K$. Because the sensitivities with respect to $K$ are relatively high, one must pick highly stable resistances for the $K$-determining elements. The circuits become very $Q$-sensitive for high values of $Q$. This is particularly true of the band-pass case, where the sensitivities of $Q$ with respect to $K$ and the two resistors contains $Q$ itself. For high-pass and low-pass filtering, where one often works with a relatively low $Q$ filter, this is not a real problem; if one is interested in a high $Q$ band-pass filter this can be a serious limitation.

## F. INFINITE GAIN STATE VARIABLE FILTERS

The infinite gain state variable filter (IGSVF) is shown in Fig.3.43, where the notation that we developed for analog computers is employed. For the generalized circuit shown the voltage transfer function becomes

$$J(s) = \frac{a_o + a_1 s + \ldots + a_{n-1}s^{n-1} + a_n s^n}{b_o + b_1 s + \ldots + b_{n-1}s^{n-1} + b_n s_n} \tag{355}$$

A second-order IGSVF is shown in Fig. 3.44.

As we will see, the IGSVF generally provides less $Q$ sensitivity to changes in the component values than any of the other realizations. This is accomplished at the cost of two additional amplifiers, and at one time this was a major

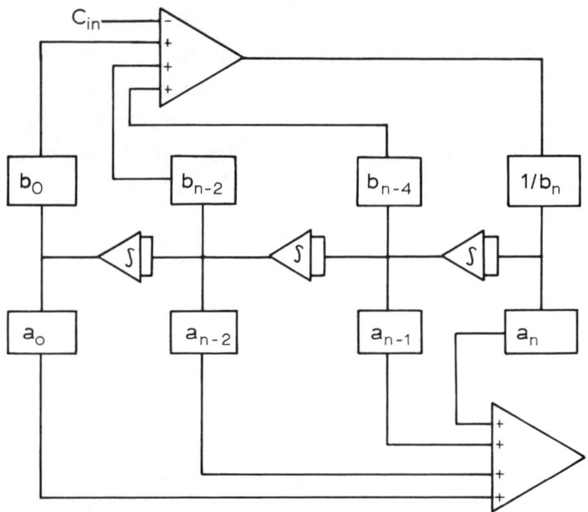


**Fig. 3.43.** The infinite gain state variable filter, a general block form representation.

consideration, not only because of the additional cost of the amplifiers but also from the viewpoint of the additional board space required to house the extra amplifiers and the additional resistors (six total as opposed to an average of three for the other implementations). Technology has changed this argument with the availability of four high-quality OAs in one standard 14-pin dip package which occupies 0.3 X 0.7 of board space. In addition, a number of manufacturers are marketing modules, some of which are on the aforementioned package, which contain not only the three amplifiers in Fig. 3.44 and an additional uncommited amplifier, but also most of the components that are also shown in the figure. Some forms bring out connections for $C_1$ and $C_2$ and either $R_3$ or $R_4$, whereas others include an integrated capacitor pair and bring out $R_2$,$R_1$ and $R_3$.

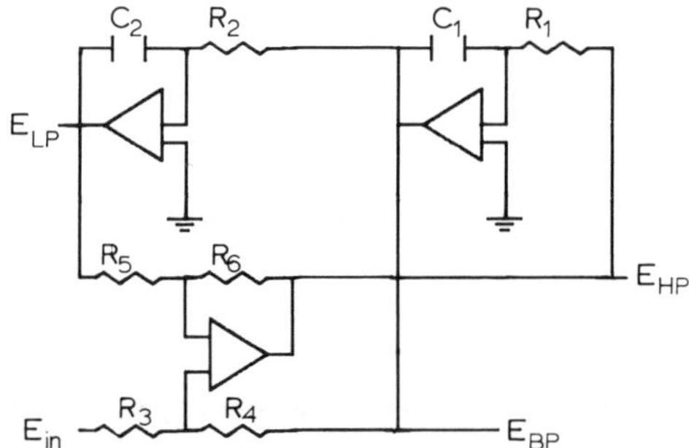


**Fig. 3.44.** A second-degree state variable filter, shown as the actual circuit should be. Note that there are outputs of low-pass, high-pass, and band-pass filters.

In either of these cases one has the advantage thay many of the temperature-sensitive components are integrated onto the same substrate and thus track thermally. All the previous discussion about the advantages pertaining to this approach apply here. The circuit also has the advantage that one simultaneously has a low-pass, high-pass, and band-pass filter, all of which are tuned at the same frequency. The transfer functions for the second-order form of Fig. 3.44 are as follows:

Low-pass: $$H(s) = \frac{\frac{1}{R_1R_2C_1C_2}\left(\frac{1 + R_6/R_5}{1 + R_3/R_4}\right)}{s^2 + s\left[\frac{1}{R_1C_1}\left(\frac{1 + R_6R_5}{1 + R_4/R_3}\right)\right] + \frac{R_6}{R_5}\left(\frac{1}{R_1R_2C_1C_2}\right)} \tag{356}$$

High-pass: $$H(s) = \frac{s^2\left[\frac{1 + R_6/R_5}{1 + R_3/R_4}\right]}{s^2 + s\left[\frac{1}{R_1C_1}\left(\frac{1 + R_6/R_5}{1 + R_4/R_3}\right)\right] + \frac{R_6}{R_5}\left(\frac{1}{R_1R_2C_1C_2}\right)} \tag{357}$$

Band-pass: $$H(s) = \frac{-s\frac{1}{R_4C_4}\left(\frac{1 + R_6/R_5}{1 + R_3/R_4}\right)}{-s^2 + s\left[\frac{1}{R_1C_1}\left(\frac{1 + R_6/R_5}{1 + R_4/R_3}\right)\right] + \frac{R_6}{R_5}\left(\frac{1}{R_1R_2C_1C_1}\right)} \tag{358}$$

For all three outputs the $\omega_o$ and $\alpha$ are the same:

$$\omega_o = \left(\frac{R_6}{R_5R_1C_1R_2C_2}\right)^{1/2} \tag{359}$$

$$\alpha = \frac{1 + R_6/R_5}{1 + R_4/R_3}\left(\frac{R_5}{R_6}\frac{R_2C_2}{R_1C_1}\right)^{1/2} \tag{360}$$

but obviously the gains are somewhat different:

Low-Pass: $$H_o = \frac{1 + R_5/R_6}{1 + R_3/R_4} \tag{361}$$

High-pass: $$H_o = \frac{1 + R_6R_5}{1 + R_3/R_4} \tag{362}$$

Band-Pass: $$H_o = \frac{R_4}{R_3} \tag{363}$$

The phase and group delays are the same as in the previous networks of each type. The sensitivities for a number of parameters are also identical:

$$S_{R_5}^{\omega_o} = S_{R_1}^{\omega_o} = S_{R_2}^{\omega_o} = S_{C_1}^{\omega_o} = S_{C_2}^{\omega_o} = \frac{-1}{2} = -S_{R_4}^{\omega_o} \tag{364}$$

For the high-pass and low-pass filters the following sensitivities are also identical:

$$S_{R_2}^{\alpha} = S_{C_3}^{\alpha} = \tfrac{1}{2} = -S_{R_1}^{\alpha} = -S_{C_1}^{\alpha} \tag{365}$$

$$S^{\alpha}_{R_6} = \frac{-1}{2} + \frac{R_6/R_5}{R_1C_1\alpha\omega_o(1 + R_4/R_3)} = -S^{\alpha}_{R_5} \tag{366}$$

$$S^{\alpha}_{R_3} = \frac{1}{1 + R_3/R_4} = -S^{\alpha}_{R_4} \tag{367}$$

$$S^{H_o}_{R_3} = -S^{H_o}_{R_4} = \frac{-1}{1 + R_4/R_3} \tag{368}$$

The remaining important sensitivity parameter is nearly identical:

Low-pass:
$$S^{H_o}_{R_5} = -S^{H_o}_{R_6} = \left(\frac{R_5/R_6}{1 + R_3/R_4}\right) \tag{369}$$

High-pass:
$$S^{H_o}_{R_5} = -S^{H_o}_{R_6} = \frac{1}{H_o}\left(\frac{R_6/R_5}{1 + R_3/R_4}\right) \tag{370}$$

The parameters for the band-pass filter are also nearly identical to those for the high-pass and low-pass filters, the primary differences being in stating the differences with respect to changes in $Q$ rather than $\alpha$, although the form of $S^{H_o}_{R_3}$ is distinctly different:

$$S^{Q}_{R_1} = S^{Q}_{C_1} = +\tfrac{1}{2} \tag{371}$$

$$S^{Q}_{R_2} = S^{Q}_{C_2} = \tfrac{1}{2} \tag{372}$$

$$S^{Q}_{R_6} = S^{Q}_{R_6} = \frac{1}{2} - \frac{R_6/R_5}{R_1C_1\alpha\omega_o(1 + R_4/R_3)} \tag{373}$$

$$S^{Q}_{R_4} = -S^{Q}_{R_3} = \frac{1}{1 + R_3/R_4} \tag{374}$$

$$S^{H_o}_{R_3} = -1 = -S^{H_o}_{R_4} \tag{374}$$

Practical implementations of this filter normally start with the initial assumption that $C_1 = C_2 = C$ as well as $R_5 = R_6 = R$. One also usually sets the nominal value of $R_3 = R$ as well. One then calculates the values for $R_1$ and $R_2$ from $C$ and the desired central frequency:

$$R_1 = R_2 = \frac{1}{\omega_o C} \tag{376}$$

The nominal value for $R_4$ is then calculated from

$$R_4 = R_3(2Q - 1) \tag{377}$$

Note that rational values for $R_4$ are possible only for $Q{>}2$. This circuit works for $Q$ values of up to 50 (i.e. $\alpha \leqslant 0.02$) and is probably the best bet for very sharp band-pass filters. The increasing availability of the integrated modules will probably lead to the increased use of this circuit even for high-pass or low-pass only circuits because of the space savings and simplicity of use. Tuning is accomplished by trimming or changing $R_1$ and $R_2$ (or $C_1$ and $C_2$) simultaneously, and $Q$ (or $\alpha$) can be independently adjusted by means of $R_4$. It should be noted, however, that $H_o$ depends on $R_4$, but in filter applications this is normally not a major consideration.

## G. THE NEGATIVE IMMITANCE CONVERTER OR GYRATOR CIRCUIT

The OA circuit in Fig. 3.45*a* is called an ideal current inversion negative immitance converter (INIC) or gyrator circuit. The behavior of this circuit is fairly easy to derive for an ideal amplifier. If a current $I_1$ flows through $Z_1$, then an identical current $I_1$ must flow through the normal feedback resistance $R_1$ owing to the voltage drop from the voltage at the amplifier output, here termed $E_a$, to the voltage at the negative input $E_i$. The amplifier adjusts $E_a$, so ideally there is no potential difference between the two inputs and as a result the voltage drop across the resistor $R_2$ is the same as that across $R_1$ and the current through $R_2$ is therefore $I_2 = -R_2/R_1 I_1$. Mathematically stated,

$$\frac{E_a}{R + Z_2} = \frac{E_0}{Z_2} \tag{378}$$

$$\frac{E_1 - E_0}{Z_1} + \frac{E_a - E_0}{R_k} = 0 \tag{379}$$

Solving for $E_0$, which we reiterate is the potential at the summing points in this case,

$$E_0 = \frac{-E_1 K Z_2}{A_1 - K Z_2} \tag{380}$$

Inspection of this equation leads to the interesting conclusion that this circuit in effect subtracts an impedance, or at least the effect of that impedance, from the impedance in the input leg of the amplifier, or it "gyrates" in $s$ space one of the impedances. We return to this feature in a later section. For now, we note only that one can construct second-order filters by this method. The only difficulty with the resulting filter is that it has a high output impedance since one takes the signal from the positive input of the amplifier rather than its output, and as a result one usually has to buffer the signal with a voltage follower. As a result,

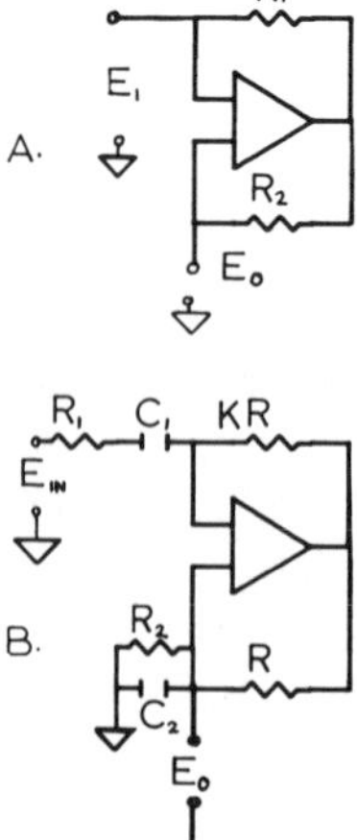


**Fig. 3.45.** The OA realization of a current inversion negative immitance convertor. (*a*) Basic converter; (*b*) band-pass filter.

one normally encounters only the band-pass filter implementation of the INIC filter.

### 1. Band-Pass INIC Filters

A band-pass filter can be constructed by employing the circuit in Fig. 3.45*b*. Substituting the resulting values of $R$ and $C$ into $Z_1$ and $Z_2$ and rearrangement leads to the voltage transfer function:

$$H(s) = \frac{-Ks/R_1C_2}{s^2 + s(1/R_1C_1 + 1/R_2C_2 - K/R_1C_2) + 1/R_1C_1R_2C_2} \tag{381}$$

The network parameters for this circuit are given by

$$H_o = \frac{K}{C_2/C_1 + R_1/R_2 - K} \tag{382}$$

$$Q = \frac{1}{\alpha} = \frac{1}{\sqrt{R_1C_1/R_2C_2} + \sqrt{R_2C_2/R_1C_1} - K\sqrt{R_2C_1/R_1C_2}} \tag{383}$$

$$\omega_o = \left(\frac{1}{R_1C_1R_2C_2}\right)^{1/2} \tag{384}$$

$$\phi = \pi + \phi_{BP} \tag{355}$$

$$\tau = \tau_{BP} \tag{356}$$

and the sensitivities that result are

$$S_K^{H_o} = 1 + H_o \tag{387}$$

$$S_{R_1}^{H_o} = \frac{-R_1/R_2}{C_2/C_1 + R_1/R_2 - K} = -S_{R_2}^{H_o} \tag{388}$$

$$S_{C_1}^{H_o} = \frac{C_2/C_1}{C_2/C_1 + R_1/R_2 - K} = -S_{C_2}^{H_o} \tag{389}$$

$$S_{R_1}^{Q} = \frac{Q}{\omega_o R_1}\left[\frac{1}{C_1} - \frac{K}{C_2}\right] - \frac{1}{2} \tag{390}$$

$$S_{R_2}^{Q} = \left[\frac{Q}{\omega_o R_2 C_2}\right] - \frac{1}{2} \tag{391}$$

$$S_{C_2}^{Q} = \frac{Q}{\omega_o C_2}\left(\frac{1}{R_2} - \frac{K}{R_1}\right) - \frac{1}{2} \tag{392}$$

$$S_K^{Q} = \frac{QK}{\omega_o R_1 C_2} \tag{393}$$

$$S_{R_1}^{\omega_o} = S_{R_2}^{\omega_o} = S_{C_1}^{\omega_o} = S_{C_2}^{\omega_o} = -\frac{1}{2} \tag{394}$$

Selection of component values normally proceeded from selection of $C_1 = C_2 = C$ and $R_1 = R_2 = R'$. One then selects a value for $K$ based upon

$$K = \frac{2-1}{Q} \tag{395}$$

$$R' = \frac{1}{\omega_o C} \tag{396}$$

The value for the $K$-setting resistors are relatively unimportant, although $R$ is usually chosen to be some convenient value in the range of 10–30$K$. The sensitivity equations for the generalized form look fairly formidable, but with the component selection above they become quite tractable:

$$S_K^Q = 2Q - 1, \quad Q = \frac{1}{2-k} \tag{397}$$

$$S_K^{H_o} = 2Q, \qquad H_o = 2Q - 1 = \frac{K}{2-K} \tag{398}$$

$$\omega_o = \frac{1}{RC} \tag{399}$$

$$S_{R_2}^Q = S_{C_1}^Q = S_{R_1}^Q = S_{C_2}^Q = Q - \tfrac{1}{2} \tag{400}$$

$$S_{C_1}^{H_o} = S_{R_2}^{H_o} = -S_{R_4}^{H_o} = -S_{C_2}^{H_o} = Q \tag{401}$$

Trimming is accomplished by simultaneous variation of $R_1$ and $R_2$, which alters $\omega_o$ while leaving $Q$ and $H_o$ unchanged. Trimming for $Q$ is accomplished by altering $K$, which does not effect the central frequency but does alter $H_o$ by an amount proportion to twice the $Q$ chose, but again one normally is not too concerned by changes in the value of $H_o$ in a band-pass filter.

## 2. GYRATORS

The INIC that we have discussed is in effect a gyrator circuit; that is, it moves the transform of a circuit element from one quadrant of the complex plane to another. We discuss briefly a second realization of the gyrator principle which very closely approximates an ideal gyrator. The circuit in Fig. 3.46 is a two-amplifier realization of a grounded load gyrator (16). The circuit draws a current $I_1$ from any voltage source $E_1$ and the equivalent impedance is defined as

$$Z_e = \frac{E_1}{I_1} = \frac{Z_1 Z_3 Z_5}{Z_2 Z_4} \tag{402}$$

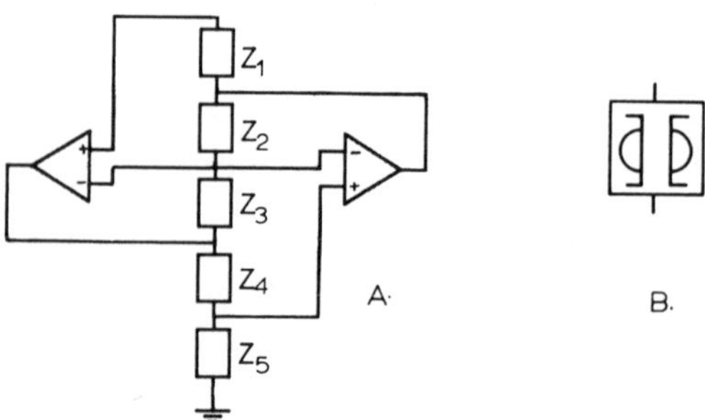


**Fig. 3.46.** A two-OA five-element implementation of a gyrator. ($a$) Basic circuit; ($b$) general symbol.

where the components are either resistors or capacitors. This result can be derived by remembering that the potentials at the inverting and noninverting inputs are identical within the capabilities of the amplifiers and by noting that the currents through elements $Z_2$ and $Z_3$ are identical, as are the currents through $Z_4$ and $Z_5$.

There are several interesting configurations of this circuit that we now examine briefly. If either $Z_2$ or $Z_4$ is a capacitor and the remaining four are resistors, then we have, for example, $Z_e = s(C_2R_1R_3R_5/R_4) = Ls$ where $L$ is the effective inductance, which is determined by the ratio of the other components. Within the current and potential handling capabilities of the OAs and the ideality of the other passive components (particularly the $Q$ of the capacitor), the equivalent inductance is ideal. One can also fabricate large value capacitances that are as close to ideal as the other components allow. For the case where $Z_1$ is a capacitor and the remaining components are resistors, $Z_e = (R_3R_5/R_2R_4C_1s)$. The cases where $Z_3$ or $Z_5$ are the capacitances is formally the same. It is also possible to fabricate components that otherwise do not exist in nature. For example, replacing two of the three odd-numbered elements leads to an effective impedance $Z_e = (R_5/R_2R_4C_1C_3s^2)$ which has been given the name "frequency-dependent negative resistance," or FDNR (16), because in frequency space, its representation is

$$\frac{-R_5}{\omega^2C_1R_2R_4C_3} \tag{403}$$

The version with the two capacitances in the even positions has the form $-\omega^2C_2C_4R_1R_2R_5$ in frequency space.

The use of gyrator-derived components as the grounded load part of a $T$ network is an obvious application, and the limitation that the final impedance must be grounded is not as restrictive as it may seem because two gyrators can share a common termination resistance (Fig. 3.47). Unfortunately, this does not mean that one can simply connect $Z_5$ of a gyrator element to the output and the far end of $Z_5$ to the inverting input in the feedback loop of an OA. In this case, the output would see an effective impedance of $Z_3Z_4Z_5/Z_2Z_4$ but the current flow to the summing point would be given by $E_{out}/Z_5$, and it is this current flow that determines the output functionality.

The more classical use of this form of gyrator is in emulating $LC$ filter circuits, either by the fairly straightforward implementation of an inductor by this method or by an interesting conceptual variation (16). In the variation one designs the circuit using resistors, capacitors, and inductors and then implements the circuit by replacing capacitors by FDNR, resistors with capacitors,

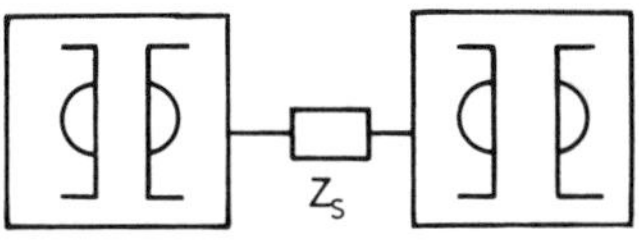


**Fig. 3.47.** Implementing floating load gyrators.

and inductors with resistors; that is, each component is replaced by one that is $1/s$ times the original component transform. The transform of the resulting circuit is identical to that of the original model.

One important feature of gyrator circuits is that they are remarkably insensitive to amplifier parameters. Stable $Q$s of better than 1000 can be constructed with quad-operational amplifiers having gain of only 40 dB of open-loop gain, at the frequency of interest.

## H. THE EFFECT OF REAL AMPLIFIERS ON FILTER PERFORMANCE

Throughout this section, we have tried to stress the fact that one needs about 80 dB loop gain in order for the circuits to behave in the "simple" manner derived. This is extremely important, as is demonstrated shortly; this is not the only place where the limitations of a real amplifier need to be considered.

Low-pass filters can suffer from the intrusion of reality, in a fairly interesting manner, which has nothing to do with loop-gain problems. For example, consider the VCVS low-pass filter that we have just treated. For direct current, that is, nominally 0 Hz, the circuit simplifies to that shown in Fig. 3.48, where the capacitors have been eliminated because they represent infinite impedances at the frequencies interest. The interior amplifier is assumed to be ideal and the parameters of interest are represented by the noise sources, the two resistances $R_a$ and $R_b$ are used to set $K = 1 + R_b/R_a$, and the new resistor $R_c$ is included as a compensation, as we will show.

Rigorous solution of the system leads to the following results:

$$E_0 = \frac{1}{R_a(1 + 1/A) + R_b/A} \{I_{B2}(R_1 + R_2)(R_a + R_b) \\ -I_{B1}[R_c(R_a + R_b) + R_aR_b - V_{os}(R_a + R_b)]\} \tag{404}$$

If we let $A \to \infty$,

$$E_0 = I_{b2}(R_1 + R_2)K - I_{b2}(R_{cK} + R_b) - V_{os}K \tag{405}$$

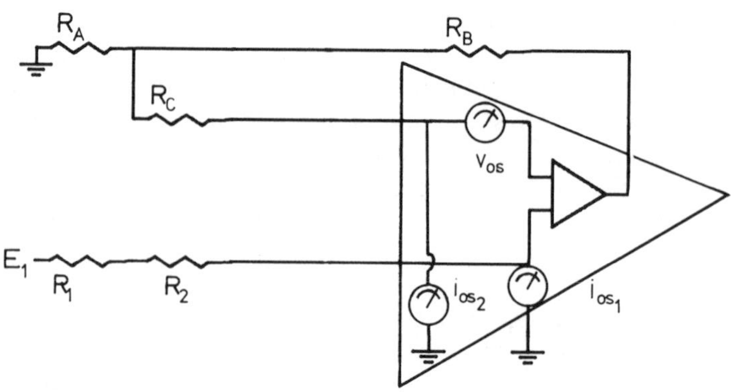


Fig. 3.48. Model for the dc behavior of a VCVS implementation of a low-pass filter.

where $I_{b1}$ and $I_{b2}$ are the bias currents at the positive and negative inputs, respectively, and $V_{os}$ is the voltage offset. $R_c$ is used to compensate for the bias currents and if the two are essentially identical and $R_c$ is chosen as

$$R_c = R_1 + R_2 - \frac{R_b}{K} \tag{406}$$

then the bias current effect can be essentially eliminated. If the bias currents are very low, as in FET amplifiers, or the impedances are low, then one may eliminate $R_c$. Since the bias currents and drifts are temperature sensitive, they can lead to time-varying components (because the internal temperature of the amplifier can change) at the output, introducing a significant error in some cases where low signal strengths are coupled with large and time-varying high-frequency components.

The effects of bias currents and noise on the other implementations of the low-pass filter are carried out in the same manner and are not treated further. For high-pass and band-pass filters the noise sources, both current and voltage, are the dominant nongain factors. One normally worries about voltage noise if high-impedance components are involved, as is often the case with filters. Then the current noise source can become at least as important as the voltage noise source. As we have previously mentioned, these noise sources are inherent in the construction of OAs and are a problem that one must live with. There are some amplifiers available that are classified as low noise amplifiers and they should be considered when very low strength signals must be handled. However, they are significantly more expensive than other amplifiers and quite often they possess far less gain and a lower unity gain crossover frequency than comparably priced "noisy" amplifiers.

The loop gain is the most serious problem normally encountered in filters, particularly high-pass or band-pass implementations, with the latter being the most susceptible, particularly in high $Q$ implementations. We treat only the infinite gain multiple feedback filter because it is the most common filter and the procedure for the others is essentially the same. In equations 252 and 253, we developed the general response for this type of filter with and without the amplifier gain factor included. We can now write

$$H'(s) = \frac{H(s)}{1 + 1/A(s)[\,1 - H(s)(1 + Y_2/Y_1)]} \tag{407}$$

where $H(s)$ is the voltage transfer function for an infinite gain amplifier. Defining the feedback gain or feedback ratio $\beta(s)$ by

$$\beta(s) = \frac{1}{1 - H(s)(1 + Y_2/Y_1)} \tag{408}$$

we can rewrite the true transfer function as

$$H'(s) = H(s)\left(1 - \frac{1}{1 + A(s)\beta(s)}\right) \tag{409}$$

Note that the feedback ratio is the feedback between the output and negative input.

We can now write the error due to finite loop gain as

$$E(s) = \frac{-H(s)}{1 + A(s)\beta(s)} \tag{410}$$

which is completely general for any filter of the infinite gain multiple feedback type. The phase of $E(s)$ is not the phase error of the filter but rather the phase of the error.

Most people, can't look at an equation of this type and gain any kind of feeling for the kind of problems that might ensue, except to make the obvious observations that when $A(s)\beta(s)$ becomes unity there is big trouble. In order to gain an appreciation for the problem it is necessary to consider a practical example. Consider a band-pass filter with a center frequency of 10 KHz, a gain of 10, and a $Q$ of 20, which is admittedly fairly sharp but still within normal design limitations. We utilize an OA with an open-loop gain of 100 dB and a unity gain crossover frequency of 10 MHz, a fairly impressive amplifier. The resulting gain and phase plots are shown in Fig. 3.49. The gain error is essentially zero at the center frequency, but it is over 3 dB off immediately on either side of $\omega_0$, representing a gain error of 2 in the immediate region of $\omega_0$. All this really means is that the filter is not quite as sharp as the ideal implementation. The phase error, which is in excess of $-40°$, looks fairly frightening in terms of

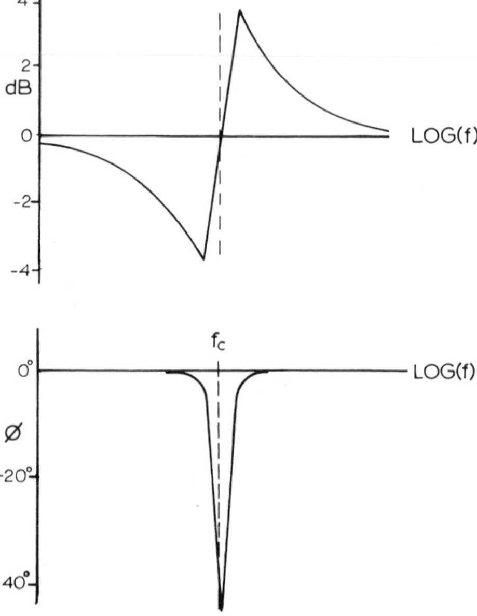


**Fig. 3.49.** Gain and phase errors for a band-pass filter arising from finite loop gain.

phase changes until one remembers that what is being plotted is the phase of the error (which is fairly small as we have just seen), not the error in the phase. If we decrease the unity gain crossover frequency, the results change dramatically. In contrast, changes of an order of magnitude in the open-loop gain do not affect this circuit, because even for an amplifier with an open-loop gain of 60 dB but a unity gain crossover frequency of 10 MHz, we have essentially the same gain in the critical region. A general rule of thumb is that for high-pass filters and band-pass filters, the unity gain crossover frequency is normally far more important than the open-loop gain, whereas the opposite is true for low-pass filters. The overall results are represented in the Bode plot portion of Fig. 3.49, which is essentially indistinguishable from a product of the amplifier gain by the loop gain of the ideal amplifier. Let us investigate the meaning of this plot in some detail. We have already seen that the gain error at low frequencies is $1/A_0$, so that if $A_0 = 100$ (40 dB) the gain error is 1%. This relationship also holds at high frequencies where the gain error is given by $1/A\,\omega t$). If we wish the gain error to be less than 1% at 10 KHz, then the separation between the closed-loop gain and the open-loop gain of the amplifier must still be 40 dB at that frequency. Let us examine what happens if the amplifier shown in Fig. 3.49 had an open-loop gain of 80 dB and a unit gain crossover frequency, also termed the gain-bandwidth product, of 1 MHz. If the circuit gain is 10, that is, 20 dB, then the limiting frequency is 1 KHz; above this frequency the gain error increases past our acceptable limit of 1%. In order to achieve a gain of 10 at 10 KHz with 1% accuracy we must use an amplifier with a unity gain crossover frequency of 10 MHz. Choosing an amplifier with a high dc gain, that is, $A_0$, but the same $f_u$ does not help. Because both amplifiers are rolling off at −6 dB per octave to unity gain at the same frequency, they have identical properties in the critical 1 KHz to 1 MHz range.

We consider only one more parameter in this section, the differential input impedance of the amplifier. We can take this term into account by inserting an admittance $Y_6$ from the negative input to common. When this term is included the only function that changes is the feedback ratio $\beta(s)$, which becomes

$$\beta(s) = \frac{1}{1 - H(s)(1 + Y_2/Y_1) + (Y_6 Y_1 Y_3)(Y_1 + Y_2 + Y_3 + Y_4)} \tag{411}$$

This can be a significant contribution for cases where the circuit impedances are high, particularly in low-pass filters, or in high-pass or band-pass filters when the capacitive component of the differential input impedance is large, as is the case in some amplifiers.

### 1. Temperature Effects

All the parameters of an OA, including the open-loop gain and differential input impedance, are temperature sensitive. Fortunately for most chemical applications, the temperature coefficient for the latter two is positive, so only in low-temperature applications need one worry about the temperature effects causing trouble. If one is designing a circuit for a low-temperature environment or experiencing difficulty with a commercial instrument that is being operated at a subnormal temperature, this is a feature that must be considered.

## I. OTHER FORMS OF FILTERS

There are a number of other forms of filters which the analytical chemist may encounter, not all of which are analog in nature. We will examine one essentially analog filter and several digital filters.

### 1. The Lock-in Amplifier

A lock-in amplifier is a singularly effective band-pass filter, which has a very high effective $Q$. Although there are various commercial implementations of the lock-in amplifier available, we can examine a simple circuit that has most of the important features. Figure 3.50 is the circuit diagram for our hypothetical lock-in amplifier. The first stage of the filter is either a high-pass or relatively broad band band-pass filter, preferably second-order or higher. The purpose of this stage is to prevent saturation of the following stage, which does most of the real filtering. The second stage, the synchronous detector, has two inputs. The first is the output of the initial filter stage and the second is a reference channel. The reference channel usually has two switch-selectable sources, either an external reference signal or an internal sinusoidal oscillator. The selected reference signal is passed through a calibrated phase shifter and then through a high gain clipper or zero crossing detector, which ideally converts the phase-shifted reference sinusoid into a square wave. The reference channel drives a solid-state switch which opens when the square wave is high and closes when the square wave is low. The simplest implementation of this switch would be an $n$-channel JFET or a MOSFET (17,18). The switch connects the series $R_1C_1$ network from the filter to the summing point of the OA. The capacitor is there to eliminate the dc effects of the filter, because the ratio of $R_0/R_1$ is normally perhaps 10. The behavior of this circuit can best be understood by considering its effect on four sine waves, one in phase with the reference and with the same frequency $f_0$, the second also at $f_0$ but 90° out of phase with the reference, and the third and fourth at frequencies of $0.75f_0$ and $1.5f_0$, respectively. The way these frequencies are altered by the synchronous detector is shown in Fig. 3.51. The output of the

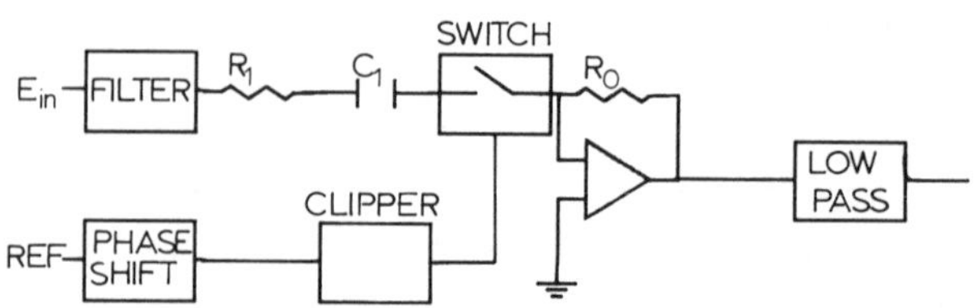


**Fig. 3.50.** Block diagram of a lock-in amplifier.

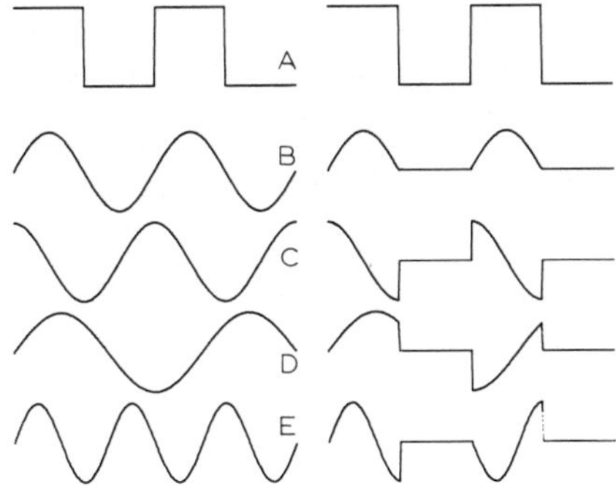


**Fig. 3.51.** The effect of synchronous demodulation on sine waves.

synchronous filter is then passed through a low-frequency cutoff low-pass filter with high dc gain. In order to understand what this final stage does to the rather strange looking wave form from the synchronous detector, we must think about the low-pass filter in a slightly different way than we have before. The dc component and any frequencies below the cutoff frequency are passed through with constant gain, whereas those above the cutoff are integrated. This is equivalent to taking the average value of the input wave form. The average value of the phase-shifted $f_0$ component is obviously zero so that the detector is phase sensitive; that is, it discriminates with respect to the phase angle between signal and reference. The signal due to the in-phase $f_0$ component is amplified by an amount $2(R_0/R_1)(G_{dc}/4)$, where $G_{dc}$ is the gain of the final low-pass filter. The components that are at $0.75f_0$ and $1.5f_0$ are attenuated by the integration action of the final low-pass filter section because the chopping action of the synchronous detector has effectively introduced higher-frequency components into the signal. The bandwidth of the effective filter formed is $1/4\tau$ for a single pole low-pass filter and $1/8\tau$ for a double pole low-pass filter, where $\tau$ is the time constant of the final filter stage. Typical time constants can be 10–100 sec; therefore a 1 KHz modulation frequency can have the effective $Q$ as high as $10^5$ or $10^6$, which is a value totally unapproachable for any rational number of sections of band-pass active filters. The one penalty invoked is that the input signal amplitude must change slowly with respect to the time constant of the final stage active filter, so that for rapidly changing signals, one must decrease the time constant and thus decrease the $Q$ of the filter. For very slowly changing signals one can abstract useful information from signals where the $S/N$ ratio is 1/1000.

A slight modification of this approach is to substitute the synchronous detector with an analog multiplier. The output signal is now full-wave rectified but the arguments proceed as before. In either approach, the square wave that is the final reference must be perfectly symmetrical, that is, the high and low times must be equal, or "leakage" occurs. This puts some fairly stringent requirements on the circuitry which transforms the reference sine wave into a square wave.

A second modification eliminates the square-wave generation step and multiplies the sine wave from the reference channel directly. Simple trigonometry predicts that the output will in general be

$$\sin(x)*\sin(y) = \tfrac{1}{2}[\cos(x - y) - \cos(x + y)] \qquad (412)$$

which contains only time-varying components unless the two signals are at the same frequency.

For the special case where $x = y$, we have

$$\sin_2(x) = \tfrac{1}{2}[1 - \cos(2x)] \qquad (413)$$

## 2. Digital Filters

Analytical instruments are no longer simply analog instruments; many systems are analog-digital hybrids with built in mini- or microcomputers. In still other cases, the data acquired is post-processed by a computer. As a result, the field of digital filtering has become important. This subject is treated in more detail in Chapter 8, but we feel it is worthwhile to treat the subject briefly here to compare the properties of digital filters to the more familiar analog filters. We must reiterate again that any filter distorts the "good data" as well as removes unwanted "noise" components if the cutoff frequencies are not chosen carefully. Most chemists accept this fact when dealing with analog filters, but too often it is overlooked in dealing with digital filters. There is nothing magic about digital filters; they have their strengths and weaknesses just like analog filters, but because of their nature some of the weaknesses are easier to overlook.

In digital signal processing, the one basic law that cannot be ignored is that one must sample at twice the frequency of the highest frequency component in the signal (19). This is well understood among Fourier transform workers, but a lot of the rest of us seem to ignore this feature. This means that if data are taken at 0.01 sec intervals, the highest frequency component that should be in the signal is 50 Hz. If we ignore this restriction when utilizing digital filtering, we are asking for trouble and the odds are extremely high that we are going to get it. This also implies that an analog low-pass filter must proceed the digitization device to eliminate high frequency components.

The first nonregression form of digital filter that really caught the attention of the analytical community was the Savitzky-Golay (S-G) filter (20). This filter has a deceptively simple algorithm. It utilizes weighted fractions of the $n$ points on either side of the point of interest to compute a filtered value for the central point:

$$y_k = \sum_{i=-n}^{n} C_i \, Y_{k+i} \qquad (414)$$

The terms $C_i$ are constants, which are repeatedly used as the filter slides across the data array. From the definition it can be seen that only an odd number of points can be used in the filter and that one must lose the $n$ initial and $n$ final points in the data set. One interesting property of the S-G filter is that one can use it to compute smoothed derivatives of different orders by changing the values of the constants.

Usually the problems that people experience when using the S-G filter arise from not really understanding what the filter is doing in terms of the frequency-gain relationship. Betty and Horlick (21) have investigated the frequency response of the 7- and 21-point S-G filters, including the derivative filters as well as the two low-pass forms. Figure 3.52 shows the results of an extension of their work to cover the low-pass forms of the four most commonly used low-pass filters, the 7-, 9-, 11-, and 21-point filters, and compares these results with the filtering action of a single and double pole analog filter.

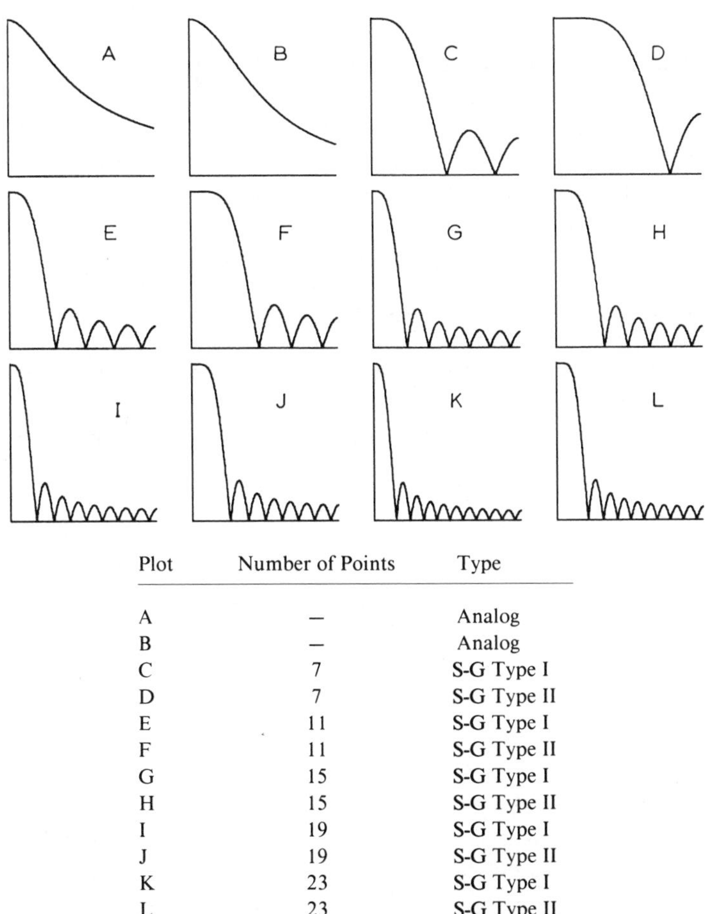


| Plot | Number of Points | Type |
|---|---|---|
| A | — | Analog |
| B | — | Analog |
| C | 7 | S-G Type I |
| D | 7 | S-G Type II |
| E | 11 | S-G Type I |
| F | 11 | S-G Type II |
| G | 15 | S-G Type I |
| H | 15 | S-G Type II |
| I | 19 | S-G Type I |
| J | 19 | S-G Type II |
| K | 23 | S-G Type I |
| L | 23 | S-G Type II |

**Fig. 3.52.** The gain characteristics of some digital and some analog low-pass filters; the digital filters are the Savitzky-Golay filter of type I or II. (Note: the axes are absolute value of gain vs. frequency.)

## IX. REAL COMPONENTS

So far we have neglected the nonideality of resistors and capacitors. All these features must be considered in order to successfully design an OA circuit or to understand the limitations of a commercial instrument.

### A. REAL RESISTORS

We have briefly mentioned some of the important properties of resistors, but we must review them and add other factors. The important parameters to consider in understanding the behavior of a resistor in a circuit are its precision, power rating, temperature coefficient, stability, series inductance, and capacitance.

#### 1. Precision

The initial accuracy rating of the resistor at a specified temperature, normally 25°C, is under zero current flow. For an $x$% resistor, 99% of a randomly chosen sample falls within $\pm x$% of the nominal value. Precision ratings are 20, 10, 5, 2, 1, 0.1, and 0.05%, although some military devices have other rating levels. The price of a resistor strongly depends upon its precision rating.

#### 2. Temperature Coefficient

The resistivity of the materials used to construct a real resistor is temperature dependent. The temperature coefficient is an expression of the rate of change of resistance with temperature normally expressed in parts per million per degree centigrade. If the resistance of the unit at 25°C is $R_0$ and the temperature coefficient is $r°$, then the resistance at the temperature $T$ in degrees centigrade is

$$R_t = R_0 \frac{1 + r°(T - 25°)}{10^6} \tag{415}$$

Normally the value of $r°$ is positive, that is, a positive temperature coefficient or increasing resistance with increasing temperature, although some special devices are available with negative coefficients. Obviously the smaller the value or $r°$, the more desirable the resistor, because it will be more stable with respect to temperature changes.

#### 3. Power Rating

The resistance is rated for the maximum power that can safely be dissipated by the device. Most resistors are of the $\frac{1}{4}$, $\frac{1}{2}$, and 1 W variety, although $\frac{1}{8}$ and $\frac{1}{10}$ W versions are sometimes encountered, particularly in the high precision types. Higher power resistances such as 2, 3, 5, and 10 are normally found only in power supplies or in the power stage of some booster circuits. One normally tends to use the lowest power rated unit, which is within the maximum expected dissipation for the circuit element. This is not always advisable because of self-heating. The power dissipated by the resistor is converted to heat, which causes a local rise in temperature in the resistance over and above the temperature of the environment. The ability of a resistor to dissipate heat into its environment is directly proportional to its surface area, and the higher the wattage rating, the greater the surface area available for dissipation. (This is only true when considering resistances, which are essentially identical except for wattage ratings.) A $\frac{1}{2}$ W resistor running at its rating limit is distinctly uncomfortable to touch, whereas a 1 W resistor dissipating $\frac{1}{2}$ W is significantly cooler. If the change in resistance due to the temperature rise is important, one should consider using a higher wattage resistor to help dissipate the heat (or heat sink the resistance to a metallic surface such as a printed circuit ground plane). If the change in resistance is not important in the particular application, one must still consider heating of the circuit board (a discolored printed circuit board is an indication of an under-specified resistor), which can lead to delamination of the copper conductor or heating of adjacent components. Even when these are not important considerations, one should consider stability.

### 4. Stability

This parameter is fairly difficult to pin down. It refers to the long-term changes in the resistance of an element. The stability is usually specified in terms of the change in ppm/1000 hr. at some rated load. The aging of a resistor is a function of several processes. No-load processes, such as breakdown of the environmental barrier and adsorption of moisture, generally tend to lower the resistance. The decomposition of the resistive element due to heating (the higher the power dissipation, the greater the change) may either increase or decrease the resistance, depending upon the resistive material.

### 5. Thermal Cycling Effects

The application of abrupt power pulses causing temperature steps can lead to changes due to compression or fracturing of the interior element. Again, the change can be either positive or negative. The stability rating of a resistor, when it can be found, may indicate the relative stability of one type relative to another, but because its testing bears no resemblance to a particular application, the numbers found should only be considered as guides. The one thing that is certain is that the real resistance changes from its initial value, and if absolute resistance is critical, one should consider provision for trimming the resistor to correct for aging.

### 6. Series Inductance and Capacitance

As mentioned above, all real resistors have an inductive and resistive component which is normally most important for high resistance values. Wire-wound resistors and potentiometers (i.e., those made from wire coils), even those called "noninductively wound," have an inherently higher inductance than other types.

Practical resistors are constructed using a number of different technologies, carbon composition, carbon film, metal film, cermet (a metal ceramic composition), ceramic, silicon nitride, conductive plastic, and metal wire being the most common, but by no means the only types. Table 3.II attempts to summarize briefly some of the major properties of some of the most common technologies. There will always be some particular resistance made for a particular purpose that violates the guidelines of the table; for instance, there *are* some 2 W metal film resistors made for very special purposes, but they are not generally available.

One method of coping with the changes in real resistances with time, temperature, and usage is to take advantage of the fact that in *most*, but not all, applications, it is not the absolute resistance that matters, but the ratio of several resistances that is important. If one can guarantee that the individual resistors have the same history and environment, to the microscopic scale, then they should change essentially identically. This seemingly improbable task is routinely accomplished now by depositing a number of carbon, metal, or cermet film resistors on a single highly insulating but thermally conductive substrate. Normally the accuracy or precision of any individual resistor may be 2% or 1%, but the ratio accuracy is commonly 0.1% or better. Also, although the temperature coefficient of any individual resistor may be as high as 100 ppm, the mismatch between resistors can often be as good as 2 ppm. An even more interesting development is the availability of multiple resistors and from one to three single turn trimmer potentiometers, all on a single substrate and in a single package. These packages can allow extremely accurate initial circuit accuracy and excellent performance stability (ratio tracking accuracy) as a function of time and still allow trimming for recalibration after several years.

## B. REAL CAPACITORS

An ideal capacitor has zero leakage current; that is, a charge stored in it stays unchanged indefinitely. It has zero power dissipation and no inductance, is totally indifferent to the polarity and magnitude of the voltage across its terminals, and is of course, totally unaffected by time and temperature. By now the reader should not be surprised to discover that no capacitor can meet any one of these specifications, let alone all of them.

### 1. Leakage Current and Insulation Resistance

For all practical purposes these are two aspects of the same phenomena. A real capacitance loses charge owing to leakage currents flowing within the capacitance; a reasonable model is to assume a capacitance with ideal charge storage properties that is connected in parallel with a resistance, the insulation resistance. One commonly sees this parameter expressed in terms of either megohms per farad (or microfarad) or as the product in seconds. The greater the value, the better the capacitor, at least in terms of this parameter.

TABLE 3.II

Resistors

| Type | Power rating (W) | Temperature range (°C) | Range of resistances (Ω) | Temperature coefficient ($R \times 10^{-6}, K^{-1}$) | Tolerances (%) | Maximum voltage (V) | Application |
|---|---|---|---|---|---|---|---|
| Metal film | 1/20 | −55 to 125 | 10 to $2.0 \times 10^5$ | −100 to 25 | 0.1, 0.5, 1.0 | | Precision resistor. |
| | 1/10 | −55 to 125 | 10 to $3.0 \times 10^5$ | −100 to 25 | 0.1, 0.5, 1.0 | | May be used over |
| | 1/8 | −55 to 125 | 25 to $1.5 \times 10^6$ | −100 to 25 | 0.1, 0.5, 1.0 | | a broad bandwidth |
| | 1/4 | −55 to 125 | 25 to $3.0 \times 10^6$ | −100 to 25 | 0.1, 0.5, 1.0 | | |
| | 1/2 | −55 to 125 | 10 to $4.0 \times 10^6$ | −100 to 25 | 0.1, 0.5, 1.0 | 700 | |
| | 1 | −55 to 125 | 25 to $4.0 \times 10^{-6}$ | −100 to 25 | 0.1, 0.5, 1.0 | 1000 | |
| | 2 | −55 to 125 | 100 to $6.0 \times 10^6$ | −100 to 25 | 0.1, 0.5, 1.0 | 1000 | |
| Carbon film | 1/10 | −55 to 125 | 1.0 to $4 \times 10^5$ | −250 to 500 | 1, 2, 5, 10 | | Common resistor. |
| | 1/8 | −55 to 125 | 1.0 to $3 \times 10^5$ | −1500 to 250 | 1, 2, 5, 10 | 150 | Broad bandwidth |
| | 1/4 | −55 to 125 | 1.0 to $5.0 \times 10^6$ | −1500 to 250 | 1, 2, 5, 10 | 250 | |
| | 1/2 | −55 to 125 | 1.0 to $1.0 \times 10^7$ | −1500 to 250 | 1, 2, 5, 10 | 350 | |
| | 1 | −55 to 125 | 1.0 to $1.5 \times 10^7$ | −1500 to 250 | 1, 2, 5, 10 | 500 | |
| | 2 | −55 to 125 | 2.0 to $1.0 \times 10^8$ | −1500 to 250 | 1, 2, 5, 10 | 1500 | |
| Carbon composition | 1/8 | −55 to 125 | 2.7 to $2.0 \times 10^7$ | −5000 to 5000 | 5, 10, 20 | | Garden-variety |
| | 1/4 | −55 to 125 | 2.7 to $2.0 \times 10^7$ | −5000 to 5000 | 5, 10, 20 | 250 | resistor. Broad |
| | 1/2 | −55 to 125 | 2.7 to $2.0 \times 10^7$ | −5000 to 5000 | 5, 10, 20 | 350 | bandwidth |
| | 1 | −55 to 125 | 2.7 to $2.0 \times 10^7$ | −5000 to 5000 | 5, 10, 20 | 500 | |
| | 2 | −55 to 125 | 10 to $2.0 \times 10^7$ | −5000 to 5000 | 5, 10, 20 | 1500 | |
| Cermat film | 1/8 | To 70 | 4.7 to $1.5 \times 10^5$ | −200 to 100 | 1, 2, 5 | | Semi-precision resistor. |
| | 1/4 | To 70 | 4.7 to $1.6 \times 10^5$ | −200 to 100 | 1, 2, 5 | | Can work around the |
| | 1/2 | To 70 | 4.7 to $1.0 \times 10^5$ | −200 to 100 | 1, 2, 5 | | "High" temperature |
| | 1 | To 70 | 10 to $1.0 \times 10^6$ | −200 to 100 | 1, 2, 5 | | coefficient |
| | Up to 115 | To 70 | 10 to $1.0 \times 10^6$ | −200 to 100 | 1, 2, 5 | | |
| Wire wound | 1/8 | −45 to 85 | 1.0 to $3 \times 10^5$ | +20 | 0.05, 0.1, | | Precision resistor. |
| | 1/4 | −45 to 85 | 1.0 to $4.5 \times 10^5$ | +10 | 0.5, 1.0 | | Inductance |
| | 1/2 | −45 to 85 | 1.0 to $1.2 \times 10^6$ | +10 | 0.01, 0.05, | | Limits high frequent, |
| | 1 | −45 to 85 | 0.5 to $1.0 \times 10^5$ | +20 to +100 | 0.1, 0.5, 1.0 | | applications |
| | 2 | −45 to 85 | 0.5 to $2.7 \times 10^5$ | +20 to +100 | 0.01, 0.05, | | |
| | Up to 250 | −45 to 85 | 0.5 to $2.7 \times 10^5$ | +20 to +100 | 0.1, 0.5, 1.0 | | |
| | | | | | 0.1, 0.5, 1.0 | | |
| | | | | | 0.1, 0.5, 1.0 | | |
| | | | | | 0.2, 0.5, | | |
| | | | | | 1.0, 3.0 | | |

### 2. Power Dissipation

An ideal capacitor produces a 90° phase shift between the voltage impressed across a capacitor and the current flowing through it. A real capacitor produces a somewhat smaller phase shift that can be viewed as being due to a finite parallel resistance.

This property is evaluated at some specific frequency, commonly 1 KHz, and is expressed in a number of ways, all of which relate to the phase angle between the current and voltage. The *power factor* is the *sine* of the angle by which the current flowing through the capacitor fails to be 90° out of phase with the voltage impressed across it. The *tangent* of this angle is called the *dissipation factor*. The ratio of the capacitive reactance to the equivalent series resistance (the reciprocal of the dissipation factor) is termed the $Q$ of the capacitor. For high-quality capacitors used in signal filtering the power factor, dissipation factor, and phase angle difference expressed in radians are so small that they are essentially identical. The equivalent series resistance $R_s$ can be determined from the relationship $R_s$ = power factor/$2\pi fC$. This parameter is important in integrators and filter circuits.

### 3. Soakage

An ideal capacitor that has been charged up to a given voltage and then briefly short circuited exhibits 0 V between the terminals when the short circuit is removed. A real capacitor, particularly electrolytics, exhibits a subsequent voltage rise to some fraction of the original impressed voltage. This is termed soakage and is due to slow rearrangements of the dipoles within the capacitor. One has not recovered all of the stored charge in the initial shorting and the dipoles slowly yield up the residual charge. The lower the soakage value quoted for a capacitor, the more capable it is of following rapid charge change requirements.

### 4. Voltage Polarity and Magnitude

All capacitors have a maximum rated voltage which should not be exceeded for any appreciable amount of time (what amounts to an appreciable amount of time is often an interesting question for which it is hard to get an answer). In general, a capacitor fails catastrophically, as a short circuit usually, when the maximum voltage rating is exceeded. Some capacitors inherently are nonpolar; that is, they are unaffected by the direction or sign of the electric field gradient whereas others, normally termed electrolytic capacitors, survive only if the correct polarity conventions are observed. With polar or electrolytic capacitors, an electrochemical reaction takes place if the polarity is reversed, and the capacitor fails destructively, permanently, and spectacularly as a short circuit. The only justification for polar capacitors is that their volume efficiency (farad/$cm^3$) is high and their cost (farad/monetary unit) is low relative to nonpolar types. A 10 $\mu$F at 25 V tantalum electrolytic capacitor may be as small as a

TABLE 3.III

Capacitors

| Type | Dielectric material | Dielectric constant, $K_a$ | Range of capacitance | Temperature Coefficient (ppm/°C) | Temperature range (°C) | Maximum voltage (V) | Insulation resistance (25°C, M-F) | Tolerance (%) | Power factor | Typical $Q$ |
|---|---|---|---|---|---|---|---|---|---|---|
| Variable | Air | 1.001 | 5–500 pF | | | 500 | | 1, 2, 5, 10, 20 | | 1500 (1 MHz) |
| Ceramic | Barium titanate | 200–16,000 | 10 pF–1 F | −30 to +30 | −55 to 125 | 6000 | 200,000 | 5, 10, 20 | $5 \times 10^{-4}$ $20 \times 10^{-4}$ | 1500 (1 KHz) |
| Oil | Paper in oil | 2.2 | 0.01–1.0 F | | −55 to 125 | 10,000 | 2,000 10 (85°C) | 5, 10, 20 | | 250 (1 MHz) |
| Mica | Mica | 5.4 | | 0–70 | −55 to 125 | 10,000 | | 1, 2, 5 | $1 \times 10^{-4}$ $-7 \times 10^{-4}$ | 600 (1 KHz) |
| Mylar (film)[a] | Mylar (polyester) | 3.0–4.5 | 0.001–1.0 F | +250 | −55 to 125 | 1,000 | 100,000 | 1, 2, 5, 10, 20 | $8 \times 10^{-4}$ $-14 \times 10^{-4}$ | 100 (1 KHz) |
| Teflon (film) | teflon | 2.1 | 100 pF–10F | −250 | −60 to 150 | 1,000 | 300,000 | 1, 2, 5, 10, 20 | $0.5 \times 10^{-4}$ $-1.5 \times 10^{-4}$ | |
| Polystyrene (film) | Polystyrene | 2.5 | 100 pF–10 F | −50 to 100 | −60 to 85 | 1,000 | 200,000 | 1, 2, 5, 10, 20 | $1 \times 10^{-4}$ $-2 \times 10^{-4}$ | 2000 (1 KHz) |
| Polycarbonate[2] (film) | Polycarbonate | 3.2 | 100 pF–20 F | Nonmontonic | −50 to 100 | 1,000 | 12,000 | 5, 10, 20 | $30 \times 10^{-4}$ $-50 \times 10^{-4}$ | 500 (1 KHz) |
| Polypropylene (film) | Polypropylene | 2.1 | 0.001–1.0 F | 300 | −55 to 105 | 500 | 250,000 | 1, 2, 5, 10, 20 | | 2500 (1 KHz) |
| Porcelain monolithic | Porcelain | | 0.1–1000 pF | 70–110 | −40 to 120 | 400 | 1,000,000 | 0.1 pf, 0.25 pf, 0.5pf 1, 2, 5, 10, 20 | | |
| Tantalum electrolytic | Tantalum oxide | 27.6 | 0.19–300 F | | −55 to 85 | 50 | | 5, 10, 20 | | |
| Aluminum electroyltic | Aluminum oxide | 8.4 | 0.1–100,000 F | | −40 to 85 | 300 | | 10, 20 | | |

[a] $Q$ is a non-linear function of frequency.

pencil eraser, an aluminum electrolytic of the same rating the size of the first joint of one's little finger, and a film and foil, nonpolar capacitor of the same rating half the size of a clenched fist. A pure, nonpolar ceramic of this rating cannot presently be built, although progress in stacked ceramics indicates that one, perhaps four, times the volume of the tantalum may be possible.

### 5. Surge Currents

Electrolytic capacitors used in power supplies have a maximum surge current rating, the maximum instantaneous current that the capacitor can tolerate. Actually, all capacitors have this limitation, but only in power supplies is one likely to encounter a circumstance where this rating limit can be exceeded without first exceeding the voltage rating.

### 6. Precision, Stability, and Temperature Coefficients

As in the case of resistors, a capacitor is rated at 25°C under specified load conditions and its value changes with time, temperature, and history. The precision ratings of capacitors depend strongly on the construction technology employed ranging from + 80%, −20% for ceramics down to 1% or less for some special film capacitors.

### 7. Construction and Properties of Real Capacitors

There are a number of technologies employed in building commercial capacitors, and the properties of the resulting product depend more strongly on the technology than is the case for resistors. The properties of some of the most important types found in analytical instruments in the low voltage sections are listed in Table 3. III.

## X. OPERATIONAL AMPLIFIERS IN CONTROL SYSTEMS

Control systems can be broken down into two broad catagories, open-loop and closed-loop control systems. A normal kitchen faucet is an open-loop control system: one turns the faucet and the internal valve controls the flow of the water. If the pressure drops, the flow rate decreases unless the operator again turns the handle. A closed-loop system is one in which the state of the system is monitored and corrective action is taken if the measured response does not match the programmed value. An example of this type of system is the more sophisticated shower temperature controller often found today. In this type of system if the head pressure of the hot water line rises or the head pressure of the cold water line drops, the valve on the hot water side is closed a proportional amount to prevent scalding. Some more sophisticated versions also try to prevent frostbite as well. We concentrate here on closed-loop control systems.

## A. AUTOTITRATOR

An autotitrator is essentially a closed-loop system composed of a pH meter, sensing and decision electronics, and a burette controller. There are several variations on the theme; we discuss the simplest, the dead stop autotitrator, and a somewhat more sophisticated version with anticipation. Under this heading we also discuss simple pH meters.

### 1. pH Meters

The pH meter is essentially a specialized voltmeter. The potential developed by an ion selective electrode (ISE) - reference electrode (RE) pair in solution is given by

$$E_c = E_t^{\circ\prime} - \frac{2.303RT}{n_t F}[\text{Log}(A_i)] - E_r^{\circ} - \frac{2.303RT}{n_r F}\log(A_r) \tag{416}$$

where $E_t^{\circ\prime}$ and $E_r^{\circ}$ are the standard potentials of the test and reference electrodes, respectively (for a glass electrode, this is not strictly a standard potential), and $A_i$ and $A_r$ are the activities of the controlling ions for the test and reference electrodes. We can rearrange this to give

$$E_c = E_t^{\circ\prime} - \frac{0.059}{n_t}\frac{T}{298}\left(\log(A_i) + \frac{n_t}{n_r}G_r\right) \tag{417}$$

where $G_r$ is the log activity of the reference ion and we have taken into account the nominal 25°C (approximate 298°K) laboratory temperature. Many commercial pH meters are constructed from a single OA but for simplicity in understanding the circuit, we use more. Figure 3.53 shows a pH meter that allows us to take into account the $n$ and the $T$ dependency of the response and to null out the effect of $E^{\circ}$, the lumped standard potentials, and sets the meter movement to some arbitrary position when a calibration solution is being monitored. Two points should be noted. The test electrode (normally a glass electrode for pH work) is shown connected to ground and a shield is shown surrounding it. This is because the glass electrode is a very high impedance device and hence very susceptible to noise pickup. The reference electrode

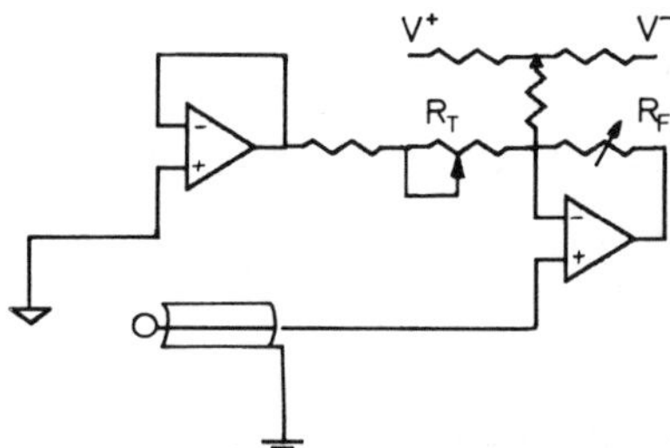


**Fig. 3.53.** A simple pH meter with an arrangement that allows compensation for the offset voltage ($P_C$) and a temperature setting ($R_T$). Note that the input signal comes from a coaxial cable which limits the amount of noise that is picked up.

connects to the input of a voltage follower amplifier so that no current flows through the reference. This is critical because any current that flows through the reference *must* come from the test electrode. The arrangement shown allows one to change the offset adjust (or calibrate) control and the temperature compensation control without interaction between them. The temperature-dependent term arising from $G_r$ is not explicitly adjustable in this configuration, because the activity of the reference ion is normally close to the unity. This introduces only a slight temperature-dependent error unless the temperature is greatly different from 298°K and the term can be taken care of by the calibrate control.

## 2. Controller Circuit

The voltage output of the pH meter (either the one shown or the voltage output available on most commercial pH meters) is connected to $R_1$ of OA2. The output of OA1 passes through a resistor $R_Z$ and then through a zener diode-diode bridge combination which limits the maximum output to ±10 V at the bridge. This voltage is dropped to ground by a potentiometer, the tap of which connects to the positive input of OA2. This circuit is a voltage comparator with adjustable hysteresis. The momentary contact pushbutton switch, PSW1, and resistor $R_2$ to the +15 V supply allow us to place a net positive potential on the negative input of OA1 relative to the potential at the positive input. The output of OA2 goes to its negative bound in response to this condition, and the zener diode-diode bridge combination limits the voltage to −10 V. This circuitry is incorporated to produce known conditions for the saturation voltage output of the amplifier. The potential at the positive input is now $-10X$, where $X$ is the voltage fraction picked off from the potentiometer tap at ranges from zero to unity. As long as the potential from the pH meter is more positive than $-10X$, the output of OA2 remains low negative, but when the potential becomes negative with respect to $-10X$, then the output of OA2 swings positive, driving the potential at the positive input to $+10X$, locking the output of OA2 into

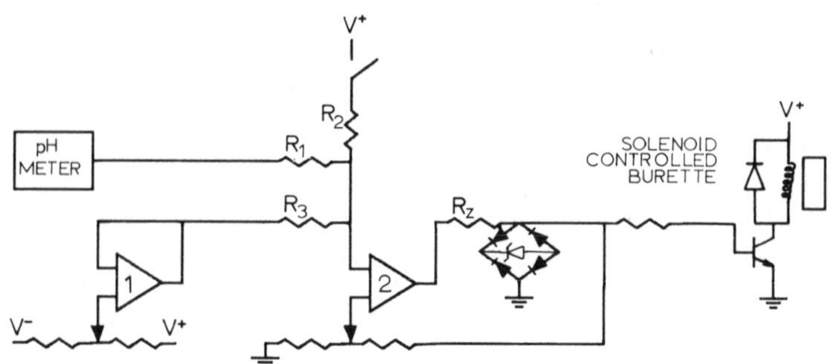


**Fig. 3.54.** An autotitrator of the bang-bang type.

positive saturation. The output of OA2 also drives the base of a *NPN* transistor which has the solenoid of a titrant control valve in its collector to power supply connection. When the output of OA2 is positive, the transistor conducts and opens up the titrant flow path; when the output of OA2 goes negative, the transistor is turned off and the titrant flow is halted. As long as the set point for the end of the titration is such that the potential for the pH meter is a negative voltage, the circuit performs properly, allowing titration to flow until the set point is reached and stopping the titration at that point. If the direction of the voltage change is incorrect, an inverter can be placed between the pH meter and the controller.

The difficulty with this simple circuit is that it produces erroneous results. The problem arises from the nature of titrations and the mixing delay of the solutions. Titrant entering the cell does not instantaneously disperse, so that local high and low concentration regions exist. If the sensor electrode is placed so that it is in the direct path of titrant entry, then the sensor senses a solution condition that is "ahead" of the general solution conditions. If turbulent flow exists, then little "packets" of titrant reach the sensor, producing in essence a "noisy" signal, and the first time the controller sees a potential that represents a past end point condition, the system locks up. One solution is to slow the response of the controller, and conceptually the simplest way to accomplish this is to insert a low-pass filter between the pH meter and the resistance $R_1$ of OA1. The filter should be a noninverting filter with a time constant of about 1 sec to allow all fluctuations to die out in solution. One can determine the correct time constant empirically be inserting a drop of titrant into an unbuffered cell and observing the results of an oscilloscope, then multiply by about 3 for safety. One can slightly modify this circuit to eliminate the low-pass filter by connecting the tap of the potentiometer to the positive input of OA2 via a resistance $R_4$ of, say, 10MΩ and connecting a capacitor from the positive input to ground. The potential at the positive input then lags behind the output and so premature lockup does not occur. The resistance $R_4$ can be made variable to set the correct time constant.

### 3. A Proportional Controller Autotitrator

Even the modified form of the circuit produces incorrect results because of the on-off nature of the control. One sometimes sees circuits of this type called "bang-bang" circuits (Fig. 3.54). A better approach is to have the rate of titrant addition be proportional to the difference between the average value of the instantaneous solution composition and the end point composition. This type of autotitrator is a member of the class of controllers known as proportional controllers. A simple implementation of this method is shown in Fig. 3.55. The output of the pH meter is added to the complement of the reference or set point value by OA2 and the output of this amplifier goes to a voltage-to-frequency converter. The VFC produces a pulse train whose frequency is directly proportional to the voltage at its input. The output of VFC drives a stepping motor

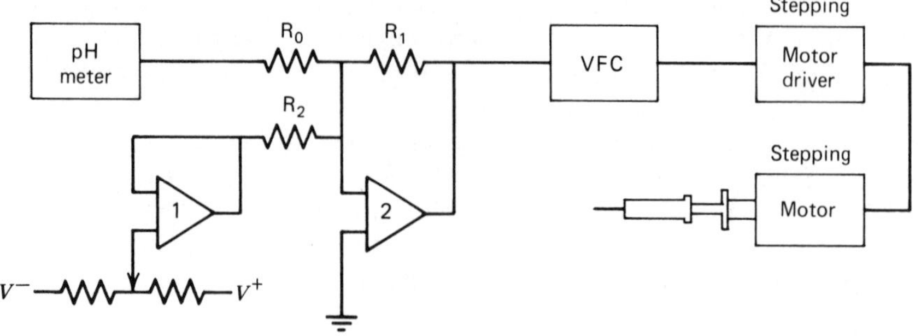


**Fig. 3.55.** A block diagram of a proportional controller autotitrator.

that is connected via a lead screw to a syringe containing the titrant. A stepping motor makes some fixed fractional rotation every time a step is applied to it. Common stepping motors require from 4 to 80 steps per revolution and step rates from zero to several thousand per second are normal. With this kind of system the rate of titrant addition is proportional to the "distance" from the end of the titration. Depending upon the sensitivity of the VFC, it may or may not be necessary to have a low-pass filter in the circuit. If required, this can simply be accomplished by placing a capacitor in the feedback loop of the summing amplifier. One other property of this type of autotitrator is that it can function as a pH-stat. Many enzymatic and simple chemical reactions produce a titratable ion as a product. One can use the rate of titrant addition to maintain a constant species activity as a measure of the reaction rate and hence the concentration of catalysts (including enzyme concentration). The output of the summing amplifier is a direct voltage representation of the rate of titrant addition in this circuit. If one is interested in the total amount of titrant added, as in a typical titration, one can use a digital counter to count the pulses to the stepping motor, or for an analog representation, one can use a long-term integrator on the output of the adder amplifier and obtain the same information. The VFC can either be a commercial unit (costs for a perfectly adequate commercial unit are now below $10), or one may be constructed from OAs.

### 4. A Voltage-to-Frequency Converter

There are a number of circuits that will produce a pulse train whose frequency is proportional to the input voltage. The critical parameters in a VFC are the maximum frequency of oscillation and the linearity of the response. For our application, the linearity is unimportant if we are going to measure the number of pulses required or the burette piston travel as the indicator of the titrant delivered, or if the pulse rate is the measure of the rate of formation of titratable species in solution. However, if we are going to integrate the output of the adder amplifier for our measure of titrant added, or are going to use that signal directly for our measure of formation rate, then linearity becomes important. When linearity is in excess of 1% of full scale, that is, when maximum deviation from a straight line drawn from zero to full scale exceeds 1% of full scale, then with the present state of the art in VFCs, it is wiser to employ a commercial VFC. In general, one can purchase VFCs with linearities of 0.1% or better for less than one can construct them.

An inexpensive VFC that is linear to better than 1% is shown in Fig. 3.56. The circuit accepts only unipolar, positive signals. A positive signal at the input to the integrator produces a negative-going ramp. The voltage at the negative input of OA2 is compared by the OA to the voltage at its positive input. The positive input voltage is clamped by the silicon diode to ground to less than 0.6 V or to $\frac{2}{3}E_s$ (negative saturation voltage). When the voltage at the inverting input becomes more negative than the voltage at the noninverting input, the output of OA2 goes to positive saturation. The potential at the input of the $p$-channel JFET (an MPF 102) is clamped by the diode to +0.6 V, which brings the drain-source resistance to about 400Ω, discharging the capacitor and returning the output of OA1 to nearly 0 V. On discharge, the potential at the inverting input of OA2 is pulled to about +0.7 V by the resistor to the positive power supply $V+$, which is a potential that is more positive than the +0.6 V at the noninverting input of OA2, so that the output of OA2 goes to negative saturation again. This turns off the JFET and the integration cycle proceeds again. The time $t_u$ required for the integrator to reach the trip point is

$$t_u = \frac{R_1 C_0}{E_i - E_{os1}} \left[ \frac{30}{27} \left( \frac{2}{3} E_{os1} - E_{os2} \right) - \frac{27}{3} V^+ \right] \tag{418}$$

where $E_{os1}$ and $E_{os2}$ are the offset voltages of OA1 and OA2, respectively. The input bias current errors can be neglected if one is using cheap FET input amplifiers such as the CA3140. The obvious factors affecting accuracy are changes in the input offset voltages, which can be trimmed to nearly zero, and changes in the saturation voltage of OA2. The latter error could be minimized by using a zener diode with a low temperature coefficient, but these alterations are not worth the effort because they typically introduce errors of only about

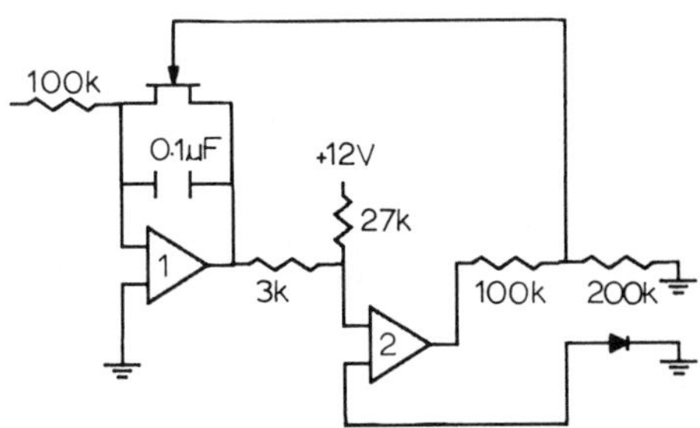


**Fig. 3.56.** An inexpensive voltage-to-frequency converter.

0.5%. There are circuit diagrams available of highly linear VFCs (12), but we view them as being of primarily academic interest. If one needs more linearity, the most logical solution is to utilize one of the commercial units.

### 5. Autotitrators with Zero Volume Change

One difficulty with the preceding system for pH-stat or, more generally, concentration-stat work, is that the additional titrant changes the solution volume as the experiment proceeds. This approach has some secondary advantages. One can utilize titrants such as bromine, which are at best difficult to work with by more traditional methods. The standardization of titrants is, at least in principle, eliminated since the amount of species generated is directly proportional to the charge utilized. Finally, one has eliminated the mechanical linkages in the system. Since the mechanical components are generally the most unreliable, the performance of the system should be improved.

Figure 3.57 shows a pH-stat designed by Adams et al. (22). The pH meter section is significantly different from the one we used in the preceding example, because of some new requirements of the circuit. One of the historical difficulties in this type of instrument has been in electrically isolating the pH measurement process for the generation process and in minimizing noise pickup, which can cause the system to oscillate near pH null. Amplifiers OA1, OA2, and OA3 form a very high input impedance differential subtractor amplifier. The solution ground electrode (a platinum flag) is required to establish a ground reference for the differential amplifier section. Any common mode noise signal present "in" the solution is eliminated by this configuration. Amplifier OA4 is essentially the "standardize" control and OA5 produces what is in effect the error signal. The signal is low-pass filtered by $C_1$ in parallel with the feedback resistance $R_6$ and $P_3$ which set the sensitivity. The isolation amplifier, OA6, is central to the success of the instrument. An isolation OA is used when it is necessary to isolate two portions of electronic circuitry from one another as in this case, or when electrodes are connected to human patients as in the case of medical electronics. Basically they are two self-contained OAs with separate power supply inputs. The two amplifiers are coupled together, either by an isolation transformer or by light-emitting diode-photodiode pairs. For the transformer method the signal from the first amplifier is modulated prior to the transformer inputs and the transformer output is filtered to reconstruct the original wave form. In either case, a high degree of isolation between the two sides results. In some systems there is a built-in dc-dc converter so that only one power supply is required.

The switch SW1 selects the on signal, the output of the isolation amplifier or one of four calibrated voltage sources, which is amplified by R10/R9 by OA7. This signal drives amplifiers OA8-OA9, which act as a constant current source. OA9 is a unity gain noninverting current booster. The voltage at the noninverting input of OA8 also appears at the inverting input which causes a current flow through resistor $R_m$. Because no current flows from the noninverting input, in must flow through the cell between the auxiliary and working electrodes.

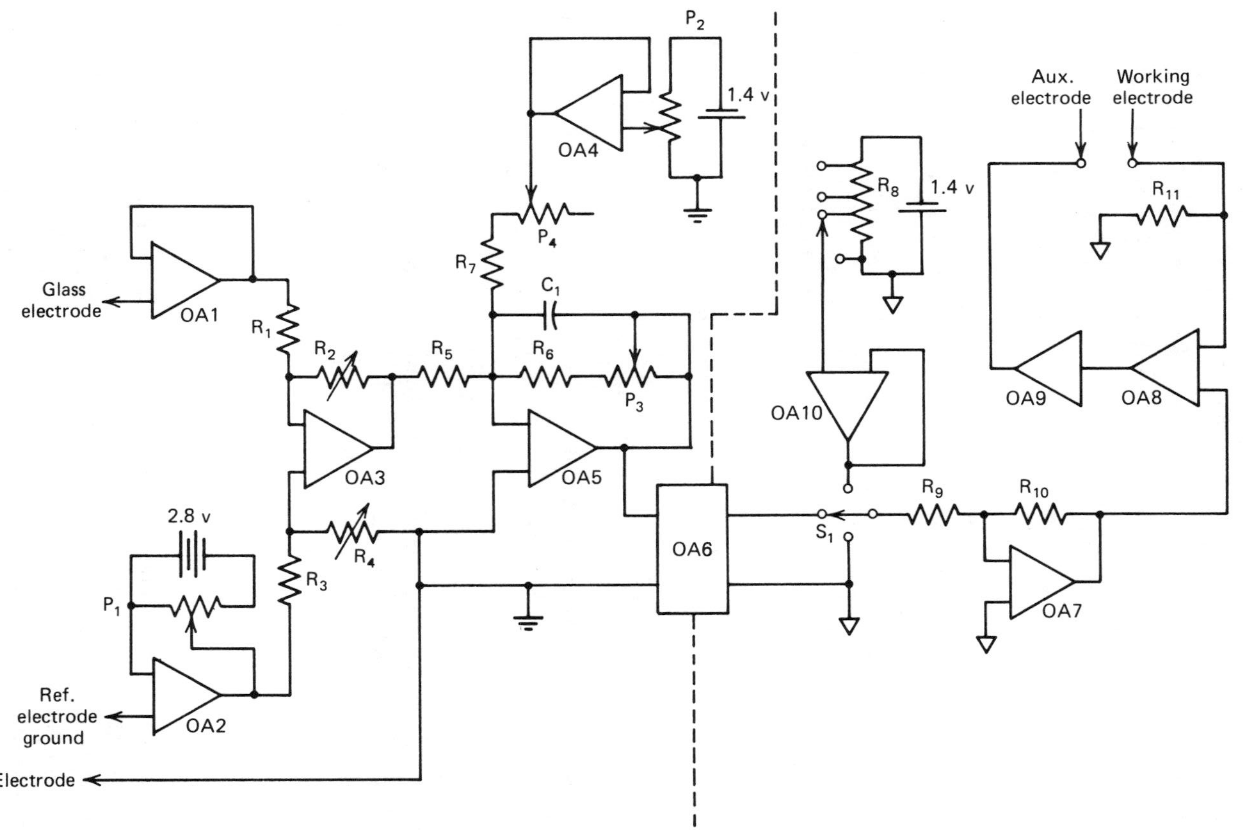


**Fig. 3.57.** A Coulometric ph-stat (see text explanation).

The overall performance is analyzed in the following manner. The electrolysis current $i_c$ is determined by

$$i_c = \frac{E_{oa7}}{R_{11}} \tag{419}$$

If OA6 is operating in unity gain noninverting mode (the amplifier employed in the original paper can be programmed for gains of +1 to +100), then the cell current in Laplace space is

$$\overline{I}_c = \frac{R_6 R_{10}}{R_{11} R_9}\left[\frac{\overline{E}_{oa}}{R_7} + \frac{\overline{E}_{oa}}{R_5}\right] \Big/ C(s + a) \tag{420}$$

where $a$ is the time constant of the feedback loop of OA5. Since the output voltage of OA4 does not change during the experiment, we can rewrite this as

$$\overline{I}_c = \frac{R_6 R_{10}}{R_m R_9}\,[\overline{E}_{oa4}/R_7 s + \overline{E}_{oa3}/R_5]\,/C(s + a) \tag{421}$$

The output of OA3 is a familiar problem, being simply a differential subtractor circuit:

$$E_{oa3} = -E_{oa1}\left(\frac{R_2}{R_1}\right) + E_{oa2}\left(\frac{R_4}{R_3}\right)B \tag{422}$$

$$B = \frac{1 + R_2/R_1}{1 + R_4/R_3} \tag{423}$$

The signals from the two electrodes are

$$E_{oa1} = \Phi_1 + \theta\,\text{pH} + E(i)_1 \tag{424}$$

$$E_{oa2} = \Phi_2 + E(i)_2 \tag{425}$$

where $\Phi_1$ is the constant related to the asymmetry potential and $\Phi_2$ is the potential of the reference electrode and $\theta$ is typically 59 mV/decade. $E(i)_1$ and $E(i)_2$ are common mode voltages with respect to current ground that were empirically determined to be a function of the electrolysis current. The output of OA3 is thus

$$E_{oa3} = -\left[\Phi_1\frac{R_2}{R_1} - \Phi_2\frac{R_{4B}}{R_3}\right] - \left[E(i)_1\left(\frac{R_2}{R_1}\right) - E(i)_2\left(\frac{R_4}{R_3}\right)B\right] - \theta\text{pH}\left(\frac{R_2}{R_1}\right) \tag{426}$$

As long as the common mode voltages depend on the electrolysis current and, hence, on time, then by adjusting $R_2$ and $R_4$ one can eliminate the common mode component from $E_{oa3}$. The signals due to $\Phi_1$ and $\Phi_2$ are also time independent on the time scale of any rational experiment at constant temperature, so that one can write

$$E_{oa3} = -\,\theta\,\text{pH}\frac{R_2}{R_1} + M \tag{427}$$

where $M$ is the constant.

$$M = \Phi_1 \frac{R_2}{R_1} - \Phi_2 \frac{R_4}{R_3} \tag{428}$$

The relationship to the cell current now becomes

$$\overline{I}_c = \frac{R_{10}}{R_{11}R_9}\left(\frac{E_{oa4}}{SR_7} - \frac{M}{SR_5} - \frac{M}{R_{5s}} - \frac{\theta \text{pH} R_2}{R_1}\right) \Big/ (s + a)C \tag{429}$$

The constant terms containing $E_{oa4}$ and $M$ can be lumped into one constant which is set at the beginning of the experiment, so that the net $I_c$ is zero at the desired pH:

$$\overline{I}_c = \frac{R_{10}}{R_{11}R_9C(s + a)}\left(\frac{P}{s} - \frac{\theta R_2}{R_1}\text{pH}\right) \tag{430}$$

The circuit is designed so that the working electrode generates $OH^-$ when the pH is too low (i.e., $H^+$ when the pH is too high). The opposite reaction or some other similar reaction occurs at the auxiliary electrode, but it is isolated from the rest of the solution by an ion exchange membrane. The rate of change of pH with respect to time is given by

$$\frac{d\,\text{pH}}{dt} = \frac{-i_c(t)}{\epsilon \pi V F} \tag{431}$$

where $V$ is the solution volume in liters, $F$ the Faraday, $\pi$ the buffer capacity of the solution, and $\epsilon$ the current efficiency (the fraction of electrons that actually produces either $OH^-$ or $H^+$ ions; ideally $\epsilon$ is unity and this value is closely approached in a number of media). In Laplace space this relationship becomes

$$s\Delta\text{pH} - \Delta\text{pH}_{t=0} = -\frac{\overline{I}_c}{\epsilon \pi V F} \tag{432}$$

We could at this point substitute our other expression for $\overline{I}_c$ into this expression and equate

$$\theta \frac{R_2}{R_1}\Delta\text{pH} = \frac{P}{s} - \theta R_2\overline{\text{pH}} \tag{433}$$

solving for the functionality of pH as the original authors in effect did. This is not quite correct. In the original paper, it was demonstrated (22) that the response time of OA3 to a pulse of acid or base is $0.9 \pm 0.1$ sec. This means that for a concentration impulse of titrant, the response has the form $1 - e^{-t/b}$ where $b$ is 0.9 sec for the particular cell geometry they employed. This feature can be incorporated into the above equation by adding the delay operator $1/(s + \tau)$or

$$\overline{I}_c = \frac{R_{10}R_6\text{pH}}{R_9R_{11}C(s + a)(s + \tau)} = -\frac{N\,\text{pH}}{(s + a)(s + \tau)} \tag{434}$$

We can now make the substitution and find

$$s\,\Delta\overline{\mathrm{pH}} - \mathrm{pH}_{t=0} = -\frac{N\,\Delta\mathrm{pH}}{VF(s+a)(s+\tau)} = -\frac{R\,\Delta\mathrm{pH}}{(s+a)(s+\tau)} \tag{435}$$

$$\overline{\mathrm{pH}} = \frac{(s+a)(s+\tau)}{s(s+a)(s+\tau)+R} \tag{436}$$

We can now express the denominator in terms of the roots of the cubic equation to give

$$\overline{\mathrm{pH}} = \frac{(s+a)(s+\tau)\mathrm{pH}_{t=0}}{(s+d_1)(s+d_2)(s+d_3)} \tag{437}$$

To return to the time plane requires finding the roots of the cubic equation and is practical only with numerical methods using a computer. Although it is trivial to write the form of the solution in time, we do not do so because of the complexity of the expression. As an alternative to the use of a computer to solve the roots, one can cheat a little and come up with an expression that preserves the essential features of the system, but that is more tractable.

The time constant of the low-pass filter, 0.1 sec, in OA5 is small with respect to the time constant of the cell-OA1-OA2-OA3 combination, and we can with reasonable safety neglect its effect. Making this approximation the equation in Laplace space simplifies to

$$s\,\Delta\overline{\mathrm{pH}} - \mathrm{pH}_{t=0} = \frac{D_a\mathrm{pH}}{(s+\tau)} \tag{438}$$

$$\Delta\overline{\mathrm{pH}} = \Delta\mathrm{pH}_{t=0}\,\frac{(s+\tau)}{s(s+\tau)+D_a} \tag{439}$$

This transform is difficult to find in transform tables, but the following transform is available:

$$\mathcal{L}^{-1}\,\frac{s+a}{s(s+b)} = \frac{a}{b} + \left(1-\frac{a}{b}\right)e^{-bt} \tag{440}$$

There is also the translation property

$$\mathcal{L}^{-1}g(s+d) = e^{-dt}F(t) \tag{441}$$

and we can recast our equation to take advantage of these relationships:

$$\frac{s+\tau}{s(s+\tau)+D_a} = \frac{s+\tau}{s^2+\tau s+D_a} = \frac{s+\tau}{(s+x)(s+y)} \tag{442}$$

$$\mathcal{L}^{-1}\,\frac{s+h}{(s+x)(s+y)} = e^{-yt}\,\mathcal{L}^{-1}\,\frac{(s+h-y)}{(s+x-y)(s+y-y)} = e^{-yt}\,\mathcal{L}^{-1}\,\frac{(s+h-y)}{s(s+x-y)} \tag{443}$$

Identifying $-y$ with $a$ and $x - y$ with $b$, we have

$$\mathcal{L}^{-1}\frac{s+\tau}{s(s+\tau)+D_a} = e^{-yt}\left\{\frac{(\tau-y)}{x-y} + \left[1 - \frac{(\tau-y)}{x-y)}\right]e^{-(x-y)t}\right\} \tag{444}$$

$$= e^{-yt}\frac{(\tau-y)}{x-y} + \frac{(x-y-\tau+y)}{x-y}e^{-xt} = \frac{1}{x-y}\{(\tau-y)e^{ty} + (x-\tau)e^{-xt}\} \tag{445}$$

We must now insert the values for the roots $x$ and $y$ in the equation:

$$x = \frac{\tau + \sqrt{\tau^2 - 4D_a}}{2},\ y = \frac{\tau - \sqrt{\tau^2 - 4D_a}}{2} \tag{446}$$

$$x - y = \tau^2 - 4D_a \tag{447}$$

$$(\tau - y) = \frac{2\tau - \tau + \sqrt{\tau^2 - 4D_a}}{2} = x \tag{448}$$

$$(x - \tau) = \frac{\tau - 2\tau + \sqrt{\tau^2 - 4D_a}}{2} = -y \tag{449}$$

It simplifies the discussion somewhat to define a new parameter $u$ such that

$$u = \sqrt{\tau^2 - 4D_a} = \sqrt{\tau^2 - \frac{4R_{10}R_6R_2\epsilon\theta}{R_1R_9R_{11}\beta VF}} \tag{450}$$

The expression for the instantaneous pH of the solution is

$$\Delta\text{pH} = \text{pH}_{t=0} + \left(\frac{1}{2u}\right)[(\tau+u)e^{-(\tau-u)t/2} - (\tau-u)e^{-(\tau+u)t/2}] \tag{451}$$

$$= \text{pH}_{t=0} + \frac{e^{-t/2}}{2u}[(\tau+u)e^{ut/2} - (\tau-u)e^{-ut/2}] \tag{452}$$

There are three special cases of this result, depending upon the value of $u$, hence on the ratio of the time constant of the cell-pH meter system to the gain of the electronic circuit $Da$. When $\tau^2$ is greater than $4Da$, the value of $u$ is real and the above expression is an adequate representation. When $\tau^2 = 4Da$, the value of $u$ is zero and the expression is inappropriate. When $\tau^2 < 4Da$, the value of $u$ is complex and although the expression is correct, it is not as informative as it might be.

For the case where $u$ is complex, we can define a new variable

$$jv = \sqrt{(-1)\,[4Da - \tau^2]} = j\sqrt{4Da - \tau^2} \tag{453}$$

$$\Delta\text{pH} = \Delta\text{pH}_{t=0} + \frac{e^{-\tau t/2}}{2jv}[\tau(e^{+jvt/2} + e^{-jvt/2}) + jv(e^{jvt/2} - e^{-jvt/2})] \tag{454}$$

and using the complex definitions of sin $z$ and cos $z$, this becomes

$$\Delta\text{pH} = \text{pH}_{t=0} + e^{-\tau t/2}\left(\frac{\tau}{v}\sin(vt/2) + \cos(vt/2)\right) \tag{455}$$

or recast in terms of a phase angle $\phi$,

$$\Delta \text{pH} = \text{pH}_{t=0}\sqrt{\frac{\tau^2 + y^2}{v^2}}\, e^{-\tau t/2}\sin(vt/2 + \phi),\ \phi = \arctan(v/\tau) \tag{456}$$

This predicts a sinusoidal oscillation of frequency $vt/2$ damped by the exponential decay $\exp(-\tau t/2)$.

The special case where $u$ is zero can best be treated by reexamining the Laplace space representation,

$$\lim_{u \to 0} x = y = \frac{\tau}{2} \tag{457}$$

$$\lim_{u \to 0} \frac{s + h}{(s + x)(s + y)} = \frac{s + \tau}{(s + \tau/2)^2} \tag{458}$$

$$\mathcal{L}^{-1}\frac{s + \tau}{(s + \tau/2)^2} = e^{-t/2}\mathcal{L}^{-1}\frac{s + \tau/2}{s^2} = e^{-\tau t/2}\left(1 + \frac{\tau t}{2}\right) \tag{459}$$

so that the expression for the pH change in this case becomes

$$\text{pH} = \text{pH}_{t=0} e^{-\tau t/2}\left(1 + \frac{\tau t}{2}\right) \tag{460}$$

The response of the cell is slow for the cases where $u$ is real and reaches an optimal response for $u = 0$. For cases where $u$ is complex, ringing occurs. In servo control nomenclature these responses are termed overdamped, critically damped, and underdamped. In one solves the more complex set of equations which takes into account the delay introduced by the capacitor in the feedback of OA5 using $a = 0.1$ and $\tau = 0.9$, the resulting computer-generated response curves are essentially identical to those for the same values of $Da$ with the simpler system.

## B. POTENTIOSTATS AS CONTROL SYSTEMS

One of the more interesting problems in control systems is the electrochemical potentiostat. The problem appears trivial enough; one must control the potential at the interface between the working electrode and the solution and monitor the current that flows as a result of the potential impressed. The equivalent circuit for a three-electrode electrochemical cell is shown in Fig. 3.58, and a simple three-amplifier potentiostat circuit is shown in Fig. 3.59. The critical features of the cell are $C_{d1}$, (the double layer capacitance of the electrode solution interface), which can be 50 $\mu\text{F}\cdot cm^2$ or more, the solution resistance between the electrode interface and the point in solution where the reference electrode intersects the electric field gradient between auxiliary and working electrodes, symbolized by $\Omega$, and the impedance of the electrochemical process, $Z_f$, assumed to be in parallel with the double layer capacitance. We can lump these three terms into a single impedance $Z_e$ and briefly investigate how a

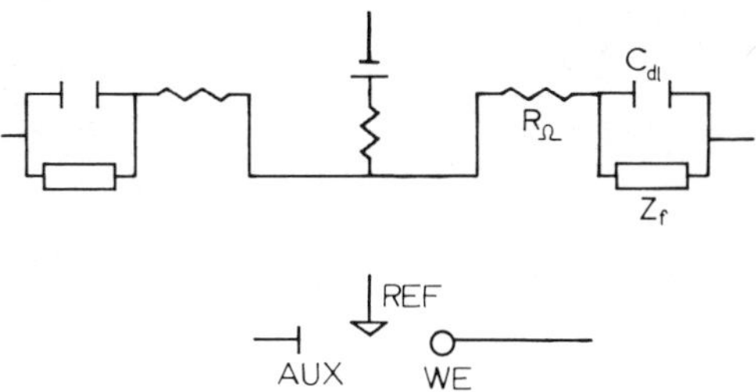


**Fig. 3.58.** Equivalent-circuit for an electrochemical cell.

potentiostat *should* work. The inverting input of the current transducer amplifier OA3 is ideally at ground potential so that to a first approximation we can assume that $Z_e$ is grounded. Then a potential applied at $R_1$ forces the output of OA1 to swing until the output of the reference electrode follower; amplifier OA2 produces a voltage so the current flows through resistors $R_1$ and $R_2$ are identical:

$$E_{oa2} = -E_1\left(\frac{R_2}{R_1}\right) \tag{461}$$

This means that the potentiostat controls the potential at the input of OA2, that is, at the reference electrode. Because the current drawn by OA2 is negligibly small, there is no potential drop across the internal resistance of the reference electrode itself, or any salt bridge connecting it to the solution proper.

This immediately presents a problem. The potential that we really wish to control is the potential on the other side of the resistance Ω. Unless the resistance or the current is vanishingly small (the former is unlikely and the latter is uninteresting by definition), there is an $iR$ drop error in our potential control, which is time dependent because the current is a function of time and potential. The answer to this problem lies in some form of dynamic $iR$ compensation scheme. There are a number of possible approaches and we briefly investigate three of them shown in Fig. 3.60.

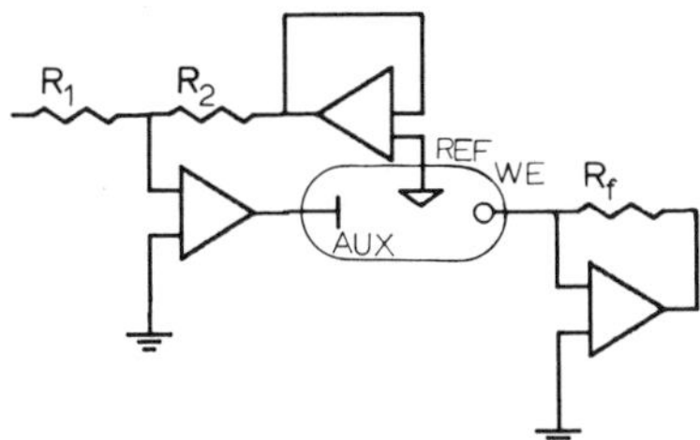


**Fig. 3.59.** A three OA implementation of a potentiostat with the reaction at the working electrode (WE) being monitored by a current follower. The chemistry is induced by the current being supplied by the auxiliary electrode (AUX) with the compensation for $iR$ drop being supplied by the follower (F) via the reference electrode.

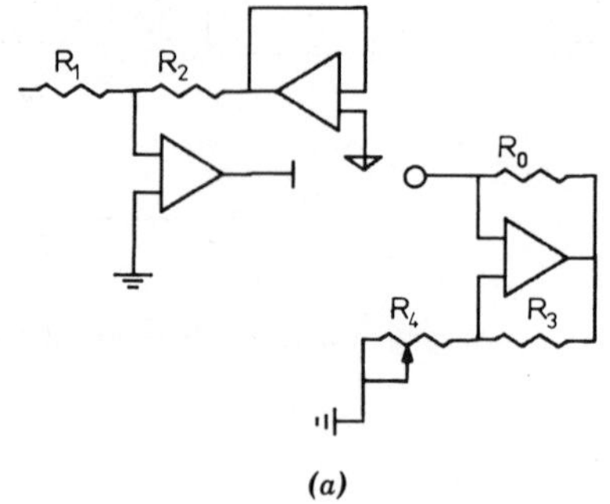


*(a)*

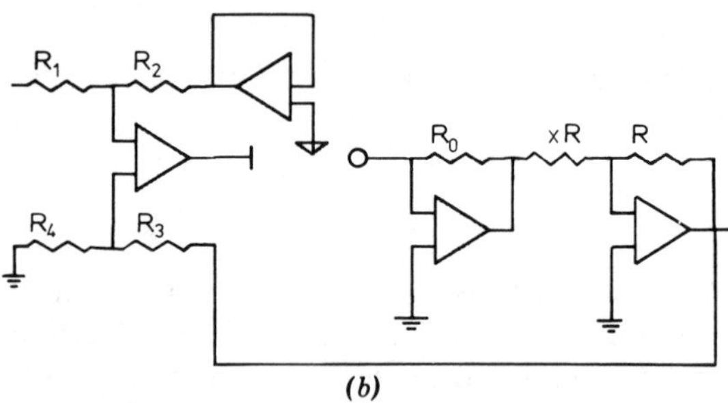


*(b)*

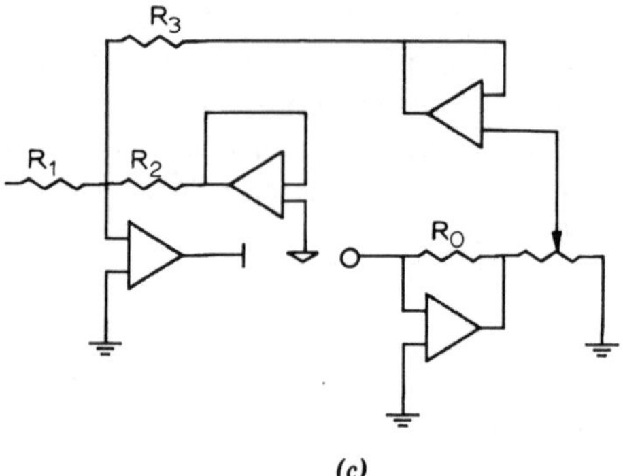


*(c)*

**Fig. 3.60.** Three Methods of IR compensation. (*a*) Negative immittance converter compensation circuit; (*b*) feedback compensator; (*c*) feedback compensator with the potentiometer setting independent of $R_3$. The last two circuits leave no residual resistance term in the response.

The first method leaves the circuitry around OA1 and OA2 alone and modifies the circuitry of OA3 into a negative immittance converter circuit, similar to the case we studied in the section on filters. We can write the equation for the output of OA3 in Laplace space as

$$\overline{E}_{oa3} = \overline{E}_r\left(\frac{R_0}{\overline{Z}_e}\right)\frac{(R_3 + R_4)\overline{Z}_e}{\overline{Z}_e(R_3 + R_4) - R_4(\overline{Z}_e + R_0)} \tag{462}$$

where $E_r$ is the potential at the reference electrode. If we make $R_3 = \theta R_0$ and $Z_e = \Omega + Z_i$ where $Z_i$ is the interfacial impedance (the parallel combination of the double layer capacitance and the faradaic impedance $Z_f$), then this becomes

$$\overline{E}_{oa3} = \overline{E}_r\frac{R_0(R_4 + \theta R_0)}{\overline{Z}_e(\theta R_0 + R_4) - \overline{Z}_e R_4 - R_4 R_0} = \frac{(R_4 + \theta R_0)\overline{E}_r}{(\overline{Z}_e\theta - R_4)} \tag{463}$$

$$= \frac{(R_4 + \theta R_0)\overline{E}_r}{\theta\overline{Z}_i + (\theta\Omega - R_4)} = \frac{(R_0 + R_4/\theta)\overline{E}_r}{Z_i + (\Omega - R_4/\theta)} \tag{464}$$

If we set $R_4/\theta = \Omega$ then the output voltage becomes

$$E_{oa3} = \frac{E_r(\Omega + R_0)}{Z_i} = \frac{E_r R_0(1 + \Omega/R_0)}{Z_i} \tag{465}$$

which differs from the ideal response $E_r R_0/Z_i$ by the constant factor $1 + \Omega/R_0$. Normally the resistance $R_0$ is much larger than the solution resistance, which contributes a negligible error. Even when the error is not negligible, it is not time dependent so that the effect of solution resistance has been transformed from a time-dependent error into a time-independent error. At this point, it should be obvious that this circuit overcomes the effect of the series resistance by emulating a negative resistance of the same magnitude. This is accomplished by moving the noninverting, hence moving the inverting, input off of ground, by an amount proportional to the current flow, thus cancelling out the *iR* drop through the solution resistance. The preferred implementation of this circuit is to have $R_3$ and $R_4$ be matched pairs and the resistance $R_4$ a noninductive potentiometer, so that one can set the compensation at one resistance level and not require readjustment as the transducer gain is changed. For reasons that we discuss later, it is not possible to compensate perfectly for all of the solution resistance. When $R_4/\theta = \Omega$ the system oscillates.

The other two implementations involve modification of the circuitry surrounding OA1. The circuit of Fig. 3.60*b* is described by

$$E_{oas3} = \frac{E_i(R_2/R_1)R_f}{Z_i + \Omega - \dfrac{(R_1 + R_2)R_3 R_f X}{(R_3 + R_4)R_1}} \tag{466}$$

If we set $R_1 + R_2 = R_3 + R_4$, then we can cancel out the effect of the solution resistance by making $(R_3R_f/R_1)X = \Omega$. This requires one more amplifier than the first solution and it is more difficult to arrange the circuit so that one does not have to retune the compensation circuitry when $R_f$ changes, but it leaves no residual resistance term in the response. This circuit also oscillates if one attempts to completely tune out the effect of solution resistance.

The final compensation scheme that we consider is shown in Fig. 3.60. The response of the system is given by

$$E_{oa3} = \frac{-E_1(R_2R_f/R_1)}{Z_1 + \Omega - XR_2(R_f/R_3)} \tag{467}$$

Ideally, one should make $R_f$ and $R_3$ matched resistor pairs so that as the value of $R_f$ changes, the compensation setting is not altered. It is possible to eliminate the follower amplifier $OA4$ from the circuit, but then the setting of the potentiometer for constant compensation is no longer independent of the value of $R_3$. This circuit shares the formal advantages of the preceding circuit, but it too oscillates when one attempts to compensate perfectly for the solution resistance.

## 1. Potentiostat Stability

The problem lies in the nature of the interfacial impedance $Z_i$ and in our neglect of the nonideality of the amplifiers, particularly OA3. We can appreciate the problem by considering only the effect of $Z_e$ on OA3 without attempting to $i$R-compensate the potentiostat. If we ignore the effect of $Z_f$, then the current transducer circuit and $Z_e$ look like a high-pass filter, where the pass-band gain is given by $R_f/\Omega$ and the corner frequency $F_c$ is given by $1/2\ \Omega\ C_{dl}$. If we decrease $\Omega$ toward zero, then $F_c$ must eventually increase to the extent that the closed-loop gain +6 dB/octave linear portion intersects the -6 dB/octave roll of the OA open-loop gain and oscillation occurs. This says that for any given cell condition (solution resistance and double layer capacitance), there is some value of the feedback resistance for the transducer resistor that cannot be exceeded without oscillation occuring. When we include the effect of the faradaic impedance, the situation improves somewhat. The theoretical response of a faradaic process to a potential excitation is a nonlinear function, but if we restrict our attention to small amplitude excitations of a reversible charge transfer process, then the response to a sinusoidal excitation is given by (23)

$$I(\omega t) = EM\sqrt{\omega \sin\left((\omega t + \frac{\pi}{4}\right)} \tag{468}$$

where $M$ is a function of a number of parameters such as the number of electrons transferred, the diffusion coefficient of the species, and the dC potential about which the sinusoidal excitation is based. The exact form of $M$, though important for a detailed study of the response of an electrochemical cell, is irrelevant for our purposes here, except to note that $M$ reaches a maximum at the normal polarographic half-wave potential and falls fairly abruptly to zero on either side of this potential. The presence of charge-transfer kinetics or solution chemical kinetics greatly alters the simple picture, but in most of these cases the frequency dependency is less than $\sqrt{\omega}$ so that the form of $Z_f$ looks more resistive and, in terms of our interest, more stabilizing. It is not difficult to show that the current-voltage relationship above is equivalent to the Laplace plane representation

$$Z_f = \frac{1}{M\sqrt{s}} \tag{469}$$

and that the form of $Z_e$ is given in Laplace space by

$$Z_e = \Omega + \frac{1}{Cs + M\sqrt{s}} \tag{470}$$

The ideal gain $G(s)$ of the simple current transducer with feedback resistance $R$ is given by

$$G(s) = \frac{R}{\Omega + \dfrac{1}{Cs + M\sqrt{s}}} = \frac{R(Cs + M\sqrt{s})}{\Omega(Cs + M\sqrt{s}) + 1} \tag{471}$$

We convert to frequency space by noting that

$$s = \frac{1-j}{\sqrt{2\omega}} \tag{472}$$

$$G(\omega t) = \frac{R}{\Omega}\left(j\omega C + \frac{(1-j)M}{\sqrt{2\omega}}\right)\frac{1}{\dfrac{1}{\Omega} + j\omega C + \dfrac{(1\text{-}j)M}{\sqrt{2\omega}}} \tag{473}$$

Rather than forming a phase magnitude solution for this expression, we have chosen to evaluate it numerically, and the results are displayed in Fig. 3.61 for a double layer capacitance of 25 $\mu\text{F}/cm^2$. The results use some fairly nominal electrochemical parameters and assume one electron transfer and a fairly typical electrode area for a hanging mercury or dropping mercury electrode of 0.035 $cm^2$. The concentration ranges from $10^{-2}$ to $10^{-7}$ moles/$\ell$. and from 250 to 0 Ω. Several features are apparent from this figure: (1) at low frequency the faradaic component is dominant, but the slope of the gain is negative so that except for concerns about sufficient loop gain, this does not affect stability; (2) at high frequency the double layer capacitance dominates the response; and (3) the stability of the cell is essentially the stability of a high-pass filter circuit composed of Ω, $C_{dl}$, and $R_0$.

The response of an electrochemical cell to a potential step contains a faradaic component which decays as $1/\sqrt{t}$, which is consistent with our $\sqrt{s}$ model for the impedance, and the peak current response for potential ramps is proportional to the square root of the sweep rate, which is also consistent. We mention this because of its implications for stability. One tends to choose the transducer feedback resistor $R_f$ so that the sweep voltametric peak current or current

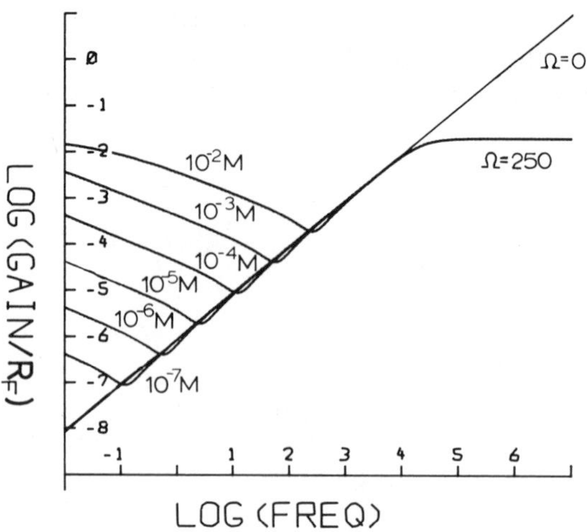


**Fig. 3.61.** Frequency response of a reversible electrochemical cell as a function of solution resistance and concentration of faradaic species.

transient some fixed time into a potential step, is near full scale. The current and electrochemical species concentration are directly proportional so that as one decreases the concentration, one increases the $R_f$ while working at fixed time or sweep rate. The value of $R_f$ required for a fixed amplitude response is equal to the square root of the experimental duration. Our Bode analysis of the system indicates that the greater $R_f$ becomes, the more likely oscillation is to occur, assuming $\Omega$ stays constant. The lower the solution resistance becomes, either because of cell geometry or solution condition changes, or because of *iR* compensation, the lower the value of $R_f$ must be to prevent oscillations. The obvious solution to this problem is to insert an additional zero into the transform, usually by placing a capacitor in the feedback loop of the transducer amplifier. Ideally this capacitance should be chosen so that the closed-loop gain flattens out about one decade prior to its intersection with the open-loop gain. This is not simple to implement optimally, since the high-frequency closed-loop gain is a function of $R_f$ and $C_{dl}$ (remember that for most systems the high-frequency response is essentially independent of the faradaic process). To further complicate matters, $C_{dl}$ is a potential dependent parameter, but fortunately the change over the accessible region usually is less than an order of magnitude, although this is the reason that a transducer may be stable at one working electrode potential and not another.

This inherent instability of the current follower configuration, as this circuit is known, has led to measurement schemes in which the current flow in a cell is determined by monitoring the voltage from across a calibrated resistor. There are basically two places where the resistor can be placed, in a series with the working electrode with the other end grounded, and between the output of the potentiostat amplifier and the auxiliary electrode. Each configuration has its own inherent difficulties; both decrease the total dynamic range of the system by the measured voltage drop, but this is usually a fairly negligible consideration. In the grounded resistor configuration, one has in effect added an additional solution resistance to the system, increasing that error term. In the other configuration the resistance does not add to the effective solution resistance, but the amplifier must be operated in differential configuration and great care in setting up the circuit necessary.

Even if the transducer amplifier does not oscillate, the potentiostat itself often breaks into oscillation. The reason for this behavior can be understood by considering a fairly detailed analysis of OA1 and OA2 in Fig. 3.62. The output of OA2 is actually related to the potential at the reference electrode $E_r$ by

$$\overline{E}_{oa2} = \frac{\overline{E}_r}{1 + 1/A_2(s)} \tag{474}$$

where $\overline{A}_2(s)$ is the open-loop gain of OA2. The potential at the reference electrode is given by

$$\overline{E}_r = \overline{E}_{oa1} \frac{\overline{Z}_e}{(\overline{Z}_e + \overline{Z}_a)} \tag{475}$$

where $Z_a$ is the auxiliary electrode analog of the impedance of the working electrode $Z_e$. If we represent the potential at the noninverting input of OA1 by $\overline{e}_a$ and remember that $\overline{E}_{oa1} = -\overline{e}_a\overline{A}_1(s)$, then by combining these relationships with the sum of the currents at the noninverting input of OA1,

$$\frac{\overline{E}_{oa1} - \overline{e}_a}{R_1} + \frac{\overline{E}_{oa2} - \overline{e}_a}{R_2} = 0 \tag{476}$$

we can derive the relationship for the output of OA1 in response to an input signal $E_1$:

$$\overline{E}_{oa1} = \frac{-\overline{E}_1 (R_2/R_1)(\overline{Z}_a + \overline{Z}_e)/\overline{Z}_e}{1\Big/\left(1 + \dfrac{1}{A_2(s)}\right) + \dfrac{1}{A_1(s)} \dfrac{R_1}{Z_e} [(\overline{Z}_a + \overline{Z}_e)/(R_2 + R_1)]} \tag{477}$$

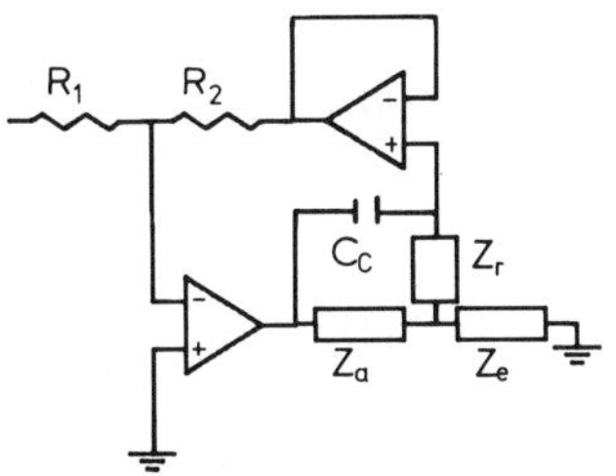


**Fig. 3.62.** Equivalent model for a potentiostat with grounded working electrode. *Az, Ze*, and *Zr* are the respective impedances of the auxiliary working response electrodes.

There are two interrelated criterian for instability in this equation: either the numerator of the equation intersects the open-loop gain of OA1 at more than the permissible angle or the denominator goes to zero at some frequency. The frequency response of the numerator is determined by $1 + \overline{Z}_a/\overline{Z}_e$ and, as we have already seen, the high-frequency response of the working electrode is controlled essentially totally by solution resistance and the double layer capacitance, so that for either electrode the high-frequency impedance is essentially given by

$$\overline{Z}_i = \frac{R_i(s + 1/R_iC_i)}{s} \tag{478}$$

where $R_i$ and $C_i$ are the solution resistance from electrode to the reference electrode and $C_i$ is the double layer capacitance of the electrode. The critical portion of the numerator thus reduces to

$$\frac{R_2}{R_1}\left(\frac{1 + \overline{Z}_a}{\overline{Z}_e}\right) = \left[1 + \frac{R_a(s + 1/R_aC_a)}{R_e(s + 1/R_eC_e)}\right]\frac{R_2}{R_1} \tag{479}$$

The only way that we can run into trouble is if the function is increasing at 6 dB/octave at the frequency of intersection with the open-loop gain, which can occur only if $1/R_eC_e$ occurs at higher frequency than $1/R_aC_a$. In fact, this is the usual situation. The double layer capacitance is directly proportional to the electrode area and the auxiliary electrode usually is larger than the working one in order to obtain the desired electric field properties. Further, the solution resistance between auxiliary and reference is usually greater than or equal to the resistance from reference to working to minimize *iR* drop effects.

Instability occurs only if $1/R_aC_a$ occurs at a frequency where the rising +6 dB/octave portion can intersect the open-loop gain and if $1/R_eC_e$ occurs at a frequency higher than the extrapolated intersection frequency (i.e., if the −6 dB rolloff has begun to cancel the +6 dB from the $Z_a$ contribution, there is be instability). When this form of instability occurs, one must add a pole to the gain; there are two places where this can be accomplished. One can add a capacitor from the output to the inverting input of OA1 or add a capacitor from the output of OA1 to the noninverting input of OA2, effectively providing a high-frequency short circuit to the impedance $Z_a$. The latter method is the preferred approach for reasons that are left up to the reader to reason out. This is the explanation for the rather mysterious capacitance that one normally finds between auxiliary and reference electrodes of potentiostats.

There are still some unpleasant possibilities in the denominator of the expression for the potentiostat. If we express the denominator in terms of the DC open-loop gains $A_i$ and −3 dB points $a_i$ of the two amplifiers and denote the denominator by $D$, then

$$D = \frac{(s + a_1)}{A_1a_1}\left(1 + \frac{R_2}{R_1}\right)\left(1 + \frac{Z_a}{Z_e}\right) + \left(\frac{1}{1 + (s + a_2)/A_2a_2}\right) \tag{480}$$

We make the substitution $G = 1 + R_2/R_1$ and assume that $Z_a/Z_e$ is significantly greater than unity (this is reasonable because we are worrying about the high-frequency portion of the response). As in the previous discussion we assume that the faradaic contribution can be ignored for the high-frequency portion of the response. We make one further assumption, which is normally employed in discussing loop stability.

$$\frac{(s + a_i)}{A_i a_i} = \frac{s}{A_i a_i} \tag{481}$$

This merely states that we are working at frequencies above the corner frequency, or −3 dB point of the amplifier; that is, the −6 dB/octave rolloff has begun, or we wouldn't be worrying about this contribution to instability. With all these assumptions our expression for $D$ is still fairly general:

$$D = \left(\frac{G}{A_1 a_1}\right) s \left(\frac{Z_a}{Z_e}\right) + \frac{A_2 a_2}{s} \tag{482}$$

Inserting the $RC$ model for the electrode impedances leads to

$$D = \frac{G}{A_1 a_2} \frac{(R_a C_a s+1)}{(R_e C_e s+1)} s + \frac{A_2 a_2}{s} \tag{483}$$

Assuming that we are worrying about frequency regions above the $RC$ time constants of the electrodes leads to

$$D = \frac{G R_a C_a}{A_1 a_1 R_e C_e} s + \frac{A_2 a_2}{s} \tag{484}$$

In frequency space this becomes

$$D(\omega t) = \frac{j\, G R_a C_a}{A_1 a_1 R_e C_e} \left( \omega^2 - \frac{A_2 A_1 a_2 a_1 R_e C_e}{G R_a C_a} \right) \tag{485}$$

which indicates that oscillation occurs at the frequency

$$\omega_0 = \frac{A_2 A_1 a_2 a_1 R_e C_e R_1}{(R_2 + R_1/(R_2 + R_1) R_a C_a} \tag{486}$$

Actually the oscillation will not occur as long as $\omega_0$ is greater than the unit gain crossover frequency of the slowest amplifier.

## C. CHRONOCOULOMETRY

The electrochemical technique of chronocoulometry involves applying a potential step to an electrode and integrating the current and monitoring the charge-time transient. This technique places some fairly stringent requirements on the amplifiers in the system (Fig. 3.63). A typical implementation uses a current follower amplifier, OA1, to convert current to voltage by means of feedback resistor $R_f$ and then integrates the response with OA2 using input

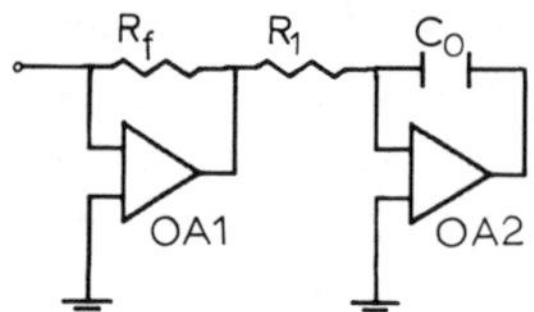


**Fig. 3.63.** A simple chronocoulometry circuit; OA1 is the current transducer, converting current to voltage, and OA2 is an integrator.

resistance $R_1$ and feedback capacitance $C_0$. The current from the cell is composed of two components, the transient current ($I_t$) due to the single or often double potential step excitation of the cell, and a background current $I_b$ due to slow electrolysis of solvent and supporting electrolyte or impurities. This latter current does not usually depend strongly on potential or time.

To a first approximation, $I_b$ can be assumed to be the same at the step potential as at the rest or pre-step potential, and can be considered an error current. It is convenient to consider the other potential and current nonidealities as error currents as well. When we do this, the total current seen by the integrator OA2 is given by

$$Ic = (I_b + I_t)\frac{R_f}{R_1} + \frac{E_{os1} - E_{os2}}{R_1} - \frac{I_1 R_f - I_2}{R_1} \tag{487}$$

where $E_{osi}$ and $I_i^-$ are the offset voltage and negative input error currents of the $i$th amplifier. The transient current from the cell depends strongly on the nature of the electrochemical process under study, but we can approximate average behavior by assuming that it has $1/\sqrt{t}$ behavior (i.e., that the system behaves as a reversible charge-transfer system), and that we can ignore the effect of the solution resistance. This also ignores the double layer charging current which would ideally be an impulse function, but actually is essentially an exponentially decaying function. Both contributions have infinite magnitude at time $t = 0_+$, which is obviously unrealizable with finite amplifiers.

Therefore, we are faced with three choices: (1) let the transducer OA1 saturate for some time period early in the transient; (2) limit the value of $E_{oa1}$ to some value below saturation by means of a clipping circuit; or (3) dynamically change the gain of the transducer during the transient. The first solution is obviously the simplest to implement and unfortunately the most common in practice. It suffers from the obvious disadvantage that an OA in saturation does not behave as an OA; the tracking between the two inputs is lost and the output potential bears little resemblance to the current at its input.

In this connection, it should be noted that the resistive voltage drop method of monitoring the cell current is not subject to this loss of control of the cell potential, although the monitored voltage still does not reflect the current transient during the saturation period. In addition, an OA when driven to saturation takes a finite amount of time to return to normal operational behavior after the overload condition is removed. At one time this period could be measured in seconds; it is still several milliseconds for some forms of modern amplifiers, although most integrated circuit amplifiers recover in microseconds or less. Placing crossed silicon diodes from the inverting input to ground minimizes the loss of control. The second approach is the most foolproof in that the amplifier never saturates, but the data during the clipping response time are again totally unrelated to the current flow through the cell.

There are a number of ways to implement clipping circuits. The simplest is to place crossed zener diodes from the output to the inverting input. When the voltage exceeds the zener breakdown voltage, the diodes conduct and limit the output voltage of the amplifier to the zener potential. This and essentially all other clipping schemes are simply circuitry placed in parallel with the normal feedback components. A second method involves placing a resistive divider from output to each power supply. A silicon diode is placed from the negative input to the summing point with the polarity such that when the output reaches the potential, which is the clipping potential, the diode becomes forward biased and goes into conduction. The book by Graeme et al. (12) contains an excellent discussion of these and other clipping schemes.

The third method is in a sense a variation on the second. During an overload condition the circuit places a smaller resistance in parallel with the normal $R_f$. When the output voltage decays to some predetermined set point, the circuit switches the resistance back out. The simplest implementation of this circuit is a FET in parallel with the $R_f$ and a variable time delay circuit to turn the FET off. One observes transients and adjusts the time delay until the FET turns off again at a point when the transducer output is just below saturation potential. There are also schemes where the potential is sensed and the switching is done automatically, but these are beyond the scope of this discussion.

The obvious advantage of this form of overload protection is that at least in principle, the total transient can be reconstructred knowing the ratio of the two resistances involves. As a matter of pure practicality, this method also by definition saturates the transducer amplifier momentarily, since the current requirement at the beginning of the transient is normally beyond the capabality of the amplifier. The ultimate form would be a scheme, in which the amplifier itself does not deliver the current at the beginning of the experiment, but only takes control when the transient has decayed to within its capabalities. In addition, the scheme should allow the reconstruction of the transient during the overload condition. We are not aware of the publication of such a circuit at this time.

To maintain some generality, we assume that the response from OA1 for a time period $\theta$ represents an overload condition and at times after $\theta$ the faradaic signal decays as $G/\sqrt{t}$. We further assume that the integration time is $10{,}000\theta$, so the overload represents a small portion of the transient. At time $\theta$, the faradaic portion of the response is about 25% of the total signal, the rest being the double layer charging transient. We can get a rough picture of the response by imagining that the faradaic impedance is in parallel with the double layer and solution resistance, rather than just the double layer capacitance. In this case, the double layer transient is $E/\Omega e^{-tr/\Omega Cdl}$, which predicts a maximum double layer charging current of $E/\Omega$. The faradaic current is given by $nFAC \sqrt{D/\pi t}$ where $A$ is the electrode area, $D$ the diffusion coefficient, $n$ the number of electrons transferred, and $C$ the species concentration in moles/$cm^3$. Picking some fairly typical values' $A$=0.04 $cm^2$, $D$ = 3.1 x $10^{-6}$, $C$ = $10^{-6}$ moles/$cm^3$, and $n$ = 2 electrons, we find that the current is 8 x $10^{-6}$ $\sec^{1/2}/\sqrt{t}$.

The current due to the faradaic process decays to 1% of its $\theta$ value at the end of the integration and because of this, the error currents can become important. In a typical experiment the current does not typically decay much below the microampere level, and so if the OAs have subnanoampere input error currents (the typical current level for FET input amplifiers is 0.1 nA), they normally do not contribute significantly to the total error. The sources that are most important are those related to the difference in the offset voltages of the two amplifiers. If the two amplifiers are 1 mV different and the integrator resistor is $10K$ $\Omega$ (these are not unreasonable parameters for short time experiments), then the error current due to this source is 0.1 MA, which can be a significant error. The traditional approach to the problem is to utilize amplifiers that have highly stable input offset voltage characteristics and to zero out the integrator drift prior to the experiment. One should also check the stability of the zero frequently between transients.

We will digress momentarily and examine the requirements for an integrator amplifier. Ignoring the offset voltage and current error terms, we concentrate on the effects of finite gain. The voltage transfer functions of an integrator circuit, taking into account the gain properties of the amplifier, are as follows:

$$G(s) = \frac{E_o(s)}{E_1(s)} = \frac{-\overline{Z}_f/\overline{Z}_1}{1 + 1/A_e} \tag{488}$$

$$G(s) = \frac{-(1/R_1C_os)}{1 + \cfrac{1}{\cfrac{A_oa_o}{s + a_o}\left[\cfrac{1}{1 + 1/R_1C_0s}\right]}} \tag{489}$$

$$G(s) = \frac{-bA_oa_o}{s^2 + [a_o(A_o + 1) + b]s + a_ob}, \quad b = \frac{1}{R_1C_0} \tag{490}$$

Solving the quadratic equation in the denominator for the two roots leads to

$$s = \frac{-[a_o(A_o + 1) + b] + \sqrt{[a_o(A_o + 1) + b]^2 - 4a_ob}}{2} \tag{491}$$

For any realistic amplifier $A_o \gg 1$, so that we can replace $a_o(A_o + 1)$ by $a_oA_o$ with very little error. When this substitution is made, we find that one of the roots is zero, so that

$$G(s) = \frac{-A_oa_ob}{s(s + a_oA_o + b)} = \frac{-1}{R_1C_0 + 1/A_oa_o}\left[\frac{1}{s} - \frac{1}{s + a_oA_o + b}\right] \tag{492}$$

As long as the time constant of the integrator is large compared to the reciprocal of the unity gain crossover frequency, which it must be, or the amplifier characteristics totally govern the response, then

$$G(s) = \frac{-1}{R_1 C_0}\left[\frac{1}{s} - \frac{1}{s + a_0 A_0 + b}\right] \quad (493)$$

It is usefuly to evaluate this response for both the square root of time signal expected from the electrochemical cell and for a simple step function. For the step function we have

$$E_0(t) = \mathcal{L}^{-1} G(s)\frac{E}{s} = \frac{-E}{R_1 C_0}\left[t + \frac{1}{A_0 a_0 + 1/R_1 C_0}(1 - e^{-(A_0 a_0 + R_1 C_0)t})\right] \quad (494)$$

If we investigate the short time behavior of the response by expansion of the exponential term, we discover that the short time response parallels the ideal response, but that the linear portion is offset an amount $1/f_c$ sec, where $f_c$ is the unity gain crossover frequency as shown in Fig. 3.64. The integrator gives an erroneous extrapolation to zero time with an intercept of $-1/f_c R_1 C_0$, not zero, and that it will not approach linear behavior for about $3/f_c$ sec. At long times the behavior can be shown to be

$$E_0(t) = EA_0(1 - e^{t/A_0 R_1 C_0}) \quad (495)$$

Although this is perhaps surprising at first glance, it is perfectly logical if one considers that the very low frequency response is totally dominated by the low-pass filter type response of the operational amplifier. These two limits place some constraints upon the characteristics of a useful integrator amplifier. The value of $1/f_c R_1 C_0$ must be negligible, compared to any intercept information one wished to obtain from the integration, and $1/f_c$ must be short compared to the time of the first useful data point. By considering the small argument

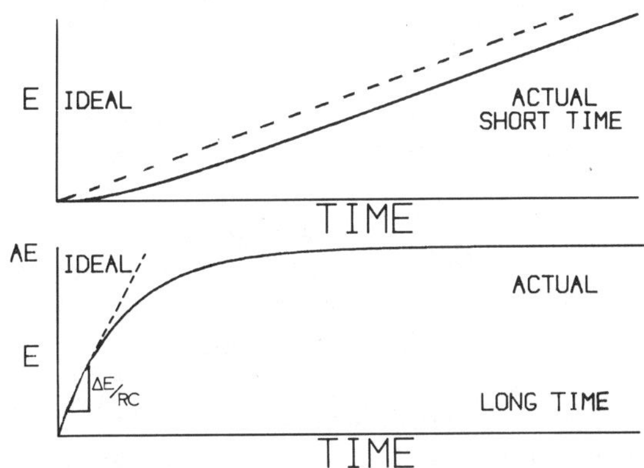


**Fig. 3.64.** The response of a real integrator to a step input function.

expansion for an exponential $e^x = 1 + x + x^2/2 + \ldots$, we can see that $t/A_oR_1C_0$ must be small compared to maximum expected experimental nonlineariy of the integrated data.

The response for the square root of time excitation is generated by using the transform space representation $E_1(s) = MR_f/\sqrt{s}$, where $R_f$ is the current transducer feedback resistance. Applying this excitation to the amplifier response transform yields the time plane representation:

$$E_0(t) = \frac{MR_f}{R_1C_0}\left(2\sqrt{t} - e^{-x^2}\int_0^x e^{v^2}\,dv\right) \qquad (496)$$

$$x^2 = \left(a_oA_o + \left(\frac{1}{R_1C_0}\right)\right)t \qquad (497)$$

The integral known as Dawson's integral (10) has an initial value of zero and a maximum value of 0.92 at $x = 0192$, and for $x \geqslant 2$ decays essentially as $1/2x$. The short time response of the integrator is shown in Fig. 3.65; by comparing these results with those for the step excitation, it can be seen that the electrochemical signal places more stress on the ideality of the short time response of the integrator by about a factor of 10. The long time behavior for the response is analogous to the previous in that at long times the low-pass feature dominates and the response fails the simple model.

It is important to note that the error currents seen by the integrator depend upon the voltage offset of both the transducer and the integrator, and that simply using a highly stable (nominally a chopper type) amplifier in the integrator itself is not sufficient; both amplifiers must be matched. One way of circumventing this problem is to use a feedback circuit to stabilize the effects of voltage drift of the transducer-integrator pair. The circuit in Fig. 3.66 is one implementation of such a circuit. The double pole, single throw switch shown is normally a pair of JFETs (24) and in the initial application, these were driven from a digital signal from a computer. When the switches are closed, the capacitor at OA3 is connected to ground and the circuit on OA3 is effectively an inverter with large gain. Any potential at the output of OA1 is integrated by OA2, and the output voltage is amplified with the sign inverted. This is fed back as a current to OA1, resulting in a decrease in the output at OA1. The stable state of this circuit is for the output of OA2 to be very near 0 V as long as the total effective error currents in OA2 are less than $E_{fs3}(R_f/R_cR_1)$, where $E_{fs3}$ is the full scale output voltage of OA3. The charge on $C_0$ is the average voltage required to maintain compensation. When the switch is opened, that voltage appears at the output of OA3 and the average value of the correction signal is applied to the transducer, OA1. The transient experiment is initiated and as long as voltage drifts of OA3, owing to its error currents, are small with respect to the compensation potential and the error currents at OA2 remain constant, then the drift in the output of OA2 owing to error currents are suppressed during the experiment. In one implementation of this circuit, we have seen less than 10 mV/sec drifts with an integrator time constant of $10^{-5}$ sec, using three FET input amplifiers which sell for \$0.35 each in lots of 100.

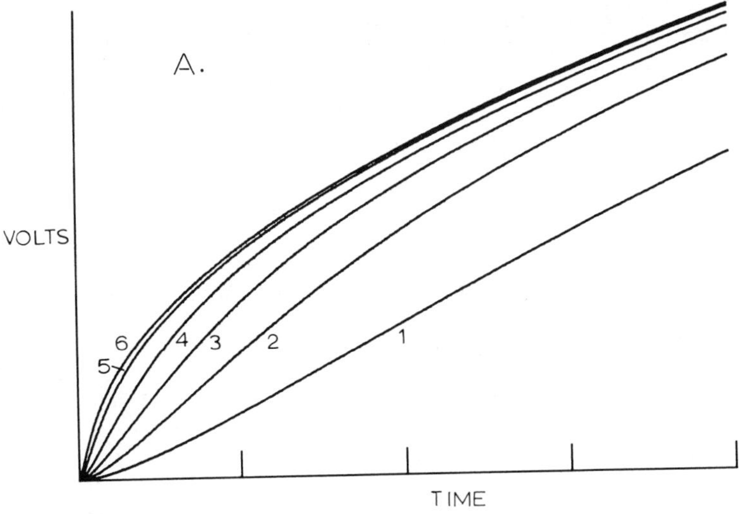


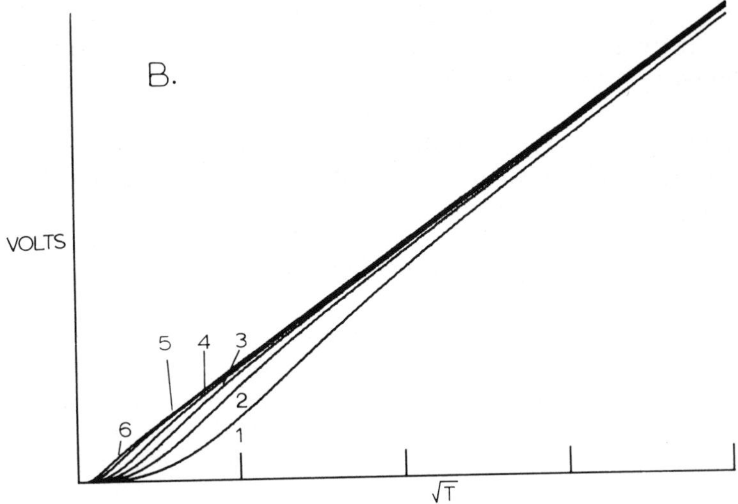


**Fig. 3.65.** The response of a real integrator to a $1/\sqrt{t}$ input function. (*a*) Short time response vs. time. (*b*) Long time response vs. $\sqrt{t}$. (1) *fu* = 20 kHz. (2) *fu* = 50 kHz. (3) *fu* = 100 kHz. (4) *fu* = 200 kHz. (5) *fu* = kHz. (6) *fu* = 1 MHz.

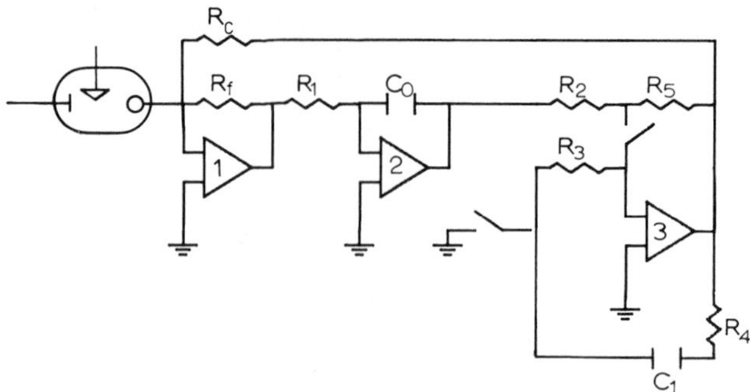


**Fig. 3.66.** A compensation scheme for current and voltage offset effects for chronocoulometry. As noted in the text, this circuit allows one to relax the requirements on the OA as comp and with the circuit in Fig. 3.63.

This circuit, with the feedback capacitance of OA2 replaced by a resistance, has also been used to suppress the background current in differential pulse polarography. Using this method, background currents 1000 times the magnitude of the pulse current have been successfully suppressed.

A detailed study of the response characteristics of this circuit is beyond the scope of this chapter, but a few details are worth noting. Since this is a closed-loop circuit, the possibility for oscillation exists. Analysis is complicated by the discontinuous nature of the signal from OA2. There is general be a discontinuity when the switch opens and the output changes from an instantaneous correction signal to an average value of that signal. Also, there is certainly a discontinuity when the switch closes at the end of the potential transient to the cell. As long as the time constant of the integrator is greater than that of the compensator, the circuit remains stable, assuming the compensator, OA3, is itself stable. When OA2 is acting as an inverter circuit, the probability of oscillation is greater. Most of the stability problems can be traced to the compensation circuit itself. The resistance $R_3$ is employed to isolate the inverting input to OA2 from ground while the switch is closed and a current $E_{os2}/R_3$ flows as a result. This means that the signal corrected for will be erroneous by an amount proportional to the ratio of this current to the current through $R_2$. This signal source disappears and is replaced by a simple offset voltage $E_{os2}$ when the switch opens. This error can be minimized by occasional offset adjustments on OA3 and by making $R_3$ large.

A far less obvious source of error and instability arises from the switch itself. ALL FETs exhibit the phenomena of charge injection as they switch states in either direction. In our application, this acts to produce a voltage spike when the switch closes and a voltage step when the switch opens, as viewed at the output of OA3. For some discrete transistors, such as the MPF-102, the magnitude of the charge is usually not sufficient to cause difficulty, but for some integrated circuits with built-in level changers, the magnitude can be sufficient to cause severe stability problems. One circuit that has proved quite tame with respect to the problem of charge injection is the 4016 or 4066, which are CMOS analog transmission gates. With proper biasing, they can handle ± 5 V signals quite nicely, containing six switches in a single package selling for well under a dollar.

## XI. CONCLUSION

Since Deford's early applications of the OA in chemical instrumentation, these components have become the heart of most analog circuitry. Because of these "dirty little pieces of silicon," the implementation of most circuits has become easier. As we have seen, the whole field of active filters has progressed to the point where a VCVS filter can be constructed from a single chip and some passive elements.

There has been an explosion in the number of types of OAs. This has been a boon to those that design chemical instrumentation. Although this choice is nice to have, unless one knows how to make this decision, one could end up using an expensive OA that's not suited for the application. One can fall into this trap not because of designer's ignorance but because of jargon that allows the manufacturers to hide the weaknesses of their product. (At least, this is where we cast the blame.)

Table 3.IV lists some common characteristics that must be used in selecting an amplifier for a particular purpose. A rather easy example is the amplifier that should be used to impedance-match a glass electrode. Assuming we are interested in static readings, which is realistic because most methods of altering solution pH operate on a time scale of tens of milliseconds or longer, we can ignore the frequency response.

Realizing that the frequency response isn't a problem, we then turn to the real problem of this measurement—how much current can be drawn across a glass electrode before a significant voltage drop is produced? Given that the electrode if interest may have a resistance of up to 100 MΩ, and assuming we want to make readings with an error of 0.01 mV ($10^{-5}$V), the using Ohm's law we see that only 100 fA ($10^{-13}$V) can be drawn through the electrode before it alters the readings. In Table 3.IV, the only OA that meets this specification is the AD515L.

Because in this case the acceptable current levels are so low, one must be especially paranoid about the actual circuit layout. When we implemented this circuit (25), we had to take a number of precautions. These were (1) encasing the electrometer in a grounded mini-box, (2) running the signal from the electrode to the box's lead on coaxial cable, and (3) making the wire from the tip of the coaxial input connector as short as possible. These all helped to guard the signal from broadcast noise.

The electrochemical problem of chronocoulometry discussed in the preceding section provides one further example of amplifier selection. We consider only the current transducer and integrator stages in this discussion. We work

TABLE 3. IV

Comparison of Some Operational Amplifiers

| Property | Units | 741 | 108 | 308 | 3140 | LF351A | AD515 | BB3554 | AD288 | ICL7605 | HA2900 | LF352A | AD50 | AD51 | LH003C |
|---|---|---|---|---|---|---|---|---|---|---|---|---|---|---|---|
| Open-loop Gain | dB | 100 | 110 | 110 | 100 | 100 | 86 | 100 | 100 | 100 | 160 | 100 | 90 | 96 | 85 |
| Input offset | mV | 6 | 0.5 | 0.5 | 15 | 20 | 1 | 1 | 1 | 0.002 | .02 | 30 | 2.0 | 2.0 | 3 |
| Input bias | nA | 500 | 2.0 | 7.0 | 0.05 | 0.2 | 0.0003 | 0.100 | 0.05 | 1.5 | 0.05 | 0.03 | 2.0 | 2.0 | 2000 |
| Unity gain | MHz | 1 | 1 | 1 | 4.5 | 3.0 | 0.35 | 90 | — | — | — | 0.14 | 70 | 56 | 30 |
| Full power band | KHz | 10 | 10 | 10 | 100 | 100 | 5 | $16 \times 10^3$ | 3.5 | 0.01 | 2.5 | 25 | 8 MHz | — | |
| Slew rate | V/sec | 0.5 | 0.5 | 0.5 | 9 | 13 | 0.3 | 1000 | 0.1 | 0.5 | — | 1.0 | 500 | 400 | 30 |
| Rise time 10–90% | sec | — | 80 | 80 | 0.08 | — | — | 0.05 | — | — | — | — | — | — | — |
| Rated load | mA | 5 | 2 | 2 | 10 | 2.0 | — | 100 | — | — | — | — | 100 | 100 | 120 |
| Input Impedance | M | 2 | 70 | 40 | 1500 | $10^6$ | $1^{-7}$ | $10^5$ | $10^2$ | $10^6$ | — | $2 \times 10^6$ | 1000 | 1000 | 100 |
| $dE_{os}/dT$ | V/°C | — | 5 | 7 | — | 10 | 50 | 50 | 2 | 0.02 | .05 | 600 | 50 | 50 | 3 |
| CMRR (IMRR) | dB | 80 | 110 | 110 | 90 | 80 | 80 | 78 | 100 | 7 | 120 | 100 | 60 | 60 | 90 |
| Supply current | mA | 2.8 | 0.6 | 0.8 | 5 | 2.8 | 1.5 | — | — | 100 | — | 1.2 | 40 | 40 | ±20 |
| Voltage swing | ±V | 10 | 13 | 13 | 13 | 15 | 10 | 10 | — | 5.0 | — | 12 | 15 | 15 | — |
| Cost (approximate) | $ | 0.50 | 3 | 1.5 | 1 | 1.25 | 25 | 165 | — | — | — | 10 | 100 | 100 | 20 |
| Type of OA | | General purpose | High perfor-mance | High perfor-mance | BIFET | BIFET | FET electro-meter | Fast settling | Trans-form isola-lation | CH2 ampli-fier | Chopper | BIFET instrumen-tation | Wide-band differ-ential FET | Wide-band differ-ential FET | Wide-band |

with both the two-amplifier implementation (Fig. 3.63) and the three-amplifier version of the circuit (Fig. 3.66) to demonstrate the changes in the amplifiers' requirements owing to the addition amplifier.

The two-amplifier circuit was evaluated in terms of its static error contributions in equation 469. If we assume (1) that no current flows through the cell prior to the application of the potential step, and (2) that the resulting step is large enough then the current transient from the $n$ electron reduction is given by:

$$E_2 = \frac{2nFAC^*R_f}{R_1C_1}\sqrt{\frac{Dt}{\sqrt{\pi}}}\left[1 + \frac{L_1R_f - L_2R_1 - E_{os}}{2nFAC^*R_f}\sqrt{\frac{\pi t}{D}}\right] \tag{498}$$

where $C^*$ is the electroactive species concentration, $D$ is the diffusion coefficient, and $E_{os}$ is the sum of the offset voltages.

We can evaluate the static requirements for these amplifiers by the following procedure. After the passive elements have been chosen, the output of the amplifier should be between 2 and 10 V at the end of the experiment. The equation's second term is the absolute error; typical settings for the maximum permissible values are 0.5% for purely analog and 0.1% for a circuit interfaced to a computer, for doing the acquisition and reduction of the data. Investigating the latter case with the limitation that the experiment's duration is 10 or less, we can rewrite the equation as

$$0.001 = \frac{L_1R_f - L_2R_1 + E_{os}}{R_1C_1} \tag{499}$$

For economic and space reasons, we limit $C_1$ to 2 $\mu$F (the low leakage capacitors required are expensive and physically oversized—polystyrene, Teflon, or polycarbonate film capacitor should be used). This means that our total error budget for $L_1(1 - R_f/R_1)$, $i_2$ –, and $E_{os}/R_1$ is only 2 nA. We dare not assume that the sign cancellation will help us. Therefore, our estimate has to be on the conservative side; thus the choice is made to budget 400 pA to each of the error terms.

The largest value for $R_f$ one can utilize without resorting to Thevenin T networks and still not have to be concerned with the resistor's inductance is 10 M$\Omega$ (use metal film resistors), leading us to the problem of combined offset voltages of the two amplifiers being 4 mV. One is not free to distribute this error randomly, because any offset voltage in the transducer amplifier acts as an error in the potential control accuracy of the potentiostat's circuit. This allows us to make the decision that the transducer amplifier should have a maximum voltage offset of 1 mv, budgeting 3 mv for the integrator.

The input current for both the transducer and the integrator is 400 pA or less. If we can set up our time constant and transducer gain, so that the resistance ratio is always unity or less, we can use amplifiers with 400 pA input current for both. It's important to remember that unless one is willing to trim the voltage offset' every few hours it's not smart to rely on the offset voltage trim because it contributes as much as 10% of the untrimmed error, in drift, throughout the day.

Determination of the dynamic requirements for the two amplifiers is more complicated. We can simplify the problem somewhat by treating each amplifiers separately. The transducer amplifier should ideally be treated by modeling the input impedance as a series resistance (uncompensated solution resistance Ω) connected to a parallel combination of the solution interface double layer capacitance $C_{dl}$ and faradaic impedance $Z_f$, as shown in Fig. 3.58. The resulting effective impedance in Laplace space is used to construct the gain equation. In turn, this is inserted into the response equation which includes the *DC* open-loop gain $A_O$ and the 3 dB frequency $a$ as shown in equation 488. If we do this, we end up with a Laplace space representation which requires evaluation of a cubic equation in $\sqrt{s}$. In actual practice, this is not necessary for most reasonable electrochemical systems.

As we have seen in the preceding section on the potentiostat, the dominant term at high frequency arises from the combination of Ω and $C_{dl}$ (see Fig. 3.62). We can feel relatively safe in constructing a model of the very short time response which contains only these terms. The application of a potential step of amplitude $\Delta E$ to this model results in a current flow given by

$$i = \frac{E}{\Omega}(1 - e^{-t\Omega C_{dl}}) \qquad (500)$$

Then the current demands on the amplifier are evaluated by specifying the values of $E$ and Ω. In chronocoulometry the step size required to achieve effectively zero concentration at the electrode of a solution species is roughly 0.300/nV, with reference to the $E^{\circ}$ value of the couple in question, where $n$ is the number of electrons involved. A reasonable minimum value of the uncompensated solution resistance is about 5 Ω (this is unrealistically small for nonaqueous solvent systems and fairly optimistic even for aqueous ones).

The peak current demand is thus 0.06/n for a maximum of 60 mA. The value of $C_{dl}$ depends upon the nature of the electrolyte-solvent system, the solution potential, and the area of the electrode, but a fairly reasonable value for $C_{dl}$ is 40 $\mu F/cm^2$. A common value for the area of the electrodes used in experiments of this sort is about 0.03–0.05$cm^2$, giving an effective capacitance of from 1 to 2 $\mu$F. If we do not wish to saturate the transducer amplifier for any appreciable amount of time the maximum value of $R_f$ is 200 Ω (or more generally $R_f$=12Ω/0.3 assuming a maximum swing of 12 V). The Bode plot for this type of model is given in Fig. 3.61, where the maximum closed-loop gain is 200/5 or 40 with the high pass region beginning at about 16 KHz. To ensure stability, the amplifier must have an open-loop gain of at least 40 at 160 KHz, that is, the unity gain crossover frequency $f_u = aA_o > 10$ MHz.

At this point, it is wise to evaluate just what kind of current flow one can expect from the faradaic process and what the magnitude of the output voltage from the transducer amplifier will be. If we pick some reasonable values for this type of experiment (i.e., $C^* = 0.001$ moles/liter and $D = 3 \times 10^{-6} cm^2$/sec), we find that the output voltage is 1 mA/electron transferred. At 75 $\mu$sec into the step the output voltage is 0.200 V, whereas at 10 sec into the step, the voltage has dropped to about 0.5 mV.

We can now turn our attention to the integrator amplifier short time requirements. In general, the data analysis performed involves some form of short time extrapolation of the charge versus the square root of time in order to determine the double layer charge. Either as an item of intrinsic interest in its own right or to remove its effect, if the objective is to examine the time functionality of the faradaic response, in order to determine some kinetic parameter.

We have already examined this case and the results were given in equation 496. There are two main features of this expression. First, it can be shown that the intercept of the extrapolation versus $\sqrt{t}$ is $-2[(aA_0 + 1/R_1C_1)\pi]^{1/2}$, not zero. Second, the time required for the expression to become linear in $\sqrt{t}$ to some acceptable fractional error $\epsilon$ is given by $t = 0.5/\epsilon[(aA_0 + 1/R_1C_1\pi]^{1/2}$. If we require that the double layer charge determined by the experiment is to be accurate to within 0.1%, then $aA_0$ must be at least 2 MHz. We can evaluate the slope accuracy requirement by noting that it will require about 75 $\mu$sec for the double layer current to decay to 0.5% of the faradaic current level which leads to a 1 MHz requirement for 0.5% slope accuracy. A 0.1% slope accuracy requirement once again requires a unity gain crossover frequency of about 2 MHz.

We can now summarize our requirements. The transducer amplifier must have an offset voltage of less than 0.5 mV, a current compliance of at least 60 mA, MHz unity gain crossover frequency, and an input current of 400 pA. The integrator amplifier requires an offset voltage of about 3 mV, 2 MHz unity gain crossover frequency, and 400 pA input bias current, but does not need more than 10 mA current compliance. The integrator amplifier can be a fairly inexpensive LF351A, but the transducer amplifier has very stringent requirements and an amplifier such as an AD50 or AD51 is necessary, both of which are in the $100 price range.

The two-amplifier implementation places some stringent requirements on the amplifiers, particularly upon the transducer amplifier. In contrast the three-amplifier configuration, (Fig. 3.66) allows us to relax these requirements somewhat. This form requires only that the DC terms be relatively *constant* during the step itself, because the additional amplifier compensates for these terms.

The new amplifier does not need particularly good high-frequency response because it need only follow the output of amplifier 3 between potential step experiments, so a unity gain crossover frequency of 0.5 MHz is more than adequate. The pre-experimental output voltage from amplifier 2, $E_2$, is $IR_c/X$, where $X$ is the DC gain of amplifier 3 and $I$ is the total current flow through the working electrode from all sources plus the total of the error currents and $E_{os}/R_1$. A typical value for $X$ is 1000, so $E_2$ is on the order of 10 mV. The offset voltage of amplifier 3 should be on the order of 10% of this or 1 mV. The accuracy of the compensation during the step requires that the output of amplifier 3 remain constant during this time period. The only term that can influence this voltage is current flowing through the hold capacitor to the negative input.

This error should be held to less than 100 mV in the 10 second period of the experiment, so the bias current required for a 1 $\mu$F compensation capacitance is 1 nA. With these requirements, a good-quality FET amplifier such as the LF351A or even a LM308A would be adequate.

We can now open up the error budget for amplifiers 1 and 2 to allow almost 1 $\mu$A of total current expressed error. We can use an LF351A for the integrator, relaxing our transducer requirements to 1 $\mu$A of bias current while keeping the rest of the requirements the same. This now allows us to use a far less expensive amplifier such as the LH0003C, costing about one-fifth as much as the AD50 that was required in the two-amplifier form.

This chapter was designed to acquaint the reader with some of the main principles of analog circuitry, to point out the mental processes and mathematical tools required, and to analyze a circuit from the standpoint of either understanding an existing circuit or successfully designing a new circuit. Except in the case of analog simulations or filter design, one can usually rough out a circuit without considering the transform representation, using only the crudest Maxwell demon conceptualization. Once the rough circuit is in hand, then increasingly detailed models are used, primarily in order to determine the requirements for the physical amplifiers and to guard against the possible effects of parasitic elements.

Although a detailed description of good circuit board design practice is beyond the scope of this chapter, several simple guidelines can be given that will eliminate most of the common problems.

1. Use only good-quality fiber glass-based boards, *NOT phenolic*.
2. Use separate grounds for different current levels. The positive input of an inverting amplifier should not be attached, even through a resistor, to a ground which is carrying more than a few microamperes of current. Our normal practice is to use several separate ground runs which come together at a single point where the DC power enters the board. In some circuits four or five different isolated ground paths are employed.
3. Bypass the power terminals of each amplifier with a 0.1 MF ceramic capacitor to power ground located as close as possible to the amplifier.
4. Keep connections to the negative input of amplifiers as short as possible to avoid antenna effects and current leakage problems.
5. Use guard rings around critical high impedance points, but beware of the fact that you may be introducing parasitic capacitances.
6. Avoid the use of mechanical switches wherever possible.
7. If you must use mechanical switches in a circuit, try to use switches that physically mount to the circuit board rather than making connections to the switch via wires.
8. Use shielded cable any time that the bare input of an OA is brought off of a circuit board. If it is a positive input, drive the shield with the output of the amplifier. If it is a negative input, it is usually best to ground the shield.

## ACKNOWLEDGMENT

The authors would like to acknowledge the following people for their help: Jeri Smith, for her time and patience in typing this manuscript; Dwight Blubaugh, for some error checking and programming the Savitzky-Golay smoothing routine; and Eileen Birch for her expert draftsmanship.

## REFERENCES

1. Paynter, H. M., *A Palimpsest on the Electronic Analog Art*, G.A. Philbrick Researchers, Inc., Dedham, Mass., 1965.
2. Ragazzini, J. R., R. H. Randall, and F. A. Russel, "Analysis of Problems in Dynamics by Electronic Circuits", *Proc. I.R.E.*, **35** 520 (May 1947).
3. Deford, D. D., mimeographed notes, Northwestern University, Evanston, Il.
4. Booman, G. L., *Anal. Chem.*, **29**, 213 (1957).
5. Kelley, M. T., D. J. Fisher, and N. C. Jones, *Anal. Chem.*, **31**, 1475 (1959).
6. *Ibid.*, **32**, 1262 (1960).
7. Reilley, C. N., *J. Chem. Educ.*, **39**, A853, (1962).
8. Konn, G. A., and T. M. Konn, *Mathematical Handbook for Scientists and Engineers*, McGraw-Hill, New York, 1968.
9. *Handbook of Mathematical Tables*, Chemical Rubber Company, Cleveland, Ohio, 1964.
10. Abranowitz, M., and I. A. Stegun, *Handbook of Mathematical Functions*, National Bureau of Standards Applied Mathematics Series #55, U.S. Government Printing Office, Washington, D.C., 1964.
11. Spiegal, M. R., *Theory of Problems of Laplace Transforms*, Schaum, New York, 1965.
12. Graeme, J. G., G. E. Tobey, and L. P Huelsman, *Operational Amplifiers, Design and Appliances*, McGraw-Hill, New York, 1971.
13. Nyquist, H., *Am. Inst. Electr. Eng.*, **47**, 214 (1928).
14. Goldman, S., *Laplace Transform Theory and Electrical Transients*, Dover, New York, 1966.
15. Fox, H. W., *Master Op-Amp Applications Handbook*, TAB Books, Blue Ridge, Pa., 1978.
16. Lynch, T. H., *Electronics*, **115**, 21 (July 1977).
17. Diefenburger, A. T., *Principles of Electronic Instrumentation*, Saunders, Philadelphia, 1979.
18. Malmstadt, H. V., C. R. Enke, and S. R. Crouch, *Electronics and Instrumentation for Scientists*, Bejamin/Cummings, Reading, Mass., 1981.
19. Bekey, G. A., and W. J. Korplus, *Hybrid Computation, Wiley, New York, 1968.*
20. Savitzky, A., and M. Golay, *Anal. Chem.*, **36**, 1627 (1964).
21. Betty, K. R., and G. Horlick, *Anal. Chem.*, **49**, 351 (1977).
22. Adams, R. E., S. R. Besto, and P. W. Carr, *Anal. Chem.*, **48**, 1989 (1976).
23. Smith, D. E., in A. J. Bard, Ed., *Electroanalytical Chemistry*, Vol. 1, Dekker, New York, 1966.
24. Woodward, W. S., T. H. Ridgway, and C. N. Reilley, *Anal. Chem.*, **46**, 1151 (1974).
25. Brimmer, S. P., T. H. Ridgway, N. Radic, and H. B. Mark. Chem. Biomed. and Environ. Inst. 12(3) 171 (1982).

Part I
Section E

Chapter 4

# TRANSDUCERS

By Galen W. Ewing, *Las Vegas, New Mexico*

**Contents**

## I. INTRODUCTION

A transducer is a device that serves as an interface between two different domains with respect to energy or information or both. Those transducers that convert energy from one form to another, at the same time effecting the transfer of information, are the principal subject of this chapter. Thus silicon photocells are discussed in connection with their ability to monitor or measure a photon flux, but not when they serve (under the designation "solar cells") as a power source.

The unit to be given chief attention is the complete device that is normally available as a module for incorporation into laboratory instruments. We describe, for example, a photomultiplier tube but not the innumerable instruments that include photomultipliers. On the other hand, our discussion is not restricted to the properties of the photocathode itself, even though this is the seat of the actual transfer of information from optical to electrical signals.

It is characteristic of transducers, as used in laboratory instruments, that they are frequently called upon to operate close to their sensitivity limit, there the signal-to-noise (*S/N*) ratio is minimal; under such conditions the transducer is said to be *noise limited*. Hence some discussion of noise is essential to the proper understanding and use of transducers.

Almost invariably (in our context), transducers convert their information into electrical signals. Hence the primary classification of transducers must be according to the energy domain that is the *source* of the information. A secondary classification corresponds to the electrical parameter that forms the output of the transducer. Thus within the categories of pressure transducers and temperature transducers, we can distinguish species in which the signal is delivered in the form of a voltage, others in which a change of resistance carries the desired information, and so on.

In the treatment that follows, the primary classifications are taken up in turn for the comparative discussion of the alternative devices in each. Later in the chapter there is some consideration of electronic circuitry, arranged according to the output characteristics.

## II. TEMPERATURE TRANSDUCERS

The coverage of this section is not intended to include the principles of thermometry, which are considered elsewhere in the Treatise (11), but to concentrate on those temperature-sensitive devices that produce directly or indirectly an electrical signal. These transducers can be classified as mechanical or electrical, in reference to the property of which the temperature variation is measured.

### A. MECHANICAL TEMPERATURE TRANSDUCERS

To this category belong those devices that depend on the coefficient of thermal expansion of some working material, which may be gas, liquid, or solid.

#### 1. The Expansion of a Gas

The gas thermometer has played a fundamental role in the development of thermodynamics, but is seldom utilized as a practical measuring instrument; therefore it is not discussed here.

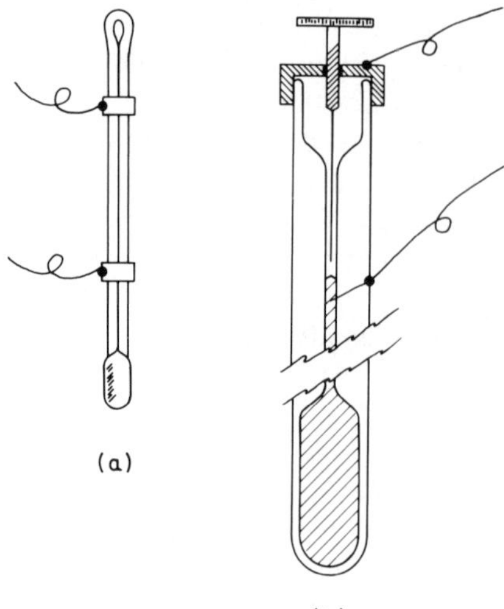


**Fig. 4.1.** Electrically contacting mercury-in-glass thermoregulators. (*a*) Fixed, (*b*) adjustment.

### 2. The Expansion of a Liquid

The familiar mercury-in-glass (or alcohol-in-glass) thermometer is the chief example of this class. The sensitivity depends on the differential expansion between the liquid and its container; hence a liquid with a large coefficient confined in a vessel made of a rigid solid with a coefficient nearly zero is optimal. The speed of response is determined by the heat capacity and thermal conductivities of both materials, as well as by the surface area of the transducer exposed to the temperature to be sensed.

Liquid expansion units can be utilized for temperature control by providing metallic electrodes to make contact with a mercury thread (Fig. 4.1). It is possible to obtain a continuous electrical readout from a mercury thermometer through a resistive or capacitive device to sense the height of the mercury thread, but the available resolution is severely limited.

### 3. The Expansion of a Solid

The differential expansion of two metals, welded together as a bimetallic strip, causes a flexure of the strip with change in temperature. This is widely used in inexpensive and yet reliable thermoregulators and time-delay relays. The effect can also be made to produce rotary motion in a dial indicator. A continuously variable electrical output can be obtained by means of a secondary transducer to detect mechanical motion; several of these are described in Section V.A.3.

### 4. The Quartz Crystal Thermometer

The piezoelectric resonant frequency of a plate of crystalline quartz has an easily measurable, reproducible temperature coefficient, expressible as a power series. Plates can be cut from the crystal in such an orientation that the coefficients of the second and higher powers are zero or negligible (4). This means that the frequency varies linearly with temperature. Hewlett-Packard has designed a thermometer based on this transducer, linear to within 0.05% over the range −40 to +230°C with $10^{-4}$ K resolution, and a time constant of about 1 sec.

## B. ELECTRICAL TEMPERATURE TRANSDUCERS

### 1. Thermocouples

Any pair of unlike electrical conductors placed in contact develop a potential across the interface, known as a contact potential. This potential generally varies with temperature. As with any source of potential difference, there is no way of measuring a single contact potential; a complete circuit must necessarily contain two junctions. A voltage observation at open circuit provides a measure of the temperature difference between the two junctions. Conventionally, the reference junction is held at the ice point so that calibration on the Celsius scale is facilitated. For convenience, particularly in recording instruments, the ice-point junction is often replaced by a junction at or above room temperature, together with a source of constant potential adjusted to simulate the ice temperature. Such surrogate "cold junctions" must be adjusted separately for each pair of thermocouple materials.

For general thermometry, couples of copper/constantan, iron/constantan, Chromel/Alumel, or platinum/platinum + 10% rhodium are commonly employed. These differ principally in respect to the temperature range over which they can be used. All are somewhat nonlinear, but over short spans the nonlinearity is usually negligible. Figure 4.2 shows the values of $dE/dT$ (called the *thermoelectric power**) as a function of temperature, for a number of conductors (14).

If a current is allowed to flow spontaneously in the thermocouple circuit, heat is absorbed at the hot junction and liberated at the cold, the *Peltier effect*. This is a reversible process that can be intensified by forcing a greater current to flow in the same sense as the spontaneous current, and reversed by a reversal of the current. The circuit can then act as a heat pump.

The magnitude of the open-circuit voltage developed by a thermocouple is given by

$$E = P(\alpha, \beta)\,\Delta T \tag{4.1}$$

where $P(\alpha, \beta)$ is the thermoelectric power of conductor $\alpha$ relative to $\beta$, usually expressed in microvolts per Kelvin, and $\Delta T$ is the difference of temperature between the two junctions, in Kelvins. If a current flows, $\Delta T$ is reduced by the action of the Peltier effect, but if the resistance of the circuit is of the order of 100 Ω or greater, the heat loss from this source is negligible.

*A misnomer, for this quantity does not have the dimensions of power.

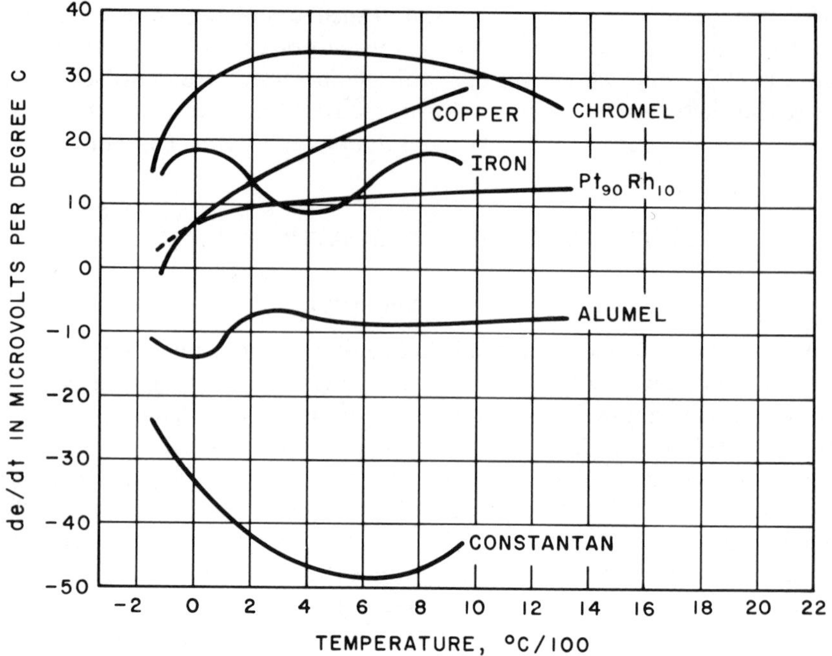


**Fig. 4.2.** Thermoelectric power of various thermocouple materials against platinum. (From reference 14, by permission of Leeds & Northrup Company.)

Kollie et al. (36) have noted errors of up to 150% in temperature readings from Chromel/Alumel thermocouples, induced by a magnetic field impressed upon the thermocouple during its measurement. The cause has been identified as the Ettinghausen-Nernst effect, the theory of which is discussed in the reference.

In order to optimize a thermocouple for use as a heat pump, the resistance must be made as small as possible, and a large current impressed from an external source (1–10 A at voltages up to 36 for a typical assembly). The conductors are most commonly fabricated of the semiconductor material $Bi_2Te_3$ or its solid solutions with $Sb_2Te_3$ or $Sb_2Se_3$, heavily doped with $n$- and $p$-type additives. A common form of couple is shown in Fig. 4.3; many such modules can be operated electrically in series while in parallel thermally.

### 2. Temperature-sensitive Resistors

All elemental metals and most alloys have positive temperature coefficients of resistance. The coefficients range from essentially zero for certain alloys such as constantan and manganin up to 0.006 $K^{-1}$ for nickel. Platinum, often used in precision resistance thermometry, has a coefficient of 0.003 $K^{-1}$.

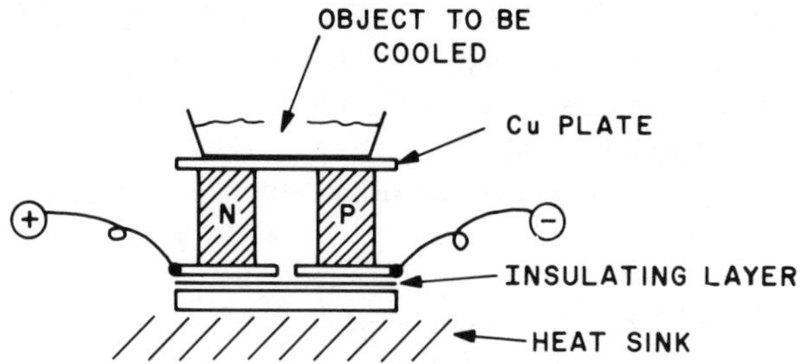


**Fig. 4.3.** A single-junction thermoelement heat pump.

The resistance of semiconductors is expected on theoretical grounds to follow an exponential relation:

$$R \propto \exp\left(\frac{\Delta E}{rkT}\right) \tag{4.2}$$

where $\Delta E$ is one-half of the forbidden energy gap between valence and conduction bands, $r$ is a constant with the value 1 when the number of majority carriers (holes or electrons) is less than the number of acceptor (impurity) sites, and 2 when this number is exceeded, $k$ is the Boltzmann constant, and $T$ is the Kelvin temperature (27). In actuality this relation is obeyed only over restricted regions. Figure 4.4 shows the curve obtained for silicon doped with phospho-

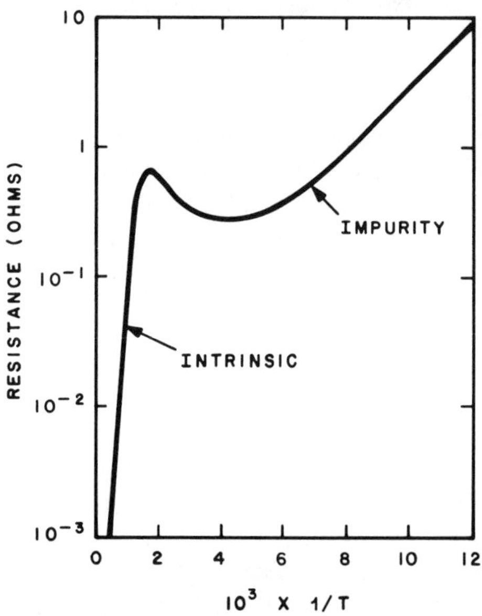


**Fig. 4.4.** The resistance-temperature characteristic of silicon doped with phosphorus. (Adapted from reference 20.)

rus (20); equation 4.2 is followed in the two linear regions, but these are separated by a region of negative slope. For the majority of semiconductors the temperature range of greatest practical interest lies in the intermediate region of negative temperature coefficient.

Semiconductor units designed to utilize this temperature characteristic, called *thermistors* (7), are fabricated by compression or sintering of mixed oxides of transition metals. They are available with room-temperature resistances from 1 Ω to 10 MΩ and coefficients from –0.03 to –0.06 $K^{-1}$.

Self-heating of resistive components owing to the current used to measure them must always be taken into consideration. This is particularly true with thermistors because of their negative coefficients; unless precautions are taken, small units burn out, for they tend to draw increasing current as the temperature is raised by the $I^2R$ heating.

### 3. Temperature-sensitive Semiconductor Junctions

The current $I$ flowing across a semiconductor junction follows the law

$$I = I_s\left[\exp\left(\frac{qV}{kT}\right) - 1\right] \tag{4.3}$$

where $q$ is the electronic charge and $V$ is the potential appearing across the junction. $I_s$, the saturation current, is a function of the geometry of the junction and of the concentration of majority and minority carriers. ($I_s$ is the constant, very small, reverse current flowing when $V$ is more negative than about –0.1 V.) Taking logarithms and rearranging gives an expression for $V$:

$$V = \left(\frac{kT}{q}\right)\ln\left(\frac{I + I_s}{I_s}\right) \tag{4.4}$$

The logarithmic term is constant if the current is unchanged, and $V$ becomes a linear function of temperature. For large currents, variation of $I_s$ must be considered, and the temperature dependence follows the relation

$$\frac{dV}{dT} = \frac{V}{T} - \frac{kT}{q}\left(\frac{1}{I_s}\cdot\frac{dI_s}{dT}\right) \tag{4.5}$$

The second term usually prevails, whereas the effect of the first term is to reduce $dE/dT$ slightly at low temperatures. When the first term is small, the coefficient is of the order of –3 mV/K (26).

This relation is manifest in any *pn* junction, operated at constant current, whether isolated in a diode or forming part of a transistor. In the latter case, the amplification of the transistor can be used to increase the sensitivity. Such a device, incorporating a transistor mounted in a probe, can form a very convenient thermometer over a range of about 100 K above and below room temperature. One commercial unit, with sensitivity adjustable from 10 to 360 mV/K, claims 0.1 K reproducibility, and nonlinearity not exceeding 0.125 K over the range –100 to +150°C (58).

Zener diodes show a temperature dependence of the characteristic reverse voltage drop, the zener voltage. The coefficient is nearly zero at about 5 or 6 V, increasingly negative below this, where the breakdown follows the zener mechanism, and positive at higher voltages, the avalanche region. For a 10 V zener, the coefficient is about +6 mV/K.

## III. DETECTORS OF RADIANT ENERGY AND CHARGED PARTICLES

Many types of transducers respond in varying degrees to incident energy delivered by electromagnetic waves or by a beam of electrons or ions. Hence detectors for these phenomena are conveniently considered together.

### A. VISUAL DETECTION

The human eye as a transducer can hardly be surpassed for certain types of measurements. Of particular importance are those applications involving matching the intensities of two beams of light of the same spectral composition, as in the half-shade polarimeter, or the Duboscq comparator (19). The eye has a tremendous range of sensitivity, but cannot be used to estimate absolute intensities with any useful degree of reliability. Because visual detection has gone out of fashion, we do not explore it further.

### B. PHOTOGRAPHY

Photographic materials composed of silver halide particles embedded in gelatin are inherently sensitive to radiant energy throughout the X-ray region, in the ultraviolet and visible from about 250 to 500 nm, and to charged particles [ions or electrons (9)] if they have sufficient kinetic energy. The lower limit in the ultraviolet is due to the absorption characteristics of gelatin. Sensitivity can be extended down to about 50 nm by coating the silver emulsion with a thin layer of anthracene or other fluorescent material, which serves to convert the short wavelength ultraviolet to a (roughly) proportional flux of visible radiation capable of penetrating the gelatin to activate the silver salts. The far ultraviolet can be recorded on *Schumann* plates, in which the emulsion is very thin, containing minimal quantities of gelatin. These plates have been utilized as far as 1 nm. The fragile Schumann emulsion must be carefully protected from abrasion.

The upper wavelength limit for photography can be increased to about 650 nm ("panchromatic" films) and as far as 1200 nm ("hypersensitized" films) by incorporating certain dyes.

The usual photographic material for recording X-radiation ("X-ray film") is coated with emulsion on both sides of the support, thus increasing the absorption of this penetrating radiation. Morimoto and Uyeda (49) have reported the results of tests on a large number of commercially available X-ray films with respect to such characteristics as speed, granularity, homogeneity, aging effects, and sensitivity, toward both X-rays of various wavelengths and darkroom safelights.

Photographic detection in strictly analytical applications is largely becoming obsolete. It continues to be useful in emission spectrography, especially for qualitative survey spectra. It is useful in certain types of X-ray diffraction, where the geometric relations of various diffracted beams can readily be established on a photographic plate or film. It is also utilized in high-resolution mass spectrometry with the Mattauch-Herzog design of instruments.

In addition, photographic detection finds application in *radioautography* to determine the distribution of radioactive substances on a surface.

The details of photographic processing are covered elsewhere in the Treatise (61). For quantitative purposes, the developed negative is examined in a densitometer, an instrument with a lamp and photocell, for measuring the "optical density," that is, the light-stopping ability of the silver deposits on the film or plate. There are so many variables in connection with photographic processing, any of which may affect calibration, that quantitative photography is generally unattractive. Its greatest advantage is its sensitivity; since the response is cumulative over a controllable exposure time, integrated energy is measured rather than the power in the beam of radiation.

## C. ELECTRON EMISSION PHOTOTRANSDUCERS

The several devices classed under this heading depend for their action upon the relase of electrons from a photocathode; they differ with respect to internal amplification and the mode of collection of the electrons.

### 1. The Photocathode

According to the classical photoelectric effect, electrons are ejected from a metallic surface by incident photons, provided the energy available is great enough to overcome the surface energy barrier (the *work function*). The quantum efficiency of this process (the number of electrons emitted per incident photon) is very low, generally less than 1%. Much more efficient (up to 20 or 30%) is the analogous effect in semiconductor photocathodes, and hence these materials are used almost exclusively in practical devices.

The effect can be understood by reference to Fig. 4.5. As in any semiconductor, electrons can exist in two energy domains, the lower or *valence* band, and the higher *conduction* band, separated by a forbidden gap. Electrons can be promoted from valence to conduction band by receiving energy as heat or as radiation. To be emitted from the surface of the solid into a vacuum, the electron must receive sufficient energy to overcome the added barrier known as the *electron affinity*, the semiconductor analog of the work function in metals. The Fermi level indicated in the figure serves as a reference, since it must indicate the same energy in the solid and in the vacuum. The energy required to liberate an electron is the sum of the gap energy, $E_g$, and the electron affinity,

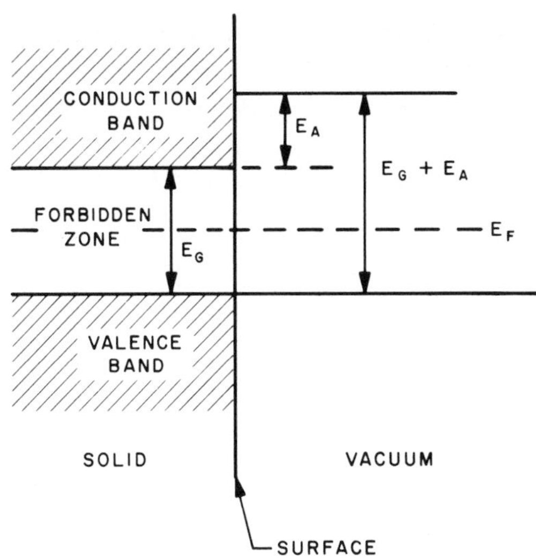


**Fig. 4.5.** Energy relations at the surface of a metal in a vacuum.

$E_A$. This is typically much less than the work function in a metal. The quantum efficiency is increased also by the ease with which electrons can move within the crystal, as compared to the situation in a metal where there are large numbers of electrons to offer collisions.

The spectral response of photoemissive tubes primarily depends on the nature of the cathode surface. Table 4.I lists most of the currently useful cathode types; the data have been collected from the literature of many manufacturers and hence may not be strictly comparable. Information on the inherent sensitivity of cathodes is usually not available; the tabulated values include the filtering effect of the tube envelope or window. Those cathodes given an "S" number are fairly uniform, industry-wide; the others are more likely to vary slightly from one manufacturer to another, the result of proprietary differences in the technique of manufacture.

It will be noted that the last several entries in the table show sensitivity only in the far ultraviolet. These "solar-blind" cathodes can be used to detect and measure low intensities of ultraviolet, even in the presence of full sunlight illumination. An ordinary candle flame gives sufficient ultraviolet to activate a relay connected to a phototube that has this type of cathode.

The cathode sensitivity is specified either in terms of photocurrent per unit luminous flux ($\mu$A/lm, the *luminous sensitivity*) or in current per unit incident power (mA/W, the *radiant sensitivity*). Radiant sensitivity must be specified for a particular wavelength, usually the point of maximum response. Luminous sensitivity is calculated in comparison with the standard CIE* luminance, that of a black body at a color temperature of 2854 K, easily approximated by a

*CIE = Commision Internationale d'Éclairage.

tungsten filament lamp. The relation between these measures of sensitivity can be derived from first principles as follows:

The current composed of photoelectrons leaving the cathode, $I_k$, is given by

$$I_k = \frac{P\lambda e Q(\lambda)}{hc} \tag{4.6}$$

where $P$ is the radiant power at wavelength $\lambda$ incident on the cathode, $e$ is the electronic charge, $h$ is Planck's constant, $c$ is the velocity of radiant energy in vacuo, and $Q(\lambda)$ is the quantum yield. The radiant cathode sensitivity, $E(\lambda)$, then becomes

$$E(\lambda) = \frac{I_k}{P} = \frac{\lambda e Q(\lambda)}{hc} \tag{4.7}$$

TABLE 4.I

Composition and Properties of Various Photocathodes

| Nominal Composition | Response designation[a] | Window[b] | Wavelength Range[c] (nm) | Sensitivity: Luminous (μA/lm) | Sensitivity: Radiant[d] (mA/W) | Quantum eff.[d] (%) |
|---|---|---|---|---|---|---|
| Ag.O.Cs[e] | S-1 | L | 400–800–1200 | 25 | 2 | 0.36 |
| $Cs_3Sb$ | S-4 | L | 300–400–650 | 40 | 42 | 13 |
| $Cs_3Sb$[f] | S-11 | L | 310–440–650 | 60 | 48 | 14 |
| $Cs_3Sb$[g] | S-17 | L | 300–490–650 | 125 | 83 | 21 |
| $Cs_3Sb$ | S-5 | U | 185–340–650 | 40 | 50 | 18 |
| $Cs_3Sb$[f] | S-21 | U | 185–440–650 | 30 | 23 | 7 |
| $Cs_3Sb$ | S-19 | S | 160–330–650 | 40 | 64 | 24 |
| $Cs_3Sb$[f] | S-13 | S | 160–440–650 | 60 | 48 | 14 |
| Ag.Bi.O.Cs[e,f] | S-10 | L | 300–450–750 | 40 | 20 | 6 |
| $Na_2KSb$[f] | S-24 | B | –380– | 32 | 64 | 23 |
| $K_2CsSb$[f] | — | B | 250–400–650 | 80 | 89 | 28 |
| $K_2CsSb$ | — | L | 300–400–650 | — | 80 | 25 |
| $CsNa_2KSb$[f] | — | S | 160–420–850 | 150 | 64 | 19 |
| $CsNa_2KSb$[f] | S-20 | L | 300–420–800 | 300 | 89 | 21 |
| $CsNa_2KSb$ | — | B | 400–575–920 | 265 | 45 | 10 |
| GaAs | — | U | 200–450–940 | 250 | 37 | 10 |
| GaAsP | — | A | 200–450–770 | 200 | 61 | 17 |
| CsTe | — | S | 160–220–320 | — | 28 | 15 |
| CsTe | — | F | 110–130–340 | 150 | 48 | 17 |
| CsI[f] | — | F | 110–120–200 | — | 22 | 20 |
| CuI[f] | — | F | 110–130–180 | — | 17 | 16 |
| KBr | — | F | 110–120–160 | — | 20 | 20 |
| RbI | — | F | –120– | — | 20 | 20 |
| CuBe | — | — | – 70–130 | — | — | 17 |

[a] Standards established by the Electronic Industries Association (EIA).

[b] A = alumina (sapphire), B = borosilicate; F = lithium fluoride; L = lime glass; S = vitreous silica; U = ultraviolet-transmitting glass.

[c] First and third numbers denote useful range, second number is wavelength of maximum response.

[d] At wavelength of maximum response.

[e] Dots indicate unspecified stoichiometry.

[f] Semitransparent cathode; others are opaque.

[g] On a reflecting surface.

For the CIE standard source, with emission $W(\lambda)$ in watts per unit wavelength, the total luminous flux, $L$, is

$$L = K \int_0^\infty W(\lambda) y(\lambda)\, d\lambda \qquad (4.8)$$

where $y(\lambda)$ is the *luminosity coefficient*, also specified by CIE, and $k$ is the luminous efficiency of radiation at the response peak of the CIE standard eye (555 nm). The cathode current produced by the standard source is

$$I_k^\circ = \int_0^\infty E(\lambda) W(\lambda)\, d\lambda = \frac{e}{hc} \int_0^\infty Q(\lambda) W(\lambda) \lambda\, d\lambda \qquad (4.9)$$

so the luminous cathode sensitivity, $S$, is given by the ratio of equations 4.9 to 4.8:

$$S = \frac{e \int_0^\infty Q(\lambda) W(\lambda) \lambda\, d\lambda}{hcK \int_0^\infty y(\lambda) W(\lambda)\, d\lambda} \qquad (4.10)$$

This equation can be solved by iterative techniques using tabulated values of $W(\lambda)$, $y(\lambda)$, and $k$, together with values of $Q(\lambda)$ determined experimentally for a given cathode. The luminous sensitivity is applicable only in the visual region where the standard eye is defined. It is not specified for those devices that are not active in this region.

Photoemissive devices are plagued with *dark current*. By this term is meant current carried by electrons leaving the cathode by any mechanism other than photoelectric. The major part of the dark current is thermal in origin, for thermionic emission is favored by the same low-energy barrier that favors easy photoemission. This portion of the dark current can be eliminated by cooling the phototube to about −20°C for antimony-alkali types. (Cooling is not as effective for the Ag.O.Cs cathode, which has a much larger dark current to begin with.) Solid $CO_2$ or Peltier cooling is usually sufficient.* The residual dark current that is not affected by cooling is due partly to the natural radioactivity of the surroundings and to cosmic radiation. Traces of $^{40}K$ in the glass envelope and in the cathode itself are one possible source of interference. There appear to be other minor causes of dark current, not completely understood.

Dark current is proportional to the area of the photocathode, and thus can be diminished by designing a tube with a small or masked cathode.

*Excessive cooling, with many cathode materials, degrades the long-wavelength response (10).

## 2. Photoemissive Vacuum Diodes

The simplest photoemissive device is composed of a photocathode, as described in the preceding section, and a collecting anode, in a transparent evacuated envelope. Conventionally the cathode is given a concave semicylindrical shape, and the anode is a thin rod or wire along the axis of the cylinder. Because the sensitivity of the cathode may vary slightly from point to point on its surface, it is mandatory to maintain the geometry constant if precisely reproducible results are to be obtained. This extends to the position of any shadow cast on the cathode by the anode wire. It is usually desirable to defocus the radiation so that it covers the entire cathode; this is sometimes accomplished with a diffusing plate of opal glass, but the spectral response may be altered somewhat.

The interelectrode capacitance is small, of the order of 1 or 2 pF. If the measuring circuit includes a resistive load of 1 MΩ, a typical value, the *RC* time constant is 1 μsec, which indicates a maximum frequency response of about 1 MHz.

Figure 4.6 is a family of characteristic current-voltage curves for a vacuum diode at various levels of illumination. Within the steeply curved region below about 10–20 V the field is insufficient to ensure that all electrons will be captured by the anode. Above this, all the photoelectrons do reach the anode, and further increase in voltage has almost no effect. The remaining slope is due largely to decrease of the energy barrier at the surface. The vacuum is never perfect, and ionization of residual gas also contributes to this rise in current. Curve *a* of Fig. 4.7 demonstrates the linearity of current as a function of illumination in a vacuum diode operated at a fixed voltage.

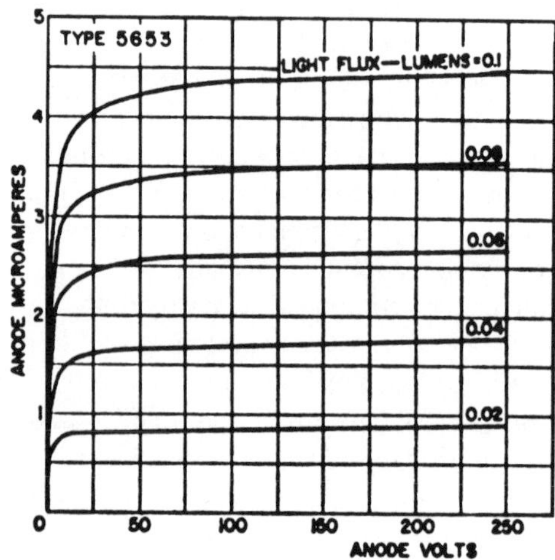


Fig. 4.6. Current-voltage characteristics of a vacuum photodiode. (Courtesy of RCA Corporation.)

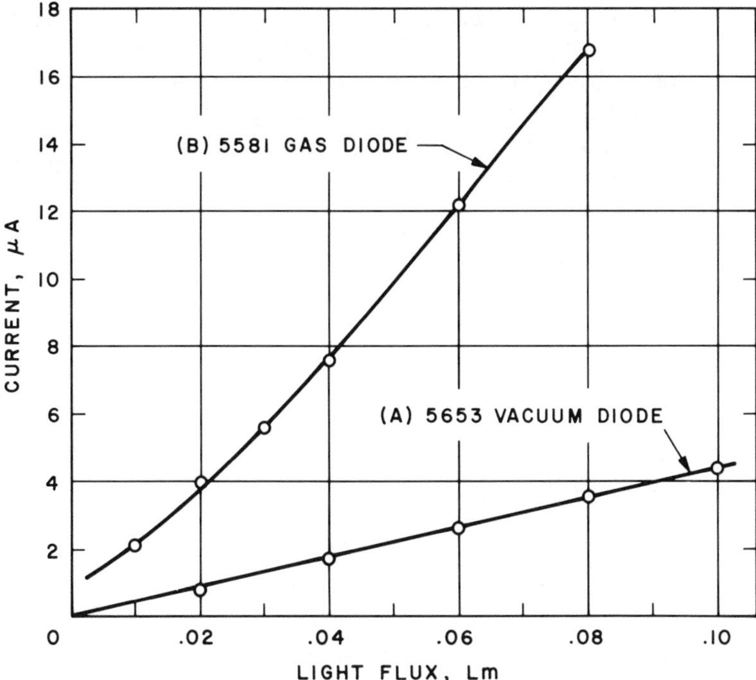


**Fig. 4.7.** Current-light flux relations of vacuum and gas-filled photodiodes. Anode potential 100 V.

Diodes are fabricated in many forms, requiring various standard sockets or clips for mountings. They may be designed to receive radiation restricted to a narrow solid angle or to be nearly nondirectional. Twin diode units are available for applications involving balanced circuitry.

Vacuum diodes are used mostly for applications where relatively high levels of illumination are to be measured. Their output can easily be amplified to increase overall sensitivity.

### 3. Photoemissive Gas Diodes

If a diode such as those described above is filled with argon or other inert gas to a pressure of about 0.1 torr, increased sensitivity results, owing to secondary ionization of gas atoms. (At voltages lower than the ionization potential of the gas, secondary ionization cannot occur, and the response of the gas diode is essentially identical to that of its vacuum counterpart.)

The current flowing through a gas phototube at wavelength $\lambda$ is given by the relation (40):

$$I_\lambda = I_k\left(\frac{\exp(\alpha d)}{1 - \gamma[\exp(\alpha d) - 1]}\right) \tag{4.11}$$

where $I_k$ is the initiating photocurrent (equation 4.6), $\alpha$ (the "first Townsend coefficient") represents the number of ions formed per unit path distance, $\gamma$ (the "second Townsend coefficient") is the number of secondary electrons formed per impacting positive ion, other minor effects being included, and $d$ is the path distance, anode to cathode.

At low levels of illumination, $\gamma$ can be assumed to be negligible, so that equation 4.11 becomes

$$I_\lambda = I_k \exp(\alpha d) \tag{4.12}$$

Since $I_k$ is directly proportional to the monochromatic radiant power $P$, and since $\alpha$ increases linearly with the voltage $V$ across the tube, we can write for the gain, $G$:

$$G = \frac{I_\lambda}{I_k} = \exp(kV) \tag{4.13}$$

Where $k$ is a proportionality constant incorporating for convenience the dimension $d$. Thus the greater the voltage across the tube, the greater the gain. Unfortunately the voltage cannot be increased much above 200 V (17) without giving rise to a glow discharge within the tube. Practical tubes are usually operated at 90 V, and show a gain of about 20.

At a fixed level of illumination, $I_k$ is a constant; hence by equation 4.13, $I_\lambda \propto \exp(kV)$, or

$$\log I_\lambda = k'V \tag{4.14}$$

in which $k'$ is a new constant. Hence the current-voltage characteristic, plotted as log $I$ against $V$, is linear. It is customary, however, to plot current directly, as in Fig. 4.8. Curve $b$ in Fig. 4.7 shows that linearity for a gas diode is poorer than

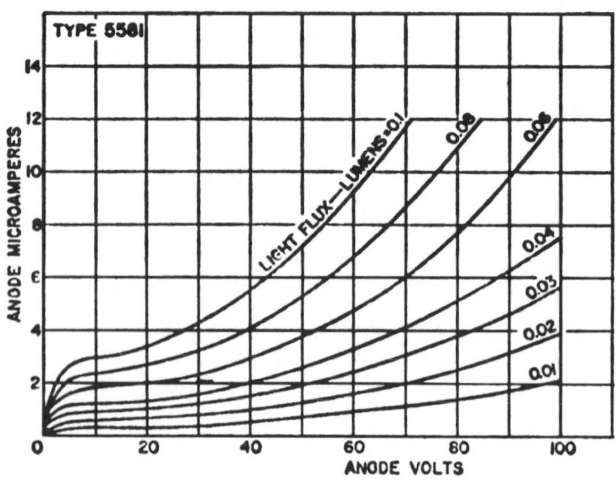


**Fig. 4.8.** Current-voltage plot for a gas-filled photodiode. (Courtesy of RCA Corporation.)

for its vacuum counterpart. The tubes selected for illustration in Figs. 4.6–4.8 are identical in all respects other than the gas content. The increase in sensitivity due to the presence of the gas is clearly evident.

Gas diodes are widely used in sound-on-film reproduction and in relay applications. In precise instrumental work they have been little used. The maximum frequency response is about 10 kHz.

## 4. Photomultipliers

By far the most widely used detector for low-level quantitative measurements in the ultraviolet and visible regions is the photomultiplier tube.* In this device, electrons from a photocathode are accelerated by an electric field and caused to hit the first of a series of electrodes called *dynodes*. The acceleration provides the electrons with enough energy to eject several secondary electrons from the dynode surface. These, in turn, are accelerated and impinge upon the second dynode, where further secondary emission occurs, and so on through many stages.

If an average of $\delta$ electrons are emitted at each of $n$ dynodes for every incident electron, it is easily seen that the current gain $G$ will be

$$G = \delta^n \tag{4.15}$$

Thus for a tube with 10 dynodes ($n = 10$) and a dynode multiplication factor $\delta = 4$, the gain is $4^{10} \simeq 10^6$. If the maximum permitted anode current is 1 mA, then the photocurrent at the cathode of this tube cannot be permitted to exceed 1 nA. But this is only the *maximum*; one can easily measure currents 1000 times less. This simple calculation suggests the utility of photomultipliers for extremely low light levels. To measure higher illumination, the gain can be reduced simply by reducing the voltages applied to the dynodes.

The structure of a photomultiplier can be broken down into three parts: the photocathode, the electron-multiplier, and the anode. As used in photomultipliers, both opaque and semitransparent cathodes have advantages. In the semitransparent type, the emissive material is coated directly on the inside of the faceplate. This gives more efficient collection of light from an extended source, often required in scintillation counting. Such tubes are regularly available with nominal faceplate diameters up to 125 mm; special ones have been made as large as 400 mm in diameter. Smaller tubes may have either type of cathode structure. The popular 1P series (1P21, visible only; 1P22, extended red; 1P28, ultraviolet-visible) use a side window and an opaque cathode.

The multiplier portion of the tube contains the dynode assembly. The dynodes are generally surfaced with $Cs_3Sb$ or similar material with a low energy barrier for emission of electrons. The mechanical structure takes several forms, as illustrated in Fig. 4.9. In the designs at *a* and *b*, the dynodes are so shaped as to focus the electrons from one stage onto the next, thereby equalizing their

*More correctly called a *multiplier phototube* or a *photoelectron multiplier*.

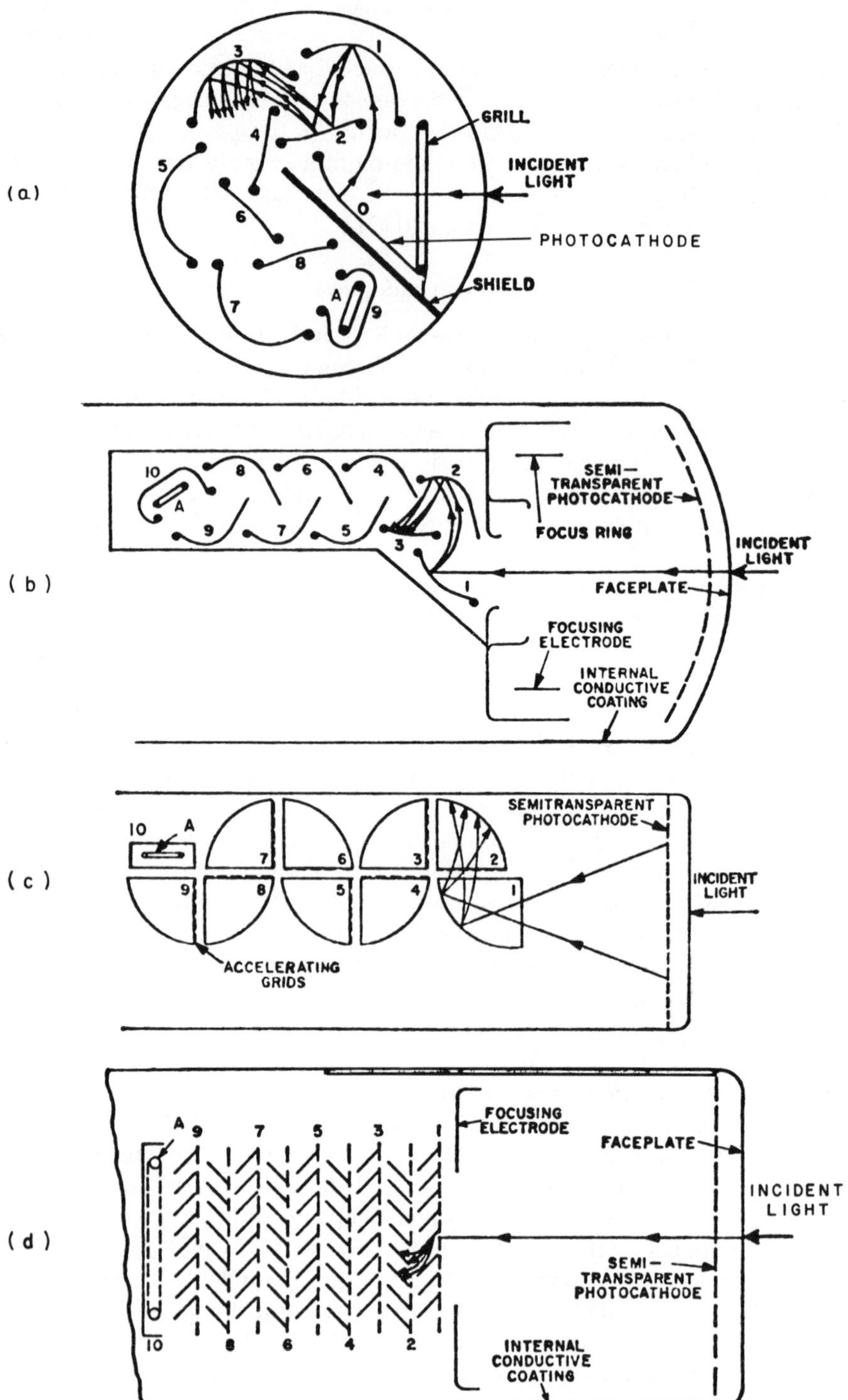


**Fig. 4.9.** Structures of several types of photomultiplier tubes: (*a*) squirrel cage, with side windows; (*b*) linear; (*c*) box-and-grid; (*d*) venetian blind. In each, *A* denotes the anode; the dynodes are numbered consecutively. (From reference 54 by permission of RCA Corporation.)

transit times through the tube. The field strengths at the dynode surfaces are rather high. These two properties give the greatest speed of response of any photomultiplier. The form at *b* is superior in speed to *a*, but also more expensive. Type *a* is appropriate where requirements in general are less stringent, and where cost and size are important. The unfocused "venetian blind" structure (*d*) shows the best stability with respect to gain, in that slight inequalities between voltages of successive dynodes have little effect; hence the smaller and more convenient, but less stable, carbon resistors can be used in the voltage divider establishing the dynode potentials. The large area of the first dynode makes for easier matching with a large area photocathode. The response speed is somewhat less than in the focused structures, perhaps 10 nsec transit time compared to 2 nsec for type *b*. This is caused largely by the weaker electrostatic field at the vanes of the dynodes. The "box-and-grid" form (*c*) is similar in properties to the venetian blind type, but shows slightly higher gain per stage for the same voltage, offset by slightly lower gain stability.

Special design considerations enter into the relation between the first stage of a photomultiplier and its cathode. In large cathode types, focusing electrodes must be included, to shape the electric field so as to optimize collection of photoelectrons. The collection efficiency typically lies between 85 and 98%. In most cases the focus electrodes are connected internally, rather than brought out to separate base pins.

Uniform transit time is important in the measurement of *pulses* of radiation, as in scintillation and photon counting. Tubes with spherically convex cathodes, hence faceplates, are superior in this respect to those with planar cathodes (Fig. 4.10) because, in the latter, electrons from the edge of the cathode must travel appreciably further than those from the center. Flat cathodes, on the other hand, are easier to couple to a scintillation crystal, but this difficulty can be overcome by the use of a fiber-optic faceplate.

The anode of a photomultiplier usually consists of a grid of fine wires within a concave last dynode, as indicated in Fig. 4.9, *a* to *d*. Most of the electrons from the penultimate dynode pass through the grid and only electrons from the final dynode are collected thereon.

The dynodes in a photomultiplier must be provided with successively more positive potentials relative to the cathode. This is conveniently done by means of a resistive voltage divider, the simplest form of which is shown in Fig. 4.11*a*. The high voltage is such as to give 100–150 V per stage. Either end of the chain may be grounded, the choice being whether it is less inconvenient to mount the tube with the anode 1–2 kV above the ground or with the cathode that amount below ground. With small tubes that are provided with integral bases, the positive side is usually grounded so that the amplifier connected to the anode need not be floated at a high potential.

In this circuit (Fig. 4.11*a*) the anode current flows through a load resistor, $R_L$, to ground, and an amplifier senses the voltage drop across this resistor. There are two major drawbacks to this arrangement, both causing nonlinear response. The first is that the potential of the anode relative to ground changes with the photocurrent, owing to the drop in $R_L$. This can be remedied by the

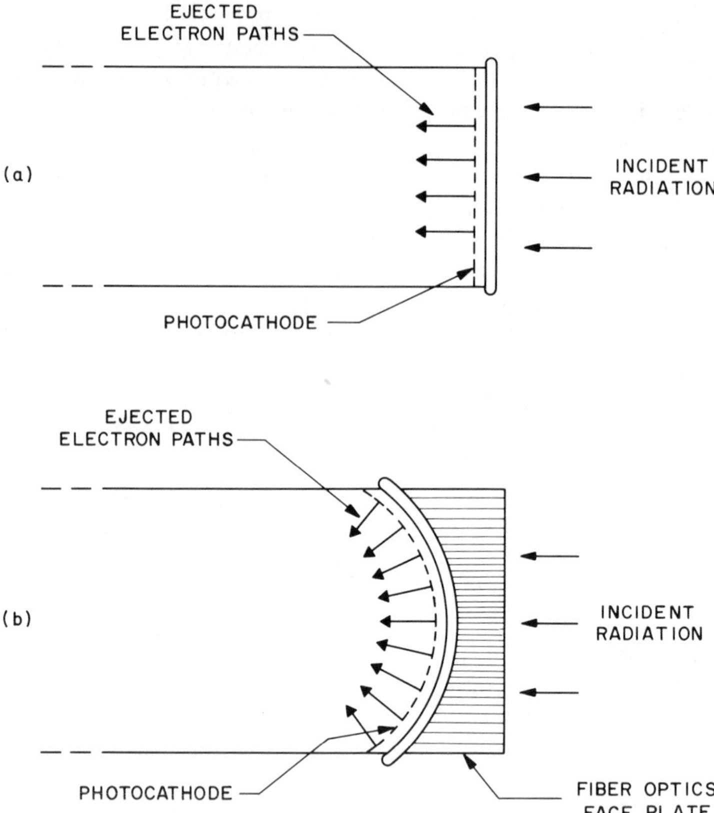


**Fig. 4.10.** Comparison of electron paths from flat and spherical photocathodes.

modification in Fig. 4.11*b*, wherein the anode is connected directly to the summing junction of a high-gain operational amplifier, acting as a current-to-voltage converter.* The load resistor $R_L$ now becomes the feedback element for the amplifier. Thus the anode is forced to remain at virtual ground, no matter how large the photocurrent may be (59).

The other fault in the original circuit lies in the ease with which the voltage divider string is loaded by the photocurrent. About three-quarters of the anode current is supplied by the last dynode (supposing $n$ to be 4); hence this current must pass through the entire resistor chain *except* $R_{Dn}$, and this current, of course, varies with the illumination. Hence the voltage drop across the resistors changes, and not all equally, with the quantity being measured. This can be overcome in the previous circuits only by making the divider resistance small enough that the current through the chain (including $R_{Dn}$) is at least 10 times greater than the maximum anode current, preferably more. This means that the

*A complete description for such a converter is given in reference 59.

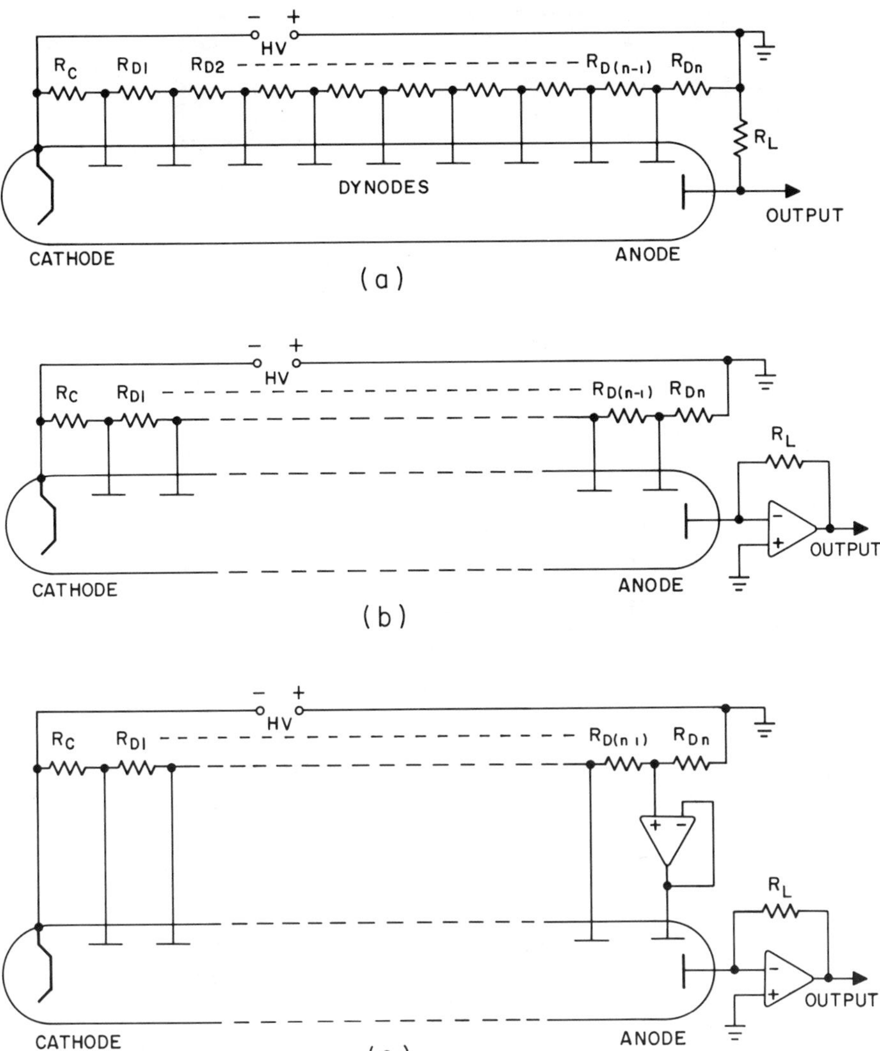


**Fig. 4.11.** Electronic circuits for photomultiplier tubes: (*a*) conventional; (*b*) with operational amplifier as current-to-voltage converter; (*c*) with voltage-follower amplifier to avoid overloading the resistor chain.

high-voltage power supply must be capable of generating 10 or more times the "useful" power, perhaps 20 W rather than 1 W. Considering that the voltage must be highly regulated, this increased wattage can be expensive. The circuit of Fig. 4.11*c* obviates this difficulty by the use of an operational amplifier connected as a voltage follower to act as a buffer between the voltage divider and the last dynode. This is usually sufficient, but a similar amplifier could be used to buffer the next dynode if needed.

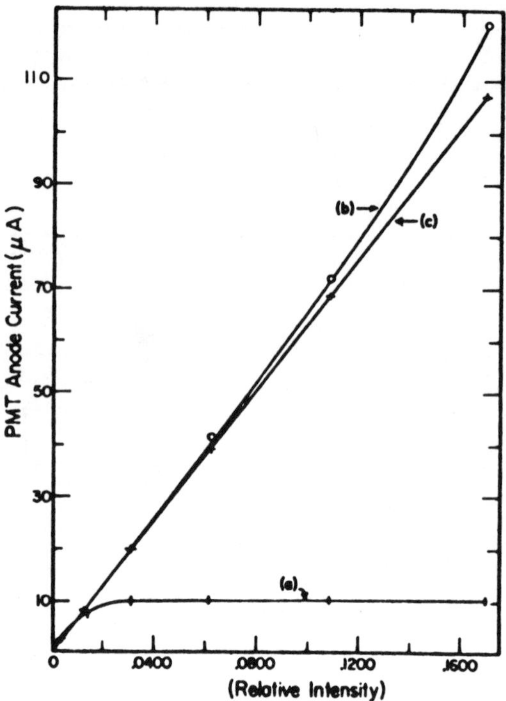


**Fig. 4.12.** Relative response of the same photomultiplier in the three circuits of Figs. 4.11. (From reference 59 by permission of *Analytical Chemistry*.)

Fig. 4.12 illustrates the dramatic improvement in linearity and dynamic range resulting from these two circuit modifications (59).

For use with pulsed light, it is desirable to bypass the last few resistors in the chain with capacitors to ground, as in Fig. 4.13*a*. Another modification, frequently recommended, is the insertion of a zener diode in place of the first resistor, also shown in Fig. 4.13*a*. This maintains the voltage drop between the cathode and the first dynode constant, ensuring that the focusing requirements continue to be met. This is especially necessary if the gain is to be varied by altering the overall voltage.

To permit a photomultiplier to pass maximal anode current, the voltage to the last few dynodes must be increased so as to overcome space-charge effects. This is accomplished by the use of a tapered rather than a uniform divider.

Special considerations regarding the use of photomultipliers in applications requiring very fast response (in the nanosecond range) have been discussed in detail by Lytle (45). Precautions necessary for highest precision work are reported by Mavrodineanu (48).

A photomultiplier can be operated at constant anode current rather than at constant potential (22,67). This can be effected by connecting it as the feedback

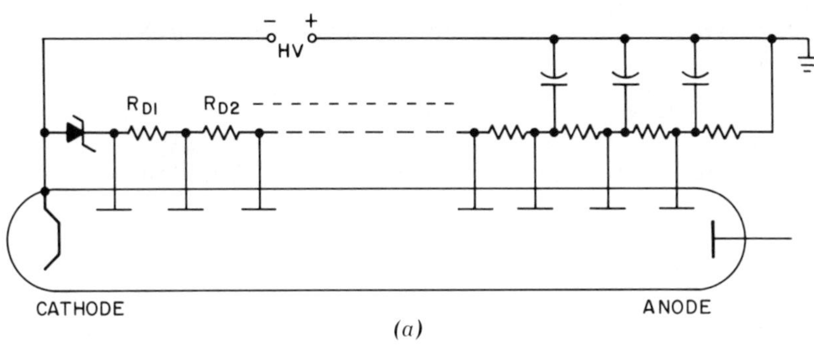


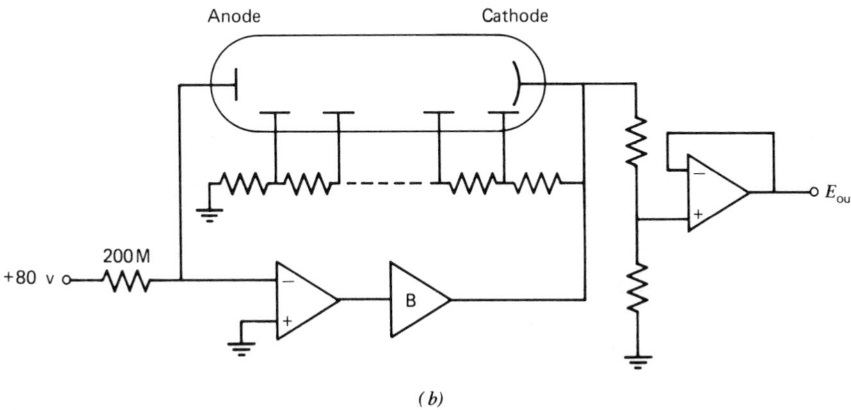


**Fig. 4.13.** Photomultiplier circuit modified for (*a*) pulsed radiation and (*b*) constant-current operation B represents a voltage-booster amplifier.

element around an operational amplifier, as in Fig. 4.13*b*. With the values shown, the anode is forced to carry a current of 0.40 $\mu$A, no matter what the level of illumination may be at the cathode. The amplifier controls the potential applied to the dynode string, adjusting it continuously to give the multiplication factor required to match the cathode and anode currents. The cathode voltage can be shown to be, under these conditions, a logarithmic function of the power in the incident radiation. This arrangement is particularly advantageous in spectrophotometric applications, because its output, when properly referenced, is directly proportional to the absorbance of a test solution.

### 5. Noise Considerations in Photodiodes and Photomultipliers

Noise can arise from many sources in vacuum photodetectors. The following can be identified; the corresponding formulas are taken from an RCA publication that gives the details of their derivation (54).

1. *Photon noise*, due to the statistical fluctuations in the arrival of photons at the cathode:

$$(S/N)_P = \sqrt{\bar{n}_P} \tag{4.16}$$

where $\bar{n}_P$ is the average number of photons incident per unit time.

2. *Photocathode noise*, based on the statistics of emission of electrons:

$$(S/N)_{PC} = \sqrt{Q\bar{n}_P} \tag{4.17}$$

3. *Dynode noise*, arising from the statistical emission of electrons at the first dynode:

$$(S/N)_D = \sqrt{\delta_1} \tag{4.18}$$

4. *Overall tube noise* at the anode, assuming all stages are identical:

$$(S/N)_a = \sqrt{Q\bar{n}_P\,(\delta - 1)/\delta} \simeq \sqrt{Q\bar{n}_P}\ (\text{for } \delta \gg 1) \tag{4.19}$$

Other forms of noise, not readily predictable, result from random radioactive events and the residual gas ionization previously mentioned. If too great currents are taken through the anode and last dynode, some photons may be produced at these sites. In certain photomultipliers, such as the box-and-grid type, a line-of-sight path may exist from anode to cathode, and this may result in positive optical feedback with attendant instability. This normally happens only if the tube is operated far beyond its ratings (62).

Ingle and Crouch (32) have published an interesting comparison between photomultipliers (PMT) and photodiode-amplifier (PD) combinations, in terms of their $S/N$ ratios. Their analysis, which includes both shot and Johnson (resistance) noise, shows that the $S/N$ ratios for the two types of detector are given respectively by

$$(S/N)_{\text{PMT}} = \frac{GR_L I_k}{\sqrt{1.6 \times 10^{-20}\, R_L + 6.3 \times 10^{-19}\, R_L^2 G^2 I_T}} \tag{4.20}$$

$$(S/N)_{\text{PD}} = \frac{R_L I_k}{\sqrt{1.6 \times 10^{-20} R_L + 3.2 \times 10^{-19}\, R_L^2 I_T}} \tag{4.21}$$

where $R_L$ is the load (or amplifier feedback) resistor, $I_T$ is the total current, including the photocathode current, $I_k$, and the dark current. If the photocurrent is large enough to render the dark current negligible, equation 4.20 reduces to

$$(S/N)_{\text{PMT}} \simeq 1.3 \times 10^9 \sqrt{I_k} \tag{4.22}$$

which means that under these conditions the photomultiplier is shot-noise limited; that is, the $S/N$ ratio is determined by the photocurrent. Similarly, if the signal voltage in a diode $E_{S,\text{PD}} = R_L I_T < 500\ \mu V$, equation 4.21 simplifies to

$$(S/N)_{\text{PD}} \simeq 7.9 \times 10^9\, I_k \sqrt{R_L} \tag{4.23}$$

so that the diode is Johnson-noise limited (the $S/N$ ratio is determined by the resistance). However, if $R_L I_T > 5\ V$, equation 4.21 becomes

$$(S/N)_{\mathrm{PD}} \simeq 1.8 \times 10^9 \sqrt{I_k} \tag{4.24}$$

and the diode becomes shot-noise limited.

Under equal conditions of illumination, and with identical cathodes, the ratio of $S/N$ ratios (from equations 4.20 and 4.21) is

$$\frac{(S/N)_{\mathrm{PMT}}}{(S/N)_{\mathrm{PD}}} = G\sqrt{\frac{1 + 20\, R_L I_T}{1 + 39\, R_L G^2 I_T}} \tag{4.25}$$

in which $G$ is the internal current gain of the photomultiplier. Manipulation of this equation shows that the $S/N$ ratio for the PMT is greater than that of the PD provided that $R_L I_T < 50$ mV. (Because simplifying assumptions were involved in the derivation, the 50 mV figure is not exact.) This relation can serve as a guide to determine whether a PMT or a PD combination is best fitted to a specific application. If the signal voltage from the diode system is greater than about 50 mV, or if the photomultiplier output (in millivolts) is greater than 50 times $G$, the selection of the diode rather than the multiplier is indicated.

## 6. Channel Multipliers

A channel electron multiplier consists of a glass tube, perhaps 100 mm long by 1 mm inside diameter, coated on the inner surface with a semiconducting material (Fig. 4.14). A potential of 2–3 kV is impressed across the ends of the resistive coating. In use, electrons enter the negative end of the tube at an angle and impinge on the wall with sufficient energy to cause multiple secondary emission. The emitted electrons are also accelerated, and strike the tube wall further toward the positive end, producing more secondaries, and so on down the tube. An anode is provided at one end and a conventional photocathode at the other. The resistive tube combines the functions of the dynode chain with its external voltage divider as they appear in conventional multipliers.

In practice the tube is curved into an arc or loose spiral in order to avoid interference from positive ions. In a straight tube a positive ion formed near the anode (by collision of an electron with a residual gas molecule) is accelerated back toward the cathode, and may strike the wall with enough energy to emit electrons, causing a spurious pulse. The curved tube prevents such ions from moving far enough to gain the required energy for secondary emission.

The gain of the channel multiplier depends upon the impressed voltage, which is often as great as $10^7$ or higher. At about $10^8$, the device saturates by space-charge limitation. Beyond this point it gives pulses of uniform magnitude, unrelated to the energy of the primary electron or photon, in a manner reminiscent of the Geiger counter. Below saturation, proportionality suffers somewhat due to the spread in lengths of path taken by different electrons. Channel multipliers have unusually low noise and low background counts, mostly

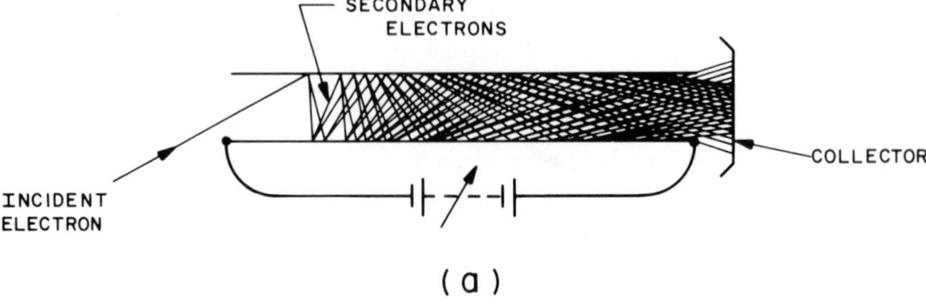


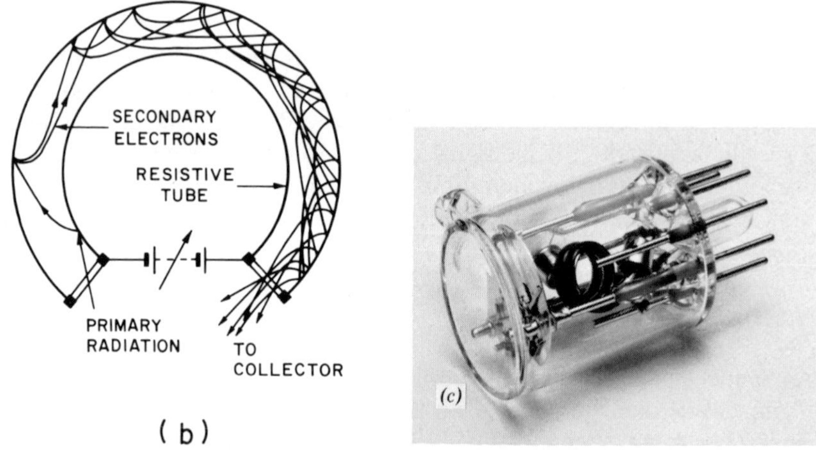


**Fig. 4.14.** Channel electron multipliers. (*a*) Linear; (*b*) curved; (*c*) photograph of Galileo model 7500. [(*c*) courtesy of Galileo Electro-Optics Corporation.]

because of their small physical size. The background can be as low as a few counts per minute.

Channel multipliers can be used in either a pulse counting or an analog mode. They are economical of space and power requirements and hence find applications in space technology as well as in the laboratory.

## D. CHARGE-SEPARATION DETECTORS

A number of types of radiation detectors depend on the action of the incident energy to cause ionization in a fluid medium, or by an analogous process, electron-hole pair production in a semiconductor. Detection of nuclear and related events by means of cloud chambers and bubble chambers are not included here.

### 1. Gas Ionization Detectors

Detectors in this category are widely used in the measurement of X- and gamma-radiation and of charged bodies such as alpha and beta particles. Their applicability in the vacuum ultraviolet region is not as well known (15).

The detection characteristics depend on the geometry of the electrodes used to detect the ions, the nature and pressure of the gas, and the input parameters of the associated electronics. A typical unit consists of a cylindrical metallic cathode with an axial wire as anode, the whole enclosed in a glass envelope. A thin window may or may not be provided for the entrance of radiation. Figure 4.15 shows such a unit. Because of the cylindrical geometry, the potential gradient is much steeper near the central anode than near the cathode.

Figure 4.16 shows schematically the relative magnitude of the response to two types of radiation of different energy content, as a function of impressed voltage. The more energetic radiation produces a greater number of ions in the gas, because the energy expended for each ionizing event is essentially constant.

At zero voltage all ions that are formed recombine, and no current results. As the voltage is increased (within region $A$), there is no change in the number of ions produced, but as they are accelerated by the field, an increasing proportion of them are collected by the electrodes, producing a measurable current. At a field of about 100 V/cm, essentially all ions are collected, only a negligible fraction remaining. At this point a plateau is seen (region $B$), so that small changes in voltage are without effect. The device operating in this mode is called an *ionization chamber*. The current resulting from a single pair of ions or even from all the ions formed by a single entering X-ray photon is too small to be observed, but for a relatively intense beam of radiation, good results can be obtained. The sensitivity to X-rays can be maximized by the use of a gas containing atoms with high atomic weight, such as methyl bromide or Freon, rather than air.

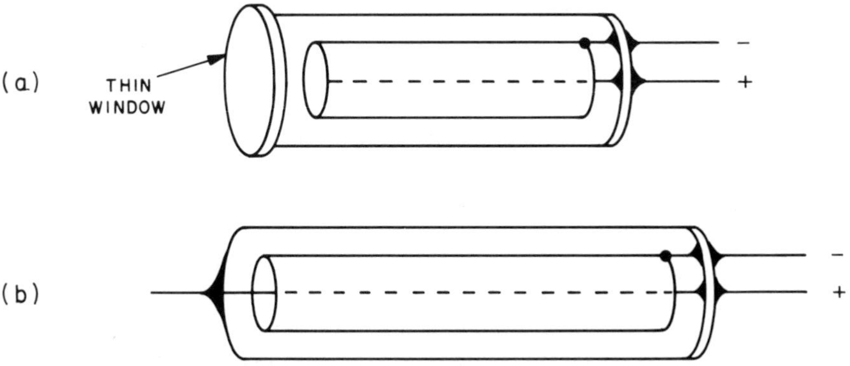


**Fig. 4.15.** Gas-ionization tubes for penetrating radiations. (*a*) End-window type for alpha and beta particles; (*b*) windowless, for gamma- and X-radiation.

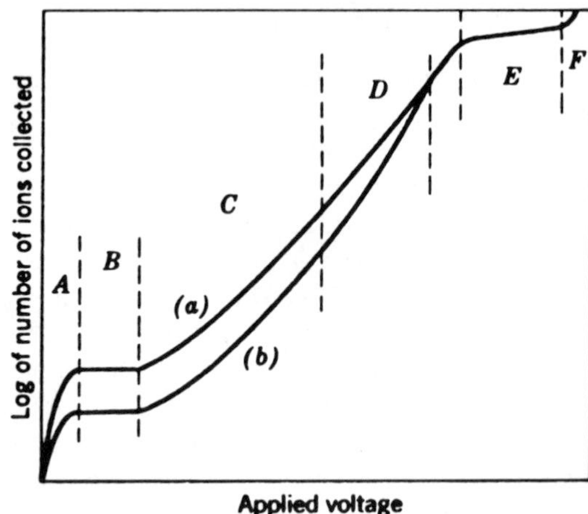


**Fig. 4.16.** The number of ions collected in a gas-ionization tube as a function of applied voltage, for radiations of two energy contents.

As the potential is raised beyond the plateau (region *C*), additional ions are formed by collision, increasing the response attributable to each individual ionizing event, and thus increasing the sensitivity. The current produced by a single entering photon is now easily observed as a pulse. As shown in Fig. 4.17, the response, now interpreted as pulse height, is proportional to the energy of the ionizing photon (38). When utilized in this mode, the device is known as a *proportional gas counter*. Since no plateau is present in this section of Fig. 4.16, the applied voltage must be very closely regulated if pulse-height proportionality is to be preserved. However, if only the *number* of pulses per unit time is required, more latitude in potential can be tolerated.

Further increase in potential brings about a saturation condition (region *E*), in which all pulses are of equal magnitude, regardless of the photon energy. This is the *Geiger* region. Beyond this (region *F*), the counter passes into a continuous discharge.

Gas-ionization detectors used in these several modes show considerable variation in operating characteristics, particularly recovery speed. It is this property that determines the maximum rate at which photons can be received and counted without excessive error. The recovery time of an ionization chamber of average dimensions is between 5 and 10 msec, giving a maximum count rate of 100–200 $sec^{-1}$. If the chamber is operated in the time-averaged or dc manner, the rate can be somewhat higher, but the precision of measurement is less. A proportional gas counter is much faster. Each electron (out of a total of perhaps 400) formed by the passage of a single X-ray photon through the gas produces an avalanche of secondary ions, all of which combine to form a pulse lasting on the order of 1 $\mu$sec. Hence the maximum rate countable is about $10^6$ $sec^{-1}$. In

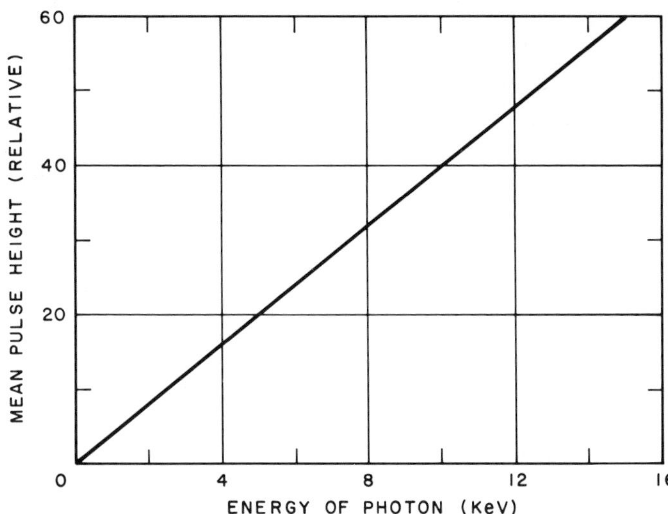


**Fig. 4.17.** Linear pulse-height relation in a proportional counter.

the Geiger region, even though the accelerating field is greater, the maximum rate is only about 500 $sec^{-1}$, the slowness being due to positive space-charge effects. Geiger counters require the addition of a quenching agent to the fill gas to prevent the emission of electrons from the cathode as a result of bombardment by positive ions. A small percentage of a halogen is effective. So also is an organic vapor, such as an alcohol, but this has limited life.

It is unlikely that any one ionization device can be optimized for all these possible operating regions. The most widely used for general detection and measurement is the proportional counter. The Geiger counter is convenient for use in portable survey meters, because the associated electronics can be much simpler.

## 2. Semiconductor Detectors

Semiconducting crystalline materials can respond to radiation by a number of mechanisms in addition to the surface emission discussed previously. The picture is complicated by the fact that more than one process may be operating concurrently or may be induced in a single device by suitable biasing.

### a. Photoconductive Cells

Perhaps the simplest of the semi-conductor photosensors is formed of a thin polycrystalline layer of cadmium sulfide or one of a few similar compounds, activated with a trace of impurity such as copper. The cell is usually made by sublimation of the desired compound in the presence of a low pressure of

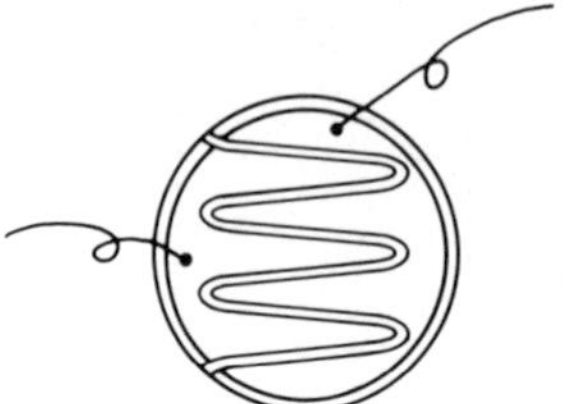

**Fig. 4.18.** Photoconductive cell. The sensitive material covers the sinuous gap between two metallic conductors.

oxygen. In one form, the compound is deposited on a metal plate or foil, and a transparent film of silver deposited over it. In another form the active material is deposited through a mask onto a ceramic substrate in such a way as to cover a serpentine gap between two metal electrodes (Fig. 4.18). The cell is usually protected with a glass window.

The conductance of the semiconductor is increased by the action of light. The mechanism involves the formation of electron-hole pairs, with promotion from valence to conduction band (cf. Fig. 4.5). The working relation for a photoconductive cell of this type is (18)

$$R = R_0\Phi^{m(\Phi)} \tag{4.26}$$

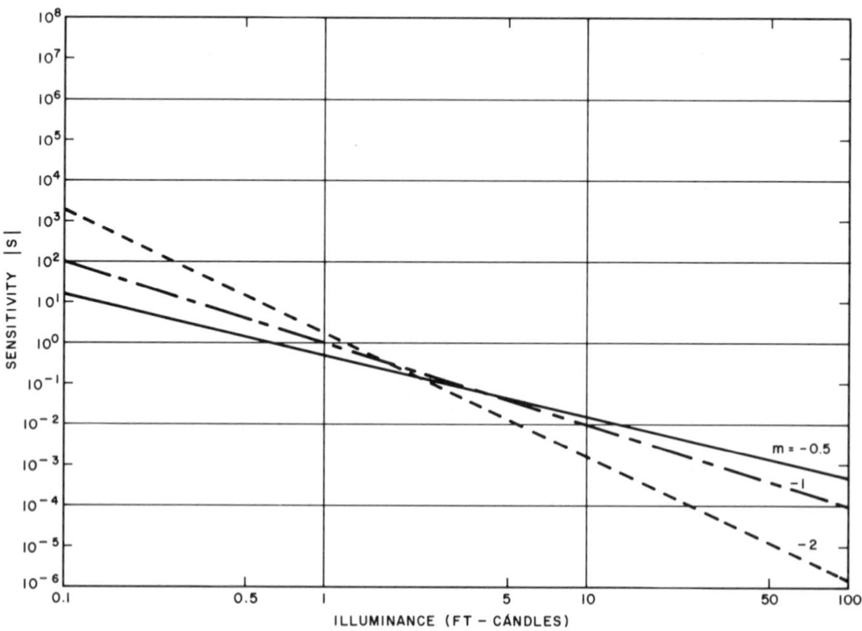


**Fig. 4.19.** The sensitivity of three typical photoconductive cells as a function of illuminance, with the value of $m$ as a parameter.

where $R$ is the observed resistance, $R_0$ is the dark resistance, and $\Phi$ is the illuminance (i.e., the radiant flux, expressed in lux or footcandles). The exponent $m$ varies from one photosensitive material to another; it is nearly constant for a given preparation of CdS, with a value between −0.4 and −3.0. For CdSe, $m$ departs from constancy as a function of $\Phi$.

Differentiation of equation 4.26 gives a measure of sensitivity, $S$:

$$|S| = \frac{1}{R_0} \cdot \left.\frac{\partial R}{\partial \Phi}\right|_m = m\Phi^{m-1} \tag{4.27}$$

This quantity, plotted against $\Phi$, gives the family of curves depicted in Fig. 4.19, which demonstrate that the sensitivity is greater for low illuminance than for high, and that small values of $m$ favor the response for more intense illuminance.

Figure 4.20 shows the spectral response curves for representative photoconductive cells (and for a silicon photodiode, to be dealt with in the next section). Some variation is to be expected for photoconductive cells prepared from the same semiconductor but with differing trace impurities and manufacturing procedures. The curve marked Cd(S,Se) is typical of devices with mixed crystals of CdS and CdSe; the wavelength of maximum response varies with the composition.

Photoconductive cells for the visible region are widely employed in relay and photometric applications. In a typical absorptiometer using a pair of CdS cells in a Wheatstone bridge configuration, it has been shown (18) that the output voltage, $E$, is related to the transmittance, $T$, of a solution by the formula

$$E = K \cdot \left(\frac{T-1}{T+1}\right) \tag{4.28}$$

where $K$ is a constant of the apparatus.

### b. Photodiodes

At a junction between $n$- and $p$-type semiconductors, the boundaries of the valence and conductance bands show an abrupt change in level (Fig. 4.21). The Fermi levels, $E_F$, must be equal if the system is in equilibrium, but if an external potential difference is impressed across the junction, the Fermi level becomes higher on the negative side than on the positive.

Consider first the situation in Fig. 4.21*a*. Incident radiation converts bonding electrons into free electron-hole pairs in the conduction band. If this occurs near the junction on the $p$ side, the electrons tend to "roll downhill" to the lower energy state on the $n$ side; similarly, holes formed on the $n$ side tend to move in the opposite direction. These processes continue until the potential builds up sufficiently to prevent further separation of charges. This corresponds to a dynamic equilibrium in which electrons and holes recombine at the same rate that they are formed. The rate of generation, and hence the illuminance, is measured by the electrical current flowing across the junction if the diode is

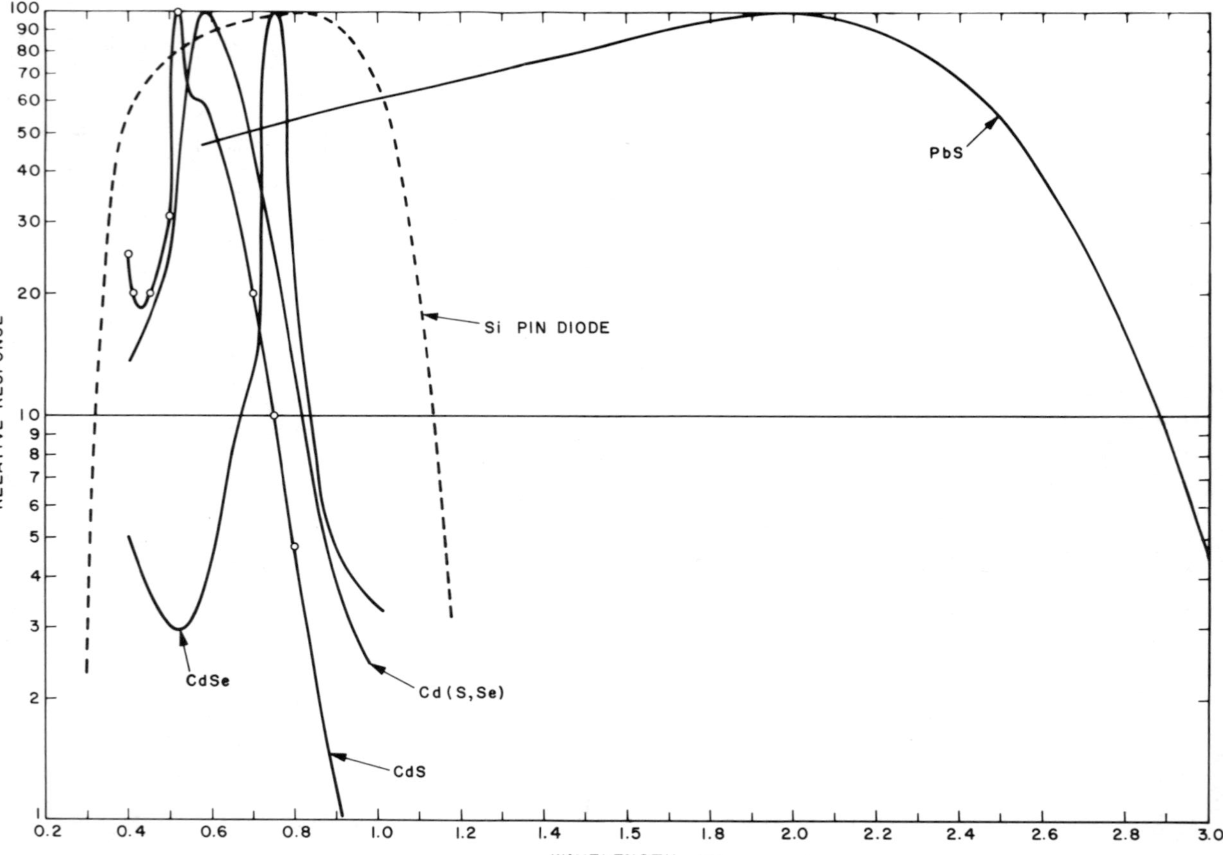


**Fig. 4.20.** The spectral sensitivity of various photoconductive materials, each normalized at 100 for maximum response.

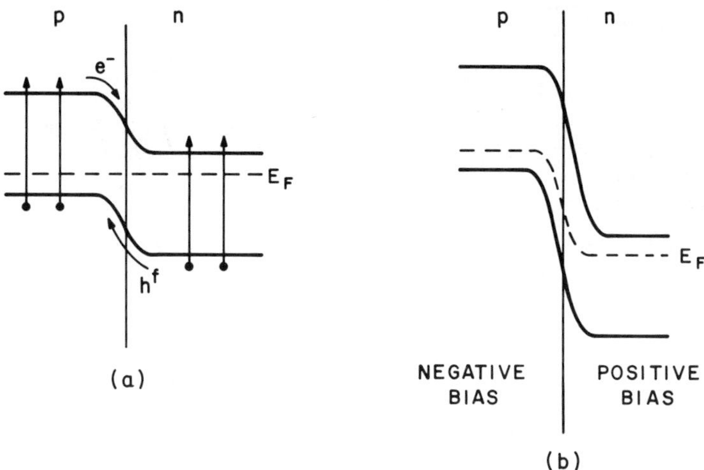


**Fig. 4.21.** Energy levels at a *pn* semiconductor junction. (*a*) Without bias; (*b*) with reverse bias. $E_F$ represents the Fermi levels.

externally short-circuited, or by the potential developed across it on open circuit. This is the *photovoltaic effect*.

If the diode is reverse biased (Fig. 5.21*b*), the applied field assists the process just described. Open-circuit measurement is of course no longer applicable. The current flowing is determined by the number of incident photons of sufficient energy, and hence is the same whether bias is applied or not. The speed of response is considerably greater in this mode, because of the increased field.

The characteristic curves of a typical photodiode, describing both modes of operation, are shown in Fig. 4.22. Note that the curve corresponding to zero illuminance is simply the familiar diode characteristic, showing a very small leakage current for negative voltage bias, passing through the origin, then increasing exponentially with positive (forward) bias. The family of curves for successive illuminance values can be measured usefully in three modes: (1) at zero current, that is, along the positive x axis, giving a response closely proportional to the logarithm of the illuminance; (2) under zero voltage, that is, along the negative current axis; or (3) at some fixed reverse bias voltage, measuring current under no-load conditions, as for example along the line marked $V_B$. The latter two modes give linear plots of current against illuminance. Sometimes a resistive load is employed and a corresponding load-line drawn on the graph ($L$); this markedly reduces the available dynamic range, and is not recommended practice.

The third mode, above, is subject to the inconvenience of a finite dark current that must be taken into account, but it permits obtaining a larger signal output than do the other modes. If connection to an operational amplifier is

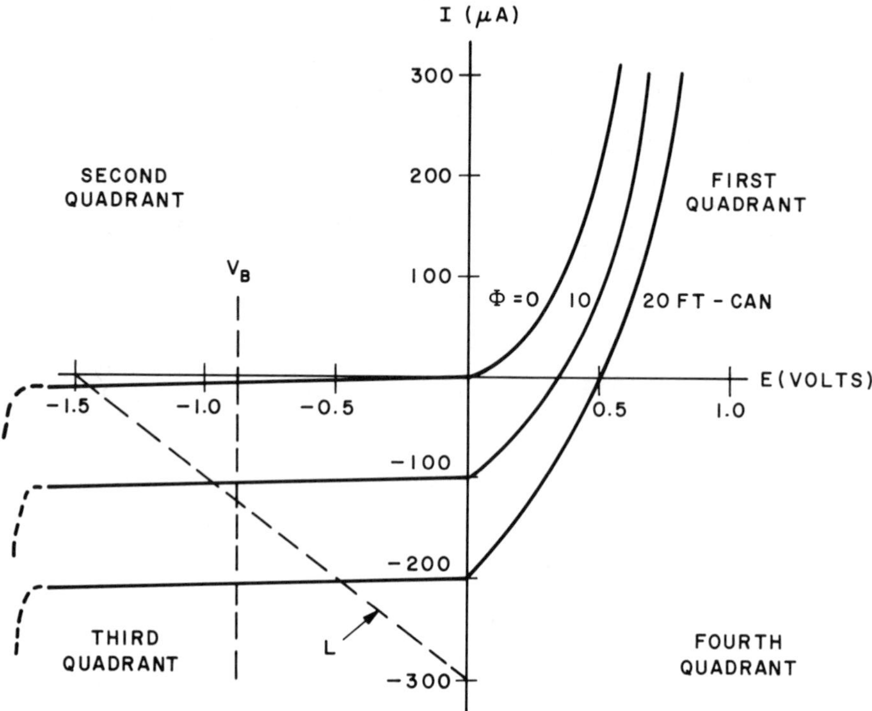


**Fig. 4.22.** Current-voltage curves for a photodiode at three levels of illumination.

contemplated, either of the first two modes should be utilized, according to preference for logarithmic or linear presentation.

Modern photodiodes are almost universally made of silicon. The relative spectral sensitivity is given in Fig. 4.20. Older cells of polycrystalline selenium or copper oxide, called *barrier-layer cells*, are always operated in the fourth quadrant (Fig. 4.22). They are less sensitive than silicon cells, but are still widely used in filter photometers and other applications because of their smaller cost.

c. LITHIUM-DRIFTED DETECTORS

The most successful semiconductor detectors for X- and gamma-ray measurements are the lithium-drifted silicon and germanium [Si(Li) and Ge(Li)*] devices. The requirement of such a detector is a *pn* junction, reverse-biased so that a depletion region exists between the *p* and *n* material. The radiation to be detected must be absorbed within the depleted region, where it produces hole and electron pairs. For penetrating radiation the depletion region must be

*"Silly" and "jelly" detectors, according to Liebhafsky.

relatively thick (1 or 2 cm), and the medium must contain atoms of high atomic weight, so that the absorption of the radiation is maximized. The latter requirement is better met with germanium than with silicon.

The incorporation of lithium results in a greater depletion depth by the following mechanism (see Fig. 4.23): a junction is produced near the surface of a block of *p*-type germanium. Metallic lithium is then diffused through the thin *n* region into the bulk of the material. Here the lithium atoms act as donors, pairing with the excess holes and leaving the region neutral, without excess holes or electrons. The neutral area is known variously by the names "depleted," "intrinsic," or "compensated." As soon as the diffusion process has progressed far enough, the device is cooled with liquid nitrogen, since the lithium has such a large diffusion coefficient that it would soon move out of the desired region if kept at room temperature. The device must be maintained at the reduced temperature for its entire life, whether in use or not. Si(Li) detectors are similar, but not as efficient in absorbing penetrating radiation.

These detectors are fast enough to count up to the limits commonly set by the associated electronics, $10^6$ sec$^{-1}$ or better. Their particular importance lies in the greatly increased energy resolution. Figure 4.24 shows the spectrum of $^{60}$Co gamma-rays plotted as number of counts against energy of photons, as determined with a NaI(Tl) scintillator and with a Ge(Li) detector. This excellent resolution has made possible the very successful energy-dispersive X- and gamma-ray spectrometers that now rival the classical wavelength-dispersive instruments.

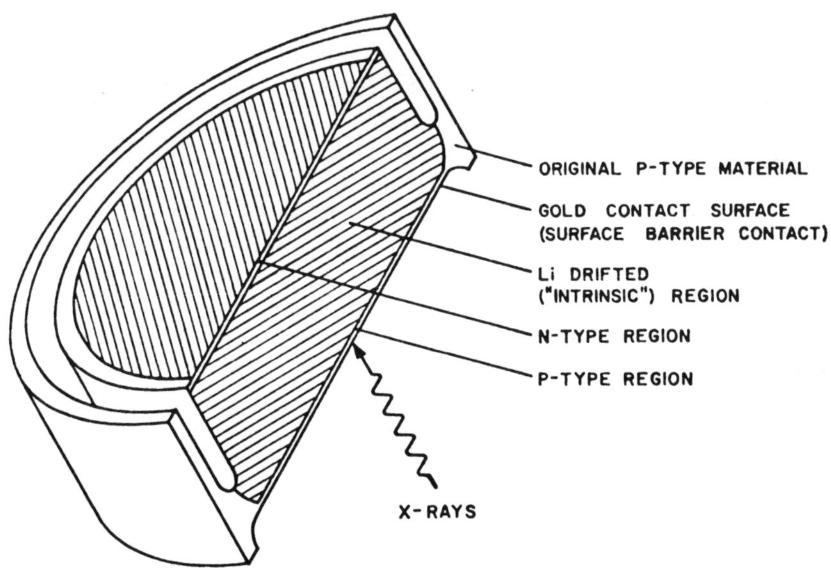


**Fig. 4.23.** Cutaway drawing of a lithium-drifted silicon detector. (Courtesy of Kevex Corporation.)

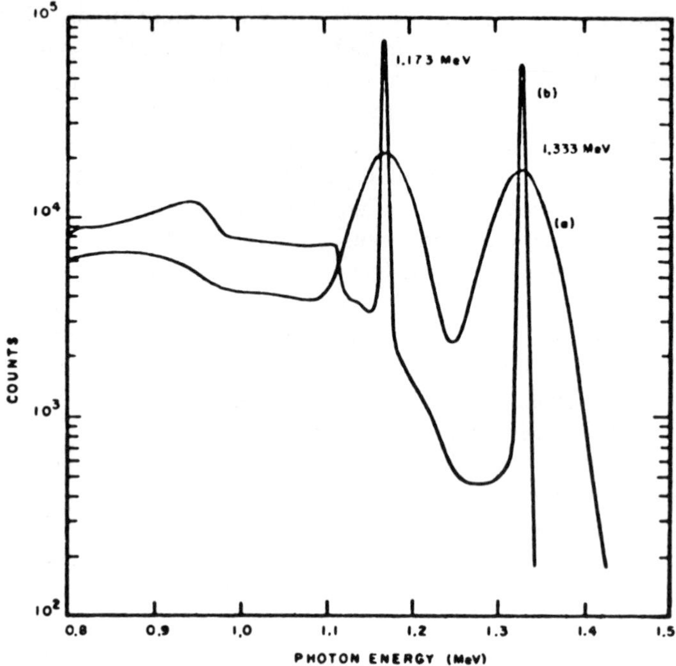


**Fig. 4.24.** Energy spectrum of $^{60}Co$ gamma radiation as determined with (*a*) a NaI(Tl) scintillator and (*b*) a Ge(Li) detector. (From reference 64 by permission of Plenum Publishing Corporation.)

Semiconductor detectors, including lithium-drifted and comparable types, as used with energetic radiations, have been discussed in detail by Goulding and Stone (25).

d. PHOTOTRANSISTORS

Since *pn* junctions are inherent in transistors, these also can be used as phototransducers. The optical properties of phototransistors, either bipolar or field effect, are those of the photodiode. The electrical characteristics, including gain, are determined by the basic transistor. In some types, the usual electrical connection to the control element of the transistor (base or gate) is used for proper biasing, whereas in others it is omitted. Phototransistors are often used in switching operations, including readers for punched cards or tape. They have not yet found wide application in chemical photometry, largely because of their rather limited dynamic range.

## E. SCINTILLATION DETECTORS

Certain materials have the property of emitting a tiny flash of fluorescent light in the visible or near ultraviolet upon the absorption of a photon or particle of energetic radiation. Such materials, called *scintillators*, can be either solid or liquid. A block of the solid, or a container of the liquid, in optical contact with a photomultiplier tube forms a highly useful detector. As Fig. 4.24 demonstrates, the scintillation detector can be used as a proportional counter; the luminosity of the flashes is proportional to the energy of the incident radiation.

A typical liquid scintillator consists of a solution of a polynuclear organic compound in toluene or similar solvent. Some useful compounds are anthracene, *p*-terphenyl, 2,5-diphenyloxazole (PPO), $\alpha$-naphthylphenyloxazole (NPO), and phenylbiphenyloxadiazole (PBD). Of these, PPO is the most efficient, but its radiation is in the ultraviolet, hence subject to absorption by toluene. This difficulty can be overcome by admixture of a second fluorescent substance that can translate the ultraviolet scintillations into the visible region. A suitable compound is 1,4-bis-2-(5-phenyloxazoyl)benzene (POPOP) or its dimethyl derivative. A recommended solution contains 5 g PPO and 0.3 g dimethyl-POPOP per liter of toluene.

Liquid scintillators are most valuable as detectors of low-energy beta radiations, such as those characteristic of the widely used tracer isotopes $^3$H (tritium),

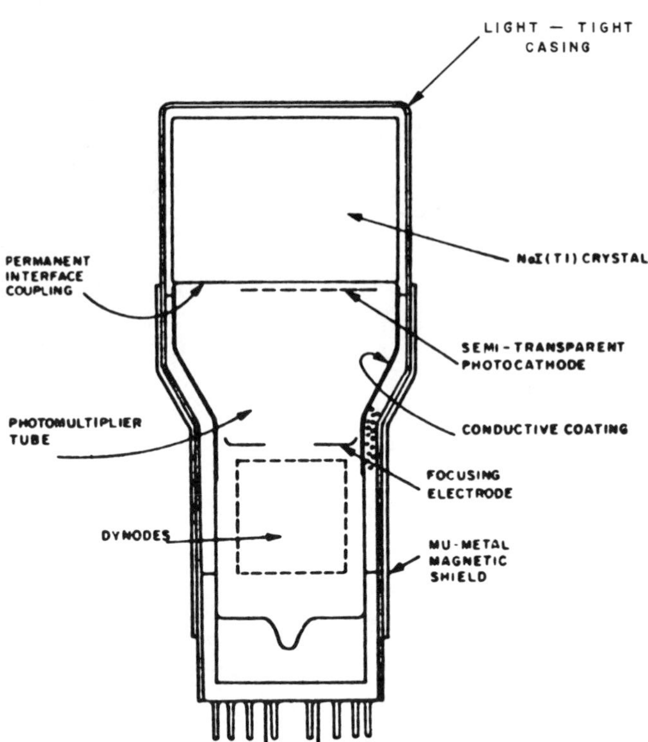


**Fig. 4.25.** Photomultiplier coupled to a NaI(Tl) crystal scintillator. (After reference 64 by permission of Plenum Publishing Corporation.)

TABLE 4.II

Characteristics of Representative Scintillation Phosphors (64)

| Material | Wavelength of maximum emission (nm) | Decay constant ($\mu$sec) | Relative pulse height |
|---|---|---|---|
| Inorganic crystals | | | |
| NaI(Tl) | 410 | 0.25 | 210 |
| CsI(Tl) | 420–570 | 1.1 | 55 |
| KI(Tl) | 410 | 1.0 | ~50 |
| LiI(Eu) | 440 | 1.4 | 74 |
| Organic crystals | | | |
| Anthracene | 440 | 0.032 | 100 |
| *trans*-Stilbene | 410 | 0.006 | 60 |
| Plastic phosphors | 350–450 | 0.003–0.005 | 23–48 |
| Liquid phosphors | 355–450 | 0.002–0.008 | 26–49 |

$^{14}C$, and $^{35}S$. Often a compound in which a tracer is incorporated can be added directly to the scintillator liquid, thus ensuring maximum efficiency in conversion of emitted particles to visible scintillations.

The principal solid material used as a scintillator is sodium iodide activated with a trace of thallium, NaI(Tl). Figure 4.25 shows the construction of a typical detector assembly with a block of NaI(Tl), perhaps 75 mm in diameter by 75 mm high, mounted in contact with the faceplate of a photomultiplier tube of equal diameter (64). The protective sheath surrounding the crystal is easily penetrated by X- or gamma-rays. Table 4.II gives some pertinent properties of a selection of scintillators (64).

A scintillation detector, including the photomultiplier, is compatible with the same power supply and signal-processing electronics as a gas-proportional counter, so that in many instrument systems, X-ray spectrometers, for example, they are directly interchangeable. The maximum counting rate of various scintillation detectors can be deduced from Table 4.II. It is of the order of $10^6$ sec$^{-1}$ for inorganic phosphors, and as great as $5 \times 10^8$ sec$^{-1}$ for organic liquids.

## F. INFRARED DETECTORS

It is customary to express the sensitivity of infrared detectors in terms of the *detectivity D*, defined by the relation (33)

$$D = \frac{S/N}{\Phi} \tag{5.29}$$

where $S$ is the signal developed as a result of incident radiant flux $\Phi$, and $N$ is the spurious signal produced by noise in the detector. $S$ and $N$ must be expressed in the same units, commonly either volts or amperes. The radiant flux is often given in watts, but may be in photons per second. The units of $D$ are the reciprocal of those employed for $\Phi$.

The value of $D$ in many cases varies inversely as the square root of the product of the area $A$ of the receptor and the frequency bandwidth, $\Delta f$. Therefore a quantity denoted by $D^*$ ("D-star"), the *specific detectivity*, is widely employed as a figure of merit for infrared detectors. If the area is measured in square centimeters, the bandwidth in hertz, and the flux in watts, then units of $D^*$ are in $\text{cm} \cdot \text{Hz}^{1/2}/\text{W}$. The bandwidth enters because of its limitation on random noise.† $D^*$ is not strictly applicable to detectors that have a different area dependence or in which the predominant noise follows a different law.

Infrared transducers belong to two major classifications; thermal and quantum (photon) detectors. In general, quantum detectors are applicable only over restricted frequency or wavelength ranges, whereas thermal detectors are limited in wavelength primarily by external features, such as the transparency of window materials. Special considerations pertinent to far-infrared detectors have been fully described by Kimmit (35).

## 1. Thermal Detectors

The devices in this category depend upon the heating effect of infrared. The radiation falls upon a target, often made of a small piece of gold foil coated with "gold black," a finely divided form of the element prepared by sublimation or electrodeposition. The target size is typically about $1 \times 0.5 \times 0.0002$ mm. It is mounted in thermal contact with the temperature sensor, and the entire assembly enclosed in a vacuum chamber behind a window of infrared-transmitting material.

The speed of response, which determines the maximum rate at which radiation can be chopped, is controlled by the time constant $R_T C_T$, where $R_T$ is the thermal resistance (reciprocal of thermal conductance) and $C_T$ is the heat capacity of the target with its associated structures. ($R_T$ and $C_T$ play analogous roles to resistance and capacitance in the time constant of an electrical circuit.) It is desirable to keep the heat capacity low, which calls for the smallest possible mass to be heated by the radiation. The thermal resistance must effect a compromise between the low value desired to minimize the time constant and the higher value required for thermal insulation from the surroundings, so that the heat is not dissipated before it has a chance to develop a maximum temperature. Time constants of 25–30 msec are typical, so that chopping of radiation is limited to 15 or 20 Hz (33).

### a. Thermocouples

Thermocouples for infrared detection are usually made of bismuth and silver, or of nickel and gold. The preferred construction utilizes two identical

†The specific detectivity is often quoted in the form $D^*(\lambda, f, \Delta f)$. For example, $D^* = 2 \times 10^9$ (2, 1000, 1) designates a value of $D^*$ measured at a wavelength of 2 $\mu$m, a chopping frequency of 1000 Hz, and an electronic bandwidth of 1 Hz. Instead of the wavelength, the kelvin temperature of a black-body source is sometimes given as the first parameter; alternatively the word "peak" is inserted, meaning that the measurement was made at the wavelength of greatest sensitivity.

targets placed close to each other in the same ambient environment, one exposed to the radiation, the other shielded from it. The two thermocouple junctions are cemented or welded to the targets. One common physical form is sketched in Fig. 4.26*a*. The thermocouple wires are as thin as possible to minimize heat conduction, but nevertheless usually serve as supports for the tiny gold targets. The potential produced, being proportional to the *difference* between the temperatures of the two targets, does not reflect changes in the ambient. In some detectors, particularly those designed as reference standards, several junctions are wired in series, with alternate junctions in contact with the two targets. This unit, called a *thermopile*, gives, with $n$ pairs of junctions, a voltage $n$ times that of a single thermocouple.

In another form of thermocouple detector, thin films of the two selected elements are deposited on a ceramic wafer, one over the other, through masks (Fig. 4.26*b*). The area where the two conductors overlap forms the target. This type is well fitted for the detection of radiation from a laser, which might be powerful enough to destroy the delicate structure described in the preceding paragraph.

b. Bolometers

A bolometer is a form of resistance thermometer adaptable to infrared detection. It can consist of a thin wire or film of nickel or other metal, usually

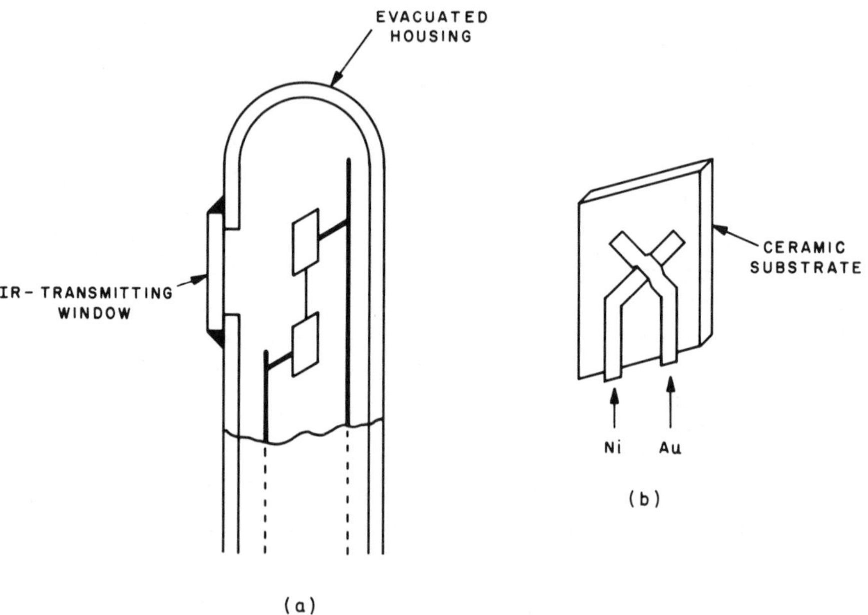


**Fig. 4.26.** Thermocouple infrared detectors: (*a*) conventional; (*b*) vacuum-deposited on a ceramic chip.

combining the functions of temperature sensor and radiation target, or it can be a thermistor affixed to a gold target.

Room-temperature bolometers have little advantage over thermocouples, and are seldom used in contemporary spectrophotometry. There is, however, an advantage in cooling a bolometer to cryogenic temperatures. The advantage comes from two sources: first is the drastic reduction in noise, and second is the reduction nearly to zero of the heat capacity of the detector, making possible the use of more massive components (63).

### c. Pneumatic Detectors

If the target is immersed in a small isolated volume of a gas, a rise in pressure of the gas results from increased radiation on the target. This principle has been incorporated in the *Golay detector* (24)(Fig. 4.27). Pressure changes in the gas (usually xenon) cause corresponding changes in the radius of curvature of a small flexible mirrored diaphragm, and this in turn alters the sharpness of focus of a secondary optical system. In the secondary system, visible radiation passes through a grid consisting of parallel bars and spaces of equal width. An image of the grid is formed on a portion of the same grid after reflection from the flexible mirror in its null position, so that the image of each open region falls on an opening, and light is transmitted to a photocell. Any defocusing or shift in the position of the image reduces the transmitted power. A fine gas leak is provided to equalize slowly the pressure difference across the diaphragm; hence the detector responds only to fluctuating (chopped) radiation. Recently designed Golay detectors utilize a light-emitting diode (LED) and phototransistor in the optical system, replacing the former incandescent lamp and vacuum phototube. The Golay detector gives flat response over the entire visible and infrared spectrum, subject only to limitation in the transmission of the window material. Its frequency (chopping) range is typically 0.1–30 Hz.

Pneumatic detectors other than the Golay type are used in certain nondispersive infrared analyzers. Here, too, an oscillating pressure caused by the absorption of a modulated beam of radiation produces corresponding motion of a

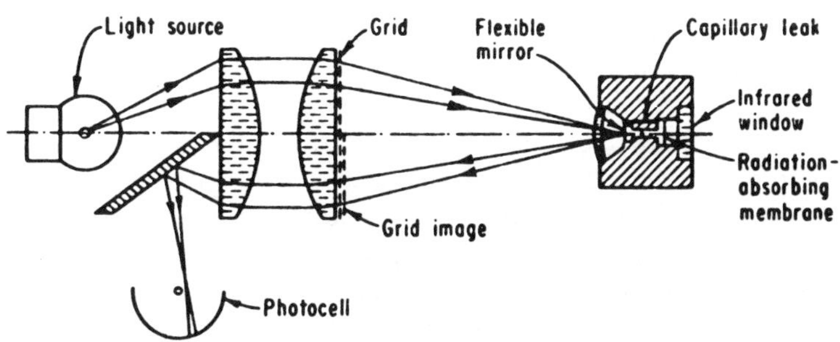


**Fig. 4.27.** Golay pneumatic detector. (From reference 33 by permission of McGraw-Hill Book Company.

diaphragm. The metallized diaphragm is made to be one plate of a capacitor, the other plate being fixed in position. An electrical current receives the same modulation by action of the vibrating capacitor.

These devices can be considered to be two-step transducers, converting information-carrying energy from a radiant form to pressure variations, and then to an electrical signal.

#### d. Pyroelectric Detectors

A number of chemical compounds form nonconducting crystals that are electrically polarized. Normally this polarization is not observable externally because stray charges picked up on the crystal surfaces neutralize the internal field. However, the resulting charge distribution cannot readjust itself quickly, so that any rapid changes in the state of polarization are reflected in easily detected potential variations, even though slow changes do not result in significant signals. Such a rapid change in polarization can be caused by a sudden temperature change, the basis for its use in infrared detection. The effect is evaluated in terms of the *pyroelectric coefficient* $p$ (units $C/cm^2\,K$) which is of the order of $10^{-8}$ to $10^{-9}$ for typical materials. The most thoroughly studied pyroelectric crystal is triglycine sulfate (TGS); others include several of the titanates, niobates, and tantalates of lithium, barium, and strontium.

A TGS detector can be fabricated from a cleaved cube of the crystal provided with metal contacts on the two cleavage faces (60). The radiation to be measured can be received either through one of the clear faces or through one of the semitransparent electrodes. This high-impedance transducer is best mounted in close proximity to a field-effect transistor to act as an impedance transformer.

It can be shown on theoretical grounds (51) that the pyroelectric detector should be significantly superior to other thermal infrared detectors with respect to detectivity, speed of response, and spectral sensitivity. Since this prediction appears to be borne out in practice, it seems likely that this device will displace all competitors both for spectrometric and laser applications throughout the entire infrared region.

### 2. Quantum Infrared Detectors

Detectors in this classification include both photovoltaic and photoconductive devices. For service in the infrared, most of these require cooling to the temperature of liquid nitrogen or helium in order to depopulate the higher vibrational levels of the detecting material. Lead sulfide and lead selenide, as photoconductive detectors in the near infrared, are exceptions, because they give adequate response for routine spectrophotometry at room temperature.

Intrinsic semiconductors can be prepared with energy gaps $E_g$ between valence and conduction bands anywhere from nearly zero to about 1.6 eV (41). These materials respond to radiation up to a long-wave limit given by

$$\lambda_m = \frac{hc}{E_g} = \frac{1.24}{E_g} \qquad (4.30)$$

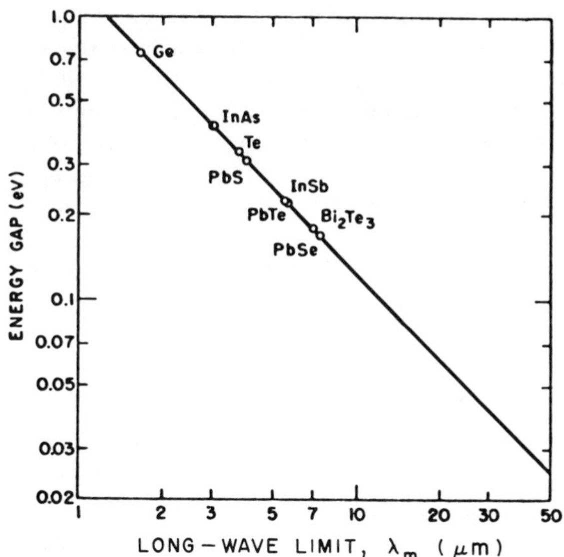


**Fig. 4.28.** The long-wave infrared limit as a function of energy gap for a number of semiconductors. (From reference 41 by permission of Academic Press.)

TABLE 4.III

Typical Infrared Detectors[a]

| Material | Type[b] | Operating temperature (K) | Wavelength range ($\mu$m) | $D^*$ (cm Hz$^{1/2}$/W) |
|---|---|---|---|---|
| Thermocouple | Th | 300 | 0.1–100 | $3 \times 10^8$ |
| Thermistor | Th | 300 | 0.1–100 | $4 \times 10^8$ |
| Golay | Th | 300 | 0.4–1000 | — |
| TGS | PE | 300 | 0.1–1000 | $7 \times 10^8$ |
| HgCdTe | PV | 77 | 8–14 | $2 \times 10^9$ |
| PbSnTe | PV | 77 | 8–14 | $3 \times 10^9$ |
| InSb | PV | 77 | 1–5.5 | $9 \times 10^{10}$ |
| InSb | PC | 77 | 2–6 | $2 \times 10^{10}$ |
| InSb | PC | 4 | 300–8000 | $2 \times 10^{10}$ |
| Si | PV | 300 | 0.4–1.1 | $5 \times 10^{13}$ |
| Ge | PV | 300 | 0.8–1.7 | $3 \times 10^{10}$ |
| Ge(Cu) | PC | 4 | 5–30 | $4 \times 10^{10}$ |
| GaAs | PC | 4 | 100–350 | — |
| PbS | PC | 300 | 0.8–3.5 | $2 \times 10^{10}$ |
| PbSe | PC | 300 | 1–5 | $1 \times 10^9$ |

[a] Compiled from manufacturers' literature.

[b] PC = photoconductive, PE = pyroelectric, PV = photovoltaic, Th = thermal.

in which the numerical constant applies if $\lambda_m$ is taken in micrometers and $E_g$ in electron-volts. The detectivity falls off linearly with wavelength below the cutoff, so it is advantageous to utilize a detector as near cutoff as possible. Figure 4.28 shows a plot of equation 4.30 for a number of semiconductor materials (41). The mixed tellurides of mercury and cadmium and of tin and lead, with the general formulas $Hg_xCd_{1-x}Te$ and $Pb_xSn_{1-x}Te$, have been studied extensively. It is possible to prepare materials with any cutoff wavelengths over rather wide ranges by varying $x$ (41).

Table 4.III gives comparative data for representative infrared detectors of the chief types. The data are not necessarily definitive, because not all were determined under the same conditions.

### G. ELECTRICAL COLLECTION OF CHARGED PARTICLES

An alternative method for the detection and measurement of beams of ions or electrons is by their direct collection on an insulated receptor connected to an electrometer amplifier. The receptor is often cup-shaped, and is known as a Faraday cup. The shape is designed to minimize loss of charge by secondary emission. A more elaborate version greatly increases sensitivity by making use of the secondary emission, as discussed in Section III.C.4. Both types are widely used in mass spectrometry and elsewhere.

## IV. DETECTORS OF NEUTRAL SPECIES

### A. GAS DETECTORS

Many transducers in this category are familiar as detectors in gas chromatography. The requirement, basically, is sensitivity to some property of individual molecules that is markedly different in magnitude for species likely to be encountered as samples, compared with a carrier gas. In general these are compound transducers, in the sense that the ultimate conversion of information from the chemical system to an electrical signal takes place at a device that can be treated as a separate transducer in its own right. For example, the thermal conductivity detector depends for its action on a differential measurement of temperature by observation of the corresponding change in electrical resistance of a wire filament or a thermistor (Section II.B.2). The feature that distinguishes this detector from other thermometric systems is the means for producing temperature differences that correlate with chemical composition of a gas stream.

In this section various types of detectors are described briefly, but for considerations of their applicability to chemical systems, the reader is referred to the sections of the Treatise dealing with chromatography.

#### 1. Thermal Conductivity Transducers

The thermal conductivity of helium at a specified temperature is considerably greater than that of all gases for which this property has been reported,

TABLE 4.IV

Thermal Conductivities of Some Gases[a] (Cal · $sec^{-1}$ · $cm^{-1}$ · $K^{-1} \times 10^6$) at 26.7°C

| | |
|---|---|
| Hydrogen | 446.32 |
| Helium | 360.36 |
| Deuterium | 334.74 |
| Neon | 115.71 |
| Methane | 81.83 |
| Oxygen | 63.64 |
| Nitrogen | 62.40 |
| Carbon dioxide | 39.67 |
| Ethanol | 34.71 |

[a] *Handbook of Chemistry and Physics*, 60th ed., CRC Press, Boca Raton, Fla., 1980, p. E-2.

excepting only hydrogen and deuterium (see Table 4.IV). The usual detector has four resistive elements exposed to gases in pairs, two to the reference (carrier) gas, two to the gas containing possible samples. Figure 4.29 shows two such elements in a typical assembly. The configuration shown for the right-hand element depends on diffusion from the gas stream and hence has a longer time constant than the flow-through geometry of the left-hand element. This arrangement gives an optimum trade-off between the high sensitivity with concomitant dependence on flow rate of the flow-through unit and the greater stability associated with the longer time constant of the diffusion unit. In the complete four-element array, two diffusion cells for carrier gas are connected electrically as opposite arms of a Wheatstone bridge, with the flow-through effluent cells in the remaining two arms. The presence of sample components with less thermal

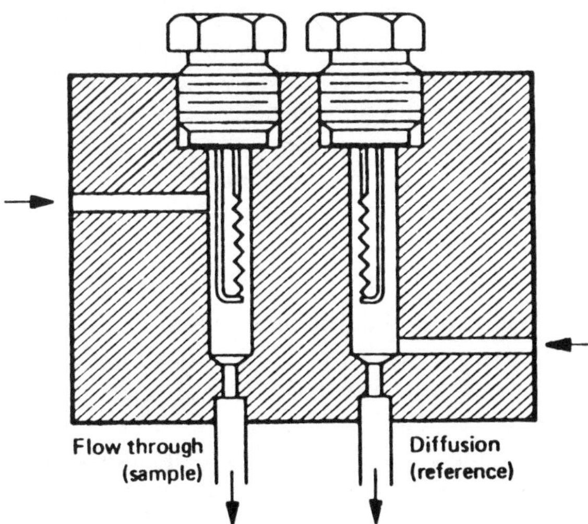


**Fig. 4.29.** Differential thermal conductance gas detector. (Courtesy of Gow-Mac Instrument Company.)

conductivity than the carrier results in reduced cooling of the corresponding resistive elements and hence an imbalance in the bridge circuit.

The theory of the thermal conductivity detector is more involved than one might expect for a device so simple conceptually, and is too lengthy to include in this chapter. There are several excellent treatments in the literature (12,34,37).

### 2. Electron-Capture Transducers

Gas detectors of this class can be considered as ionization chambers similar to those described previously (Section III.D.1), except that rather than being provided with an unchanging gas in order to observe incoming radiation, they are excited by a fixed radiation flux to permit studies of the nature of the entering gas. The excitation employed is usually the beta radiation from $^{3}H$ (250 mCi) or $^{63}Ni$ (10 mCi). Tritium is immobilized as a superficial compound on a titanium foil.

The carrier gas ($N_2$ or Ar) is ionized by beta bombardment, and the ions collected by a field that is weak enough not to cause secondary ionization by collision. Hence in the absence of a sample, a constant current of the order of $10^{-8}$ A flows through the detector. The presence of sample constituents with a high affinity for electrons (large capture cross-section) reduces the current. Electron capture cells can be operated in pairs, one with carrier gas only, the other with column effluent, allowing the standing current to be compensated. Details of theory and practical design can be found in the literature (43,44,46).

### 3. Flame-Ionization Transducers

This detector depends on the formation of ions in a hydrogen-air flame when sample components (primarily organic) are present. A typical unit has a burner orifice of 0.5 mm, a flow of column effluent plus added hydrogen of 25 $cm^3$/min, and air at 500 $cm^3$/min. A ring or cylindrical collector electrode is located 0.5–1 cm above the orifice, and is polarized at +400 V, the orifice being at ground potential. There appear to be differences of opinion about the mechanism of formation of ions in the flame; the problem has been discussed by Blades (5). Hydrocarbons give a response proportional to the number of carbon atoms over a considerable range. For a general review, see reference 23.

### 4. Flame-Photometric Transducers

This detector is similar to the flame-ionization type in its use of a hydrogen-air flame, though better results are obtained with a fuel-rich flame, whereas the ionization detector uses the reverse ratio. This detector is used primarily for the detection of organic compounds containing sulfur or phosphorus (8), and has been modified successfully for the detection of organometallic compounds (1).

Whereas the hydrogen-air flame is nearly nonluminous, characteristic radiation is produced in the region immediately *above* the flame. For sulfur, a band at 394 nm is observed, for phosphorus at 526 nm. Detection is by means of a photomultiplier with suitable filters.

## 5. Gas Density Transducers

Superficially this type of detector resembles the thermal conductivity cell, differing only in the configuration of the channels for gas flow. The principle of operation, however, is quite distinct. In its modern form (50) (Fig. 4.30), this device consists of three parallel tubes, oriented vertically, connected at top and bottom. The reference (carrier) gas, which can be nitrogen, enters at the center of the first tube (A), the column effluent at the center of the second (C), and all gases exit at the center of the third tube (E). The hot-wire or thermistor elements are located in the horizontal connecting tubes between the first and second vertical members (at B and B′), and in this application act as anemometers. The action of the detector is based on the relative effect of the earth's gravitational field on components of the effluent that are heavier or lighter than the carrier gas. Those that are more dense tend to take the lower path to the exit, whereas the less dense take the upper route. In the first case, increased gas flow in C-D′-E relative to that in C-D-E results in reduced flow of carrier past the lower detector (B′) and increased flow at B. Hence the element at B is cooled more effectively by the flowing gas than is that at B′, and the resistance changes accordingly. It is to be noted that the sample components never come in contact with the resistive sensors.

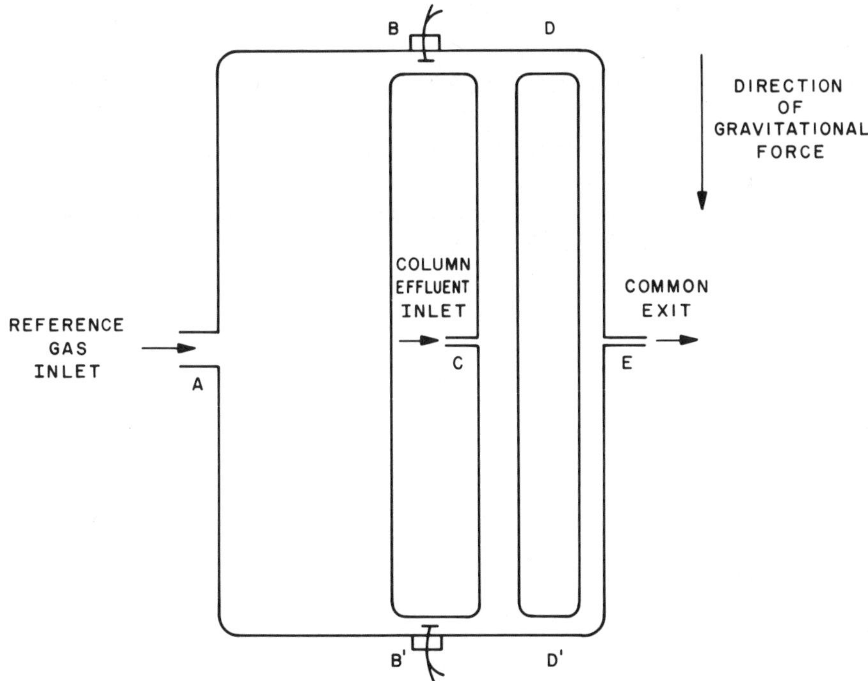


**Fig. 4.30.** Gas density balance. The widths of the various gas channels are scaled according to reference 50.

The response of this transducer depends solely on the relative densities of the carrier and effluent gases, and because the molecular weight of a near-perfect gas is proportional to its density, the device can provide an unequivocal measure of the molecular weight. Vermont and Guillemin (66) have pointed out that because of its dependence on determinable factors, the gas density transducer is well suited for calibration of other gas detectors.

### B. NEUTRON DETECTORS

Since neutrons are uncharged, they cannot interact with matter via coulombic forces, hence cannot cause ionization or pair formation as do energetic charged particles. They can, however, interact with atomic nuclei by several distinct mechanisms, hence these can be utilized in their detection (21).

Fast neutrons can be detected by the recoil protons ejected when such neutrons collide with hydrogen atoms in paraffin or other hydrogenous material. The ratio of the number of protons ejected to the number of incident neutrons is roughly proportional to the kinetic energy of the neutrons. For 1 MeV neutrons, the ratio is about 7 in $10^4$.

Thermal neutrons are readily detected by the $^{10}B(n,\alpha)^7Li$ reaction. An ionization chamber or proportional gas counter lined with boron or filled with $BF_3$ gas is convenient. The nuclide $^{10}B$ is about 19.6% abundant in the natural element; enrichment of course increases detection sensitivity. Fission and neutron-capture activation processes can also be used in neutron detection.

## V. TRANSDUCERS FOR THE MEASUREMENT OF PRESSURE

Confusion often results from the many units for expressing pressure. The accepted unit in the SI system is the pascal (Pa), equal to 1 newton per square meter; its general use is strongly recommended. A list of some pertinent conversion factors is given in Table 4.V.

TABLE 4.V

Pressure Conversion Factors

| To convert from pascals to designated unit, multiply by: | Unit | To convert from designated unit to pascals, multiply by: |
|---|---|---|
| $9.86923 \times 10^{-6}$ | atm | $1.01325 \times 10^{5}$ |
| $1.00000 \times 10^{-5}$ | bar | $1.00000 \times 10^{5}$ |
| $1.45037 \times 10^{-4}$ | $lb/in^2$(psi) | $6.89476 \times 10^{3}$ |
| $3.34562 \times 10^{-4}$ | ft $H_2O$ (4°C) | $2.98898 \times 10^{3}$ |
| $4.01474 \times 10^{-5}$ | in. $H_2O$ (4°C) | $2.49082 \times 10^{2}$ |
| $7.50062 \times 10^{-3}$ | mm Hg (0°C) (torr) | $1.33322 \times 10^{2}$ |
| $1.00000 \times 10^{1}$ | $dyne/cm^2$ | $1.00000 \times 10^{-1}$ |
| $7.50062 \times 10^{2}$ | $\mu$m Hg (0°C) (microns) | $1.33322 \times 10^{-3}$ |

Many types of transducers, usually called *pressure gauges*, are required in order to permit measurements of both static and dynamic pressures over the vast range from near-perfect vacuum to many thousands of atmospheres. For pressures greater than a few tenths of an atmosphere, the forces involved are generally great enough that the direct mechanical effects of pressure can be used in its measurement. Below this, pressure measurement must depend on molecular properties of a gas in contact with the transducer.

## A. FORCE TRANSDUCERS

Pressure gauges of this classification include a variety of devices in which a displacement of a solid or liquid component is directly produced by the pressure (or differential pressure) to be measured.

### 1. Liquid Manometers

In this category fall the various types of manometers employing mercury, water, or oil in a transparent tube. Readout is most commonly a visual measurement of the difference in height of two columns by reference to a millimeter scale. Electrical readings can be made, as with liquid-in-glass thermometers, by a capacitive or optical device to sense the position of the meniscus.

The *McLeod gauge* is a manually operated instrument with which a measured volume of low-pressure gas is isolated and compressed until its pressure is great enough to be read on a mercury manometer. It is inherently incapable of continuous operation, and hence is not commonly automated or given electrical readout. It is useful as providing a means of absolute calibration of other gauges, and is also used for measurements on closed systems in which the pressure is expected to be nearly constant once a desired vacuum is attained.

### 2. Mechanical Pressure Indicators

Included here are devices based on the flexure of a diaphragm or aneroid capsule in response to a pressure differential (30), in which the motion is transferred to a pointer or recorder by mechanical means. An example is the familiar *Bourdon gauge* (Fig. 4.31). In this device the pressure to be measured is transmitted to the interior of a flattened metallic tube that has been curved to a circular or spiral shape. An increase in pressure tends to straighten the tube, whereas a decrease accentuates the curvature. The resulting motion of the tip, through a system of levers and gears, turns the indicating pointer.

Figure 4.31*b* shows a multiple-turn Bourdon tube made of vitreous silica (chosen for is desirable elastic qualities) and provided with an optical lever for highest sensitivity measurements. The manufacturer specifies precision as close as 10 ppm; full-scale spans range from 35 kPa to 3.5 MPa (5 to 500 psi). Related gauges utilize various types of metal bellows or capsules to perform the same function as the Bourdon tube (30).

Gauges of these and related kinds are usually connected directly to the fluid under test. If this is impracticable because of possible corrosion or fouling, an

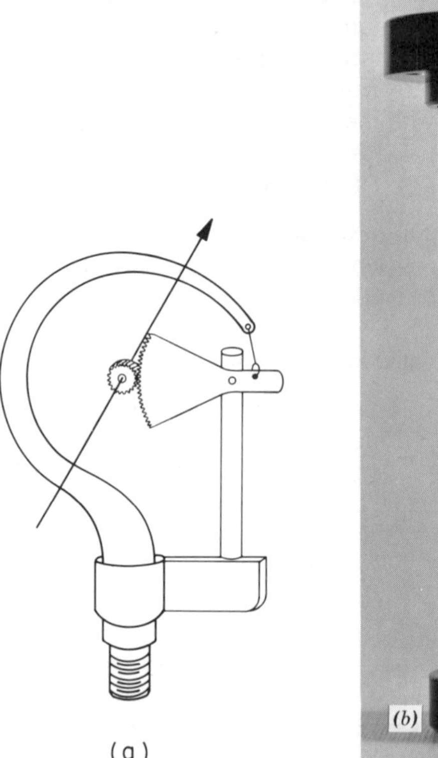


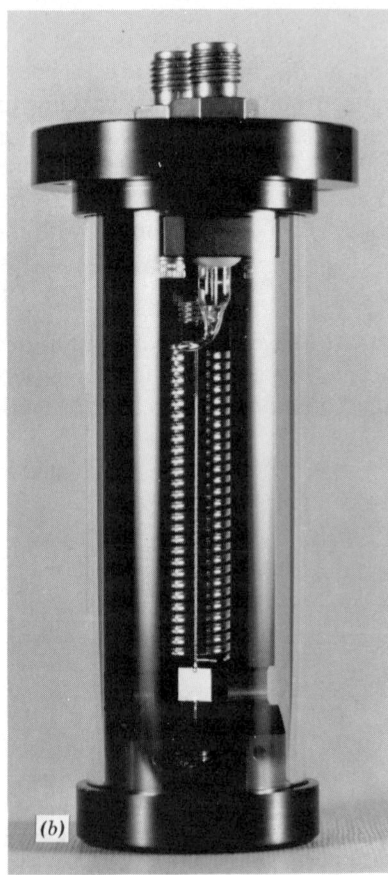


**Fig. 4.31.** Bourdon-tube pressure gauges. (*a*) Rough type; (*b*) highly sensitive vitreous silica gauge. [(*b*) Courtesy of Texas Instruments, Inc.]

intermediate fluid can be employed. This is commonly water or oil, filling the sensor but isolated from the system by a slack diaphragm so that pressure is transmitted without contact of the two fluids.

### 3. Diaphragm Gauges with Electrical Readout (3)

The motion generated in a Bourdon or diaphragm gauge can be made to control a variable resistor, capacitor or inductor, thereby producing an electrical signal functionally related to the magnitude of the pressure. A variable resistor is rather unsatisfactory; the wiper, actuated by the diaphragm, must press firmly against the resistive windings in order to give reliable readings, but this increases the friction, so that the system is suitable only for measuring rather large pressure changes, and even then with only poor precision.

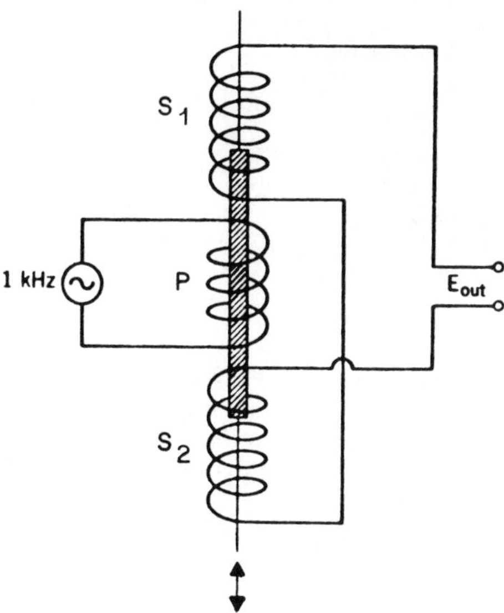


Fig. 4.32. Linear differential transformer.

### a. Inductive Readout

A linear differential transformer (LDT) provides a friction-less method of observing the motion of a diaphragm (Fig. 4.32). The LDT is provided with three windings: a primary, $P$, and two secondaries, $S$. The latter are connected in phase opposition, and mounted symmetrically on either side of the primary. An iron slug rigidly attached to the diaphragm moves axially through the coils. In the quiescent condition the slug contributes equally to the inductive coupling of $S_1$ and $S_2$ to the primary, but with departure from the central position, it causes proportional inequality in such coupling. The primary is energized with a constant alternating voltage, so that the output is null at zero displacement, where the ac currents induced in $S_1$ and $S_2$ exactly cancel each other. A suitable phase-sensitive ac detection circuit develops a signal with magnitude proportional to the difference in pressure on opposite sides of the diaphragm, with sign indicating which side has the higher pressure.

Another type of electromagnetic transducer takes the form of a conventional microphone. A small coil of wire rigidly attached to the moving diaphragm in the field of a permanent magnet produces a voltage proportional to the time derivative of the pressure.

Still another variety, known as a variable reluctance transducer, depends on the variation of the magnetic flux in an iron core under constant excitation, as a segment of the core is moved within an air gap (Fig. 4.33).

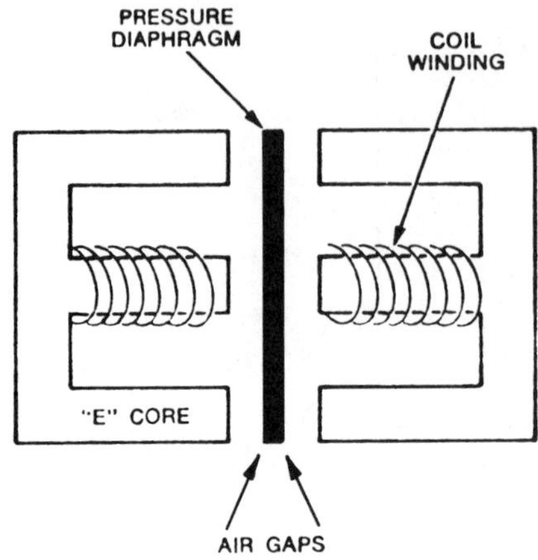


**Fig. 4.33.** Variable reluctance pressure gauge. The steel diaphragm flexes with a pressure differential, altering the magnetic paths. (From reference 3 by permission of Bell & Howell.)

### b. Capacitive Readout

In this type of electrical transducer the metallic diaphragm constitutes one plate of a capacitor or one of a differential pair of capacitors, as in Fig. 4.34. There are a number of alternative circuit arrangements possible with a capacitive transducer. In one method, an alternating voltage in the audio-frequency range is impressed across the two plates of Fig. 4.34*b*, and a differential amplifier measures the difference in the voltages between the two plates and the grounded diaphragm; the signal is proportional to the linear displacement of the diaphragm, but somewhat nonlinear with respect to the pressure. Loriot and Moran (42) have described a careful calibration check of a commercial gauge of this type in the pressure range of $2.6 \times 10^{-2}$ to $6.5 \times 10^{-4}$ Pa ($2 \times 10^{-4}$ to $5 \times 10^{-6}$torr). They estimated the overall accuracy to be within about 2%.

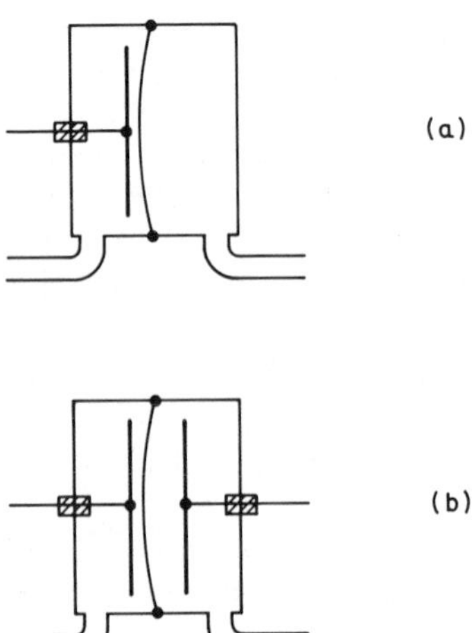


**Fig. 4.34.** Capacitance gauges. (*a*) Single-ended; (*b*) differential.

In a second measuring technique, the capacitor (Fig. 4.34*a*) is made one element of an *LC* tuned circuit controlling the frequency of an electronic oscillator. For static pressures, the frequency is given by $f = 1/(2\pi\sqrt{LC})$ where $L$ is the inductance of the circuit (external to the pressure gauge) and $C$ is the capacitance of the gauge. If the pressure is itself alternating, then the frequency of the oscillator varies around a central quiescent value, and the modulation carries the desired information. The latter arrangement is more nearly free of the effects of slow oscillator drift, difficult to overcome in static measurements. This method is competitive in sensitivity in spite of the square-root capacitance dependence, because of the great precision possible in the measurement of frequency by counting techniques.

### c. Strain-Gauge Readout

A strain gauge consists of an array of fine wires or metal foil either mounted under tension between supports or bonded to a solid surface, in such a manner that a slight dimensional change (within the elastic limit) results from any motion or flexure of the support. Such an alteration of dimensions produces a reversible change in electrical resistance. See reference 39 for details.

Both bonded and unbonded strain gauges are used in diaphragm pressure gauges. For greatest sensitivity and convenience, four identical resistive units are so mounted that two of them sense the motion of the diaphragm whereas the other two do not, all four sharing the same ambient temperature. A Wheatstone bridge circuit then gives the maximum response with complete temperature compensation.

### d. Magnetoresistive Transducers

The imposition of a magnetic field on a semiconductor alters its resistance in a reversible manner (28). This effect has been used in a pressure transducer. A silicon chip is attached mechanically to a diaphragm, and so positioned relative to a permanent magnet that motion of the diaphragm produces a corresponding change in the field sensed and hence in the resistance. Errors of 1–2% in a pressure range of 0–700 kPa (0–100 psi) can be expected.

### e. Piezoresistive Transducers

It has been found that a mechanical stress applied uniaxially to a silicon or germanium crystal produces a large anisotropic change in its resistivity (29). A pressure sensor based on this property has been fabricated, utilizing the manufacturing techniques developed for integrated electronic circuits (68). Figure 4.35 is an enlarged photograph of the sensor together with its associated electronic amplifiers, all mounted on a ceramic substrate. The back surface of the silicon chip bearing the sensor has been etched away, so that it is only 0.025 mm in thickness. It forms one face of a permanently evacuated capsule, the other wall of which is much thicker. Hence the thin silicon chip itself becomes the pressure-sensitive diaphragm. It is reported to give a linear response from about 2 to 200 kPa (0.02–2 atm), with a maximum error of about 2% of full scale.

**Fig. 4.35.** Semiconductor pressure gauge. The rectangular chip to the right is the pressure sensor, the two smaller rectangles are operational amplifiers, the larger circular opening is the vacuum connection, and the nine irregularly shaped areas at the top are laser-trimmed thick-film resistors. The assembly, less than 1 in.$^2$ in area, is shown without its hermetic cover. (Courtesy of National Semiconductor Corporation.)

## 4. Piezoelectric Transducers

A number of anisotropic crystals are piezoelectric; that is, the strain produced by a suitably oriented stress is accompanied by the appearance of a potential difference that can be sensed by electrodes affixed to the surface. These can serve as the active element in pressure transducers, particularly for applications where a fast response time is essential, such as shock-tube studies. Typically, a disk several millimeters in diameter is mounted flush with the wall of the shock tube, its surface, covered with vacuum-deposited metal, serving as one electrode. The crystal must be backed with a carefully designed inertial mass to keep it in place while not confusing its response by reflection of acoustic waves. A good discussion has been published by Ragland and Cullen (53), who selected lead metaniobate as preferable to several other materials, including quartz. They found the transducer to be linear over the range 700–7000 kPa.

### 5. The Piezo-Junction Effect

A localized stress applied by a point pressure contact on a semiconductor in the vicinity of a *pn* junction produces a marked change in its electrical characteristics (55,56). A force of 0.03 N (~3 g–wt) increases the current through a germanium diode, for a particular voltage, by a factor of about 10. The effective capacitance is also altered.

This effect can be utilized to best advantage by applying the stress to the base-emitter junction of a bipolar transistor. The current gain of the transistor ($\beta$ or $h_{FE}$) is reduced by as much as three orders of magnitude. Pressure-sensitive transistors, called *piezo-transistors*, are available in several models, with full-scale ranges from 0.100 to 20.0 psi (70–140 Pa). The linearity and hysteresis errors are typically not greater than 0.5%. This tiny transducer has been reported in various pressure measuring applications in chemical instrumentation (52).

## B. PRESSURE TRANSDUCERS BASED ON MOLECULAR PROPERTIES

For measurements of pressures appreciably below atmospheric, a number of types of vacuum gauges are available. Because these are based on molecular properties of gases, their response cannot be assumed to be uniform for different working gases. Hence in most cases calibration must be in terms of a specified gas, such as dry air.

### 1. Thermal Conductance Gauges

The thermal conductivity of a gas, mentioned previously (Section IV.A.1.) in connection with a detector for gas chromatography, can also be utilized for pressure measurement. The kinetic molecular theory predicts, and experiment verifies, that the thermal conductivity of a gas is pressure-dependent only at pressures low enough that the mean free path of the molecules is greater than the distance between the heat source and sink. Under these conditions, the thermal conductance is directly proportional to the pressure. The lower limit is determined by sensitivity considerations and depends on the method of measurement.

The earliest embodiment of this principle was the Pirani gauge, in which a metal filament is mounted in a glass envelope to be sealed to the vacuum system under test. This filament, and an identical one permanently sealed in high vacuum (for temperature compensation), are connected as two arms of a Wheatstone bridge fed with either constant voltage or constant current.

Similar devices are the thermocouple gauge, in which the change of temperature of the filament is measured by a thermo-junction, and the thermistor gauge, where the filament is replaced by a small bead thermistor to take advantage of its high temperature coefficient. The latter has been reported useful over nearly six decades of pressure, down to 0.1 Pa ($10^{-3}$torr). Improved range and sensitivity can be attained at the expense of more involved electronics by operating the gauge as a null indicator in a feedback system. The gauge is held at constant temperature and the power required is taken as a measure of pressure.

### 2. Gas-Ionization Vacuum Gauges

The number of ions formed in a gas by a constant flux of ionizing radiation is pressure dependent, and several important vacuum gauges are based on this relation. The response is actually related to the molecular density of the gas (the number of molecules per unit volume), rather than to the pressure, which fact must always be taken into account in applications where temperature varies.

#### a. Bayard-Alpert and Related Gauges

The ionization gauge, as originally designed, used the geometry of Fig. 4.36*a*, similar to a triode amplifier tube. The grid is made positive (100–200 V) relative to the filamentary cathode and the "plate," or ion collector, some 20 V less positive than the grid. Thermionic electrons from the cathode are accelerated toward the grid, but most of them pass through, decelerate, and reverse direction, then oscillate around the grid wires several times until finally captured. This trajectory gives enhanced probability of collisions with gas molecules. Positive ions formed by such collisions in the region between grid and collector are accelerated to the collector, constituting an ion current, detectable with a sensitive galvanometer or an electrometer amplifier. One fault of this gauge is due to soft X-rays, produced by electron bombardment of the grid and incident upon the collector. Photoelectrons ejected from the collector by this X-radiation are indistinguishable from ion current, and limit the low-pressure response.

This difficulty is greatly reduced in the Bayard-Alpert design (2) (Fig. 4.36*b*), in which the filament is placed outside the spiral grid, and the ion collector becomes an axial wire. The action is the same, but only a very small fraction of the X-rays are intercepted by the collector. This gauge can measure pressures from about 1 mPa to 10 nPa ($10^{-5}$–$10^{-10}$torr). It is usually operated at a constant current, in which case the collector current is proportional to the pressure.

#### b. Alpha-Particle Ionization Gauges

An alternative source of ionizing radiation is available in various radioisotopes. Alpha radiation is more effective at producing ions in a small volume of gas than is either beta or gamma. One of the most convenient sources, with a half-life of 21 yr, is $^{210}$Pb ("radium–D") in secular equilibrium with its second daughter, $^{210}$Po (13).

The measuring cell must be designed so that the inner dimensions are considerably less than the range of the alpha particles at the pressures to be encountered. If this were not the case, the particles would be completely absorbed within the gas, and the number of ion pairs would be nearly independent of pressure.

Alpha ionization gauges have the advantage of not being damaged by exposure to atmospheric pressure, a condition that would cause a hot filament gauge to burn out.

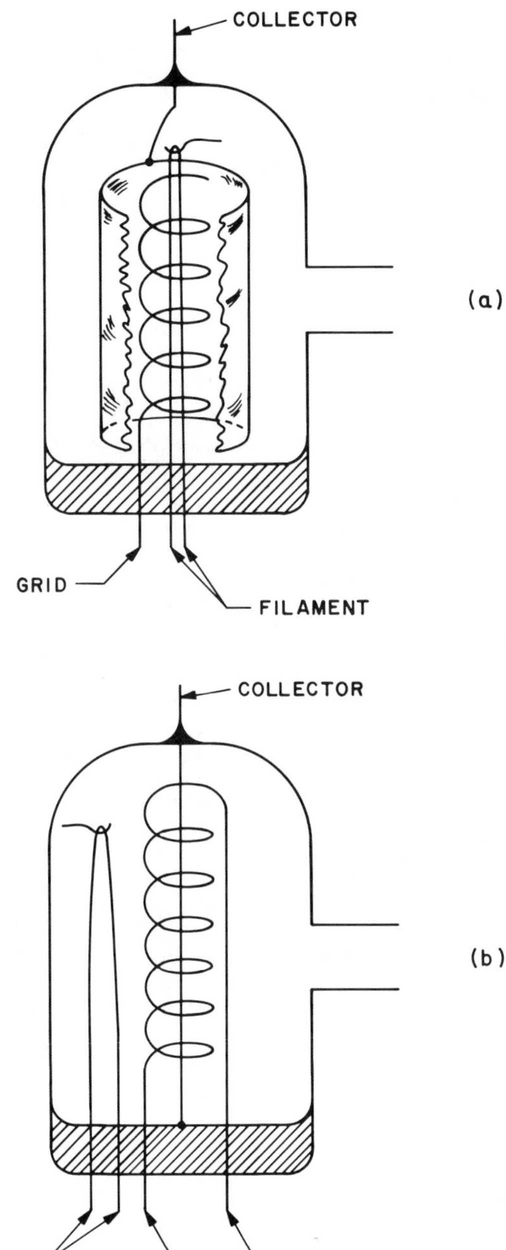


**Fig. 4.36.** Ionization vacuum gauges. (*a*) Concentric type; (*b*) Bayard-Alpert style.

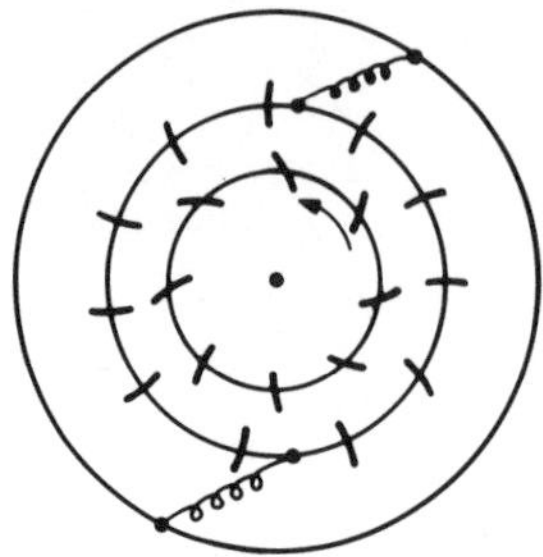

**Fig. 4.37.** Plan of squirrel-cage molecular vacuum gauge. The pointer or recording mechanism is driven by the outer of the vane-carrying rings. (Courtesy of General Electric Company.)

### 3. Molecular Impact Gauges.

If a velocity can somehow be imparted to the molecules of a gas, and the molecules subsequently made to impinge on the surface of a movable vane, momentum is transferred to the vane in proportion to the number of molecular impacts. Gauges making use of this principle are independent of the nature of the gas, in contradistinction to ionization types. Two transducers for vacuum measurement have been designed on this principle.

#### a. The Squirrel-Cage Molecular Gauge

This device (31) utilizes two sets of vanes (Fig. 4.37) in an orientation reminiscent of a turbine. The inner set, rotated by a small synchronous motor, strikes gas molecules, throwing them against the vanes of the outer set. The latter vanes are mounted on a movable frame that is restrained by a spring, and hence cause a deflection in an attached pointer in proportion to the molecular density of the gas. The instrument is calibrated to read from about 0.005 to 20 torr (0.5–2 Pa).

#### b. The Knudsen Gauge

In this gauge, energy is imparted to the gas molecules from a pair of heaters (H in Fig. 4.38), so disposed as to produce a torque in the vane. This vane is mounted on a quartz fiber suspension with a small mirror for optical readout. This gauge is useful down to the order of $10^{-5}$ Pa ($10^{-7}$ torr).

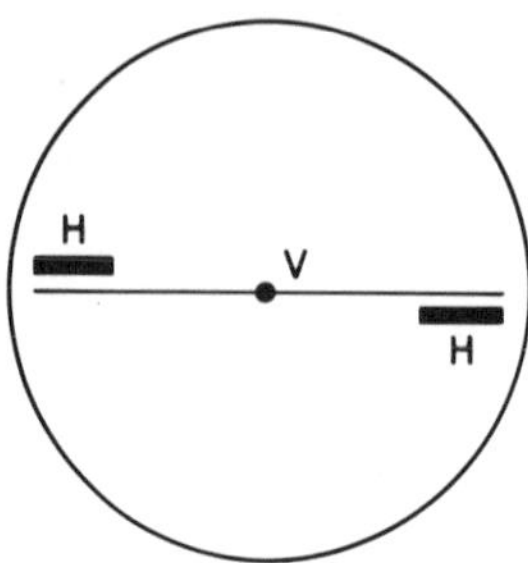


**Fig. 4.38.** Plan of the Knudsen gauge. The vane *V* can turn about its center suspension. Heaters are designated by H.

#### 4. Residual Gas Analyzers

Some of the smaller types of mass spectrometers are useful for determining simultaneously the partial pressures of all the components in a mixture of gases. These are widely used, under the generic term "residual gas analyzers," to monitor extensive vacuum systems in industry. They have far greater diagnostic value than the vacuum gauges described in the preceding paragraphs. For example, it may be possible to determine the source of the low pressure gases encountered, since the mass spectrum of air leaking into the system is readily distinguishable from that of products from the pyrolysis of vacuum pump oil or other sources.

The most commonly used devices are the quadrupole and the ion-cyclotron resonance types of mass analyzers. These cannot be described in detail here (57).

## VI. TRANSDUCERS AS ELECTRICAL COMPONENTS

Up to this point we have categorized transducers according to the physical quantity to be measured. It is also worthwhile to consider them in terms of the several electrical properties that change in magnitude with the input signal. From this standpoint, a transducer can be classed as one of two types, according to whether it acts as a source of electrical energy (which we may call an *active* transducer), or requires an external source (*passive*).

### A. ACTIVE TRANSDUCERS

Under this heading one can differentiate between devices in which the desired information is delivered in the form of a *potential*, and those in which it is carried by a *current*. In the first class fall the several piezoelectric transducers and pyroelectric radiation detectors. The second class includes transducers based on the production of ions and on the photoelectric emission of electrons. Several transducers can be used in either voltage or current modes.

#### 1. Potential-generating Transducers

It is essential that devices in this class be measured with a circuit of impedance much greater than the output impedance of the transducer, for otherwise the signal is downgraded by loading effects. This requirement can be troublesome with high-impedance transducers, such as piezoelectric crystals. The problem can best be solved by means of an impedance conversion stage, such as an emitter-follower transistor amplifier, a source-follower field-effect transistor, or an operational amplifier in the voltage-follower configuration (Fig. 4.39). Each of these circuits accepts a voltage signal impressed on its control element and produces a nearly equal output voltage across the load resistance, $R_L$. In the transistor circuits (*a* and *b*), the output is slightly lower than the input, but bears a constant ratio to it, typically 0.95–0.99. In the operational amplifier (*c*) this

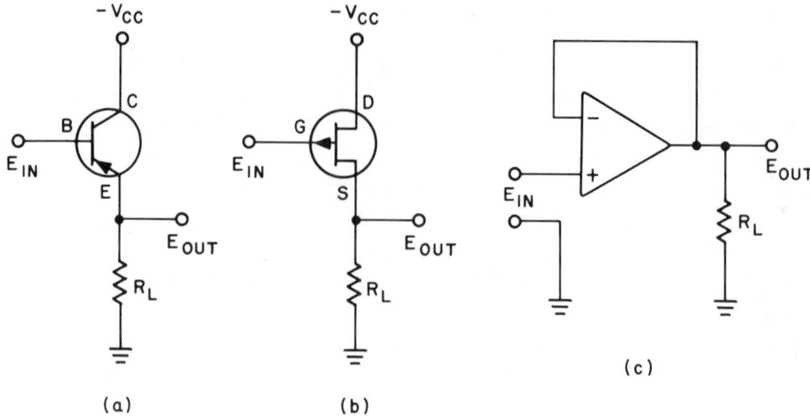


**Fig. 4.39.** Impedance converting or voltage-following circuits: (*a*) bipolar *pnp* transistor; (*b*) *p*-channel field-effect transistor; (*c*) operational amplifier.

drop is compensated, so that the input and output voltages are equal. Certain amplifier models are optimized for this application, and can show current gains (equivalent to input/output impedance ratios) as high as $10^9$.

If the signal to be measured represents small potential changes in the presence of a relatively large constant voltage, then a differential amplifier for impedance conversion is called for. This is often called an *instrumentation amplifier*. Such an amplifier can be obtained as a complete modular unit, or it can be created by suitable interconnection of several operational amplifiers. The rejection of common mode voltages can be as high as $10^6$ to 1; that is, 1 $\mu$V of signal can be discerned for each volt of the potential present equally at the two inputs.

For measurements of small dc potentials in low resistance circuits, the classical potentiometer is unexcelled with respect to accuracy, but this is only attainable at the expense of time and convenience. Digital electronic voltmeters give immediate readout, and are usually provided with computer-compatible coded output. These are available with residual uncertainties from 1 part in 1000 to 1 in $10^5$.

## 2. Current-generating Transducers

Those transducers in which electrons or ions are liberated or created in a direct functional relation to the property being measured are best treated as current sources. If the rate of production of charged particles is small enough that their effect can be observed individually, then the best electronic support is generally by digital (counting) techniques. This has long been applied to radioactivity measurements, and more recently to photon counting (47).

The alternative to counting is the use of an *electrometer* or *charge amplifier*. For this purpose an amplifier with a very low effective input impedance is

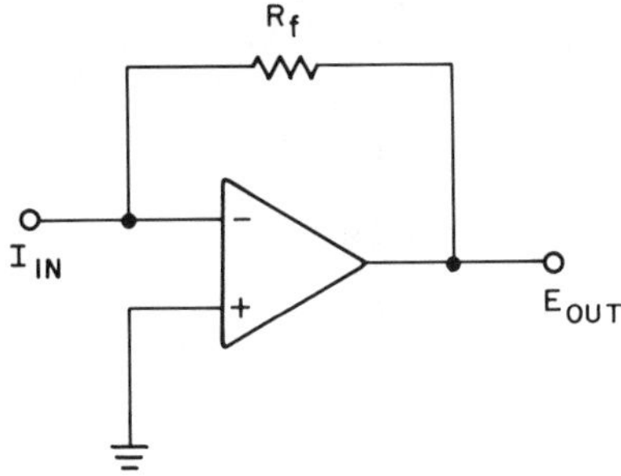


**Fig. 4.40.** Current-to-voltage converter using an operational amplifier.

required, a condition fulfilled by an operational amplifier with a feedback resistor but no input resistor (Fig. 4.40; see also Fig.4.11).*

For somewhat higher level currents, the circuit of Fig. 4.40 is still applicable, but the specifications of the amplifier can be relaxed. The designations "electrometer" and "charge amplifier" are no longer appropriate; the general term for the circuit is *current-to-voltage converter*.

### 3. Intermediate Types of Active Transducers

Thermocouples and junction photocells are low impedance devices that can be treated as either current or voltage sources. The information transduced may appear in rather different forms, however. This was mentioned in connection with the photodiode (Section III.D.2.b. and Fig. 4.22). In the case of the thermocouple (Section II.B.1), Peltier heating and cooling of the junctions occurs if current is allowed to flow, thus modifying the temperature one is trying to measure.

## B. PASSIVE TRANSDUCERS

### 1. Resistive Transducers

The conventional Wheatstone bridge (16) is by far the most commonly employed circuit for the measurement of resistive transducers. It may be operated in a continuously balanced mode, or its off-balance voltage or current can be monitored. In the first case, the resistance of the unknown arm ($R_x$, Fig. 4.41) is given by the familiar relation:

$$R_x = \frac{R_2 R_3}{R_1} \tag{4.31}$$

If the bridge is balanced by a servo system, there will necessarily be a time lag in

*Note than an operational amplifier for this service should have a high intrinsic input impedance and low bias current, but that the amplifier with its associated network, as shown in Fig. 4.40, has a vanishingly low input impedance because of the virtual ground condition.

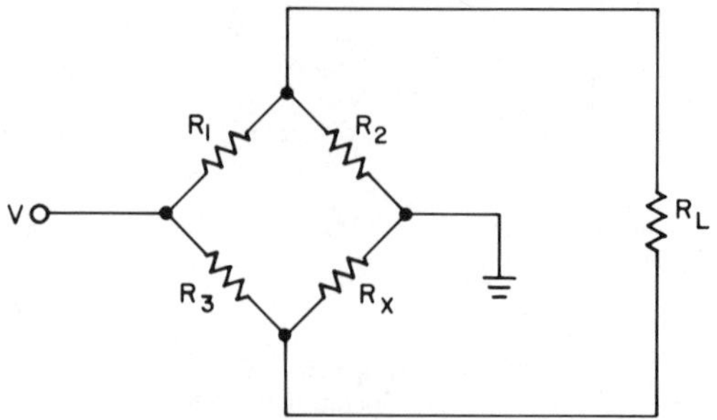


**Fig. 4.41.** Wheatstone bridge circuit.

the establishment of this relation, and this may be significant if the resistance is changing rapidly.

If the bridge is operated in an unbalanced mode, it can readily be shown (18) that the current $I_L$ through the load $R_L$ is given by:

$$I_L = \frac{V_T}{R_T + R_L} \tag{4.32}$$

where $V_T$ and $R_T$ are the Thévenin equivalent voltage and resistance, namely (for $R_1 = R_2$),

$$R_T = \frac{R_1}{2} + \frac{R_3 R_x}{R_3 + R_x} \tag{4.33}$$

and

$$V_T = V\left(\frac{1}{2} - \frac{R_3}{R_3 + R_x}\right) \tag{4.34}$$

These equations show that the signal delivered by an unbalanced bridge cannot be a linear function of the variable resistance; nevertheless it finds many areas of usefulness.

A potentially troublesome point with the Wheatstone bridge lies in its grounding. One would like to reference both the bridge power source and the amplifier to a common ground level, but this cannot be done in the simple bridge. If the bridge can be operated on alternating current, the problem can be solved by the use of either inductive or capacitive coupling between bridge and amplifier. If direct current must be used, then a possible solution is to utilize a differential amplifier with high common mode rejection, such as an instrumentation amplifier (16).

Another approach to resistance measurement is the use of an operational amplifier in the configuration of Fig. 4.42, called a "pseudo-bridge" (6). It can be shown that, if the open-loop gain of the amplifier is high, the output is given by the relation

$$E_{\text{out}} = V\left(\frac{R_2 R_3 - R_1 R_x}{R_2 (R_1 + R_3)}\right) \tag{4.35}$$

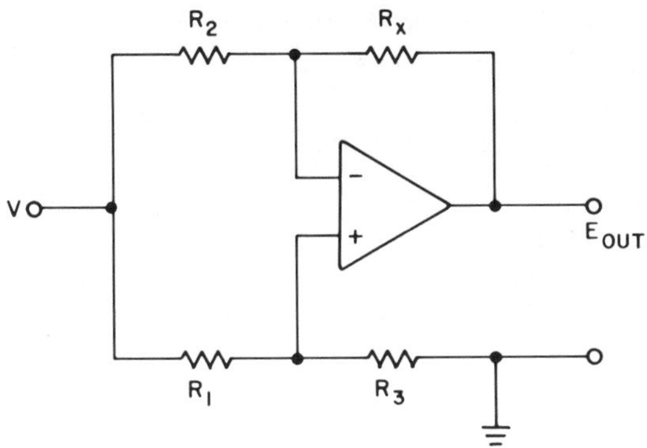


Fig. 4.42. Pseudo-Wheatstone bridge.

If the resistors are adjusted to null the output, then the quantity on the right must be zero, and

$$R_x = \frac{R_2 R_3}{R_1} \tag{4.36}$$

which is identical to the balanced bridge equation (equation 4.31).

When operated unbalanced, with $R_1$ and $R_2$ set equal, equation 4.35 predicts a straight-line relation between $E_{\text{out}}$ and $R_x$:

$$R_x = R_3 - \frac{E_{\text{out}}}{V}(R_1 + R_3) \tag{4.37}$$

The advantages of the pseudo-bridge are, first, the linear relation of equation 4.37, which can extend over several orders of magnitude, and second, the economy of electronic components as compared with the true bridge coupled to an instrumentation amplifier.

It should be realized that there is no fundamental reason why a bridge (or equivalent) must be selected for the measurement of resistance. Ohm's law can be applied directly, either by impressing a constant potential across the resistance and measuring the current, or maintaining a constant current and measuring the resulting potential drop. The latter method gives a linear response, whereas the former yields a reciprocal relation. Both circuits are commonly used in ohmmeters.

## 2. Capacitive Transducers

It was mentioned (in Section V.A.3.b) that a capacitive transducer can be measured either by an ac bridge or by incorporating it into an oscillator circuit and monitoring the resonant frequency. A recent study (65) points out that capacitive bridges are capable of greater precision but less sensitivity as compared with an oscillator method; one of the best commercial bridges has a sensitivity of about 1 part in $10^8$ against about 4 parts in $10^{10}$ for an oscillator.

The accuracy of the same bridge is a few parts in $10^6$, whereas the oscillator requires calibration for each measurement.

### 3. Inductive Transducers

Except for high-frequency circuitry, inductive devices tend to be bulky and awkward to shield adequately against stray alternating magnetic fields, so they are little used as transducers. The linear differential transformer and certain forms of dynamic or variable reluctance microphones are described in Section V.A.3.a.

## C. MODULATION TECHNIQUES

Special emphasis should be given to the advantages of modulation in connection with transducer electronics. In the present context this means impressing the desired information onto a relatively high-frequency alternating current, usually in the form of amplitude modulation.

The carrier alternating current may be generated in a local oscillator (or for convenience extracted from the power line at the prevailing 50 or 60 Hz), or it may arise from periodic interruption of the excitation to the transducer. Examples of the latter type occur widely in optical instruments where the light beam is chopped by a rotating shutter or equivalent.

The chief advantage of such techniques is a substantial increase in the $S/N$ ratio. This arises because the electronic amplifier can be ac-coupled in such a way as to discriminate against slow drifts such as might be caused by changes in ambient conditions, and against that class of noise that shows a reciprocal dependence on the frequency. A device designed to tune to the exact frequency of the carrier is termed a *lock-in amplifier*.

A related method of reducing noise originating from powerline pickup is to insert at an appropriate point in the information-processing electronics an integrator with a time constant ($RC$) equal to the period of the line voltage (Fig. 4.43). For 60 Hz line frequency, $RC = 1/60$ sec, so, for example, if $C = 1.00\ \mu F$, then $R$ should be 16.7 kΩ. This results in averaging out any variations that occur at a repetition rate of $60n$ Hz, where $n$ is an integer designating

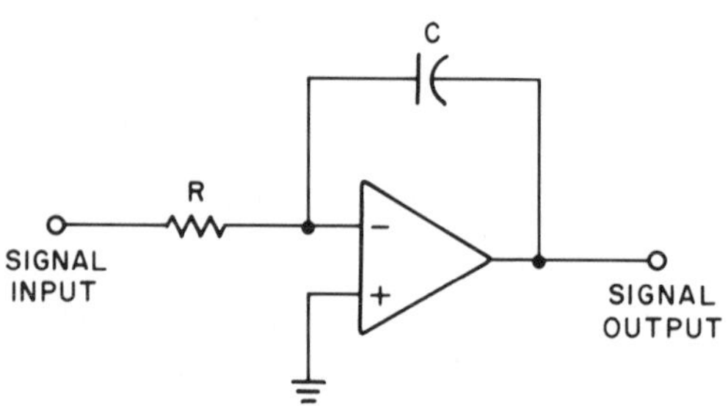


**Fig. 4.43.** Operational amplifier integrator.

harmonics of 60 Hz. If the carrier frequency is well removed from all these harmonics, it passes through the integrator nearly unchanged.

## REFERENCES


1. Aue, W. A., and H. H. Hill, Jr., *Anal. Chem.*, 45, 729 (1973).
2. Bayard, R. T., and D. Alpert, *Rev. Sci. Instrum.*, 21, 571 (1950).
3. Bell & Howell, CEC Instruments Div., Pasadena, *The Bell & Howell Pressure Transducer Handbook*, 1974.
4. Benjaminson, A. *Hewlett-Packard J.*, 16(7), 1 (1965); D. L. Hammond, *ibid.*, p. 3.
5. Blades, A. T., *J. Chromatogr. Sci.*, 11, 251 (1973).
6. Borchardt, I. G., and L. R. Holland, *Rev. Sci. Instr.*, 46, 67 (1975).
7. Boucher, E. A., *J. Chem. Educ.*, 44, A935 (1967).
8. Brody, S. S., and J. E. Chaney, *J. Gas Chromatogr.*, 4, 42 (1966).
9. Cocks, G. G., "Electron Microscopy," Chapter 73, in I. M. Kolthoff, P. J. Elving, and E. B. Sandell, Eds., *Treatise on Analytical Chemistry*, Part I, Vol. 6, Wiley-Interscience, New York, 1965, p. 4113.
10. Cole, M., and D. Ryer, *Electro-Opt. Syst. Des.*, 4, 16 (June, 1972).
11. Corruccini, R. J., "Principles of thermometry," Chapter 87 in I. M. Kolthoff, P. J. Elving, and E. B. Sandell, Eds., *Treatise on Analytical Chemistry*, Part I, Vol. 8, Wiley-Interscience, New York, 1968, p. 4937.
12. Dal Nogare, S., and R. S. Juvet, Jr., *Gas-Liquid Chromatography*, Wiley-Interscience, New York, 1962, pp. 189–214.
13. Deisler, P. F., Jr., K. W. McHenry, Jr., and R. H. Wilhelm, *Anal. Chem.*, 27, 1366 (1955).
14. Dike, P. H., *Thermoelectric Thermometry*, Leeds & Northrup, Philadelphia, 1954.
15. Ederer, D. L., and P. Dhez, *Rev. Sci. Instrum.*, 46, 144 (1975).
16. Ewing, G. W., *J. Chem. Educ.*, 52, A239 (1975).
17. Ewing, G. W., and J. P. Lucania, unpublished observations (1972).
18. Ewing, G. W., and M. G. Young, *Anal. Lett.*, *1*, 565 (1968).
19. Fortune, W. B., "Color Comparators," Chapter 3 in M. G. Mellon Ed., *Analytical Absorption Spectroscopy*, Wiley, New York, 1950.
20. Friedberg, S. A., in H. C. Wolfe, Ed., *Temperature, its Measurement and Control in Science and Industry*, Vol. 2, Reinhold, New York, 1955, Chap. 20.
21. Friedlander, G., J. W. Kennedy, E. S. Macias, and J. M. Miller, *Nuclear and Radiochemistry*, 3rd ed., Wiley-Interscience, New York, 1981, p. 233.
22. Gilford, S. R., D. E. Gregg, O. W. Shadle, T. B. Ferguson, and L. A. Marzetta, *Rev. Sci. Instrum.*, 24, 696 (1953).
23. Giuffrida, L., in I. I. Domsky and J. A. Perry, Eds., *Recent Advances in Gas Chromatography*, Dekker, New York, 1971, pp. 125–135.
24. Golay, M. J. E., *Rev. Sci. Instrum.*, 18, 357 (1947).
25. Goulding, F. S., and Y. Stone, *Science*, 170, 280 (1970).
26. Gray, P. E., D. DeWitt, and A. R. Boothroyd, *SEEC Notes I: PEM: Physical Electronics and Circuit Models of Transistors*, Wiley, New York, 1962, pp. 3–24ff.
27. Hannay, N. B., *Semiconductors*, Reinhold, New York, 1959, p. 35.
28. Ref. 27, p. 364.
29. Ref. 27, p. 366.

30. Holzbock, W. G. *Instruments for Measurement and Control*, 2nd ed., Reinhold, New York, 1962, pp. 71ff.

31. Ref. 30, p. 80.

32. Ingle, J. D., and S. R. Crouch, *Anal. Chem.*, 43, 1331 (1971); ibid., 44, 1709 (1972).

33. Jamieson, J. A., R. H. McFee, G. N. Plass, R. H. Grube, and R. G. Richards, *Infrared Physics and Engineering*, McGraw-Hill, New York, 1963, Chap. 5.

34. Johns, T., and A. C. Stapp, *J. Chromatogr. Sci.*, 11, 234 (1973).

35. Kimmitt, M. F., *Far Infrared Techniques*, Pion, London, 1970, pp. 60ff.

36. Kollie, T. G., R. L. Anderson, J. L. Horton, and M. J. Roberts, *Rev. Sci. Instrum.*, 48, 501 (1977).

37. Lawson, A. E., Jr., and J. M. Miller, *J. Gas Chromatogr.*, 4, 273 (1966).

38. Liebhafsky, H. A., H. G. Pfeiffer, E. H. Winslow, and P. D. Zemany, *X-ray Absorption and Emission in Analytical Chemistry*, Wiley, New York, 1960, Chap. 2.

39. Lion, K. S., *Elements of Electrical and Electronic Instrumentation*, McGraw-Hill, New York, 1975, pp. 47–50.

40. Loeb, L. B., *Basic Processes of Gaseous Electronics*, University of California Press, 1960.

41. Long, D., and J. L. Schmit, in *Infrared Detectors*, Vol. 5 of R. K. Willardson and A. C. Beer, Eds., *Semiconductors and Semimetals*, Academic, New York, 1970, Chap. 5.

42. Loriot, G., and T. Moran, *Rev. Sci. Instrum.*, 46, 140 (1975).

43. Lovelock, J. E., *Anal. Chem.*, 35, 474 (1963).

44. Lovelock, J. E. G. R. Shoemake, and A. Zlatkis, *Anal. Chem.*, 36, 1410 (1964).

45. Lytle, F. E., *Anal. Chem.*, 46, 545A, 817A (1974).

46. Maggs, R. J., P. L. Joynes, A. J. Davies, and J. E. Lovelock *Anal. Chem.*, 43, 1966 (1971).

47. Malmstadt, H. V., M. L. Franklin, and G. Horlick, *Anal. Chem.*, 44(8), 63A (1972).

48. Mavrodineanu, R., "An Accurate Spectrophotometer for Measuring the Transmittance of Solid and Liquid Materials," in R. Mavrodineanu. J. I. Schultz, and O. Menis. Eds., *Accuracy in Spectrophotometry and Luminescence Measurements*, *NBS Spec. Publ. 378*, Washington, D. C., 1973, p. 31.

49. Morimoto, H., and R. Uyeda, *Acta Crystallogr., 16, 1107 (1963)*.

50. Nerheim, A. G., *Anal. Chem.*, 35, 1640 (1963).

51. Putley, E. H., in *Infrared Detectors*, Vol. 5 of R. K. Willardson and A. C. Beer, Eds., *Semiconductors and Semimetals*, Academic, New York, 1970, Chap. 6.

52. Quinn, E. J., and T. Posipanko, *Rev. Sci. Instrum.*., 41, 475 (1970).

53. Ragland, K. W., and R. E. Cullen, *Rev. Sci. Instrum.*, 38, 740 (1967).

54. RCA Corporation, *RCA Photomultiplier Manual*, Harrison, N.J., 1970.

55. Rindner, W., *J. Appl. Phys.*, 33, 2479 (1962).

56. Rindner, W., and I. Braun, *J. Appl. Phys.*, 34, 1958 (1963).

57. Roboz, J., *Introduction to Mass Spectrometry*, Wiley-Interscience, New York, 1968.

58. Ruehle, R. A., *Electronics*, 48(6), 127 (1975).

59. Santini, R.E., and H. L. Pardue, *Anal. Chem.*, 42, 706 (1970).

60. Sarjeant, W. J., Z. Kucerovsky, D. Rumbold, and E. Brannen, *Rev. Sci. Instrum.*, 41, 1890 (1970).

61. Scribner, B. F., and M. Margoshes, "Emission Spectrography," Chapter 64 in I. M. Kolthoff, P. J. Elving, and E. B. Sandell, Eds., *Treatise on Analytical Chemistry*, Part I, Vol. 6, Wiley-Interscience, New York, 1965, p. 3347.

62. Sharpe, J., *Photoelectric Cells and Photomultipliers*, EMI Electronics, Ltd., via Gencom Div., Varian/EMI, Plainview, New York, 1961.

63. Stewart, J. E., *Infrared Spectroscopy: Experimental Methods and Techniques*, Dekker, New York, 1970, Chap. 11.

64. Steyn, J. J., and S. S. Nargolwalla, "Detectors," in J. Krugers, Ed., *Instrumentation in Applied Nuclear Chemistry*, Plenum Press, New York, 1973, pp. 116–117.

65. Van Degrift, C. T., *Rev. Sci. Instrum.*, 46, 599 (1975).

66. Vermont, J., and C. L. Guillemin, *Anal. Chem.*, 45, 775 (1973).

67. Wood, W. A., and S. R. Gilford, *Anal. Biochem.*, 2, 589 (1961).

68. Zias, A. R., and W. F. J. Hare, *Electronics*, 45(12), 83 (1972).

Part I
Section E

Chapter 5

# AUTOMATION: INSTRUMENTATION FOR ANALYSIS SYSTEMS

By Marvin Margoshes and Donald A. Burns,
*Technicon Corporation, Tarrytown, New York*

## Contents

## I. INTRODUCTION

The subject of this chapter is the automation of "wet" chemical analyses. We exclude automated instruments that do not significantly apply chemical reactions, such as on-line industrial analyzers that infer chemical composition from spectra or refractive index. We also exclude many automated devices in the laboratory that do not carry out what is traditionally known as "wet chemistry," though they do perform chemical analyses. This group of devices would, for example, include most chromatographs, pH meters, and emission, infrared, and X-ray spectrometers.

We include in this chapter devices that combine the sample with one or more reagents to effect processes that have traditionally been carried out in test tubes, beakers, and flasks. We first describe the general principles of operation of the analyzers, followed by descriptions of individual unit processes, with some specific applications described for illustrative purposes.

Emphasis is given to instrumentation and devices in which the entire analytical process is automated, from sample to result. The reader should be aware, however, that it may often be more effective to automate only a part of the analysis, with the other parts to be done manually.

Most of our emphasis is on segmented continuous-flow (CF) methods of analysis. This is partly because of our greater familiarity with this form of chemical automation, and partly because this type of analyzer has been in use longer than the discrete analyzers and has been more widely applied.

The CF and discrete analyzers have been used fairly extensively in chemical analyses other than clinical chemistry, though biomedical applications have predominated. Centrifugal analyzers have had little if any use outside of clinical chemistry. In this chapter, however, we de-emphasize clinical chemistry. The clinical applications of automated analyzers have been described rather well elsewhere (43,76,77), while the industrial uses have been less extensively treated.

Continuous-flow analyzers were first developed for use in clinical chemistry (58). Since the AutoAnalyzer system was introduced by the Technicon Instruments Corporation in 1957, the number and variety of analyses done in clinical chemistry laboratories have vastly increased. Most authorities agree that automation has itself been a prime factor in generating this increase, simply because the manually performed analyses were expensive and unreliable and the results were often not obtained in time to guide the physician. Most analyses of clinical specimens are now done by automated wet chemistry, and physical methods (spectroscopy, etc.) have come to dominate some other areas of chemical analysis, leading to the common observation that "the chemistry has gone out of analytical chemistry."

One of the reasons that the majority of automated chemical analyzers are found only in clinical chemistry laboratories is that little information on the apparatus has been made available in analytical chemistry textbooks and journals. Therefore, the capabilities of these devices are not appreciated by most analytical chemists. We hope in this chapter to provide a source of information. In order to make the chapter as generally useful as possible, we emphasize the principles employed and the types of instruments and modules that are available. The reader will be able to determine the usefulness of the equipment and procedures for his individual requirements, with the aid of some illustrative applications. For access to more detailed information, the reader should consult a bibliography on continuous flow published by the Technicon Corporation (67) and the book by Furman (34).

## II. BENEFITS AND DRAWBACKS OF AUTOMATING

We give here a summary of advantages that are usually sought in automating chemical analyses, and the reasons that automation may sometimes be impractical or unwise.

### A. COST

Economy is the most common objective of automation, in analytical chemistry as in other fields. The potential for savings in personnel costs is so obvious that no further discussion seems necessary. In the early days of clinical laboratory automation, savings in reagent costs were often the prime economic motive. Some of the automated instruments require much less sample and reagent than do manual methods. The savings can be quite significant when either the sample or the reagents are expensive.

Many chemists think that there can be no important economic advantage to automating analyses unless the workload is large enough to occupy the instrument fully. This is not true. Automated instruments are often able to analyze in an hour or two as many samples as several technicians can handle in a full day. In such cases, the instrument pays for itself even though it may sit idle most of the time.

### B. NEED FOR CONTINUOUS OPERATION

Many industrial processes require continuous analysis of the materials, the product, or the intermediate process stream. The cost of manual analyses becomes high, since technicians must be on hand at all times. The savings from automation can be quite large in such a case, even when the number of samples analyzed is small. If, for instance, the process requires analyses once an hour around the clock, an automated analyzer can be cost effective even though it is greatly underutilized.

### C. HAZARDS

Some analytical methods employ dangerous reagents, and the sample itself may be dangerous. Considerable care must then be exercised. It is true that "familiarity breeds contempt," and chemists and technicians may in time ignore safety precautions that were established when the method was first set up. At times, the analyses must be done in a hazardous environment (such as in an explosives plant) and it is important to reduce the number of people at risk in case of an accident. Automated equipment offers clear safety advantages in all these cases.

### D. SKILL LEVEL AND MONOTONY

It may happen that an analysis can be done well manually only by a skilled, well-motivated operator. Repetitive tasks are unattractive to most skilled persons. The result is often loss of interest followed by frequent mistakes, rapid turnover of personnel, and high personnel costs. Automation may then be the only way to get the analyses done reliably.

## E. IMPROVED PRECISION

Automated devices are commonly more precise than humans in repeated operations, including the steps of a chemical analysis. Automation may often be the only economical way to obtain precise analyses.

## F. MORE RAPID ANSWERS

Automated methods are often able to provide analytical results with less delay than by manual methods. When the method includes a slow reaction, the automated method need not wait for equilibrium to be reached because in automation the reaction time before measurement of the product can be controlled. If the workload is irregular, an automated method that is capable of high throughput can prevent a backlog from being built up, thus assuring rapid results.

## G. FIT TO OTHER AUTOMATION

The analytical laboratory increasingly exists within an automated environment. The analytical result must often be directed to a computer for calculation of results, report preparation, or process control. Manual methods may not fit well within an automated environment.

## H. RELIABILITY

Automated systems are less likely than humans to make gross errors such as interchanging samples, using the wrong size pipette, or misreading a scale on a recorder, meter, or burette. Automated devices do not strike. A well-designed and well-made device, if properly maintained, may lose fewer days per year from breakdowns than do humans from illness.

## I. WORKLOAD

The most common reason for automating is undoubtedly the requirement that a laboratory perform many analyses each day, especially when the heavy workload is expected to continue indefinitely. It should certainly be stressed, however, that a large workload is not an essential prerequisite to automation. An automated analyzer is often fully justified for one or more of the reasons listed above, even though it may be active only a few hours each day.

## J. DRAWBACKS

Anyone planning to automate an analytical method should be aware of certain difficulties that are often encountered. Some of the difficulties can be avoided with care. When they cannot be avoided, they should be anticipated in the planning, so that their effects will not come as a late surprise.

### 1. Processes That Are Difficult to Automate

It may not be feasible to automate the existing method. There are certain steps in manual analyses that are rather difficult to automate. Conventional gravimetric methods offer one example. Steps such as filtration, burning off the filter paper, and drying the residue to constant weight are difficult to automate. In nearly any case, a new analytical method can be developed that makes use only of easily automated steps. The extra cost of developing and validating the new method, when added to the normal costs of automating, might make the change to automation too expensive.

### 2. Legal Restrictions

The manual method of analysis may be written into laws or regulations. Even when the automated method follows the manual technique fairly closely, it can be extremely slow and costly to have the new method made official. Examples of this sort of limitation on automation (and of innovation in general) occur often in the food and pharmaceutical industries.

### 3. Human Factors

Human factors should not be ignored when automation is considered. There is a transition period when automation is introduced, during which personnel are trained in the new apparatus and techniques, new working schedules are put into practice, and start-up problems are resolved. It is important to keep in mind that the automated method may be subjected to conscious or unconscious sabotage. Laboratory personnel, even skilled scientists, often have an emotional attachment to the familiar. They can resist change to new procedures in subtle ways, often without deliberation or planning. Some "accidents" may actually be deliberate. The laboratory personnel may feel that their jobs are threatened by automation, and they may respond by taking effective steps to "prove" that the new equipment is faulty.

## III. CATEGORIES OF AUTOMATED ANALYZERS

Automated analyzers can be broadly divided into CF and batch analyzers.

In a CF system, the samples and reagents flow along a glass or plastic tube where the analytical steps are carried out. In batch analyzers, discrete containers hold the sample and the reagents, and all the analytical steps take place in the individual containers. In some analyzers, the containers may be grouped into an arrangement for carrying groups of samples through the steps in parallel. One example of this is the centrifugal analyzer, which analyzes 2–30 individual samples in a special centrifuge rotor.

Many automated analyzers combine features of continuous-flow and batch analyzers. Often, a batch analyzer passes the sample through a flow-through cuvette for the final colorimetric measurement. Less often, a CF analyzer may incorporate one analytical step that is performed in individual sample containers. This usually is done when a slow chemical reaction requires an incubation time of half an hour or longer.

## A. CONTINUOUS-FLOW ANALYZERS

It is a sound general principle that automation is often best accomplished by departing greatly from manual methods. Continuous-flow analysis (CFA) is a good example of this principle. The traditional methods of analysis carry each sample through the analytical procedure in one or more individual containers. In CFA, individual samples are introduced into a flowing stream of reagent in a tube. The stream may be mixed with other reagents, it may be heated or cooled, and several other procedures can be performed on the flowing stream. Finally, the samples pass through or past a detector for quantitation of the analyte.

The fluid flow through a tube is laminar, moving with a velocity that decreases hyperbolically with the radial distance from the axis of the tube. The fluid on the axis of the tube moves at twice the average velocity, and there is a thin stagnant layer at the tube surface. The flow pattern changes to a turbulent profile, in which the fluid velocity changes less along the tube radius, when the Reynolds number is larger than 2000–3000. Reynolds numbers high enough for the flow to be turbulent can be reached only with quite large liquid flows (e.g., 100–150 ml/min of water in a 1 mm tube), and this is not practical in an analytical system.

The flow pattern can be altered by an irregularity in the tube, such as indentations in the wall, curved sections of tubing, constrictions or enlargements of the tube diameter, or objects placed within the tube. A rigorous treatment of all possible cases cannot be done briefly, so we limit ourselves here to the relatively simple case of a straight tube.

All CFA systems operate under laminar flow conditions, so that the liquid at the center flows with a velocity twice the average fluid velocity. The flow decreases hyperbolically toward the wall of the tube, becoming stagnant exactly at the wall. When a sample is inserted into a flowing stream, different portions of the sample move downstream at different velocities, depending on where each portion is located radially. The effect is to disperse the sample along the axis of the tube. This axial spreading determines the rate at which samples are analyzed, because CFA requires that sample-to-sample mixing be eliminated or at least that it be reduced to an acceptable level.

While the sample-to-sample mixing must be prevented, mixing of the sample with reagent is obviously essential. In a flowing stream, as in a test tube or beaker, mixing by diffusion is too slow a process to be practical. Some means must be provided to speed up the mixing process. A stirrer, such as can be used in a beaker, is usually not practical in a flowing stream. Mixing is usually done by disturbing the laminar flow pattern. Ideally, the same disturbance should limit the axial spreading of the sample.

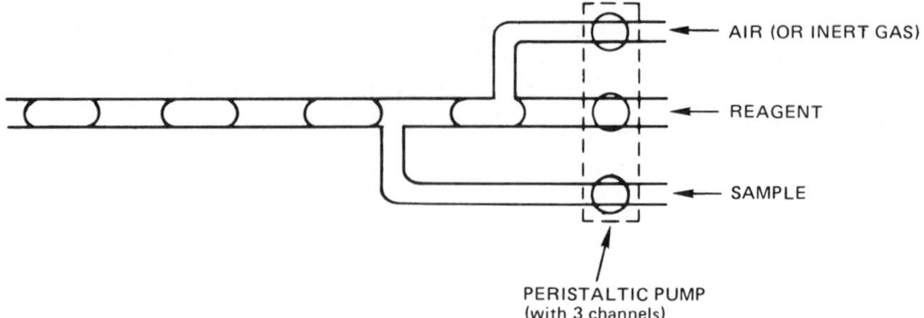


**Fig. 5.1.** Simplest manifold for segmented CF system.

## B. SEGMENTED CONTINUOUS-FLOW ANALYSIS

Skeggs (58) introduced the concept of segmentation in CF analysis. In this technique, air bubbles are added to the flowing stream at regular intervals (e.g., every 2 sec). Each bubble is large enough to fill the diameter of the tube. The stream is now made up of alternating segments of liquid and air, as shown in Fig. 5.1. An inert gas can be used in place of air if the reaction is affected by oxygen, or segments of a liquid immiscible with the sample and reagent can be substituted. Sample can be added to a segmented stream in a T-fitting, which is also shown in the figure. Each liquid segment then contains some reagent and some sample, the proportion between the two liquids being determined by the rates at which they are pumped into the tube. It is convenient to use for this purpose a peristaltic pump that can accept two or more tubes. The relative flows of the sample and reagents are determined by the sizes of the tubing used for each in the pump. In the same pump, larger tubes produce higher flows. An increase or decrease in the motor speed in the pump affects the flows in all the tubes proportionately, and the relative amounts of sample and reagent are not changed.

Another tube in the same pump can be used to pump the air to form the bubbles. Control of the bubble frequency is best done by a mechanism, called an air bar, that normally clamps the air tube but releases it at the desired intervals.

Figure 5.2 shows the flow pattern in a liquid segment. The arrows in the

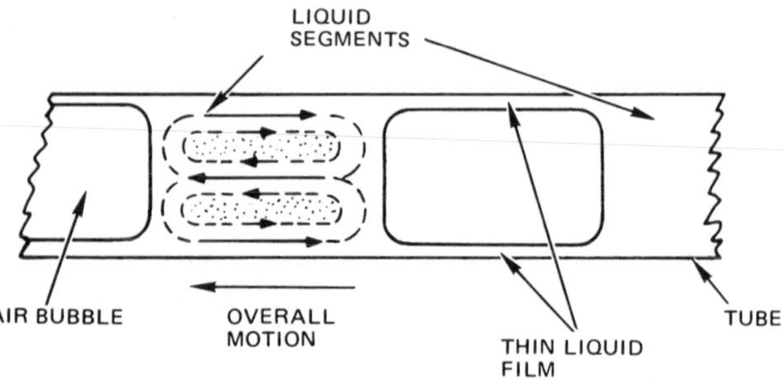


**Fig. 5.2.** Flow pattern within a single liquid segment.

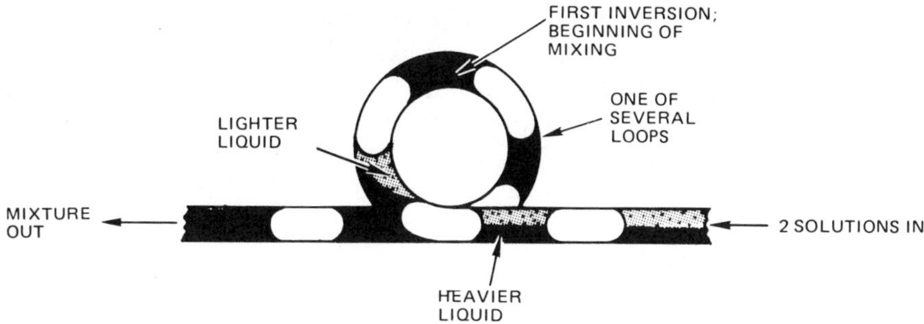


**Fig. 5.3.** Mixing coil for CF system.

figure show the fluid motion relative to the movement of the segment. As the segment moves along the tube, liquid at or near the tube axis can move freely, but liquid at the tube walls is held back. The liquid near the walls does not move backward along the tube; it only moves backward relative to the segment. This flow pattern contributes to a rapid mixing within the segment. The mixing can be increased by flowing the segmented stream through a coil, as in Fig. 5.3. If two liquids in a segment have a different density, the heavier liquid moves to the back of a segment in an ascending part of the tube, and to the front in a descending part. It is much like repeatedly inverting a stoppered flask to mix the contents.

### C. CARRYOVER, OR WASH

As the air bubbles move along the conduit, they never actually touch the wall because a thin film of liquid remains after each segment. The thickness of this film is a function of surface tension, viscosity, and flow rate. Its presence gives rise to carryover, for each segment leaves some material in its trailing film to be picked up by, and mixed with, the following segment.

In a segmented CF analyzer, a sample is initially spread out among several successive segments. For instance, if the sample is adjusted for 60 samples/hr (1 min/sample) at a sample-to-wash ratio of 2 : 1, the probe aspirates sample for 40 sec and wash solution for 20 sec. At a typical bubble rate of 30/min, the stream consists of 20 segments that contain sample plus reagent, followed by 10 segments that contain wash liquid plus reagent, then by another series of 20 segments containing the next sample, and so on.

The leading segment of each group that constitutes an individual sample becomes diluted with wash liquid as it moves along the tube. The next segment in turn picks up from the first a smaller quantity of the wash liquid, the third segment receives still a smaller quantity, and so forth. Somewhere down the sequence of segments, the dilution with wash liquid has become negligibly small, and the proportioning of reagent and sample are for practical purposes the same as when the liquids had first combined.

Similarly, the first intersample segment, containing the wash liquid, receives some sample from the segment preceding it. The second segment becomes contaminated to a smaller extent, the third segment is less contaminated, and so on. After some number of segments, contamination by the preceding sample has become negligibly small, and the first segment of the succeeding sample receives only reagent and wash liquid from the segment before it.

Figure 5.4*a* shows the distribution of concentration with time as the segments flow through or past a detector. In this example, the air bubbles are assumed to have been removed from the flowing stream shortly before the detector, and enough mixing has taken place to smooth out the changes in concentration between individual segments. After the dip corresponding to the wash liquid, there is a rising curve where the concentration of analyte increases, then a region of steady state where succeeding segments all have the same composition, then a falling region where the concentration of analyte decreases.

The number of samples that can be processed per hour through a CF analyzer may be increased by shortening both the sampling and wash periods. The sampling period can be shortened until the curve just reaches the point shown as A in Fig. 5.4*b*. Sampling for a longer period would merely prolong the steady state. The wash period can be reduced by not forcing the concentration of analyte to return fully to the base line between samples. Reaching base line between samples is less important than reaching base line before the concentration maximum of the next sample. Figure 5.4*c* shows the shape of the concentration-time curve when the sampling times are reduced, with a dotted line indicating that the material remaining from the first sample returns to base line before the peak of the second sample.

The shape of the rise and fall curves can be described in most cases by exponentials (61,70,71,74), and the concentration does not actually return to base line in a finite time. The rate of drop in concentration after each sample has the colloquial name "wash." The residue from one sample that remains to contribute to the peak of the next sample is the "carryover." Carryover is sometimes thought of as a problem that is peculiar to CF analyzers. In fact, it is common to all analytical systems, manual or automated, in which the same apparatus components contact more than one sample.

The carryover can be reduced to any desired low level simply by increasing the wash time or the sampling time. Increasing either time decreases the number of samples that can be processed per day. It is a practical question to evaluate the trade-off between carryover and sampling speed. The main parameters to be considered are the wash characteristics of the analytical system, the acceptable error due to carryover, and the sample-to-sample variability.

If all samples were identical, any amount of carryover could be accepted. Imagine a sequence of samples, all of the same composition, passing through a CF analyzer with 10% carryover. Each peak after the first would be increased by exactly 10%, and a simple correction would leave no residual error. If, however, all the samples except one had identical concentrations, $C$, but that one had a concentration $2C$, the sample following the elevated one would have its peak increased by 10% of $2C$, or $0.2C$, instead of the expected increase of $0.1C$. Subtracting a constant carryover from that peak would leave a significant error.

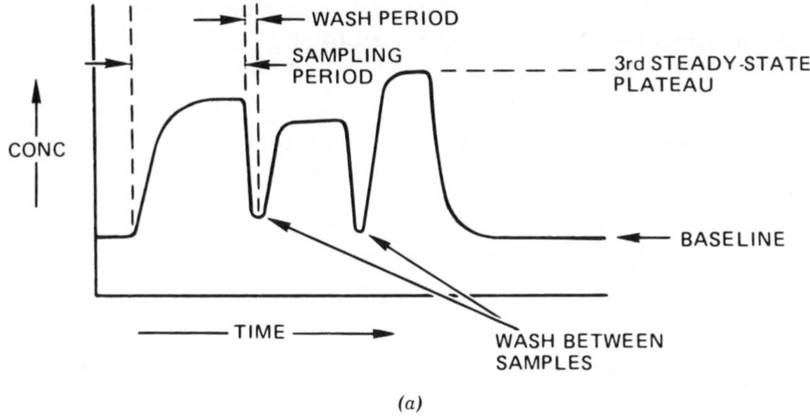


*(a)*

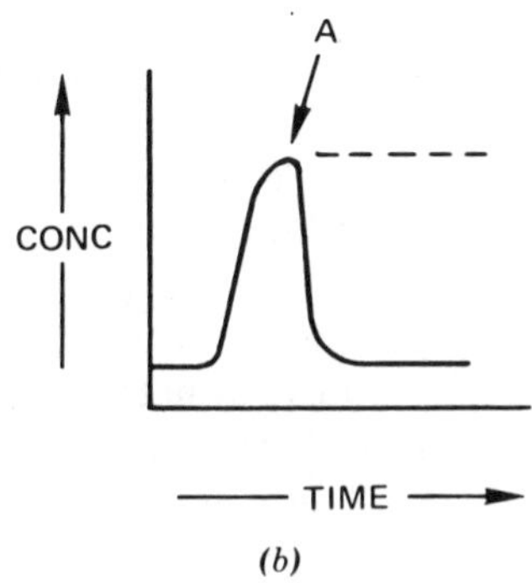


*(b)*

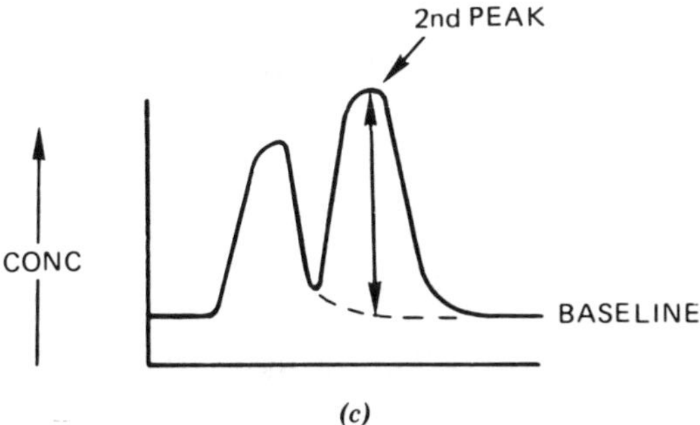


*(c)*

**Fig. 5.4.** Tracings of concentration change versus time: (*a*) washin, washout, and steady state; (*b*) minimum sampling time (plateau would begin at A); (*c*) minimum wash time (second peak over base line).

There are techniques for carryover correction that enable a significant increase in sampling rate. These techniques are discussed later in this chapter (Section VI.C).

## D. UNSEGMENTED CONTINUOUS-FLOW (FLOW INJECTION) ANALYSIS

One year after Skeggs' first publication on segmented CF analysis, Spackman et al. (63) described another type of CF analyzer which they developed for the reaction of the effluent from the column of an ion exchange amino acid analyzer with ninhydrin reagent. The system combined the column effluent, at 0.5 ml/min, with 0.25 ml/min of ninhydrin reagent and flowed the mixture through nearly 3 m of narrow (0.7 mm i.d.) tubing. Spackman et al. observed that peak spreading was excessive with large i.d. tubes at these flow rates and the required 15 min residence time for the reaction to give adequate color development. They pointed out that a 0.5 ml volume of effluent, after mixing with half that volume of reagent, would occupy a length of more than 150 cm in the narrow tube. Some peak spreading occurred, but at an acceptable level.

The unsegmented-flow system has found continued use in amino acid analyzers, where segmented flow has also been applied (30,31). It has recently been rediscovered (49,56,65) for more general purposes, and has been given the name flow injection analysis, or FIA. The name FIA comes from the practice in the newer form of the apparatus of introducing the sample into the flowing reagent stream as a discrete bolus, by means of a liquid chromatography valve (65) or by rapid injection through the side of the tubing (75).

Eliminating the segmenting air bubble from the stream simplifies the apparatus in a minor way, because an air bar is not required. One loses the twin advantages of segmentation, which are to control peak spreading and to enhance mixing of the sample with reagent. Peak spreading in FIA is not controlled. Velocity in a straight, round conduit under laminar flow conditions is given by $v = 2V\,[1 - (r/R)^2]$, where $v$ is the velocity at radius $r$, $V$ is the average velocity, and $R$ is the radius. Clearly, $v = 2V$ at the axis ($r = 0$) and $v = 0$ at the wall ($r = R$). The material at the center of the tube reaches the detector in half the time for the peak to reach there. This band spreading becomes worse with increasing residence time, and there is a simple relation between the minimum reaction time and the maximum number of samples that can be processed per hour. Forming the conduit into a tight coil causes some radial mixing, as does radial diffusion, and this reduces the band spreading but not by enough to make a major difference in the maximum sample throughput. If the reaction between sample and reagent is essentially instantaneous, FIA occasionally permits the analysis of more samples per hour than has been achieved until now by segmented flow. If the reaction is not that rapid, segmented flow analyzers have higher throughputs.

Mixing in unsegmented flow is not well understood. Some mixing of the sample bolus with the reagent stream comes from diffusion across the interface between sample and reagent. Any unevenness in the tube can also promote mixing. The flow lines are altered in a curved tube compared to a straight tube, because the liquid that passes around the outside of the curve must travel a greater distance. For this reason, it is helpful to have the tube in the form of a tight coil.

Residence times in unsegmented flow analyzers are usually short, a few seconds, and it is unlikely that complete mixing occurs in this time. Also, when the sample is introduced as a discrete bolus, the proportioning of sample and reagent can never become equal at all positions along the bolus. There must always be a larger proportion of sample at the center of the bolus than at the leading or trailing portions. In a few cases, it may be possible to use this to advantage to effect a titration. In other cases, the variation in proportioning is a drawback. Under unfavorable conditions, there may be an inadequate amount of reagent at the center of the bolus to react completely with the sample.

Perhaps the most serious drawback of FIA is that relatively large amounts of sample and reagent must be consumed for each analysis, or else tube diameters must be made quite small and pumping pressures must increase. It has been shown (46) that pumping pressures in FIA vary inversely with the fourth power of the sample size, if the ratio of sample to reagent volume is kept constant. (The increase of pressure can be reduced if the quantity of sample is reduced compared to the reagent volume, but this leads to poorer sensitivity.) The pumping pressure also increases linearly with residence time of the sample, which is a further limitation when the reaction is not extremely fast.

There are stratagems to reduce reagent consumption and increase residence time in unsegmented flow analysis. One is to inject boluses of sample and reagent into separate streams that then merge. It is apparent that this technique gives unreliable results unless the timing of the merger of the two streams can be kept rather precise. A longer reaction time can be achieved by intermittent flow, in which the stream is stopped for periods of time. Both these techniques introduce added complexity to the apparatus, and in our opinion they are poor substitutes for segmented flow methods to achieve the same goals.

All these limitations of unsegmented flow analyzers are well known to the practitioners of liquid chromatography, and they are described in the literature (25). The fact that the limitations are not known to many other analytical chemists is a reflection of the increasing degree of specialization in our profession.

## E. DISCRETE ANALYZERS

The primary purpose of this chapter is to introduce the reader to automated analysis by continuous flow. Only a brief summary of discrete analyzers is given.

A discrete analyzer is any device that automates chemical analysis and in which the sample and reagents react in a vessel rather than in a flowing solution in a tube. A discrete analyzer may pass successive samples through tubing in one or more steps of the analysis, usually to load the sample into the reaction container or to bring the reacted solution into a colorimeter or to another detector.

The container in which the reaction takes place is generally shaped like a test tube or a beaker. A conveyor mechanism (endless belt, turntable, etc.) moves the container past several stations where various operations are performed. Examples include sample addition, reagent addition, mixing, temperature-controlled incubation, and detection. In some instruments, the container forms a cuvette and the measurement of color development is made as the container moves past a reading station. In other instruments a probe is dipped into the solution and the sample is pumped into a cuvette.

The container may then pass other stations, where it is emptied of the sample and reagents and is washed and dried before being returned to the point where the next sample and reagents are added. Alternatively, the container is discarded after one use, in which case there must be some mechanism that dispenses new containers, or the user has to load new containers manually.

In one form of discrete analyzer, the disposable cuvette comes from the manufacturer with a measured quantity of reagent. The reagent is often in the form of a powder or a pellet that is dissolved by adding a measured amount of solvent just before the container is used. It is generally more expensive to purchase the container with preloaded reagent than to add a reagent solution to empty containers. Moreover, the preloaded containers leave the user more dependent on the manufacturer. On the other hand, there is an important convenience factor. Automated analyzers that use disposable cuvettes can be designed to be relatively undemanding of operator skill. The operator need only select the container that is needed for a particular determination and place it in the analyzer with a sample. The operator need not prepare solutions in advance, and there need be no concern that a solution prepared some time earlier has gone bad.

## F. CENTRIFUGAL ANALYZERS

The centrifugal analyzer (4) is a unique type of discrete analyzer. It employs a special centrifuge rotor that has from 2 to 30 or more sets of chambers. Each set has three chambers that are arranged radially and connected by a channel. The outer chamber in each set is equipped with windows so that the absorbance or fluorescence signal can be measured on the solution in the chamber.

The operation of the centrifugal analyzer is indicated in Fig. 5.5. Prior to the start of the run, the chambers closest to the center of the rotor are filled with measured amounts of the samples, and the chambers next furthest from the center are filled with measured amounts of reagent. (Special rotors allow sample or reagent to be added later, when the rotor is spinning. In the interest of brevity, only the simpler design is discussed here. For a more complete description of the centrifugal analyzers and for newer developments, see references 21, 48, and 72.)

The loaded rotor is then placed in the analyzer, and the motor is started.

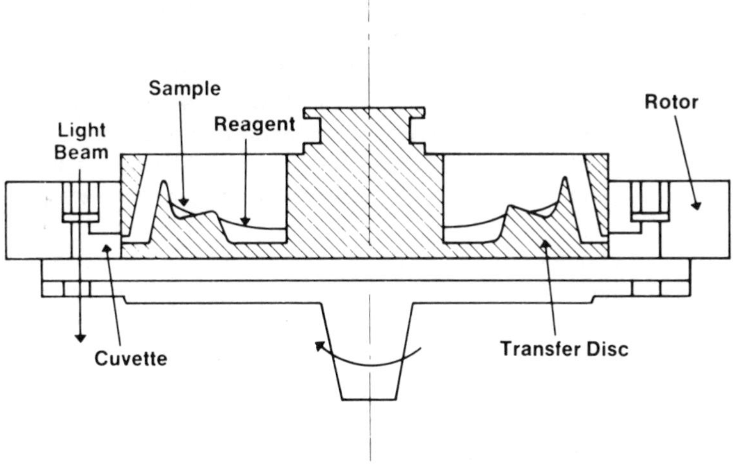


**Fig. 5.5.** Diagram of basic centrifugal analyzer. (Courtesy Union Carbide Corp.)

Centrifugal force moves the reagents into the sample chambers and the mixtures into the outermost chambers. As the rotor continues to spin, the outer chambers go by a measuring point where a light beam passes through the chamber to a detector. A new measurement of transmission is obtained for each cell with each rotation of the rotor. The centrifugal analyzer includes a computer that interprets the readings and derives the rate of each reaction and from that the concentration of the analyte in each sample.

The special advantage of the centrifugal analyzer is that it permits the measurement of rapid reactions. With good design, the first transmission measurement can be made within milliseconds after the reactants mix. New transmission measurements are made each time the rotor makes one revolution. A second advantage is that several samples are analyzed at the same time, so that one instrument can carry out a large number of analyses per hour. The limitation on analysis rate is usually the rate at which rotors can be cleaned and reloaded.

## G. OTHER KINETIC ANALYZERS

Some automated analyzers are expressly designed for reaction rate methods in which conditions are established so that the rate is proportional to the concentration of the analyte. The centrifugal analyzer is an excellent example. It provides a rapid mixing of sample and reagents and repeated measurement of absorbance over a period of time. (The centrifugal analyzer can also be used for assays other than kinetic methods.) Batch analyzers are also often used in a kinetic mode. The reagent and sample are mixed just before the cell is placed in the colorimeter, or sometimes the instrument is designed so that a final addition of sample or reagent can take place with the cell already in place for measurement.

Flow-through colorimeter cells can also be used. In the most common form, a reaction mixture is pumped in, the pump is stopped while the reaction rate is measured, and the following solution then displaces the one for which the measurement has been completed. A special version of this scheme for segmented flow analyzers (10,51) includes provision to divert the flow around the cell during measurement. A diversion valve is needed because of the springy nature of air-segmented streams; it is difficult to start or stop such streams quickly.

Another kinetic method for CF analyzers flows the same stream through two or more colorimeter cells in series, with a length of tubing between the cells to give a fixed delay. In this way, the absorbance is measured at two or more time intervals. For this method to work well, the instrument should be built so that the bubbles can be passed through the flow cell. This requires "electronic debubbling" circuits that inhibit absorbance readings while a bubble is in the flow cell. In unsegmented CF analyzers or in segmented flow analyzers that require debubbling of the stream before the colorimeter cell, the peak absorbance declines between measurements even if there is no reaction, because of a change in the shape of the peak as the solution moves between flow cells.

The advantage of kinetic analysis is that there is an inherent correction for sample blank absorbance, since that blank does not usually change with time. However, there can be competing reactions that cause the absorbance to change with time, one of the components of the reaction mixture may become depleted, or some other circumstance may affect the relation between reaction rate and analyte concentration. When such problems are expected, it is advisable to introduce a check to ensure that the rate of reaction is not changing with time. The simplest check method is to make three readings, equally spaced in time. Each pair of readings gives one measure of the rate, and the two should agree within the measurement tolerance if the rate is constant during the analysis. Some kinetic analyzers make use of more absorbance readings and more complex checks on constancy of rate.

Kinetic analyzers are most frequently used in clinical chemistry. The initial application was the measurement of enzyme activity in serum; the amount of certain enzymes in the serum is of diagnostic value for many medical conditions (6,24). Kinetic analyzers can also be used to determine compounds such as glucose that are enzyme substrates (6,38), and for other substances when the analytical reaction can be made to take place at a convenient rate. Kinetic methods are less often used for industrial analyses.

## IV. MULTICHANNEL ANALYZERS

In the common case that two or more determinations are required on each of a series of samples, it is advantageous to develop automated analyzers that do all the determinations on a single instrument. This consolidation saves the costs of dividing the sample among several instruments and later combining results from the individual instruments into a single report. The multichannel analyzer usually costs less than a number of single channel analyzers.

Two or more CF analyzers are conveniently combined in a single instrument, such as the one in Fig. 5.6. The AutoAnalyzer II pump can accommodate as many as 24 pump tubes, so a single pump can serve two, three, or even four analytical channels. A single sampler can serve an almost unlimited number of CF channels; the tube from the sampler is connected to as many T-fittings as there are analytical channels, and an aliquot of the sample is withdrawn at each fitting. Additional consolidation may be achieved by using a two-pen or three-pen recorder so that two or three analysis results are shown on a single chart. A computer or special analog or digital computation circuits can sense the colorimeter reading for each channel, convert the reading to concentration by reference to stored calibration data, and print the results on a report form.

Analyzers have been built expressly for multichannel application. The first instrument of the kind was the SMA 12/30 system; it performed 12 determinations on each of 30 samples/h. The instrument was soon modified to process 60 samples/h (SMA 12/60 system) and for six parallel determinations (SMA 6/60

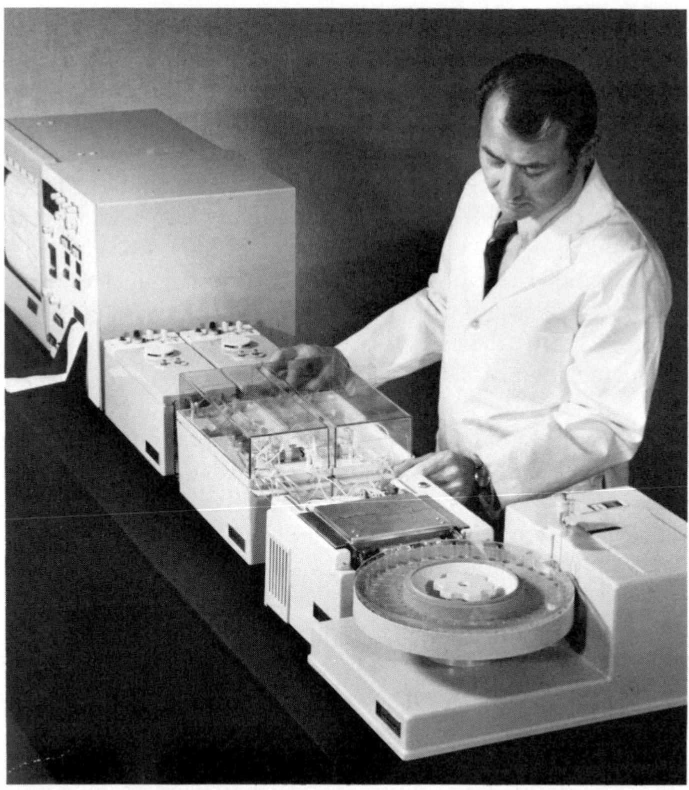

**Fig. 5.6.** Dual-channel AutoAnalyzer II (courtesy Technicon Corp).

**Fig. 5.7.** Technicon SMA 18/60 (Sequential multiple analyzer for 18 analytes at 60/hr).

system). Figure 5.7 shows an SMA 18/60 system, in which a 12-channel and a 6-channel instrument share a common sampler.

One problem in running this type of instrument is that the individual analytical channels have to be adjusted so that the several analytical results for one sample are shown on the chart recorder in the correct time sequence, or are available to the data processing circuit at the correct time. The adjustment procedure is known as "phasing." It is ordinarily done by adding to or reducing the length of tubing before each colorimeter cell, so that the elapsed time from sampler to colorimeter cell for each channel is such that all channels for a given sample arrive at a detector within one cycle time and are evenly spaced for proper display. Thus on the SMA 12/60 system, arrival times are 5 sec apart so that all 12 channels can be displayed in 60 sec.

The most modern SMA systems include a computer. One immediate gain of including the computer is that it eliminates the need for phasing. The computer can store each of the analytical results until all the data have been assembled for a given sample, and then print out the complete analysis report. Other operations that the computer can perform are discussed in Section VIII.

SMA systems with computers are the SMAC system (39), a 20-channel instrument, and the SMA II system, which can have up to 18 channels. The computer in the SMA II system does a mathematical correction for carryover, which makes it possible to speed up the analysis rate from 60 to 90 samples/hr. The SMAC system has an advanced CF hydraulic system that enables the analysis of 150 samples/h without carryover correction.

Multichannel analyzers may also be built on the discrete principle. The most straightforward method is to mount several parallel and independent one-channel analyzers on a common chassis, with a mechanism to transport the sample past each sample intake in turn. Other analyzers achieve some commonality, such as providing for a single sample intake mechanism to dispense aliquots of sample to each analyzer channel.

An advantage that discrete multichannel analyzers have compared to CF systems is that the discrete analyzers can more easily be built to provide selectivity. The SMA systems must aliquot sample and reagents to each channel for each sample, whether or not that particular analysis is needed on that sample. The batch analyzers usually do not have this limitation. The operator can control the machine by pressing buttons or throwing switches, or by some other means, so that unwanted analyses are not done. The pumps for dispensing sample and reagents are turned on only when that sample is being processed for the desired analysis. The benefits that come from selective multichannel analysis are that less sample and less reagent may be consumed. However, the savings may be less than the cost of the hardware to make the analyzer selective.

The SMA systems need fairly small quantities of sample. The SMAC system can do 20 determinations on as little as $\frac{1}{4}$ ml of serum. It is only rarely that the laboratory must get along with less than that volume. Selective analyzers may use less reagent than do instruments that do all determinations on each sample, but they may be more costly by virtue of the extra mechanisms that make them selective. The equipment costs and reagent costs may balance out rather evenly. Either selective or nonselective analyzers may prove to be the more economical for a given application. The cost analysis must take into account equipment and reagent costs, and other factors such as the average number of determinations required for each sample. If the reagents are unstable and large amounts are generally discarded unused, there is no saving on reagents in a selective analyzer.

The Technicon Monitor IV instrument is designed expressly for industrial applications. It is for use in or outside of the laboratory, and it can be left unattended for up to a week. The Monitor IV instrument is most often applied to the measurement of impurities in air or water. Special sample intake devices are available for these purposes, including a sparger to extract impurities from air, in-line filters to remove solids from water, and sampling valves to enable samples from several sources to be directed to the one analyzer.

## V. UNIT OPERATIONS IN AUTOMATED ANALYSIS

If one's first encounter with an automated wet-chemical analyzer involves something like a multichannel clinical analyzer (perhaps one of Technicon's SMA 12/60 or SMAC instruments), he or she is likely to be overwhelmed by its apparent complexity. This is no more surprising a reaction than when the average nontechnical person peers under the chassis of a color television set while the knowledgeable repairman searches out a faulty module. In both cases, the seemingly complicated circuit (whether hydraulic or electric) can be reduced to a series of far simpler "unit" operations. In this section, we consider several of these simple tasks.

In any analytical procedure, certain basic operations are nearly always included in the total analysis. For example, we always require a sample that is in a suitable form for analysis. If the sample is a liquid, it may only be necessary to move a defined volume of it into the analytical system. If the initial concentration is too high, the sample must be diluted; if it is too low, a concentration step is in order. Solid samples may require disintegration in order to achieve dissolution of the active ingredient, followed by removal of insoluble constituents by filtration or decantation, or the removal of soluble constituents by dialysis, extraction, or distillation. Each of these procedures is an example of a "unit operation" which can be considered part of the sample preparation step. We deal with a few of the more important ones individually.

## A. SAMPLING

When the sample is a liquid, one may only be required to transfer an aliquot of it from one container into another. Automated aliquoting is accomplished in discrete (batch) systems by either of two standard methods, a reversible peristaltic pump or a pair of mechanized syringes. In both cases a movable probe is provided which serves the dual purpose of aspirating a small volume of sample from one vessel, and then dispensing the sample with a selected volume of diluent into a second vessel in which a reaction and/or incubation takes place.

The peristaltic pump version is diagrammed in Fig. 5.8. With the probe in position A, the pump operates left-to-right and aliquots a small portion of the sample from its container (frequently on a turntable holding many sample cups). The probe then moves to position B and the pump operates right-to-left to dispense the sample and a defined volume of diluent into a reaction vessel. The pump may be driven by a stepping motor, and the number of steps (operator selectable) dictates the volumes moved. Typically one might aspirate 0.1 ml and dilute it with 0.9 ml to achieve a 1 : 10 dilution. Needless to say, the sample volume is sufficiently small so as not to travel any appreciable distance

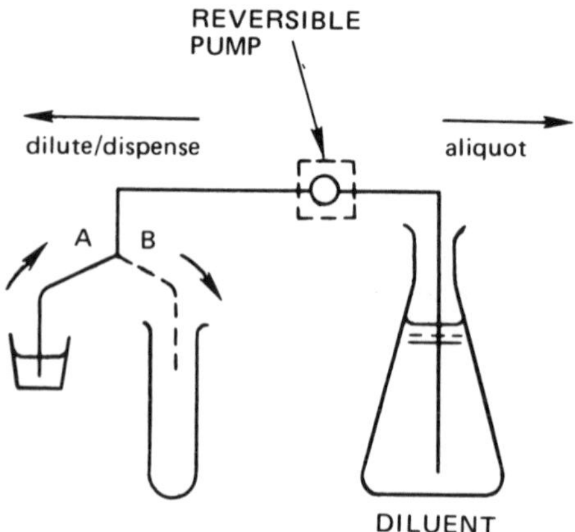


**Fig. 5.8.** Discrete liquid sampler (diluter/dispenser)—peristaltic pump version.

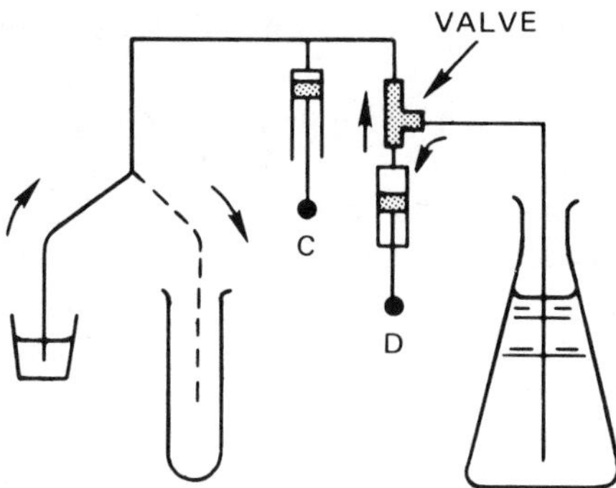


**Fig. 5.9.** Discrete liquid sampler—syringe version.

from the end of the probe—certainly never back through the pump and into the diluent container. Various manufacturers offer a variety of features such as hand-held probes, combination probe/turntable assemblies, provision for washing the probe between samples, and sometimes the aspiration of a small bubble of air to separate the sample physically from the diluent.

Figure 5.9 depicts a two-syringe version of an automatic diluter/dispenser. With the probe in the sample cup, syringe C draws in a defined volume of sample while syringe D is likewise charging itself with diluent. The valve prevents syringe D from affecting the sample volume. When the probe is over the reaction vessel, both syringes empty, dispensing sample and diluent through the probe. The syringes can be driven electrically, pneumatically, or manually. A great many instruments are available; a major reference (45) lists 27 diluters and 49 pipetters.

Figure 5.10 shows the hydraulic arrangement for sampling and diluting simultaneously in the CF technique. The sample probe alternates between successive cups in the sampler's tray and the wash vessel, the relative times in each position defining the sample/wash ratio. The air-segmented diluent and the sample are mixed in coil M as described previously. When reaction times

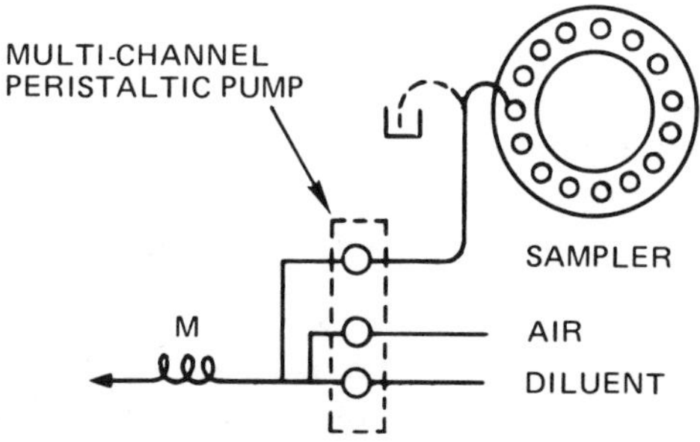


**Fig. 5.10.** Continuous liquid sampler.

are very short (e.g., 1 min between mixing and detection), air segmentation *may* not be required.

Aliquoting is accomplished in CF systems by choosing the correct combination of tube size and pumping time. Since the multichannel proportioning pump employs rollers of constant linear velocity, it follows that pump tube inner diameter controls the pumping rate. The sampler's probe aspirates sample from its container for as long as its adjustable cam permits. Sampling rates are typically between 10 and 150/h, with sample-to-wash ratios varying between one and nine (but special circumstances may require values outside these ranges). Thus a pump tube flow-rated at 1.20 ml/min aspirating a sample from a sampler set for 60/h and a sample/wash ratio of 2/1 would aliquot 0.8 ml of sample/test. Between samples the system would aspirate 0.4 ml of wash liquid.

There are alternatives to peristaltic pumping. Bochinski (8) has proposed a programmed fluid sampling and analysis apparatus in which a portion of a continuously flowing liquid sample and various reagents are defined in tall, narrow cylinders (by adjustable overflow points) and transferred by gas pressure into a reaction vessel, the bottom portion of which contains a built-in photometer for quantitation. Boasting "no moving parts," the system is extremely simple and useful for batch analyses not requiring on-line mixing, extraction, or washing of fluid-containing lines.

When a sample probe moves from one sample cup to another, there is *always* some material transferred to (and therefore contaminating) the second cup. Sample probes must be washed in some manner to prevent this carryover, and many schemes have been devised to accomplish this washing. For example, in the Technicon systems the probe makes an intermediate stop in a wash vessel through which there is a continuous flow of fresh wash liquid. A more vigorous washing was proposed by Ohlin (50) wherein the probe passed vertically through a funnel-shaped collar that was only slightly larger than the probe itself. Suction was applied to the annular space to cause wash liquid flowing around the probe to flush away any loosely adhering sample deposits on the probe's outer surface.

## B. SAMPLE PREPARATION

Solid samples require somewhat different techniques, but the end result is generally to place the solid into solution so that it can be treated as a liquid. Easily dissolved materials may be disintegrated or triturated rapidly. Some materials may require digestion.

### 1. Disintegration or Trituration

Tablets and similar samples can be disintegrated and dispersed in either aqueous or organic solvents in devices such as the Technicon SOLIDprep module. Liquid volume is adjustable from 25 to 125 ml, and a ceramic rotor and stator produce a grinding action at one or two of six selectable speeds. A carousel on top of the instrument provides for the unattended preparation of up to 20 samples, at cycle times variable over the range of 10–40 samples/h. Capsules, pills, and so forth that are analyzed on a single sample basis need not be weighed, whereas powders, food samples, and tissues, for example, must be preweighed in order to properly report results.

Although the SOLIDprep module is equipped with jets designed to wash thoroughly the contents of the sample cups quantitatively into the homogenizing vessel, these jets are sometimes ineffective with samples such as lettuce leaves. An alternate method of handling such samples is to roll them up in an inert material such as a 4 in. square of aluminum foil, forming a capsule-shaped sample which then falls easily out of the sample cup and even precludes the need for wash jets. This scheme has been proposed for use with food samples being analyzed for pesticides, and the metal foil is completely disintegrated and is unlikely to present any problems to the valving system of the instrument.

Bartels et al. (5) have developed an automatic mortar and pestle that wet-grinds solid samples. It is one component of an automated batch analyzer, which also includes a centrifuge to remove solid particles from the dissolved samples.

### 2. Dissolution

Dissolution studies on pharmaceutical tablets are well known to analytical chemists, and several methods have been proposed to simulate the physicochemical release of the active material from the composite tablet. Hanson's dissolution apparatus and the Technicon SASDRA* system are among the available instruments that can analyze the contents of such devices as the rotating basket and generate a graph of the time dependence of the release of the active ingredient. When very small amounts of solution are withdrawn from the dissolution vessel, no significant change in concentration from the true value results from this volume change. If appreciable volumes are withdrawn, however, the operator must compensate either hydraulically or mathematically.

Burns (9) has devised a solid sampling device for monitoring a parameter in a flowing solid. A bead-chain was drawn continuously through a tube whose outside diameter was just slightly larger than the bead diameter, thus trapping solid material in the spaces between the beads. Powders were dispensed continuously into a magnetically mixed vessel where any contained water was extracted into absolute ethanol containing the moisture indicator $CoCl_2$. The color change at 660 nm was related to water content.

### 3. Digestion

Digestion of samples prior to analysis is sometimes required, as with protein-containing materials prior to total Kjeldahl nitrogen determinations. The classical technique involves prolonged heating with concentrated acid with peroxide as a catalyst, followed by neutralization and distillation of ammonia from

* SASDRA = sample acquisition system for dissolution rate analysis.

alkaline solution into standard acid. Back-titration gives the answer, if there are no losses and if digestion is complete. Continuous digestion has been performed in a rotating helix in which the reaction mixture is carried along by the spiral into the heated area where the digestion actually takes place. Evolved gases must be removed continuously and scrubbed from the air, and the liquid must be cooled and removed from the spiral before it reaches the end of the tube. The chemistry that follows is straightforward, but the cost, complexity, and limited sample-handling capacity of the continuous digestor has made it not popular.

A somewhat simpler means of performing certain digestions is found in single-purpose monitors, in which a tightly wound quartz coil surrounds a high intensity UV lamp. This approach is used by Technicon* for TOC (total organic carbon) monitoring in industrial streams.

Frequently a trade-off is perceived in which some operations, although they could be fully automated, are done manually in the interest of cost savings. Accordingly, batch-type digesters are available that accommodate 20 or 40 tubes at a time. Following this kind of parallel off-line treatment, the actual analysis is done serially (and automatically) in a conventional CFA system. One should never lose sight of the fact that, depending upon workload and the availability of personnel, some operations may be cost-effective only when performed manually.

Erni and Muller (29) have used the continuous digester (helix) in an optimization study of the simultaneous analysis of nitrogen and total phosphorus. Continuous digestion has also been used in the analysis of such pesticide residues as Dasanit (3) at 20 samples/h with accuracies of ±1% for various formulations of this insecticide.

## C. MIXING

In batch systems, mixing is generally accomplished by small stirrers that descend into each reaction vessel for a predetermined time, then rise up, still spinning, and fling material onto the inside surface of the vessel to avoid contaminating the next sample. Occasionally, mixing is done by external agitation of the entire vessel, and this is surely safer than any rotating probe within the reaction mixture. Another alternative uses a gentle stream of air for mixing, either directed obliquely at the surface to impart a rotating motion within the liquid, or actually bubbled through the liquid to create convection currents within the vessel.

In segmented CF systems, the end-over-end tumbling action produced by the helical coil provides for efficient mixing of each segment, as long as there is a density difference between the two liquids which are to be mixed. The flow hydrodynamics of a segmented stream also promotes mixing. This is described in Section III.B.

*Technicon Industrial Method #535-78IM, Feb. 1, 1978, revised March 12, 1981; Technicon Instruments Corp., Tarrytown, New York.

## D. WEIGHING AND PIPETTING

Weighing still seems to be largely a manual operation. There are, however, at least three approaches to its automation. Solids such as tablets are nearly always analyzed on a "per tablet" basis, but solids such as powders must be wieghed in some fashion so that analyses may be performed on a weight basis. (1) Samples can be placed on a balance, manually or automatically, and the weight of (for example) one scoop of the sample can be noted, again either manually or automatically, so that any associated data handling modules can incorporate the weight into the final calculations. Several companies today offer automatic balances with digital data output appropriately coded for direct computer entry. (2) Powdered samples can be "sprinkled" slowly onto a balance until a preset weight is obtained. This essentially makes every sample weigh the same, so calculations of compositions are simplified.

(3) Docherty (26) incorporated automatic weighing into a procedure for analyzing fertilizer. Samples were taken continuously from the manufacturing plant's conveyor belt, subdivided, weighed, and transferred to an analytical system, all without human intervention.

An early CHN analyzer included an electronic-null balance, which weighed the sample as it was placed in a tared combustion boat. Adjusting the potentiometer of the balance simultaneously varied an input to an analog dividing circuit, thereby providing weight-based answers. In another approach, the sample is weighed on-line and this weight is used to determine the volume of diluent in which the sample is dissolved or dispersed. The sample concentrations are always constant.

Methods for pipetting in automated apparatus have already been shown in Figs. 5.8 and 5.9. Renoe et al. (54) built a module for an automated analyzer that pipettes samples and reagents by weight into reaction vessels on a turntable.

## E. GROUP SEPARATIONS

Here we discuss several methods by which major separations may be performed. These separations generally precede more specific chemical reactions, and their purpose is to provide a sample cleanup to simplify subsequent operations. More often than not, these are *physical* separations and do not involve any chemistry per se.

### 1. Dialysis and Ultrafiltration

Dialysis is a well-recognized method for removing unwanted large molecules. It is performed routinely in many clinical analyzers that use segmented CFA, even those operating as fast as 150 samples/h. Although dialysis has been used for the simultaneous dilution of a sample, that is seldom its primary purpose and only its separation feature is addressed here. The simplest dialyzers for continuous on-line sample treatment consist of a semicircular groove machined into a planar surface, its mirror image machined into a second planar surface, and the two surfaces properly aligned with a piece of dialysis membrane sandwiched between the two surfaces. When liquid streams flow on each side of the separating membrane, small molecules in either stream pass (to some degree) into the other stream, but large molecules do not. Excess "wash" is minimized by employing segmented streams on both sides of the membrane.

Early dialyzers were submerged in constant temperature water baths, but current designs operate in ambient air. Channels can vary in length from about 3 to 24 in., residence time being selected as needed for the particular analytical method. Type "C" membranes (cellulose) are most common, although other materials may be employed when speed must be increased or when gases must be accommodated.

Ultrafiltration may be regarded as a special case of dialysis in which pressure is developed by restricting the outlet of one channel in order to increase the flow rate of that channel into the opposite one. In practice, one might pump 1.2 ml/min into the donor side and permit only 0.8 ml/min to flow out the exit, thereby forcing 0.4 ml/min to flow across the membrane—much more than would normally cross. This technique would be used when concentration is a potential problem, and one could ill afford to settle for the relatively low percentage of analyte that normally dialyzes into the recipient stream.

Dialysis or ultrafiltration can be used to remove particulates in addition to its usual value in removing macromolecules. This is particularly useful when aqueous samples are being treated for HPLC analysis. For example, fermentation broths generally contain particles that require such measures as passage through a Millipore-type filter in order to protect an HPLC column. But when sensitivity is not a problem, and when the analyte is dialyzable, then the usual cellophane membrane can provide for the simultaneous operations of dilution and filtration. An example is the determination of cephalosporins by the method of Miller and Neuss (47) preceded by AutoAnalyzer sampler, proportioning pump, and dialysis unit. Unmeasured, untreated beers were diluted and pumped through the donor side of the dialyzer and the recipient stream was debubbled and passed into the loop of an automatic HPLC injection valve. A simple timer maintained synchronization between the sampler and the valve.

### 2. Extraction

Extraction is an especially useful operation which is at best tedious when done manually, and which can be automated with relative ease. Extractions generally begin with an aqueous solution containing the analyte and contaminants. Usually, the analyte is isolated by extracting it into an organic solvent. The contaminants are sometimes extracted into the organic phase, leaving the analyte in the aqueous solution. Frequently, pH plays an important role, and the original solution must be buffered at the optimum pH for the extraction. The organic solvent (or mixture of solvents) may be either heavier or lighter than the aqueous solution, so the phase separator that follows the extraction coil may take different forms. The extraction takes place in a mixing coil (which may be either glass or plastic) where the two phases are forced into intimate contact with one another for a sufficient time to permit the analyte to distribute itself between the two phases, approaching the theoretical distribution coefficients. By proper choice of solvent and pH, this can be nearly 100% efficient.

Adler et al. (2) have described in some detail various considerations required in the relatively difficult extraction of body fluids such as blood, plasma, and urine. Precautions must be taken to avoid the formation of emulsions, because once formed, they are difficult to cope with. Unless there are special reasons for not doing so, the phase separator should be constructed of a material which is wet by the phase one wishes to save, whereas the extraction coil should be made of material which is wet by the other phase. Thus when urine is to be extracted by chloroform, a glass (hydrophilic) coil would be followed by a polypropylene (hydrophobic) phase separator.

A further discussion of phase separators is given below.

### 3. Filtration and Centrifugation

Treatments such as disintegration usually produce some insoluble debris that must be removed prior to the actual analysis. The manual steps of decantation or filtration have been automated in CFA, and many methods employ these schemes. If the unwanted particles are especially dense, one can simply permit their settling in the SOLIDprep's homogenization chamber prior to pumping out the supernatant solution. When density difference is insufficient, the suspension can be pumped out and subjected to continuous, on-line filtration in a module which uses long rolls of 1 in.-wide filter paper in continuous motion to present constantly a clean surface and avoid sample cross contamination. Different grades of filter paper can be employed, including one that is treated to render it hydrophobic and thereby capable of retaining traces of water as well as suspended particles, and passing only an organic solvent (which, presumably, has dissolved the analyte).

Burns and Lazer (19) have shown that it may be advantageous to remove protein by chemical precipitation rather than by dialysis. Peak spreading was reduced from 12 to 3 min by replacing the dialysis module with a continuous filter. In most analyses, however, the use of the dialyzer doesn't impart such poor "wash" into a system; note, for instance, that the SMAC instrument operates at 150 samples/h and uses dialysis in several channels.

Particle separation by centrifugation is rarely done in automated analyzers, but it is possible. The apparatus of Bartels et al. (5) incorporates a centrifuge, with provision for washing and steam-cleaning the basket between samples.

### 4. Decantation

Decantation can be implemented in CF systems (when density difference is sufficient) by providing a relatively long path in a horizontal plane (i.e., no mixing coils) to permit particulates in each segment to group themselves at the bottom trailing edge of each segment. A suitably shaped T-fitting can then direct the particles (such as agglutinated blood cells) to waste, while retaining the segmented supernatant for subsequent treatment in the system. The long path prior to the fitting need not be straight; it can be a large diameter coil (e.g., 4–6 in.) of three or four rotations, essentially in a horizontal plane, so that individual segments travel neither up nor down.

Smith and Wilde (60) have described a method of continuous decantation to achieve the separation generally required in immunoassays. Their apparatus consisted of T-shaped tubes oriented to produce flow directions 45° to the vertical.

## 5. Distillation

A little-used but very important analytical operation is the separation of components by a difference in boiling point. All heating coils discussed previously have been operated at temperatures somewhat below the boiling points of all components of the stream passing through. But if the temperature of an aqueous stream is raised to somewhat more than 100°C, the liquid vaporizes and steam emerges.

Reference to Fig. 5.11 will help in the following description. Phenol (b.p. = 182°C) has an appreciable vapor pressure at 150°C, and small quantities of it distill with water from aqueous solution at that temperature, thereby separating it from higher-boiling constituents that might otherwise react with color-forming reagents. The system shown operates at slightly reduced pressure. The 3.42 ml/min that is pumped into the heated coil contains both liquid and segmenting air. By pumping out 4.02 ml/min, the pressure is reduced and the air expands into the slight vacuum, which promotes vaporization at a lower temperature. Nonvolatile liquids emerging from the heated coil are continuously removed to waste (W) by the line labeled "nondistillables." Volatiles that survive the first (insulated) cylinder and reach the second (cooled) one condense

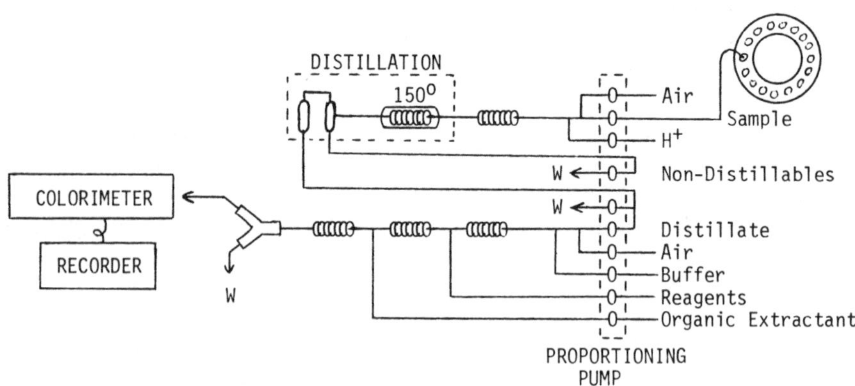


**Fig. 5.11.** Flow diagram (manifold) for automated analysis of phenol.

and are continuously removed for analysis. A colored complex (formed by reaction with alkaline ferricyanide and 4-aminoantipyrine) is extracted into isobutyly alcohol and measured at 460 nm (69).

Freistad (33) has employed continuous distillation to remove excess solvents in the automated analysis of several nitro-containing pesticides. Ott et al. (52) used continuous steam distillation in a fluorometric method for determining the pesticide Carbaryl.

### 6. Washing

In the manual practice of analytical chemistry, one frequently washes a precipitate by mixing, centrifugation, and decantation. But materials such as precipitates can be washed successfully only if there is a means of later separating the solid from the liquid. Although centrifugation doesn't lend itself very neatly to automation in CF systems, there are at least two schemes wherein a solid can be immobilized for a thorough washing.

Cohen and Stern (22) described an automated radioimmunoassay in which a magnetic field is the key to separation. Antibodies are bound to particles that contain magnetic material and a slurry of this solid phase is used as a reagent. After this solid phase medium is allowed to incubate with competing antigens in the sample and with the radio-label, the entire mixture is pumped through a tube which has been placed between the poles of an electromagnet. When the magnet is "on," that portion of antigen which is bound to the antibody is held by the magnet, and the unbound antigen (and everything else in the analytical stream) is washed away by the continuously flowing stream. After separation has taken place, the magnet is turned "off" and the bound labeled portion is valved to the flow-through cell of a scintillation detector. This technique is described further, with diagrams, in Section V.K.7.

Snyder et al. (62) described another immunochemical reaction for animal diseases such as hog cholera. In this automated enzyme-labeled antibody (ELA) test, plastic test tubes are coated on the inside with an antigen. Antibodies, if present in the serum of test animals, bind to the antigen on the inner surface of the test tubes. Washing then removes all excess serum. An enzyme-labeled conjugate is then added, and it binds to the antibodies *if* they were present in the serum under test. Again, everything not bound to this "sandwich" (plastic-antigen-antibody-conjugate-enzyme) is washed away by repeatedly filling and draining the test tubes. Finally, a substrate is introduced (one chosen to be split by the enzyme) and its colored product can be measured in a colorimeter and is an indirect measure of the level of antibodies in the serum.

## F. PHASE SEPARATIONS

In CF systems we are nearly always dealing with two phases, liquid and gas. Ocasionally, we also have a third phase, solids. Here we deal with the separation of phases on a continuous basis.

### 1. Liquid/Gas Separation

The separation of a liquid and a gas (usually air) is the simplest kind of phase separation. It is really nothing more than debubbling the usual segmented stream, and this has been adequately covered in an earlier section.

A not-so-obvious example, though, is the use of a gas-dialysis membrane to remove a portion of the gas between liquid segments. In the analysis of plant effluents for cyanide (68), the sample is acidified with a mixture of phosphoric and hypophosphorous acids and exposed to uv radiation to convert metallocyanides into simple cyanides. The hydrogen cyanide formed mixes with the air that was used to segment the reaction mixture, and some of this hydrogen cyanide diffuses through the gas permeable membrane in a dialyzer module and is absorbed in the recipient stream. Further reaction produces a colored product that is determined by colorimetry at 570 nm.

### 2. Liquid/Liquid Separation

The separation of two liquids is possible when they are immiscible, and this operation is performed every time an extraction is done. This has been covered in part above (see Section V.E.2). Frequently, when an organic extraction is required, one has already started with an air-segmented stream. One may choose to remove the air or leave it in, thus having either two or three phases.

Phase separators vary in geometry depending upon which phase is to be saved for subsequent operations, and whether or not this saved phase is to remain segmented, be debubbled and resegmented, and so on. To accommodate all the different possibilities, including stream-splitting, phase separators may have from three to six openings. Also, tube inner diameters may vary from 1 to more than 2 mm, construction materials may be hydrophylic or hydrophobic, and phase separators may or may not contain Teflon inserts for "directing" the organic phase along the correct path.

Although very high percentages of materials may be transferred from one phase to another, the overall efficiency of each separation is seldom over 80%. The reason for this will become clear. If one were to adjust carefully a phase separator and its in-flowing and out-flowing pump tubes for 100% removal of the desired phase, and the system were to change (for any of several reasons) in the wrong direction, then some portion of the unwanted phase would appear where it should not. This could negate the whole purpose of performing the extraction in the first place. Consequently, one generally settles for 70–80% of the desired phase, losing some of it with the unwanted phase, and allowing for long-term changes in the flow patterns with no danger of getting the wrong phase. For a single extraction, a 75% efficiency is acceptable, but for three consecutive phase separations of 75% each, one recovers only 42% of the starting material (*if* distribution coefficients were essentially 1.0 for each transfer).

Sansur et al. (57) employed as many as four phase separators in series, and still operated at 40 samples/h. Morphine, a drug of abuse, can be analyzed fluorometrically only after it is hydrolyzed (to release the bound form from its conjugate) and "cleaned up" chemically by extracting in and out of organic solvents at different pH values. With proper attention to fine details of hydraulic fittings, "wash" need not be a problem, even when a total of 40 pump tubes are required to analyze for a single parameter.

Finally, recall that "hydrophobic" filter paper is available which can continuously separate aqueous and organic phases by passing the latter and retaining the former along with particulates. Kawase et al. (42) used a phase separator with a membrane of tetrafluoroethylene polymer.

### 3. Liquid/Solid Separation

This has been adequately covered in preceding sections on filtration, decantation, and washing.

## G. TEMPERATURE CONTROL

Some reactions in which temperature is not an important factor can be performed at ambient conditions. At other times responses would be affected so seriously that temperature control is mandatory.

Constant temperature can be maintained in several ways. Reaction coils may be embedded in jackets through which water is circulated, or they may be encased in a solid heat-sink compound (paste) with a self-contained heater and thermostat. Large heated metal blocks with drilled holes are used to hold digestion tubes for Kjeldahl nitrogen analyses. Air baths have been employed where it is deemed inconvenient to use liquids for heat transfer. Solid heaters generally require considerable time to reach operating temperature, whereas the temperature of air can be changed rapidly (as in chromatography where oven temperature is sometimes changed during the analysis for improved resolution). Electronics can now give the analyst any desired degree of control he is willing to pay for, from $\pm 0.005°$ for kinetic enzyme analyses and bioassays to $\pm 5°$ for others.

## H. INCUBATION

When reactions are relatively short (a minute or two), there is no objectionable loss of resolution as the multisegmented samples wind their way through coils and tubes. But when reactions require tens of minutes or a few hours, the "wash" problem which goes with continuous flow makes some analyses nearly impossible. Biological reactions present the worst problems.

Burns and Hansen (18) mixed CF with batch processing (and have in fact alternated these techniques) to take advantage of what each has to offer. Fermentation broths were sampled continuously, filtered batchwise, pumped continuously and synchronously dispensed into individual vessels for discrete incubation, and finally transferred serially by continuous flow to a colorimeter for analysis. It was obvious to the authors that a 3-h incubation period precluded the use of a long delay coil, unless extremely low analytical rates could be tolerated. The instrument described could operate at 40–60 samples/h, depending upon the incubation period required (3 or 2 hr, respectively).

The storage coil (59) offers another way to reach incubation times of many hours. Successive segmented solutions of samples and reagents are pumped one after another into a very long coil of tubing. The pumping is stopped just as the first sample reaches the end of the coil. The ends of the coil may then be clamped shut, and the coil can be removed and placed in an incubator. When the incubation is complete, the coil can be connected into an analyzer, where the samples are then pumped out of the coil in sequence.

Just as some reactions are very *temperature*-dependent, others are extremely *time*-dependent. Kinetic enzyme reactions, as the term implies, require the analyst to note the *rate* of reaction in contrast to an end-point determination. This has been handled in at least three ways in continuous flow, and a comment or two about each scheme is in order.

### 1. Multiple Flow Cells

Assuming that three points are adequate to define the slope of a curve and establish that it is linear, one can pass a *reacting* mixture through three flow cells, serially, separated by 2 min each. The change in absorbance from cell 1 to cell 2 should equal the change from cell 2 to cell 3. If it does, the rate is known to be linear and can be easily determined from either or both slopes. Here the incubation time is not especially long, but it must be precise.

### 2. Stopped Flow

With appropriate valving, one can employ "programmed debubbling" and stop the flow through a flow cell when its light path is bubblefree. Burns (12) has shown that during a 1-min period at constant temperature, clinically significant enzymes can be determined by displaying the derivative of the absorbance change. This same principle (with unsegmented streams) is the basis of several popular kinetic enzyme analyzers, and stopped flow methods are frequently used for research on reaction kinetics.

### 3. Reversed Flow

An incubation gradient can be generated in such a manner as to provide an incubation time that is *twice* the cycle time. It is only necessary to synchronize a four-way valve with a sampler in such a manner that the flow through a coil is reversed once each cycle. This takes advantages of the "first in, last out" principle, permitting a 1 min cycle time to produce a 2 min incubation time for each sample. Burns (11) has shown that this can operate at 150 samples/h (i.e., the 24-sec cycle time provides a 48-sec incubation period of which about 30 sec is used to define the rate of the reaction).

## I. CONCENTRATION IMPROVEMENT

In many analytical procedures there is more analyte than is actually required, and a dilution may be in order. But frequently the concentration is too low and some means must be provided to increase it prior to detection. Here are two approaches to the problem.

### 1. Extraction

When the analyte is transferred from one phase to another (e.g., from aqueous to organic), nearly any ratio of the two solvents can be chosen. If the ratio is 1 : 1 and the distribution coefficients are favorable, there may be no net change in concentration. But if the contents of two volumes of aqueous solution are extracted into one volume or organic solvent, there is a twofold concentration improvement. In a CF system, flow rates of 2 ml/min of sample-containing aqueous media and 1 ml/min of appropriate organic solvent can accomplish this same 2x concentration improvement. Accordingly, one can extract into ever-decreasing volumes and achieve an ever-increasing concentration. It is not unreasonable to expect a gain of an order of magnitude during a triple extraction.

### 2. Evaporation

Decreasing the volume of a solution of nonvolatile analyte is another obvious method of obtaining a concentration improvement. Evaporators of all kinds are available to the analyst, each designed for a specific purpose. Rotary evaporators provide greatly increased surface area and low pressure to achieve rapid removal of solvent for individual samples, and heated blocks (sometimes with the help of a stream of nitrogen) provide for the simultaneous evaporation of solvents from multiple tubes.

When one reads in the literature "...evaporate to dryness and take up in...," he or she assumes that all hope for a fully automated system has gone, and a manual step must surely be required. This is quite frequently a need in the HPLC realm, where the best solvent for extracting a material from its native matrix may not be the best solvent for introducing the sample into a column. This solvent exchange that is called for can be performed on an EDM module. Diagramed in Fig. 5.12, the EDM module consists of an inert matrix of circular cross section, wound around two pulleys, and threaded through a glass evaporator tube. The evaporator tube has a sidearm going to vacuum, so there is a continuous flow of air through the tube, over the surface of the inert matrix. The sample (in the original solvent) is pumped onto the matrix at a point where the fast-moving air causes it to flow in a sheath stream over the surface of the matrix. About halfway down the evaporator tube, all the solvent is gone, and the residue is carried out of the tube to the take-up station where it is redissolved in a second solvent.

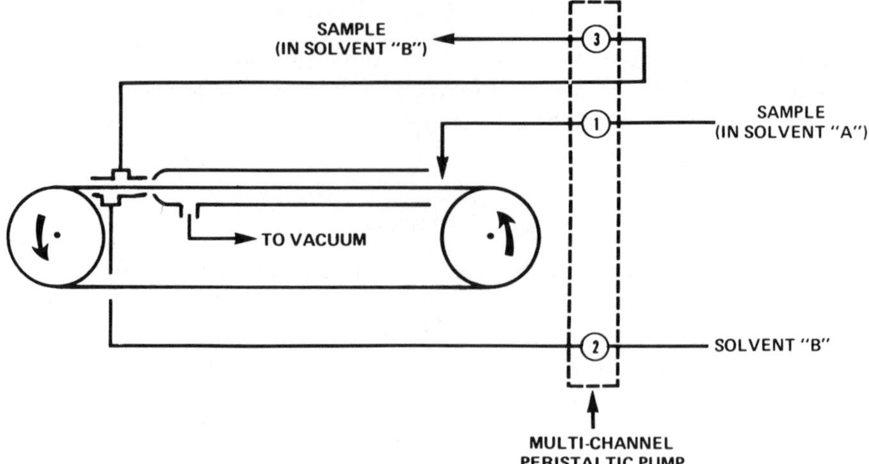


**Fig. 5.12.** Evaporation to dryness module (EDM) for continuous solvent removal.

The EDM module can also be used to concentrate trace levels of substances so they can be analyzed without special detectors. By stopping the matrix for 3 min and letting the solution evaporate on the right-hand end of the matrix, one can build up a sizable residue. The matrix can then be restarted and passed through the take-up station in 0.3 min. If the solvent flow rates are the same, then a 10-fold concentration improvement is obtained. By combining higher solvent ratios with the stopping of the matrix, samples can be concentrated even more. This should prove particularly useful to pesticide analysts where trace levels are generally encountered.

Provision has been made on the EDM module to heat the air that is drawn into the evaporator tube, to substitute an inert gas for air, to filter the air to remove UV-absorbing components, and to protect the sample from light by using its opaque cover.

## J. CHROMATOGRAPHY

Although the principle is the same in both cases, chromatography is employed both as an analytical tool and as a cleanup procedure prior to analysis. As an analyzer, it must of course be combined with a suitable detector (UV, fluorescence, electrochemical, etc.). Also, considerable effort must usually be expended in getting a sample ready to go into the chromatograph. But once a sample is properly prepared, automatic samplers and injection valves can provide for unattended analyses, complete with report generation when that option has been purchased.

The use of gel permeation chromatography (GPC) as a means of cleaning up samples for pesticide residue analysis was introduced by Stalling et al. (64) and automated that same year by Tindle and Stalling (73). Applied to fish lipids, it allowed unattended processing of up to 23 samples with good reproducibility

(CV's = 5%) and low carryover (less than 1%). The 23 sample loops were loaded manually with 5 ml each of tissue extracts, and 24 h later the tubes in a fraction collector contained the appropriate fractions from the chromatographic column, ready for analysis by whatever method was appropriate.

Florisil column cleanup has been a popular method of preparing samples for GC analysis. Stimac (66) modified a standard AOAC procedure, eliminating the need for further cleanup of many chlorinated pesticides. Getz et al. (35) used a column packed with a carbonaceous resin (Ambersorb XE-340$^{TM}$) to clean up samples prior to analysis via thin layer chromatography (TLC), GC, or HPLC. Material for analysis was placed in a SOLIDprep module, and the resulting slurry was filtered with the continuous filter module. A system of valves then permitted loading samples onto a column, eluting cleaned up samples from the column into a fraction collector, or regenerating the column. The instrumentation was used with soil samples containing several pesticides and their metabolites.

Sometimes analytes exist at such low levels that a concentration step is required before analysis. Chromatography can come to the rescue on occasion. Euston and Baker (32) addressed this problem for trace levels of pollutants in water and proposed a time programming scheme for on-column concentration as part of an automated HPLC analysis. In reverse phase chromatography water is normally used as the weaker component of the mobile phase. But if the water contains an analyte that is held by the column packing, then the sample itself becomes the mobile phase and its analyte is concentrated on the top of the column. After an appropriate volume has passed through the column, the mobile phase is switched to one that will permit separation and elution from the column, onto an analytical column for finer separation and detection.

As a unit operation, then, chromatography may be considered both as a cleanup procedure and as a means of separating components (which may have already been cleaned up by alternate schemes) prior to analysis.

## K. DETECTION

The most common means of detection in automated analysis is probably colorimetry, that is, measuring absorption of a solution in the visible region of the spectrum. One can broaden the spectrum to include uv analysis, make the method more selective by employing fluorometry, or incorporate alternate techniques such as flame photometry, electrochemical detection, or thermal analysis. We shall touch upon each of these, using examples from the more than 500 methods in current use by industry and involving continuous flow principles.

### 1. Colorimetry

The determination of cyanide in an industrial plant effluent is the example to be used for colorimetry. The sample is acidified and exposed to uv radiation to convert metallocyanides into simple cyanides. The hydrogen cyanide formed is separated from the reaction matrix by passage of the stream over a gas perme-

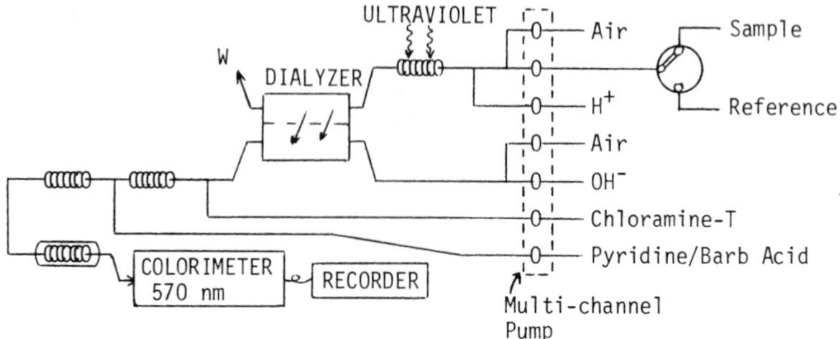


**Fig. 5.13.** Flow diagram for automated cyanide analysis in a process stream.

able membrane. The liberated hydrogen cyanide is absorbed in dilute alkali and converted to cyanogen chloride which reacts with pyridine and barbituric acid to yield a reddish-violet colored complex. The colored material is measured at 570 nm in a 15 mm flow cell, giving measurements in the range of 0–1.0 mg/liter. The flow diagram (Fig. 5.13) shows (in addition to the colorimeter) the hydraulic arrangement incorporating a UV digestor and a dialyzer module with a gas membrane. The wavelength is determined by interference filters.

## 2. Ultraviolet Spectrophotometry

An automated method for dissolution analysis of erythromycin in pharmaceutical tablets is handled by differential spectrophotometry (i.e. a blank is run simultaneously for correction). A filtered solution of the sample is equally divided between a sample stream and a blank stream. The blank stream is first treated with acid, then proceeds as with the sample stream. The UV absorption is measured at 236 nm (4 nm bandwidth) in a 15 mm flow cell. Analyses are performed in the range 8–400 $\mu$g/ml. Although not shown in the diagram (Fig. 5.14), the sample actually comes from a SASDRA module which simulta-

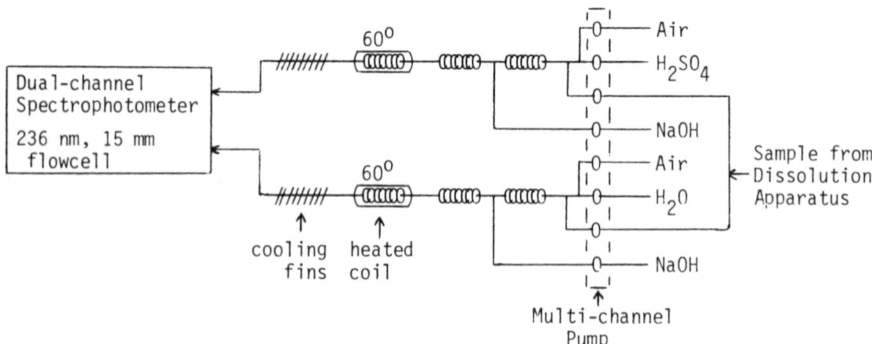


**Fig. 5.14.** Flow diagram for automated erythromycin analysis using a second channel for blank subtraction.

neously removes samples from six dissolution vessels and a standard vessel, then analyzes them serially, repeating this process at, for example, 15 min intervals. Although colorimeters can use a relatively inexpensive light source (ranging from an automobile headlamp to a tungsten/halogen lamp), uv instruments require either a hydrogen or a deuterium lamp. Also, interference filters are usually adequate in the visible, but a grating or prism monochromator is needed in the UV. This means far more expensive power supplies and quartz optics; hence costs are higher.

### 3. Fluorometry

The example of this technique is particularly interesting because it incorporates (in addition to the fluorometer) an on-line cleanup procedure. Roy and Buccafuri (55), who developed this totally automated method for calcium pantothenate, used separate on-stream slurries of magnesium trisilicate and Dowex 50W-X4 resin for sample purification. The magnesium salt removes riboflavin from the sample stream, which is then filtered with the Continuous Filter module. The filtrate is mixed with the Dowex resin which removes $\beta$-alanine, which reacts with *o*-phthalaldehyde at pH 9.3 in the presence of a reducing agent to form a highly fluorescent compound. The fluorogen is excited at 340 nm and emits at 455 nm. The multivitamin tablets or capsules may contain 0.5–25 mg/dose. The analytical rate with tablets is 20/h, as depicted in Fig. 5.15.

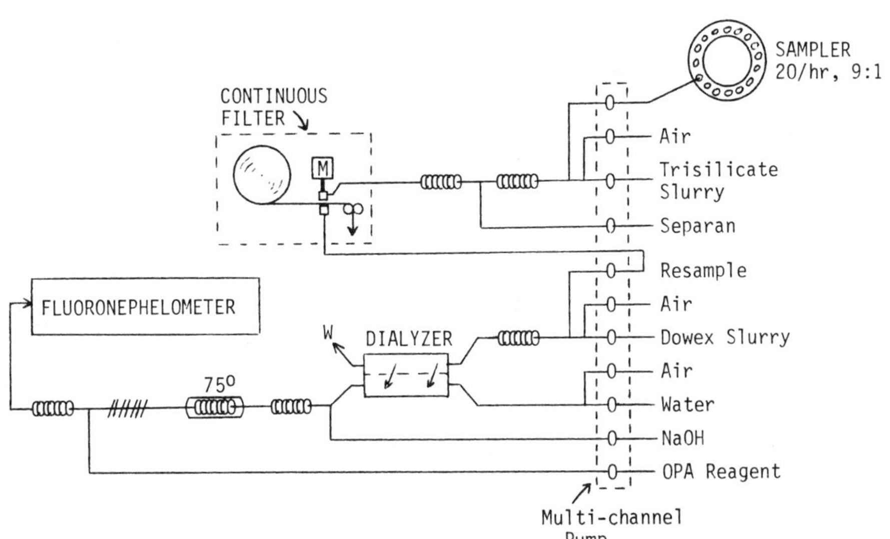


**Fig. 5.15.** Flow diagram for automated analysis of the vitamin calcium pantothenate.

### 4. Flame Spectroscopy

It is often useful to combine automated analysis modules with flame photometry or atomic absorption apparatus. Graham (37) reviewed several uses of automated sample cleanup for atomic absorption. They included dilution, filtration or dialysis to remove particles or large molecules, acid digestion to eliminate organic compounds, and solvent extraction. The last is especially useful, because it can separate and concentrate the analyte and because analytical sensitivity in flames is often improved when the sample is in an organic solvent.

Gouldon and Brooksbank (36) devised an interesting automated method to determine arsenic, antimony, and selenium in waters, using atomic absorption in a tube burner. In this method, the samples are mixed with a solution of hydrochloric acid, stannous chloride, and potassium iodide. The stream is segmented with an inert gas (argon) in order to maintain nonoxidizing conditions. The stream passes through a first heating coil, it is combined with a second stream bearing a suspension of aluminum particles, and it then passes through a second heating coil. The gaseous stribine, arsine, and hydrogen selenide are stripped in a special column with a flow of argon, which passes into the tube burner. The analytical system can process 40 samples/h.

Automated methods are equally suitable for nonflame atomic-absorption analysis, as has been demonstrated for the determination of mercury (28,40,44).

### 5. Electrochemistry

Virtually any electrochemical method can be applied in automated analysis. Voltammetry in flow cells (hydrodynamic voltammetry) has become an important field of scientific research in recent years, and it has also found significant practical application; one example is the use of flow-through voltammetric detectors in liquid chromatography. In this section, we emphasize the applications of electrochemistry in automated continuous flow analyzers, where the electrochemical procedure is combined with one or more of the other unit operations.

Although colorimetry or spectrophotometry usually measures the bulk composition of the solution, most electrochemical measurements respond to the composition of the solution at the surface of the electrode. The response time of the electrode and the size of the response for a given solution composition are both improved when the solution is well stirred. It is therefore important to design the electrode and the measurement cell to promote good stirring near the electrode surface. The air bubbles in segmented flow streams promote good stirring, and it is often possible to pass the bubbles through the electrode cell.

Several manufacturers make analyzers with flow-through ion sensitive electrodes, especially for electrolytes in serum. A description of the flow-through electrodes in the SMAC analyzer has been given (53). A feature of these electrodes is that the bubbled stream is passed through the electrodes. Prior to this development, it was believed that passing air bubbles through the electrodes would cause excessive electrical noise, but techniques have been developed that circumvent the problem (53). The passing air bubble does not wipe dry the surface of the electrode or of the tubing between the ion sensitive electrode and the reference electrode. The remaining film of liquid is sufficient to assure electrical connection between the sensing and reference electrodes at all times. The use of a differential amplifier that is grounded to the flowing stream and an active electrical filter also aid in minimizing electrical noise, including the small voltages generated by rubbing of the flow tubing in the peristaltic pump.

An automated polarographic analyzer developed by Cullen et al. (23) for pharmaceutical tablets incorporates several novel design features. These investigators needed to scan a range of 900 mV at a rate of 5 mV/sec. They could have stopped the flow of solution in the polarographic cell for the required period, but they chose instead to maintain the sample flow, which made it necessary to prolong the sampling time for a total of four minutes to assure adequate washin and washout time and a steady state for 3 min. The solution was adequately deoxygenated by segmenting the flowing stream with bubbles of nitrogen. Another feature of their system is that it incorporates a pair of storage coils between the SOLIDprep sampler and the rest of the analytical system to speed up the system throughput. The polarographic system draws its sample from one of the storage coils while the sampler fills the other. The result is to double effectively the sample throughput rate because the polarographic system does not have to be idle while the sampler is cleaning itself and is grinding and dissolving the next tablet.

Some electrochemical cells require a larger surface volume than can be achieved in a tube, and packed-bed systems can be used in these cases. A segmented stream should not be passed through a packed bed, for that would fragment the air bubbles. Peak spreading in a packed bed reactor can be minimized by proper design (25), but a residence time longer than several seconds may be possible only at the expense of pumping pressures higher than peristaltic pumps can produce. Nevertheless, packed-bed electrochemical cells can be versatile components of a CF analyzer, as was shown by Blaedel and Strohl (7). They showed that it is possible with their cell to quantitatively oxidize Ce(III) to Ce(IV) for use as a reagent. Another application was to reduce Fe(III) to Fe(II) or U(VI) to U(IV) followed by titration with permanganate. By controlling the conditions of the analysis, they could measure iron selectively in the presence of uranium or the two elements together. The packed-bed cell was also used to separate metal ions from one another, either by selective deposition or by electrodepositing all of the metal ions and selectively stripping off one or more.

### 6. Thermal

Potassium ions react with sodium perchlorate to form a precipitate of potassium perchlorate. The reaction is exothermic, and the resulting change in enthalpy is proportional to the potassium content of the sample. Measurements can be made of this heat of reaction with a thermometric detector. In the

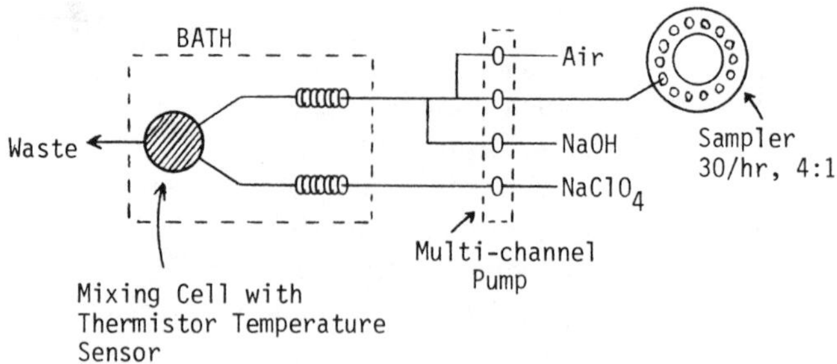


**Fig. 5.16.** Flow diagram for automated thermal analysis of potassium.

automated method portrayed in Fig. 5.16 the flow cell is thermostated above ambient temperature, and the detector is in thermal contact with the reaction mixture in the magnetically stirred flow cell. Sodium hydroxide is added to mask interference from ammonium ions (present in fertilizers, which are often analyzed by this method). Samples can be run at 30/h in the range 6–26% $K_2O$.

## 7. Radioactivity

Radioimmunoassays (RIA) combine the immune reaction with the detection of a radioactive species. The immune reaction is an extremely sensitive one in which antigens (Ag) and antibodies (Ab) combine in a manner that permits the measurement of either one. One way of detecting trace levels of an Ag is to dilute the unknown level with a known level of a radio-labeled Ag, let both compete for an Ab, then separate the free from the bound form. These are the steps that must be performed:

1. Pipette sample, standard, radio-label, and Ab.
2. Incubate.
3. Separate free and bound forms.
4. Count radioactive bound Ag with scintillation detector.
5. Calculate isotope dilution (permitting the analysis of very low levels of the unlabeled Ag).

Manually, this may be accomplished batchwise in test tubes, and the separation step can be very time-consuming. However, automation has been achieved by wrapping magnetic particles in a material to which the Ab can be attached. One can then add this magnetic Ab solid phase to a mixture of unknown sample and known label, allowing sufficient time in mixing coils for the required incubation of the competing Ag's.

The actual separation can then be accomplished by electromagnets placed around the transmission tubing, as shown in Fig. 5.17. The solid phase (after combining with the competing Ag's) is retained by the magnetic field, whereas unbound Ag passes to waste (Fig. 5.17*a*). After separation has taken place, the

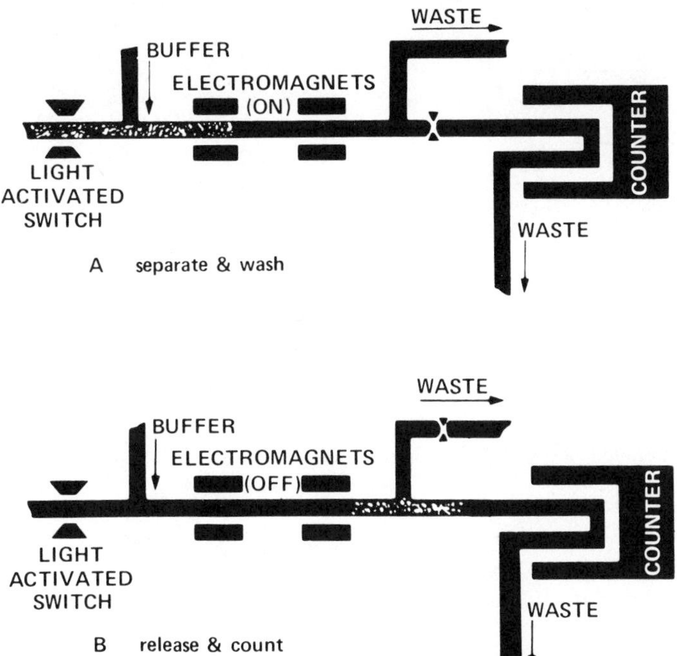


**Fig. 5.17.** Flow diagram for automated radioimmunoassay with a magnetic separation step. (*a*) Bound antigen held by electromagnet; (*b*) antigen released for passage through scintillation counter.

bound portion is released and flows through the scintillation detector for counting (Fig. 5.17*b*).

The automated RIA can operate at 30 samples/h, without human intervention. Cohen and Stern (22) have described its use for thyroxine ($T_4$) and a host of other immunoassays such as triiodothyronine, cortisol, and digoxin.

## L. COMBINING UNIT OPERATIONS

This section coordinates (where it has not already been done) the various unit operations required for the total automation of specific analyses. Table 5.1 summarizes the different unit operations that are employed in the 12 examples. The full automation of the determination of seven of the analytes has been adequately covered in Section V.K. (detection), and another (phenol) has been covered in Section V.E.5. (distillation). The remaining four are addressed here.

The analysis of theophylline in serum may be the most comprehensive combination of unit operations in this chapter. Here, the versatility of HPLC has been combined with several modules of the AutoAnalyzer system to handle therapeutic levels of drugs in biological fluids at rates of analysis much faster

TABLE 5.I

Application of Unit Operations to Selected Analyses

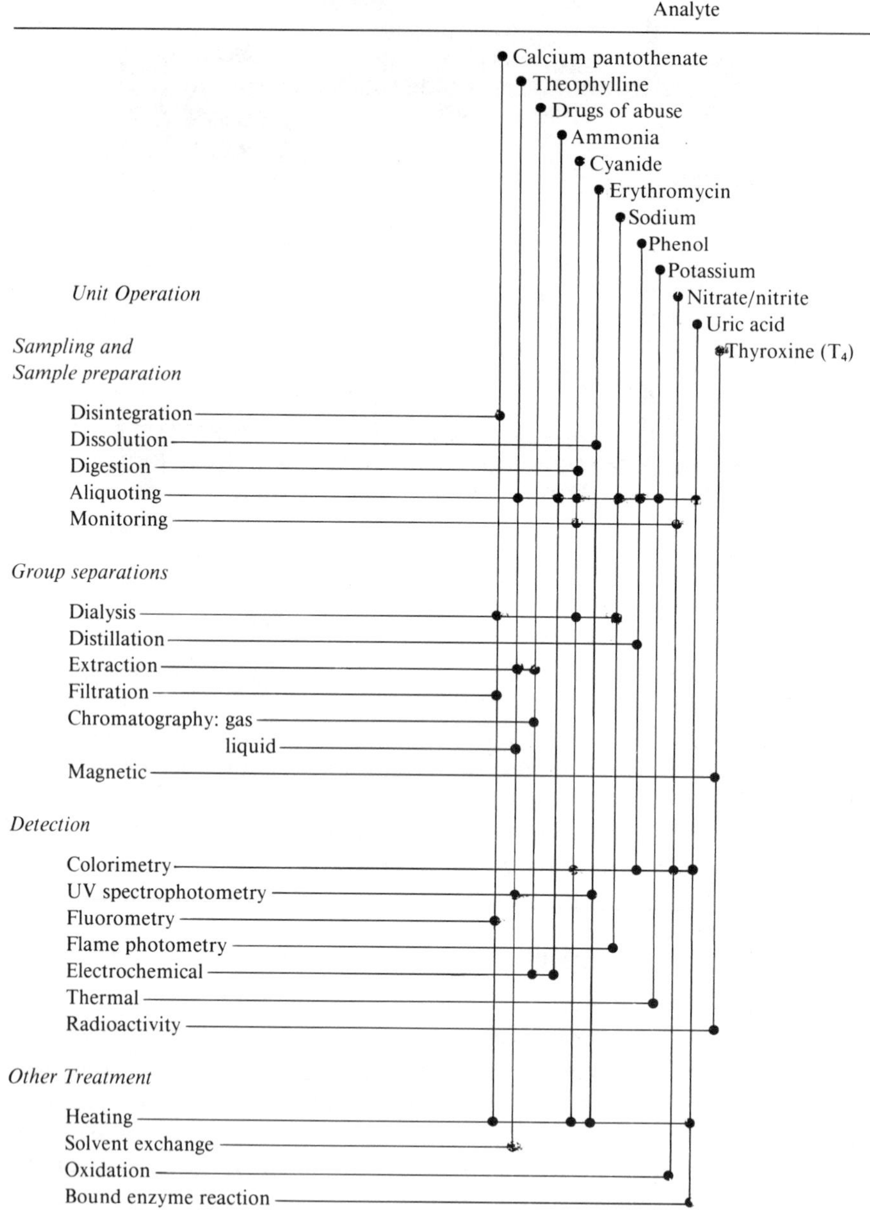

| Unit Operation | Calcium pantothenate | Theophylline | Drugs of abuse | Ammonia | Cyanide | Erythromycin | Sodium | Phenol | Potassium | Nitrate/nitrite | Uric acid | Thyroxine ($T_4$) |
|---|---|---|---|---|---|---|---|---|---|---|---|---|
| *Sampling and Sample preparation* | | | | | | | | | | | | |
| Disintegration | • | | | | | | | | | | | |
| Dissolution | | | | | | • | | | | | | |
| Digestion | | | | | • | | | | | | | |
| Aliquoting | | • | | • | • | | • | • | • | | • | |
| Monitoring | | | | | • | | | | | • | | |
| *Group separations* | | | | | | | | | | | | |
| Dialysis | • | | | | • | | • | | | | | |
| Distillation | | | | | | | | • | | | | |
| Extraction | | • | • | | | | | | | | | |
| Filtration | • | | | | | | | | | | | |
| Chromatography: gas | | | • | | | | | | | | | |
| liquid | | • | | | | | | | | | | |
| Magnetic | | | | | | | | | | | | • |
| *Detection* | | | | | | | | | | | | |
| Colorimetry | | | | | • | | | • | | • | • | |
| UV spectrophotometry | | • | | | | • | | | | | | |
| Fluorometry | • | | | | | | | | | | | |
| Flame photometry | | | | | | | • | | | | | |
| Electrochemical | | | • | • | | | | | | | | |
| Thermal | | | | | | | | | • | | | |
| Radioactivity | | | | | | | | | | | | • |
| *Other Treatment* | | | | | | | | | | | | |
| Heating | • | | | | • | • | | | | | • | |
| Solvent exchange | | • | | | | | | | | | | |
| Oxidation | | | | | | | | | | • | | |
| Bound enzyme reaction | | | | | | | | | | | • | |

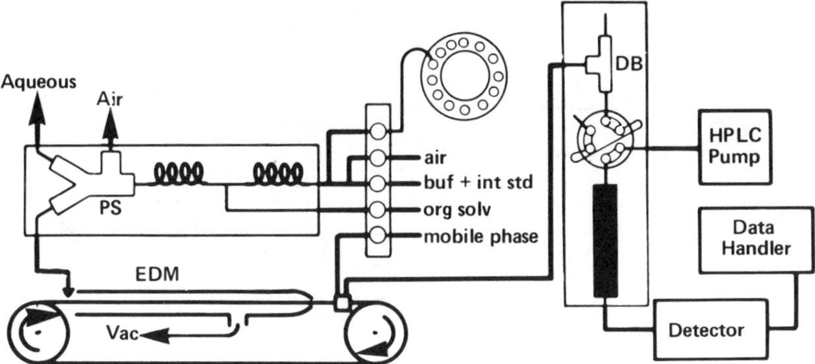


**Fig. 5.18.** Fully automated sample treatment for liquid chromatography (FAST-LC): untreated sample is aspirated, diluted, buffered, mixed with internal standard, extracted, transferred to a second solvent, and injected into a chromatographic column without any operator involvement.

than manual procedures and with absolutely no operator intervention. The analyzer has been described (13,16,17) and evaluated (1,27) elsewhere, and a configuration for theophylline is shown in Fig. 5.18. On the left are modules for automated sample preparation: sampler, pump (for adding buffer, internal standard, and organic solvent), cartridge for extraction and phase separation following deproteinization and the EDM module for solvent exchange and concentration. The right-hand portion of the diagram shows the HPLC pump, automatic injection valve, column, UV detector, and data handler. It operates at 20–30 samples/h, using only 50 $\mu$l of sample, and determines concentrations as low as 1.0 $\mu$g/ml with RSDs under 1%.

Drugs of abuse in the urine of suspected addicts have been measured by total automation in a combination of the AutoAnalyzer and a three-channel gas chromatograph (20). The general requirements for an instrument to analyze for traces of amphetamines, barbiturates, and narcotic alkaloids were (1) separate GC columns for each drug group, so no compromises would have to be made, (2) isothermal mode for fastest operation, and (3) a clean sample, so columns would have an acceptable lifetime. In operation, urine samples (unmeasured and untreated) were aspirated from a single sampler and divided into two streams, where appropriate cleanup cartridges handled acidic and basic drugs separately. The stream from the acidic drug cartridge went to one GC column, whereas the stream from the basic drug cartridge was further divided for injection into two different CG columns. All three columns ran in parallel. The two cleanup cartridges were identical, except for pH. The barbiturate cartridge is shown in Fig. 5.19, wherein a multichannel peristaltic pump directs fluids through mixing coils, extraction coils, and phase separators. Detection was possible at a few micrograms per milliliter following a triple extraction to separate interfering substances. A special syringe permitted programmed injection into the appropriate GC column.

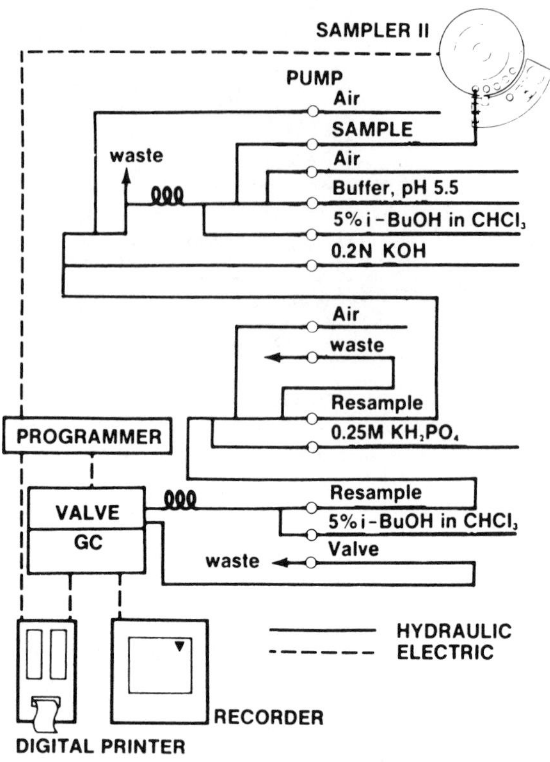


**Fig. 5.19.** Totally automated gas chromatograph for the analysis of drugs of abuse in body fluids (includes sample preparation).

The next example has been selected to demonstrate (1) the sampling of a gas (all other examples have dealt with liquids or solids), (2) an on-line oxidation column, and (3) continuous monitoring. Figure 5.20 shows an automated system for the determination of $NO + NO_2$ in ambient air based on a method for determining nitrogen dioxide. Prior to absorption, a solid oxidant consisting of $CrO_3$ on firebrick converts NO in the air sample to $NO_2$. A reagent blank is eliminated by using a differential flow system in which the absorbing reagent flows through the reference cell, providing a continuous blank for the sample cell. This monitor responds $2\frac{1}{4}$ min after a step change in input concentration, and reaches 90% of the new concentration 4 min later. Its full scale response is 0.15 ppm with its oxidation tube operating at 95% efficiency.

The final example of a combination of unit operations to achieve automation involves a bound enzyme inside a coil to effect a continuous catalytic conversion without consuming an otherwise expensive reagent. As seen in Fig. 5.21, the sample is diluted with sodium borate solution, segmented with air, and allowed to enter the donor side of the dialyzer. The recipient stream is an air-segmented mixture of sodium borate and hydroxylamine hydrochloride; when

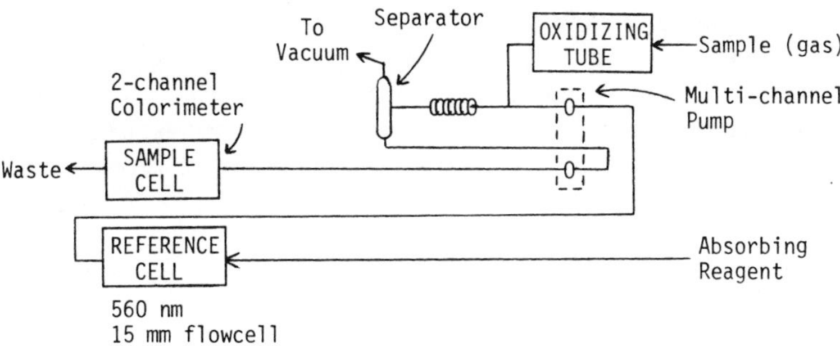


**Fig. 5.20.** Flow diagram for a $NO/NO_2$ analyzer.

it leaves the dialyzer, it contains the dialyzed uric acid. Sodium tungstate is then added, mixed, and allowed to react with phosphotungstic acid in a heating bath maintained at 50–52°C. The resulting blue color is measured colorimetrically at 660 nm in a 50 mm flow cell. Sample blanks are determined by the on-line treatment of the sample solution (prior to dialysis) with immobilized uricase enzyme in a nylon coil. The uricase selectively destroys the uric acid present in the sample and leaves the interfering substances intact. The difference in sample and blank readings reflects the concentration of uric acid present in the original sample. This procedure has been proposed as a chemical index for ascertaining the presence and the degree of insect infestation in a variety of food products, since uric acid is excreted by insects as one of the end products of nitrogen metabolism. It is effective in the range 20–1000 μg/g of sample, and can operate at 30 samples/h.

Before ending this section on the combination of unit operations, it should be pointed out that occasionally the sample preparation (cleanup) phase and the separation and/or analysis phase do not operate in mutually convenient time

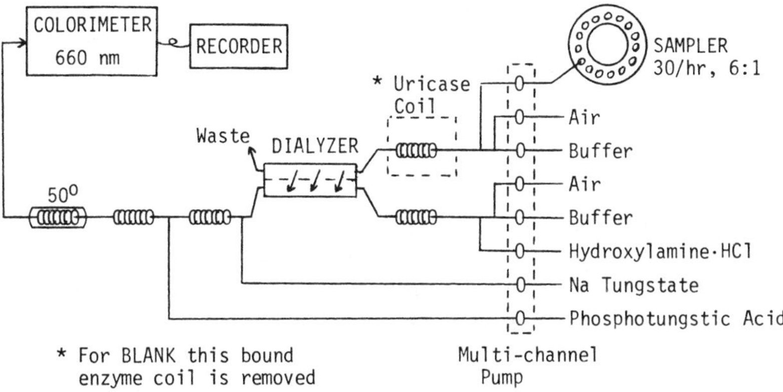


**Fig. 5.21.** Analyzer used to determine insect infestation in foods using a bound enzyme coil in place of an expensive reagent.

frames. For example, it may require only 2–3 min/sample to get it ready for injection into a chromatograph, whereas the subsequent chromatographic separation and detection may require 10–15 min/sample. Clearly, one should consider an off-line sample preparation and temporary storage as the first phase of analysis, to be followed by a slower running analytical phase (perhaps overnight). This subject—automated sample preparation—has been reviewed elsewhere (15).

The ways in which these unit operations can be combined to provide fully automated systems is limited only by the operator's imagination (or lack of it!).

## VI. DATA HANDLING

During the last several years the treatment of signals from most types of analytical equipment has become very sophisticated. The days of having to translate a meter reading into meaningful concentration units are nearly gone. With the ever-decreasing cost of microcomputers and their peripherals, many instruments today have data-handling features built into them. We consider here several aspects of data presentation.

### A. RECORDERS AND INTEGRATORS

While a meter gives a momentary and instantaneous presentation of data (perhaps in terms of a voltage or current, sometimes in terms of concentration), it is more common to employ a strip-chart recorder that can provide both a visual indication of signal strength at that moment and a permanent record of the history of the changing signal. Generally this recorder tracing is of a peak and valley form, where the peak heights (or areas) represent concentrations of successive samples. In the AutoAnalyzer system, for example, peak heights are essentially proportional to sample concentration, but in many commercial chromatographs (both liquid and gas) the peak area is integrated as a measure of concentration.

Although integrators do not ordinarily require a recorder, some versions operate from a recorder's retransmitting slide wire or are otherwise mechanically linked to the recorder. Most operators would be reluctant to part with their recorders since the tracing is a type of insurance against integrator failure, permitting data to be retrieved manually if required. Even more important to many operators is the system diagnostics possible by studying the recorder tracing of a poorly running system. A well trained operator can examine peak shapes and troubleshoot with relative ease. The operator can, in fact, learn much about the system hydraulics, electronics, and chemistry with but a glance at several minutes of recorder tracings.

Many integrators have a built-in printer that produces numerical (and sometimes even alpha-numeric) data on adding-machine-size paper. Depending on the price one is willing to pay, he or she may obtain only raw data, or scaled numbers in correct concentration units, sometimes with sample identification, and occasionally with percent of total area (as a post-run printout, when all data are available).

## B. MICROPROCESSORS AND COMPUTERS

The line between microprocessors and computers is at best nebulous. Strictly speaking, a computer is a reasonably complete collection of hardware and software which can do "number crunching" and produce a hard copy (final report). A microprocessor could be limited to control functions only, operating valves and motors without actually producing any answers. But in today's world of sophisticated analytical instruments, the self-contained microprocessor generally handles both control functions and data handling.

The power of the microprocessor within an instrument is nowhere more evident than in modern chromatographs. Properly prepared samples can be injected automatically, their chromatograms examined by elegant software, the chromatographic conditions (solvent composition, column temperature, mobile phase delivery rate, etc.) changed automatically, and optimum conditions established without operator intervention.

This same kind of control and data handling capability has recently been added to CF systems (14). A three-unit addition to an AutoAnalyzer system includes the microprocessor/controller, a so-called AutoValve module (for reagent switching during automatic start-up or shutdown), and a terminal for instrument/operator dialogue. The control functions include monitoring the base line until it is acceptably flat, then enabling the sampler to begin providing samples to the system. As peaks emerge from the detector and are recorded, their shapes are evaluated for evidence of air spikes, short samples, out of pre-set limits, and so forth. Acceptable peaks are quantitated according to a set of standards that are analyzed along with the samples, and poor peaks are flagged with appropriate error designations. As the run progresses, tentative answers are printed in real time, with corrections for carryover if that has been requested. When the run is completed (e.g., one tray of 40 samples analyzed for up to four components each), final answers are printed with additional corrections for base line drift and any gain changes that may have occurred (e.g., changes in sensitivity due to reagent degradation with time). At the end of the analysis session, the AutoValve module can be programmed to switch from reagents to a wash solution for a defined period, then to air so the system can be shut down clean and dry. Alternately, it may go into an intermittent mode in which reagents are pumped much more slowly to conserve them, while maintaining an ever-ready condition for immediate restart without having to wait for system equilibration.

## C. CALIBRATION AND ERROR CORRECTION

When an analytical system is linear (i.e., when the detector output is truly proportional to analyte concentration), calibration can be as simple as processing a single standard. Gain can be adjusted (automatically or manually) so that output is zero for the base line and numerically equal to the concentration of the standard when its peak is being quantitated. Frequently it is risky to put so much weight on a single standard, however, so several standards are processed and a least squares fit of a straight line is made through their points (peak heights or peak areas). All except the very smallest microprocessors are capable of performing these calculations with relative ease.

For nonlinear systems, one can approximate the curve with a series of straight lines, a so-called piecewise linear fit. This scheme produces exact answers at the calibration points, but not midway between these points. If the curve is known to be quadratic, then a least squares fit of a second-degree polynomial equation, although a bit more cumbersome to handle, is a preferred method for calibration. Similarly, exponentials can be accommodated with appropriate equations. Typically, an operator might choose a five- or six-point standard curve, sometimes using duplicate standards, and all real-time answers are based upon this curve. One or more of these standards (generally the highest concentration) can be rerun at the end of the group of samples, and the entire standard curve recalculated on the basis of this new value (which may have changed due to sample evaporation or deterioration of any component of the reaction mixture). The final printout is then based upon an interpolation of sample readouts between (presumed) linearly changing standards.

The phenomenon of carryover can be ignored when it is known to be very small, or it can be defined and used in error-correction algorithms when faster rates of analysis are required and/or higher precision is required. Consider a series of three samples: a high level followed by two low levels, designated $H$, $L$, and $L'$. Assuming the system exhibits carryover, the $L$ is higher than its theoretical value owing to carryover from $H$, and $L'$, is higher owing to carryover from $L$. Both $L$ and $L'$ have some true value, $X$, which is unknown to us.

$L'$ is equal to $X$ plus some fraction of $L$ (equation 1), and $L$ is equal to $X$ plus some fraction of $H$ (equation 2). If we now subtract equation 1 from equation 2, we have an expression for the difference between $L$ and $L'$ (equation 3), which can be rearranged to provide a definition of the carryover factor, $f$ (equation 4).

$$L' = X + f(L) \tag{1}$$

$$L = X + f(H) \tag{2}$$

$$L - L' = f(H - L) \tag{3}$$

$$f = \frac{L - L'}{H - L} \tag{4}$$

This factor can then be used to produce corrected answers, in real time, by applying equation 5

$$C_c = \frac{C_u - f(C_p)}{1 - f} \tag{5}$$

where $C_c$ is the concentration (corrected), $C_u$ is the concentration (uncorrected), $f$ is the carryover factor from equation 4, and $C_p$ is the concentration of the preceding sample. The $(1 - f)$ term in the denominator corrects for that portion of the $C_c$ sample that will be lost to the subsequent sample by carryover. When corrections for carryover, base line drift, and sensitivity change are incorporated into a continuous flow system, throughput can often be increased by 50% while retaining precision as low as 0.5% relative standard deviation. The system has been evaluated by Kane et al. (41).

## VII. WHEN TO AUTOMATE (AND WHEN NOT TO)

The most obvious reasons for automation are listed in Table 5.II. The list is not exhaustive, but it does reflect the main motivations for automation; the savings of time, personnel, and materials, and the improvement in analytical accuracy.

Many questions must be answered before automation can be justified. Some important questions are posed in Table 5.III. We often presume that unless

TABLE 5.II

Reasons for Automating

1. Saves time
   a. of people involved
   b. by providing answers sooner
   c. by increasing sample throughput
2. Reduces or eliminates personnel who would
   a. prepare samples
   b. operate equipment
   c. tabulate data
   d. interpret results
   e. write reports
3. Saves money by
   a. reducing time and personnel (as above)
   b. using less reagents
   c. using less apparatus
4. Increases accuracy by
   a. eliminating human errors
   b. treating all standards and samples identically, thus avoiding human bias

TABLE 5.III

Questions to be Answered

1. Is the work load sufficiently large?
2. Will automation be
   a. faster?
   b. more accurate?
   c. more economical?
3. Does automation imply computerization?
   a. self-contained or central?
   b. on-line or off-line?
4. Should one automate *all* operations, or leave some for the operator?

TABLE 5.IV
Disadvantages of Automation

1. May be very expensive to implement.
2. May be very time-consuming to develop
3. May be too rigid and nonversatile (fixed cycle)
4. One faulty part may put entire system down, especially if there is no override
5. Intermediate results may not be readily available
6. May not comply with "standard" method
7. Personnel attitude is frequently negative
8. Suppliers' quality control must be excellent, and expendables used in automated procedures cannot be changed significantly

hundreds of samples must be analyzed each day, automation is not warranted. But is has been determined that some programs of automation have paid for themselves in less than 1 year, even with only a dozen analyses per day. The final decision must be made on the basis of a careful financial analysis of the situation.

Assuming that all the questions are answered satisfactorily and that one is ready to undertake a program of automation, there are some potential disadvantages that still must be considered; these are given in Table 5.IV. With proper planning and attention to detail, many of these disadvantages can be designed out of the system.

The message is simple: automation has been shown time and again to be technically and economically beneficial, but careful planning by knowledgeable people is essential before implementation.

## ACKNOWLEDGEMENT

AutoAnalyzer, AutoValve, EDM, Monitor IV, SASDRA, SMA, SMAC, and SOLIDprep are trademarks of the Technicon Instruments Corporation.

## REFERENCES

1. Abdou, H. M., F. M. Russo-Alesi, and V. Fernandez, *Pharm. Tech.*, **5**(3), 40 (1981).
2. Adler, H. J., D. A. Burns, and L. R. Snyder, "Continuous Extraction of Body Fluid Samples Including Whole Blood, Plasma, and Urine," in *Advances in Automated Analysis, 1972 Technicon International Congress*, Vol. 9, Mediad, Tarrytown, N.Y., 1973, p. 81.
3. Anderson, C. A., in G. Zweig, Ed., *Pesticides and Plant Growth Regulators*, Vol. VII, Academic, New York, 1973, Chap. 10.
4. Anderson, N. G., *Science*, **166**, 317 (1969).
5. Bartels, H., R. D. Werder, W. Schürman, and R. W. Arndt, *J. Automatic Chem.*, **1**, 28 (1978).
6. Bergmeyer, H. U., Ed., *Methods of Enzymatic Analysis*, 2nd Eng. ed., Academic, New York, 1974.

7. Blaedel, W. J., and J. H. Strohl, *Anal Chem*, **36**, 1245 (1964).
8. Bochinski, J. H., U.S. Pat. No. 4,025,311 (May 24, 1977).
9. Burns, D. A., "Continuous Monitoring of Moisture in a Flowing Solid Using the AutoAnalyzer and a New Sampling Device," in *Automation in Analytical Chemistry, Technicon Symposia 1965*, Mediad, New York, 1966, p. 193.
10. Burns, D. A., U.S. Pat. No. 3,921,439 (Nov. 25, 1975).
11. Burns, D. A., "Kinetic Enzyme Analysis by Flow-Reversal in a Continuous-Flow System," presented at the American Chemical Society National Meeting, New York City, April 8, 1976; U.S. Patent No. 3,876,344, (April 8, 1975).
12. Burns, D. A., U.S. Pat. No. 3,921,439 (Nov. 25, 1975).
13. Burns, D. A., *Res./Dev.* **28**(4), 22 (1977).
14. Burns, D. A., "A Modern & Versatile Microcomputer for Control & Data Handling of AutoAnalyzer II Continuous Flow Analytical Systems," Abstr. 653, *Pittsburgh Conference on Analytical Chemistry and Applied Spectroscopy*, Atlantic City, N.J., March 10–14, 1980.
15. Burns, D. A., *Anal Chem*, **53**, 1403A (1981).
16. Burns, D. A., *Pharm. Tech.*, **5**(3), 53 (1981).
17. Burns, D. A., J. I. Fernandez, J. R. Gant, and A. L. Pietrantonio, *Amer. Lab.* **11**(10), 79 (1979).
18. Burns, D. A., and G. D. Hansen, *Ann. N.Y. Acad. Sci.*, **153**, 541 (1968).
19. Burns, D. A., and L. Lazer, "Automated Analysis of Polynucleotide Phosporylase Employing the Continuous Filter," in *Automation in Analytical Chemistry, Technicon Symposia 1965*, Mediad, New York, 1966, p. 42.
20. Burns, D. A., L. R. Snyder, and H. J. Adler, "Total Automation of the Gas Chromatograph by Combination with the AutoAnalyzer," in *Advances in Automated Analysis, 1972 Technicon International Congress*, Vol. 6, Mediad, Tarrytown, N.Y., 1973, p. 23.
21. Burtis, C. A., T. O. Tiffany, and C. D. Scott, "The Use of a Centrifugal Fast Analyzer for Biochemical and Immunological Analysis," in D. Glick, Ed., *Methods of Biochemical Analysis,* Vol. 23, Wiley, New York, 1976, pp. 189–248.
22. Cohen, E., and M. Stern, "The New Technicon Automated Radioimmunoassay System," in *Advances in Automated Analysis, Technicon International Congress 1976*, Vol. 1, Mediad, Tarrytown, N.Y., 1977, p. 232.
23. Cullen, L. F., M. P. Brindle, and G. J. Papariello, *J. Pharm. Sci.*, **62**, 1708 (1973).
24. Curtius, H. C., and M. Roth, Eds., *Clinical Biochemistry. Principles and Methods*, Vol. II, Walter de Gruyter, Berlin and New York, 1974, pp. 1142–1304.
25. Deelder, R. S., M. G. F. Kroll, A. J. B. Beeren, and J. H. M. van den Berg, *J. Chromatogr.*, **149**, 669 (1978).
26. Docherty, A. C., "Automatic Sampling and Analysis of Compound Fertilizers," in *Automation in Analytical Chemistry, Technicon Symposia 1967*, Vol. 1, Mediad, White Plains, N.Y., 1968, p. 265.
27. Dolan, J. W., Sj. van der Wal, S. J. Bannister, and L. R. Snyder, *Clin. Chem.*, **26**, 871 (1980).
28. El-Awady, A. A., R. B. Miller, and M. J. Carter, *Anal Chem.*, **48**, 110 (1976).
29. Erni, P. E., and H. R. Muller, *Anal. Chim. Acta*, **103**, 189 (1978).
30. Ertinghausen, G., H. J. Adler, A. S. Reichler, and N. Kinnard, "Fully Automated High-Speed Ion-Exchange Chromatography of Amino Acids," in *Advances in Automated Analysis, Technicon International Congress, 1969*, Vol. I, Mediad, White Plains, N.Y., 1970, p. 333.
31. Ertinghausen, G., H. J. Adler, and A. S. Reichler, *J. Chromatogr.*, **42**, 355 (1969).
32. Euston, C. B., and D. R. Baker, *Amer. Lab.*, **9**(3), 91 (1979).
33. Freistad, H. O., "Automated Colorimetric Determination of Residues of Parathion and Similar Compounds in Plant Extracts," *IUPAC 2nd Int'l. Congress Pest. Chem.*, Tel Aviv, Israel, 1971.
34. Furman, W. B., *Continuous Flow Analysis, Theory and Practice*, Dekker, New York, 1976.

35. Getz, M. E., G. W. Hanes, and K. R. Hill, "Progress in the Automation of Extraction and Cleanup Procedures for Detecting Trace Amounts of Pesticides in Environmental Samples," in *Trace Organic Analysis: New Frontiers in Analytical Chemistry. 9th Materials Research Symposium*," Natl. Bur. Stds. Spec. Publ. 519, 1979, p. 345.
36. Gouldon, P. D., and P. Brooksbank, *Anal. Chem.*, **46**, 1431 (1974).
37. Graham., T. F., *Am. Lab.*, **6**(9), 77 (1974).
38. Guilbault, G. G., *Handbook of Enzymatic Methods of Analysis*, Dekker, New York, 1976.
39. Isreeli, J., and W. Smythe, "SMAC: The Third Generation Sequential Multiple Analyzer," in *Advances in Automated Analysis, 1972 Technicon International Congress*, Vol. 1, Mediad, Tarrytown, N.Y., 1973, p. 13.
40. Jirka, A. M., and M. J. Carter, *Anal. Chem.*, **50**, 91 (1978).
41. Kane, P. F., B. R. Bennett, and S. Gulik, J. Assoc. Off. Anal. Chem. **64**, 1322 (1981).
42. Kawase, J., A. Nakae, and M. Yamanake, *Anal. Chem.*, **51**, 1640 (1979).
43. Kessler, G., "An Automated System of Analysis," in S. Frankel, S. Reitman, and A. C. Sonnenwirth, Eds., *Gradwohl's Clinical Laboratory Methods and Diagnosis*, Vol. I, 7th ed., C. V. Mosby, St. Louis, 1970, pp. 323–372.
44. Koirtyohann, S. R., and M. Khalil, *Anal. Chem.*, **48**, 136 (1976).
45. "LabGuide," *Anal. Chem.*, **52**(10), (1980).
46. Margoshes, M., *Anal. Chem.*, **49**, 17 (1977).
47. Miller, R. D., and N. Neuss, *J. Antibodies*, **29**, 902 (1976).
48. Mrochek, J. E., C. A. Burtis, W. F. Johnson, M. L. Bauer, D. G. Lakomy, R. K. Genung, and C. D. Scott, *Clin. Chem.*, **23**, 1416 (1977).
49. Nagy, G., Zs. Feher, and E. Pungor, *Anal. Chim. Acta*, **52**, 47 (1970).
50. Ohlin, L. E., U.S. Pat. No. 3,552,212 (Jan. 5, 1971).
51. Olansky, A. S., and S. N. Deming, *Clin. Chem.*, **24**, 2115 (1978).
52. Ott, D. E., M. Ittig, and H. O Freistad, *J. Assoc. Off. Anal. Chem.*, **54**, 160 (1971).
53. Rao, J. K., M. H. Pelavin, and S. Morgenstern, "SMAC: High-Speed Continuous Flow, Ion-Selective Electrodes for Sodium and Potassium: Theory and Design", in *Advances in Automated Analysis, 1972 Technicon International Congress*, Vol. 1, Mediad, Tarrytown, N.Y., 1973, p. 33.
54. Renoe, B. W., K. R. O'Keefe, and H. V. Malmstadt, *Anal. Chem.*, **48**, 661 (1976).
55. Roy, R. B., and A. Buccafuri, *J. Assoc. Off. Anal. Chem.*, **61**, 720 (1978).
56. Ruzicka, J., and E. Hansen, Anal. Chim. Acta, **78**, 17 (1975).
57. Sansur, M., H. J. Adler, and D. A. Burns, "An Automated Fluorometric Method for Determination of Total Morphine in Urine," in *Advances in Automated Analysis, 1972 Technicon International Congress*, Vol. 6, Mediad, Tarrytown, N.Y., 1973, p. 3.
58. Skeggs, L. T., Jr., *Amer. J. Clin. Pathol.*, **28**, 311 (1957).
59. Skeggs, L. T., U.S. Pat. No. 3,097,927, July 16, 1963.
60. Smith, D. P., and C. E. Wilde, *Clin. Chim. Acta*, , **78**, 351 (1977).
61. Snyder, L. R., J. Levine, R. Stoy, and A. Conetta, *Anal. Chem.*, **48**, 942A (1976).
62. Snyder, M. L., D. R. Downing, and W. C. Stewart, "A Semiautomated Enzyme-Labelled Antibody Test for Hog Cholera," in *Advances in Automated Analysis, Technicon International Congress 1976*, Vol. 2, Mediad, Tarrytown, N.Y., 1977, p. 326.
63. Spackman, D. H., W. H. Stein, and S. Moore, *Anal. Chem.*, **30**, 1190 (1958).
64. Stalling, D. L., R. C. Tindle, and J. L. Johnson, *J. Assoc. Off. Anal. Chem.*, **55**, 32 (1972).
65. Stewart, K. K., G. R. Belcher, and P. E. Hare, *Anal. Biochem.*, **70**, 167 (1976).
66. Stimac, R. M., *J. Assoc. Off. Anal. Chem.*, **62**, 85 (1979).

67. *Technicon AutoAnalyzer Bibliography 1956–1967; Technicon Bibliography 1967–1973; Technicon Bibliography Supplement No. 1, 1974; Supplement No. 2, 1975: Supplement No. 3, 1975; Technicon Bibliography Papers Nos. 2540–2702, 8026–9200*, Technicon Corp., Tarrytown, New York.

68. Technicon Industrial Method No. 301–73WM, Technicon Industrial Systems, Tarrytown, New York.

69. Technicon Industrial Method No. 352–74W, Technicon Industrial Systems, Tarrytown, New York.

70. Thiers, R. E., R. R. Cole, and W. J. Kirsch, *Clin. Chem.*, **13**, 451 (1967).

71. Thiers, R. E., A. H. Reed, and K. Delander, *Clin. Chem.*, **17**, 42 (1971).

72. Tiffany, T. O., *CRC Critical Reviews in Clinical Laboratory Sciences*, **5**, 129 (1974).

73. Tindle, R. C., and D. L. Stalling, *Anal. Chem.*, **44**, 1968 (1972).

74. Walker, W. H. C., "Theoretical Aspects," in W. B. Furman, Ed., *Continuous Flow Analysis, Theory and Practices*, Dekker, New York, 1976.

75. White, V. R., and J. M. Fitzgerald, *Anal. Chem.*, **47**, 903 (1975).

76. White, W. L., M. M. Erickson, and S. C. Stevens, *Analytical Automation*, 2nd ed., C. V. Mosby, St. Louis, 1972.

77. White, W. L., M. N. Erickson, and S. C. Stevens, *Chemistry for the Clinical Laboratory*, 4th ed., C. V. Mosby, St. Louis, 1976, Chap. 3, pp. 36–91.

Part I
Section E

Chapter 6

# AUTOMATION: INSTRUMENTATION FOR PROCESS MONITORING

BY KENNETH W. GARDINER, *Applied Science Program, University of California, Riverside, California*

**Contents**

# I. INTRODUCTION

## A. BACKGROUND AND RATIONALE

Process monitoring is primarily the application of instrumental analytical methods for the control of manufacturing processes and product quality. It is, in effect, a direct attempt to bring into the plant the usefulness of laboratory-type analyses to provide on-line, real-time measurements in support of plant operations. However, in recent years the control of the generation and disposal of process waste products has become an increasingly critical factor. The instrumental analytical procedures for this requirement must now be included in any treatment of the process monitoring field.

Early in the development and growth of process industries, monitoring techniques were principally concerned with temperature, pressure, flow, refractive index, and other physical property measurements. The required instrumentation was comparatively simple and posed no serious problems in design, construction, or applicability. It was soon recognized, however, that the determination of chemical composition was of even greater importance. Consequently, classical quantitative and qualitative chemical analyses on specimens of the flowing process stream were performed in plant control laboratories. Even under the best of conditions, time intervals between sampling and the availability of test results were generally too great to permit optimum control. Furthermore, depending upon the frequency of sampling, it was quite possible for an undesirable change in stream composition to have occurred without having been detected. This was essentially the state of the process monitoring field prior to World War II.

The tremendous production demands made on process industries during the war had a profound impact on all forms of technology. Not the least of those affected was the science of instrumental measurements. Impressive advances in electronics and instrument technology in the 1940s resulted in the rapid translation of many classical analytical capabilities into equivalent scientific instrument systems. However, in contrast to the rapid development of laboratory instruments, effective on-line and in-plant process analyzers appeared more slowly. This condition has persisted in spite of the established need for performing on-line measurements.

To distinguish between laboratory-type measurements and process monitoring, one must first consider those differences dictated by the conditions under which each is used. Listed in Table 6.I are the major features that differentiate process monitors and process monitoring from conventional analytical instruments and their uses.

Although basic measuring principles may be common to both laboratory and process instruments, the marked differences in design and appearance result from the requirement of satisfying the criteria given in Table 6.I.

At this point it is appropriate to emphasize the need and value of process monitoring. From an operating standpoint, the economic desirability of controlling the process to maximize material usage and standardize product quality is certainly a justification. Unquestionably, economic considerations prompted the original development of continuous on-line analyzers. The ability to detect and correct potentially dangerous operating trends, or outright process malfunc-

TABLE 6.I

Characteristics that Distinguish Process Monitors from Laboratory-type Analyzers

| |
|---|
| Process monitors must do the following: |
| Operate in "hostile" environments. |
| Work with nonideal ("dirty") samples and difficult sampling conditions. |
| Have long-term operating stability. |
| Require a minimum of operator attention and maintenance. |
| A process monitor is generally a single component analyzer with a fixed working range. |
| The high costs of installation and sampling for process monitoring are factors. |

tions, is obviously a desirable feature. The alarm capability of an on-line monitor is often considered one of its most valuable functions. A timely alarm permits manual corrections to be made in the operation before more serious problems can develop. In more sophisticated systems, the alarm signal may be used to activate automatic controls which can effect necessary process corrections without the need of human intervention.

The value of the "real-time" analysis feature of process monitoring is emphasized if one considers the waste discharge problems of various process industries. Recently, this particular aspect has received a great deal of attention as part of the emerging public concern over the quality of our environment. The ability to monitor for the presence of harmful pollutants is mandatory so that corrective actions can be initiated *before* large amounts have been released. From an abatement standpoint it does little good to analyze the environment *after* the damage has been done. Unmonitored process discharges are always a threat because of the large volumes and high flow rates characteristic of most process waste streams. Thus source monitoring on a continuous and real-time basis is a prerequisite for effective industrial pollution control.

The immediate detection of harmful vapors in process areas has also become a subject of increased attention. This has been prompted by the enforcement of the Federal Government's Occupational Safety and Health Act (OSHA). The analyses in this case involve samples that are considerably cleaner than a plant's waste effluent. Nevertheless, the monitors that will be used must be the process type because of the nature of the surroundings in which the measurements will be made. In many respects, "in-plant" use implies virtually the same operating conditions as "on-line" use as far as the instrumentation is concerned.

To a certain extent the same qualification applies in the case of conventional, or outdoor, atmospheric monitoring and the analyzers employed. Rugged operating and environmental conditions more than offset the advantages gained by working with the comparatively clean samples associated with atmospheric analyses. There is no question that considerable impetus has been given to the development of many new analytical instruments by the passage of governmental regulations in the environmental quality area. For this presentation it is assumed that most environmental analyses have more in common with process monitoring than with conventional laboratory instrumental measurements. This assumption is based primarily on the applicability of the criteria given in Table 6.I.

The lack of suitable commercial process monitors has from time to time induced a number of the larger process industries to develop and build their own analysis systems. Chemical and petroleum refining companies are prime examples. It has often been this activity that has produced many of the small private entrepreneurial ventures that engage in the manufacture of process monitors. In other cases, process companies have licensed or sold their developments to well-established, commercial instrument manufacturers. The latter, in turn, have funded additional development and application efforts which have greatly extended the usefulness of the original systems. These same large instrument manufacturers have, of course, developed a number of their own process and field monitors that have had wide acceptance. The combination of the custom nature of many process applications and the large financial investment associated with such work has unquestionably deterred the more rapid development of the process monitoring field.

## B. FACTORS INFLUENCING THE CHOICE OF MONITORING SYSTEM

Typically, the choice of the measuring system to be used in any given process situation is generally dictated by a limited number of practical considerations. From an economic standpoint, plant personnel look primarily at the capital investment and cost-benefit aspects associated with the monitoring system under consideration. These include not only the initial capital outlay for systems hardware but also the cost of installation, including delivery of required services to the monitoring site. An estimate of the anticipated maintenance is an equally important item.

The monitoring site environment often requires that suitable protective housing be constructed. Rigorous plant building codes may dictate that the housing be substantial. As a result, it is usually expensive. Often the cost of providing and servicing a suitable location for a specific analyzer can exceed the instrument's sales price by a substantial factor. Consequently, many manufacturers specifically design their monitors to be enclosed in protective metal cases or explosion-proof cast metal condulets. These enclosures are mounted directly on a structural member at the monitoring site. The major portion of the instrument electronics and the necessary control panel may be remotely located in an appropriate control room some distance from the monitoring area. On-line analyzers that are completely contained in protective enclosures, however, do pose the problem of providing ready on-site access to instrument operating elements and controls.

Another aspect in the choice of a monitoring system is often the degree of developed user acceptability enjoyed by the particular unit under consideration. Concomitant with this are the reputation and established technical ability of the manufacturer and vendor of the system. Unlike standard laboratory instruments, process monitors tend to have a large element of custom design and single-purpose use built into them. A system that measures compound A in company X's process may be found incapable of measuring the same material to an acceptable degree of accuracy in company Y's process. The variable and complex makeup of most process streams generally requires that a certain amount of customer application work may have to be performed by the instrument manufacturer. Industrial process compositions are often proprietary and the application work performed may be confidential and thus of no value as far as other potential customers for the monitoring system are concerned. Consequently, market acceptance depends to a large extent upon the instrument manufacturer's ability to satisfy the specific needs of a substantial number of knowledgeable users. Cooperative application work involving both customer and vendor is the surest way to establish this kind of reputation. Unfortunately, the security policies of potential users, as previously indicated, have prevented the extensive employment of this approach.

In addition to cost and acceptability considerations, the selection of a monitoring system is influenced by the specificity and sensitivity required to satisfy the desired measurement. Considering that process samples are usually far from ideal, compromises on specificity may be prudent, if not mandatory, in certain cases. The problem of monitoring the "phenol content" of a process stream is representative. The "phenols" present consist of regular phenol, cresols, and xylenols. The individual identification and measurement of each of the phenolic compounds present in order to determine the total phenol concentration would require a time-consuming and costly procedure. The use of a monitoring system providing a "composite" analysis for all phenol-type compounds in a single determination would be preferable. A photometric on-line technique was, in effect, developed for this purpose (92). The method is based on the fact that phenols in aqueous solution all exhibit a similar shift in their uv absorption spectra when going from an acidic to an alkaline condition. By adding a strong base to the continuously flowing sample, a portion of which has been previously acidified and used as a zero fluid, the phenolate ion for all the species is produced, and an increased absorption is observed at 289 nm. A previous calibration with known amounts of phenol present provides the relationship between observed absorbance and "total phenol" content. Differences in individual absorbtivities are fortunately not very great.

The realistic assessment of the accuracy and precision that is needed is also essential in the selection of a monitoring system. Unlike laboratory analyses involving ideal samples, a process stream determination may not require or justify an equally high degree of accuracy for a given measurement. To apply a maximum accuracy method to a nonideal system that operates adequately within a ±5–10% range is certainly not desirable from a cost-benefit standpoint. Unfortunately, the judicious matching of monitoring system performance with actual need is often given too little attention in practice.

## II. THE BASIC ELEMENTS OF A PROCESS MONITORING SYSTEM

The concept of process monitoring can best be presented from a systems point of view (Fig. 6.1). As shown, the major subsystems include those associated with sampling, sample treatment, the measurement, and data output. Depending upon the degree of automation desired, the data output system may consist of a simple analog signal display or, for a totally automated system, a computer-based subsystem for data reduction and closed-loop process control. The following sections present a more detailed description of the main elements that make up a complete process monitoring system.

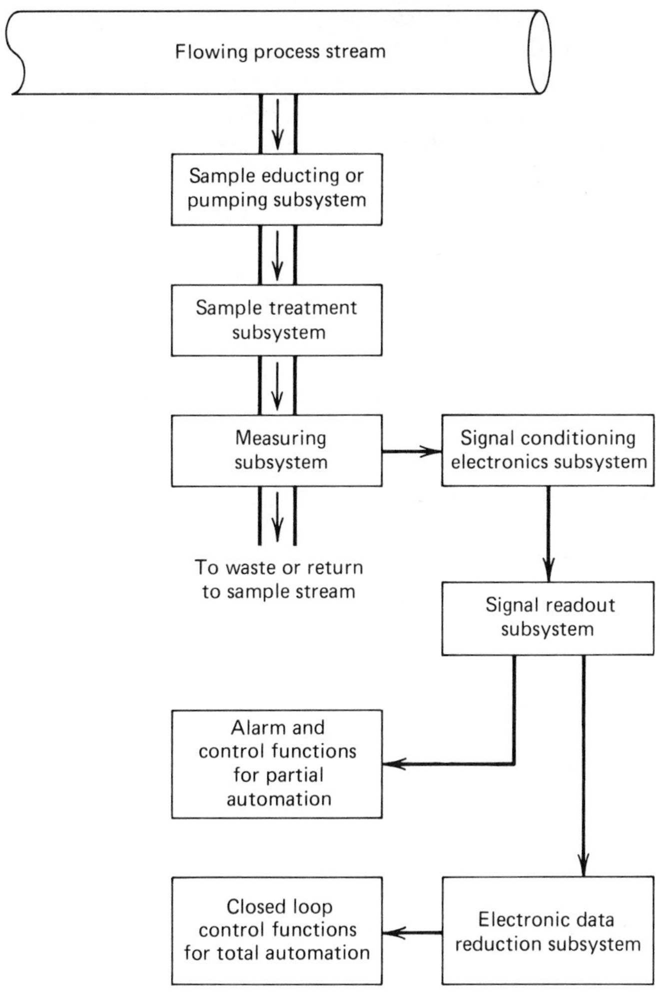


**Fig. 6.1.** Schematic representation of the systems concept for process monitoring.

## A. SAMPLING CONSIDERATIONS

The delivery to the measuring instrument of a representative and measurable sample from the process stream is perhaps the most critical step in the entire process monitoring procedure. First, the actual sampling point must be located to assure that sample representativeness and homogeneity are achieved on a continuous basis. Phase separations or other discontinuities in the stream if reflected in the test sample result in erroneous signals that can affect alarms and control functions. This danger exists whether the measurement is being made directly in the main flowing stream or an aliquot is being diverted to the measuring instrument.

If the process stream is operating under pressure, a pressure reducing subsystem must be used to aid in removing the sample for measurement. If the stream is hot, the sampling subsystem, the sample transfer lines, and all other portions of the monitoring system coming in contact with the sample must be heated to prevent condensation or solidification and blockage, depending on whether the sample is a gas or liquid. If the main stream contains an appreciable amount of suspended matter, a filtering system is usually incorporated in the sampling subsystem. For continuous use, dual filters are preferred so that one filter element can be cleaned by a back-flushing procedure while the other, which has just been cleaned, is alternately treating the sample stream. In many heated process gas samples, water vapor can often be present as an impurity. If a room-temperature measurement is called for, the test stream must be cooled and the condensed water vapor removed by a trap or demister.

If the process stream pressure is low, then the test sample must be removed by mechanical pumping or some other form of educting to provide sufficient

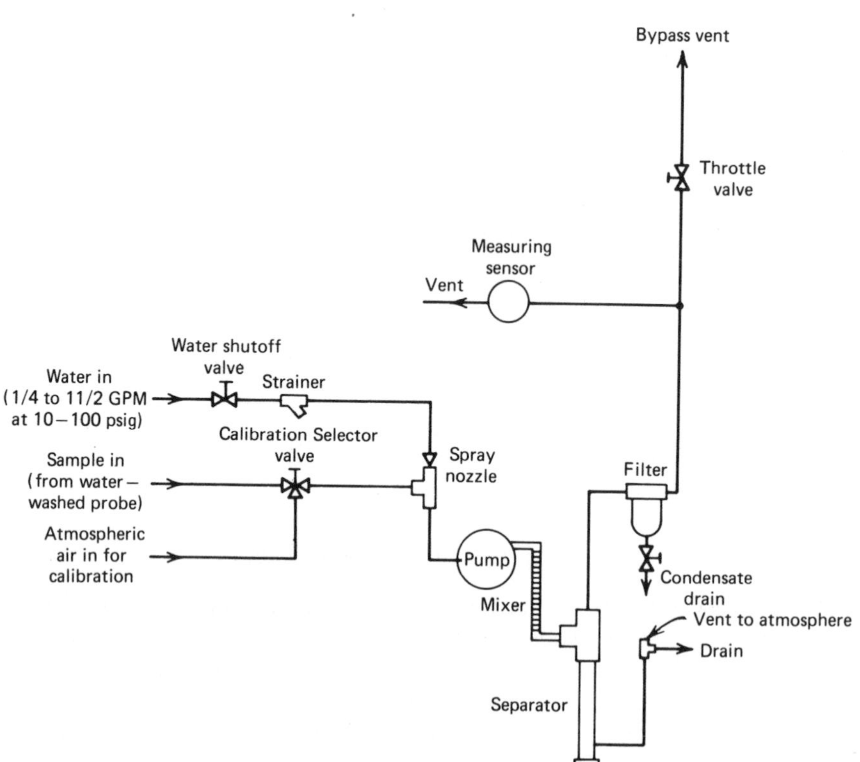


**Fig. 6.2.** Schematic drawing of a pump and sample washing system for sampling particulate-laden gas streams. Courtesy of Teledyne Analytical Instruments.

pressure to deliver the specimen to the measuring instrument (e.g., cf. 106). A system for pumping particularly dirty gas streams is shown schematically in Fig. 6.2. The peristaltic pump used is most effective for those situations involving very high loadings of particulates (129). Normally, such streams would rapidly plug conventional mechanical pumps or water or steam educting samplers. Gases or specimens from fluidized-bed catalytic cracking processes employed in the petroleum industry are appropriate examples.

In the system in Fig. 6.2, a water-washed probe with the aid of the pump extracts the sample, cools it, and delivers it to an on-line monitor (Fig. 6.3). A gravity separation trap is used to recover the gas sample, whereas the liquid portion containing all the particulate matter is diverted to waste. Continuous internal washing of the sample probe and attendant tubing not only serves to

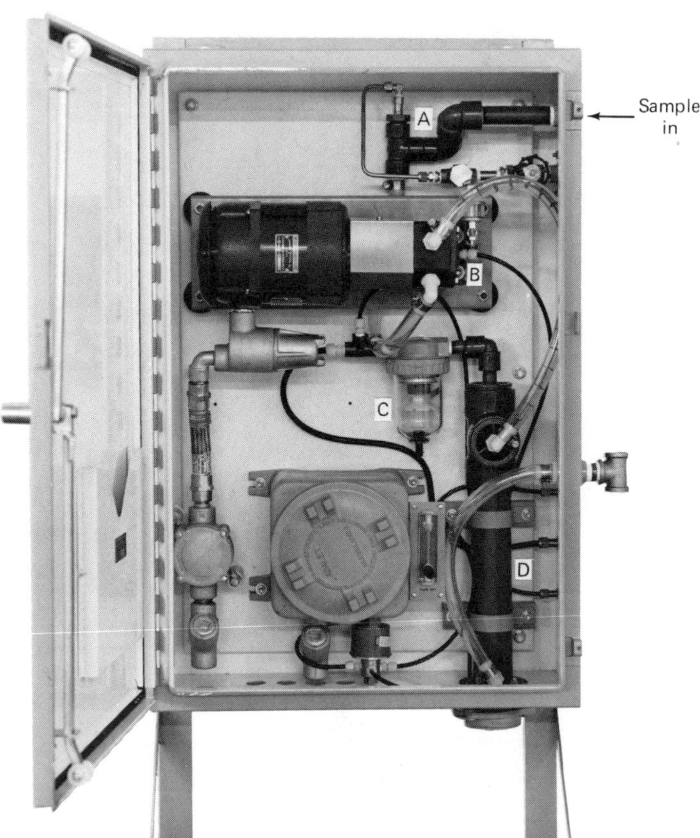


**Fig. 6.3.** Process monitor for flue gas oxygen analysis showing (A) spray wash for incoming sample, (B) slurry pump, (C) demister, (D) gas-liquid separator, and (E) oxygen monitor in protective condulet. Courtesy of Teledyne Analytical Instruments.

keep the lines open but provides a means of removing any corrosive gases, such as $SO_2$, that may be present. The use of reinforced polymeric materials wherever possible also helps to minimize corrosion problems. For example, the peristaltic pump's flexible inner liner is constructed of heavy gauge Neoprene and the portion of the pump housing in contact with the sample is reinforced polyvinyl chloride. Special attention must be given to corrosion possibilities in many process monitoring applications (37).

## B. SAMPLE TREATMENT

Although the water cooling and washing action of a sampling system such as that described in the immediately preceding section does, in fact, provide a degree of sampling conditioning, there are often cases where a much more extensive treatment is required. The treatment system shown in Fig. 6.4 is part

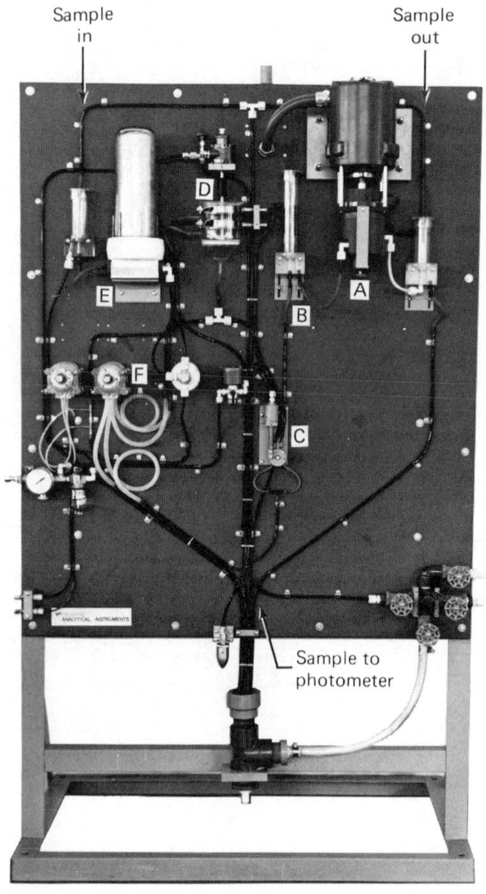


**Fig. 6.4.** Sample treating system for processing refinery waste water prior to photometric anaylsis. A, Homogenizer; B, sample debubbler; C, selector valve for streams to photometer; D, air sparging unit; E, oil drop filter; F, peristaltic pumps. Courtesy Teledyne Analytical Instruments.

of an oil-in-water monitor for continuously analyzing refinery waste water discharges (112). The oil to be measured is dissolved and dispersed in the water along with other non-oil hydrocarbons and dirt. To obtain the measurement, a differential, dual wavelength, uv absorption photometric technique is applied. The method involves alternately viewing a continuously flowing homogenized raw sample and a similar stream with all of the oil, and only the oil, removed. The difference in the two absorbances observed represents only the oil content, because the effect of other uv absorbing hydrocarbons that may be present has been compensated by the "zeroing" step.

Total homogenization of the sample is provided on a continuous basis by the mechanical blender located in the right-hand portion of the system as shown pictorially in Fig. 6.4 and schematically in Fig. 6.5. In the left-hand part of the system, the oil is being removed from a portion of the flowing sample by first filtering and then air sparging. The two separately treated streams are alternately directed through the measuring photometer by a clock-controlled valve. The effect of the dirt in the homogenized sample is also compensated by the

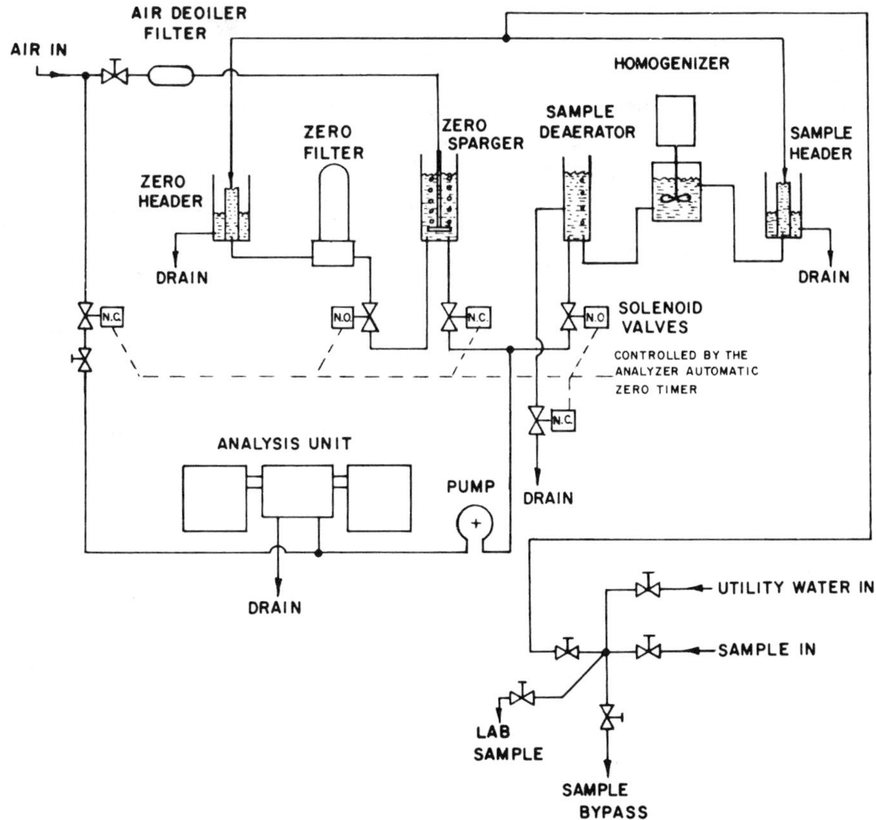


**Fig. 6.5.** Schematic drawing of sample treating system for on-line monitoring of oil in refinery waste water. Courtesy of Teledyne Analytical Instruments.

dual wavelength, ratioing principle employed. Depending upon the amount of dirt in the stream, this particular sample-treating system is capable of operating continuously for extended time periods with a minimum of operator attention. Many other forms of sample conditioning are often mentioned in the commercial literature of process instrument manufacturers.

## C. THE MEASURING SUBSYSTEM

The heart of any process monitoring system is, of course, the actual measuring instrument itself. Examples of some of the many types of analyzers used are given in a following portion of this presentation. In brief, virtually all known methods of analytical instrument sensing and transduction have been tried, as far as process monitoring is concerned. However, measurements based on photometric, separatory, and electrochemical principles have survived as probably the most widely used. The fairly demanding requirements of process and field applications (Table 6.I) are obviously limiting factors that rule out the use of many sophisticated instrumental methods.

The often encountered abrasive and corrosive nature of process streams encourages the use of systems having as little direct contact as possible between the stream and the measuring system's active elements. Thus flow-through photometers have a special appeal. The sample is confined in a sample cell which in turn can be isolated from the sensitive measuring transducer, the radiant energy source, and their attendant electronics. The sample cell itself constitutes no particular problem. It is generally constructed of chemically resistant materials and can be easily removed and disassembled when cleaning of the cell windows is required.

Measuring instruments employing separatory principles are applied primarily to the monitoring of gas streams, although procedures for liquid streams are becoming increasingly important. The physical separation of complex gas mixtures with chromatographic columns provides the petroleum refining industry, for example, with one of its most widely used forms of process monitoring. A more detailed discussion of this method is given in Section III-G. As an analytical technique, gas chromatography uniquely combines a most effective means of sample pretreatment, that is, the complete separation of all the components present, with a choice of several integral measuring devices. A wide selection of separatory columns and the flexibility allowed in choosing a detector make gas chromatography a most versatile measuring system.

Liquid chromatography is an increasingly important form of process analysis, particularly for the drug manufacturing and food industries. Although gas chromatographs are used as true on-line monitors, the column separation of liquid samples under high pressure is currently more of a laboratory-style procedure.

The most widely used electrochemical measuring systems employ potentiometric, polarographic, or galvanic sensors. All are characterized by the incorporation of a thin membrane, either glass or some polymeric material, to isolate the sensor's active electrodes and electrolyte solution from the sample being analyzed.

A detailed discussion of the various forms of transduction provided by analytical detectors can be found in Chapter 4.

## D. SIGNAL READOUT SUBSYSTEMS

Process monitor signals are commonly presented in the dc millivolt range. However, many instrument manufacturers offer the customer a choice of either milliampere or voltage outputs, with the latter designed to facilitate computer interfacing. Because of the value of being able to observe trends in the process stream being monitored, strip or circular paper chart potentiometric recorders are often used as signal readout devices. A 12 or 24 hr circular chart provides the plant operator with the entire record at one glance. Strip charts do have the disadvantage of visibly exposing only a limited section of the record at any given time, and one must remove and unwind the chart roll in order to inspect the entire trace.

Milliammeters and millivoltmeters constitute the major form of nonrecording, analog signal presentation, with digital displays gradually becoming more widely used. The optical meter relay is a special version of readout meter that provides an internal switching system to activate alarms or control circuits. Solid-state electronic switching elements are also widely used for the same purpose. The conversion of analog signals to digital outputs permits the use of electronic data processing and tape storage. The many advantages in economy of storage and ease of future accessibility of the recorded data when in tape form are becoming increasingly evident. The methods employed to amplify and convert the analog signals from detectors are covered in this volume in Chapters 1–4. Computer methods for data reduction that apply to many process monitoring systems are covered in Chapters 7–9.

## E. SPECIAL CONSIDERATIONS IN THE CONSTRUCTION OF PROCESS MONITORING SYSTEMS

The constant threat of corrosive atmospheres and the generally hostile environment associated with process monitoring require that special precautions be observed when designing and installing a monitoring system (e.g., cf. 37). Sensitive electronic, optical, and mechanical components must be housed in rugged and durable compartments provided in the instrument enclosure itself. The enclosure is generally constructed of heavy-gauge steel sheet. Safety requirements in many operating areas often demand that all electrical and electronic devices be contained in explosion-proof condulets. The condulets in turn may then be mounted in a weather-resistant metal case to provide additional protection (Fig. 6.6). The elimination of possible explosion hazards and internal corrosion from atmospheric causes can be assured by continuously purging the instrument enclosure with dry nitrogen.

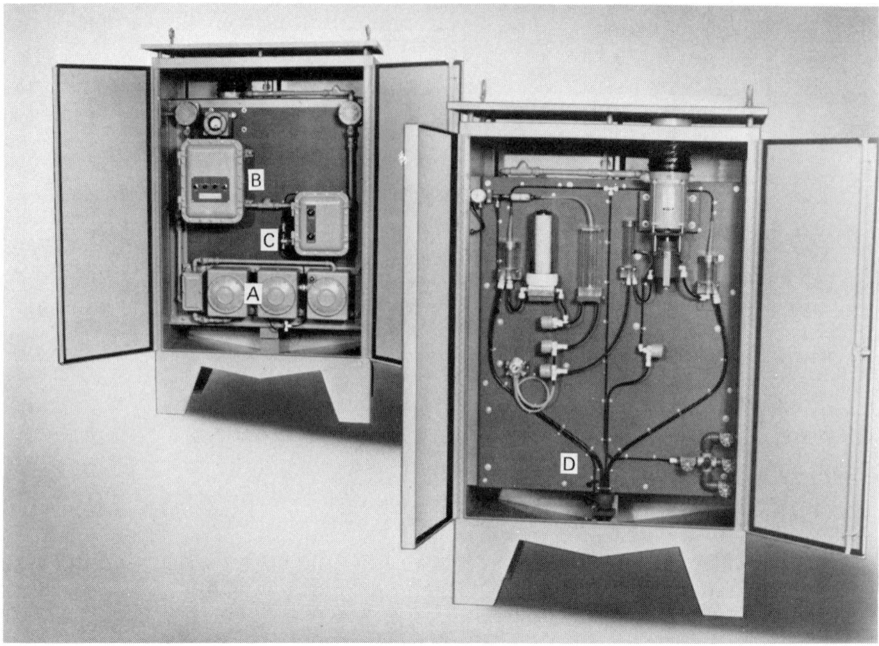


**Fig. 6.6.** Process photometer showing explosion-proof condulets and entire system mounted in protective steel enclosure. A, Process photometer; B, photometer electronics and controls; C, sample selecting system; D, sample treating system. Courtesy of Teledyne Analytical Instruments.

As previously stated, the increasing use of polymeric materials for construction purposes, including sample lines, mounting plates, valves, and pumps, attests to the seriousness of the corrosion problem. Reinforced fiber glass enclosures have been advantageously employed in particularly bad environments. Special precautions regarding the corrosion of soldered connections and switch contacts in electronic assemblies must also be observed. Generally speaking, almost every element of an on-line or in-plant monitoring system may be vulnerable to some form of physical or environmental damage. Also included in this category are the unwanted effects of vibration, sudden mechanical shock, and thermal shock.

## F. THE CALIBRATION OF PROCESS MONITORING SYSTEMS

A major obstacle to achieving the optimum use of a process monitoring system is more often than not the inability to calibrate the method. The usually encountered complex and variable nature of the sample stream to be measured constitutes the main problem. Although the species to be determined can usually be obtained in pure form, the preparation of standard samples that duplicate process conditions is quite another matter. The instrument's response to the species of interest in an ideal matrix must often be accepted as a suitable form of calibration. This does leave undetermined the extent of the error that may be introduced by possible matrix interferences. One practical way to confirm a monitoring system's accuracy is to conduct over a period of time simultaneous analyses with a standard reference method and then compare the two records. This does require a considerable amount of batch sampling in order to make certain that all possible stream variations have been encountered.

The proprietary nature of most industrial processes makes prior calibration of a monitoring system by the instrument manufacturer a most difficult if not impossible task. Samples supplied by the customer for this purpose are often not completely representative of the process stream itself. A cooperative in-plant effort that permits the instrument manufacturer to adapt the system directly to the actual process stream is a far more effective way to assure ultimate satisfaction.

## III. TYPES OF PROCESS MONITORS

The complete classification of all available process monitors poses a difficult problem because of the great number of different measuring systems that have been and are employed. In part, the problem arises from the extensive amount of customizing required of monitoring systems to satisfy users' needs. Although there are many "standard" models that can be systematically cataloged, it is their virtually unlimited modifications that defy classification. Consequently, it is possible to present here only selected examples of those monitors that are probably the most widely used or that utilize a particularly appropriate form of measurement. The majority of the systems chosen are concerned with the determination of chemical composition and structure. Therefore, the many instruments employed for temperature, pressure, flow, viscosity, vapor pressure, melting point, cloud point, and a host of other physical property-based measurements are not discussed. Also, more sophisticated and complex measuring systems such as those represented by magnetic mass spectrometry, nuclear magnetic resonance, and Raman spectrometry are not included, although it can be pointed out that they are employed for certain process measurements. It is felt that, in spite of the usage claims, such analyzers cannot be considered as typical candidates for on-stream or in-plant classification. The cost and technological problems associated with attempting to reduce such systems to satisfy the rugged requirements characteristic of process monitoring are not particularly encouraging at this time.

Marginal examples of system complexity for acceptance in this classification include instruments for measuring total organic carbon contents of waste water effluents and gas chromatography/mass spectrometers for trace organic determinations. Aside from these, the monitoring systems described in the following sections are extensively used either as direct on-line measuring systems or for providing in-plant analytical capabilities.

A comment must be made about the availability of the literature appropriate to the subject matter covered here. It is an unfortunate fact that in spite of the recognized importance of process monitoring there is still a lack of available standard text and reference books devoted to this subject. Isolated contributions covering specific monitoring developments are of course to be found in various parts of the scientific literature. The best sources of specific and detailed information are the instrument manufacturers' brochures, data sheets, and operator instruction manuals. Because such literature is usually not included in regular reference libraries, anyone wanting information must contact the appropriate manufacturer and request the desired manuals. As an aid in this respect, a liberal use of references to manufacturing sources is available in this chapter for those interested in a further study of the systems presented. Attention is also called to the list of supplementary readings included at the end of this chapter.

## A. PHYSICAL PROPERTY-BASED MONITORS

As stated above, the wide diversity of measuring techniques employed in process monitoring makes simple classifications impossible. For example, there are many determinations made that are based on a specific physical property or parameter of a given material, such as its electrical or thermal conductance. Although there is a very large number of instrumental systems available, it is possible to present here only a limited selection of such physical property-based monitors. It is felt that the following choices are representative of some of the more widely used systems.

### 1. Thermal Conductivity Analyzers

Differences in the heat conducting properties of gases provide a simple means for monitoring process gas streams (38). For example, thermal conductivity analyzers are used to control the quality of various commercial grades of nitrogen and oxygen produced by air liquefaction processes. The level of hydrogen in the continuous gas phase hydrogenation of unsaturated hydrocarbons is another typical measurement that can be made by the thermal conduction method (27).

Thermal conductivity has a drawback in that it provides a nonselective form of measurement. All gases present in a sample contribute to the magnitude of the reading observed; hence the method is limited to the determination of a single component in either a binary or a pseudo-binary mixture. In the latter case, the component-of-interest's thermal conduction varies with its change in concentration whereas the conductivity contribution of the matrix remains constant.

The measuring principle is based on the detection of the change in ohmic resistance of a small, heated, platinum resistance element, or thermistor bead, when a process gas stream of varying composition passes over it. The change in resistance is measured with a Wheatstone bridge circuit and the electrical analog output is related to the composition of the stream through a preestablished calibration procedure.

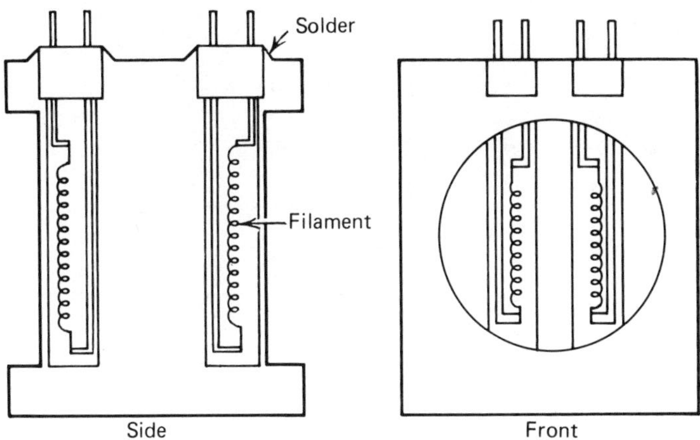


**Fig. 6.7.** Dual element thermal conductivity detector. Courtesy of Teledyne Analytical Instruments.

In practice, differential measurements are generally employed to reduce matrix gas effects and interferences from background temperature fluctuations. A reference gas of constant composition is simultaneously monitored along with the process stream in a dual channel detector (Fig. 6.7). The net difference between the two thermal conduction readings yields a more accurate measurement of the component of interest. The measuring elements are recessed in cavities so the sample and reference gases contact them only by diffusion. In this manner, varying heat losses resulting from changes in flow rate are minimized. Commercial instruments permit monitoring from a low parts per million level of hydrogen to percentages of carbon dioxide or methane in air (85,100). The sensitivity achieved in any given case depends upon the magnitude of the thermal conductivity difference between the two components involved.

### 2. Piezoelectric Sorption Hygrometer

The observed change in vibrating frequency of an hygroscopically coated, quartz crystal oscillator is utilized in monitoring for the presence of water vapor in process gas streams (80,133,81,82). The measurement is made with two coated crystal oscillators operated in tandem (Fig. 6.8). Water vapor is sorbed and then desorbed on each crystal by alternately switching the sample stream and a dry reference gas between the two every 30 sec. While one crystal is absorbing water and decreasing in frequency, the other is drying and increasing in frequency. The detected changes in frequency are electronically compared and the resulting differential is related to the stream's moisture content by prior calibration. Because desorption is a slow process, the 30 sec switching cycle is required in order that decreasing changes in moisture content can be detected with reasonable accuracy.

The system is effective for moisture contents in the range 1–25,000 ppm by volume and an accuracy of ±10% is claimed for readings between 1 and

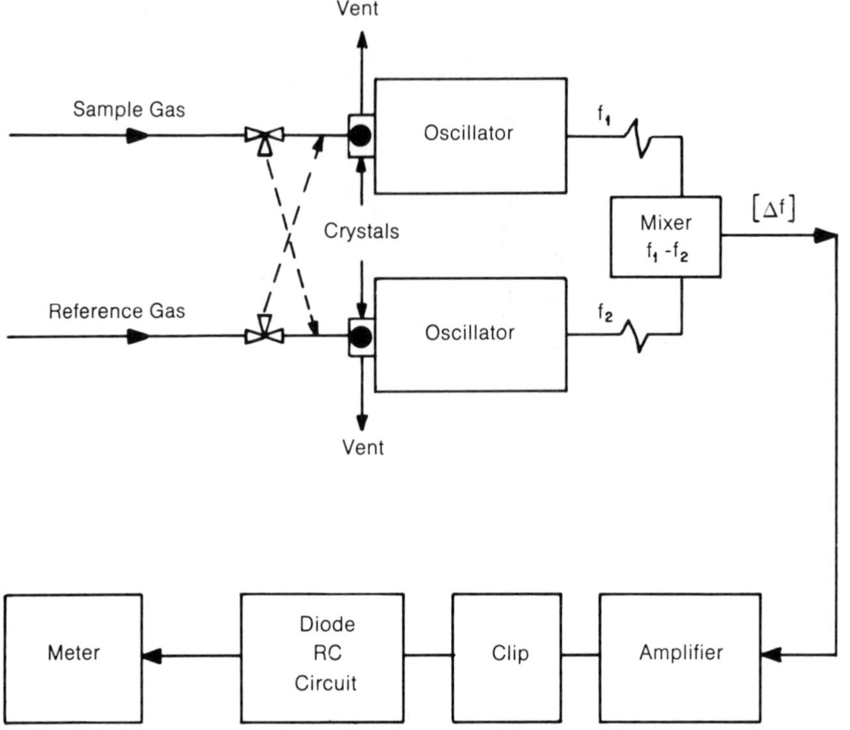


**Fig. 6.8.** Schematic representation of a piezoelectric sorption hygrometer for process moisture measurements. Courtesy of Du Pont Instrument Products Division.

2500 ppm (45). The method has been successfully applied to hydrogen, ethylene, various fluorocarbons, and certain natural gas stream monitoring. The deposition of dirt or the absorption of components other than water vapor on the crystal surface cannot be tolerated. The instrumentation for on-line use is compact and readily adaptable to meet in-plant, explosion-proof requirements (Fig. 6.9).

### 3. Ionic Conductance and Dielectric Property Measuring Instruments

#### a. Solution Conductivity Monitors

One of the oldest forms of process monitoring is that based on the electrical conductance of aqueous solutions as a function of electrolyte concentration (86). Like thermal conduction in gases, ionic conduction in solutions is nonspecific. Thus accurate quantitative determinations can be made only when the test sample is truly binary in composition and contains a single conducting species that varies in concentration. If the electrolyte fraction is multicomponent, then the system is considered pseudo-binary if all components but the test species remain constant. A differential measuring technique would then be employed. Typical process applications include the monitoring of the purity of

**Fig. 6.9.** Explosion-proof version of a piezoelectric sorption hygrometer for process monitoring. Courtesy Du Pont Instrument Products Division.

water from ion exchange columns, demineralizers, and distillation columns, as well as the strengths (from ppm to percentages) of acid, caustic, and a variety of inorganic salt solutions.

Solution conductivity is measured with a conductivity cell consisting of two flat, parallel, metal electrodes mounted a few centimeters apart in a rigid plastic or glass cell body. Platinum or platinum-coated metals are the electrode materials most often employed to eliminate the possibility of sample corrosion problems (87). The electrodes are energized from a low voltage, alternating current source to avoid current-limiting polarization effects. Because conductivity is, by definition, the reciprocal of solution resistance, a balanced bridge circuit is the measuring system. A sample's observed conductance is related to electrolyte concentration by calibration with known solutions of the conducting species. Measurements are made either by immersing a conductivity probe directly into the sample stream or by diverting a continuously flowing portion of the stream to a remotely positioned, flow-through cell. Maintenance of electrode cleanliness is essential and constitutes one of the method's major drawbacks, along with its nonspecificity, in process applications.

### b. Capacitance Measurements

The polar character of a limited number of molecular species serves as a means for their continuous detection. For example, water is determinable from the parts per million to the percentage range when present in a wide variety of liquid hydrocarbons by virtue of this property. The principle is not confined to the measurement of liquid samples alone. It has been extensively employed in industry to monitor moisture in a number of granular solids such as tobacco, cereals, meal, polymers, and coal, as well as in fabrics and paper (65). Like conductance measurements, the method is nonselective and in addition is prone to many interferences. The resulting uncertainties in its reliability when applied to nonbinary and nonideal samples have limited the method's general applicability and acceptance.

A capacitance measurement consists of detecting with an ac bridge the capacitance variations of a pair of fixed electrodes as the test material is passed between them. Electrode geometry and sample density are critical parameters, with uncontrolled variations in the latter often contributing the major source of error. The equipment required for accurate work is expensive, and the operator attention required for optimum performance is not an inconsequential factor. The capacitance principle has probably been most effectively used in industry as a "nonanalytical" tool for detecting liquid levels in tanks and liquid-liquid interfaces in chemical reactors.

### 4. Beta Radiation Devices

Beta ray absorption is principally used as a means for determining sheet and film thickness in the metal rolling, paper forming, and plastic industries. A carbon-14 radioisotope serves as the beta source and an ionization chamber is the detector (69). These two elements are positioned so the moving film passes between them. The measured attenuation of the incident radiation is directly related to sample thickness by calibration with standard film samples.

As another example, the particulate content of stack gas discharges can be monitored by beta ray absorption. The measurement consists of the collection of the particulates on a fixed portion of a continuous filter tape by filtration for a set period of time. After the filtration period, the tape is advanced into the radiometric section of the monitor. The portion of the tape containing the collected particulates is positioned between the beta source and the detector (Fig. 6.10). Because of the relatively low energy of the source used, the measurement is unaffected by particulate chemical composition. A linear response is claimed for filtrates ranging to 80 mg in weight distributed over a 9 $cm^2$ surface area of the tape. Approximately 12 measurements can be made per hour, depending upon sample stream conditions, with an overall accuracy of ±5%, as compared with a reference gravimetric value (84).

### 5. X-Ray Monitoring Systems

X-Ray absorption and secondary X-ray emission measurements are both utilized as process monitoring methods. X-Ray absorption is applied chiefly to the inspection of cast and molded articles as a means of detecting internal flaws. Such chemical determinations as the continuous measurement of sulfur in

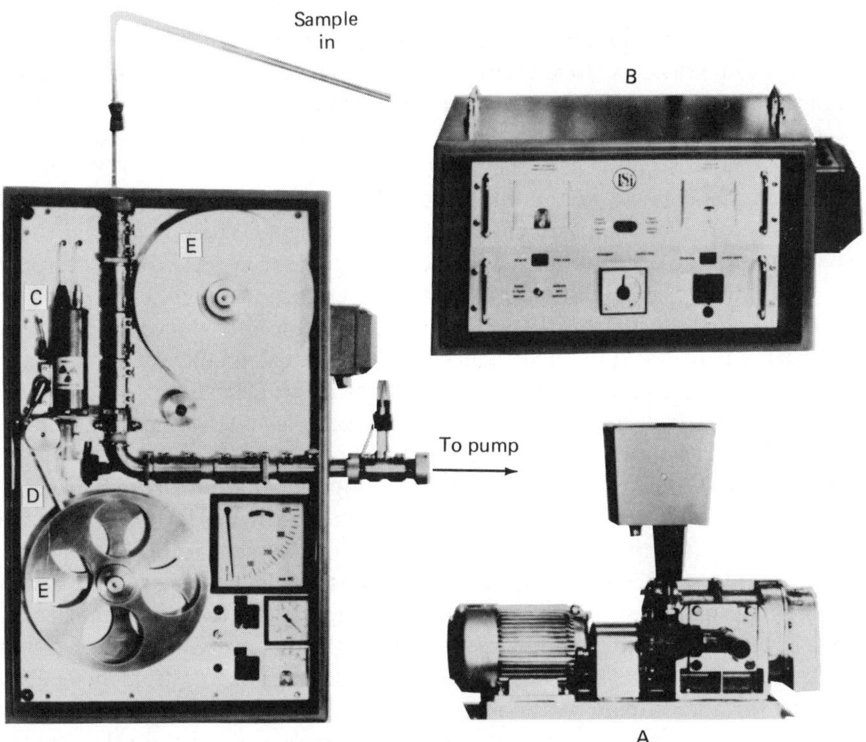


**Fig. 6.10.** Beta ray absorption analyzer for monitoring particulates in stack gas. A, sampling pump; B, electronic systems and controls; C, radiometric gauge; D, paper tape for sample collection; E, paper tape storage reels. Courtesy of Lear Siegler, Inc.

process hydrocarbon streams have also been accomplished by X-ray absorption (28). In general, absorption methods are not extensively used for process compositional analyses.

On the other hand, X-ray emission or fluorescence methods do have a fairly widespread industrial acceptance, particularly in the metal coating and alloy industries. The determination of plating weights of corrosion-resistant metals on steel sheets is a typical example. For liquid samples, the monitoring of lead tetraethyl additions to gasoline is performed with X-ray fluorescent equipment.

X-ray fluorescence analyzers consist of an X-ray source, a sample holder, a diffraction crystal monochromator to resolve the characteristic secondary emitted radiations from the sample, and an ionization counter as detector (116). Continuous flow-through systems are available, although many industrial applications are batch determinations.

### 6. Paramagnetic Oxygen Analyzers

Oxygen is unique in that its paramagnetic property is much greater than that of any other normally encountered gas. This characteristic has been utilized industrially, for example, to monitor the oxygen content in combustion processes, in inert gas treating operations, and for a large number of oxygen-sensitive chemical processes. The paramagnetic oxygen analyzer (110) consists of a small diamagnetic glass dumbbell suspended in a flow-through cavity by a torsionally stressed, quartz fiber suspension. A magnetic field is applied to the cavity containing the dumbbell. When an oxygen-bearing gas is passed through the cavity, the dumbbell is deflected by an amount proportional to the paramagnetic influence exerted on the magnetic field by the introduced oxygen. The displacement of the dumbbell, which is proportional to the stream's oxygen content, is detected either by optical means or by an electrostatic null balance method (13). Prior calibration with reference samples relates the displacement observed directly to the oxygen concentration.

Depending upon the "torsional stress" imparted to the fiber suspension, dumbbell arrangements can be constructed with sensitivities for oxygen ranging from the parts per million to high percentages. The fragile nature of the suspensions has always been a disadvantage, because sudden shocks or excessive vibrations usually result in a broken fiber. Recently, the use of metal suspensions has made available paramagnetic systems that are much more rugged than the quartz fiber units (119). Because of the suspension's extremely delicate balance, sample streams cannot contain material that might condense or deposit on the dumbbell. Significant errors in accuracy result from the slightest changes in this respect. It might also be noted that nitric oxide, because of its paramagnetic nature, constitutes a possible interference.

An alternate version of the paramagnetic principle utilizes the "magnetic wind" concept (88,101). Thermomagnetic oxygen analyzers contain a dual channel, flow-through, thermal conductivity cell (Section III.A.1). A strong magnetic field is applied to one of the cavities containing a sensing element. Oxygen molecules in the flowing stream are preferentially attracted into the imposed magnetic field. The oxygen entering the bottom of the cavity flows upward over the heated sensing element. As it becomes warm, it loses its paramagnetic character. A convection current is created as the warmer molecules are displaced by the incoming cooler oxygen molecules of the sample. Cooling of the thermal sensing element by the convection current results in a change in resistance that is measured by a bridge circuit. The nonmagnetically influenced sensor constitutes the compensating reference arm of the measuring bridge.

### 7. Ionization-Based Monitors

The electrical conductance of ionized gases provides a very sensitive means for detecting and determining a number of organic compounds. Like many other forms of instrumental sensing, the technique is not specific unless it is employed as a detector for a separatory system, such as a gas chromatograph. Ionization of gases can be achieved in a number of ways. For example, electrical discharges, beta rays, and ultraviolet light are typical means of excitation but are of limited industrial applicability.

Thermal excitation probably serves as the major ionizing source for process monitoring purposes. A small hydrogen flame provides sufficient energy to ionize the carbon atoms of most hydrocarbon molecules (39,90). The mechanism accounting for this phenomenon is not clearly understood at present. Systems employing flame ionization detection are used to monitor the purity of inert gas streams, the effectiveness of hydrocarbon scrubbers, combustion efficiencies, and a wide variety of additional industrial and safety applications. The method, being nonspecific, provides a "total carbon count" of all the hydrocarbons present if the stream is of mixed composition. For quantitative measurements of mixtures or unknown compounds, the determined values must be reported in terms of concentrations "equivalent to" some reference material. In other words, the magnitude of the ion current observed is directly proportional to the total number of carbon atoms entering the flame per unit time. In turn, this can be stated as being equivalent to the introduction of so many molecules of methane, hexane, or some other selected reference compound. For streams containing a known hydrocarbon, direct calibration and determinations are, of course, possible.

In a typical flame ionization detector, the sample stream is continuously blended with the hydrogen fuel at the base of the burner (Fig. 6.11). Combustion is supported by the surrounding air atmosphere which must be hydrocarbon-free. The ion collection cathode is positioned slightly above, and surrounding, the tip of the small flame. A dc potential of 500 V is impressed between the

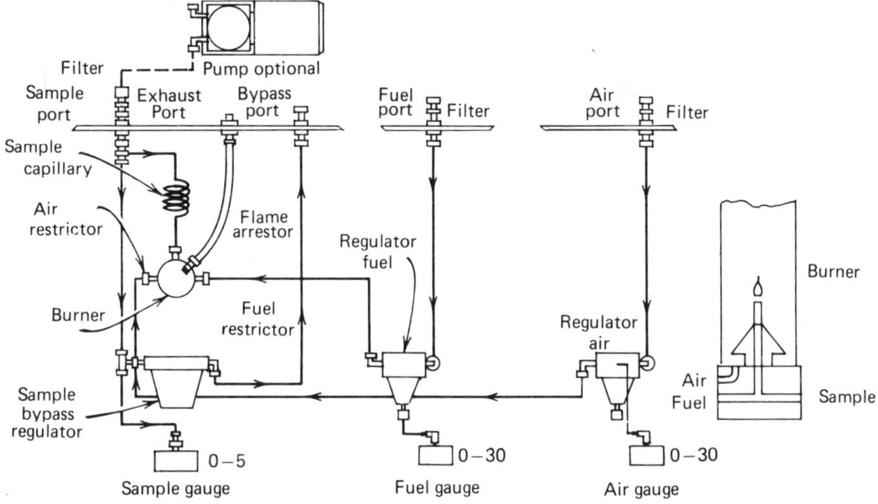


**Fig. 6.11.** Flame ionization detector flow schematic. Courtesy of Beckman Instruments, Inc.

collector electrode and the base of the flame. Carbon ions generated in the flame cause a small current of the order of $10^{-10}$–$10^{-12}$ A to flow in the collector electrode circuit. Electrometer-type amplification generates a linear analog signal that is directly related to the hydrocarbon content.

Sensitivities in the order of parts per billion to low parts per million are readily achieved with available commercial equipment (14,104,121). In practice, the technique is generally limited to quantitative measurements below 500 ppm because of charge saturation effects produced by higher hydrocarbon concentrations. The substitution of atoms such as oxygen, nitrogen, sulfur, or the halogens in the hydrocarbon molecule has a noticeable effect on flame response. Dirt particles interfere by causing substantial changes in the flame's noise level. In spite of these limitations, flame ionization detectors continue to be extremely valuable monitors, particularly in the air pollution field.

### 8. Gas Combustibles Analyzers

The concentration of hydrocarbon gases in combustion discharges, air-fuel mixtures, and various process streams generally are at higher levels than are suitable for flame ionization detection. Such determinations are possible by measuring the combustible characteristics of the compounds. The combustible measurement, like thermal conductivity (Section III.A.1), uses a heated resistance element as the sensor. A major difference, however, is that a combustibles detector has a catalytically active surface. This permits a low temperature oxidation, or burning, of the materials to be measured and thereby prolongs the useful life of the detector. The resistance change of the sensor, produced by the combustion process on its surface, is measured by a bridge circuit, and the signal is related to hydrocarbon content by calibration with standard gas mixtures. Industrial measurements are generally in the 0.5–10% range.

Flow control and matrix variations of the sample stream are critical parameters, and the dual sensor concept, as described for thermal conductivity, must be used to compensate for these. The reference sensor, not being catalytically active, is sensitive to all thermal effects except those generated by the "combustibles" process.

Both small portable analyzers and on-stream systems are extensively used to monitor boiler stack discharges for pollution and burner efficiency purposes (102,122). Often these analyzers consist of both a "combustibles" unit and an oxygen detector, for the two measurements together are more informative about the state of the combustion process (122). A pump or educting system is used to draw the sample from the stack, as described in Section II.A.

## B. GENERAL OPTICAL ANALYZERS

The refraction and scattering of light are optical phenomena of long-standing value to process monitoring, although neither provides much in the way of analytical specificity. This shortcoming, at least for refractive index measurements, is offset by the comparative simplicity of the method (77). Light scattering, however, represents a far more complex situation and, except for direct turbidity measurements, is a very difficult technique to apply to nonideal samples. Refractive index measurements are mainly used for liquid stream monitoring, whereas light scattering is primarily employed for measuring the particulate loading of both gaseous and liquid samples.

Both methods permit the complete isolation of the sample from any of the critical parts of the measuring instrument, such as the light source or the sensor. The measurements are nondestructive and can be made directly in pipes or reaction vessels, thus avoiding the expense of sampling systems.

### 1. Refractive Index Monitors

A schematic representation of an on-line refractive index monitoring system is given in Fig. 6.12. The flowing process stream is in direct contact with a spinel prism window which serves as the "second medium" needed to produce the

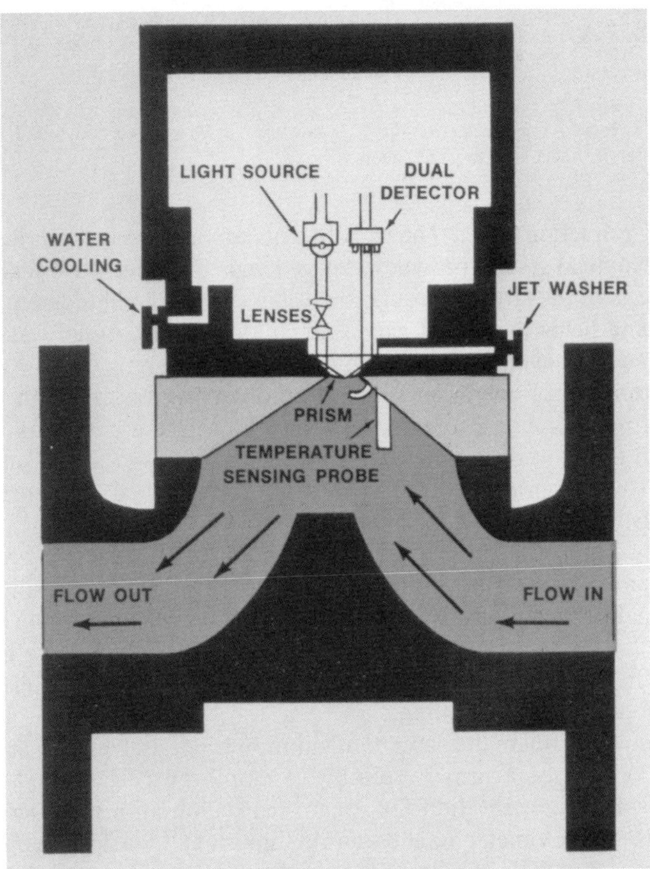


**Fig. 6.12.** Representation of an on-line process refractometer system. Courtesy of Anacon, Inc.

**Fig. 6.13.** Process refractometer for direct in-line use. A, Weir valve body; B, refractometer; C, instrument service parts. Courtesy of Anacon, Inc.

necessary refraction effect. The use of a regular valve body modified to contain the entire optical system provides a rugged and compact instrument (Fig. 6.13). The passage of the stream through the valve sweeps the liquid across the prism window and helps to prevent excessive buildup of dirt. Small washing jets for periodic additional cleaning are also included.

The principal of operation is based on observing variations in the "critical angle" of refraction as a function of change in solution concentration (5). The alignment of the source image with the reflecting faces of the prism interface is selected so the refracted beam is directed onto a dual element, solid-state, photoresistive detector. The light beam moves back and forth across the measuring detector as the critical angle of refraction changes with variations in stream composition. The varying resistive output of the measuring detector is ratioed against the output of the reference detector, which is totally irradiated at all times by the refracted light beam. Thus variations in source intensity are compensated, and the observed output signal can be accurately related to solution composition by calibration.

In those cases where the valve configuration is not usable, such as for samples in reaction vessels, a probe assembly is employed (Fig. 6.14) (5). If greater sensitivity and compensation for stream matrix variations are desired, then a differential refractometer is chosen (6). Important industrial refractive index measurements include the "solids" content of beverages and juices, dissolved solids in boiler condensates and process water, the purity of styrene, and the control of various fractionating columns.

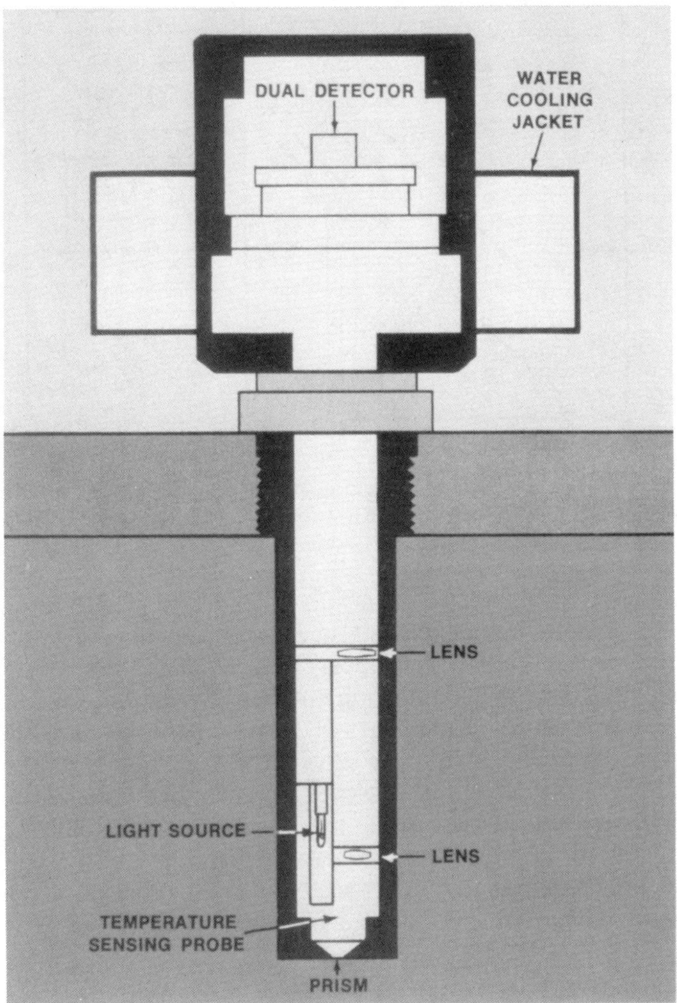


**Fig. 6.14.** Schematic representation of a probe-type process refractometer. Courtesy of Anacon, Inc.

## 2. Monitors Based on the Scattering of Light

Particulate matter in both gaseous and liquid process streams is detected by observing the intensity of the light scattered at a right angle from an incident beam directed through a sample. The observed variations in intensity are related by calibration to the particle concentration. Particle size, and to a certain extent particle shape, affects the reading. Unfortunately, these influences severely restrict the quantitative usefulness of the principle. A typical nonquantitative application involves detecting the appearance of suspended solids such as boiler scale in recycled boiler feed waters.

Light scattering instrumentation consists of a white light source, an elemen-

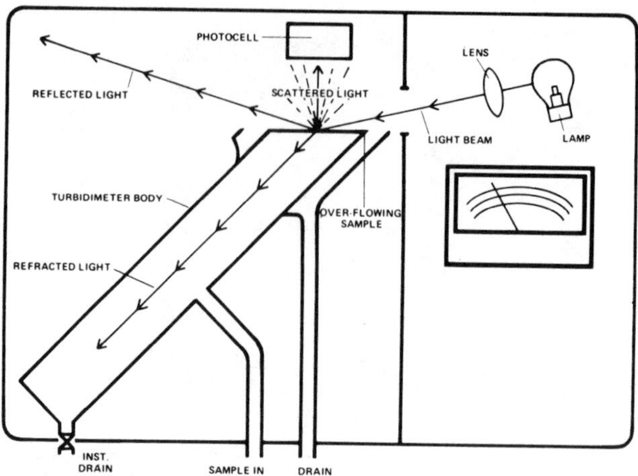


**Fig. 6.15.** Schematic drawing representing a suspended solids monitoring system for use with extremely turbid process samples. Courtesy of Hach Chemical Company.

tary focusing and collimating system, and a photodetector such as a photoresistive cell or photomultiplier. Flow-through cells are commonly employed to contain the sample. Window cleanliness can be a more or less serious problem, depending upon the amount and nature of the "dirt" in the sample stream. To avoid this drawback, the sampling and measurement technique depicted in Fig. 6.15 has been devised (56). As shown, the sample does not contact any critical portion of the measurement system. Because only a limited depth at the surface is involved in the scattering process, very turbid samples have been effectively treated this way. When the measurement must be made in a pipe or vessel, an insertion probe monitor can be used (7). Suspended solids in the concentration range of 0–250,000 ppm are measurable by these techniques.

## C. PHOTOMETRIC ANALYZERS

The high degree of chemical specificity achieved with spectrophotometric techniques makes them a most widely used and effective form of process monitoring. Absorption measurements in the range of 220 nm to approximately 7 $\mu$m account for many important determinations in the chemical, petrochemical, petroleum, and food processing industries. Innumerable laboratory studies utilizing uv, visible, near ir, and ir spectrophotometry provided the base for the development of on-line methods needed for monitoring those process streams whose complex composition defied analysis by other techniques. In addition, the analytical sensitivity of the ultraviolet region and the versatility of colormetric procedures are important assets in establishing reliable methods for environmental monitoring. The fact that the sample is observed in a confining cell allows photometric methods to be applied to corrosive and hot samples as well as flowing streams under high pressure.

A process photometer generally measures just one species in the stream; thus comparatively simple nondispersive optical systems are best suited for the process environment. The needed monochromaticity is supplied by nondispersive filters ranging in type from colored glass elements to highly selective interference filters. The engineering development and the packaging of process photometric systems are examples of some of the more noteworthy achievements of the process instrument industry.

The following sections contain descriptions of representative, commonly encountered process monitoring photometers. Limited space prevents the inclusion of all the different models in use. The operating principles of the examples given do, however, encompass the current state of the art for on-line process photometry.

### 1. Ultraviolet and Visible Light Photometers

On-line process photometers utilizing uv light sources are extensively employed for the measurement or organic compounds containing conjugated double bonds such as benzene, toluene, and other unsaturated ring molecules. Also, industrially important gases such as chlorine, bromine, ozone, hydrogen sulfide, and sulfur dioxide have measurable uv absorptions. Certainly, one of the more convenient aspects of uv absorption methods is that aqueous samples can be viewed directly, because the commonly used wavelengths are unaffected by water.

Visible region measurements are basically concerned with the relationships between observed colors and the chemical composition of the sample. Other industrial applications include the continuous monitoring of product color simply for quality control purposes. The measurement of the color of oils in terms of ASTM or Saybolt color units is a typical example. Many cosmetics and consumable colored liquids and food products are manufactured under close scrutiny by a color-measuring photometer serving as part of the on-line control system.

The generation of a color by the addition of a selected chemical reactant to the process sample is another form of photmetric colorimetry. Specific molecular species can be isolated and identified in complex aqueous mixtures in this manner. Otherwise colorless acid gases such as sulfur dioxide and nitric oxide can be trapped in water and converted to characteristically colored species when reacted with certain reagents.

Because of their basic similarities in design and construction, uv and visible photometers are considered as one and the same from a systems standpoint. A single instrument with a conventional incandescent source and a vacuum pototube or photomultiplier detector can make measurements from the near uv to the red region of the spectrum. All that need be changed to do this are the optical filters. For industrially important measurements in the uv range of 220–400 nm, a better source of uv light such as an hydrogen, mercury vapor, or quartz-iodine lamp would be used.

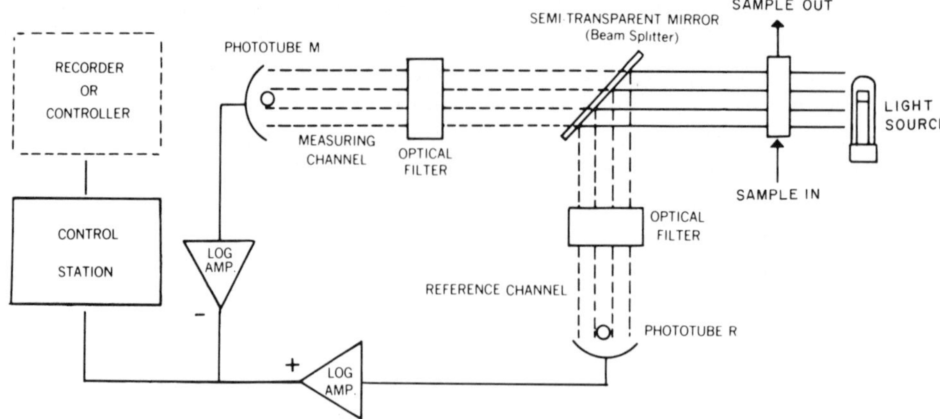


**Fig. 6.16.** Diagram of an uv-visible region process photometer of the fixed beam, or unmodulated, light source type. Courtesy of Du Pont Instrument Products Division.

The most frequently used, commercially available photometric units for process monitoring appear in two somewhat different physical configurations. These are referrred to as being either the fixed-beam type or the modulated-beam type. As is implied, the fixed-beam system employes unmodulated light (Fig. 6.16), whereas the modulated configuration uses a rotating filter wheel to bring two selected wavelength filters alternately into the beam emerging from the sample cell (Fig. 6.17).

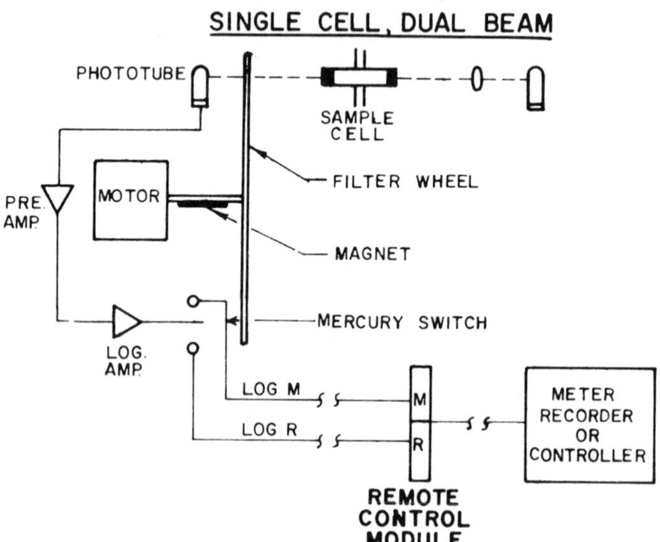


**Fig. 6.17.** Schematic representation of an uv-visible region process photometer system employing the modulated light beam principle. The filter wheel contains the two interference filters needed to provide the dual wavelength ratioing measurement. Courtesy of Teledyne Analytical Instruments.

In an unmodulated beam photometer, collimated source energy is directed through the sample cell and, after being attenuated, is separated by interference filters into two selected wavelengths for final measurement (46). Dual photodetectors provide a ratio between the outputs at the two selected wavelengths. These are chosen so that one falls in a characteristic absorption region of the component of interest and the other, or reference wavelength, is at some value not sensitive to concentration changes of that component. Thus the signal generated by the reference wavelength is used to compensate for interferences caused by dirt in the sample, dirty cell windows, or fluctuations in the light source. Changing the cell path length can provide an expansion of the measuring range capability from parts per million to high percentages for many compounds.

The modular design of the photometer provides protection for the source and detector assemblies and permits considerable flexibility in the length and type of sample cell that can be used. Generally, the sample is brought continuously from the process to the monitor and is passed through a flow-through cell. The modualr concept does, however, permit true on-line measurements to be made by the use of windows in the process pipe itself. The source and detector units are then mounted at opposite windows. This system has been effectively adapted for the measurement of $SO_2$ and the oxides of nitrogen in industrial stack effluents (47,49), as well as for a variety of other on-line determinations (e.g.,46).

In the modulated, or chopped, beam system, a single flow-through sample cell is also used (Fig. 6.17) (123). However, unlike the fixed beam unit, the light emerging from the sample cell atlernately passes through the two selected optical filters as the filter wheel rotates at approximately 1800 rpm. The resulting pulsed beams are directed to a single phototube. The two components of the photodetector's pulsed output are electronically isolated and converted into two direct current, logarithmic signals that are subtracted, or "ratioed," to provide a compensated final signal. This unit is also assembled in modular form for on-line use in either safe or hazardous areas (Fig. 6.18).

Both types of photometers are capable of handling either gaseous or liquid samples on a continuous basis. Cells with quartz windows have been used at working pressures up to 400 psi. Sample cells are constructed so they can be easily replaced or disassembled for periodic cleaning of the windows if required.

Performance characteristics for these types of photometers are outstanding and include such operating specifications as less than 1% noise, signal drift of less than 1% per day, ±1% full scale accuracy, a measuring range of .02–3.0 absorbance units, and a reproducibility of no worse than ±2% of the full scale value.

A somewhat different arrangement for direct in-stream photometric measurements is shown in Fig. 6.19 (74). The probe, containing folded light path, is inserted directly into a stack for the measurement of $SO_2$ or $NO_x$. A collimated light beam is directed from the source through a protective window in the mounting flange to a reflecting mirror at the end of the probe. The gaseous stream permeates the probe through small ports and enters the beam path. The

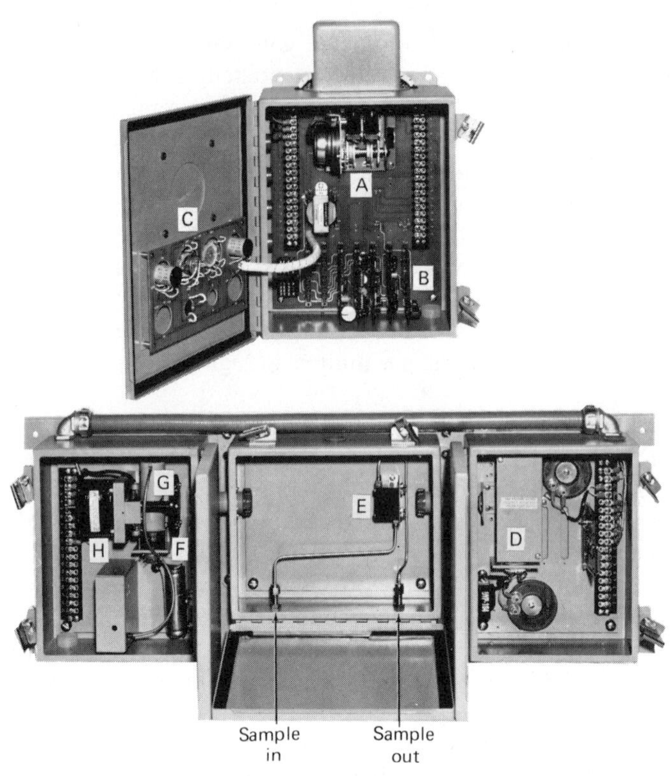


**Fig. 6.18.** Modulated beam process photometer. A, Motor-driven cam system; B, electronic systems; C, control panel; D, uv light source; E, flow-through cell; F, filter wheel; G, photodetector; H, filter wheel drive. Courtesy of Teledyne Analytical Instruments.

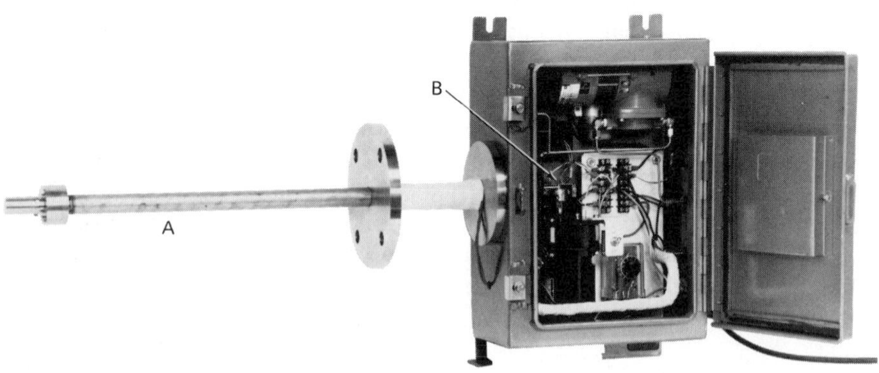


**Fig. 6.19.** Probe-type, modulated beam, uv absorption photometer for direct in-sample measurements. A, Insertion probe; B, filter optics and photodetector system. Courtesy of ITT Barton Process Instruments and Controls.

attenuated beam is directed back into the externally located measuring system. Dual wavelength, modulated beam optics are employed to provide a ratio reading proportional to the absorbing species' concentration.

### 2. Near Infrared Photometers

The near ir region of the electromagnetic spectrum covers the approximate range of 0.8–3 $\mu$m in wavelength. This region is used sparingly for analyses when compared with the popularity of either the uv-visible or the ir regions. Nevertheless, useful near ir absorption frequencies are exhibited by such compounds as alcohols, amines, certain aromatics, and by a number of olefinic compounds. The most frequent application of near ir photometry, however, involves the determination of water in solid and liquid samples. The water molecule exhibits two prominent absorption bands centered at 1.9 and 1.4 $\mu$m, respectively. The relative intensity to the bands, coupled with the fact that very few other compounds absorb strongly at these wavelengths, provides a unique means for the specific determination of moisture.

Industrially important applications include the measurement of moisture in cellophane and paper sheets, wood chips, plastic molding powders, and dried products. Typical examples involving liquid process streams might include the measurement of water in a wide range of aromatic hydrocarbons, alcohols, amines, ketones, chlorinated hydrocarbons, organic acids, olefins, and industrial oils. The near ir method does not provide the measuring sensitivity achieved with uv or visible region methods. Hence typical working concentrations for the moisture determination tend to be from hundreds of parts per million and low percentages, depending upon the matrix solution, to as high as 80%.

Moisture in solids is measured with a "reflection-type," near ir photometer (Fig. 6.20). A tungsten filament lamp provides adequate energy for working in

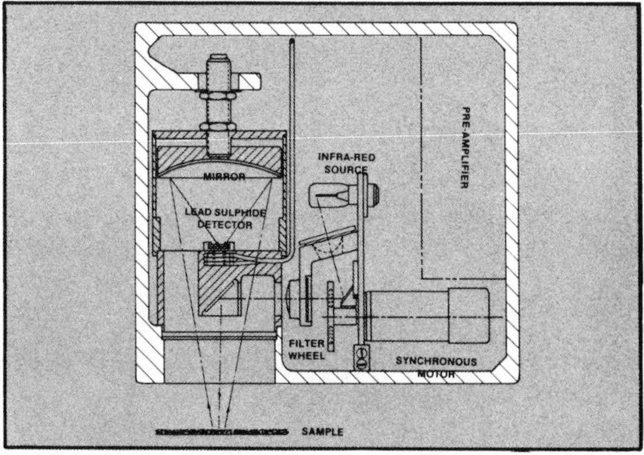


**Fig. 6.20.** Schematic diagram of a "reflection-type" near ir process photometer for continuously monitoring the moisture content of solids. Courtesy of Anacon, Inc.

the 1–2 $\mu$m region. The detectors most commonly employed are the solid-state variety, such as lead sulfide or lead selenide. Nondispersive optics with dual wavelength interference-type filters as monochromators and use of the modulated beam principle make up the rest of the photometric system.

When the method is applied, the incident radiation is directed by appropriate focusing and reflecting elements onto the moving sample stream. Limited penetration of the beam at the surface of the sample results in a measurable absorption by the water in the sample. The attenuated reflected beam is focused on the detector whose output is related to moisture content by calibration with known samples. An accuracy of the order of ±1% of the full scale reading is generally claimed for most applications (8).

For liquid streams, moisture is determined with a nondispersive, absorption-type photometer (Fig. 6.21). The system is the modulated beam type and employs two wavelengths for ratioing purposes. The interference filters and detector used are identical to those for the previously described reflection unit. Cell path lengths of 0.01–4.0 in. are common and all windows are preferably of quartz, although sapphire is recommended if sample pressures are high, that is, in the 300–500 psi range (124). The water peak at 1.9 $\mu$m is normally selected as

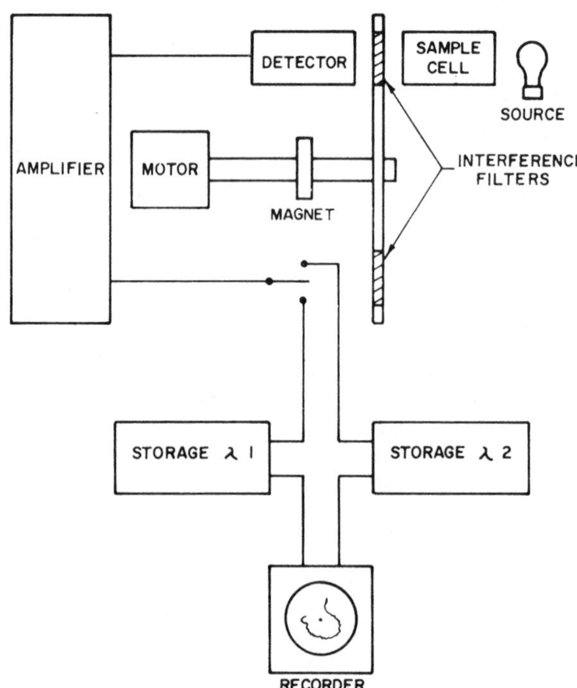


**Fig. 6.21.** Schematic drawing of a nondispersive, modulated beam, near ir process photometer system. Measurements are made by absorption. Courtesy of Teledyne Analytical Instruments.

**Fig. 6.22.** On-line process near ir photometer for the continuous monitoring of gas and liquid streams. A, Electronic systems; B, ir source; C, filter wheel and photodetector system; D, location of flow-through cell. Courtesy of Teledyne Analytical Instruments.

the measuring wavelength. A suitable reference value must be chosen at some point where absorption is not significantly influenced by changes in water concentration. Generally, wavelengths in the 1 $\mu$m range of the spectrum are best suited for this purpose.

The photometer is designed so that most of the electronics are in a protective enclosure isolated from the sample cell compartment (Fig. 6.22). The two-module configuration shown is not so convenient to use as is the three-module, uv-visible region system previously described. Short path lengths between source, sample cell, and detector are essential when working in the infrared regions because interference from surrounding atmospheric moisture can cause sizable measuring errors, particularly when working at low water concentrations.

### 3. Infrared Photometers

Absorption measurements in the infrared region of the spectrum extending from approximately 2.5 to 7.0 $\mu$m are well suited for a variety of process gas applications. The monitoring of carbon dioxide in ammonia (32) and ethylene oxide production, in diving chamber and submarine atmospheres, in furnace blanketing and heat treating operations, in medical respiratory applications, and in fuel combustion studies are typical examples. Infrared absorption is not limited to the measurement of carbon dioxide; sulfur dioxide (25), water vapor, carbon monoxide, methane, ethylene, nitric oxide (25), and ammonia are also measurable. A wide working range of concentrations, extending from the low parts per million to high percentages depending upon the absorbance of the particular material of interest, is not unusual.

Like the previously described photometers, ir process monitors are almost exclusively of the nondispersive filter type (77). However, there is a significant difference in that some versions do not employ interference filter optics for spectral isolation purposes. The alternative used, designated the "positive filter" method, utilizes a dual beam, two cell configuration with a pneumatic detector that is sensitized with a sample of the specific compound to be measured (Fig. 6.23) (15). The same general principle with a slightly different configuration of the pneumatic dectector is also widely employed (105).

In use, the radiant energy from two adjacent sources, usually hot Nichrome wire elements, is chopped by a rotating sectored disc and then directed through two cells mounted in tandem. One cell contains a nonabsorbing reference gas to compensate for background interferences while the process stream sample flows through the other cell. The pneumatic detector is divided into two compartments by a flexible diaphragm, one compartment for each beam. The diaphragm supports one plate of a capacitor that is an integral part of a sensitive electronic measuring circuit. Thus any pulsation of the diaphragm caused by unequal absorptions of the two beams by the gas in the detector is converted into a corresponding pulsed output signal. The amplitude of this signal is proportional to concentration changes in the flowing sample. By "sensitizing" both compartments of the detector with the species of interest, high specificity and sensitivity are achieved. Selected optical filters ae placed between the sources and the cells to refine further the selectivity of the method.

Sample cell lengths are variable and range from $\frac{1}{4}$ in. to as much as 15 in., thus providing a working concentration range of ppm to high percentages for many of the materials previously cited. Being compactly assembled, the instruments are readily mounted in explosion-proof and weatherproof enclosures for in-plant installation (Fig. 6.24).

The more conventional interference filter types of infrared photometers are also used for monitoring industrial gas and liquid samples (9). In operating principle and design, they closely resemble the modulated beam photometers previously described. A solid-state photoconductive detector is generally used, however, for effective detection.

A unique portable infrared unit (Fig. 6.25) has found wide acceptance for spot check monitoring in a number of process gas applications (132). A variable-path gas cell makes wide range of absorbance measurements possible. The

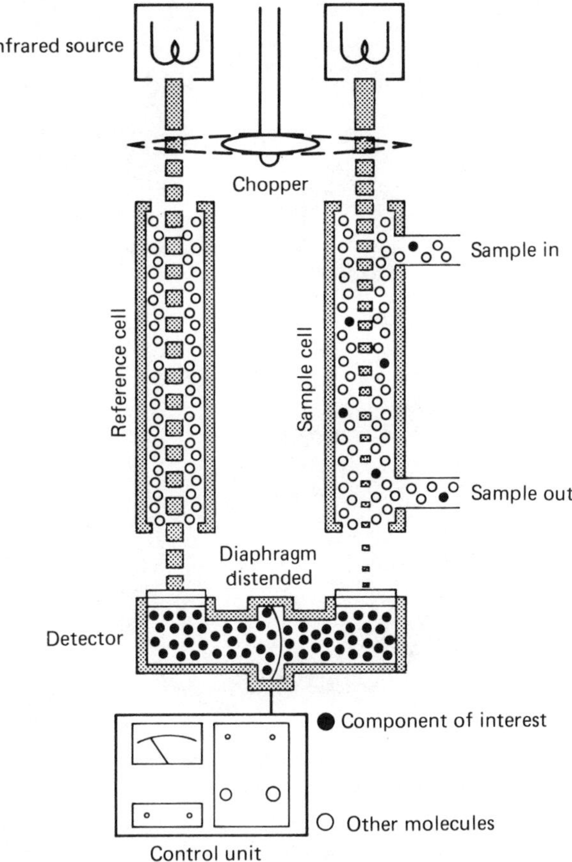


**Fig. 6.23.** Flow diagram showing main elements of a process ir photometric system for continuously monitoring gas streams. Courtesy of Beckman Instruments, Inc.

basic unit is of the modulated beam, interference filter, dual wavelength variety and uses a lithium tantalate, solid state ir detector (Fig. 6.26). The optical filter system can be expanded, however, to provide more than two fixed wavelengths for continuous multicomponent analyses. Another version of the basic unit contains a variable filter providing wavelengths from 2.5 to 14.5 $\mu$m. Variable path lengths ranging from $\frac{3}{4}$ to 20 m can be literally "dialed" into the gas cell which is less than 1 m in length. Exceptional analytical sensitivity, with measurable minimum concentrations for $SO_2$, CO, $CCl_4$, $NH_3$, acetone, and trichloroethylene being reported in the tenths to hundredths of a part per million, is a feature of this system (132).

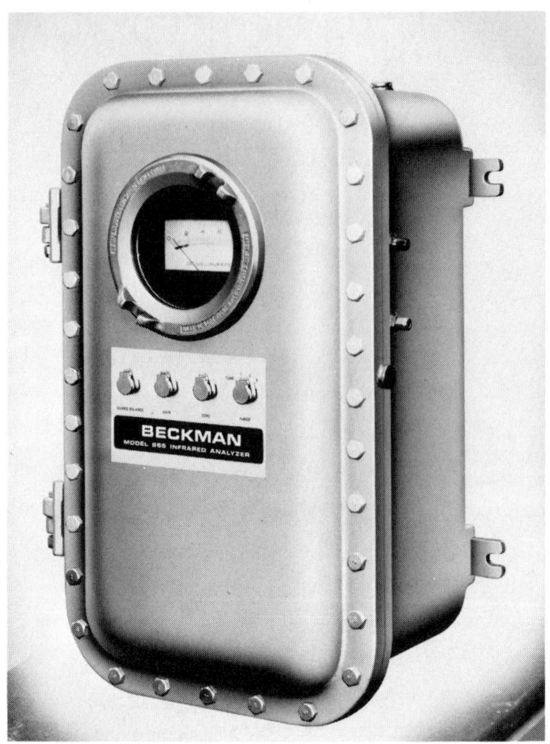


**Fig. 6.24.** Process ir photometer mounted in explosion-proof enclosure for on-line monitoring in a hazardous area. Courtesy of Beckman Instruments, Inc.

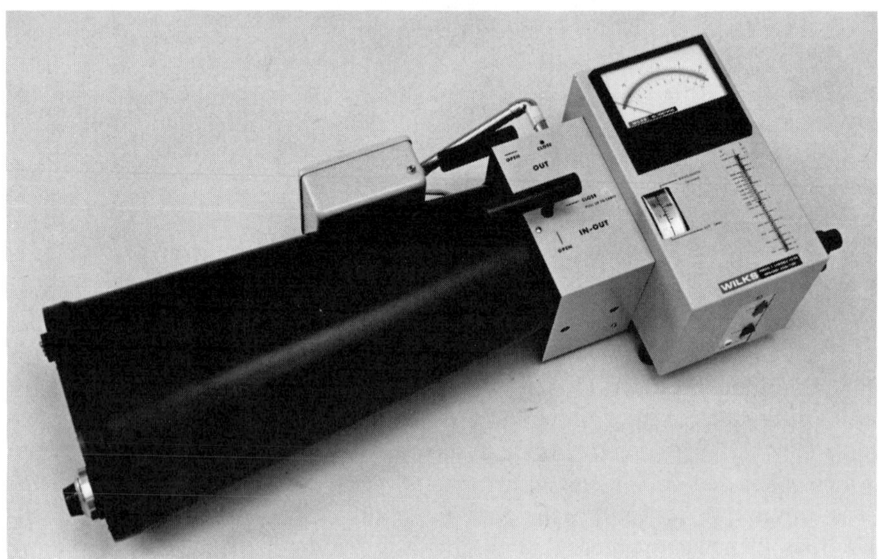


**Fig. 6.25.** Portable ir photometer for field analyses of gas and liquid samples. Shown with variable path length cell at left. Courtesy of Wilks Scientific Corporation.

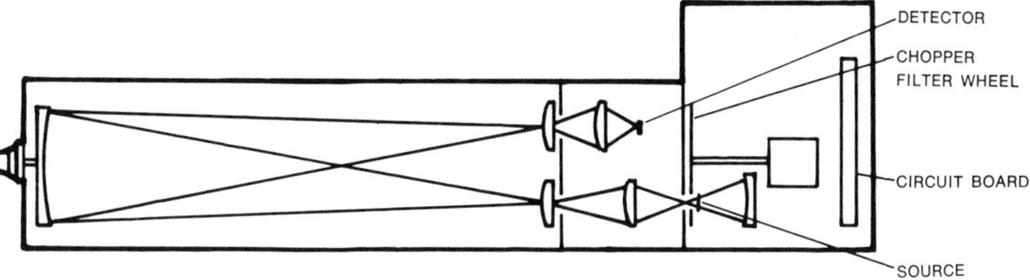


**Fig. 6.26.** Schematic diagram of portable ir photometer for field applications. Courtesy of Wilks Scientific Corporation.

### 4. Fluorescence Measuring Monitors

The use of an high intensity uv light source to induce the gas phase fluorescence of $SO_2$ and NO is a fairly recent addition to process analysis technology. The emergence of the method is a direct result of the need for a continuous analyzer for source and ambient monitoring. The highly selective nature and inherent sensitivity provided are features that make the technique attractive for other process applications. An unfortunate weakness, however, is that samples must be free of particulates and excess water vapor if reasonable accuracy is to be realized. To achieve this condition with a stack gas sample without losing an appreciable amount of the $SO_2$ is a formidable challenge. The measurable range for $SO_2$ is from 1.0 to 5000 ppm with a minimum detectable concentration of 1.0 ppm.

The optical arrangement employed for fluorescence measurements is not unlike that used for the monitoring of particulates by light scattering (Section III.B.2). As shown in Figure 6.27, the uv light source irradiates the flowing gas sample in the cell compartment causing fluorescence (3). The intensity of the emitted radiation is measured by a photomultiplier detector positioned at 90° to the path of the incident beam. Corning and Wratten filters are used in both the incident and emitted beams to provide some measure of wavelength selectivity. Calibration with known reference samples gives the relationship between detector output and the concentration of the fluorescing species in the process sample.

A system (Figure 6.28) using pulsed uv light as the irradiating source has been developed for the specific fluorometric measurement of ambient and stack sulfur dioxide (127). Interference filter optics screen out unwanted radiations from both the source and the irradiated sample. Pulsed uv light in the 200 nm range is reported to be optimum for inducing $SO_2$ fluorescence free of most interferences, such as air quenching.

A fluorescence photometer incorporating a zinc metal vapor excitation source is also proposed for ambient and source monitoring of NO (107). This

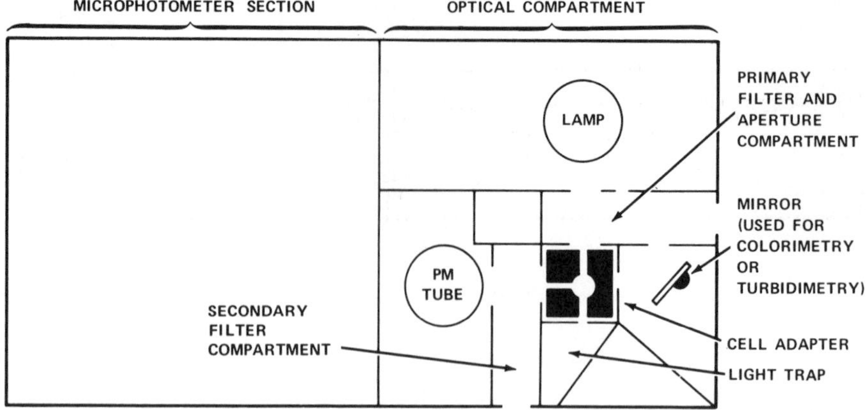


**Fig. 6.27.** Representation of the optical arrangement for the continuous fluorophotometric monitoring of flowing liquid streams. Flow-through cell is located in cell adapter section. Courtesy of American Instrument Company, Division of Travenol Laboratories, Inc.

particular type of lamp has a high energy output at 214 nm. Aside from the source, the optical arrangement used is similar to that already described. The system is reported to be capable of measuring nitric oxide concentrations from 0.04 ppm to 300 ppm in pure nitrogen gas samples. However, it has not been definitely established as yet that similar sensitivities are realized in industrial stack samples.

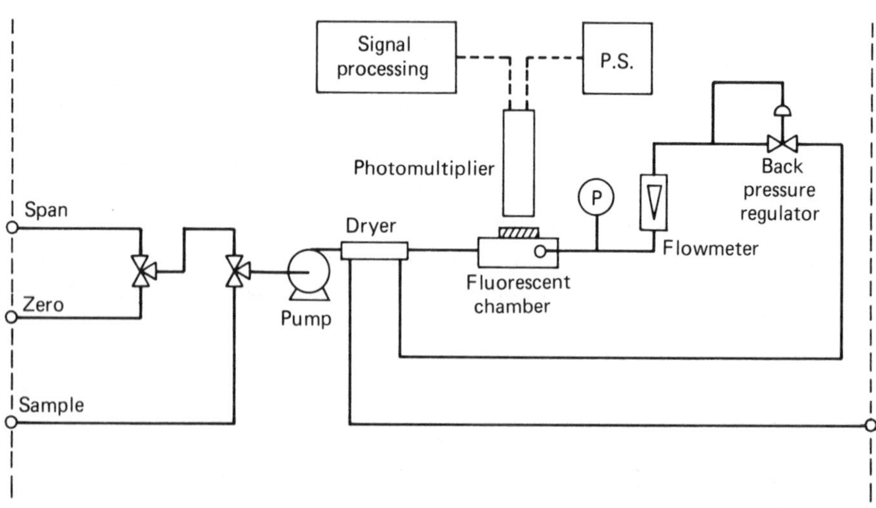


**Fig. 6.28.** Flow schematic of a pulsed, uv source fluorophotometer system for continuously monitoring ambient and stack $SO_2$ (U.S. Pats. 3,826,920 and 3,845,309). Courtesy of Thermo Electron Corporation.

## 5. Chemiluminescence Analyzers

The reactivity of ozone with certain compounds producing excited species that emit measurable light is the principle employed in chemiluminescent monitoring systems. Ozone itself can be determined this way, as can nitric oxide (34,53). For ozone determinations, ethylene gas is the reactant that is excited and chemiluminesces. In the presence of an excess of ethylene, the emitted light intensity is linearly proportional to the ozone concentration. To make the measurement, a sample of the ozone-bearing gas is drawn into a small reaction cell where it is mixed with an ethylene-air mixture (Fig. 6.29). The resulting chemiluminescence in the 300–600 nm region is measured with a photomultiplier detector for concentrations of ozone ranging from 10 ppb to 10 ppm (96).

The reaction between ozone and nitric oxide (N0) produces an unstable and light-emitting species of nitrogen dioxide ($NO_2$). This reaction, like the ethylene reaction, can be used to determine ozone, but it is primarily applied for the measurement of NO and mixtures of NO and $NO_2$, referred to as $NO_x$ (34,53). To achieve the latter, the $NO_2$ mixture is catalytically reduced to ozone-reactive

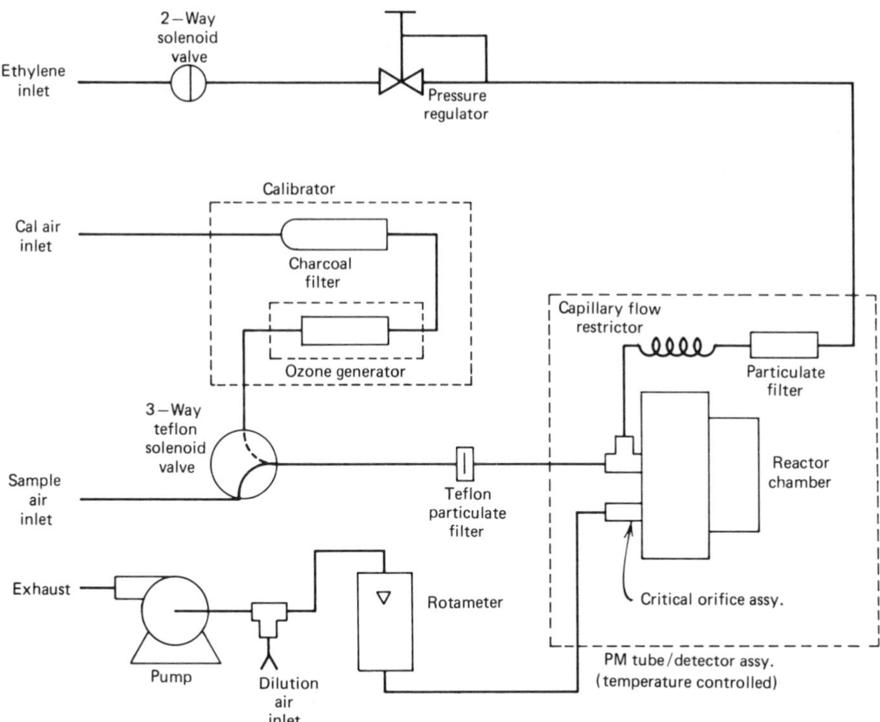


**Fig. 6.29.** Schematic diagram of a system for the continuous chemiluminescent photometric measurement of ozone. Courtesy of Meloy Laboratories, Inc.

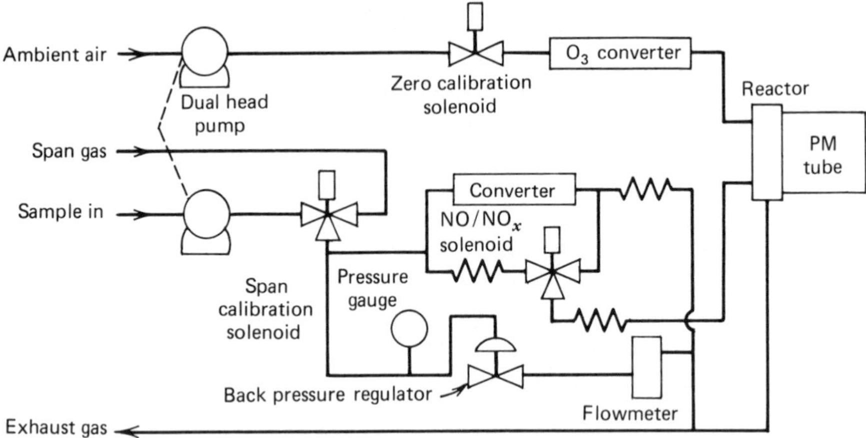


**Fig. 6.30.** Schematic representation of a continuous monitoring system based on the photometric measurement of the induced chemiluminescence of nitric oxide. Courtesy of Beckman Instruments, Inc.

NO. A difference measurement of before and after the conversion provides the determination of all the nitrogen oxides present. A flow schematic of an instrumental system for these determinations is given in Fig. 6.30 (16). Several other commercially available chemiluminescent analyzers (e.g., 2,97) utilize essentially the same procedure as that illustrated in Fig. 6.30. Useful working ranges for available monitors are approximately 0.01 ppm to as much as 2000 ppm for the various oxides mentioned. Studies have revealed that the reliability of the results obtained by the chemiluminescent principle for ambient air monitoring is comparable to that achieved with the Federal Reference Method for $NO_2$ (12).

## D. ELECTROCHEMICAL MONITORING SYSTEMS

Electrochemical process monitors are essentially utilized for two basic forms of measurement. One involves detecting the potential change resulting from variations in an ionic species' concentration in an aqueous solution. The other is the measurement of the flow of current in a pair of immersed electrodes as a function of the concentration of oxidizable or reducible species in the test sample. Unquestionable, the most familiar and widely used method in the first category is the determination of solution pH, or hydrogen ion concentration. The analyses of such electrochemically active gases as $O_2$, $SO_2$, $H_2O$ vapor, and NO are measurements of significant industrial importance and are representative of the second group. A susceptibility to interferences and a sensitivity to the condition of the sample are unfortunately restrictive features common to most process-type electrochemical monitors. For example, electrode surface cleanliness is a critical consideration and with dirty process streams the maintenance of the sensors is a major factor. The following samples include a limited selection of some of the more well-established and commonly employed electrochemical analyzers (117), and a few newer techniques that are gaining acceptance.

### 1. Potentiometric Sensors

a. THE MEASUREMENT OF pH

A solution's acidity or basicity is measured with a glass electrode sensitive to changes in hydrogen ion concentration. Although the glass electrode is a comparatively fragile sensor, the relative simplicity of pH-measuring instruments makes them ideal for many direct on-line monitoring and control uses (17,54). Such systems contain a working pair of electrodes consisting of the pH-sensitive element and a standard calomel reference electrode. The measuring and reference electrodes are integrally mounted in a rigid and rugged protective flow-through cell or probe assembly for process applications. Sample temperature affects pH and an electrical means for temperature compensation is an essential part of the measuring system. Changes in pH are measured in the millivolt range at very high ohmic resistance with essentially no current flow through electrodes; hence special precautions are taken in the design of electrode leads to avoid stray capacitance and current effects. This necessitates the use of very high input impedance, electrometer-type amplification. Available electrode designs and types are numerous, and pH measuring systems for virtually every kind of need are commercially available (e.g., 17).

It is completely outside the scope of this presentation to try to list even the major process applications involving pH measurements. They are seemingly endless because they occur in some form of other in practically all process industries.

b. OTHER ION SELECTIVE ELECTRODE SYSTEMS

Glass electrode technology has developed over the past years to the point where selective electrodes for a number of ionic species are now commercially produced (108). Like their pH counterpart, ion specific sensors have been adapted to innumerable on-line needs (109). Such industrially important materials as ammonia, chlorine, and various chemical oxidizers and reducers can be monitored. Copper, cyanide, lead, nitrate, and sulfide ions can also be accurately determined.

On-line analyses involving ion specific electrodes usually incorporate some form of sample pretreatment prior to measurement (Fig. 6.31). The schematic flow diagram shown represents a monitoring system for the measurement of cyanide. Reagent No. 1 is acidified EDTA used to release the metal-bound cyanide, and reagent No. 2 is a basic indicator introduced before final sensing.

To avoid giving a wrong impression, it must be pointed out that the unknowledgeable and indiscriminate application of ion specific electrodes is not advisable nor is it productive. A thorough knowledge of the test stream's chemical composition and an understanding of the possible interferences are far more

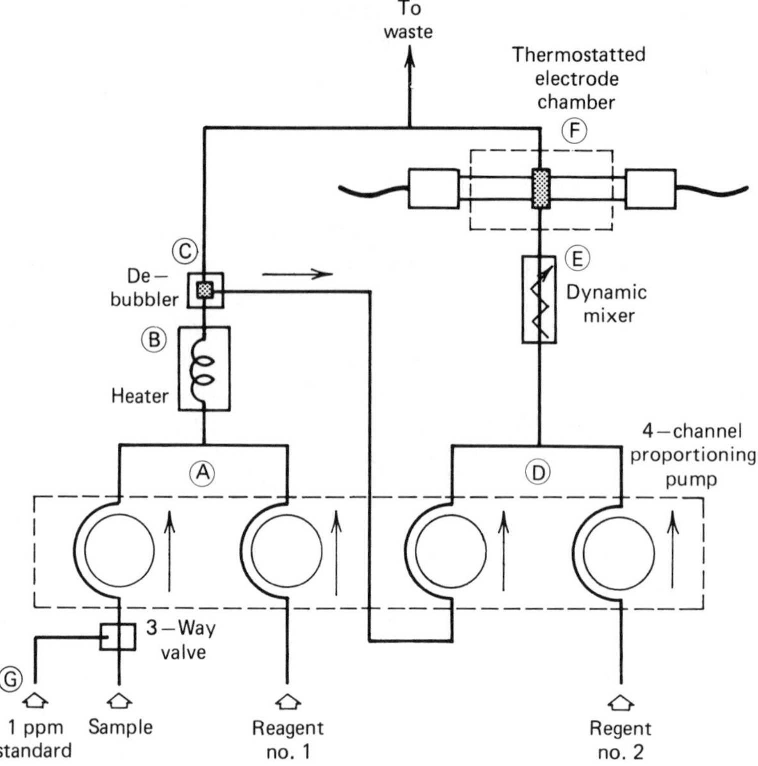


**Fig. 6.31.** Schematic flow diagram of an on-line system for the continuous ion specific electrode determination of cyanide. Courtesy of Orion Research.

critical here than is the case for determining pH. Continuing advances in the application area, however, assure that the technique will become an increasingly useful means of process monitoring.

## 2. Galvanic Sensors

The spontaneous generation of an electric current by a galvanic cell upon introduction of an electrochemically reactive gas is the basic means for several important determinations. Although oxygen is the species of major interest, galvanic sensors for the measurement of $SO_2$ and NO, for example, have been developed. A major advantage of the galvanic approach is the high degree of analytical specificity achieved with little effort or expense. The simplicity of the instrumentation needed makes the technique ideal for on-line and in-plant purposes.

### a. Microfuel Cell Detectors

A "microfuel cell" oxygen sensor utilizes the well-known electrochemical combination of a noble metal cathode, a sacrificial metal anode, and an alkaline electrolyte solution. For ease of handling, the complete system can be enclosed in a small plastic disposable container (Fig. 6.32). A very thin polymeric membrane covering the cathode allows oxygen to diffuse into the reactive electrode surface while the caustic electrolyte is held in the cell (125). Either sample from the process stream is educted and passed over the sensor mounted in a flow-through compartment or a cell-containing probe is inserted directly in the line. The spontaneous reactions that occur involve the reduction of $O_2$ molecules to hydroxyl ions at the cathode and the equivalent oxidation of lead metal to lead oxide at the anode.

The current generated, which is proportional to the oxygen partial pressure, can be measured directly with a low torque microammeter or amplified and displayed on a conventional potentiometric recorder or current meter (Fig. 6.33). A number of adaptations incorporating the microfuel cell are available for monitoring boilers and stacks for combustion efficiency, for safety applications, for various inerting and gas mixing uses, and for medical purposes (126). Linear outputs are realized for oxygen concentrations ranging from 1 to 2 ppm to 100%. A convenience of the system is that ambient air (20.9% oxygen) can be used for instrument calibration.

Several different classes of cells have been developed, as it was found that electrolyte buffering was required for measurements made in the presence of different levels of carbon dioxide. Otherwise, the cells are remarkably free of interferences from such commonly encountered species as water vapor, hydrocarbons, and particulate matter in the sample stream. Because of the aqueous electrolyte employed, working temperatures are limited to less than 125°F and

**Fig. 6.32.** Galvanic-type micro fuel cell sensor for the detection of gaseous oxygen. Cell diameter is approximately 1 in. Courtesy of Teledyne Analytical Instruments.

water cooling or washing is required for hot samples. Other galvanic $O_2$ systems, but not in the "microfuel cell" configuration, are also used industrially (89,103).

The microfuel cell design concept has also been adapted in particular to the measurement of $SO_2$ and NO in stack effluents and ambient atmospheres (128). At the present time, usage is somewhat restricted because of the number of interferences that appear to seriously affect cell performance in industrial use.

b. HIGH-TEMPERATURE, SOLID-STATE OXYGEN CELL

A unique galvanic-type sensor for oxygen operating at 850°C employs an yttria-stabilized zirconia matrix as the cell electrolyte (131). The zirconia matrix is in the form of a pencil-sized hollow tube with both inner and outer surfaces metallized by porous platinum electrodes. The sample gas is passed through the cylinder's inner bore and the outside surface is exposed to ambient air. The driving force for the cell is the partial pressure difference of oxygen outside (ambient) and inside (sample) the cylinder. Oxygen molecules on the higher partial pressure side (ambient) spontaneously gain electrons, become ions, and enter the zirconia electrolyte. At the inner electrode, oxygen molecules are formed from these ions. The potential thus generated is inversely proportional to the logarithm of the analyzed sample's oxygen partial pressure. Consequently, the system's readout of oxygen concentration is on a logarithmic scale.

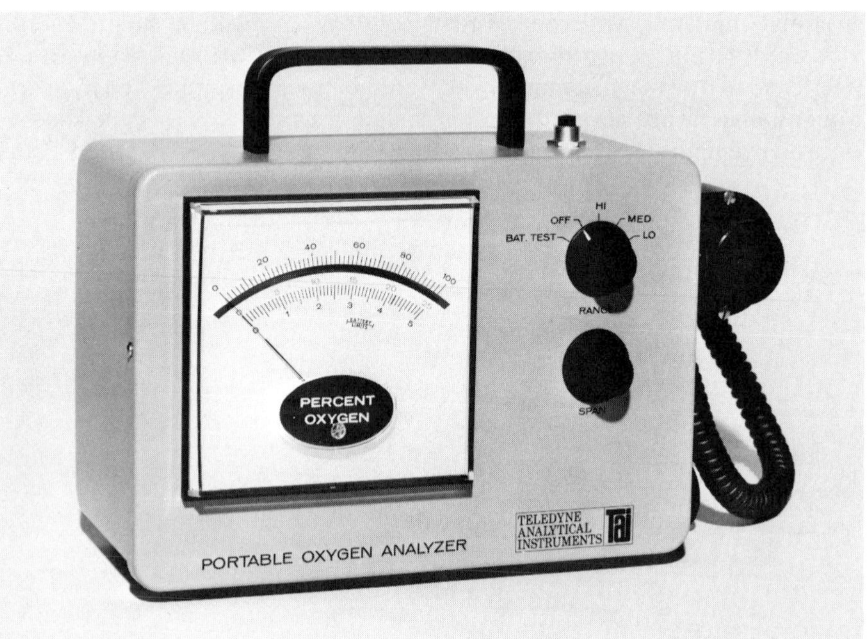


**Fig. 6.33.** Small, portable oxygen analyzer for field use employing a galvanic-type micro fuel cell sensor (located in probe at right). Courtesy of Teledyne Analytical Instruments.

Because of its high operating temperature, the cell is proposed for direct insertion in hot flue gases. Although the cost of a water-washed sample eductor or pump is eliminated, direct insertion may expose the sensor to serious interferences caused by the combustible materials that may be present. At the high operating temperature used, an oxidation of the combustibles is assured inside the cell, thus reducing the sample's oxygen content before it can be measured. Also, it is not clear what effects on accuracy result from the high moisture contents characteristic of flue gas streams.

### 3. Polarographic Systems

The use of voltage-induced electrochemical oxidations or reductions in aqueous solutions is not unknown in the process area. Dissolved chlorine, for example, is regularly determined in this manner. Conventional dropping mercury electrodes are not very practical, however, as far as on-line or in-plant uses are concerned. Solid electrodes have the disadvantage of being easily fouled by dirt or oily materials in the sample. Thus the most widely used version employing the polarographic principle is embodied in the membrane-covered electrode (33).

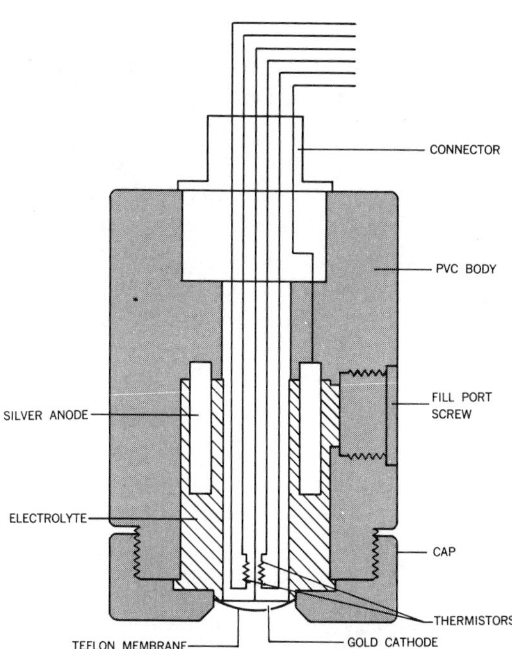


**Fig. 6.34.** Schematic diagram of a membrane-type polarographic oxygen cell. Courtesy of Beckman Instruments, Inc.

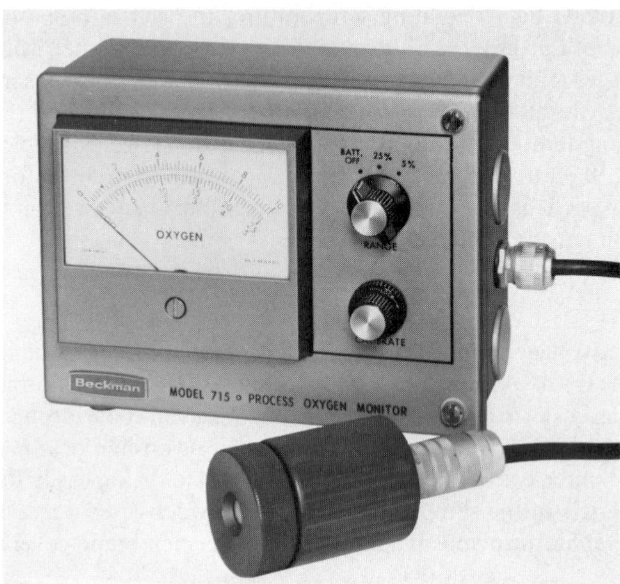


**Fig. 6.35.** Polarographic-type process oxygen monitor with portable probe. Courtesy of Beckman Instruments, Inc.

Although gaseous $SO_2$, $NO_2$, and $Cl_2$ are determinable by membrane electrode polarography, by far the most frequently employed measurement involves the determination of oxygen. The dissolved oxygen content of aqueous streams is particularly amenable to measurement by this technique (e.g., 135). A typical polarographic oxygen cell (Fig. 6.34) contains a gold cathode and a silver anode immersed in an aqueous potassium chloride solution (18). A thin, oxygen-permeable, polymeric membrane is stretched over the cathode and holds the electrolyte in the cell body. Unlike the disposable "microfuel" cell, provisions are made in this cell's construction to permit the replacement of the electrolyte solution on a periodic basis during use. Failure to do this results in incorrect readings because of the changing pH of the cell's electrolyte solution as the cell is operated. A potential of approximately 0.8 V is impressed across the electrode. Temperature compensation is mandatory and is supplied by a built-in thermistor network. Compact and rugged instrumentation is easily achieved because of the uncomplicated nature of the polarographic system (Fig. 6.35).

### 4. Coulometric Analyzers

Both primary and secondary coulometric techniques have been adapted for on-line and in-plant use. Primary coulometric monitors determine the concentration of electrochemically reactive species by 100% current-efficient electrolysis. Secondary coulometric methods involve the electrochemical generation and detection of a titrant for the species of interest. These methods are chiefly used to monitor process gas streams and ambient atmospheres for $H_2O$ vapor, $O_2$, $SO_2$, $NO_2$, and ozone, although there are many other industrial applications extant. Coulometric methods are characterized by their inherent sensitivity, with certain species being accurately measured in the parts per billion range.

#### a. Coulometric Water Vapor Monitors

The accurate determination of trace amounts of water vapor in nitrogen, argon, helium, carbon dioxide, and freon gases is an extremely demanding measurement. The advent of the coulometric, or electrolytic, moisture cell (78) reduced this determination to a routine form of process monitoring (19,48). However, the method is not universally applicable because a number of compounds interfere with the measurement. Potential users of such devices are advised to check with the manufacturer for application information prior to going on-stream.

The electrolytic moisture cell consists of a length of small-diameter glass tubing, coiled in spiral form, securely encased in a molded plastic protective cell housing. Inside the tubing are two separated fine-platinum-wire electrodes wound in a bifilar manner and firmly embedded in the glass but with their bulk exposed. The uniform space between the wires is filled with phosphoric anhydride ($P_2O_5$) produced by the electrolysis of syrupy phosphoric acid that is poured through the tubing. A 90 V dc source is impressed across the electrodes for operating purposes.

In operation, moisture in the flowing test sample is totally absorbed by the $P_2O_5$ forming a current-conducting phosphoric acid species. The impressed voltage causes electrolysis to occur and the absorbed water is completely electrolyzed to hydrogen and oxygen. The current flowing is a direct measure of the moisture concentration at a given mass flow rate. Thus under fixed sample flow conditions, the coulometric current detected can be read directly on a meter as parts per million of moisture. Water contents in the range of 1–1000 ppm are most effectively handled by this method.

Interferences in the moisture determination can occur if the stream contains appreciable amounts of hydrogen. A catalytic recombination of the hydrogen with oxygen produces additional amounts of $H_2O$ and a high reading results (35). Also, hydrocarbon vapors present in the sample stream are polymerized or decomposed by the hot phosphoric hydrate medium being electrolyzed during the cell's operation. Fouling of the electrode system results, and it must then be cleaned and regenerated to become operable again. Amines, ammonia, and alcohols also interfere because they either interact with the electrolyte or electrolyze to give spurious readings.

#### b. Coulometric Oxygen Monitors

Coulometric electrolysis has also been applied to the analysis of the trace oxygen contents in some process streams (63,79). The oxygen-bearing stream is bubbled through a liquid-immersed spongy silver cathode contained in an airtight cell. A voltage impressed across the silver cathode and a cadmium anode causes the complete reduction of all of the oxygen passing through the cell. Again an established sample flow rate results in a direct relationship between observed output current and the oxygen concentration of the incoming sample. In view of the ready availability of galvanic and polarographic sensors for this purpose, coulometric oxygen monitors have had limited applicability.

c. Electrochemical Titrators

Monitoring systems for the continuous coulometric titration of trace concentrations of ozone (93), nitrogen dioxide (20,93), and sulfur-containing compounds (75) in process gas streams and ambient atmospheres are commercially available. Also available are batch-treating instruments for nitrogen, halogen, and sulfur determinations (43). The portable nature of the instrumentation for several of these measurements is a most convenient feature (Fig. 6.36).

Typically, the analysis of atmospheric ozone utilizes an iodometric titration (30). The incoming $O_3$ reacts with a KI solution to produce $I_2$ in the cathode region of the cell (Fig. 6.37). The cathode surface is covered by a thin layer of hydrogen gas, preformed by passing a current until a self-limiting, complete polarization occurs. The appearance of $I_2$ immediately causes a depolarization of the cathode because of the reaction of $H_2$ with the $I_2$to form soluble HI. Thus a cathodic current again automatically flows to reestablish the $H_2$ film polarization condition. The cathode current observed for a fixed incoming sample flow rate is directly proportional to the sample's ozone concentration (94).

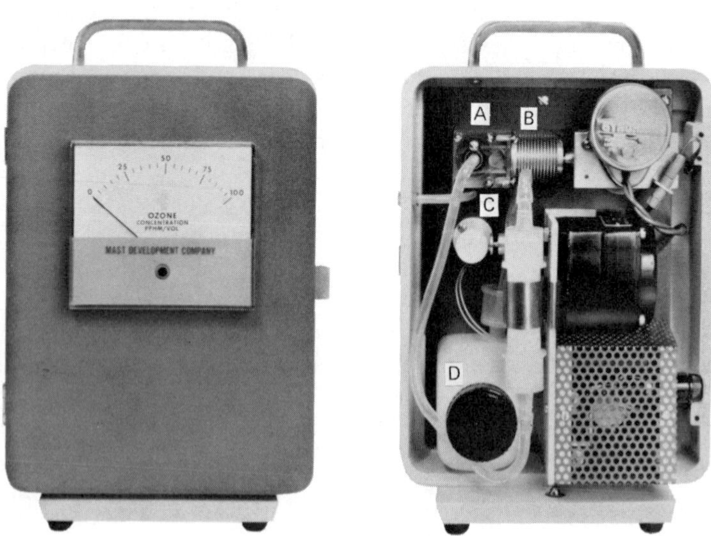


**Fig. 6.36.** Small, portable, and self-contained ozone analyzer for atmospheric monitoring. A, Measuring electrode; B, reagent solution pump; C, air sampling pump; D, reagent solution reservoir. Courtesy of Mast Development Company.

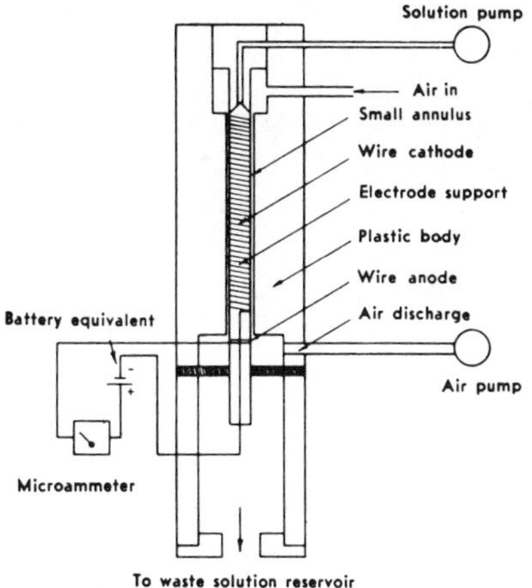


**Fig. 6.37.** Schematic representation of a microcoulomb ozone-sensing cell. Courtesy of Mast Development Company.

Nitrogen dioxide is determinable in the parts per billion to parts per million range by the same iodometric reaction (20) and with a bromometric titration (114). The coulometric generation if iodine is also employed in the electrolytic version of the Karl Fischer determination of water (11,99).

The continuous monitoring of sulfur-bearing compounds in waste gases from petroleum, petrochemical, and pulp and Kraft mill operations is accomplished with an electrolytically generated bromine titration (75). In operation, the sample is bubbled through a titration cell containing a three-electrode array immersed in an aqueous bromide solution. The following reactions occur at the three electrodes:

Sensing Cathode

$$Br_2 + 2e^- = 2\,Br^-$$

Common Anode

$$2\,Br^- - 2e^- = Br_2$$

Generating Cathode

$$2\,H^+ + 2e^- = H_2$$

Oxidizable sulfur species produce a decrease in the cell's established electrolytically generated $Br_2$ concentration. As the $Br_2$ balance is altered, the sensing cathode output is fed through a control circuit which causes more $Br_2$ to be generated at the anode to restore the initial balance. The current flow needed to reestablish the original $Br_2$ null point is directly related to the sample's sulfur content. Mercaptans, $H_2S$, $SO_2$, and various other organic sulfides are determinable at the parts per billion to low parts per million levels in this manner.

A three electrode system for the titration of $SO_2$ in an $I^-$–$I_2$ solution is also available (21). The reference electrode senses the imbalance in the cathode-anode current as the incoming $SO_2$ disturbs the initial $I^-$–$I_2$ equilibrium and a null balance circuit is used as the means of measurement.

## E. EMISSION SPECTROSCOPIC AND FLAME PHOTOMETRIC MONITORING SYSTEMS

Inorganic and metallic compounds are analyzed by emission spectroscopic methods using high thermal energy sources for sample excitation. The excitation sources most commonly used are the electrical arc or spark or an hydrogen flame. Because of the extremely high energy levels generated by arcs and sparks, the resulting emission spectra consist of a multitude of the characteristic lines of the elements present in the sample. These can be resolved only with sophisticated dispersive-type optical systems employing gratings or prisms. Consequently, emission-type spectroscopic monitors are comparatively complicated and expensive units. Although emission spectroscopy is a batch analysis method, multichannel units can provide the simultaneous readout of as many as a dozen elements per sample in less than 1 min.

A hydrogen flame, on the other hand, generates substantially less heat than does the arc or spark source; hence resulting emission spectra are far less complicated, and less sophisticated nondispersive optical systems are adequate for measurement. Flame systems have another important advantage over arc and spark units for process use in that samples can be fed rapidly and continuously into the flame from a flowing stream. Atomic absorption is a variation of the flame method and is employed when greater specificity and sensitivity are desired. There is not much evidence in the industrial literature, however, to indicate atomic absorption photometers are widely used as on-stream or in-plant monitors.

### 1. Arc and Spark Emission Monitors

Optical emission spectroscopy is chiefly used by the primary metals industry as a means for in-plant control of product quality. A number of years ago, the demand for the quick determination of as many as one to two dozen critical elements involved in the production of aluminum resulted in the development of a multi-channel, direct-reading, emission spectrometer system (60–62). The technique involves batch samples, but results are obtained in less than 1 min. Measuring capabilities from the subparts per million level to low percentages for some 48 elements in either liquid or solid samples are routine for today's instrumentation (4). Generally, a low-voltage spark is the choice of an excitation source for most process monitoring applications involving quantitative measurements.

A multichannel spectrometer utilizes a series of photomultiplier detectors in place of the usual recording photographic plate (Fig. 6.38). Each photomultiplier location corresponds to a chosen analytical wavelength for a given single element. Thus as many elements as there are usable photomultiplier positions can be determined simultaneously. The outputs of the photomultipliers are electronically processed, and the total emitted radiation for each element can be displayed as a concentration reading on a meter. Precalibration is of course required and is achieved by using certified standard reference samples contain-

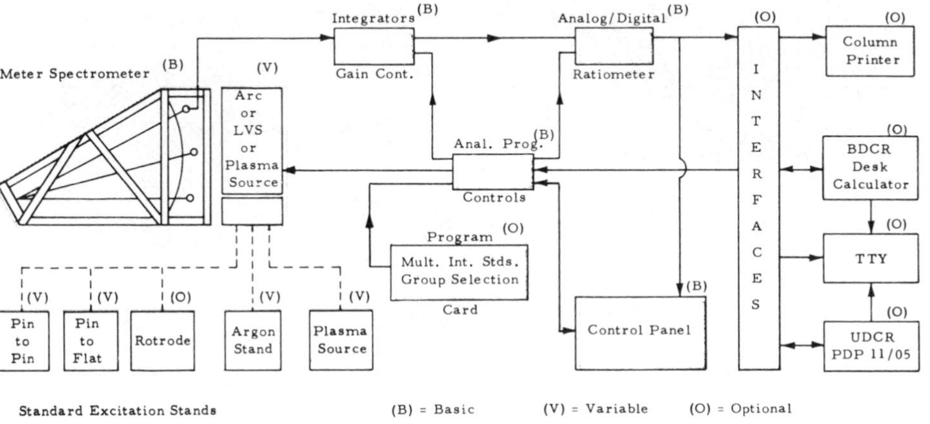


**Fig. 6.38.** Schematic diagram of an automated multichannel emission spectrometer system for in-plant monitoring purposes. Courtesy of Applied Research Laboratories, Division of Bausch and Lomb.

ing all of the elements of interest in the appropriate matrix. Systems are also available in which signal outputs are digitized and then processed by a dedicated minicomputer to provide dynamic background corrections, automatic standardizations, and matrix dilution normalizations for each batch sample examined.

## 2. Flame Photometers as Process Monitors

Flame excitation is often desirable from an optical emission standpoint because the lower excitation energy provided by a flame results in the appearance of only a few characteristic spectral lines, or bands, for each element analyzed. However, because of the lower source energy available, fewer elements can be determined than are possible with an arc or spark. The flame technique has one big advantage, though, in that it is a truly continuous method and not a batch system. Advantage has been taken of its continuous nature, and flame emission analysis has been adapted to the monitoring of the total sulfur content in process gas streams, stacks, and the atmosphere (Fig. 6.39). The measurement is possible because of the hydrogen flames excitation of sulfur's 394 nm emission band. All sulfur atoms whether in the form of $SO_2$, $SO_3$, $H_2S$, or mercaptans are determinable in this manner.

In operations, the sample is drawn through the flame with the aid of a pump. Nondispersive interference filter optics are used to screen out background light from the flame. Sensitivities at the parts per billion level are achieved, with the upper limit being about 10 ppm (98). The method may be similarly used for the determination of phosphorous compounds by measuring the emission intensity at 526 nm (98).

**Fig. 6.39.** Flame photometric monitoring system for the measurement of total sulfur content in process streams and the atmosphere. Courtesy of Meloy Laboratories, Inc.

## F. COMBUSTION-BASED ANALYZERS

The determination of the organic content in water is an important measurement, primarily because of environmental concerns about water quality. Many process industries, such as petroleum, petrochemical, and food processing, generate very large volumes of waste water that contain varying amounts of organic materials. These wastes must be disposed of, and their discharge into natural water bodies, and even municipal waste water treating systems, can seriously affect the biological activity of the receiving body. In general, instrumentation for providing reasonably rapid analyses has not been available until fairly recently (26). Prior to this time, the standard procedures for this measurement involved either long-term (2–5 days), "biological-oxygen-demand" (BOD) studies or laboratory-style, "chemical-oxygen-demand" (COD) measurements requiring several hours to complete. Instrumentation currently available for this measurement consists of batch sample devices that are still more laboratory-like in operation and construction than are most in-plant process monitors. Automatic sampling systems are employed and results can be obtained in a matter of minutes.

The total carbon content of a water sample consists of all of the carbon atoms present from both organic and inorganic materials. The latter includes dissolved carbon dioxide and dissolved and suspended inorganic carbonates. On the other hand, the total organic carbon content is the sample's total carbon content less the dissolved carbon dioxide and the inorganic carbonates. Instrumental systems for the total organic carbon measurement are therefore dual channel devices. One channel determines all the carbon present in the sample. The other provides a means for removing carbon dioxide and carbonates by adding acid and then measuring the remaining carbon present, as in the first channel. An electronic comparison of the two signals generated provides the desired organic carbon analysis.

### 1. Total Carbon Analyzers

Total carbon is determined by removing a microliter-sized sample from the process stream with an automatic sampling valve and injecting it into the heated entrance port of the instrument (Fig. 6.40). A carrier gas containing oxygen sweeps the sample into a high temperature (950°C) combustion tube packed with an oxidizing catalyst. All entering carbon emerges as carbon dioxide which is then quantitatively determined by a nondispersive, ir absorption analyzer (22,70). The water vapor from the combustion and the vaporized sample is condensed and removed from the stream prior to the ir analysis. The

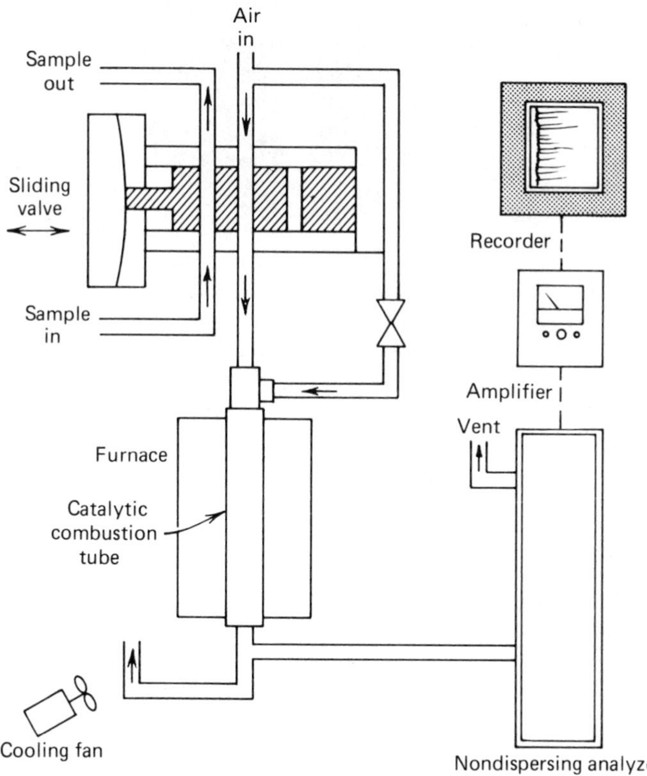


**Fig. 6.40.** Flow diagram for a process-type total carbon analyzer with an automatic sampling value. Courtesy of Beckman Instruments, Inc.

**Fig. 6.41.** Total carbon analyzer for process stream monitoring. A, Automatic sample valve; B, Combustion tube furnace; C, sample inlet manifold; D, ir analyzer. Courtesy of Beckman Instruments, Inc.

reading is presented as the total carbon content on a scale determined by prior calibration.

A drawback of such systems is that the samples must be free of particulate matter, which would cause blockage of the restricted orifices of the sampling valve or the injection port of the furnace. Filtering is not advisable because it may reduce the carbon reading if a significant part of the carbonaceous material is present in a particulate state. A substantial amount of equipment is needed to perform this measurement in process situations (Fig. 6.41).

## 2. Total Organic Carbon Monitors

The differential measurement that yields only the total organic carbon content is performed by a dual channel process (Fig. 6.42). As indicated, two

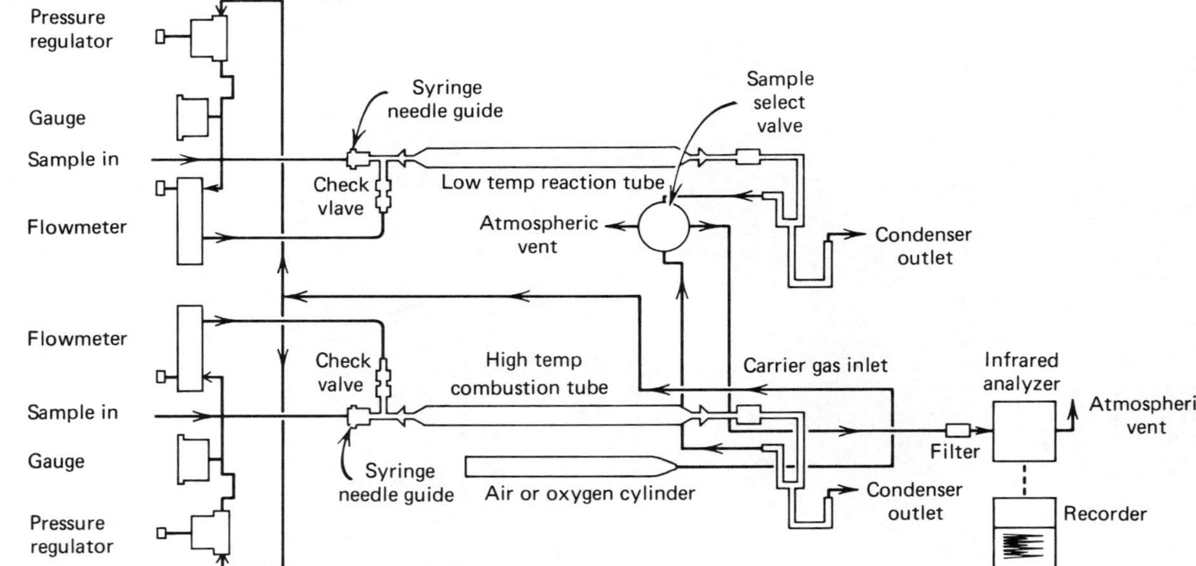


**Fig. 6.42.** Schematic representation of a total organic carbon measurement system. Courtesy of Beckman Instruments, Inc.

**Fig. 6.43.** A total organic carbon analyzer for the batch analysis of aqueous process streams. Courtesy of Beckman Instruments, Inc.

separate microliter specimens of the stream are alternately and manually injected into the combusion tubes (23). Systems employing automatic sampling are also available (71). For one portion of the sample, a low-temperature (150°C) converter tube containing a mineral acid is used to liberate the carbonates and dissolved carbon dioxide that may be present. These combined fractions are measured as described above. The other portion of the sample is treated in a high termperature tube as for the total carbon measurement. Both outputs of carbon dioxide are sequentially read with a nondispersive ir analyzer. The two readings are electronically subtracted to yield the total organic carbon value. Today's total organic carbon analyzers are manually operated devices best suited for in-plant rather than on-line use. They are more compact instruments in construction and laboratory-like in appearance (Fig. 6.43) than are the total carbon analyzers.

Another version of the total organic carbon determination utilizes a flame ionization detector rather than nondispersive ir as the measuring subsystem (Fig. 6.44). In this procedure, the syringe-injected samples are catalytically combusted at a high temperature as described above. However, the emerging water vapor is completely removed and the two dry carbon dioxide streams to which hydrogen is added are sequentially directed into a second reactor tube at

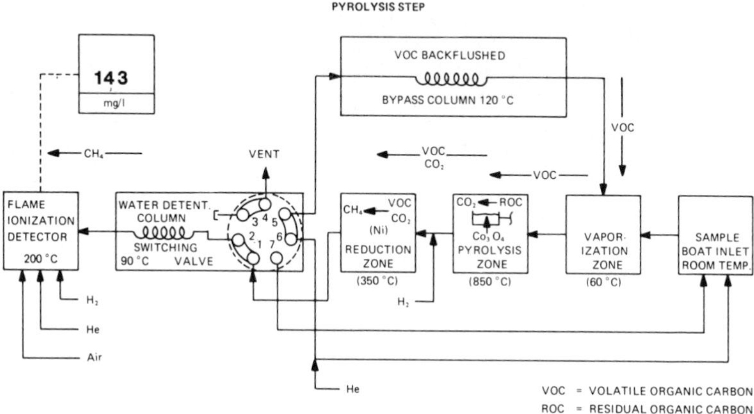


**Fig. 6.44.** Schematic flow diagram of a total organic carbon analyzer employing a flame ionization detector. Courtesy of Dohrman Division, Envirotech Corporation.

350°C. This reactor contains a reducing nickle catalyst which effects the complete methanization of the carbon dioxide. The two separate portions of the dry methane-containing carrier stream are passed through a flame ionization detector (Section III.A.5) with the output signals being related to the carbon content by a calibration plot (42,72).

Sample cleanliness restrictions, as far as particulates are concerned, are the same as those cited for the total carbon measurement. Results are obtained in approximately 5 min from the time of sample injection.

### 3. Oxygen Demand Analyzers

The biological activity of a water body can be represented by the rate at which biological degradation of the included organic material depletes the dissolved oxygen content. An excessive depletion rate can lower the oxygen content to a level that does not support the water's natural life systems, and eutrophication ultimately occurs. Thus a more rapid instrumental means of measuring the relationship between oxygen demand and hydrocarbon content than is afforded by conventional BOD tests is desirable. The following instrumental methods have been developed to satisfy this need.

A schematic flow diagram for the batchwise instrumental determination of oxygen uptake by a water sample's organic content is shown in Fig. 6.45 (55,73). A microliter-sized sample is syringe injected into a nitrogen carrier gas stream containing a known, constant low level (approximately 200 ppm) of oxygen. The vaporized sample is combusted in a heated tube and the oxygen level of the carrier stream is reduced by an amount directly proportional to the sample's carbon content. After leaving the combustion zone, the carrier stream

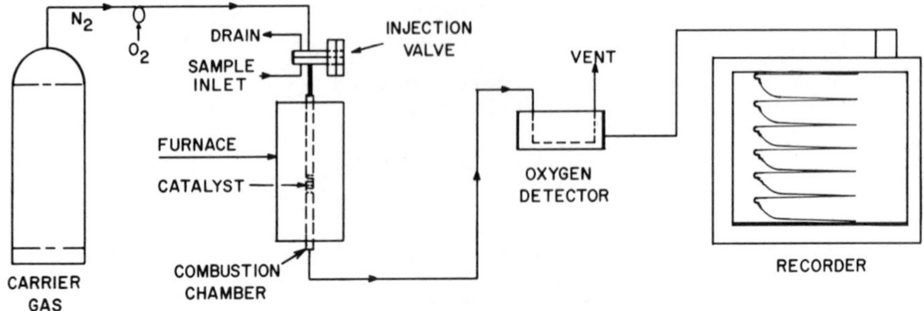


**Fig. 6.45.** Flow schematic of a total oxygen demand monitoring system. Courtesy of Ionics, Incorporated.

is dried and any other acid gases, such as sulfur dioxide formed from organic sulfur present, are trapped. The oxygen content is then accurately determined by a high temperature, solid state oxygen detector (Section III.D.2.b). Precalibration with standard potassium acid phthalate solutions provides a relationship between the oxygen cell's output and the sample's "total oxygen demand." The complete procedure takes 2–3 min (10,73).

Results obtained by this method are reported to be in excellent agreement with values obtained by standard laboratory wet chemical analyses based on the chemical oxidation (COD) procedure. A trend agreement is also seen with the classical biological oxygen demand (BOD) results (134).

A similarly rapid method employing carbon dioxide instead of oxygen as the oxidizing medium is also available (Fig. 6.46) (111,118). The carbon of the organic material is converted to carbon monoxide, as is the equivalent portion

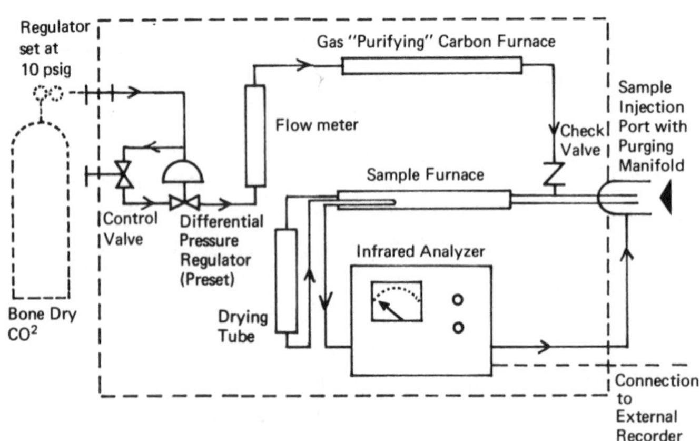


**Fig. 6.46.** Flow diagram for the instrumental measurement of the "Chemical oxygen demand" of an aqueous sample. Courtesy of Precision Scientific, A Subsidiary of GCA Corporation.

of the carbon dioxide consumed in the catalyzed combustion. The combined carbon monoxide fractions are measured by a process-type, nondispersive, infrared analyzer. Precalibration with standard solutions establishes the relationship between the carbon monoxide reading and the equivalent "chemical oxygen demand" of the sample. Typical values fall in the range of 10–300 mg $O_2$/liter with an accuracy of ±1% of the full scale reading.

## G. PROCESS GAS CHROMATOGRAPH SYSTEMS

Gas chromatography, founded on the fundamental work of James and Martin (76) and Martin and Synge (91), is a batchwise, analytical separatory process that provides both qualitative and quantitative information about the test sample (77). Although the separatory process itself involves only the gas phase, liquid process streams can also be treated. The latter need only be sufficiently volatile to be vaporized readily at modest temperatures without suffering thermal damage.

In principle, gas chromatography separates a complex mixture into its discrete components through an infinitely large number of sorption-desorption steps. These are carried out in a column packed with a suitable absorbent material. A very small sample volume (microliter to milliliter range) is injected onto the column at the receiving end and is swept through by an inert gas stream. Column dimensions are typically $\frac{1}{4}$in. i.d. and 3–20 ft in length. At the outlet of the column, the flowing carrier stream consists of a series of separated binary mixtures, or bands. Each binary mixture, or concentration band, represents one of the components in the originally injected sample. The separate bands are sequentially detected by an appropriate sensor system.

Separated components are identified by observing the time required for them to emerge from the column, as indicated by the intervals between detector signals. The physical principles applying to column separations are extensively discussed in the literature (31,36). For convenience in practice, the detector's output is usually displayed as a series of peaks on a recorder strip chart with the x axis representing a time base for determining retention times. The amplitude of each peak is taken as being proportional to the component's concentration for given test conditions. Calibration samples are employed to provide the proper relationship between observed retention time, signal amplitude, and component concentration.

In construction, conventional process gas chromatographs are comparatively uncomplicated instruments consisting of three basic subsystems. These are the sample inlet valve, the separatory column, and the detector-readout system. As a practical consequence, conventional gas chromatographs for on-stream and in-plant use are readily mounted in compact, explosion-proof or all-weather enclosures (Fig. 6.47) (24,29,66,113).

Of a gas chromatograph's three main subsystems, probably the most critical from an operating standpoint is the sample valve. Repetitive batchwise sampling of microliter volumes is a demanding requirement for any mechanical

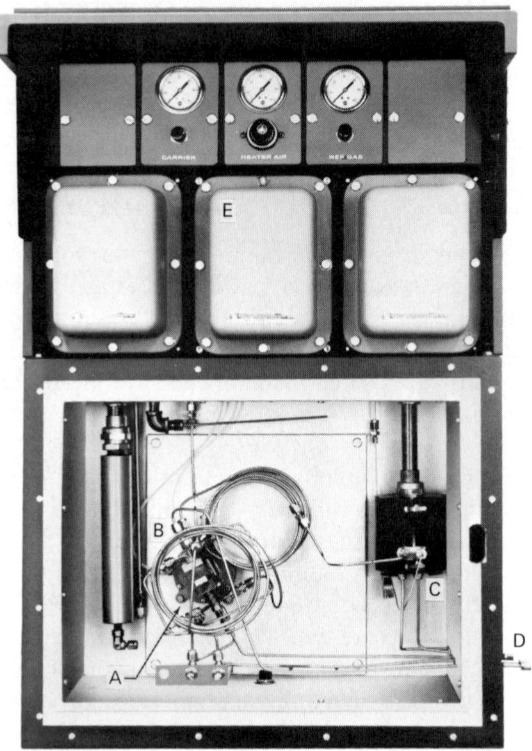


**Fig. 6.47.** Process gas chromatograph showing modularized construction and use of explosion-proof design. A, Sample valve; B, chromatograph columns; C, detector block; D, sample and support gases manifold; E, electronic and control module. Courtesy of Beckman Instruments, Inc.

sampling system. A great deal of effort has been expended in developing rotary and piston-type mechanisms for this purpose. Both motor-driven and pneumatically activated valves are extensively employed. The valve's critical element is either a rotating or sliding solid member containing several precision-machined small cavities or channels. These fixed-volume cavities serve as the means for precisely extracting a discrete volume of the process stream and introducing it onto the column. Any slight variation in the valve's sampling volume completely destroys the integrity of the quantitative determination. Consequently, there are severe limitations on the degree of sample dirtiness and corrosiveness that can be tolerated. If liquids are being sampled, a volatilization step must be introduced prior to column injection. A preheater section located between the sample valve and the column serves this purpose.

The most commonly used detectors for process units are those based on thermal conductivity (Section III.A.1) or ionization phenomena (Section III.A.7). The selection of the detector to be used depends on the types of materials to be measured and the sensitivity required. As an example, monitoring petroleum refining processes involves analyzing systems consisting of innumerable hydrocarbon species that are separated effectively by gas chromatography (130). Flame ionization, in turn, is specifically suited to detecting hydrocarbons at very low concentrations but thermal conductivity is adequate for less stringent requirements. Other types of ionization detectors, such as electron capture units, may be selected when monitoring halogenated, oxygenated, or phosphorus- or sulfur-bearing compounds in chemical process streams.

Although hardly qualifying as a process type of subsystem, a quadrople mass spectrometer detector for gas chromatography provides an unparalleled degree of analytical specificity (51,52,64). This is particularly true for the identication of trace amounts of potentially harmful organic materials of complex molecular structure that may be present in industrial and domestic waters (83,95). Long-lived pesticides and possible carcinogens, mutagens, and teratogens in chemical processing plant effluents are typical examples (137). Because of the critical nature and growing importance of such measurements from a process monitoring standpoint, gas chromatograph-mass spectrometer analyzers are included here.

The major elements of such a system are shown schematically in Fig. 6.48. An enriching step is usually employed just prior to the entrance to the mass

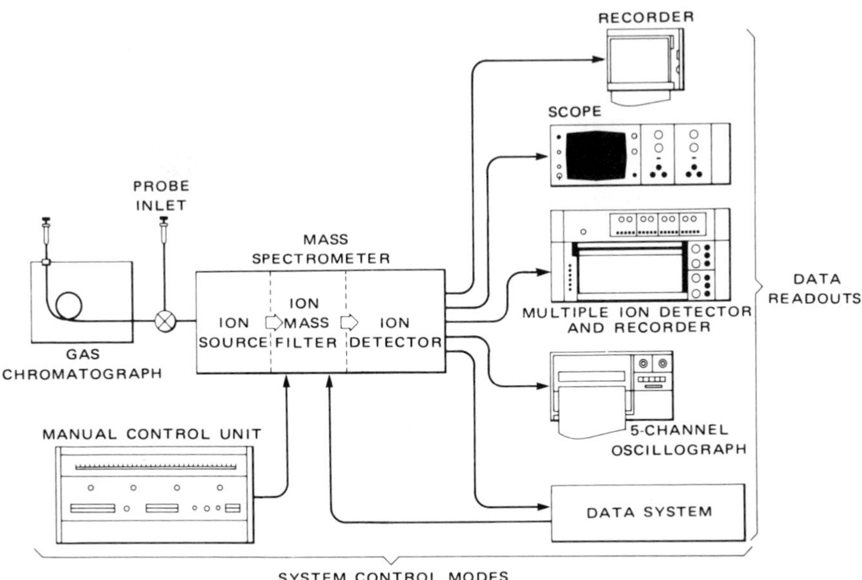


**Fig. 6.48.** Schematic drawing of a gas chromatograph-mass spectrometer system. Courtesy of Hewlett-Packard, Scientific Instruments Division.

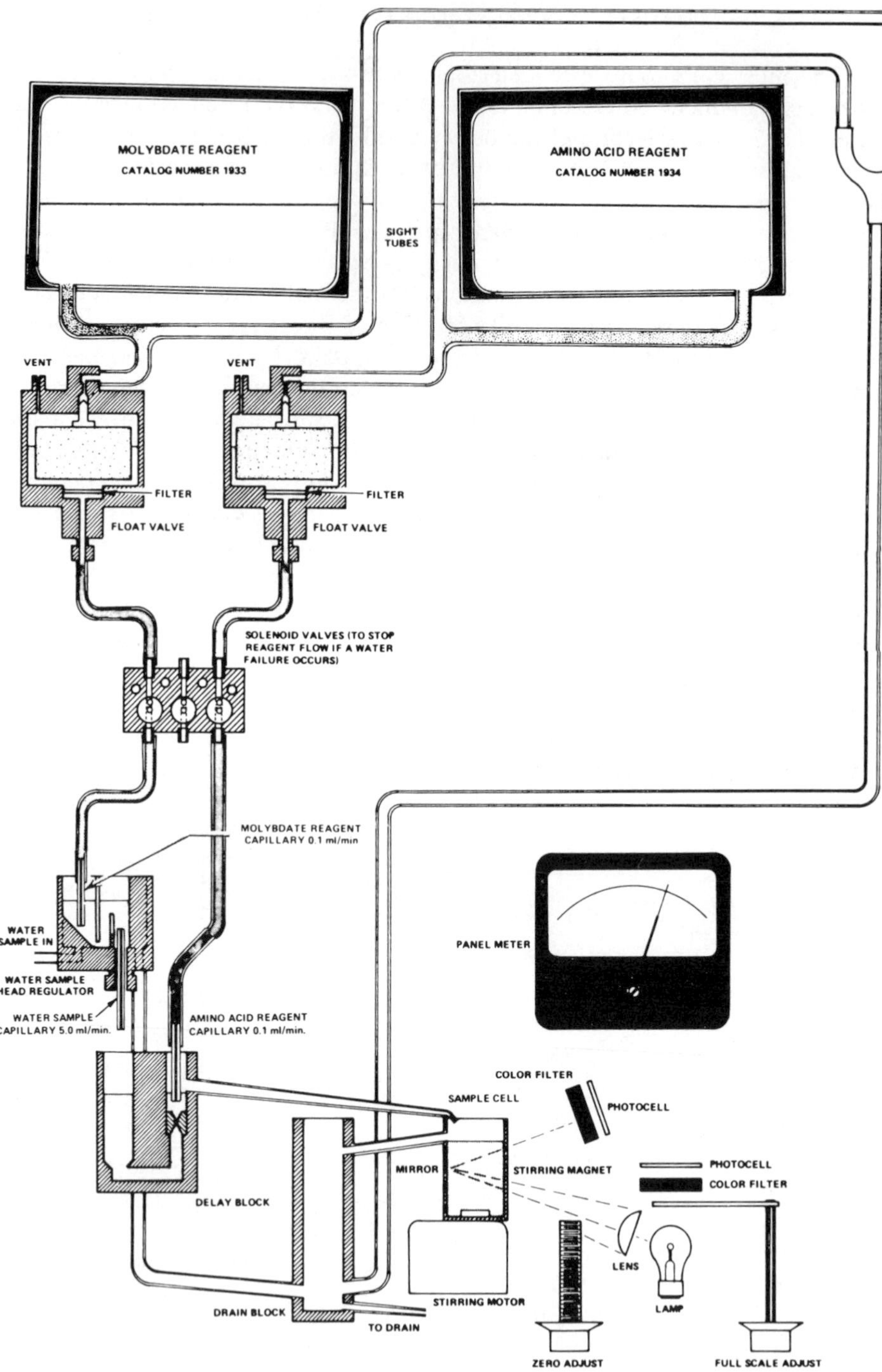


**Fig. 6.49.** Schematic representation of an automatic titration system utilizing photometric detection for the continuous monitoring of aqueous process streams. Courtesy of Hach Chemical Company.

spectrometer, because the sample consists chiefly of carrier gas and only a very small amount of the hydrocarbon species. The enriched species of interest is broken up into a number of ionized fragments by either electron impact or bombardment with an introduced reagent gas such as methane ("chemical ionization")(51). The resulting charged fragments are separated according to their various mass-to-charge ratios by electrostatic fields imposed in the mass spectrometer, or "mass filter." The separated fragments are collectively measured by an electron multiplier detector. The individualized "cracking pattern" for each compound serves as a fingerprint and provides an unambiguous means of identification. A minicomputer is a required integral part of the readout system. Quantitative accuracies of the order of ±0.3% can be achieved at the nanogram level by the chemical ionization technique (51,64).

In summary, gas chromatography is one of the most widely applied forms of industrial process monitoring. The number of individual applications of this technique is so great that it is impossible to include here even a representative listing. The reader desiring specific information should refer to the literature appropriate to his particular field of interest.

## H. ROBOT MONITORING SYSTEMS

Although the electronic-type instrument systems previously described are preferred for process monitoring, there are still a number of determinations that require classical wet chemical procedures as part of the measurement. Automated continuous methods have been developed to eliminate the time-consum-

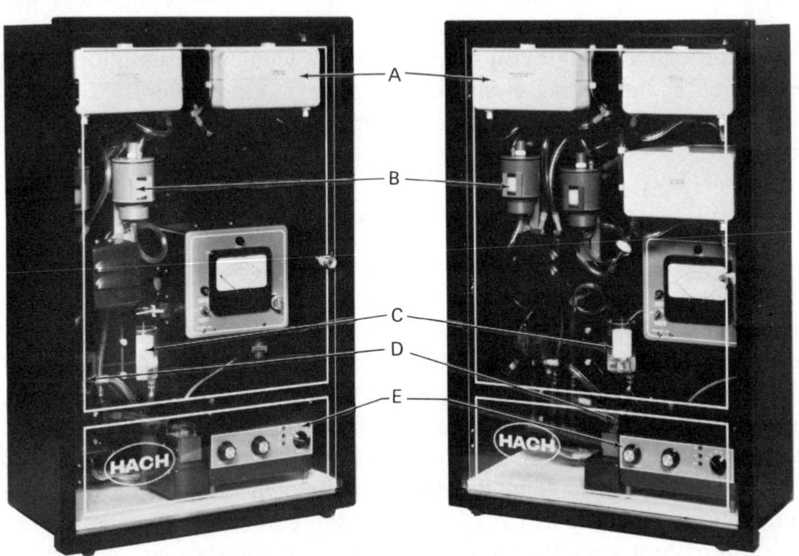


**Fig. 6.50.** Automatic photometric titrator systems. A, Reagent containers; B, reagent float valves; C, mixing cell; D, delay block; E, photometer. Courtesy of Hach Chemical Company.

ing human manipulations required for the conventional laboratory determinations (57,115,120). Such systems are treated in detail in Chapter 5.

Essentially, automated chemical analysis consists of the continuous introduction of a sample that is mixed with a metered amount of a reagent to produce a measurable color change or other form of detectible end point. Typical process applications involving this concept are the determinations of either silica (58) or phosphate (59) in a process or waste water stream (Fig. 6.49). A sample is continuously mixed with the flowing reagent stream, both being delivered to a mixing compartment at the rate of 30–50 ml/min by constant head, gravity flow. Capillaries of selected bore size determine the volumes that are delivered. The flow rates are chosen to provide a sufficient delay time of approximately 7 min for the color reaction to be completed prior to delivery of the sample to the measuring photometer. Silica and phosphate contents are measurable in the range 0.5–5.0 mg/liter with this technique. The necessary reagents are stored in containers at the top of the unit (Fig. 6.50). The photometer's photodetector output is continuously displayed on a meter reading in concentration units for the component being determined.

## IV. INSTRUMENT-COMPUTER INTERFACING FOR PROCESS MONITORING AND CONTROL

Process monitors are prolific generators of data by virtue of their continuous operation. In many cases, the circular or strip-chart analog record produced by conventional recorders is an adequate means of data presentation and storage. However, in many instances the data generated must be further processed immediately to provide qualitative and quantitative information suitable for interpretive or control purposes. Such data reduction is achieved by interfacing the monitor with a suitable computer system. A detailed presentation of instrument-computer interfacing is given in Chapter 8. Consequently, only a very brief treatment need be given here concerning the use of computers with process monitors.

### A. CONVENTIONAL COMPUTER SYSTEMS

Because the signals from process monitors are usually in analog form, an analog computer can be employed directly for data processing in many cases. Simple analog computers specifically designed for this purpose are extensively used where comparatively few numerical manipulations of the data are required (e.g., 68). Process optimization is a typical example, because this can be achieved by simple differentiation or integration of the outputs from the appropriate on-line monitors to yield necessary controller instructions. Analog computers, however, do not lend themselves well to the processing of the large amounts of information that may be needed in complex process applications. Also, more complex mathematical treatments of the data are generally not feasible by analog methods. These problems are readily overcome, however, by the use of digital computational procedures.

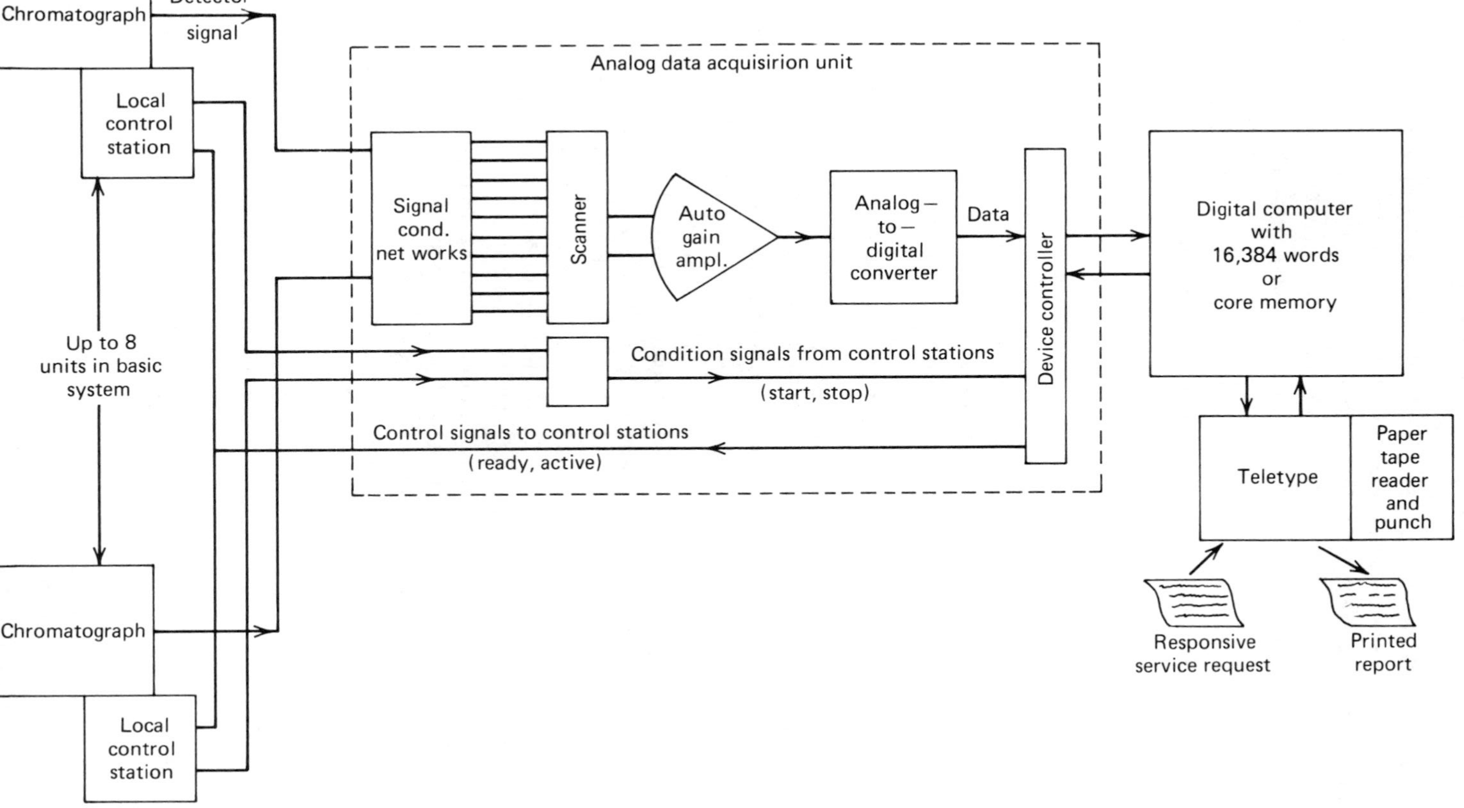


**Fig. 6.51.** Schematic diagram of a typical computer-controlled process monitoring system using gas chromatographs as the signal generators. Courtesy of Electronic Associates, Inc.

Small, compact, highly accurate digital computers well suited for process use are commercially available and offer a wide range of data-reduction capabilities. A typical system employing a digital computer for both data processing and control purposes is shown schematically in Fig. 6.51. The monitors in this case are process gas chromatographs (44,50), but they could just as well be on-line photometers or any of the other monitoring systems previously cited. Computer command dictated by the results of the data treatment are fed back into the system to the appropriate control station associated with each monitor. Printed records for archival purposes are simultaneously generated by the included print-out devices.

The variations in design of computer-based systems for process monitoring applications are virtually unlimited, and numerous examples and guidelines are to be found in the technical literature and in manufacturers' publications (e.g., 40,41,44,136). Unquestionably, the use of total automation for process control purposes will continue to grow as monitoring instruments improve in reliability and versatility and more easily adaptable computer systems are developed.

### B. PROGRAMMABLE CONTROLLERS

The recent advent of "programmable controllers" (1) provides a less expensive and sophisticated way to achieve "closed loop automation than is afforded by the use of conventional mini or micro computer systems. A complete treatment of control systems is to be found in Chapter 3, so the briefest of discussions is given here. A programmable controller is described as a one-bit computer that receives an input signal, compares it with a stored value, and responds with an appropriate output as determined by a set of simple programmed instructions. The simplicity of the programming required is such that it does not demand a knowledge of a conventional computer language such as Fortran. Thus plant operating personnel with no formal training in computer programming can readily communicate with the sytsem.

A typical example of the use of a combined monitoring-programmable controller system is the on-off activation and proportional speed of large blower fans for tunnel ventilation purposes (1). The presence and buildup of atmospheric carbon monoxide in a combined inhabited space is the concern. The appearance of carbon monoxide is determined by on-line, nondispersive, infrared analyzers. The signals from the monitors are fed to programmed controllers regulating the fan motors. Activation of the fans and the speed at which they run are determined by the preentered instructions in the controller's memory.

## V. MONITORING SYSTEMS CATEGORIZED ACCORDING TO USE

*A. The Measurement of Moisture*

| | | |
|---|---|---|
| 1. Piezoelectric Sorption Hygrometry | *Section* | III.A.2 |
| 2. Near Ir Photometer | | III.C.2 |

| | |
|---|---|
| 3. Electrochemical Titrators | III.D.4.c |
| 4. Coulometric Water Vapor Monitor | III.D.4.a |

*B. The Measurement of Oxygen*

| | |
|---|---|
| 1. Thermal Conductivity Analyzers | III.A.1 |
| 2. Paramagnetic Oxygen Analyzers | III.A.6 |
| 3. Microfuel Cell Oxygen Analyzer | III.D.2.a |
| 4. High-Temperature, Solid-State Probe | III.D.2.b |
| 5. Polarographic Monitor | III.D.3 |
| 6. Coulometric Monitor | III.D.4.b |

*C. The Measurement of Hydrocarbons*

| | |
|---|---|
| 1. Thermal Conductivity | III.A.1 |
| 2. Flame Ionization Detector | III.A.7 |
| 3. Combustibles Analyzer | III.A.8 |
| 4. Uv Photometric Analyzers | III.C.1 |
| 5. Ir Photometric Analyzers | III.C.3 |
| 6. Fluorophotometric Systems | III.C.4 |
| 7. Chemiluminescence Analyzers | III.C.5 |
| 8. Combustion-based Systems | III.F.1, 2 |
| 9. Gas Chromatograph Systems | III.G |

*D. Flue Gas Measurements*

| | |
|---|---|
| 1. Particulates by Beta Ray Absorption | III.A.4 |
| 2. Paramagnetic Oxygen Analyzers | III.A.6 |
| 3. Combustibles Analyzer | III.A.8 |
| 4. Particulate Analysis | III.B.2 |
| 5. Uv Photometric Analyzers | III.C.1 |
| 6. Chemiluminescence Analyzers | III.C.5 |
| 7. Microfuel Cell Oxygen Analyzer | III.D.2.a |
| 8. Microfuel Cell $SO_2$ and $NO_x$ Analyzers | III.D.2.a |
| 9. High-Temperature, Solid-State Probe | III.D.2.b |

*E. Water Quality Measurements*

| | |
|---|---|
| 1. Solution Conductivity | III.A.3 |
| 2. Turbidity Analyzers | III.B.2 |
| 3. Uv Photometric Analyzers | III.C.1 |
| 4. Fluorophotometric Systems | III.C.4 |

| | | |
|---|---|---|
| 5. | pH Analyzers | III.D.1.a |
| 6. | Ion Specific Electrodes | III.D.1.b |
| 7. | Polarographic Dissolved Oxygen Analyzers | III.D.3 |
| 8. | Electrochemical Titrators | III.D.4.c |
| 9. | Total Carbon and Organic Carbon Analyzers | III.F.1, 2 |
| 10. | Oxygen Demand Analyzers | III.F.3 |
| 11. | Gas Chromatograph Systems | III.G |
| 12. | Robot Chemist System | III.H |
| 13. | Autotitrimeters | III.H |

*F. The Analysis of Metals*

| | | |
|---|---|---|
| 1. | X-Ray Monitors | III.A.5 |
| 2. | Multichannel Spectrographs | III.E.1 |

*G. Trace Gas Measurements*

| | | |
|---|---|---|
| 1. | Thermal Conductivity | III.A.1 |
| 2. | Flame Ionization Detectors | III.A.7 |
| 3. | Chemiluminescence Analyzers | III.C.5 |
| 4. | Microfuel Cell Oxygen Analyzers | III.D.2.a |
| 5. | Microfuel Cell $SO_2$ and $NO_x$ Analyzers | III.D.2.a |
| 6. | High-Temperature, Solid-State Probe | III.D.2.b |
| 7. | Polarographic Oxygen Analyzer | III.D.3 |
| 8. | Coulometric Moisture Monitor | III.D.4.a |
| 9. | Coulometric Oxygen and Ozone Analyzer | III.D.4.b |
| 10. | Electrochemical Titrators | III.D.4.c |
| 11. | Coulometric $SO_2$ and $NO_2$ Analyzers | III.D.4.c |
| 12. | Flame Emission Monitor | III.E.2 |
| 13. | Gas Chromatograph Systems | III.G |

## REFERENCES

1. Abel, D., *Instr. Control Syst.*, 49 (July 1974).
2. Aero Chem Research Laboratories, Inc., Sybron Corp., Princeton N.J., Brochures TB-73-1, TB-74-1, TB-74-2.
3. American Instrument Co., Div. Travenol Laboratories, Inc., Silver Spring, Md., Catalog No. J4-7461, J4-7439, J4-7440.
4. Applied Research Laboratories, Division of Bausch and Lomb, Sunland, Calif., Brochure—*Quantometric Analyzer*.

5. Anacon, Inc., Ashland, Mass., Brochure—*Model 47 Process Analyzer*.
6. *Ibid.*, Brochure—*Model 800 Process Refractometer*.
7. *Ibid.*, Brochure—*Model 600 Suspended Solids Monitor*.
8. *Ibid.*, Brochure—*Model 106 Moisture Analyzer*.
9. *Ibid.*, Brochure—*Model 206 Infrared Analyzer*.
10. Arin, M. L., F. H. Meller, and D. L. Brown, paper presented at Instrument Society of America Meeting, Midland, Mich., (Feb. 9, 1972).
11. Barendrecht, E., *Nature*, **183**, 1181, (1959).
12. Baumgardner, R. E., T. A. Clark, and R. K. Stevens, *Environ. Sci. Technol.*, **9**, 67 (1975).
13. Beckman Instruments, Inc., Process Instruments Div., Fullerton, Calif. Bulletin 0-4016-C.
14. *Ibid.*, Bulletins 4104-C, 4115A.
15. *Ibid.*, Bulletins 4129-C, 4136A.
16. *Ibid.*, Bulletin 4133-B.
17. *Ibid.*, Bulletin 4111-B.
18. *Ibid.*, Bulletin 4091-C.
19. *Ibid.*, Bulletin 4101-A.
20. *Ibid.*, Bulletin 4118.
21. *Ibid.*, Bulletin 4097-B.
22. *Ibid.*, Bulletin 4093-A.
23. *Ibid.*, Bulletins 4082-B, 4082-D.
24. *Ibid.*, Bulletins 4135-A, 4120-A.
25. *Ibid.*, Application Data Sheets 4396, 4397.
26. *Ibid.*, Application Data Sheets 4374, 4377, 4379.
27. *Ibid.*, Application Data Sheet 4399.
28. Beerbower, A., in *Analysis Instrumentation*, III, Control Engineering Reprint 492 (1959).
29. The Bendix Corporation, Process Instruments Division, Ronceverte, W.V., Brochure and Operation Manual—*Model 6171 Process Chromatograph*.
30. Brewer, A. W., *New Sci.*, 32 (July 11, 1957).
31. Cassidy, H. G., *Fundamentals of Chromatography*, Interscience, New York, (1957).
32. Chapman, R. L, *Tech Bulletin 5037-A*, Beckman Instruments Div., Fullerton, Calif.
33. Clark, L. C., Jr., U.S. Pat. 2,913,386.
34. Clough, P. N., and B. A. Thrush, *Trans. Faraday Soc.*, **63**, 915 (1967).
35. Czuha, M. J., K. W. Gardiner, and D. T. Sawyer, *J. Electroanal. Chem.*, **4**, 51 (1962).
36. Dal Nogare, S., and R. S. Juvet, *Gas-Liquid Chromatography*, Inter-Science, New York, (1962).
37. David, F., and F. Martin, *Instr. Control Syst.*, 61 (June, 1974).
38. Daynes, H. A., *Gas Analysis by Measurement of Thermal Conductivity*, Cambridge University Press, Cambridge, England, (1933).
39. Desty, D. H., C. J. Geach, and A. Goldup in *Gas Chromatography*, R. P. W. Scott, Ed., Butterworths, London, (1960), p. 46.
40. Digital Equipment Corp., Maynard, Mass., Laboratory Data Products Application Notes 1000, 1001 (Feb. 1974), 2000.
41. Digital Equipment Corp., Marlborough, Mass., "*DEC System-10,*" *J. Appl. Res.*, **1**.
42. Dohrman Division, Envirotech Corp., Santa Clara, Calif., Bulletins SM-674-072750, SM-686-102750.
43. *Ibid.*, Brochure SM-690-0737-100.
44. Downer, W., and J. Muldoon, *EAI ProPACE, Process Chromatographs and Computers*, Electronic Associates, Inc., West Long Branch, N.J.

45. E. I. DuPont de Nemours and Co., Instrument Products Division, Wilmington, Del., Brochure A-82382, 1974.
46. *Ibid.*, Brochures A-87899, 11-73; A-95489, 11-74; A-72058; A-83851.
47. *Ibid.*, Brochures A-76674 (Bulletin 461-A); A-81723, 7-74; A-81340; A-81339; A-82899
48. *Ibid.*, Bulletin E-03456.
49. *Ibid.*, Bulletins 460/1, 463.
50. Ewing, R. C., *Oil Gas J.*, 77 (Oct. 18, 1971).
51. Finnigan Corporation, Sunnyvale, Calif. *F Series Automated GC/Ms with Integrated Data System*.
52. Ibid., *Finnigan Spectra*, **3** (1) (Jan. 1973).
53. Fontijn, A., A. J. Sabadell, and R. J. Ronco, *Anal. Chem.*, **42**, 575 (1970).
54. The Foxboro Company, Foxboro, Mass., Bulletin K-15B, Instruction Manuals M1 14-248, (1973); M1 14-249, (1973); M1 14-251, (1973); 18-227, (1971); 14-240, (1968); 14-241, (1970); 14-243, (1969).
55. Goldstein, A. L., W. E. Katz, F. H. Meller, and D. M. Murdoch in paper presented at American Chemical Society Meeting, Atlantic City, N.J. (Sept. 12, 1968).
56. Hach Chemical Company, Ames, Iowa, Bulletins 2411-4ED, 2426-6ED; Manuals 7-15-74-4ED, 9-13-74-7ED.
57. *Ibid*., Bulletin 1407-4ED; Instruction Manual for Model 1407.
58. *Ibid.*, Bulletins 651B-3ED, 1234B-4ED; Manuals for Models 651B and 1234B.
59. *Ibid.*, Brochure—*Uncomplicated Silica Analysis*, 8-1-72.
60. Hassler, M. F., E. Davidson, H. Orr, and W. H. Berry, *Mikrochim. Acta*, **1955**, 596.
61. Hassler, M. F., *J. Opt. Soc. Am.*, **41**, 870 (1951).
62. Hassler, M. F., *Spectrochim. Acta*, **6**, 69 (1953).
63. Hersch, P. A., *Am. Lab.*, 29 (Aug. 1973).
64. Hewlett-Packard Corp., Scientific Instruments Div., Palo Alto, Calif., Bulletin Series 5980A.
65. Holzbock, W. G., *Instruments for Measurement and Control*, 2nd ed., Reinhold, New York, (1962).
66. Honeywell, Inc., Process Control Div., Fort Washington, Pa., Brochure D-475.
67. Houser, E. A., *Principles of Sample Handling and Sampling Systems Design for Process Analysis*, Instrument Society of America, Pittsburgh, Pa., (1972).
68. Hybrid Controls, Inc., Lansdale, Pa., Specifications S504-3, S-504-4.
69. Industrial Nucleonics Corp., Columbus, Ohio, Brochure—*Acc Ray 510 System*.
70. Ionics, Inc., Instrument Div., Watertown, Mass., Brochure—*Model 1214 Total Carbon Analyzer*.
71. *Ibid.*, Brochure—*Model 1218 Total Organic Carbon Analyzer*.
72. *Ibid.*, Bulletin—*Model 1224 Total Organic Carbon Analyzer*; Bulletin OCA-1, 1-15-72.
73. *Ibid*., Brochure—*Model 225 Total Oxygen Demand Analyzer; Model 1236 Total Oxygen Demand Analyzer*.
74. ITT Barton, Process Instruments and Controls, Monterey Park, Calif., Bulletin 1410; Technical Information Sheet 13-G1-66-1, 7-28-72.
75. *Ibid.*, Bulletins 286-3, Gl-20-2, Gl-400-2, 402-1, 407/8-1.
76. James, A. T., and A. J. P. Martin, *Biochem. J.*, **50**, 679 (1952).
77. Karasek, F. W., *Analytical Instrumentation in Process Monitoring*, A.C.S. Short Courses. American Chemical Society, Washington, D.C., 1970.
78. Keidel, F. A., *Anal. Chem.*, **31**, 2043 (1959).
79. Keidel, F. A., *Ind. Eng. Chem.*, **52**, 490 (1960).
80. King, W. H., Jr., U.S. Pats. 3,164,000, 3,478,573.
81. King, W. H., Jr., *Anal. Chem.*, **36**, 1735 (1964).
82. King, W. H., Jr., in A. Wexler, Ed., *Humidity and Moisture, Measurement and Control in Science and Industry*, Vol. 1, Reinhold, New York, (1965).

83. Knight, J. B., *Finnigan Corporation Application Tips*, No. 14, Revision II, January 29 (1971). Finnigan Corp., Sunnyvale, Calif.
84. Lear Siegler, Inc., Englewood, Colo., Brochures: Argos/PL/001A/5/74, Argos/PL/002A/7/74.
85. Leeds and Northrup Co., North Wales, Pa., Data Sheet C3.1001-1969.
86. *Ibid.*, Brochure C2.2113-DS.
87. *Ibid.*, Brochure D2.2114-DS.
88. *Ibid.*, Manual 177208 Issue 5, Model 7803-G Analyzer.
89. Lockwood and McLorie, Inc. Horsham, Pa, Bulletins 0272, 0374.
90. Lovelock, J. E., *Anal. Chem.*, **33**, 162 (1961).
91. Martin, A. J. P., and R. L. M. Synge, *Biochem. J.*, **35**, 1358 (1941).
92. Martin, J. M., C. R. Orr, C. B. Kincannon, and J. L. Bishop, *J. Water Pollution Control Federation*, **39**, 21 (1967).
93. Mast Development Co., Davenport, Iowa, Form Nos. C-30, C-7970TCB; Manual Model 724-2; Brochure Model 724-2M.
94. Mast, G. M., and H. E. Saunders, *I.S.A. Trans.*, **1**, 325 (1962).
95. McQuire, J. M., A. L. Alford, and M. H. Carter, *Environmental Protection Technology Series*, EPA-R2-73-234, July 1973, Corvallis, Or; 97330.
96. Meloy Laboratories, Inc., Springfield, Va., Brochures Models OA325, OA350; Manual Model OA350-2.
97. *Ibid.*, Brochure Series NA500.
98. *Ibid.*, Bulletins SA160, FSA190.
99. Meyer, A. S., Jr., and C. M. Boyd, *Anal. Chem.*, **31**, 215 (1959).
100. Milton Roy Company, Hays-Republic Div., Michigan City, Ind., Publication 65:B643; Manual 6/74-E(SH)643D-3.
101. *Ibid.*, Manual Model 631 Analyzer, No. 1/70-E631/771.
102. *Ibid.*, Manual No. 10/73-E647-1,2; Data Sheets B647,B647-3.
103. *Ibid.*, Brochure Model 623.02 Analyzer.
104. Mine Safety Appliance Co., Instrument Div., Pittsburgh Pa; Bulletin No. 0714-10, 0711-1.
105. *Ibid.*, Bulletins Nos. 0705-3, 0705-20.
106. Morrow, N. L., R. S. Breif, and R. R. Bertrand, *Chem. Eng.*, 85 (Jan. 24, 1972).
107. National Bureau of Standards, *Dimensions*, 138 (June 1974). National Bureau of Standards, Washington, D.C.
108. Orion Research, Monitor Division, Cambridge, Mass., Brochure—*On-Line Analytical Systems, Series 1000 Monitors*.
109. *Ibid.*, Newsletter/Specific Ion Electrode Technology, Vol. V, No. 1 (1973).
110. Pauling, L., U.S. Pat. 2,416,344.
111. Precision Scientific, a Subsidiary of GCA Corporation, Chicago, Ill., Bulletin 644A.
112. Pust, H. W., R. E. Kreider, and K. W. Gardiner, *Analysis Instrumentation*, Vol. 9, Instrument Society of America, Pittsburgh, Pa.
113. Pye Unicam, Ltd., Cambridge, CB1 2PX, England, Brochure Series 404 Mk II.
114. Rostenbach, R. E., and R. G. Kling, *J. Air Pollut. Control Assoc.*, **12**, 459 (1962).
115. Skeggs, L. T., U.S. Pat. 2,797,149.
116. Skoog, D. A., and D. M. West, *Principles of Instrumental Analysis*, Holt, Rinehart & Winston, New York, (1971), Chap. 14, p. 349.
117. Smith, D. E., and F. H. Zimmerli, *Electrochemical Methods of Process Analysis*, Instrument Society of America, Pittsburgh, Pa., (1972).
118. Stenger, V. A. and C. E. Van Hall, *Anal. Chem.*, **39**, 206 (1967).
119. Taylor Servomex, Sybron Corp., Crowborough, Sussex, England, Manual Type OA.137 Analyzer.

120. Technicon Instruments Corp., Industrial Systems Division, Tarrytown, N.Y., Technical Publications Nos. TA1-0170-20, 2533-6-1/6-4-7.5, 2558R-10-3/6-4-7.5, 3018-2-3/R4-4-2, 3012-10-2-1W; Manuals TP0-0169-10, TN1-0169-00.

121. Teledyne Analytical Instruments, San Gabriel, Calif. Brochure—*Series 400 Analyzers*.

122. *Ibid.*, Brochures—*Model 9700, Model 102 Analyzers*.

123. *Ibid.*, Brochure—*Series 600 Process Photometers*.

124. *Ibid.*, Brochure—*Series 500 Analyzers*.

125. *Ibid.*, U.S. Pats. 3,429,796; 3,767,552.

126. *Ibid.*, Brochure—*Series 300 Analyzers*.

127. Thermo Electron Corp., Waltham, Mass., Brochure—*Pulsed Fluorescent $SO_2$ Analyzer*; U.S. Pats. 3,826,920; 3,845,309.

128. Theta Sensors, Inc., A Subsidiary of Meteorology Research, Inc., Altadena, Calif., Brochure—*Theta Sensor SP1000*.

129. Vanton Pump and Equipment Corp., Hillside, N.Y., 07205.

130. Villalobas, R., Technical Bulletin 5039, Beckman Instruments, Inc., Process Instruments Division, Fullerton, Calif. 92634.

131. Westinghouse Electric Corp., Computer and Instrumentation Div., Orrville, Ohio, 44667, Bulletin 106-101.

132. Wilks Scientific Corp., Norwalk, Conn. Bulletins M1-4-10M, 1-74; M1-5-10M, 6-73; M15-10M; 3-74; M101-10M, 2-75.

133. Williamson, J. A. and D. W. Janzen, *Analysis Instrumentation*, Vol. 10, Instrument Society of America, Pittsburgh, Pa.

134. Wood, E. D., A. E. Perry, and M. C. Hitchcock, paper presented at American Chemical Society Meeting, Houston, Tex., Feb. 26, 1970.

135. Yellow Springs Instrument Co., Scientific Div., Yellow Springs, Ohio, Brochures Model 57, 54APB, 54ARC, 51B; Items 004483 P/N A-05730-B, A-05716A, Aug. '74.

136. Ziegler, E. D. Henneberg, and G. Schomburg, *Anal. Chem.*, **42**, 51A (1970).

137. Zwick, D. and M. Benstock, *Water Wasteland, The Nader Report*, Grossman, New York, (1971).

## SUPPLEMENTARY BIBLIOGRAPHY

1. Anderson, N. A., *Instrumentation for Process Measurement*, 2nd Ed., Chilton, Radnor, Pa., (1972).
2. Brown, J. E., "*On-Stream Process Analyzers*," *Chem. Eng.*, 164 (May 6, 1968).
3. Clevett, K. J., *Handbook of Process Stream Analysis*, Halsted Press, Chichester, England, (1973).
4. Siggia, S., *Continuous Analysis of Chemical Process Systems*, Wiley, New York, (1959).

Part I
Section E

Chapter 7

# COMPUTER SYSTEMS: STRUCTURE AND DATA PROCESSING

BY CARL F. IJAMES AND CHARLES L. WILKINS, *University of California, Riverside Riverside, California*

Contents



## I. INTRODUCTION

In this chapter an overview of the important concepts of computer architecture and computer programming is presented. Furthermore, the combination of these concepts into functioning laboratory analysis systems is considered. The earlier chapters on analog and digital electronics introduced many of the basic terms and principles that are applied in this discussion. This chapter, in turn, serves a similar function for subsequent chapters on interfacing, data acquisition, and programming.

Computer technology is such a rapidly advancing field that revolutionary developments occur regularly, almost monthly. The primary thrust in hardware development is toward miniaturization, first implementing a complete central processing unit on a single integrated circuit (the microprocessor), and then increasing the processing power of that chip. Software has also been advancing, with user-friendly, interactive systems the goal. For chemists, this means that computer-controlled instrumentation is becoming the rule rather than the exception. Obviously any attempt to relate the state of the art is doomed to failure even with the shortest publication delays. With these constraints in mind this chapter emphasizes the functional organization of computer architecture, both hardware and software. The logical design of computer systems is not likely to change as rapidly and dramatically as the details of hardware implementations. On the other hand, a specific system example is useful in illustrating the concepts of hardware design, so the Motorola MC68000 microprocessor is employed when an example is necessary.

## II. INTERNAL STRUCTURE OF COMPUTERS

Figure 7.1 is a block diagram of the fundamental hardware components of a computer system. The central processing unit (CPU) performs arithmetic, logi-

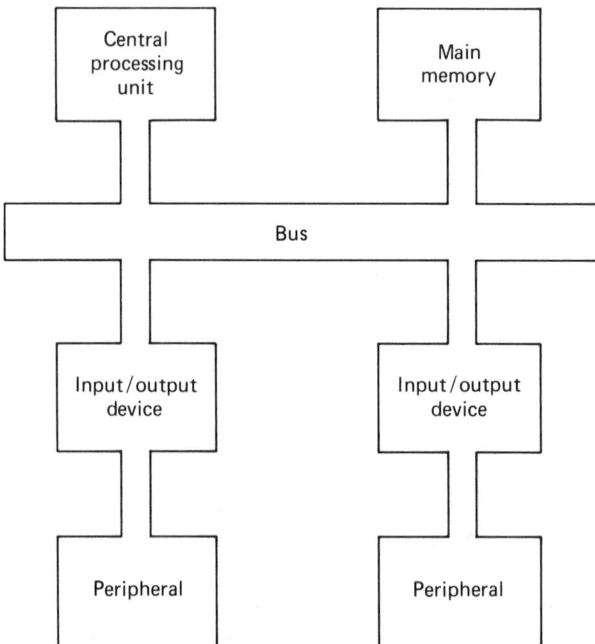


**Fig. 7.1.** Functional elements of a computer.

cal, and control operations. The main memory contains the current program and data. Input/output (I/O) devices provide communication between the peripherals and the computer. The peripherals include the operator's terminal, experiments, mass storage devices, fast printers, and other computers. The bus consists of the interconnections between each of the other components. Each of these is now discussed in more detail.

## A. CENTRAL PROCESSING UNIT

The CPU contains various registers and an arithmetic logic unit (ALU) to perform the actual numerical operations. One important characteristic of a computer is the word size. This is the number of bits (binary digits) operated upon by a single machine instruction, and is usually the number of bits contained in each memory location and register. Word sizes range from 4 bits in small microprocessors used in control applications to 80 or more in very large supercomputers. The larger the word size, the fewer the number of operations necessary to move a fixed amount of data, and therefore the less time required for the move. Conversely, the smaller the word size, the cheaper the cost of a fixed number of words of memory. Typical word sizes are 8 bits for microprocessors, 16 bits for high-performance microprocessors and most minicomputers, 20–24 bits for computers designed for signal averaging Fourier transform instruments (NMR, ir, mass spectrometry), and 32 bits for high-performance minicomputers and larger systems.

The registers in the CPU function as memory locations, and can be used to store operands for the ALU. Typically, registers operate much faster than memory so it is more time efficient to employ them in repetitive operations. A major decision in hardware systems design is the number of registers to provide. The simplest architecture is based upon a single register, the accumulator, requiring all operations to have this accumulator as the source of one operand and as the destination of the result. For example, the contents of a memory location can be added to the contents of the accumulator and the sum left in the accumulator. A more sophisticated design provides several registers, each of which can function as an accumulator. Finally, the most flexible design allows memory locations as well as registers to function as both the source and destination of any operation. Only in this design is the speed difference between register and memory operations important, since only here does the programmer have a choice.

For programs, one register is used as the program counter. This is the location in memory of the next instruction to be executed. Another register is used to contain various status flags, consisting of one bit per flag. These indicate the results of the most recent operation, and are used to control conditional branches. Commonly there are flags reporting whether the last result was positive, negative, or zero; the occurrence of an overflow; whether interrupts are enabled; and the current mode of the processor. Most processors allow interrupts; that is, they have an input line that can signal the processor to stop whatever it is currently doing and cause it to begin execution of another program. These interrupts can be enabled or disabled by setting or clearing the appropriate flag bit in the status register. Also, some processors have certain instructions that are described as privileged. These are useful in multi-user environments to prevent one user from disrupting another. In order to execute these privileged instructions the processor must be in the proper mode. The less powerful microprocessors usually do not implement this feature, since they are typically used in single-user environments. If more than one mode is available, the current mode is determined by the status register.

Finally, the CPU contains the circuitry necessary to interpret and execute each instruction as it is received from memory. These instructions include arithmetic operations, logical operations, comparisons, branches, and data movement, among others.

## B. MAIN MEMORY

The main memory of a computer is termed random access memory (RAM), because any location can be accessed in any order. Also, memory is classed as either read-write or read-only memory (ROM). The term RAM is generally used only to describe read-write memory, which is either volatile or nonvolatile. Volatile memory loses its contents when power is removed, whereas nonvolatile memory is altered only by specifically writing into a location. Originally, nonvolatile core memory was used, but the majority of RAM in use today is volatile semiconductor memory. ROM is usually programmed by the manufacturer and is used when fast, permanent storage is necessary. In a general-purpose computer most of the memory is read-write, since many different programs must be executed. On the other hand, in a dedicated control computer, only one program may be necessary. Once written, it is usually modified infrequently, so such programs are often stored in ROM. Even in the dedicated computer some read-write memory is usually present, in order to store parameters that may be varied by the operator.

The minimum memory size required in a computer system is determined by the application, whereas the maximum size possible is limited by the hardware design. Eight bit microprocessors very often use two words (16 bits) to form an address, so they have a memory address range of 0 to $2^{16} - 1$, or 65535. Memory size is usually specified in bytes, where 1 byte is 8 bits. When referring to memory sizes a common symbol is k, for kilo. One k is $2^{10}$, or 1024, so an 8 bit microprocessor, using 16 bit addressing, can address 64 kbytes of memory. Another common symbol is M, for mega, which is $2^{20}$, or 1048576. The address range of computers with word sizes larger than 8 bits depends upon the specific implementation. Some large computer systems can contain up to 64 Mbytes. Until the last few years, memory was very expensive, usually constituting a major portion of the cost of a computer system. Now, with high-density semiconductor RAM instead of core memory, up to 256 kbits may be stored upon a single integrated circuit. Memory costs are now only a small fraction of the total cost of a computer system. Small 8 bit personal computers now provide as much memory as the larger computers of only 10 years ago.

## C. INPUT/OUTPUT DEVICES

Next there are the I/O devices, which provide communication between the computer and the various peripherals. These are accessed by the CPU either by special I/O instructions and a separate I/O portion of the bus, or merely as memory addresses which do not correspond to any actual memory in the system. This last scheme is termed memory-mapped I/O, and employs the same address and data portions of the bus used to communicate with main memory. A control word is written to a location to signal the device that data transfer is requested, and then the data is written to or read from another location to effect the transfer. By using more than one memory location for the data, large blocks can be moved quickly. Also, another location is used by the device to report the status of the transfer and may signal completion of data transfer by a slower peripheral. Another way to transfer large blocks of data even more quickly is to use direct memory access, or DMA. Once the I/O device has been instructed to begin transfer by either of the above schemes, it moves data directly to or from the main memory. Although it has control of the memory, the CPU cannot access memory, so the CPU must remain idle during this time. By avoiding the necessity of the CPU fetching each data word from the device and moving it to memory the total time for the transfer is greatly reduced. I/O devices are grouped as either digital or analog, depending upon the form of the data which with they deal. The computer can only manipulate digital data directly, whereas most experiments provide analog data and require analog control signals. To perform the translation from one form to the other, converters are used: either analog-to-digital (ADC) or digital-to-analog (DAC). These are discussed in depth in other chapters.

Digital data can be transmitted either serially or in parallel. Serial communications use one signal path for each direction, and send data one bit at a time. For the computer to send and receive data requires two paths, usually consisting of three wires: send, receive, and ground. Parallel communications use a separate data path for each bit, requiring nine wires for 1 byte. The advantage of parallel data transmission is speed, since an entire word is sent in the same time required for a serial arrangement to transmit 1 bit. The disadvantage is that for long data paths, the wire cost becomes prohibitive. The bus of a computer employs parallel communications to give the greatest possible system speed. Most low-speed peripherals, such as terminals, are serial devices. One feature becoming common in new instruments is a built-in serial interface, to allow direct communication with a computer. These can now be found on instruments ranging from analytical balances to uv-visible spectrophotometers, and are another sign of the growing popularity of microprocessor control.

## D. PERIPHERALS

Lastly, there are the peripherals. In the broadest sense, these include anything connected to the computer. The three most important peripherals are mass storage units, terminals for communication with the operator, and, in laboratories, the experiments or analytical instruments.

Mass storage units provide nonvolatile, read-write data storage. The most common mass storage devices use magnetic recording on either tapes or disks. Tapes give large capacities but are relatively slow because they are serial devices, not random access. To read a standard 2400 foot magnetic tape may take as long as 20 min for an average performance tape drive. Therefore, tapes are used to archive programs and data, and for transferring data from one computer system to another.

Disks, on the other hand, allow random access at times in the range of tens to hundreds of milliseconds. Naturally, disk performance is rapidly increasing, so these times may decrease by orders of magnitude in the near future. Data is stored on disks in concentric circles called tracks. Each disk surface contains from 30 to hundreds of tracks. Also, a single disk drive may contain many disk platters, each rotating upon a common spindle. Disks are classified as either fixed or moving head, depending upon whether the read-write head is stationary over a single track or allowed to move from track to track. Fixed head disks give much faster performance because no time is wasted waiting for the head to be positioned over the correct track, but are much more complex and more expensive than moving head drives. Most disk drives are the moving head type, using one read-write head per disk surface. Disks are also classified as either removable or fixed media. If the disk platters can be transferred from one drive to another to allow transfer of data, then the disks have removable media. Finally, disks may have either rigid platters or flexible platters. The flexible platters, called floppy disks, are used in small, low-capacity drives in inexpensive home and laboratory computer systems. In these systems, the read-write head actually touches the surface of the disk, causing wear and limiting the lifetime of the media. Rigid platters are made of metal instead of thin plastic, and can rotate much faster. In these drives the heads "fly" about 50 $\mu$in. above the disk surface, at speeds of about 150 mph. The very precise positioning possible with these designs yields much higher data densities than the floppy disks.

The latest development in disk drives is the Winchester drive. These drives feature moving heads and rigid platters; they are usually fixed media. They were developed by International Business Machines, Inc., and the first model contained two platters, each with a capacity of 30 Mbytes. The drive was called the 30-30, as was the famous rifle, and so was nicknamed the Winchester. Each platter and head assembly is hermetically sealed at the factory, providing a dustfree environment and allowing much smaller head-to-platter distances. This increases the recording density, and allows at least 10 times the data capacity of similarly sized, nonsealed units while increasing the cost over that of floppy drives only by a factor of about 2. Typical capacities of disk drives range from 100 kbytes for a small floppy to more than 1000 Mbytes for a very high performance drive (5).

## E. EXAMPLE: MOTOROLA 68000

Introduced by Motorola Semiconductor, Inc. in 1979, the 68000 is now available from several sources and is finding ever wider applications in personal, laboratory, and business systems. The 68000 is a general register 16 bit machine, with many features designed to readily allow future expansion to a full 32 bit architecture. The program counter and 17 general-purpose registers are 32 bits wide, and the status register is 16 bits. The data paths internal to the chip itself are all 32 bits wide, but the data bus that communicates with the outside world is 16 bits wide. Only the lower 24 bits of the program counter are used, to provide an address range of 16 Mbytes. In future implementations the full 32 bits may be utilized, which would allow 4096 Mbytes to be addressed, more than most large computers of today. The 32 bit program counter means that the full address range is available to the user at any time without using any of the memory segmentation schemes necessary when a CPU uses more address bits than its word size. The integrated circuit is a 64 pin dual-in-line package, and operates from a single +5 V supply. The address and data lines are separate, not multiplexed, to yield greater performance. The 68000 is compatible with the existing 8 bit 6800 family of interface chips, and new, 16 bit I/O chips are being designed by Motorola (4).

There are eight data registers and nine address registers. All 17 can be used to hold addresses, but to save execution time the condition codes in the status register are affected only by operations upon the data registers. The 68000 provides 56 different instructions, providing for a variety of arithmetic, logical, data, and bit manipulation operations, and program control. Fourteen addressing modes allow extremely flexible data accessing and manipulation. With two exceptions, any data manipulation instruction can use any memory location or a register as either the source or the destination of the data, and can operate upon data widths of 8, 16, or 32 bits. The exceptions are the multiply and divide instructions, which must have one 16 bit and one 32 bit argument, and which must have a register as the destination of the result. Finally, since the 68000 uses memory-mapped I/O, any of the instructions can be used to facilitate input and output operations.

The 14 addressing modes that the 68000 provides give the programmer much flexibility. For example, data can be accessed by specifying their address as a register, by giving their absolute address in memory, by placing their address in a register or in memory, or by adding an offset to the address contained in a register. In each case, the actual address in memory must be calculated from the specified data. Internally, there are three ALU's, each 16 bits wide. One ALU serves to process the user's data, and the other two perform 32 bit address calculations. By using three ALU's in parallel, the complex address calculations for future data can be calculated simultaneously with the processing of current data. A pre-fetch queue contains the current instruction being executed, the next instruction, and the address of the instruction following that one. Since the execution time of each instruction varies, any time that the bus is idle this queue fetches information that will be necessary in the future. This decreases the wasted time, that is, time that the bus sits idle, to less than 10%. If the queue contains a conditional branch instruction, it fetches the addresses of each possible outcome, saving even more time (6).

The standard version of the 68000 uses a 12 MHz clock signal, although faster and slower versions are available. The fastest move operation is from one register to another, and requires four clock cycles, or 333.3 nsec at 12 MHz. The slowest operation is a signed divide, which requires up to 170 clock cycles, or 14.19$\mu$sec. This performance compares quite well with minicomputers of the last several years, being comparable to a Digital Equipment Corporation PDP 11/60, a common 16 bit minicomputer model. Also, this is significantly faster than other currently available 16 bit microprocessors, and certainly faster than current 8 bit systems (3).

## III. LANGUAGES AND PROGRAMMING

In the early days of laboratory automation the choice of computer systems was limited. Very few systems were available, and so the user was forced to purchase the hardware system that came the closest to meeting his needs. The software to operate the system was an afterthought, because it all had to be custom written anyway. Now there are many systems to choose from, which means that probably several manufacturers produce a computer system that will fit the hardware requirements of an experiment. Given this, the availability of software is becoming the dominant factor in the purchase decision. Certainly there always remain specialized research applications that demand a custom-designed package, but frequently yesterday's one-of-a-kind research system becomes today's off-the-shelf system, found in many laboratories. A prime example is the chromatography data system. Initially, the computing power necessary to acquire data from a chromatograph was well into the range of minicomputers, which meant that few laboratories had the necessary capability. Now an inexpensive microcomputer system, costing less than $5000 (in mid–1983), can provide complete automation of not just one, but several chromatographs: acquiring the data, controlling the instruments, plotting the data, integrating the chromatogram, and generating a complete report for the user.

The major reason for this shift in priorities is the change in relative cost of hardware and software. Until several years ago, the hardware constituted the bulk of the cost of a computer system. Software expenses were relatively inconsequential. Now, with the advances in semiconductor technology, the hardware costs have dropped dramatically. Coupled with this, the users of computers have become much more demanding of the software. Programs must now be friendly, easy to use, and idiot-proof instead of merely usable by a highly skilled programmer. The large increase in programming effort necessary to reach these goals has naturally increased the cost of the software proportionally. In current laboratory systems the software may represent several times the cost of the hardware, instead of less than one-fourth as was common earlier (7).

Although complete systems are available which relieve the user from any programming whatsoever, in many cases some knowledge of programming is helpful. Although later chapters cover programming in detail, this discussion reviews the various philosophies of languages in general. Languages are described as either low level or high level, depending upon the features and supplied functions. Generally the higher the level of a language, the fewer the number of statements necessary to implement a given algorithm. The language designations, from lowest to highest level, are assembly languages, interpretive languages, and compiled languages (2). Each of these is now discussed in some detail.

### A. ASSEMBLY LANGUAGE

The most fundamental language possible is the machine language of the computer. This consists of a sequence of binary numbers, each of which is decoded into a single machine instruction. Because memorizing a list of numbers and their associated functions is clumsy, to say the least, assembly language was developed. This substitutes a mnemonic for each possible machine instruction, and requires a program, called the assembler, to translate the human-readable mnemonics into the binary machine code that the computer can execute. Assembly language gives the user maximum flexibility and control, and also allows for the fastest possible execution times. Most early laboratory computers were programmed in assembly language, since memory was expensive and the computers were relatively slow.

The major problems with assembly language are that programming is slower than with higher level languages, program modification is difficult for anyone but the original author, debugging a program is much more difficult, documentation aids are minimal, and a detailed knowledge of the computer system is necessary. This last problem means that an assembly language programmer expert on one computer system must learn a whole new architecture if he desires to program on a different machine. Debugging involves fixing the errors in logic and typing that inevitably creep into a program. Since higher-level languages that produce compact and efficient machine code are now available for most computer systems, the use of assembly language for routine laboratory programming is rapidly diminishing. Only in cases where the advantages of speed and flexibility outweigh the problems is assembly language programming used today. This usually takes the form of efficient subroutines for data acquisition, Fourier transformation, and instrument control which are called from a program written in another language.

### B. COMPILED LANGUAGES

Because all programs must be reduced to machine language to execute, a natural next step after an assembler is a compiler. This is a program that translates a high level language into machine code. It differs from an assembler in that each program statement may generate several machine instructions, instead of the one-to-one correspondence of assembly language. High level languages are defined according to standards independent of any one machine architecture, and generally are transportable from computer to computer. All that is required is a compiler for each computer to perform the translation.

Modern compilers optimize the machine language generated so that either execution speed or memory usage is minimized, at the option of the user, and can approach the results of a skilled assembly language programmer in terms of either. One important feature of compiled languages is that small program modules can be developed independently of one another, and so used in many programs. Many libraries of these subroutines are available for specialized needs. The main disadvantage of a compiler is that the translation step is relatively slow, and must be repeated each time a program bug is discovered and fixed. Also, some form of mass storage is required to store the compiler, the source program, and the generated machine language.

## C. INTERPRETED LANGUAGES

One way to avoid the inconvenience of recompiling an entire program after any change is to use an interpreted language. This employs a program which translates a high level language into a condensed intermediate form. This intermediate form can be executed by the interpreter, but can also be used to reconstruct the original program. This means that when a change is made, only that portion of the intermediate code that is affected is retranslated. Thus changes can be made and evaluated quickly and easily. The problem is that the process of executing the intermediate code by the interpreter is not nearly as efficient as the execution of the machine code generated by a compiler, typically slower by a factor of five or so. However, no mass storage is absolutely necessary, so a minimal computer system can be constructed with the interpreter in ROM and no external disks.

Even the problem of execution speed can be addressed by using a combined approach of an interpreter and a compiler. The interpreter is used to develop a program, and then the finished program is compiled. This allows use of the best features of both modes, and is becoming more popular as the compilers become available. The only drawback is that the interpreter does not lend itself conveniently to the loading of separate modules as does the compiler, so this feature is generally not available in interpreted languages.

## D. POPULAR LANGUAGES

The most widespread language today is BASIC, an interpretive language conceived in the early 1960s to teach programming. It has become very popular with the growth of the personal computer, since it can easily be supplied in ROM without requiring disks or other mass storage devices. Because personal computers are often used in low-cost laboratory systems, it is commonly encountered by chemists. As of this writing, no industry-wide standard definition of BASIC has been adopted, which means that each manufacturer supplies his own dialect, not necessarily compatible with anyone else's dialect. Though it was developed as an interactive interpreted language to facilitate teaching, several compiled versions are now available. BASIC is a very flexible language, making it easy to learn. This flexibility is also the major disadvantage. Very little structure is required in BASIC programs, allowing what is known as "spaghetti code" to proliferate. This is a program that uses so many unnecessary branches in the logic flow that often even the original programmer is hard pressed to remember just what they all mean. Also, once these bad habits are learned it is extremely difficult to force oneself to write clear, well documented, well organized programs.

Obviously not every programmer falls into this trap, but enough have to spur the development of languages with rigid syntax and necessarily straightforward logic flow. This is called top-down, or structured programming, where each program function follows in the natural order dictated by the data processing goals of the program. For example, to find the area of a rectangle, first the lengths of the sides are entered, then the area is calculated, and then the result printed out. Although seemingly trivial, this shows the logical procession of program flow in the top-down approach. The best known structured language is PASCAL, a compiled language developed in the early 1970s for the purpose of teaching this programming style. Since it was developed for teaching and not as a general-purpose language, PASCAL has several weaknesses (notably in character manipulation). Nevertheless, it has developed a wide following, and is used in several laboratory computer systems. Most implementations of PASCAL adhere to the original definition, so many PASCAL programs can be transported from system to system.

Among scientific programmers, FORTRAN is the most widely known language. This compiled language, developed in the mid 1950s, is syntactically less flexible than BASIC but not nearly as structured as PASCAL. An industry standard FORTRAN exists, and the latest version is FORTRAN-77. This allows much greater portability from system to system than is possible with BASIC. Many libraries of FORTRAN subroutines are available for specialized mathematical and graphic functions.

## IV. INCORPORATION OF COMPUTERS IN THE LABORATORY

Now that the individual components of computer systems are familiar, the integration of these components into a useful laboratory tool is considered. Requirements in a laboratory include open- or closed-loop control; data acquisition, storage, and analysis; and general record-keeping and report generation. To meet these needs five general approaches exist, each of which has advantages and limitations. First, the simplest approach is to record the data locally and then transport it to a distant computer. This approach is termed off-line, and is rapidly becoming extinct. The next level of integration is to dedicate a microprocessor to a single experiment. Only the peripherals necessary for that one application are installed. Next, a general-purpose microcomputer is used. Enough peripherals are included to allow the computer to interface with several experiments, but due to the limited processing power of the microprocessor only one experiment can be active at any time. A logical step upward from this is to use a more powerful computer, either a high-performance microprocessor or a minicomputer, to allow all of the experiments to be active together. Finally, a networking, or distributing processing, approach can be used. Here, each equipment has its own microprocessor which can perform the control and data logging functions necessary. Each microprocessor is then connected to a central computer to perform the actual data analysis and manipulation.

## A. OFF-LINE DATA ACQUISITION

The benefits of having experimental output in a computer-readable form are many and obvious. The first approach used in laboratory automation was the off-line analysis of previously recorded data. The term off-line means that there is no direct connection between the experiment and the computer, as opposed to on-line, where such a link does exist. The data can be stored in either digital or analog form, but must be in digital form when presented to the computer. The most common off-line storage device for analog data is a magnetic tape recorder. For digital data, punched cards, paper tape, or magnetic tape can be used. With the great cost reductions in computer systems of the last few years, it has become economically favorable to include enough processing power in the experiment itself to eliminate the need for off-line data analysis in most cases. Only in those cases where no computer control is necessary, where the rate of data acquisition is very slow, or where the amount of data processing necessary is beyond the capability of the computers available locally is off-line processing used today.

## B. THE DEDICATED MICROPROCESSOR

In the late 1970s most instrument manufacturers began large-scale design and marketing of "microprocessor controlled instrumentation." By using one or more dedicated microprocessors in an instrument, complex and flexible control functions are easy to implement. For example, in gas chromatography there are several different heaters that must be controlled. The injector, the column, and the detector are all usually at different temperatures, and often the column temperature must be varied over a wide range during the course of an analysis. The analog control electronics necessary to carry out all these tasks is fairly complex and cumbersome. However, one microprocessor can easily monitor three temperature sensors and control three heaters. Furthermore, it can also monitor carrier gas flow rate, turn on and off relays to control automatic injectors and inlet splitters, start a strip-chart recorder, and still have some processing power left over. Complex instruments such as state-of-the-art mass spectrometers may use two or three microprocessors just for control, along with another computer for data acquisition and analysis. Today, the per unit cost of adding microprocessor control to an instrument at the time of manufacture may be as low as $100. To add to an existing system is much more expensive since the hardware and software must be custom designed.

The main advantage of the built-in, dedicated microprocessor is that the instrument is completely independent of any other instrument in the laboratory. Equipment failure or programming errors in this one instrument do not destroy data from another experiment. In a teaching environment, it is certain that students will make mistakes of this nature, and so this independence is of prime importance. However, if more than one such instrument exists in the laboratory, each system duplicates all the peripherals found in all the other systems. Thus terminals, printers, mass storage devices, plotters, ADCs, and so on must all be duplicated for each instrument. Generally these peripherals make up the bulk of the hardware costs in such a system. Obviously this can rapidly become very expensive.

### C. GENERAL-PURPOSE MICROPROCESSOR

Often in a laboratory there are several experiments in progress, each of which could benefit from computer control. However, the expense of developing a dedicated microprocessor system for each one may be uneconomical. One solution is to purchase one microprocessor system, with the appropriate peripherals and sufficient processing power to carry out any one of the experiments. It then can be moved from experiment to experiment as needed. This type of system may cost from 2 to 10 times what a single minimal microprocessor system costs, but if enough different experiments are planned it is still cost-effective. General-purpose microprocessor systems range in cost from under $500 to over $10000, including software for many standard experiments. Systems designed for chromatography are the most popular, and average costs are $3000–$4000.

A major advantage of the general-purpose approach is the great flexibility provided. Not only can experiments be controlled and data acquired, but the system is also available for programming, word processing, and even record-keeping. One of the fastest growing areas in laboratory computers is the laboratory information management system, or LIMS. This is an all-purpose lab notebook implemented upon a computer. Notes can be kept, sample numbers recorded, data logged, graphs prepared, and reports suitable for publication can be generated, all within this one program package. Another advantage of this approach is that the expensive peripherals are not duplicated with each experiment.

The problems with this approach lie in competition for the computer system among researchers in the same laboratory. Because it can be interfaced to only one experiment at a time, scheduling becomes necessary. Though certainly possible, this can become inconvenient. Moreover, it is much easier for one user to inadvertently destroy another's data or software. This is especially so if one or more inexperienced people use the system on a routine basis. Finally, if the single computer breaks down, all the experiments are affected. With care all these difficulties can be avoided, but the prospective users must be aware of them before selecting such a system.

## D. LARGE-SCALE COMPUTER SYSTEMS

Although microprocessors are becoming ever more powerful, some experiments have such demanding data acquisition or control requirements that a larger computer system is necessary. Essentially, this is the dedicated microprocessor approach with additional processing power included. In many cases a dedicated minicomputer is included in a large and complex instrument, such as a Fourier transform nuclear magnetic resonance (FTNMR) spectrometer.

Alternatively, it is possible to use one computer to control several experiments simultaneously, instead of sequentially or with independent computers. However, with the current pricing trends in the laboratory computer industry, this is becoming less and less cost effective. Minicomputers range from $10000 to $300000 in cost, a substantial increase over the microprocessor systems. Also, with this approach, it is even easier for one user to disrupt another's experiment. This is very difficult to implement in a training environment. A unique system such as this would require a substantial investment in custom software, again making the other alternatives much more attractive. Last, there is once again the dependence of all experiments upon one computer system.

## E. NETWORKING

The final approach, termed networking, or distributed processing, is a compromise between one all-powerful computer and several smaller systems. Here each experiment has its own computer for local control and data acquisition, but the data thus obtained can be passed on to another computer for analysis. With many instruments, for example, FTNMR, analyzing the data may require more than an order of magnitude longer than acquiring it. In order to relieve this bottleneck a second computer is used. This second system is then tailored for data manipulation, unlike the computers used in today's instruments, which were designed primarily for control and acquisition. Because this second computer system is usually much cheaper than the cost of a second instrument, it is a very effective way to increase instrument productivity. Another circumstance where networking several computers together is advantageous is in the utilization of expensive peripherals. For example, only one computer system in a laboratory may have extensive disk storage, a fast (and expensive) printer or plotter, or magnetic tape facilities for archiving or exchanging data. By linking the other computers in the laboratory to this one system, all may utilize its peripherals. Only the minimal peripherals necessary for the experiment need be attached to each of the other computers.

The method of connecting the various computers together is determined by the volume of data to be transferred. If this is modest, say, a list of peak areas and retention times from a chromatography data system, a simple serial link would be adequate. For the many thousands of data points typical in a single FTNMR free induction decay, faster parallel communications may become necessary. Also, there are several different schemes for arranging the computers in the network. The two most common arrangements are as a star and as a ring.

In the star configuration, there is one central computer that can communicate with each of the local computers. The local computers can not communicate directly with one another. The ring network places the computers in a circle. Each system can communicate only with the computers on each side of it in the ring. Messages must be passed around the ring until they arrive at the destination computer (1).

## V. CONCLUSION

This chapter discusses the logical components of computer hardware and programming. In addition, the combination of these components into useful laboratory computer systems is discussed. By careful consideration of the needs of an entire laboratory, and consideration of the available computer systems, many savings are possible with a general-purpose microprocessor or networking approach to laboratory automation.

## ACKNOWLEDGMENTS

We would like to acknowledge the support of the Shell Companies Foundation through a Shell Doctoral Fellowship to Carl Ijames and the support of the National Science Foundation under grant CHE-82-08073.

## REFERENCES

1. Dessey, R. E., *Anal. Chem.* **54**(11), 1167A (1982).
2. Dessey, R. E., *Anal. Chem.* **55**(6), 650A (1983).
3. Grappel, R., and J. Hemenway, *EDN*, 179 (April 1, 1981).
4. Scanlon, L. J., *The 68000: Principles and Programming*, Howard W. Sams, Indianapolis, 1981.
5. Solomon, L., and S. Veit, *Comput. Electron.*, **21**(8), 36 (1983).
6. Starnes, T. W., *Byte*, **8**(4), 70 (1983).
7. Wintz, P., *Digital Design*, 30 (Nov. 1980).

Part I
Section E

Chapter 8

# INTERFACING PRINCIPLES FOR LABORATORY COMPUTERS


BY SAM P. PERONE, *Lawrence Livermore Laboratories, University of California, Livermore, California*

DAVID O. JONES, *David O. Jones Company, Spokane, Washington*

H. R. BRAND, *Lawrence Livermore Laboratories, University of California, Livermore, California*


**Contents**

Digital instruments and digital computers have become important elements in modern analytical instrumentation. There are many reasons for this. Together with the advances in instrumentation of the 1950s and 1960s came the need to use electronic data processing to handle the vast amounts of raw data that could be generated. In the late 1960s and early 1970s, as a result of space-age technological advances, digital devices became so compact and inexpensive that laboratory-size computers became a reality. These minicomputers could be connected to instruments to acquire and process digitized data automatically, control experimental parameters, print reports, and so forth. Because of the heavy analytical load carried by gas chromatography (GC) in industrial laboratories, chromatographs were among the first types of instruments to be computerized on a wide scale. High-resolution mass spectrometry (MS) benefited greatly from the availability of laboratory computers, because of the need to handle large quantities of data for runs lasting only a few seconds. Also, such methods as Fourier-transform NMR and ir spectroscopy as well as GC/MS became feasible as routine analytical techniques only when instruments were developed that included dedicated minicomputers.

The 1970s also saw the development of microprocessors and microcomputers. These devices represented another couple of orders of magnitude of reduction in size and cost of the heart of digital computers—memory and central processor. This level of technology was versatile and more difficult to implement, but the size and cost advantages made it very attractive for applications that would be replicated or mass-produced. This era saw the advent of pocket calculators, electronic games, and personal computer systems. Microprocessors also replaced minicomputers in many analytical instruments and also found their way into instruments requiring only minimal computations or control (e.g., pH meters).

Many other areas of analytical instrumentation have benefited greatly from computerization, and the reader is referred to review articles describing various applications in detail (3). The objective of this chapter is to introduce the fundamental principles of on-line computer instrumentation, focusing attention on the characteristics of digital devices important for interfacing laboratory instruments to digital computers. These fundamentals include a consideration of number systems, digital logic, digital devices, analog/digital translation devices, sampling of raw data, interface design, and on-line computer operations. After the principles of interfacing devices and their components are presented, the design of interface devices for some specific functions are described. It is important to emphasize that the general principles discussed here apply as well to microcomputer and minicomputer technology.

## I. THE DIGITAL COMPUTER

Figure 8.1 provides a block diagram of the essential components for a typical digital-computer configuration. The *memory* is a component capable of storing many thousands of binary-coded (digital) *packets* of information. Each packet is composed of *n* binary digits (*bits*) and is called a *word* of information. Each

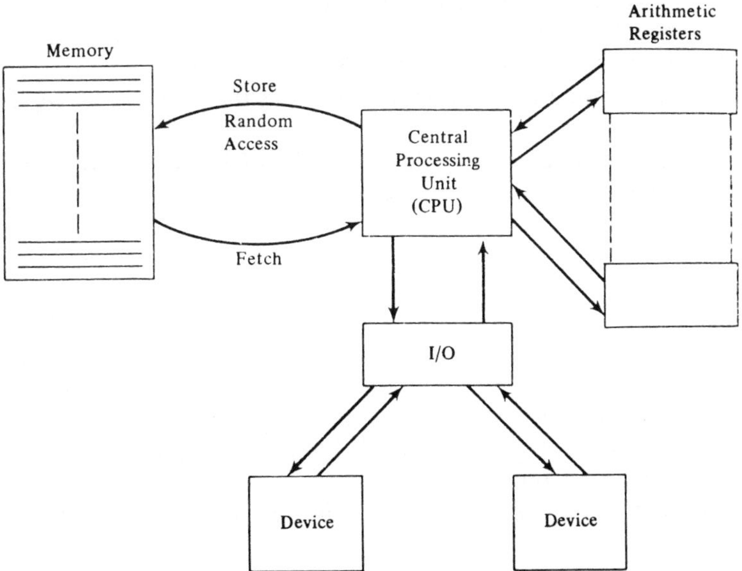


**Fig. 8.1.** Typical digital computer configuration.

of these $n$ bit words has an *address* associated with it, and the information contained within can be fetched or stored by specifying the address. The internal memory is usually randomly accessible (called random access memory or RAM) and is either a semiconductor or a magnetic-ferrite-core type (2). Three types of information are stored in memory; *data*, *instruction*, or *address*. The speed with which information is fetched or stored in RAM is the *memory access time* and is usually about 0.1–1 $\mu$sec.

The *central processing unit* (CPU) controls the overall operation of the computer. It is made up of electronic registers and logic circuits that execute the simple logical and arithmetic operations of which the computer is capable. The CPU also contains or controls registers that dictate the flow of digital values to and from RAM. When these operations are executed in appropriate sequences, the computer can accomplish complex mathematical or data processing functions. Moreover, if one provides the appropriate electronic *interface*, these simple operations can be used to control experimental systems, acquire data, or print results on a teleprinter, line printer, oscilloscope, or other peripheral devices.

The sequence of instructions to be executed by the computer is called a *program*. In actuality, the program is a set of binary-coded instructions stored in memory. The CPU fetches each instruction from memory, interprets and executes it, and then moves on to the next instruction. The CPU fetches instructions sequentially from memory, unless told to do otherwise by one of the instructions.

The *arithmetic registers* are high-speed electronic *accumulators* (ACs). That is, each is a set of $n$ electronic two-state devices (like flip-flops; see next section), which can be used to accumulate intermediate results of binary arithmetic involving $n$ bit data. Nearly all the arithmetic and logical operations of the CPU are carried out in the arithmetic registers. Binary information can be transferred to or from memory and the arithmetic registers by the execution of appropriate machine instructions.

The *I/O* (input/output) *bus* allows transfer of binary-coded information between peripheral devices and the central processing unit. The number of peripheral devices that can be connected to the I/O bus is limited primarily by the sophistication of the peripheral device-selection hardware and software (programs).

A small digital computer (of the types currently being used for laboratory automation and experimental control)* may have an instruction set that includes of the order of 50 to 100 different instructions. Each of these instructions corresponds to a specific binary coding which, when decoded by the CPU, results in the execution of a fairly simple arithmetic or logical step. Examples of some simple machine operations are binary addition of a datum in some memory location to the contents of an AC, the transfer of the contents of an AC

*Examples are the small desk-top laboratory computers manufactured by Data General Corp., Hewlett Packard Corp., Digital Equipment Corp., and others.

to a memory location (and vice versa), rotation of the binary digits of the AC contents to the left or right, and the application of logical tests such as determining whether the AC is zero, nonzero, odd, even, positive, or negative. By developing programs composed of appropriate sequences of many of these elementary operations, the most sophisticated mathematical computations can be carried out. Since the computer can execute instructions so rapidly—of the order of $10^6$ instructions per second—it can complete complex computations with fantastic speed.

## A. PROGRAMMING THE DIGITAL COMPUTER

The programming of a computer is usually accomplished with some sort of symbolic language. That is, readily recognized symbols are used to represent simple machine operations or groups of machine operations. Translating programs are supplied by the computer manufacturer to convert symbolic programs into the binary-coded machine-language programs. The simplest of these symbolic languages is the *assembly language*, where there is nearly a one-for-one conversion from symbolic statements to machine language. A program for translating these programs into machine language is called an *assembler*. The relationship between assembly and machine languages is shown in Fig. 8.2. The figure shows that the assembly language instructions, such as LDA Z (which may mean load the contents of memory location Z into the A register) must be translated into numerical machine language, the only language that the central processer can understand.

Because programming in assembly language can become very tedious, higher-level languages have been developed where single statements can be translated into large blocks of machine-language program segments. The translating program is called a *compiler*, and one such high-level language is FORTRAN. The relationship to machine language is also shown in Fig. 8.2.

Obviously, it is much simpler to prepare programs in FORTRAN than in assembly language. However, compiler-generated programs may be very inefficient in the utilization of available memory space. Moreover, speed of execution and synchronization of computations with outside events are relatively difficult to control with these programs. These considerations are particularly important for *on-line* computer applications in the laboratory; therefore, assembly-lan-

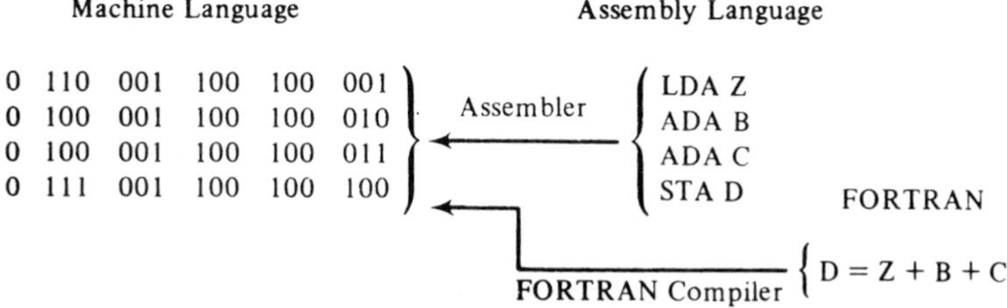


**Fig. 8.2.** Relationship of machine language to assembly language and FORTRAN.

guage programming must be used extensively so that the programmer can exercise the detailed control of computer operations required. The next section deals extensively with such programming.

## B. OFF-LINE COMPUTERS

The computer configuration with which most scientists are familiar is the *off-line* system (see Fig. 8.3). To use the computer in this configuration, the scientist typically writes a data processing program in FORTRAN or some other high-level computer language, runs the experiment(s), manually tabulates the data from the strip-chart recorder or oscilloscope trace, transfers the tabulated data to the computer by manual entries directly through a keyboard terminal or

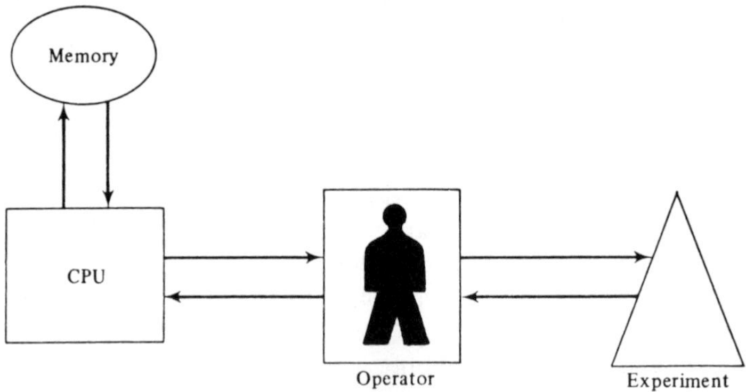


A. Off-line Computer System

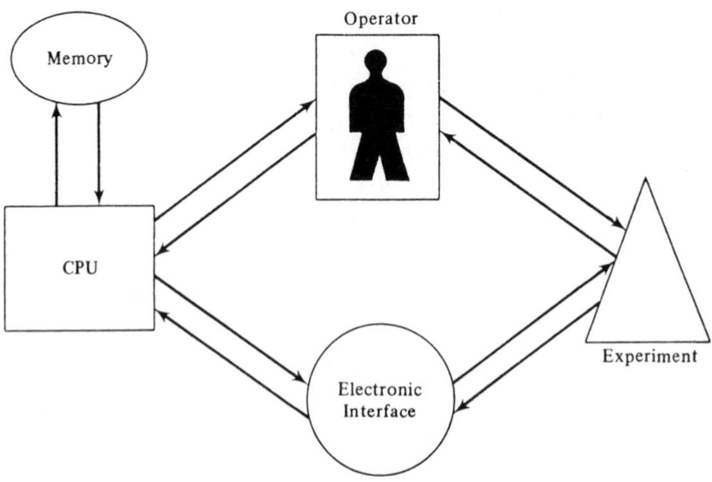


B. On-line Computer System

**Fig. 8.3A** (*a*) Off-line computer system. (*b*) On-line computer system.

indirectly via some storage medium such as punched cards, punched tape, magnetic tape, or magnetic diskettes, and then waits until the program has been executed and the results printed. Turnaround times may vary from a few minutes to a few days, depending on the capacity of the computer facility, the number of users, and the backlog and priorities of work to be processed.

### C. ON-LINE COMPUTERS

For the investigator who requires very rapid or instantaneous results from his computer system, for whatever reason, the solution may be to employ an on-line computer system. The most important distinction of this configuration is that there is a *direct* line of communication between the experiment and the computer (Fig. 8.3*b*). The line of communication is through an electronic *interface*. (This interface includes control logic, electronic elements to provide timing and synchronization, and conversion modules, such as digitizing devices, which translate real-world data into information that the digital computer recognizes.) Data are acquired under computer control or supervision, and the program for data processing is usually resident in memory to provide for very rapid completion of the computational tasks. Results may be made available to the investigator quickly by means of teleprinter display, line-printer, oscilloscope, or other forms of printout. In addition, the computer may be programmed to communicate directly with the experiment by controlling electronic or electromechanical devices, such as solid-state switches, relays, stepping motors, servomotors, or any other devices that can be activated by changes in voltage or current.

The advantages of on-line computer operation, then, include rapid turnaround of data collection and data analysis tasks, elimination of manual communication links by substituting an electronic interface, possible direct computer control of the experiment, and possible real-time interaction between the computer and the experiment. Because the computer can make computations and decisions at speeds exceeding most ordinary data-acquisition rates, it can modify the experimental conditions during the experiment. An additional advantage, of course, is that the logistic barriers of the remote computer system are eliminated.

## II. PRINCIPLES OF DIGITAL INSTRUMENTATION

### A. BINARY AND OCTAL NUMBER SYSTEMS

All information handled or generated by the central processing unit (CPU) must be binary or binary-coded machine language. This includes instructions, memory addresses, and data. Thus the small-computer user must quickly become familiar with this number system. It would be well to review here the binary number system and binary arithmetic.

The *decimal* number $369^*_{10}$ can be broken down into $3 \times 10^2 + 6 \times 10^1 + 9 \times 10^0$. Similarly, the *binary* number (in base 2) 10101 represents $1 \times 2^4 + 0 \times 2^3 + 1 \times 2^2 + 0 \times 2^1 + 1 \times 2^0 = 21_{10}$. Large binary numbers, for example, 101101110010101, are conveniently represented in a shorthand fashion for easy recall or reference. The usual shorthand is the *octal* (base 8) system. The applicability of the octal system can be seen from the binary representation of the numbers 0 to 7:

| | | | | | |
|---|---|---|---|---|---|
| 000 | 0 | 011 | 3 | 110 | 6 |
| 001 | 1 | 100 | 4 | 111 | 7 |
| 010 | 2 | 101 | 5 | | |

This sequence illustrates the normal binary counting sequence, which can be extended to an infinite number of binary digits (bits). It also shows the octal digits equivalent to all 3 bit combinations. To convert any large binary number to octal, group binary digits in groups of three, *starting at the rightmost digit*, for example,

$$\begin{matrix} ..1 & 001 & 101 \\ 1 & 1 & 5 \end{matrix} = 115_8$$

$$\begin{matrix} 101 & 101 & 110 & 010 & 101 \\ 5 & 5 & 6 & 2 & 5 \end{matrix} = 55625_8$$

$$201_8 = 010 \quad 000 \quad 001$$
$$356_8 = 011 \quad 101 \quad 110$$

Counting in octal: 0, 1, 2, 3, 4, 5, 6, 7, 10, 11, 12, 13, 14, 15, 16, 17, 20, 21,
22, 23,. . ., 75, 76, 77, 100, 101,
102,. . . ., 766, 777, 1000,
1001,. . ., 7776, 7777, 10000,. . .

Often it is necessary to convert numbers from one base to another. The above examples illustrate the ease of converting from octal to binary and vice-versa. Consider the following examples:

## B. DECIMAL-TO-BINARY CONVERSION

For example, $876_{10}$ = ? in binary?

$$\begin{array}{rl} 876 & \\ -512 & \quad 512 = 2^9 = \text{largest power of 2 to fit in } 876_{10} \\ \hline 364 & \\ -256 & \quad 256 = 2^8 = \text{largest power of 2 to fit in } 364_{10} \\ \hline 108 & \end{array}$$

*369 to the base 10.

$$
\begin{array}{rl}
-\ 64 & 64 = 2^6 = \text{largest power of 2 to fit in } 108_{10} \\
\hline
44 & \\
-\ 32 & 32 = 2^5 \\
\hline
12 & \\
-\ 8 & 8 = 2^3 \\
\hline
4 & \\
-\ 4 & 4 = 2^2 \\
\hline
0 &
\end{array}
$$

Thus the binary representation of $876_{10}$ must include $2^9$, $2^8$, $2^6$, $2^5$, $2^3$, and $2^2$, and the binary number is

$$1\quad 101\quad 101\quad 100$$

The octal equivalent of this binary number is $1554_8$.

## C. DIGITAL INFORMATION

The laboratory scientist is accustomed to seeing experimental information displayed as *analog* data. That is, data are usually made available in a *continuous signal* such as given by a strip-chart recorder or oscilloscope trace. Unfortunately, the digital computer cannot directly handle analog information. The computer must have information in a digital (usually binary-coded) format.

The differences between analog and digital information are illustrated graphically in Fig. 8.4. Whereas analog data are continuous with an infinite number of real values between any two points on a trace, digital data are discrete, with well-defined finite limits of resolution between any two points. Thus, for example, when digitizing an analog signal that varies between 0 and 15 V, if only 4 bits can be used to encode the digital information, the digital resolution is $\pm\frac{1}{2}$ V; that is, only 16 values ($2^4$) can be represented by 4 bits, allowing only 0, 1, 2, 3,..., 15 V digital values (see Fig. 8.4). If only 2 bits are used to encode the digital information, the resolution decreases even further as there are now only four digital states possible.

It should be obvious, then, that digitizing data *always* results in some *loss* of information! We put up with this only because it is a required format change in order to be able to use the powerful data handling features of the digital computer. However, it is important for the scientist developing and using computerized instrumentation to be aware of the problems inherent in this approach, and to learn how to take maximum advantage of digital instrumentation. To this end we have included the following discussions describing digital devices, digitization methods, sampling considerations, and so forth.

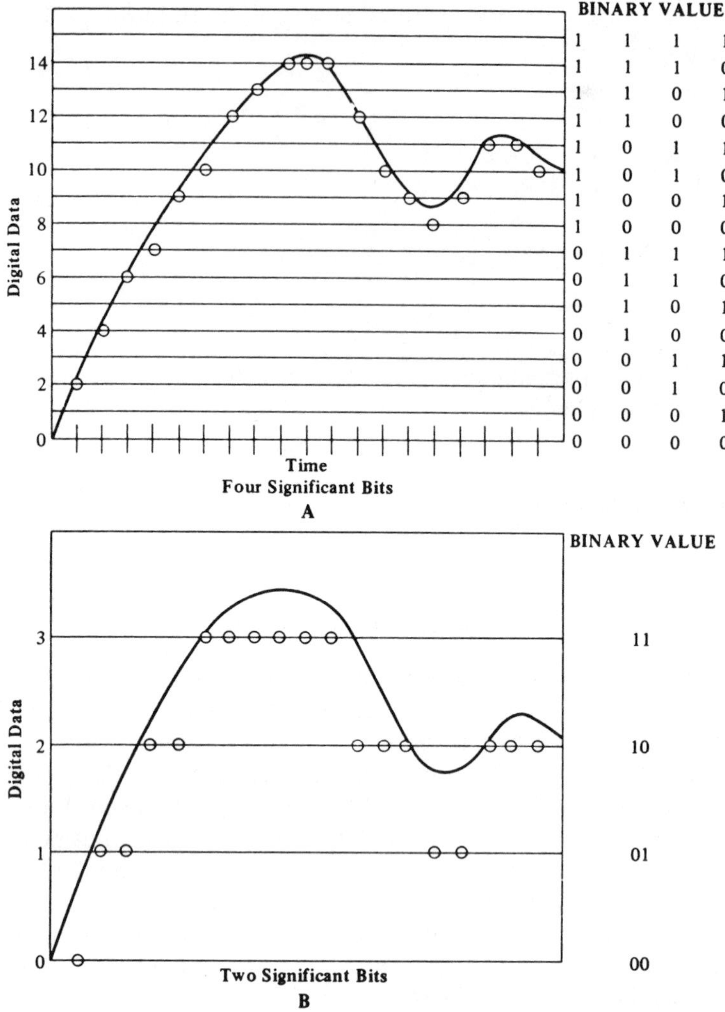


**Fig. 8.4.** Digital representation of analog data.

## D. DIGITAL LOGIC STATES

Digital logic devices are the foundation used to build digital instrumentation, including computers and interfaces. They consist of a set of electronic circuits that perform simple logic operations. They can be connected to make more complex building blocks to perform the necessary logic, storage, arithmetic, interface, and timing operations that result in digital instrumentation. Logic operations are performed by the logic states generated with various digital devices. The logic states are usually represented by the binary number system.

In the simplest case, a digital logic function can be simulated with conven-

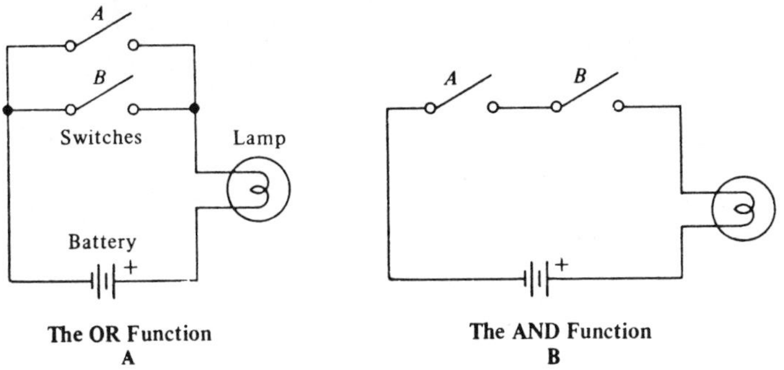


**Fig. 8.5.** Simple logic functions.

tional switches and a battery. The output states can be indicated with a light bulb. Two such simple logic circuits are shown in Fig. 8.5.

The circuit in Fig. 8.5*a* is performing the basic OR digital logic function. When both switches A and B are open, no current flows to the lamp. However, when either switch A *or* B or both are closed, current flows to light the lamp. The *output* state, that is, whether the lamp is on or off, can correspond to the numbers of the binary number system. The *on* condition can be defined as the binary number 1 and the *off* condition as the binary number 0. The *input* functions or switch positions can be defined in the same manner, with the open switch condition as a binary 0 and the closed condition as a binary 1.

A circuit for performing another basic logic function is that in Fig. 8.5*b*. In it, the two switches A and B are connected in series rather than in parallel. In this case, both switches A *and* B must be closed for the lamp to light. This is a basic digital logic circuit to perform the AND function.

The operation of the circuits presented in Fig. 8.5 can be described in terms of the binary number system in tabular form as presented in Table 8.I. The output states are listed for all combinations of the input-switch conditions. If the circuits presented in Fig. 8.5 are expanded with more input switches, similar but more complex tables can be constructed. In addition, the AND and OR

TABLE 8.I

Binary Number Representation of Logic Circuit Operation

| OR function | | | AND function | | |
|---|---|---|---|---|---|
| Input conditions | | Output condition | Input conditions | | Output condition |
| Sw. A | Sw. B | | Sw. A | Sw. B | |
| 1 | 1 | 1 | 1 | 1 | 1 |
| 1 | 0 | 1 | 1 | 0 | 0 |
| 0 | 1 | 1 | 0 | 1 | 0 |
| 0 | 0 | 0 | 0 | 0 | 0 |

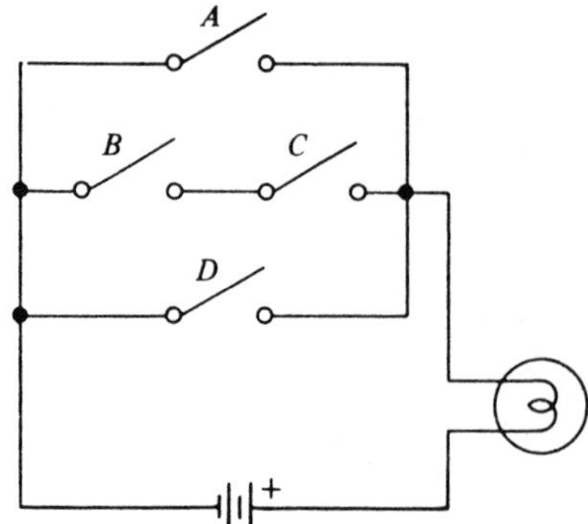


**Fig. 8.6.** Complex logic function.

circuits presented above can be connected together to perform more complex logic operations. A complex logic circuit performing both AND or OR logic is shown in Fig. 8.6.

The logic circuits presented above are useful approximations for defining the simple nature of digital logic. They are not, however, used in modern electronic digital computers. Modern electronic digital logic uses transistors to perform the switching operations and is available in *integrated-circuit* form. Integrated-circuits are miniaturized solid-state devices that may contain several complete electronic circuits, and a single integrated-circuit logic package can perform extremely complex logic functions. The logic states in integrated-circuit digital logic are usually represented by voltage levels, such as 0 and +5 V for both the inputs and outputs.

Often it is necessary to define the operation of a logic system in a *sentence* format. This can easily be done, and for the circuit in Fig. 8.6, *the logic output is equal to a binary 1 when either A OR (B AND C) OR D is equal to a binary 1.*

## E. SIMPLE LOGIC ELEMENTS: INTRODUCTION TO GATES

Modern electronic digital logic comes in several microelectronic integrated-circuit forms; discussion of integrated-circuit logic types, including electronic-circuit descriptions and a comparison of operating characteristics, is presented elsewhere (4). Suffice it to say here that several different types of logic exist with different electronic operating characteristics and requirements. They differ in their electronic circuits, not in the logic functions that they perform. The principles discussed here apply equally well to all.

There are five basic, simple logic functions from which all others are usually constructed. They are presented along with their common symbols and "truth tables" (see below) in Fig. 8.7. The actual hardware electronic device used to perform a particular basic logic function is called a *gate*. The symbols presented in Fig. 8.7 are often referred to as the *basic positive logic gate symbols*.

Of the logic functions presented in Fig. 8.7, the AND and OR gates have been considered above using switches and box diagrams. Their operation is the

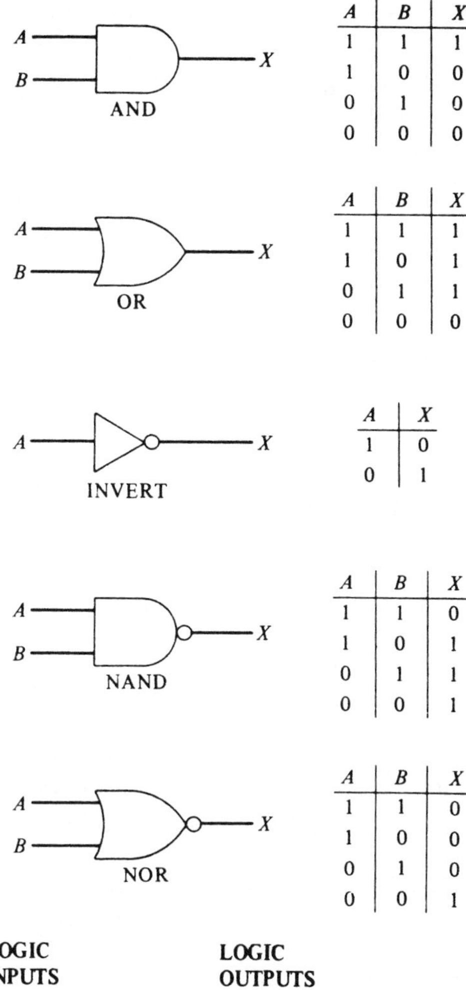


| A | B | X |
|---|---|---|
| 1 | 1 | 1 |
| 1 | 0 | 0 |
| 0 | 1 | 0 |
| 0 | 0 | 0 |

| A | B | X |
|---|---|---|
| 1 | 1 | 1 |
| 1 | 0 | 1 |
| 0 | 1 | 1 |
| 0 | 0 | 0 |

| A | X |
|---|---|
| 1 | 0 |
| 0 | 1 |

| A | B | X |
|---|---|---|
| 1 | 1 | 0 |
| 1 | 0 | 1 |
| 0 | 1 | 1 |
| 0 | 0 | 1 |

| A | B | X |
|---|---|---|
| 1 | 1 | 0 |
| 1 | 0 | 0 |
| 0 | 1 | 0 |
| 0 | 0 | 1 |

**Fig. 8.7.** Common electronic logic symbols and truth tables.

same as that already discussed. The symbols presented in the figure are used from now on.

The INVERT or NOT gate does exactly what its title suggests; that is, it negates the input. When a binary 1 is placed on the input, the output is a binary 0. In like manner, when a binary 0 is placed on the input, the output is a binary 1. The inverter is a fundamental part of two more basic gates, the NAND and NOR gates. The term NAND is derived from AND and NOT and functionally refers to an AND gate followed by an inverter (Fig. 8.8). In fact, the NAND function can be generated in such a manner. The logical operating characteristics of the NAND and NOR gates can be defined in the same manner as was

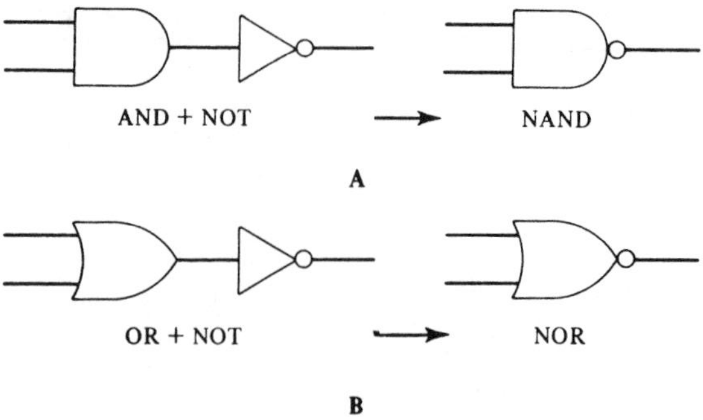

AND + NOT → NAND

A

OR + NOT → NOR

B

**Fig. 8.8.** Logical construction of NAND and NOR gates.

done for the AND and OR functions previously described. The logical operation tabulations presented in Fig. 8.7 are commonly called *truth tables*. Truth tables are by far the most commonly used road map to the operation of logic systems.

## F. BOOLEAN ALGEBRA

Boolean algebra is the algebra of logic and consists of a symbolic method of studying logical operations. We have previously discussed AND, OR, and NOT or INVERT functions. The Boolean algebra symbols for these functions are given in Table 8.II. The AND function can be represented by a dot or by placing the variables adjacent to each other. The OR function can be represented by "+." The NOT or INVERT function can be represented by a prime beside, or bar above, the variable.

A complex logic function can, of course, be written in terms of Boolean algebra. For example, $X = AB + CD + F(\overline{G} + H)$ is read, "$X = 1$ whenever $A$ AND $B = 1$ OR $C$ AND $D = 1$, OR $F = 1$ when $G = \mathrm{O}$ OR $H = 1$". Notice that $G = \mathrm{O}$ is a requirement rather than $G = 1$. This is because of the complement bar above it in the Boolean expression. It implies that $G$ must be equal to O; $\overline{G}$ is usually expressed verbally as "not $G$."

Many theorems and postulates have been developed for Boolean algebra and are necessary for the design and evaluation of complex logic systems. They are

TABLE 8.II

Symbols for Boolean Algebra Operations

| Function | Example | Symbol | Example |
|---|---|---|---|
| AND | $X = A$ AND $B$ | | $X = A \cdot B$ or $X = AB$ |
| OR | $X = A$ OR $B$ | + | $X = A + B$ |
| NOT | NOT $A$ | $\bar{}$ or′ | $\overline{A}$ or $A'$ |

not discussed here but are presented elsewhere (5). Only one theorem is presented here. It is De Morgan's theorem, which is used to complement complex logic expressions. It is necessary for an understanding of NAND and NOR gates. For example, a NAND gate is equivalent to an AND gate followed by an inverter. In other words, it is an *output-inverted* or *complemented* AND gate. Thus a Boolean expression can be written in the following manner for a NAND gate. It says that $X = 1$ when the complement of ($A$ AND $B$) = 1:

$$X = (AB)' \tag{1}$$

This is a correct expression, but it makes the construction of a complex truth table cumbersome because it requires first constructing the AND gate truth table and then complementing its output column. With the use of De Morgan's theorem, the expression can be placed in a more convenient form. De Morgan's theorem states that

$$(AB)' = \overline{A} + \overline{B} \tag{2}$$

and

$$(A + B)' = \overline{A} \cdot \overline{B} \tag{3}$$

The more general implications of De Morgan's theorem can be considered after defining two new terms, a *dual* and a *literal*. To obtain the dual of a Boolean expression, one must interchange all occurrences of a "+" and a "·" and of a 1 and a 0. A literal is defined as any single *variable* within the dual expression. For example, in the expression

$$\overline{A} \cdot B + C \tag{4}$$

whose dual is

$$\overline{A} + B \cdot C \tag{5}$$

the letters $\overline{A}$, $B$, and $C$ are all literals. *In order to complement a Boolean expression, one must complement all literals in the dual expression*. This is illustrated in Table 8.III. The application of these rules for the NAND and NOR functions leads to the statements of De Morgan's theorem presented above.

Boolean algebra is often used to interpret and generate logic diagrams. Consider the diagram presented in Fig. 8.9$a$. It has two inputs $A$ and $B$. $B$ is

TABLE 8.III
Complementing Boolean Expressions

| Expression | Dual | Complement |
|---|---|---|
| $1 \cdot A + B$ | $0 + AB$ | $\overline{A}\overline{B}$ |
| $\overline{A}B + C$ | $\overline{A} + BC$ | $A + \overline{B}\overline{C}$ |
| $A(B + C)$ | $(A + B)(A + C)$ | $(\overline{A} + \overline{B})(\overline{A} + \overline{C})$ |

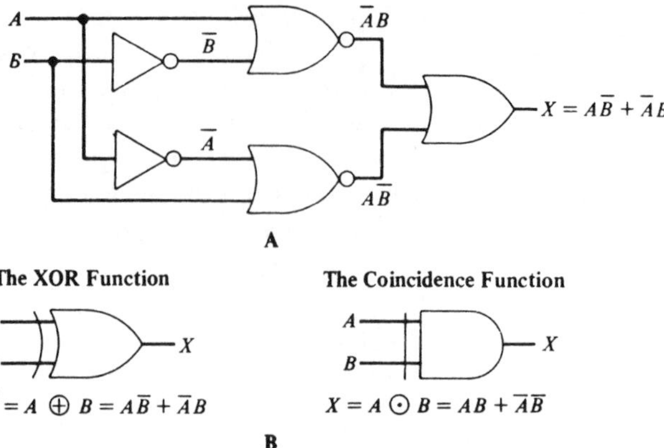


**Fig. 8.9.** Exclusive OR and coincidence gate.

inverted by the upper inverter to $\overline{B}$, and $A$ is inverted by the lower inverter to $\overline{A}$. The inputs to the upper NOR gate are then $A$ and $\overline{B}$. From the Boolean expression for a NOR gate (equation 3), the output of the upper gate is $\overline{A} \cdot B$. In a like manner, the output of the lower NOR gate is $A \cdot \overline{B}$. These outputs become inputs to the OR gate, resulting in the output expression

$$X = A \cdot \overline{B} + \overline{A} \cdot B \tag{6}$$

This particular function is called the *exclusive* OR function (XOR) and is the basis for binary arithmetic operations. It says that $X$ is true if $A$ or $B$, but not both, are true. It has a defining symbol "⊕" and is written as

$$X = A \oplus B \equiv A \cdot \overline{B} + \overline{A} \cdot B \tag{7}$$

The XOR function is available in integrated-circuit form and is designated by the symbol presented in Fig. 8.9*b*.

There is one more function that must be defined, the *coincidence* function. As the name implies, this function gives a 1 output whenever both inputs are the same, either all O's or all l's. Its symbol is also illustrated in Fig. 8.9*b*. The reader should construct its logic diagram using AND, OR, NAND, NOR, and INVERT gates in a manner similar to that just presented for the XOR function.

## G. FLIP-FLOPS

The basic device used for counting and storage operations is the *flip-flop* or *bistable multivibrator*. The flip-flop, in its many forms, can be constructed from individual gates, but it is usually purchased as a single unit in integrated-circuit form.

The most basic flip-flop is called a *reset-set* or RS flip-flop. It can be constructed from two cross-coupled NAND gates as illustrated in Fig. 8.10*a*. It has

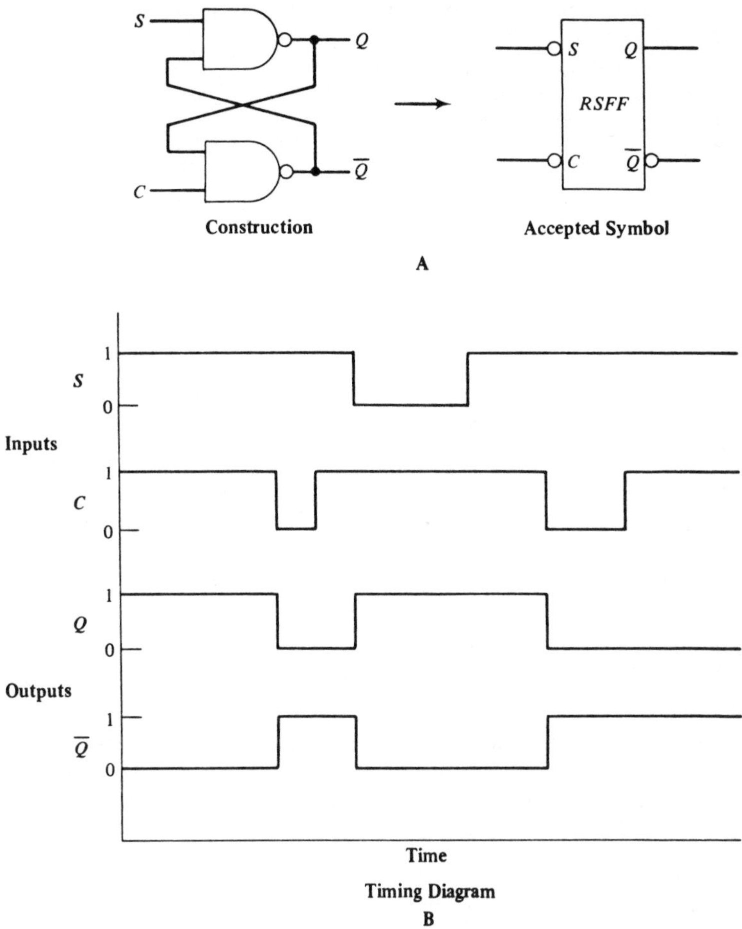


**Fig. 8.10.** The reset-set (RS) flip-flop.

two inputs, labeled $S$ for set and $C$ for clear, and two outputs, labeled $Q$ and $\overline{Q}$. The $Q$ output goes to a binary 1 and remains there when the $S$ input momentarily goes from 1 to 0. In like manner, the $\overline{Q}$ output goes to a binary 1 and remains there when the $C$ input momentarily goes from 1 to 0. Whenever $Q$ is a 1, $\overline{Q}$ is a 0; that is, they are always complementary. As a result, the C input also clears the $Q$ output to a logical 0. The RS flip-flop remains in whatever state it has been set or cleared to until its states are changed by applying negative-going pulses to one or the other of the inputs. It is thus a bistable device with two stable states that can be used for binary data storage.

A truth table for the operation of the RS flip-flop is presented in Table 8.IV. This is a somewhat different kind of table from those we have considered before in that it has time as a variable. $t_n$ is the time *before* the specified input

TABLE 8.IV

RS Flip-Flop Truth Table[a]

| | Inputs | | $t_n$ | $t_{n+1}$ |
|---|---|---|---|---|
| Case | S | C | Q | Q |
| 1 | 1 | 0 | X | 0 |
| 2 | 0 | 1 | X | 1 |
| 3 | 1 | 1 | X | NC |
| 4 | 0 | 0 | X | U |

[a] NC means the output is unchanged from its previous state; U means the output is undefined and may go to either state.

conditions have been imposed, and $t_{n+1}$ is the time *after* they have been imposed. Because $Q$ and $\overline{Q}$ always complement each other, only $Q$ is given in the table. NC means that the input is unchanged from its previous state. U means that the output in undefined and may go to either state. X means that the inputs at $t_n$ can be in any state.

The RS flip-flop is commonly used for control and storage operations in cases where two input signals occur. One signal is used to set its output and allow storage of a binary 1, whereas the other input is used to clear its output and store a binary 0. Notice in Table 8.IV that when $S = C = 0$, the output is undefined. When using the RS flip-flop as a storage element with two data inputs that always complement each other, this condition offers no hindrance to its use. However, for other operations, such as counting, where only one data input signal is provided, the RS flip-flop cannot be used directly.

A timing chart for the RS flip-flop is presented in Fig. 8.10*b*. Notice that the $Q$ output is initially set at 1 and remains so until the $C$ input goes to a binary 0. $Q$ then changes to 0 and remains there until it is again set to 1 by the $S$ input. $\overline{Q}$ is then cleared back to 0. Notice that the $\overline{Q}$ output always complements the $Q$ output state. Timing charts are useful for defining the dynamic operation of logic devices. Often they are used in place of, as well as with, truth tables.

A more versatile flip-flop than the RS type is called a *clocked* flip-flop (Fig. 8.11); it can be used for counting operations. The clocked flip-flop has direct set $S_D$ and direct clear $C_D$ inputs that operate in the same manner as the $S$ and $C$ inputs of the RS flip-flop. This is commonly called *asynchronous* operation since no timing requirements are made. For this mode of operation, the same truth table as presented in Table 8.IV can be used, by substituting $S_D$ and $C_D$ for $S$ and $C$. The clocked flip-flop, however, has another mode of operation called the *synchronous* mode. In the synchronous mode, information is entered into the flip-flop through the AND-gated $S$ and $C$ inputs ($S_1$, $S_2$ and $C_1$, $C_2$). The flip-flop does not change state, however, until a transition occurs at the *clock* or $T$ input. The truth table for the synchronous mode of operation is presented in Table 8.V. Again X indicates that an input can be in either logical state. In the truth table, $t_n$ refers to the input conditions prior to a timing pulse,

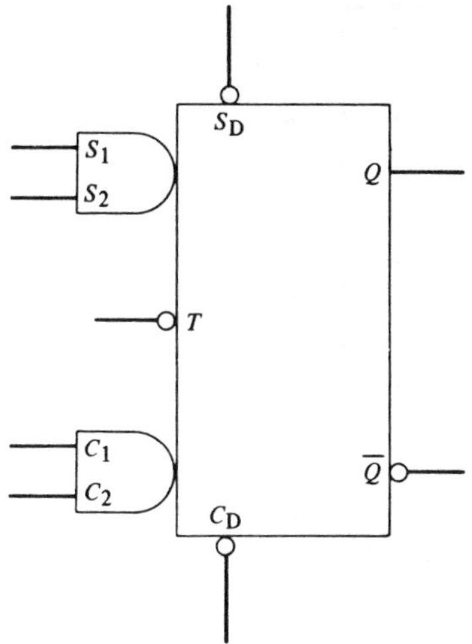


**Fig. 8.11.** The clocked flip-flop.

and $t_{n+1}$ refers to the time after a timing pulse has been applied to the $T$ input. Notice, however, that there is still an undefined state. It occurs when all gated synchronous inputs are 1. As a result, the clocked flip-flop still cannot be used for counting in this mode. If, however, $S_1$ is connected to $\overline{Q}$, $C_2$ is connected to $Q$, and $S_2$ and $C_1$ are connected to a binary 1, as is shown in Fig. 8.12*a*, a different condition exists. Because either $Q$ or $\overline{Q}$ must always equal 0 and since $Q$ is connected to $C_2$ and $\overline{Q}$ is connected to $S_1$, a condition where all gated inputs equal 1 can never be generated. As a result, there is never an undefined output state. If a pulse train is applied to the $T$ input, the flip-flop changes state on each negative-going pulse. This is called *JK operation*, and a truth table for it is presented in Table 8.VI. Because the $S_1$ and $C_2$ inputs are connected to the

TABLE 8.V
Clocked Flip-Flop Truth Table

| | $t_n$ | | | | $t_{n+1}$ |
|---|---|---|---|---|---|
| Case | $S_1$ | $S_2$ | $C_1$ | $C_2$ | $Q$ |
| 1 | 0 | X | 0 | X | NC |
| 2 | X | 0 | X | 0 | NC |
| 3 | X | 0 | 0 | X | NC |
| 4 | 0 | X | X | 0 | NC |
| 5 | 0 | X | 1 | 1 | 0 |
| 6 | X | 0 | 1 | 1 | 0 |
| 7 | 1 | 1 | 0 | X | 1 |
| 8 | 1 | 1 | X | 0 | 1 |
| 9 | 1 | 1 | 1 | 1 | U |

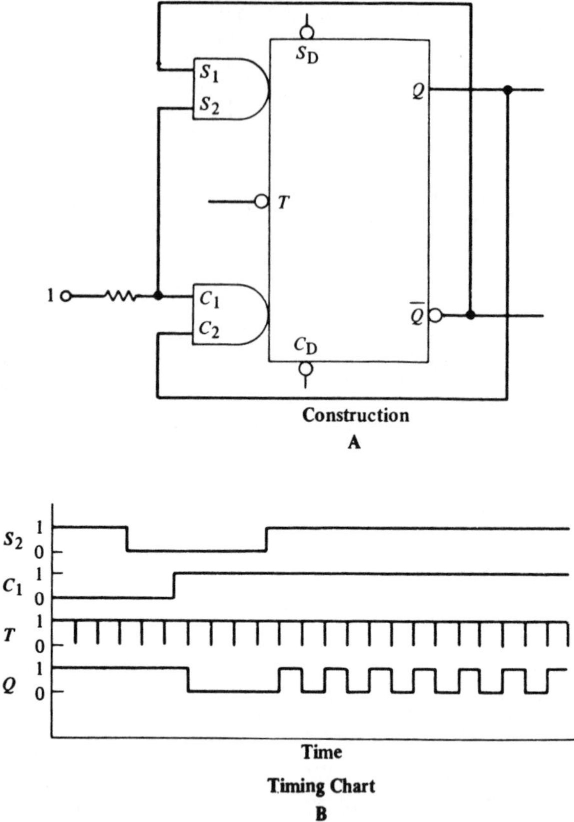


**Fig. 8.12.** *JK* operation of a clocked flip-flop.

outputs, they are not listed in the table. The timing chart is shown in Fig. 8.12*b*. Notice that when $S_2$ and $C_1$ are 1, the $Q$ output changes from 0 to 1 on every other input pulse, dividing the frequency in half. This flip-flop serves for counting functions.

TABLE 8.VI
*JK* Flip-Flop Truth Table

| $t_n$ | | $t_{n+1}$ |
|---|---|---|
| $S_2$ | $C_1$ | $Q$ |
| 0 | 0 | NC |
| 1 | 0 | 1 |
| 0 | 1 | 0 |
| 1 | 1 | $\overline{Q}_n$ (complements) |

Normally, one would not have to connect a clocked flip-flop into the *JK* mode because many integrated-circuit *JK* flip-flops are available (e.g., Fig. 8.13). Notice that the gated *S* and *C* inputs are renamed *J* and *K*. The connections from *J* and *K* to $\overline{Q}$ and *Q* are made internally and usually do not appear on the diagram.

There are, in addition to the RS, clocked, and *JK* flip-flops, other popular configurations for many varied applications of input gating, including capacitor coupling for ac-only operation.

There is, in addition, another general class of flip-flops called *master-slave* types. They have been developed to help overcome the critical nature of the timing requirements for the flip-flops already discussed. When one actually uses clocked flip-flops such as the *JK* units described above, a particular problem often occurs. It is that the output changes that result from a given clock pulse and set of input parameters applied to the flip-flop can change the input information during the life of the clock pulse. This results because of the connections between the outputs and inputs necessary for *JK* operation (Fig. 8.12*a*). Very critical timing requirements are often needed to ensure that the flip-flop does not settle in the wrong state. This may involve very careful synchronization of clock pulses and input information, and selection of the clock-pulse duty cycle.

A master-slave flip-flop is actually two flip-flops in one, with a master flip-flop that feeds data to a slave flip-flop. The resulting flip-flop configuration can be any of the types already discussed, such as clocked or *JK*. A *JK* master-slave flip-flop and clock-input wave form are presented in Fig. 8.14. For the sake of illustration, the various internal gates and connections are presented. In actual practice, the symbols for master-slave flip-flops are not distinguished from those already presented. Almost all clocked-types of flip-flops available in integrated-circuit form are of the master-slave type. *All the counter and register circuits*

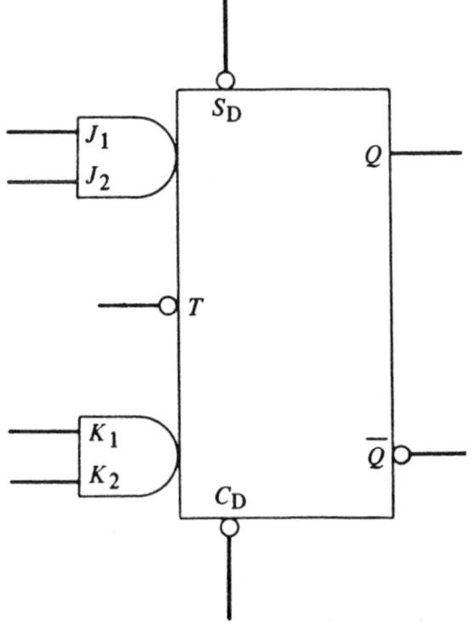


**Fig. 8.13.** The *JK* flip flop.

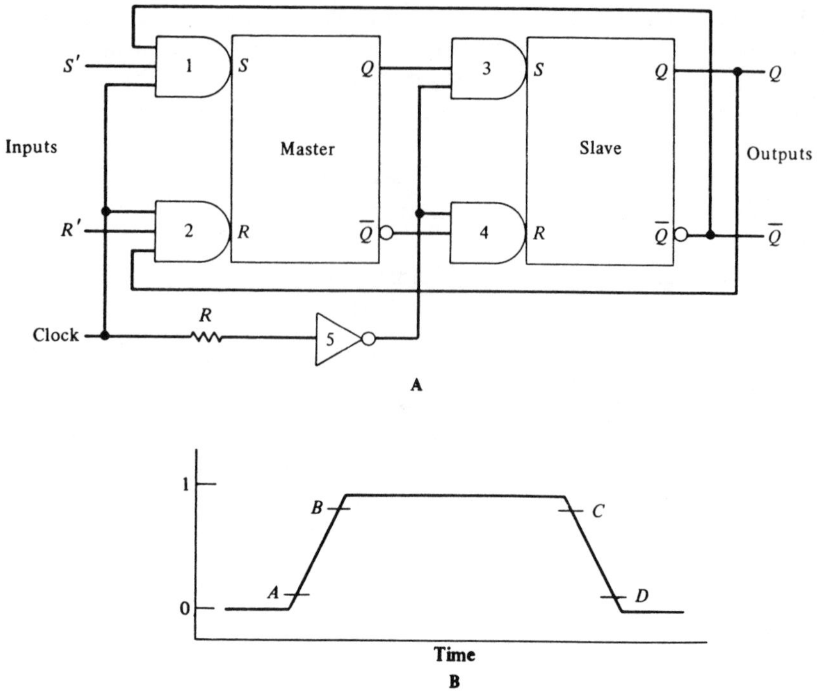


**Fig. 8.14.** The *JK* master-slave flip-flop.

*presented in the pages that follow were designed with master-slave flip-flops.*

Referring to Fig. 8.14*a*, notice that both the master and the slave are gated RS flip-flops. The clock-pulse wave form presented in Fig. 14*b* has four points on it labeled A, B, C, and D. The operation of the master-slave *JK* flip-flop can be explained as follows: as the clock pulse goes positive from 0 to 1 past point A, the slave input gates 3 and 4 become disabled, isolating the slave from the master. The state of the master prior to A is stored on the slave outputs. As the clock pulse passes point B, input gates 1 and 2 of the master are enabled, allowing data to be transferred in through the *S* and *R*inputs. As the clock passes C, gates 1 and 2 are disabled, isolating *S* and *R* from the master. As the clock pulse further falls past D, inverter 5 enables gates 3 and 4, allowing data transfer from the master to the slave, which does not change until the clock pulse is completed. The effects of changes in the slave outputs cannot reach the master inputs during a clock pulse. Master-slave flip-flop configurations thus exhibit much less critical timing requirements and much better immunity from noise.

## H. DATA-STORAGE LATCHES

One simple data-storage device is called a data *latch*. It is often used for the temporary storage of binary information. A data latch can be built with either

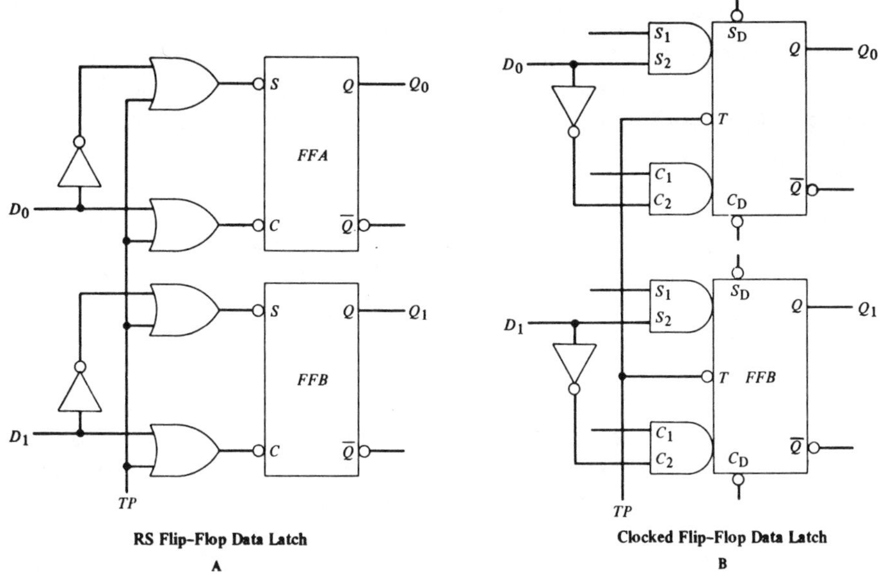


**Fig. 8.15** (*a*) RS flip-flop data latch. (*b*) Clocked flip-flop data latch.

RS or synchronous flip-flops. The one shown in Fig. 8.15*a* is constructed from RS flip-flops. When the timing-pulse (TP) input is high (a binary 1), the outputs of all the input OR gates are high. No information from the data inputs $D_0$ and $D_1$ can be transferred to the latch. However, when the TP input is a binary 0, the data inputs can transfer information to the flip-flops. For example, if a binary 1 is presented at $D_0$, the $S$ input of FFA is a binary 0, which sets $Q$ to a binary 1. Notice that the $C$ input is a binary 1. As a result, when TP is low, the $Q$ outputs follow the $D$ inputs. When TP is high, the $Q$ outputs do not change regardless of what conditions occur at the $D$ inputs. Data are thus stored in binary form until the TP input again goes momentarily to 0.

A data latch can also be built from synchronous master-slave flip-flops, as illustrated with the clocked flip-flops presented in Fig. 8.15*b*. The clocked flip-flop data latch operates in much the same way as the RS flip-flop data latch: the TP line must undergo a negative-going 1 to 0 transition each time data is transferred. However, the clocked flip-flop data latch operates only in a synchronous manner, that is, each time a negative-going clock pulse is present. This is in contrast to the RS flip-flop data latch where the outputs follow the inputs whenever the TP input is 0. A data-storage latch can be constructed from *JK* flip-flops. Latches can, of course, be constructed for any number of data bits. Integrated-circuit data latches are commonly available in 4 bit, 8 bit, and larger configurations.

## I. SHIFT REGISTERS

Shift registers are, like data latches, often used for data storage. They also have other features that make them more versatile. They can acquire and output data, in both serial and parallel modes. Also, they can be used to move data within the register while it is being stored.

The simplest shift register is the *serial* I/O type, shown in Fig. 8.16 (omitting the dashed line outputs). The 4 bit register presented is constructed from master-slave clocked flip-flops. It can, however, also be constructed from *JK* flip-flops in exactly the same manner. *It cannot be constructed from RS flip-flops*. Notice that a shorthand notation, with the AND gate symbols omitted, is used for the clocked flip-flops. The operation of the register is quite simple. The $C_D$ inputs are all tied together to provide a common "clear" line. This allows all the flip-flops to have their $Q$ outputs set to 0 simultaneously. The first stage inputs to FFA are connected in exactly the same manner as for the synchronous data latch previously presented. Successive stages have their outputs and inputs connected together to allow data to be transferred from one to the other in a serial manner. If the register is cleared, a binary 1 is presented at the data input, and a clock-pulse transition from 1 to 0 is provided on the TP input line; the binary 1 appears on the output of FFA. Notice that a binary 1 is now applied to the $S$ input of FFB. The next clock pulse transfers the binary 1 to the output of FFB, presenting it at the input of FFC. Thus data present at the data input appear at the output of FFD four clock pulses later. An $n$ clock-pulse data delay can be generated by using $n$ flip-flops.

In some applications, digital data might be received in serial but is needed in parallel for the computer or data-acquisition system. A serial-to-parallel converter can be built by providing enough clock pulses and flip-flops to fill the serial I/O shift register with data. Then the outputs of each flip-flop in the register can be read simultaneously. This is a *serial-input parallel-output* shift register and is the configuration in Fig. 8.16 with the dashed (parallel) outputs, $Q_0$ through $Q_3$.

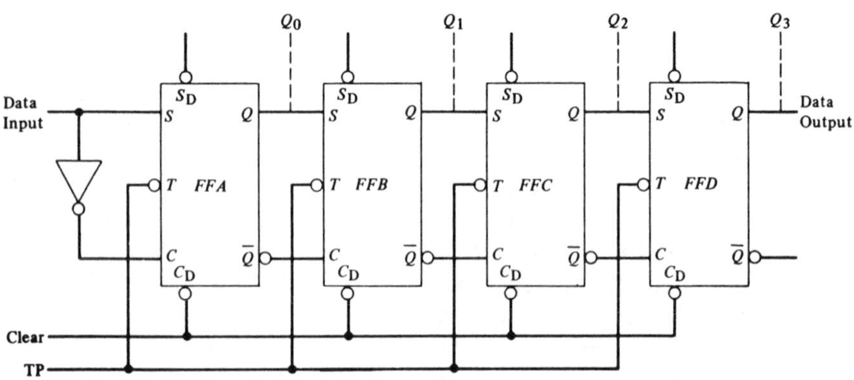


**Fig. 8.16.** Four bit shift register.

## J. ASYNCHRONOUS COUNTERS

Often one must provide counters in an interface to count events, such as the number of data points taken, the number of times data exceed a predetermined threshold, to divide down the clock frequencies, and so forth. The flip-flop used in modern integrated-circuit counters is the master-slave *JK* flip-flop. It can be used to construct two basic types of counters, asynchronous and synchronous, that count up or down in a variety of counting schemes.

The simplest counter is the *asynchronous* binary up-counter. The one presented in Fig. 8.17, (with its timing chart) counts from 0 to 15 and resets. Initially, all flip-flops in the counter are cleared to 0. When the first negative-going transition or clock pulse is presented at the count input, the output of FFA ($Q_0$) changes state from 0 to 1. When the second clock pulse occurs, it again changes the state of FFA, returning it from 1 to 0. This transition causes

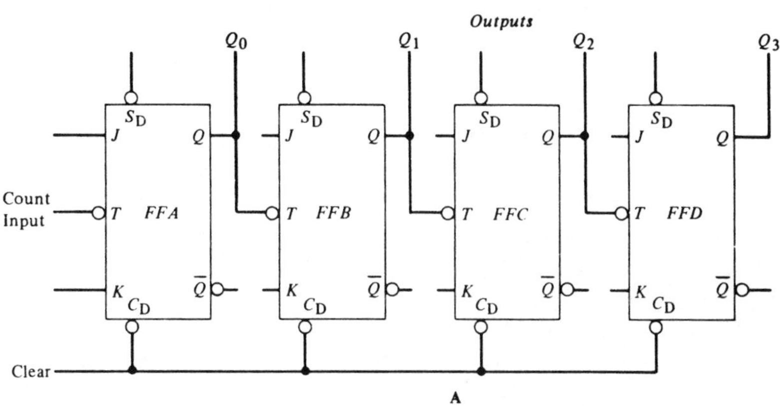


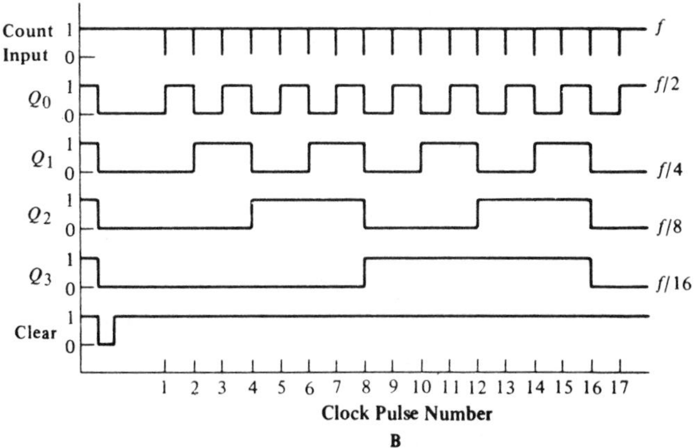


**Fig. 8.17.** Asynchronous binary counter.

FFB ($Q_1$) to change state from a binary 0 to a binary 1. When FFB returns to a binary 0, it causes FFC ($Q_2$) to change state. In the same manner, FFC causes FFD to change state. Any number of flip-flops can be connected to count to larger numbers. The asynchronous counter is often called a "ripple counter" because of the way counts ripple through or are passed along from flip-flop to flip-flop. From the timing chart, we can see that binary frequency division occurs with $Q_0$ = f/2, $Q_1$ = f/4, $Q_2$ = f/16. This is then a convenient way to divide down clock frequencies in binary orders of magnitude. Notice also that we can read the binary number output of the counter from the timing chart after any given number of clock pulses has occurred. Consider the case after 10 clock pulses. Reading down from the top of the chart at clock pulse number 10, $Q_0 = 0$, $Q_1 = 1$, $Q_2 = 0$, and $Q_3 = 1$. The binary number for 10 is thus 1010 since $Q_0$ is the least significant bit. Notice that $1010_2 = 10_{10}$.

In addition to an asynchronous binary up-counter, an asynchronous binary down-counter can also be designed. For a 4 bit counter, it counts down from 15 to 0. This is accomplished by disconnecting each T input in the up-counter form the corresponding $Q$ output and connecting it to the corresponding $\overline{Q}$ output.

Because we are accustomed to thinking in terms of the decade number system, decade counters are often used in interface systems. They not only count in powers of 10 but can be used to divide down clock frequencies in decade rather than binary steps. The *decade counter* presented in Fig. 8.18 along with its timing chart, follows the count sequence presented in Table 8.VII. The BCD (binary-coded decimal) number system is a binary representation of the decimal number system.

Notice that the timing chart is the same as the binary up-counter timing chart in Fig. 8.17*b* through clock pulse number 8. From there on, however, it differs. The chief requirement for the operation of the decade counter is that it must reset to 0 after a count of $9_{10}$ or $1001_2$. Notice in Fig. 8.18 that the $\overline{Q}$ output of FFD is connected to the $J$ input of FFB. This allows FFB to change state only

TABLE 8.VII
Binary-coded Decimal (BCD) Number System

| Decade | *BCD* |
|---|---|
| 0 | 0000 |
| 1 | 0001 |
| 2 | 0010 |
| 3 | 0011 |
| 4 | 0100 |
| 5 | 0101 |
| 6 | 0110 |
| 7 | 0111 |
| 8 | 1000 |
| 9 | 1001 |

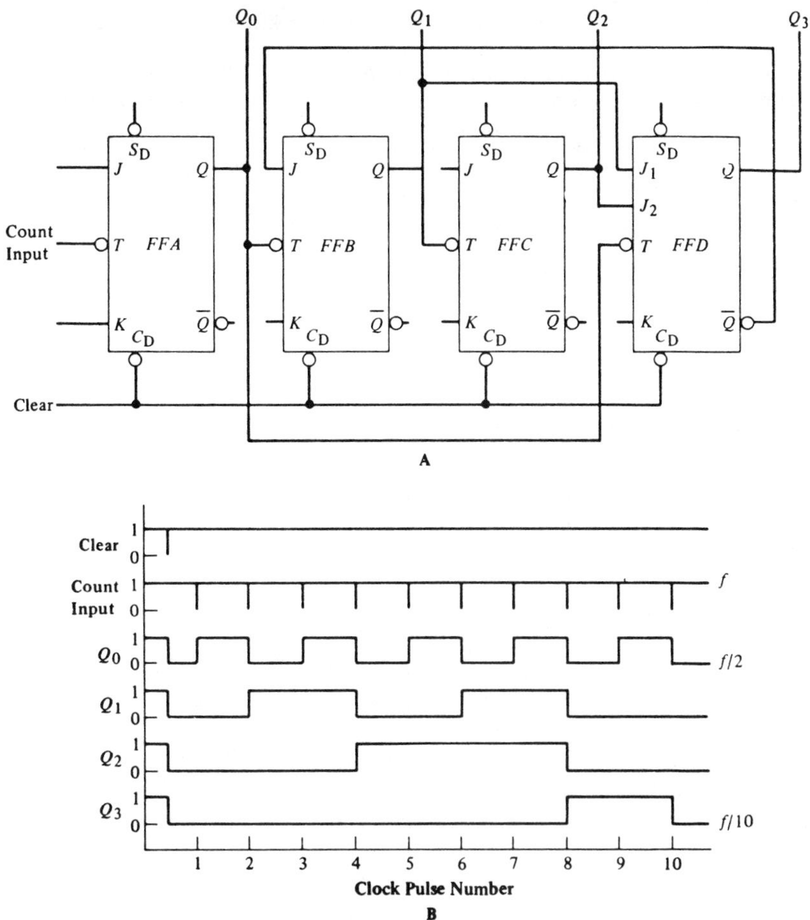


**Fig. 8.18.** Asynchronous decade counter.

when the $\overline{Q}$ output of FFD = 1. This occurs for counts 0 to 7. However, for counts 8, 9, and above, FFB is disabled. Notice also that FFD has two $J$ inputs, $J_1$ and $J_2$. They allow FFD to change state from 0 to 1 ($Q$ output) only when $Q_1 = 1$, $Q_2 = 1$, and $Q_0$ goes from 1 to 0. These conditions are only present at a count of 7 and allow the count of 8 to occur. The count transition from 9 to 0 occurring on clock pulse number 10 in Fig. 8.18*b* occurs in the following way: for a count of 9, FFB is disabled because of the connection from $\overline{Q}$ of FFD to its $J$ input. It cannot change from $Q_1 = 0$ state that it is in. Inputs $J_1$ and $J_2$ of FFD are also 0. Any 1-to-0 timing signal thus sets $Q_3$ back to 0. This can be verified by reviewing the *JK* flip-flop truth table presented in Table 8.VI. Clock pulse 10 toggles $Q_0$ of FFA from 1 to 0. $Q_0$ is connected to the T input of FFD, causing its $Q$ output also to go from 1 to 0. The counter thus resets to 0 on clock pulse 10.

## K. SYNCHRONOUS COUNTERS

Unlike asynchronous counters, where the output change of one flip-flop is applied to the clock input and thus changes the state of a succeeding flip-flop, *synchronous*-counter flip-flop outputs set up the $J$ and $K$ inputs of succeeding flip-flops so that a common clock signal can cause the proper count sequence to occur.

A synchronous binary up-counter is illustrated in Fig. 8.19*a*. Compare this

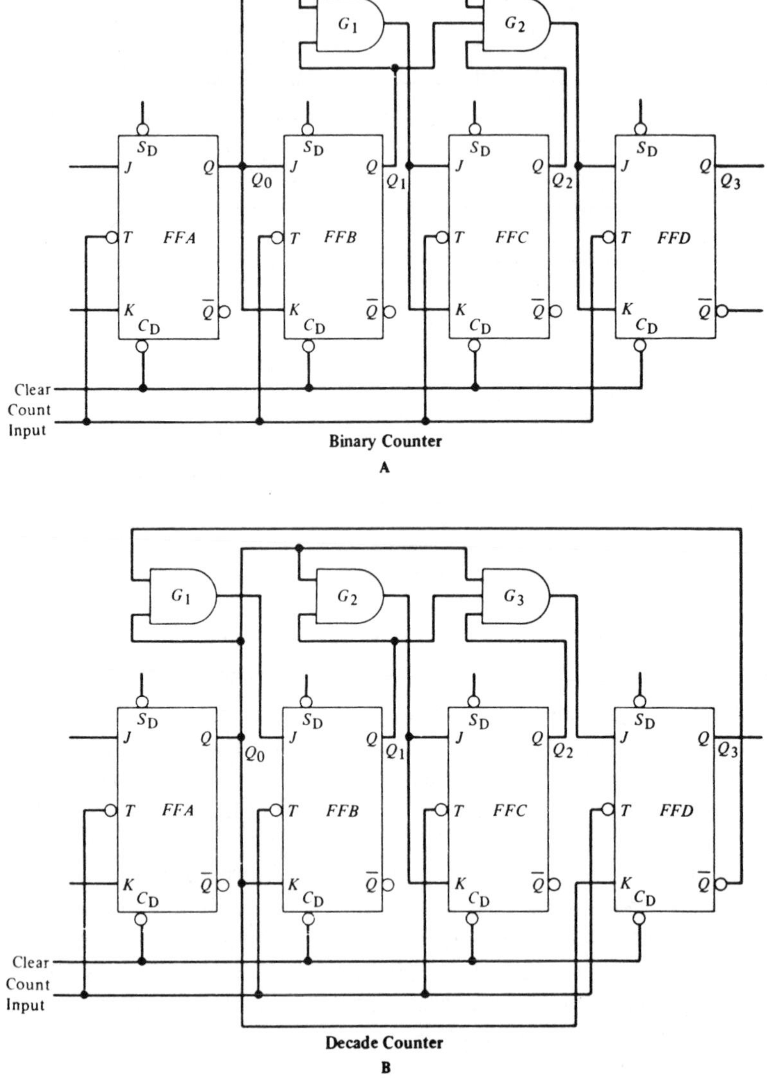


**Fig. 8.19.** Synchronous binary and decade counters.

counter with the asynchronous binary up-counter in Fig. 8.17. Notice that the synchronous counter requires external gating whereas the asynchronous counter does not. This is because the count sequence is generated by the external gates which set up the $J$ and $K$ inputs of each flip-flop. The timing chart in Fig. 8.17*b* for the asynchronous binary counter can also be used for the synchronous counter.

A synchronous *decade* up-counter is shown in Fig. 8.19*b*; it should be compared to the asynchronous decade counter in Fig. 8.18. The timing chart in Fig. 8.18*b* applies equally well to both counters. Notice again that the synchronous counter involves more complex external gating. The evaluation of the operation of the synchronous decade up-counter is the same as for the binary synchronous counter if one remembers that a flip-flop changes state only when it receives a clock pulse and when either or both of the $J$ and $K$ inputs are a logical 1.

The question arises, why one would use the more complex synchronous counters rather than the simple asynchronous counters? One important reason is that in asynchronous counters the various count sequences must ripple from one flip-flop to the next. This means that an incoming count on the first flip-flop that will eventually change the state of a flip-flop farther down the line must pass through all intermediate flip-flops. This operation takes a period of time equal to the sum of the signal *propagation delay times* of each succeeding flip-flop. The maximum count rate possible with asynchronous counters is determined by the total propagation delay of all the flip-flops in sequence. However, in synchronous counting systems, the count sequences do not have to ripple down all the flip-flops. The propagation delay for a given synchronous counter is usually no more than that for one flip-flop and one gate. The result is that the synchronous counters can operate at much higher speeds. However, as the number of stages increases in synchronous counters, so does the number of inputs to the gates used between counter stages. In fact, if one looks at a synchronous binary up-counter of 15 stages, one sees that an input gate with 14 inputs is required at the last stage. One way around this problem is to use semisynchronous operation. In this mode of operation, flip-flops are run synchronously up to perhaps eight or nine stages, and then these units are connected together asynchronously. When used in decade counters, each decade may be made up of four individual stages, all of which are synchronous internally, but each external stage or decade is connected to another decade asynchronously. This decreases the cost and complexity, while still allowing high-speed operation. The operation is not, however, at as high a speed as is a totally synchronous system.

## III. INTERFACING DEVICES

A computer is considered on-line to an experimental system when a direct electronic communication link exists between the experiment and the computer

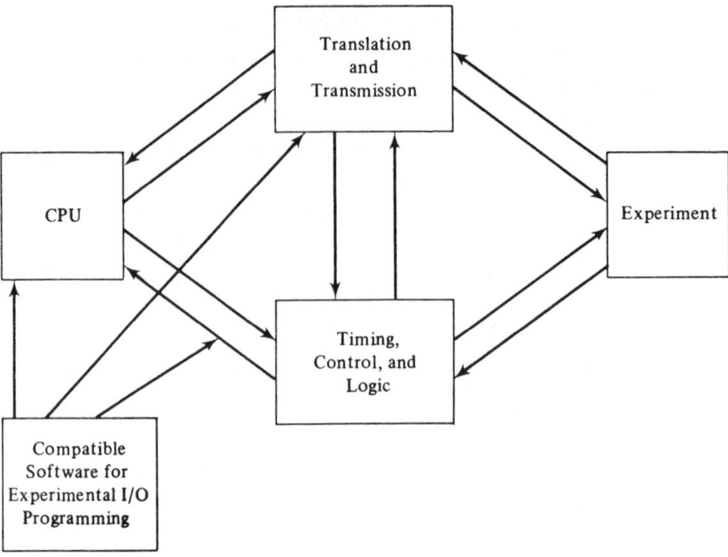


**Fig. 8.20.** On-line computer communication functions.

and, perhaps, between the computer and the experiment. Fig. 8.20 illustrates the various functions required for such communication. These functions are carried out by the following:

1. *Translation and transmission elements*. These include analog and digital hardware that allow the appropriate conversion or handling of electronic information for communication between experiment and computer. Typical elements include analog-to-digital converters, digital-to-analog converters, voltage amplifiers, current-to-voltage converters, signal conditioners, sample-and-hold amplifiers, and multiplexers.
2. *Timing, control, and logic elements*. These include such hardware as a digital clock, logic gates, flip-flops, counters, one-shots, Schmitt triggers, analog switches, and level converters.
3. *Appropriate software to drive the interface hardware*. The I/O programs are necessary and important parts of the interface. Software often provides the necessary controls needed for the operation of the electronic elements in the interface.

## A. TIMING DEVICES

Figure 8.21 illustrates the most important experimental interface function, the generation of a stable *time base*. The *time-base* generator, or *digital clock*, is generally a combination of a crystal-controlled, stable, fixed-frequency *oscillator* that emits a pulse train of very accurately known frequency, plus a *counter*

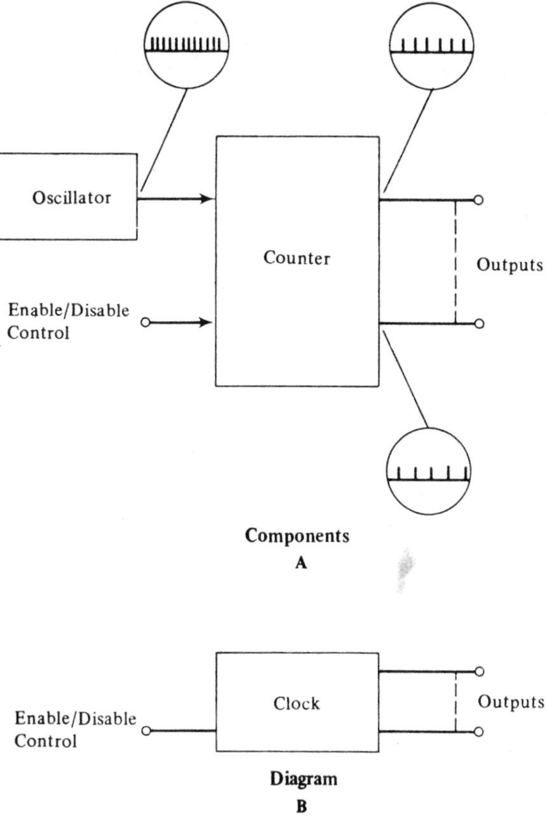


**Fig. 8.21.** Interface time-base generator.

or scaler logic section used to divide this frequency into a variety of output frequencies. The output frequency can be incrementally variable through the use of counters and can be controlled by enabling (starting) or disabling (stopping) the counter system. In subsequent discussions, we use the block diagram in Fig. 8.21*b* to represent the clock, which shows only an enable/disable input and a series of outputs.

The generation of accurately known and reproducible adjustable time delays is often required in an interface system. The *monostable multivibrator* or *one-shot* is often used for generating this type of delay function. The diagram shown in Fig. 8.22 is for a gated one-shot in that there is a series of gates that allow logic to be performed at the input of the device. The one-shot has the characteristic of remaining in a stable state until a trigger pulse is applied to the input; at this time the output changes state for a period of time called the *pulse width* (PW), which depends on the value of the resistance-capacitance ($RC$) timing network employed in the circuit. Generally, the one-shot device has a provision

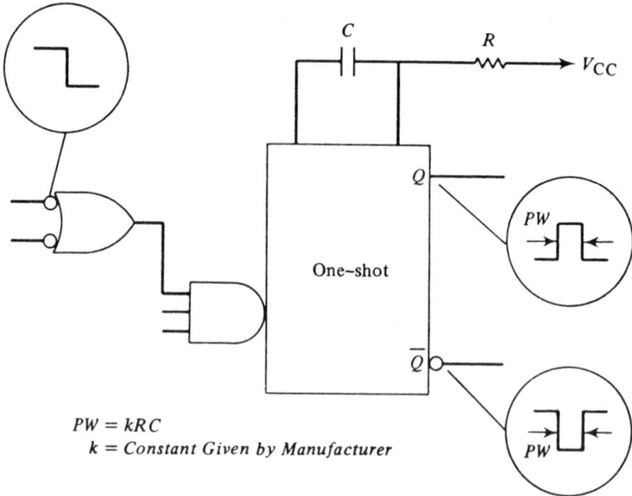


**Fig. 8.22.** Monostable multivibrator or one shot.

for attaching different resistances and capacitances. Delay times can generally be varied over wide ranges, typically from nanoseconds to many seconds or even minutes. Many integrated-circuit one-shots are commercially available.

## B. TRANSLATION ELEMENTS

### 1. Digital-To-Analog Conversion (DAC)

A *digital-to-analog converter* (DAC) is used to change digital numerical information into a continuously variable analog output. DACs are often used as control devices in chemical experiments. For example, in a typical fast-sweep polarographic experiment, a DAC can be used to provide the electrochemical-cell control-voltages and ramp functions. Because the computer can generate numbers in any sequence, nonlinear ramps can be generated, which provide a wide variety of versatile control signals.

A basic DAC application is illustrated in Fig. 8.23. The DAC takes a digital input and converts it to an analog output which consists of a series of voltage steps. The minimum step magnitude is a function of the dynamic range and resolution of the converter. For example, if the converter has a 10 V output maximum and has a resolution of 1 part in 1024, or 10 bits, the minimum voltage step on the output is about 10 mV. Notice in Fig. 8.23 that there are digital inputs and an analog output, an *enable* input at which a conversion can be started, and an *end-of-conversion* output that indicates when a conversion is complete. The actual conversion from digital numbers to an analog output is accomplished within the converter by a series of resistors and switches called a *ladder network*; the operation is discussed in detail elsewhere (6).

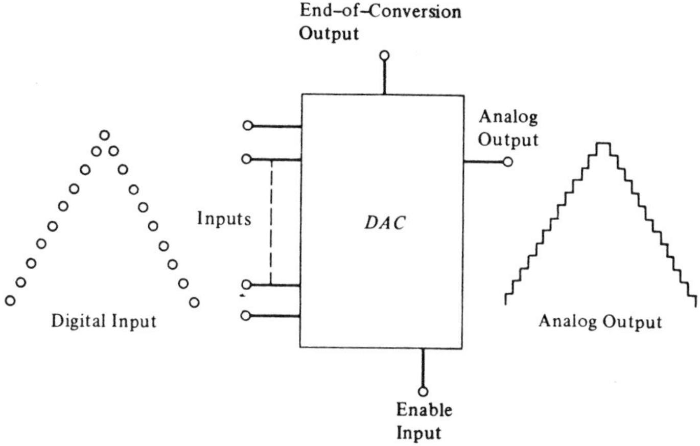


**Fig. 8.23.** Digital-to-analog conversion.

## 2. Voltage Comparators and Schmitt Triggers

The voltage *comparator* is an analog device with two inputs and an output. It is generally used to compare one voltage with another and to indicate which is larger. A voltage comparator and its resulting wave forms are shown in Fig. 8.24. Notice that there are both a reference voltage and an input signal. The comparator compares the input signal with the reference voltage. When the input signal rises above the magnitude of the reference voltage, the comparator output changes state from 1 to 0. When the input signal falls below the magnitude of the reference voltage, the output changes back from 0 to 1.

A device somewhat similar to a comparator is the *Schmitt trigger* shown in Fig. 8.25. In it, there are input, output, and upper and lower threshold terminals. Its typical wave form is also shown in Fig. 8.25. The two reference levels $V_{t+}$ and $V_{t-}$ are the upper and lower thresholds. When the input signal goes above the upper threshold, the output changes from 1 to 0. However, when it goes below the upper threshold, the output signal does not change as long as it is above the lower threshold. When it goes below the lower threshold, the output changes. In a like manner, the increasing signal does not change the output when it only goes above the lower threshold. The difference between the upper threshold and lower threshold is commonly called the *hysteresis* or *backlash level* of the Schmitt trigger. It is used to provide noise immunity for the device.

In Fig. 8.26, both Schmitt trigger and comparator outputs are given where a noisy input signal exists. Notice in the case of the comparator that as the signal oscillates above and below the comparator reference voltage the output also oscillates. Notice in the case of the Schmitt trigger that the magnitude of the noise is less than the magnitude of the hysteresis; as a result, the output of the Schmitt trigger does not have noise spikes whereas that of the comparator does. The trigger can be used to extract control signals from noisy input signals and

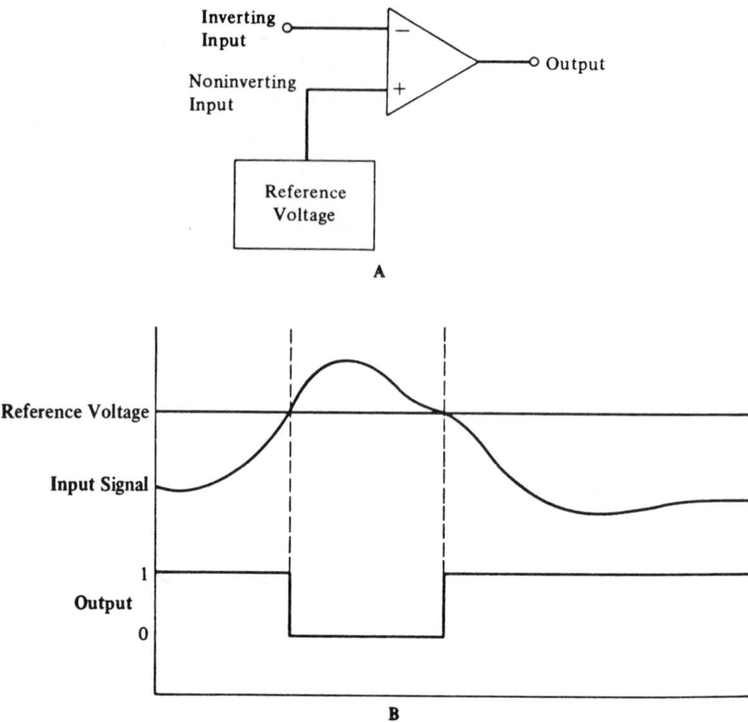


**Fig. 8.24.** Voltage comparator.

eliminate the noise by proper selection of the hysteresis level. It is often used as a level conversion device to take digital signals of one level and convert them to that of another level.

The lack of hysteresis in the comparator makes it more susceptible to noisy environments. It also makes it a more accurate switch.

### 3. Analog Multiplexers

When one has several signals that must be connected to the input of one ADC, a common device used is the *analog-signal multiplexer* (Fig. 8.27). With it, several input signals can be sequentially switched to one output. Often the input signals come from the output of sample-and-hold or track-and-hold amplifiers. Analog-signal multiplexers are a series of electronic switches with digital control inputs to allow a computer or interface to control this operation.

### 4. Analog Switches

Analog switches can generally be divided into three categories: mechanical, electromechanical, and electronic. Common toggle, slide, and rotary switches are examples of mechanical types; their contacts are opened or closed manually.

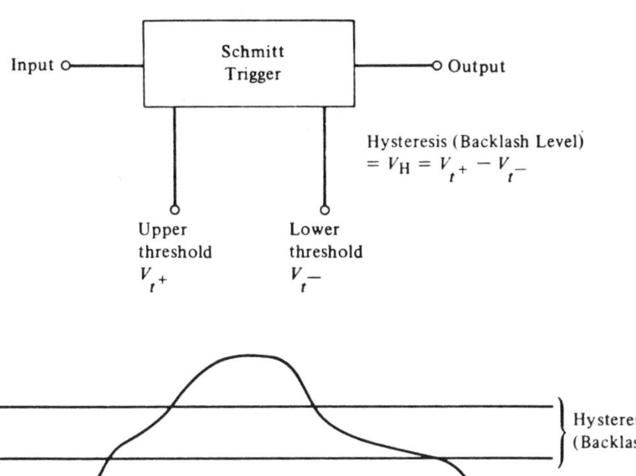


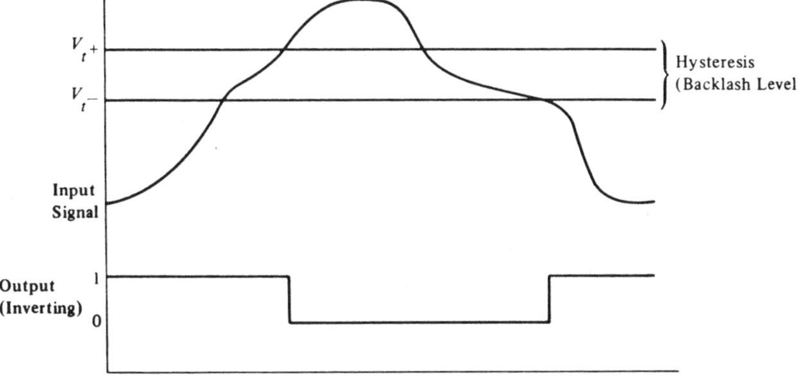


Fig. 8.25. Schmitt trigger.

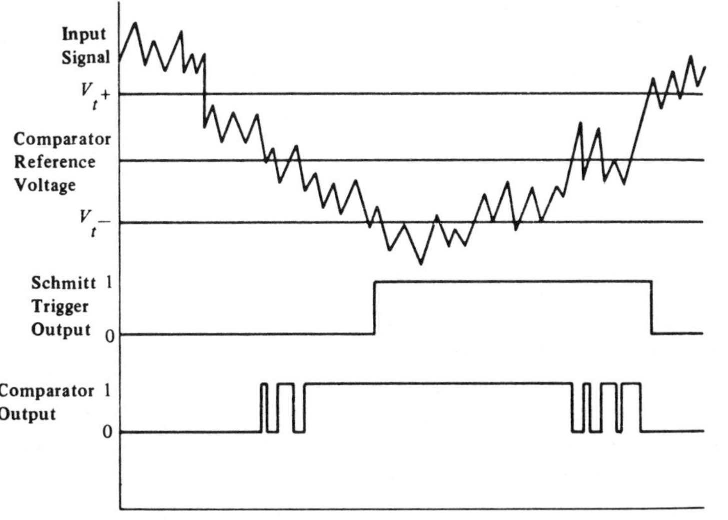


**Fig. 8.26.** Comparison of Schmitt trigger and voltage comparator for noisy signals.

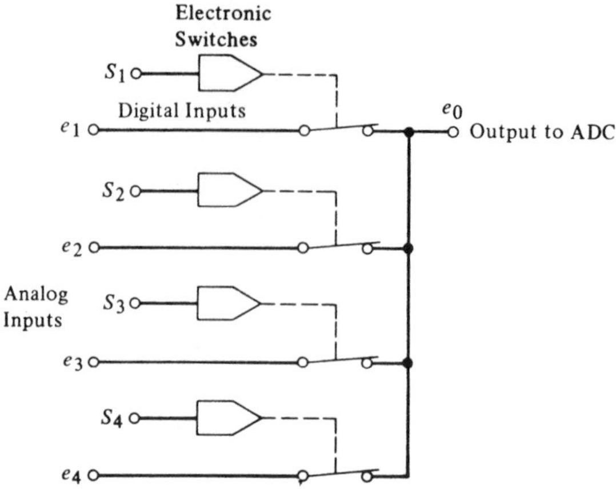


**Fig. 8.27.** Analog multiplexer.

Electromechanical switches are usually relays of one sort or another in which an electromagnetic coil is energized to open or close the contacts. The contacts may be either dry or coated with mercury. Mercury-wetted contacts usually exhibit somewhat better switching characteristics.

Electromechanical analog switches exhibit some very desirable characteristics. They have essentially zero resistance when the contacts are closed and infinite resistance when they are open. They can handle a range of many orders of magnitude of voltage and current of either polarity. They also have, as might be expected. Some undesirable characteristics. First, they have contact bounce whenever the contacts are opened or closed. This may result in significant signal noise that must be filtered out. Mercury-wetted contacts often help to lessen bounce noise, but do not eliminate it. Second, electromechanical switches are rather slow in switching; the faster ones take 1 msec or so to open or close.

Electronic analog switches are usually constructed from *junction field-effect transistors* (JFETs) or *metal oxide-silicon field-effect transistors* (MOSFETs). FETs used as switches are usually driven by transistor driving circuits that open or close them. A typical FET analog switch might be closed when a logical 0 from integrated-circuit logic is applied. FET switches have advantages over electromechanical analog switches in one principal area—speed. They have orders-of-magnitude greater switching speeds, with turn-on or turn-off times of less than 1 usec. Because they have no contacts, the do not exhibit contact bounce noise. They may, however, if not designed into a circuit correctly, generate electronic switching spikes.

Electronic analog switches also have some shortcomings. They have finite on and off resistances. JFETs usually have the smaller on or closed resistance, some types as low as 1 Ω or so. MOSFETs, on the other hand, have the higher off or open resistance, typically of the order of $10^{16}$ Ω. In other words, FET switches are not ideal switches, but rather can be considered as electronically variable resistors.

FET switches are restricted to certain ranges of voltage and current, and have polarity restrictions. Common voltage levels are of the order of 10 V or so, but some specialized units go as high as 100 V. Maximum currents are often no greater than 100 mA but may be greater. FET switches usually do not work well for signal levels less than a few millivolts because of a small inherent voltage drop, in the transistors, that may vary with operating conditions.

Generally speaking, one should use electromechanical analog switches for low-level or high-level signals that need not be switched in times less than 1 msec or so and electronic analog switches for moderate signal levels with switching times less than 1 usec or so. Both electromechanical and electronic types are available in configurations that can be driven directly from integrated-circuit digital logic.

## 5. Analog-To-Digital Conversion (ADC)

A typical *analog-to-digital converter* (ADC) is illustrated in Fig. 8.28. It consists of an analog input, digital outputs, a start conversion input, and an end-of-conversion output. The ADC changes an analog or continuous voltage from an experimental system into a series of discrete digital values so that a computer can be presented with digital data in a format that it can handle. The most common output format is a binary digital representation of the analog input. There are many types of ADCs, some fast and some slow, some that require high-level voltage inputs in the range of from one to several volts, and some that require low-level voltage, current, or resistance inputs. In this section, we discuss several types of fast converters.

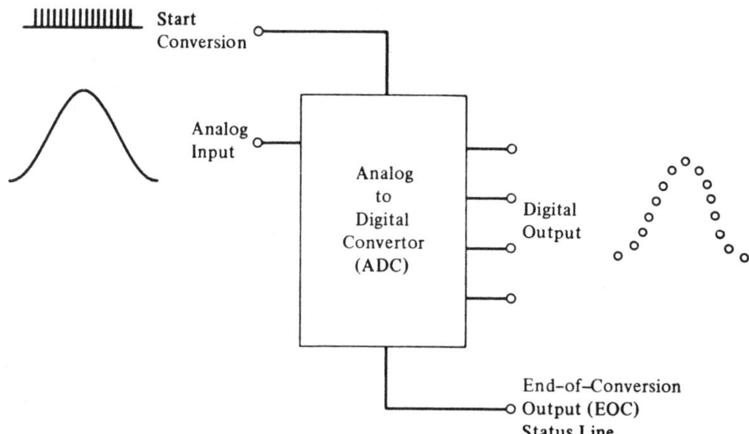


**Fig. 8.28.** Analog-to-digital conversion.

### a. Counter Converters

Probably the simplest ADC is the *counter converter* (Fig. 8.29*a*). It is usually a fast high-level converter having conversion times of less than 1 msec. This makes it capable of providing more than 1000 data points/sec. Basically, it consists of a comparator, a clock or pulse generator, a counter, and a DAC. When an analog input signal is presented to one input of the comparator and a start-of-conversion signal is presented to the counter, the counter resets to 0 and starts counting up, and presents a digital input to the DAC. The DAC in turn provides a corresponding analog output voltage to the other input of the comparator. When the counter output number reaches a magnitude that provides a voltage to the comparator (through the DAC) equal in magnitude to the analog input signal, the comparator changes state, turning off the clock and stopping the counter. The digital output representation of the analog input is read in parallel from the counter outputs. A status or end-of-conversion signal can be obtained from the output of the comparator as it changes state upon completion of a conversion. The counter converter is simple and inexpensive. The conversion time is proportional to the magnitude of the input voltage. That is, since the counter always starts counting from 0, the larger the analog input voltage, the longer it takes for the counter to count up to the value where the DAC output applied to the comparator is equal to the analog input voltage.

### b. Continuous Converters

The *continuous converter* in Fig. 8.29*b* is very similar to the counter converter in Fig. 8.29*a* with the exception that the counter used can count both up and down; the comparator, instead of turning off the clock, controls the counting direction of the counter; and the counter counts all the time. When the analog input signal voltage has been exceeded by the output of the DAC, the comparator changes the direction of the counter from up to down. When the DAC output falls below the analog input, the comparator changes the direction of the counter to count up. The up-down counter tracks the analog input voltage if it is not changing faster than the counter can follow. When a new signal is applied to the analog input, the continuous converter locks onto it and follows it. Digital outputs can be read at intervals from the output of the counter. After it locks on to a signal, the continuous converter is extremely fast.

### c. Successive-Approximation Converters

The 4-bit *successive-approximation* ADC shown in Fig. 8.30 differs from both the counter and the continuous converters in that, in place of a counter, it has a pattern generator. It consists of a comparator, a DAC, a buffer-register data latch in which the digital output is stored, the pattern generator, and some control logic. Its operation is most easily understood with reference to Fig. 8.31. All the possible number combinations generated by the pattern generator are presented from a 4 bit successive-approximation ADC. Notice that for the 4 bit conversion, it always takes four steps to completion. Unlike the counter con-

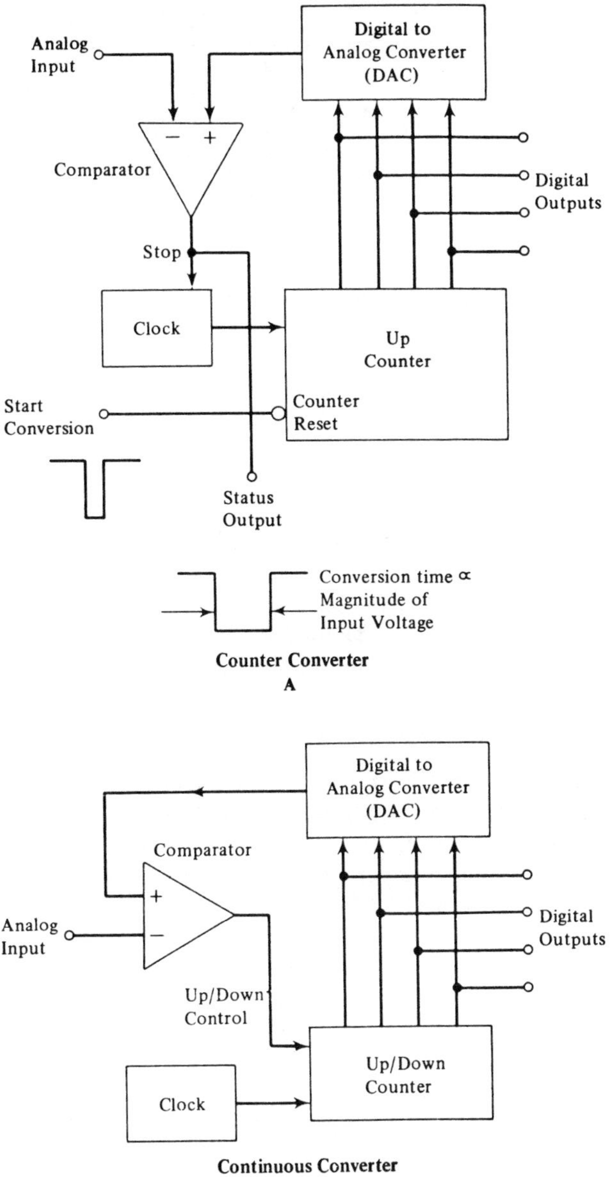


**Fig. 8.29.** Counter and continuous analog-to-digital converters.

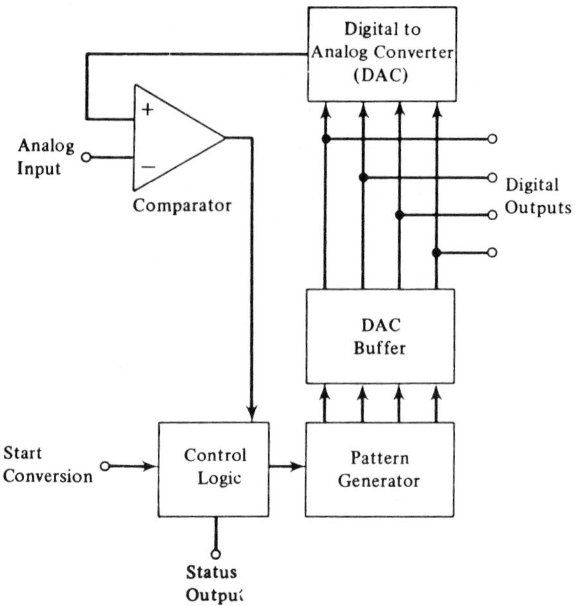


**Fig. 8.30.** Successive approximation analog-to-digital converter.

verter, where the conversion time is a function of the magnitude of the number, a successive-approximation converter always has a fixed conversion time.

## 6. Important Features of ADCs

IN selecting high-level ADCs, several criteria are important. They include the input voltage range, format of the output, resolution of the output, logic voltage levels of the output, conversion speed, control signals, and power-supply requirements. Input voltages generally range from 1 to 10 V full scale. They may be positive, negative, or bipolar.

The most commonly used ADC output format is the binary number system because of its direct compatibility with many digital computers. Other codes, usually binary-coded decimal (BCD), are sometimes used. In addition, bipolar input converters such as the ±1 or ±10 V units can have different forms of binary coding to account for the dual polarity.

The resolution and dynamic range of the output of the converter are determined by the number of data bits available. Common binary converters have 8, 10, 12, or more output bit configurations, giving resolutions of 1 part in 256, 1024, 4096, or greater, respectively. Converters using BCD output configurations often have three or four decimal digits represented, giving resolutions in the range of $10^3$ to $10^4$. Higher-resolution converters are available for applications requiring more resolution.

By far the most common output logic voltage levels are 0 and +5 V [the

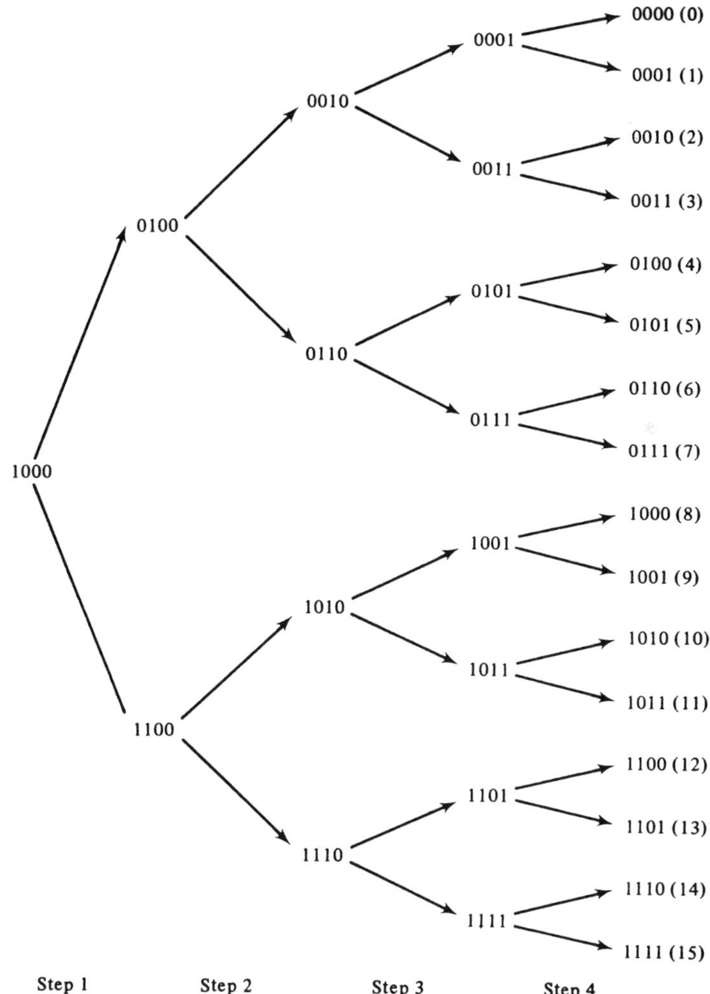


**Fig. 8.31.** Successive approximation conversion pattern.

common integrated-circuit levels for diode-transistor (DTL) and transistor-transistor (TTL) logic [4]. There are, however, different logic voltage levels used, including 0 and +12 V, 0 and −3 V, and others. The logic voltage levels should, of course, be compatible with those used on the computer or data system. If they are not, some sort of voltage-level conversion is necessary.

*Conversion time* is the time required, after a start signal, for a conversion to be completed. It determines the maximum data rate of a given converter. High converter data rates range from 10 kHz to about 10 MHz.

Control signals for fast ADC units usually consist of a start command and an end-of-conversion or status signal. Other control signals are possible. In units with output storage-buffers, controls may be available to load or store in the buffer.

Power-supply requirements, though they may seem trivial, are in fact very important. The presence or absence of an internal reference supply is a good example of this. The magnitude and stability of the general power supply are also important. ADC units with built-in voltage regulators have much less severe power-supply requirements than those with no internal regulation.

## C. ANALOG SAMPLING

A useful interface device is the *track-and-hold* amplifier. Although in practice one would generally buy a ready-built track-and-hold system, it is useful here to review its operation. A track-and-hold amplifier is shown in Fig. 8.32, along with its response curves and timing chart. An incoming signal is fed to the

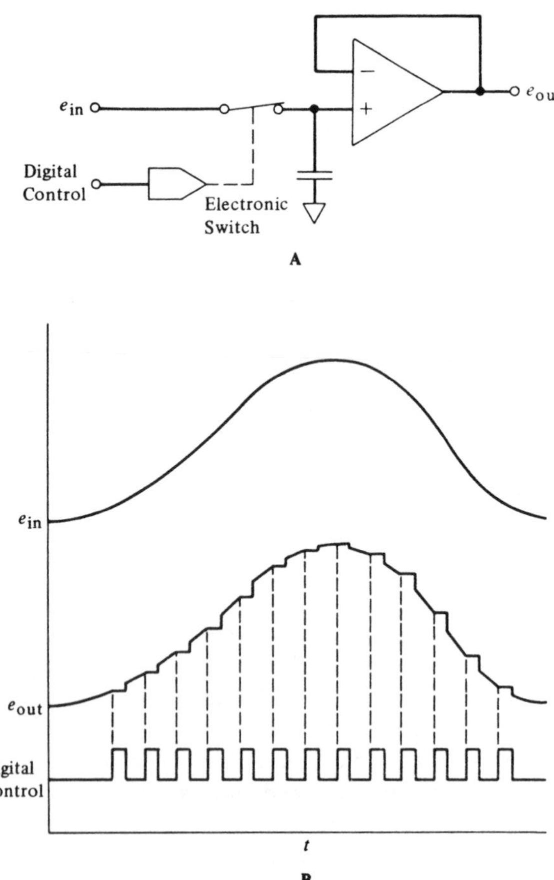


**Fig. 8.32.** Track-and-hold amplifier.

capacitor when the electronic switch is closed. In this mode, the output remains at the voltage last seen by the input capacitor. This is illustrated in Fig. 8.32*b*, where, each time the digital control signal goes to a logical 1, the electronic switch opens, and the magnitude of the voltage at the time it opened is stored on the capacitor. Notice that the output wave form results in a series of levels stored on the capacitor and read from the amplifier output.

Track-and-hold amplifiers are used to store incoming signals briefly for an ADC. Often this is necessary when several signals need to be stored simultaneously and then sequentially switched into the input of the converter. It is also useful when converting extremely short transient signals. In effect the track-and-hold amplifier reduces the aperture time, or actual time that data is taken, of the ADC to the aperture time of the amplifier switch, a condensation of at least two orders of magnitude, usually. (See further discussion below.)

## IV. DATA ACQUISITION PRINCIPLES

Using the analog and digital devices described above, one can begin to design the specific communication link, or interface, between the digital computer and experimental systems to be operated on-line. Several fundamental considerations must be kept in mind: how does the particular computer use, recognize, and interpret information from the outside world? How does the computer transmit information? What are the computer machine-language instructions available for I/0 functions?

One fundamental principle that should be emphasized here is that *the computer communicates with the outside world by recognizing binary voltage level changes at incoming terminals and producing other such changes at other (outgoing) terminals*. Thus the interface design is reduced to the problem of monitoring and interpreting voltage-level changes produced by the computer and ensuring that voltage-level changes produced by the experiment are detected and properly interpreted by the computer. Thus interface design can be completed only if both the I/0 hardware *and* software of the computer are well understood. The experimenter must also know the availability and characteristics of the hardware interface components described above.

### A. PROGRAM-CONTROLLED DATA ACQUISITION

Two kinds of data acquisition approaches can be defined. One is to operate the data acquisition programming under *interrupt control*. That is, the computer is operated in a mode where data acquisition devices are serviced upon demand. The other approach is called *program-controlled* data acquisition. This approach involves programming the computer to look for service requests from specific devices and wait for these requests if necessary. The latter approach is discussed here.

Perhaps the best way to discuss program-controlled data acquisition is to consider a specific problem. Consider the case where an experiment of a transient nature is to be conducted and data acquired during the lifetime of the experiment, for example, the monitoring of the gaseous products of an explosion. It is desired to use the computer to initiate the explosion and simultaneously initiate data acquisition. Data acquisition is desired at a constant rate of 10 kHz. When a specified total number of data has been taken, it is desired to have the computer terminate data acquisition and reset the experimental instrumentation to original conditions. Each data point is to be taken in from the ADC and stored in memory for later processing.

A schematic diagram of the computerized data-acquisition system is given in Fig. 8.33. The computer initiates the data-acquisition cycle by executing a COMMAND instruction; this terminal goes to a "1" state and is connected to the Enable input of a 10 kHz clock, the output of which enables a 10 bit ADC every 0.1 msec. Simultaneous with enabling the clock and the data acquisition process, the COMMAND output initiates the experiment. The experimental output is continuously available at the input to the ADC. Every time a conversion is completed, the ADC sets a STATUS flip-flop. When that flag goes to a 1 state, the computer can determine that a conversion has been completed and that the digitized datum has been strobed into the input buffer register.

Under program-controlled data acquisition, the computer is programmed to test the STATUS bit to determine when each conversion has been completed.

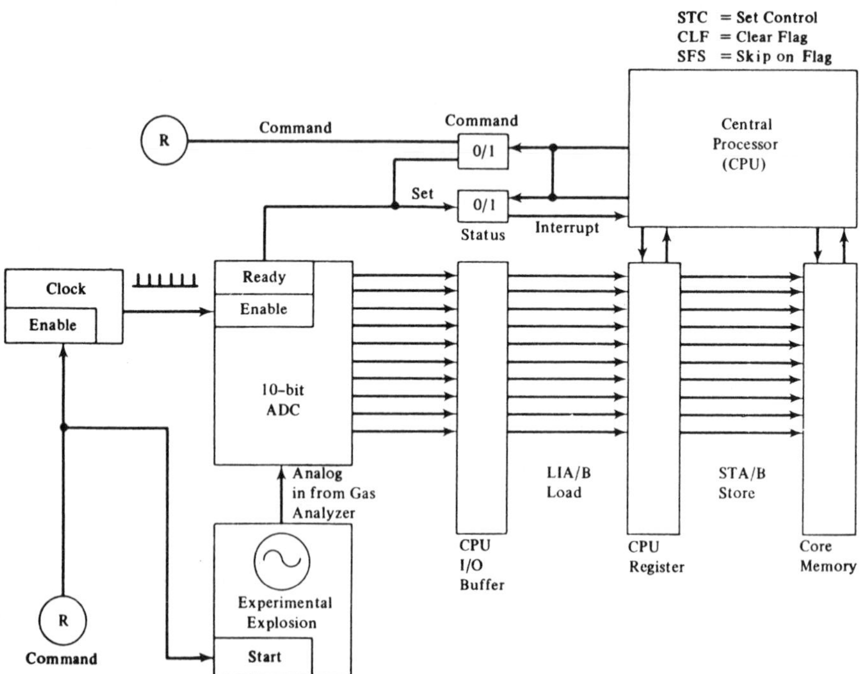


**Fig. 8.33.** Program-controlled data acquisition system.

When the computer gets a "true" answer in querying this flip-flop, it goes to a data input routine which loads the contents of the buffer register into a CPU register, clears the STATUS bit, and then stores the datum in core memory. This routine must keep track of the total number of data taken and handle sequential storage of data in a specified block of memory. When the specified total number of data has been taken, the computer terminates the data acquisition by clearing the COMMAND bit. This disables the clock and resets the experimental instrumentation to initial conditions.

### B. TIMING AND SYNCHRONIZATION IN DATA ACQUISITION

Although the need for synchronization between data acquisition operations and experimental events should be obvious, the importance of this has not been illustrated. Fig. 8.34 describes what can happen when synchronization error occurs. The trace representing the real data which starts at time zero, $t_0$, is *presumably* sampled by the data acquisition system at time points indicated by the clock pulses on the $x$ axis. Because the clock can generate pulses at an accurately known frequency $f$, the time between pulses, $1/f$, is precisely known. The program generally assumes that the first clock pulse is seen at exactly the fundamental time interval $1/f$ after $t_0$. However, this is true only if data acquisition has been synchronized exactly with the start of the experiment. If synchronization has not occured, the first clock pulse can come anywhere during that first time interval. If the first clock pulse occurs early, as shown in Fig. 8.34, and the program is not aware of the synchronization error, then the program may assume that the first datum obtained really corresponds to the time assigned to the first clock pulse on the diagram. Thus the data points taken at the $x$'s on the real data trace are effectively displaced along the time axis to the points indicated by the squares on the diagram, and the digitized wave form seen by the computer has the appearance of having been translated on the time axis. For

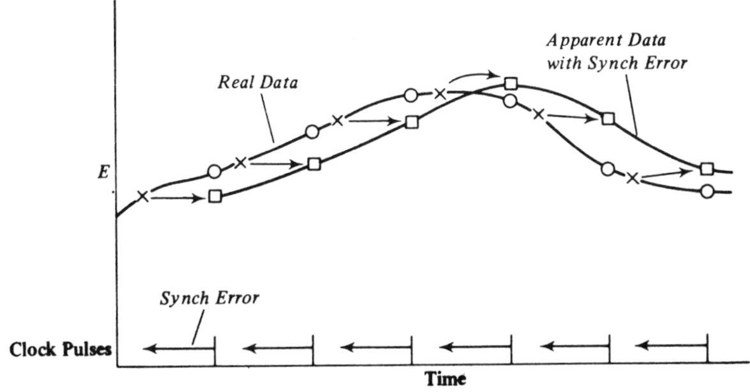


Fig. 8.34. Data acquisition timing and synchronization error.

experiments where the data density is great, an error of this sort may be insignificant; however, for most experiments, this type of error may cause severe difficulties in data processing. (An example of such an experiment is one involving ensemble signal averaging, described below.)

Generally, a crystal oscillator is used for precise timing. However, this type of clock provides a continuously available pulse train at a fixed frequency. Thus there is no way to determine when a given clock pulse occurs in real time. The uncertainty can be minimized if one selects a clock with a very high frequency and scales this down to the desired frequency range (see Fig. 8.21). The countdown logic can be initialized, enabled, or disabled. Thus the output of the scaler provides a pulse train where the uncertainty in the duration of the first time interval is no greater than the time interval, $1/f_0$, of the crystal oscillator. For example, if the basic clock rate, $f_0$, is scaled by a factor of 100, then the uncertainty in the initial scaled time interval is no greater than 1%.

The scaled clock is most valuable for establishing a time base for experiments that cannot be started at a precisely known moment by external control. If the start of a spontaneously initiated experiment can be detected electronically, this signal can be used to enable the scaler logic of the clock. Thus a time base precisely synchronized with the experiment is obtained.

If the exact frequency of a free-running clock is to be used for data acquisition timing, synchronization can be achieved by simple gating. This is demonstrated in Fig. 8.35. The output of the free-running clock is brought to one input of an AND gate. The other input can be enabled by the command output of the computer or some other source. When this input of the AND gate is conditioned true, clock pulses can get through the gate and are seen at the output. The first clock pulse that gets through the AND gate is used to initiate data acquisition and simultaneously start the experiment. Subsequent pulses are seen at

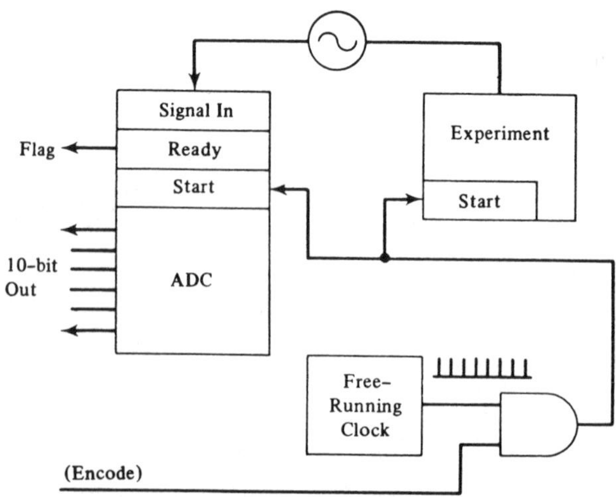


**Fig. 8.35.** Synchronization of experimental time-base generator.

exact multiples of the fundamental time interval, $1/f_0$, of the free-running clock. Thus data acquisition is exactly synchronized with the start of the experiment because the first available clock pulse was used to initiate the experiment. Note that this synchronization approach is applicable only if the experiment can be initiated externally. For an experiment that initiates spontaneously, the alternative approach of enabling scaler logic on a high-frequency clock should be used. The scaled clock is the most generally applicable and is the type implied in most illustrations here. *The laboratory computer user should be keenly aware of the synchronization limits and capabilities of the time-base generator (clock) used in his system*.

## C. ENSEMBLE-AVERAGING APPLICATION

A good example of many of the principles discussed in preceding sections, and a useful application of the digital computer for enhancing experimental measurements, is the technique of *ensemble signal averaging*. This technique can be applied in cases where experimental data are obtained with large amounts of superimposed background noise. Although many approaches can be taken to handle instrumental problems leading to noisy data, it is not always possible to eliminate noise. (For example, a situation where standard noise-elimination procedures may be inadequate is the case where the source of the noise is not in the electronics but is an inherent part of the experimental system.) When the frequency of the noise is similar to the frequency of the fundamental waveform of interest, conventional filtering techniques are not adequate. In such cases, some sort of signal-averaging approach must be used in order to extract the fundamental signal from the noise. However, two conditions must be met: first, the signal must be repeatable, and second, the noise must be random and not synchronized with the experimental output.

The ensemble signal-averaging approach involves running repetitive experiments and using the computer to acquire the digitized wave form each experiment and to sum the repetitive wave forms. When many such experimental outputs have been summed in this coherent fashion, the random noise fluctuations in the individual wave forms begin to cancel. The $S/N$ ratio, in fact, should increase proportionately to the square root of the number of averaging cycles. This approach is illustrated in Fig. 8.36.

It is extremely important in an ensemble-averaging experiment that the experimental output be synchronized exactly with the data acquisition process. If any significant fluctuation in synchronization with the time base occurs, a distortion in the extracted fundamental wave form is observed.

## D. SAMPLING OF EXPERIMENTAL DATA

### 1. Using Track-and-Hold Amplifiers

The tracking capabilities of even a high-speed ADC are relatively limited. To improve the tracking features of a data-acquisition system substantially, a track-and-hold (T/H) amplifier should precede the ADC.

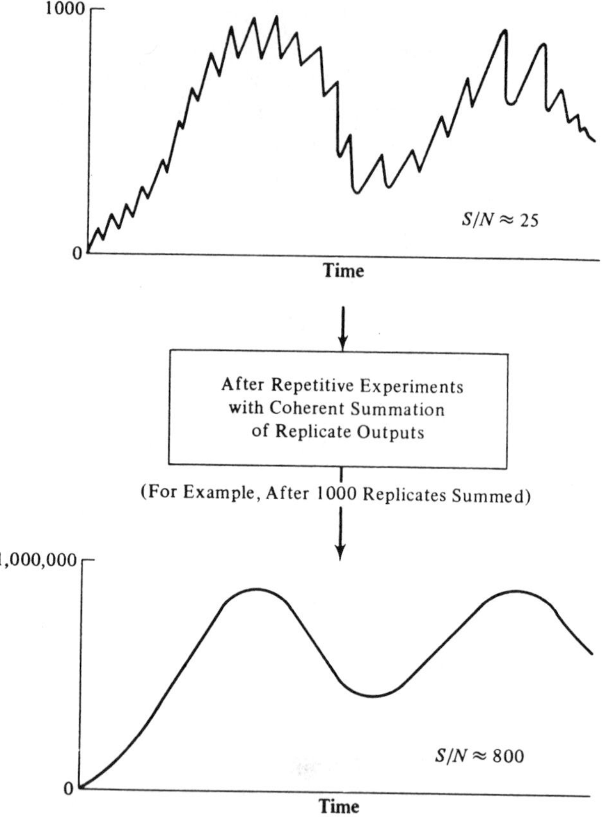


**Fig. 8.36.** Ensemble signal averaging.

During each hold period, the voltage output of the T/H amplifier remains constant so that digitization may take place (see Fig. 8.32*b*). The digital output of the ADC reflects the voltage level at the specific time corresponding to the beginning of each hold period, despite the fact that the conversion is completed at some later time. Thus the T/H amplifier allows wave form sampling with time-base precision independent of the conversion rate of the ADC. The uncertainty in the timing of the sampling (the aperture time) depends on the switch-opening time and is a characteristic of the T/H amplifier. T/H amplifiers are available commercially with aperture times of the order of 10–100 nsec. Other important characteristics include *response* and *settling* times, which refer to the amplifier's ability to follow rapidly changing signals. These features actually limit the overall acquisition or sampling rate. Sampling intervals (the time from the end of one hold period to the beginning of the next) of the order of one to several microseconds can be attained with currently available devices.

### 2. Sampling Frequency

The sampling frequency selected for data acquisition obviously should be related to the bandwidth of the sampled wave form. From information theory, the criterion for adequate sampling is that the minimum sampling frequency (*Nyquist frequency*) must be twice the bandwidth of the sampled wave form. Thus for a 100 Hz signal, the sampling frequency must be at least 200 Hz to retain the information inherent in the wave form.

This criterion is strictly applicable in such applications as the sampling of interferograms for Fourier transform analysis. However, sampling frequencies considerably greater than the Nyquist frequency should be used to allow faithful reproduction of the signal for straightforward data processing algorithms. A rule-of-thumb criterion is that the sampling frequency should be at least 10 times the bandwidth of the wave form. (Of course, the previously discussed limits imposed by ADC or T/H aperture times or amplifier response place an effective upper limit on sampling frequencies with a specified resolution.)

Oversampling of an experimental signal can cause problems mainly because of excessive memory requirements. On the other hand, undersampling can cause more serious problems. One of these is the possibility of producing signal artifacts by *aliasing*. This phenomenon can occur when sampling frequencies lower than the Nyquist frequency are used. Figure 8.37 illustrates how a 3 kHz sine wave can be aliased to a 1 kHz signal or a 158 Hz signal by using sampling frequencies of 4 kHz and 3.16 kHz, respectively.

### 3. Multiplexing

It is sometimes necessary to sample more than one experimental wave form simultaneously during a single experiment. To accomplish this, an analog multiplexer device can be used. We discuss here the configurational and sampling considerations for multiplexing analog signals.

The primary characteristic of an analog multiplexer is that it can accommodate multiple analog inputs, any one of which can be sampled through a single output channel. The selection of the input channel to be transmitted to the

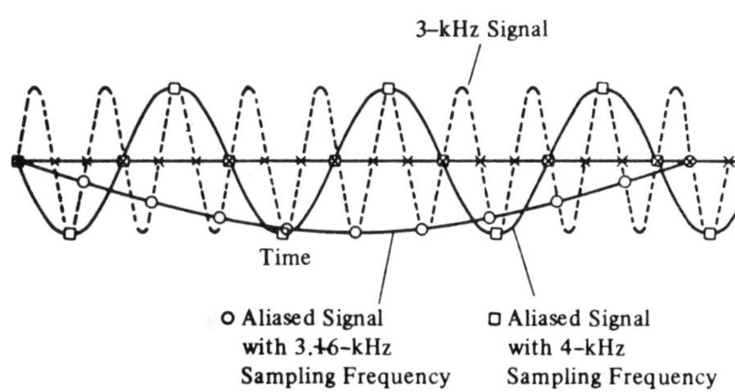


Fig. 8.37. Aliased signal phenomenon.

digitization hardware can be accomplished by the output of a binary-coded command from the computer or by generating an appropriate external sequencing code. The critical characteristic that we must consider here is the time required to switch between input channels. With solid-state analog switches, the sequential selection of analog inputs can proceed with time intervals of the order of a few microseconds or less between channels.

The reader should recognize immediately that the analog multiplexer need not provide the slow step in an overall data-acquisition process. Indeed, the slower processes are associated with the data acquisition hardware and software that follow the analog multiplexer. Another point that should be emphasized here is that it is impossible for the multiplexer to sample independent wave forms in a truly simultaneous fashion. Some finite time interval must exist between samplings of different channels. The manner in which this problem is handled depends on the need for acquiring truly simultaneous data from different channels.

Figure 8.38 illustrates two alternative configurations for multiplexing analog signals from four independent sources. In Figure 8.38*a*, the multiplexer is followed by a T/H amplifier and ADC. This configuration is used if there is no need to achieve simultaneous sampling of the four input channels. Moreover, if the total time required to complete data acquisition from the four channels is small compared to the time interval between samplings, this configuration can be used. For example, if an overall data acquisition frequency for all four channels of 10 Hz is employed and the total sampling time per channel is

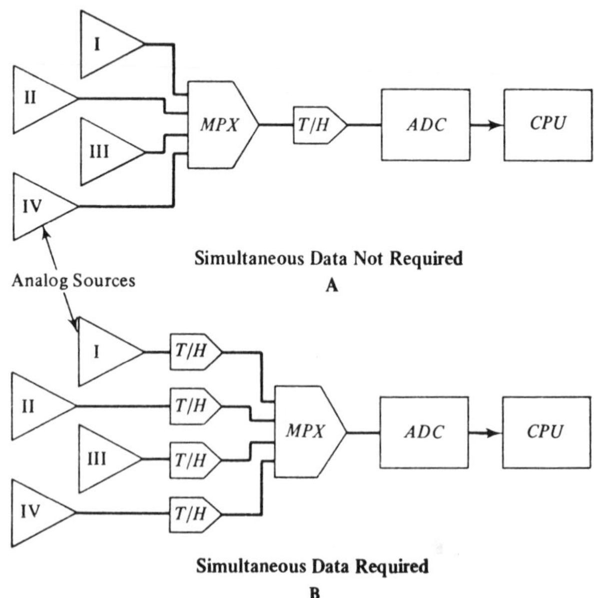


**Fig. 8.38.** Multiplexed data acquisition system.

100 μsec, the maximum *skew* of the sample data is 300 μsec. That is, the fourth channel is sampled 300 μsec after the first channel. This amount of skew can be considered negligible compared to the 100 msec time interval between samplings.

For the case where simultaneous sampling is required and the time required to sequentially sample each of the input channels is long compared to the overall data acquisition time base, the configuration shown in Fig. 8.38*b* is recommended. In this case, each analog signal is funneled through a T/H amplifier. Because all the T/H amplifiers can be gated to the *hold* mode simultaneously, the time required to sample all channels through the multiplexer is inconsequential, provided that it does not exceed the overall data acquisition time-base interval. Another consideration is the *droop* specification on the T/H amplifier. That is, the T/H amplifier must be capable of holding the analog signal without significant decay until the particular channel is sampled and the digitization complete.

## V. GENERAL CHARACTERISTICS OF MINICOMPUTER SYSTEMS

The discussions in this section and in succeeding sections are important to the reader who is interested in a general-purpose laboratory computer system. If the scientist is interested in a "turn-key" system for a particular job such as mass spectrometry, gas chromatography, or Fourier transform spectroscopy, then he or she need not be concerned with the detailed characteristics of the computer hardware. In this case, the scientist should only be interested in the overall characteristics of the turn-key system and its capability for doing the specified job.

A list of the general characteristics of minicomputer systems is provided in Table 8.VIII. Within the categories outlined in Table 8.VIII, the various commercially available minicomputers can differ widely in their characteristics. When all the minicomputer features are taken together, they determine the applicability of the system to the particular problem. It is difficult to say which of these features is of greatest importance to the user. However, as long a the

TABLE 8.VIII

General Minicomputer Charateristics

| |
|---|
| Word size |
| Processor architecture |
| I/O structure |
| Machine-language features |
| Speed (memory cycle and execution) |
| Hardware accessories included or available (EAU, DMA, priority interrupt, read-only memory, etc.) |
| Standard peripherals Available |
| Standard software available |

computer is to be used for on-line laboratory applications, among the most important features for evaluation of the system is the I/O structure, and we focus on this aspect here.

## A. COMPARISON OF I/O STRUCTURES AND INSTRUCTIONS

One general observation should be made in comparing the I/O considerations for various computer architectures. Although the channeling of information may be considerably different because of the computer architecture, the fundamental aspects of the machine language for I/O operations are surprisingly similar from one computer system to another. This is because the I/O functions can be reduced to the fundamental operations of *data transfers in and out, control commands from the computer, checking the status of external devices, and device selection*. There are basically two kinds of data transfer procedures, program controlled and interrupt controlled. With program control the computer is programmed to query periodically the status of one or more peripheral devices. Whenever a device is ready to accept or give data the program detects this "ready" status and initiates a data transfer. Whenever the device is not ready or "busy," this status can likewise be detected and no data transfer operation occurs. Alternatively, interrupt-controlled data transfers can be arranged by a user's program. Here, the computer I/O hardware is set up to allow peripheral devices to indicate their status to the CPU, causing it to suspend operation, whenever a device is "ready," and cause a peripheral service subprogram to be executed. Thus data transfers can be accomplished *on demand* with interrupt control, and more efficient use of the computer's time is accomplished.

There are some fundamental differences in I/O structure that may exist even within systems that have similar overall architecture. For many computer systems, we can consider at least two different I/O structures. One of these is the multichannel, buffered, hardware priority interrupt structure. In this case each peripheral device can be connected to a specific I/O channel which incorporates digital buffer registers for I/O data, as well as command and status check hardware for that unique channel. Several distinct I/O channels exist, each with its own channel number so that it can be addressed directly by an I/O program. Interrupt requests are handled on an automatic basis where each channel has an assigned priority implemented in hardware. Another structure is the party-line structure.

In the party-line I/O structure, a single I/O line is available, and all external peripheral devices hang in parallel on this common line, or bus. This bus provides control lines, device-selection lines, interrupt, and a status-check line. The device-selection decoding logic is provided by the device rather than the computer hardware. All devices hang in parallel on the I/O bus, and data are strobed in or out of these devices from or to the working registers by the proper decoding of control commands to the device-selector hardware associated with each device. Status checking is provided through the common status line, which can be activated through the device selection hardware by feeding back a command bit to the computer. The interrupt line can be set by any of the devices in parallel, but the computer does not know *a priori* which device has caused the interrupt. The computer must execute a prearranged sequence of status-check software operations to determine which of the devices on the party line has caused the interrupt. If several devices are capable of interrupting the system, then the interrupt service program may be quite lengthy and involved. The priority of interrupt service is obviously established by the software routine. Priority can be established simply by determining which device is checked first upon the recognition of the interrupt. Interrupts are serviced by causing the execution of a "Jump Subroutine" instruction.

Fortunately, a standardized interface bus design has gained wide acceptance recently. This is the IEEE 488 interface design, which is shown schematically in Fig. 8.39. There are 16 active signal lines, grouped into three different functional sets: data (eight lines), data byte transfer control (three lines), and general interface management (five lines). Data transfers are controlled by the second set of lines. For example, any "listening" device that is addressed can set the "ready" line when it is able to accept data. The sender or "talker" device puts data on the data bus and indicates "valid data" on the line. The listener can acknowledge successful receipt of the data on the "data accepted" line. Multiple devices can be addressed simultaneously, but the data transfer rate will be limited by the slowest "listener."

Many commercial instruments, peripheral devices, and computer systems are currently providing I/O hardware and protocol consistent with the IEEE 488 standard.

## B. MICROPROGRAMMABLE COMPUTERS

Traditionally computer systems discussed thus far all share one common feature. They have been fixed-instruction, general-purpose machines. They are general-purpose in that the instruction set and overall architecture are not designed for any specific task, but can be adapted to a wide variety of applications. They are fixed-instruction devices because the machine instructions are hardwired into the CPU and are not generally alterable by the user.

Other common characteristics include the fact that the speed of execution of machine instructions is tied to the memory cycle time and generally requires some integral number of memory cycles for completion. Moreover, the CPU is the dominant *control device* in each system.

An alternative to the FIGP system is the *microprogrammed computer*. One possible configuration is shown in Fig. 8.40. The heart of this system is the *control memory*. This is usually a *read-only-memory* (ROM) device.

As the name implies, a ROM is a memory module with an unalterable array of binary words. The contents of each location can be fetched and read but cannot be deleted or changed. ROMs can be constructed from solid-state semiconductor materials, which allow much faster memory access time. In fact, access time of less than 100 nsec are possible.

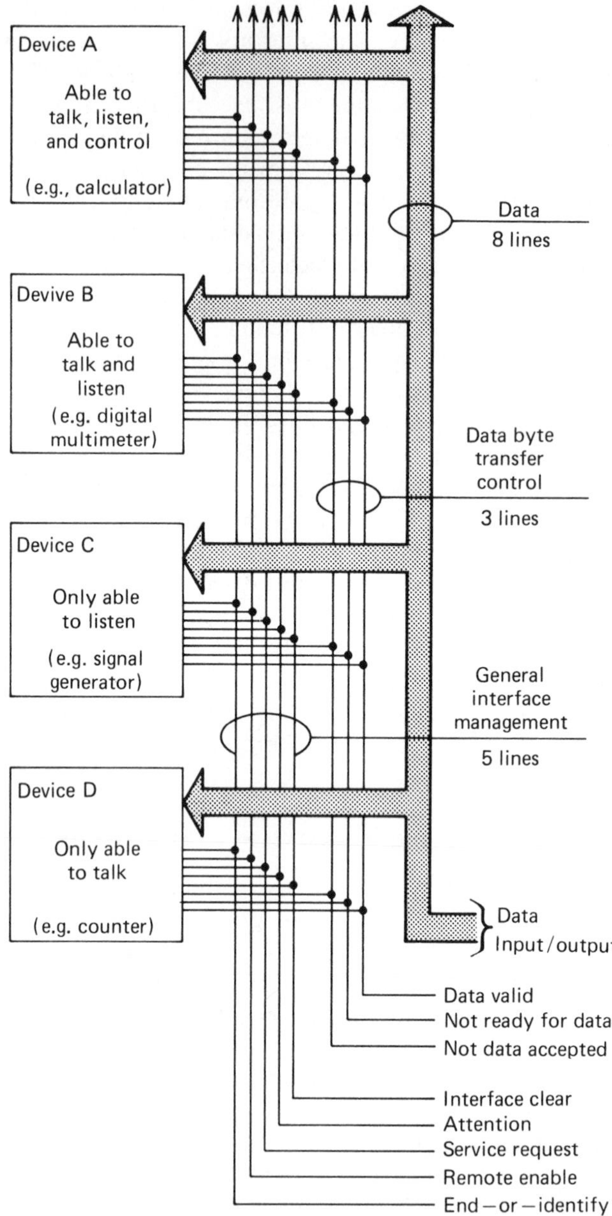


**Fig. 8.39.** HP-IB (IEEE 488).

It is the availability of high-speed ROM's which makes microprogrammable computers feasible and advantageous. With reference to Fig. 8.40, the general philosophy of the microprogrammable computer is to provide a *control device* capable of executing a limited, but very fundamental, set of "microinstruc-

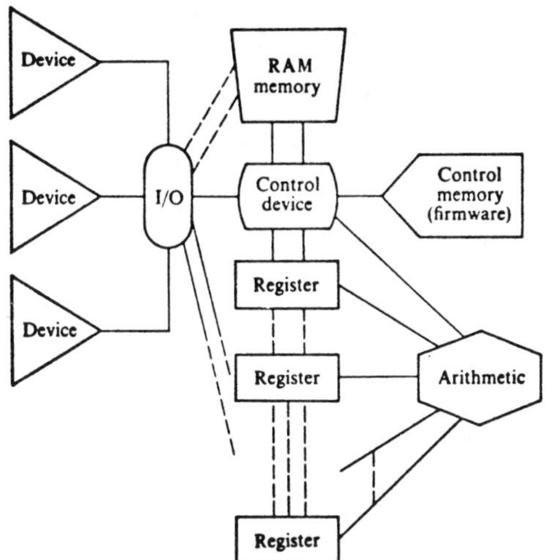


**Fig. 8.40.** Typical microprogrammable computer configuration.

tions"—such as logical AND, OR, 1 bit rotate, 1 bit shift, add, clear, complement, logical tests, I/O, register-register, and register-memory transfers. These hard-wired machine instructions (microinstructions) are normally more limited and fundamental than the instruction set provided with a FIGP computer. To generate more powerful instructions, the ROM is made to contain various subroutines that generate appropriate sequences of the microinstructions to accomplish a given algorithm. Because the ROM has such rapid access time, and because several semiconductor (high-speed) working registers are usually provided as "scratch-pad" memory, these ROM subroutines can be executed rapidly. Thus the speed of execution of many instructions can be divorced from the memory cycle time of the computer. (In fact, in some designs the control memory can operate completely independently of the normal CPU memory operations.)

Very sophisticated and complex algorithms can be implemented by the ROM subroutines (called "firmware"), and each can be called by a macro instruction defined by the user for his assembly language programming. These firmware subroutines can be executed very rapidly, but may be limited ultimately by the number of references to "slow" random access (RAM) memory required for completion.

The microprogrammable computer functions are similar to the FIGP computer, except that the user may *define his own machine-language instruction set*. That is, the nature of the programming language used depends on the firmware subroutines specified. Thus the user may choose to have macro instructions—such as MPY, SQRT, LOG, STRTADC, and DACOUT—which call specific firmware subroutines to accomplish indicated operations. This flexibility is possible because of the reasonable cost of producing an infinite variety of specific ROM's.

The advantages of microprogramming should be obvious. One is that a computer can be tailor-made, as specified by the user, for specific computational or control operations. Another is that the macro instruction set can be changed by replacing the control memory. Also, it should be obvious that the macro instruction set can provide a very high-level programming language, which is essentially the "machine language" as far as the programmer is concerned. Thus "machine language" instruction sets defined by BASIC, FORTRAN, ALGOL, and so forth are possible.

These basic principles have been utilized most effectively in conjunction with microprocessors. These are large-scale integrated-circuit devices which contain all the arithmetic, logical, input-output, and memory addressing hardware of the central processing unit of a general-purpose digital computer. These devices can be mass produced for less than $10 each and require postage-stamp type space. When combined with specialty programs in ROM they can be designed to perform sophisticated control, data collection, and data transfer functions previously handled by complex hardware devices and/or dedicated minicomputers. These devices can also form the nucleus of general-purpose "microcomputers" which perform essentially the same functions described earlier for minicomputers, except for a much lower price. Because of the generally lower microprocessor speed and less sophisticated elementary instruction set, and the general lack of extensive standard software packages, these "micro" systems are somewhat less powerful than the "minicomputer" systems. However, the differences in capabilities are quickly disappearing.

## VI. FUNDAMENTAL PRINCIPLES FOR INTERFACE DESIGN

It seems appropriate here to pull together many of the instrumental concepts discussed throughout the book relating to the design of interfacing between computer and experimental systems.

### A. SCOPE OF THE INTERFACE

It should be recognized, first of all, that the interface includes everything from the transducer or transducers used in the laboratory instrumentation to the computer software designed to communicate with that experiment. Everything in between contributes to the interface. This includes the computer I/O hardware, the signal handling devices of the analog instrumentation, translation elements, signal transmission devices, and connections to and from the computer. It should be emphasized here that it is unrealistic to consider interface design without recognizing all the components that are part of the interface, starting with the chemical instrumentation itself.

### B. INTERFACE DESIGN

In this section, we try to provide some general guidelines for interface design. The various considerations are summarized in Table 8.IX. The first of these

TABLE 8.IX
General Interface Design Considerations

| |
|---|
| CPU I/O structure |
| Primary (I/O bus features: COMMAND, STATUS CHECK, DATA IN/OUT) |
| Secondary (Standard interface devices: buffer registers, sense switches, DACs, ADCs, CLOCK, MPX, T/H, logic devices, etc.) |
| Laboratory instrumentation characteristics |
| Output features (transducer, noise, range, location, bandwidth, etc.) |
| Control requirements (start/stop, time base synchronization, experimental control parameters, etc.) |
| Processing and data handling |
| Passive (Data logging) |
| Active (real-time control) |
| Dedicated or time-shared |

steps involves a definition of what the outside world sees at the computer terminals. As described earlier, the general approach is similar for various computers. That is, each computer system must provide for the output of command bits as well as status checking and input and output of binary-coded data. Details may differ from one computer system to the next.

In addition to the primary features of the computer's I/O structure, the user must consider the characteristics of any standard interface devices incorporated into the system. For example, a computer system might include an interfaced "internal" time base generator, or "real-time clock." Other standard interface devices might include I/O buffer registers, sense switches, multiplexers, digital-to-analog converters, and analog-to-digital converters. These and other devices may be included as an integral part of a purchased computer system and it is necessary for the user to understand the required input characteristics, the specified output characteristics, and the nature of the programming required to properly operate the standard interface devices. Most importantly, the user must be aware of the limitations of applicability of the primary and secondary I/O devices. For example, he or she must be aware of whether a digital data input is buffered or unbuffered at the computer interface. The user must also know whether the computer or the external device, or both, determines whether a true flag bit causes an interrupt or not. With regard to the secondary devices, the user must be aware of the analog range limitation of the ADC or DAC devices; how the real-time clock is synchronized with external events; how the computer determines the status of sense switches; and so forth.

The second major function in interface design is to define the characteristics of the laboratory instrumentation which the user desires to place on-line to the computer. One aspect of the laboratory instrumentation that must be considered is the output characteristics. For example, what are the inherent features of the transducer element. Does it produce a current or a voltage signal related to the fundamental phenomenon of interest? Perhaps a discontinuous output is obtained such as in the discrete pulse output of photomultiplier when used for photon counting, or the pulse train output of the quartz thermometer. Also to be considered are the characteristics of the fundamental transducer output with respect to the signal frequency, amplitude, and noise background, and the analog preprocessing functions desired. For example, the signal may have to be amplified, differentiated, integrated, compared, and so forth with analog devices. Another essential consideration is the location of the fundamental instrumentation with respect to the data acquisition system. That is, over what distance must analog or digital signals be transmitted? This distance consideration must be coupled with the inherent characteristics of the signal in order to design the proper transmission elements. The user must also determine the overall nature of the experiment and associated signals with which the time base must be synchronized. The user must be aware of what inherent experimental characteristics are accessible for synchronization. For example, can the experiment be initiated at a precise moment in time or does the experiment initiate spontaneously? Finally, the required digitization features must be established. That is, the bandwidth of the experimental signal must be established in order to define an adequate data acquisition frequency. The dynamic range of the analog elements producing the experimental signal must be evaluated in order to establish the required precision of the analog-to-digital converter required. The noise rejection features of the interface hardware or data acquisition software can be determined only when the noise level of the experimental signal has been established.

The other major aspect of laboratory instrumentation to be considered for interface design includes the control requirements. For example, exactly how is an experiment initiated? Does it require a manual-mechanical operation, such as mixing, or can an electrical signal be used? Also, how is the experiment terminated? What mechanical or electronic control functions are required? Do these control functions involve the establishment of voltage, current, or wave form frequency values? What precision and accuracy of controls are desired?

A third and very important consideration in overall interface design involves the determination of the specific processing and data handling requirements. For example, the user must determine whether the on-line application involves dedicated use of the computer or perhaps time sharing with several different devices or instruments. Furthermore, the design of the interface may be considerably different, depending on whether the program involves simple data logging operations or real-time interaction with the experiment. To be more specific, if the application involves a time-shared operation, the user may very well choose to design the data acquisition around the interrupt system. If the application requires a considerable amount of real-time computer processing and interaction with the experiment, the user may have to design the interfacing to minimize the amount of real-time bookkeeping achieved by the computer. For example, a direct memory access channel may be used, or the hardware interface might be designed to include certain bookkeeping or logical functions that might ordinarily be executed by the computer in real time.

Obviously, the preceding discussion on general considerations for interface design leaves much to be desired with regard to specific systems. However, to do an adequate job of describing interface considerations for the wide variety of possible applications is beyond the scope of this chapter. It is hoped that the inexperienced reader will be able to use the preceding discussion as a guide for approaching his or her interface design problems. Some specific examples follow.

## VII. INTERFACING EXAMPLES

Two different interfacing examples are presented here for illustrative purposes. The first of these involves computer-controlled sampling of multiple ion selective electrodes in flow stream analysis (8). Figure 8.41 depicts the multiplexed potentiometer with computer and interface. The computer interface contains a software programmable clock, a 16 bit DAC, an 11 bit ADC (±10 V range) preceded by a T/H amplifier, two general-purpose I/O registers, and a digital store scope. Each register contains a status bit, control bit, 16 bits into and out of the computer. All bits are TTL compatible. A timing diagram (Fig. 8.42) displays the manner in which multiple electrodes are selected and monitored during the monitoring section of the program. After a base frequency (1 Hz) clock tick is detected, the first electrode is selected, along with its corresponding DAC value output. The time allowed for settling is 10 msec ($\tau_1$). The output is then averaged for time $\tau_2$ by summing data taken at a 10 KHz

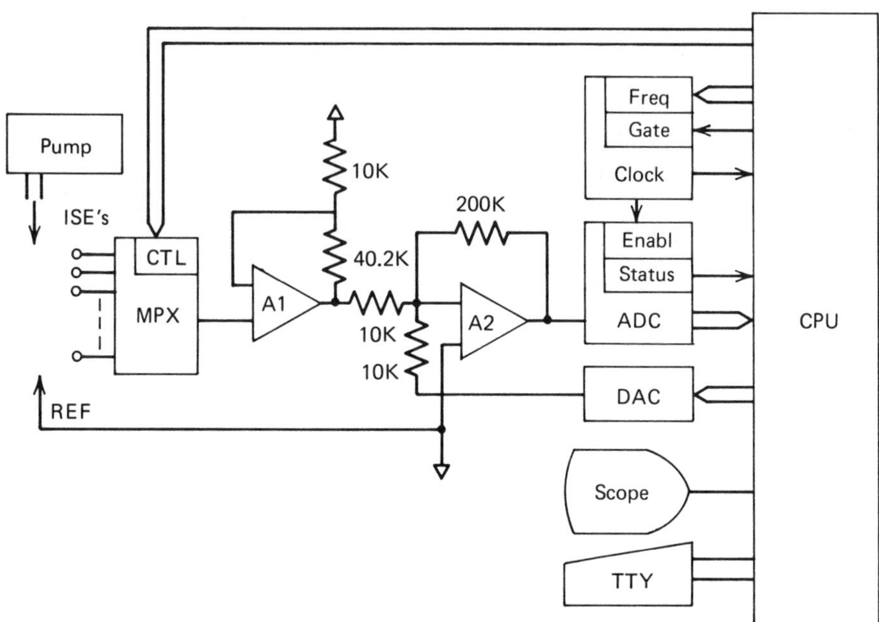


**Fig. 8.41.** Schematic diagram of multiplexed potentiometer and on-line computer system for continuous ISE monitoring.

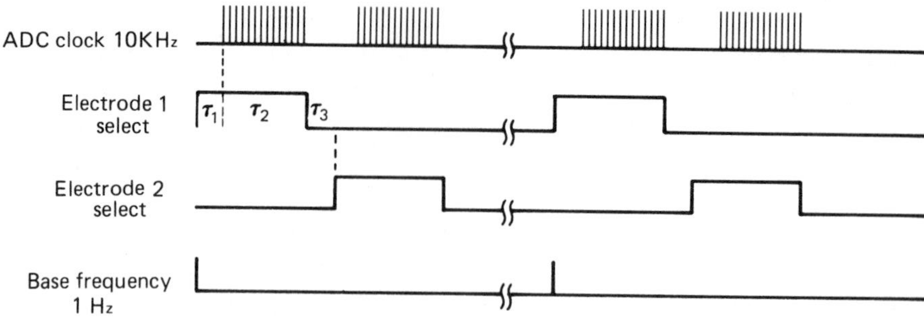


**Fig. 8.42.** Timing diagram for sampling of multiple ISE inputs to on-line computer system.

rate. ($\tau_2$ was set equal to 16.7 msec, or 1/60 sec, in order to reject possible 60 Hz pickup.) $\tau_3$ is the variable time required by BASIC to operate on the resulting value. The same procedure is followed for each succeeding electrode, with sampling functions for all electrodes completed in <1 sec.

The second example refers to computerized data acquisition of rapid transient data in flash photolysis experiments (1). The basic components of the computerized data acquisition system are shown in Fig. 8.43. The data acquisition system is composed of the following components: a 10 MHz crystal clock scaled down to various available frequencies from 500 kHz to 0.01 Hz (either a square wave or 2 $\mu$sec pulse train output was available at decade frequencies and at intervals of 1, 2, or 5 times each decade frequency); a ±10-V T/H amplifier with 10 nsec aperture time and 300 nsec settling time; and a ±10 V, 11 bit ADC which has a conversion time of 3.8 $\mu$sec. The data acquisition and auxiliary logic elements are composed entirely of TTL logic components, and are based on a general-purpose interface design published elsewhere (7).

Digital data acquisition is triggered by the negative edge of a phototransistor pulse which is activated by the light flash. The anlog signal is recorded simultaneously on an external storage oscilloscope. The phototransistor output is shaped by a Schmitt trigger to provide a 1-0-1 pulse compatible with TTL logic. An *RC* filter at the Schmitt trigger input filters out high-frequency noise and eliminates premature triggering. The Schmitt trigger pulse triggers a flip-flop which then directly initiates data acquisition and the oscilloscope trace simultaneously. A timing diagram of the sequence of events is shown in Fig. 8.44. Because the pulse of light is about 15$\mu$sec wide, there is an uncertainty in the position of a true zero time. Also, there is a 1–2 $\mu$sec delay between the start of the flash and the start of the time base, owing to the shape of the phototransistor pulse and the level needed to trigger the Schmitt trigger. For this work, time zero was arbitrarily defined as the instant that the RUN flip-flop changes state, as shown in Fig. 8.44.

These two examples of interface design do not begin to cover the many possible interfacing functions for chemical experiments. However, they do serve to illustrate that not only is the interface circuitry important, but also the

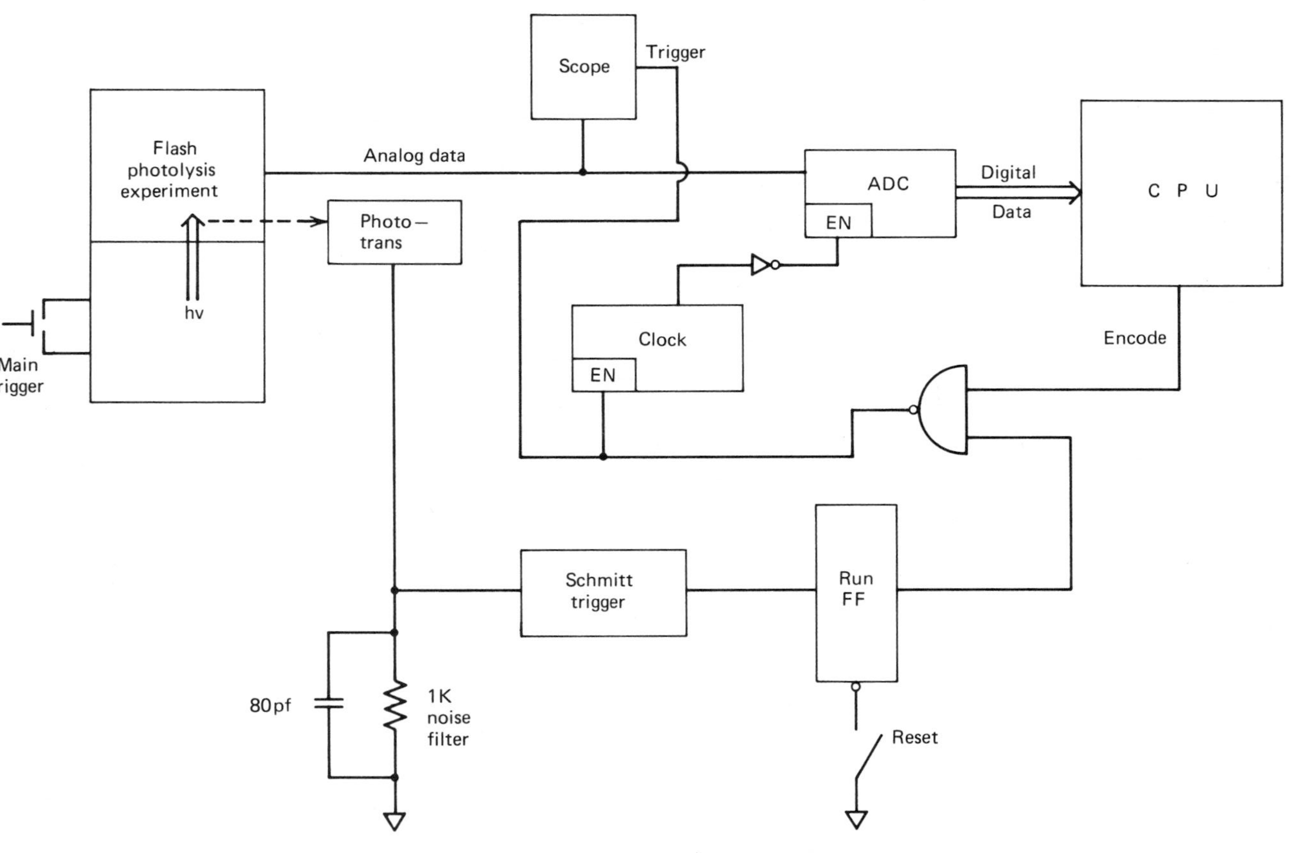


Fig. 8.43. Block diagram of data acquisition system.

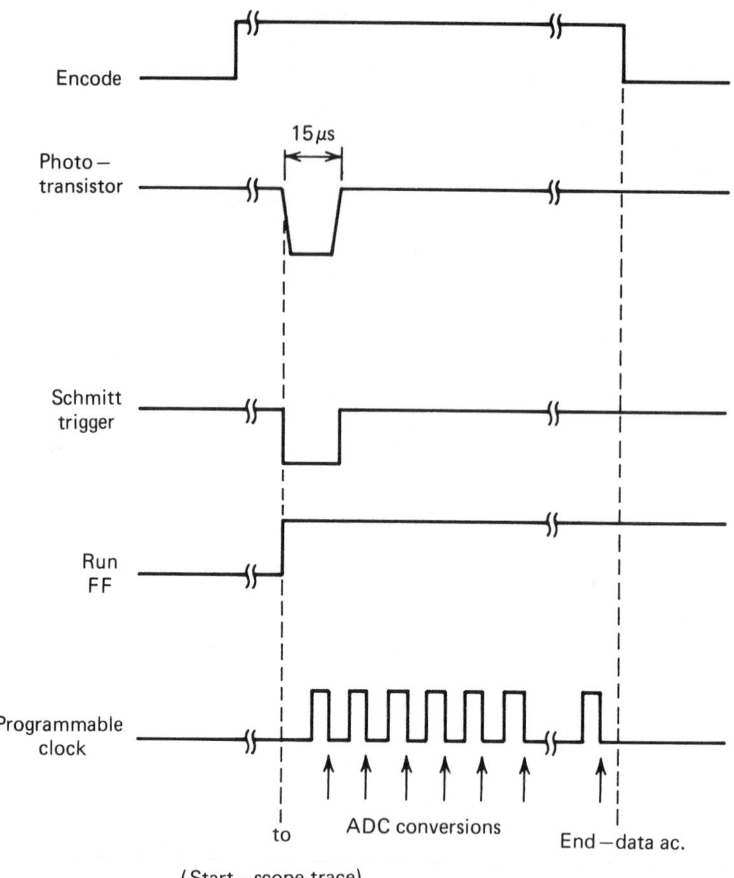


**Fig. 8.44.** Timing diagram.

associated timing diagrams are an invaluable aid to interpretation. Many other applications of computers in chemical experimentation have been summarized previously (3), and the reader is referred to that summary for literature references to a wide variety of interfacing functions.

## VIII. INTERFACING STANDARDS

In the past few years, a large number of instruments have entered the marketplace bearing one or more standard interfaces. In addition, the number and type of interfacing modules designed specifically for use with standard interfaces has greatly increased. These two facts, combined with the proliferation of commercial interfaces designed specifically to plug directly into popular computers, herald the time when most computer interfacing will be done by purchasing commercial products and plugging them together.

There are many advantages to using commercial products with standard interfaces. First, the system designer can concentrate his effort on the job of producing a viable system to acquire and process the data and need not be concerned with the details or problems related to making the electrical connections between the computer and the instruments. Second, system reliability can be improved by obtaining spare interfaces, not to mention the fact that commercial equipment constructed using modern printed circuit board techniques are typically more reliable than breadboards, wire-wrap boards, or the like. Third, data acquisition and control systems can be constructed and maintained by personnel not intimately familiar with all the details of computer interfacing. This can be a major benefit to sites that have little or no electronics support.

Standard interfaces are not yet a panacea. Some standard interfaces are unsuitable for applications requiring high throughput. Not all types of interfaces are commercially available for a particular standard, and the cost of purchasing and maintaining many different standard interfaces is often prohibitive. Because the standard interface must itself be interfaced to the host computer, it is not always possible to obtain commercially the standard interface board for a particular computer system. In addition, designing and constructing the hardware necessary to connect a special-purpose data acquisition or control system to a standard interface are often just as difficult as interfacing the special purpose hardware directly to the host computer.

In spite of these disadvantages, standard interfaces now account for a fair portion of the mini- and microcomputer business. Basically, standard interfaces come in three general types:

1. Data acquisition and control hardware directly interfaced to a standard computer bus.
2. Instrument communication buses.
3. Serial communication lines.

## A. STANDARD COMPUTER BUSES

The simplest of the three types to use is the directly interfaced hardware. What has been standardized in this interface is the computer I/O bus structure. The three examples are the Q-bus used by the LSI-11 series of microcomputers form Digital Equipment Corp., the Intel Multi-bus, and the S-100 bus. Any computer that operates on one of these buses can be interfaced to any of the data acquisition and control products available for that bus by merely configuring the interface card and plugging it in. Some of the interfaces commercially available are ADCs, DACs, timers, counters/scalars, relay control interfaces, thermocouple-specific ADCs, general-purpose digital control interfaces, and time of day clocks. Selecting a computer that is compatible with a standard bus for which the interface modules already exist can significantly reduce the cost of a system and greatly decrease the system construction time. There are also other benefits to this approach: it eliminates many of the maintenance problems associated with one-of-a-kind interfaces, it eliminates the need for an electronics design engineer, and finally, it makes the system easy to duplicate and document.

## B. STANDARD INSTRUMENT BUSES

Because of design constraints or economics, it is not always possible to use a computer system that is compatible with one of the standard computer buses. The standard instrument buses are a good choice in this case. The standard instrument buses all have one thing in common: they must all be interfaced to the host computer. At first, this may seem like a major problem. However, in most cases the computer manufacturer or an independent vendor offers an interface between the computer and the standard instrument bus. In cases where no such interface exists, use of a standard instrument bus may still prove beneficial when more than one instrument supporting the instrument bus is to be interfaced to the computer. It can also be beneficial when the only choice is between reproducing the commercial instrument and interfacing the computer to the standard instrument bus.

### 1. Standard Instrument Buses—The GPIB

The most popular standard instrument bus is the IEEE 488 standard bus shown in Fig. 8.39. This standard bus is also known as the HPIB (Hewlett Packard interface bus) and as the GPIB (general-purpose interface bus). A relatively large number of instrument manufacturers now market GPIB-compatible instruments, or instruments that can be upgraded for GPIB compatibility. Most of the GPIB instruments contain dedicated microprocessors, and are as fully functional through the GPIB as they are through the front panel.

The GPIB is a byte parallel bus with eight additional control lines. Up to 15 instruments can be interfaced to the bus at any one time. Because all the instruments are interfaced to a common bus, each is assigned a unique address in the range 0–31 so that communication with a single device is possible. One of the devices on the GPIB must be the bus controller. The remaining devices on the GPIB are termed the slaves. It is the bus controller that initiates and controls all transfers of data over the GPIB, even those transfers involving data transferred from a slave to the bus controller. Typical transfer rates over the GPIB are 5000–300,000 bytes/sec, with the maximum transfer rate being less than 1,000,0000 bytes/sec. Because the actual transfer rate between two GPIB devices is limited by the transfer rate of the slower of the two devices, care should be taken when selecting GPIB compatible instruments for applications that require high transfer rates.

Because a slave device is not allowed to initiate a data transfer to the bus controller, the GPIB was designed with a common bus control signal known as SRQ, a mnemonic for service request. It is used by slave devices when they require service of some sort, or when they need to inform the bus controller that they have completed their task. Typically, SRQ is generated by an instrument that has acquired data and is now ready to transfer that data to the host computer (bus controller). When the bus controller recognizes SRQ, it must first determine which device is requesting service. This is done using either the serial poll or the parallel poll protocol. The parallel poll is the faster of the two polling methods, but it can only be used when eight or fewer devices are capable of generating SRQ. Once the bus controller determines which device is requesting service, it initiates the proper data transfer(s) to satisfy the device.

At this point, one problem with GPIB should be mentioned. Not all manufacturers of GPIB instruments implement the entire IEEE 488 standard. This most often surfaces in the handling of SRQ, serial poll, and parallel poll. Many times a particular GPIB instrument supports serial poll and not parallel poll, or conversely. Care should therefore be exercised when selecting GPIB instruments to make sure they support all the features necessary to the intended application.

In instances where it is desirable for more than one device to be the bus controller, the GPIB makes provisions whereby bus mastership can be passed to another device on the GPIB in an orderly fashion. This feature can be used to allow multiple computers to access a single group of shared instruments. In addition, it allows for intercomputer communication as well. Some experimental computer networks have even been built around the GPIB.

A final thought about the GPIB. There is no reason that a single host computer can not be interfaced to multiple independent GPIBs. Such a scheme allows higher throughput by allowing parallel communication paths. It also allows the number of GPIB devices interfaced to a single computer to be greater than 15. Admittedly, there is an incremental cost associated with each new GPIB interface. However, this cost is often overwhelmed by the additional hardware and software costs incurred with other possible solutions.

### 2. Standard Instrument Buses—CAMAC

The second standard instrumentation bus is known as CAMAC, an acronym for computer automated measurement and control. Like the GPIB, CAMAC has been standardized by the IEEE in IEEE Standard Number 583. In addition, a number of other standards exist involving CAMAC interfaces and the like. Serial highway interfacing is covered by IEEE 595, parallel highway interfacing is covered by IEEE 596, and block transfers in CAMAC systems are covered by IEEE 683. Having multiple controllers in a single crate is covered by IEEE 675.

Before further discussion of CAMAC is possible, a number of CAMAC terms must be defined. A simple CAMAC system is composed of a CAMAC "crate" which contains "plug-ins." A CAMAC crate is composed of a chassis with power supplies and a backplane bus with 25 "stations." Each station is numbered, beginning with 1 at the left of the crate (as viewed from the front of the crate). A module is electrically associated with one station, but the module may physically span several. Station number 25 is reserved for the "crate controller." The crate controller is a special plug-in that has all the circuitry to control the CAMAC bus, or "dataway." In addition to the bused dataway signals, the crate controller also has access to 24 pairs of point-to-point signals from each of the remaining 24 stations. Each CAMAC crate must contain a crate controller.

Unlike GPIB instruments, most CAMAC instruments can not be operated without a host computer. The front panels of CAMAC modules are usually limited to gain control knobs and the like, with no mechanism for initiating or displaying data. This can be very frustrating when trying to determine whether software or hardware is the cause of a data acquisition failure. CAMAC can, however, support a complete computer system within the CAMAC crate which can replace the host computer. Such a configuration can be economical in both space and money.

A CAMAC "module" is a plug-in that can be placed in any of the first 24 stations. Available modules cover the range of GPIB hardware and such things as memory and reed relays. Like GPIB devices, modules are slaves that respond to commands issued from the controller. Unlike GPIB devices, modules are addressed by a point-to-point signal from the crate controller. The address of a module is its station number within the CAMAC crate. This has the advantages that modules do not have to contain the necessary circuitry to set and remember their address, nor do they have to compare their address with addresses coming down the bus. Indeed, on CAMAC, there is no address bus, nor is the data bus multiplexed between addresses and data as with GPIB. The disadvantages of CMAC's addressing scheme is that moving modules within the CAMAC crate necessitates changing, or informing the software in the host computer that the module has been moved and so will now respond to a new address.

Another difference between GPIB and CAMAC is the size of the data bus. CAMAC uses two 24 bit parallel buses. The "read bus" is used for all data transfers from modules to the crate controller. Similarly, the "write bus" is used for all transfers from the crate controller to the modules. In addition to these two buses, CAMAC utilizes a third parallel bus to carry the function code for the module to execute. The function code bus is a 5 bit parallel bus. It is complemented by a fourth 4 bit parallel bus carrying the station "subaddress." The subaddress is most useful with modules that contain multiple identical instruments; that is, some modules contain four ADCs, whereas others contain multiple high-speed scalars. The subaddress feature of CAMAC makes addressing each of these individual elements within a module simple and natural. In CAMAC, this collection of four buses and a few other associated control lines is known as the "dataway."

CAMAC systems can be configured as parallel systems where multiple CAMAC host computer interfaces each communicate with a different CAMAC crate, or they can be configured as a "daisy chain" serial loop (see Fig. 8.45). The parallel configuration is the faster of the two, but is limited by the constraint that the crate controller to host computer cables are limited to approximately 3 m. The daisy chain serial loop configuration is significantly slower than the parallel scheme because all information must flow serially, and must always flow in the same direction around the loop. The serial configuration does, however, allow remote placement of the CAMAC crates.

Like the GPIB, an interface is required between the host computer's bus and the CAMAC dataway. Unlike the GPIB, where the computer interface serves the purpose of controlling the bus and interfacing the host computer to

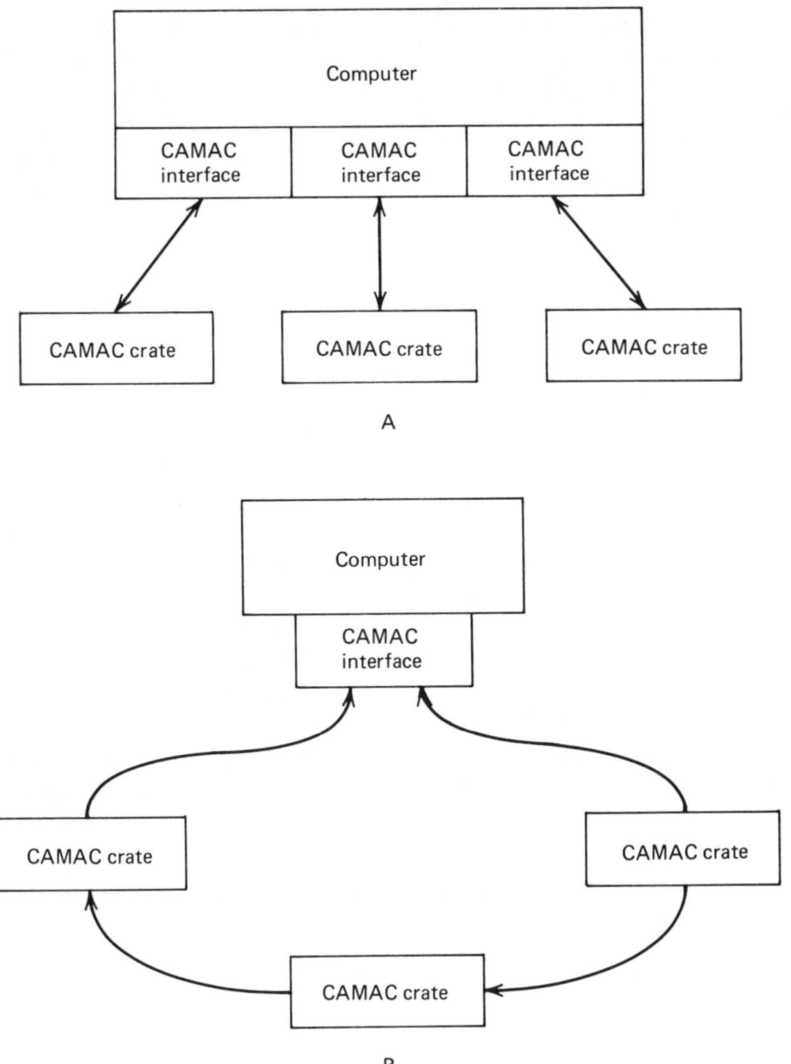


**Fig. 8.45.** (*a*) CAMAC crates interfaced in parallel. (*b*) CAMAC crates interfaced in a serial daisy chain.

the bus, CAMAC interfaces the host computer to the crate controller. Typically, CAMAC vendors market both the computer interface hardware and the CAMAC crate controller plug-in as a single item. Obviously, purchasing a CAMAC interface in this way is preferable to obtaining a computer interface and then trying to locate the compatible crate controller, or vice versa.

Because of the high parallelism in the CAMAC dataway, high throughput is typical of CAMAC. Also, because CAMAC basically originated in the nuclear physics and nuclear chemistry domain, CAMAC modules capable of very high speed operation are easily obtainable. Digitizers with local storage are available with a 1 MHz sampling rate at a resolution of 10 bits, and a 50 MHz sampling rates can be obtained when only 6 bits of resolution are required. Scalars capable of 100 MHz count rates are also available. Many high-speed CAMAC modules include solid-state memory so that acquired data can be stored locally. Later, these data can be transferred to the host computer at a more reasonable rate. Using direct memory access computer interfaces, information can be moved very rapidly between CAMAC modules and the host computer. Most modules that include local memory are designed to accommodate block transfers of part or all of their memory.

Another advantage of CAMAC for high-speed applications is directly related to modules with local memory. Because each module has its own local memory, there is no memory nor bus contention to worry about during high-speed data acquisition involving many channels of data. The only problem encountered is limited local storage of data by the modules. However, with memory chips quadrupling in capacity every 3 years or less, this soon may not be much of a problem.

Finally, like the GPIB, CAMAC provides a signaling method for modules to inform the crate controller, and thus the host computer, that they have completed their assigned task, or are ready for the next task. In CAMAC, this signal is known as Look-At-Me, or LAM for short. There are 24 point-to-point LAM lines, one for each of the first 24 stations. The crate controller typically concentrates all 24 LAMs into a single request tied to the host computer interrupt hardware. The host computer can then read the state of each of the 24 LAMs by issuing a special instruction to the crate controller, or the host can individually poll each of the modules.

### 3. Serial Communication Lines

The third type of standard instrument interface is the serial communication line. Traditionally, serial communication lines have been used to interface computer terminals to the host computer. More recently, serial communication lines have also been used to interconnect multiple computer systems into networks, and to interface computers with certain types of peripherals such as slow line printers and cassette tape drive. Currently, many instruments are commercially available with serial communication lines. Most of these are "smart instruments"; that is, they contain a dedicated micro- or mini-computer capable of being programmed by the operator, or by a host computer via the serial communication line. In most cases, the instrument's programming language is very limited, and is specifically tailored around the function of the instrument. As an example, a particular smart spectrophotometer executes programs made up of commands such as "measure," "smooth," "store," "display," "plot," "absorbance," and "transmittance." Although this programming language is quite limited by computer programming standards, it is a very natural language for acquiring and locally processing the spectrophotometric data. This very high level programming language also facilitates rapid host computer program development because the instructions that the host computer must send to control the smart instrument can all be composed and debugged manually at the smart instrument keyboard. This can eliminate many hours spent puzzling over why the data received from the smart instrument is not what was expected, or why the data was never received at all.

Serial communication lines derive their name from the fact that they transmit their information serially, one bit at a time. Serial interfaces are basically combined parallel-to-serial converters (transmitter) and serial-to-parallel converters (receiver). Data are transmitted over a serial line in little packets known as characters. A character is composed of a start bit (for synchornization purposes), 5–8 data bits, an optional parity bit, and finally 1, $1\frac{1}{2}$, or 2 stop bits. For proper transmission, both the transmitter and the receiver must agree on the number of data bits, the parity bit, if any, and the number of stop bits that comprise a character. In addition, the transmitter and receiver must also agree on the rate at which the bits are transmitted. This rate is typically given in bits per second or "baud." Although it is permissible to use any baud rate within the limits of the electrical transmission specifications for the physical transmission line, it is conventional to operate at 110, 300, 600, 1200, 2400, 4800, 9600, or occasionally, 19,200 baud. Nearly all serial interfaces come equiped with a clocking crystal and frequency division circuitry to provide these baud rates.

Another characteristic of serial interfaces is their mode of transmission. The simplest mode is known as simplex. Simplex allows transmission in only one direction and is not commonly used. The second mode is half-duplex. Half-duplex allows two-way communications, but not simultaneous two-way communications. Like simplex, half-duplex is not widely used because of its limitations. The third mode is full-duplex. Full-duplex allows simultaneous two-way communications. Nearly all computer terminals support full-duplex operation, as do smart instruments. As before, both serial interfaces must agree on the mode of communication. In addition, because transmission is done on one wire of the communication cable, and reception is done on another with full-duplex operation, care must be taken to make sure that the transmit line from the computer's serial interface is connected to the receive line of the smart instrument, and vice versa. Depending on the configuration of the respective serial interfaces, a null modem may be required.

Serial communication lines have many advantages. They are simple to construct with the modern UART (universal asynchronous receiver/transmitter) integrated circuit chip. Even better, they are widely available and quite inexpensive. Serial interfaces are very simple from the programming point of view. They can communicate with any device capable of sending and receiving serial data in the proper format. Finally, serial data can be transmitted over quite reasonable distances, allowing the smart instrument to be located away from the host computer. In fact, by using MODEMs, serial data may be transmitted over commercial telephone lines. Such transmissions are, however, limited in data rate, and noise can corrupt the transmission.

Along with their advantages, serial communication lines have two disadvantages. Serial communication lines come in four incompatible standards: RS 232 (usually RS 232C, which is also known as EIA), RS 422, RS 423, and 20 mA current loop. Luckily, 20 mA current loops are not generally used on modern equipment. In addition, RS 422 and RS 423 are relatively new and the majority of commercial serial interfaces capable of handling either or both are also capable of handling RS 232C. The popularity of RS 232C is mostly based upon the computer terminal market. Nearly all computer terminals now on the market are RS 232C compatible. Thus computer interfaces for serial communications using the RS 232C standard are easily available, and are typically quite inexpensive. The other disadvantage of serial communication lines is their limited transfer rate. RS 232C is basically only usable at rates at or below 29,200 baud. At best asynchronous serial transmission requires 2 bits of overhead for 8 bits of data. Thus RS 232C is limited to 1920 byte/sec. This is greatly slower than either the GPIB or CAMAC. The new standards, RS 422 and RS 423, are capable of much higher baud rates. However, since serial communication lines have no hardware handshake mechanism for assuring that the receiver is ready to receive the data, complicated software is required to implement a serial communications protocol. Use of a serial communications protocol to communicate with a smart instrument is, of course, limited to the protocols supported by the smart instrument. Also, at rates much above 500,000 baud, most micro- and minicomputers become strained to keep up with the incoming data when using the standard serial interfaces. Finally, RS is an acronym for recommended standard, and not all manufacturers adhere strictly to the standard, nor do they always implement the entire standard. This can lead to problems, especially when trying to connect smart instruments that utilize the full standard to the host computer's serial communication interfaces that were designed to communicate with computer terminals, because most terminals only require a subset of the standard, and the computer interfaces were designed accordingly.

## IX. CONCLUDING REMARKS

No mention has been made of the parallel interfaces available on many commercial instruments because there is no widely adopted standard. The type of connector, the number of interface lines, the format of the data, the voltage levels used to convey bits, and the timing and synchronization signal are typically different from manufacturer to manufacturer, and sometimes are different for different instruments all offered by the same manufacturer. To be sure, using standard computer parallel interfaces with some added circuitry to interface to the parallel interface port on these instruments is less complicated than interfacing to instruments with no parallel port. However, because no standard exists the advantages of standard software, direct plug compatibility, and simplicity of system construction and maintenance available with previously mentioned standards are all lost.

Also lacking from this section on interfacing standards is a discussion of the complexity of software required to support these standards. This is due to the highly variable form and function of the host computer interface to the standard. However, a few general comments on software complexity can be made, but remember that these comments will not be true for all computers. The simplest interface to support with software is the serial communication line when communications protocols are not required. The transmission or reception of characters is quite straightforward, and often the host computer's operating system's terminal support software can be used where multiple terminals are permitted. The standard computer bus interfaces run a close second to serial interfaces for ease of generating support software. Some manufacturers even supply general programs to support their product that may be used directly, or may be used as programming examples. Next in the line of increasing software complexity (and therefore difficulty) is CAMAC. The CAMAC manufacturers who market computer-to-CAMAC interfaces typically also sell the basic software necessary to support their product. The major problem with CAMAC comes in the programming of the CAMAC modules because often the documentation accompanying the modules is brief and mixes the programming information with the electrical specifications. After CAMAC comes serial communications lines requiring communications protocols. Most new protocols require forming cyclic redundancy check (CRC) values that are appended to the end of the data message and serve as an error check. In addition, the new protocols require a header to be sent before the data which often also contains its own CRC value. Finally comes the software to support the GPIB. Many computer to GPIB interface manufacturers now offer a set of GPIB communication subroutines to support their product. This has greatly reduced the software cost involved with GPIB. The downfall of GPIB is caused by the lack of standard command protocols. Each GPIB device has its own format for the commands it can execute, and these commands are typically strings of nonmnemonic ASCII characters.

## REFERENCES

1. Dahnke, K. F., S. S. Fratoni, Jr., and S. P. Perone, *Anal. Chem.*, **48**, 296 (1976).
2. Nashelsky, L., *Digital Computer Theory*, Wiley-Interscience, New York, 1966, pp. 238ff.
3. Perone, S. P., and D. O. Jones, *Digital Computers in Scientific Instrumentation*, McGraw-Hill, New York, 1973, Chap. 12 and references therein.
4. Perone, S. P., and D. O. Jones, *Digital Computers in Scientific Instrumentation*, McGraw-Hill, New York, 1973, Chaps. 7 and 8, and Appendix B.
5. Perone, S. P., and D. O. Jones, *Digital Computers in Scientific Instrumentation*, McGraw-Hill, New York, 1973, Appendix C.

6. Perone, S. P., and D. O. Jones, *Digital Computers in Scientific Instrumentation*, McGraw-Hill, New York, 1973, Chap. 8.

7. Schmidlin, E. D., in C. L. Wilkins, S. P. Perone, C. E. Klopfenstein, R. C. Williams, and D. E. Jones, Eds., *Digital Logic and Laboratory Computer Experiments*, Plenum Press, New York, 1975, Appendix F.

8. Zipper, J. J., B. Fleet, and S. P. Perone, *Anal. Chem.*, 46, 2111 (1974).

Part I
Section E

Chapter 9

# COMPUTER: PROGRAMMING

By Thomas L. Isenhour, *University of North Carolina, Chapel Hill, North Carolina*
AND
Peter C. Jurs, *Pennsylvania State University, University Park, Pennsylvania*

**Contents**

## I. INTRODUCTION

Computers are machines capable of processing information. Next to the development of language, the transportable information storage device (book) may be man's greatest invention. Surely, *Homo sapiens* becomes *Homo intellectus* with the ability to transmit knowledge down through the ages in the form of written records.

The computer is the next stage of evolution in the handling of information. Computers are dynamic books in that they can correlate information in addition to storing it. Computers, can, thereby, create new information and may eventually achieve the status of one of man's greatest inventions: language, books, and computers.

Man has been inventing computing devices for several centuries now. But only in the last 20 years has the programmable, high-speed, digital computer become well developed and widely available. We limit our discussion to that class of computing machines having the following properties:

1. Speed—tens of thousands to millions of mathematical operations per second.
2. Accuracy—5, 10, or even 20 significant figures without undue effort.
3. Programmable—controlled by a stored program which includes flexibilities such as conditional branching.

Computers are used in a great variety of ways, but the major scientific uses may be summarized as follows:

1. Computers can routinely and conveniently perform numerical calculations more rapidly and accurately than if performed by hand. (A second generation computer of the IBM 7090 class can perform approximately the equivalent of a man-life of calculations every 2 h.)
2. Computers can produce answers that would be useless if the time for manual calculations were required. (For example, the necessary calculations for correcting the trajectory of a spacecraft must sometimes be accomplished within a limited time. By the time a manual calculation could be done the answer might be only academic.)
3. Certain types of experiments may be optimized or simulated by computer calculations. (For example, a knowledge of the functionality of the variable involved often allows a computer simulation of a plant process. From such simulations the model may be optimized and the optimum plant constructed without a costly trial-and-error process of development.)
4. Computers are routinely used for information storage and the retrieval, organization, management, and presentation of large data banks. (Thousands of spectra may be searched and compared to an unknown, for example. The 10 nearest matches might then be presented in order of probability. Such operations may be accomplished in a matter of seconds or minutes with a portion of the computer's various memory devices acting as a permanent file for all the library data.)

5. Computers can be used to gather data from experiments as produced (in real time), and they can often be incorporated into experimental apparatus so that they direct the experiment. For example, computerized X-ray diffractometers are sufficiently automatic to collect data for many hours unattended.

6. Computers can be programmed to display intelligence by learning to perform tasks while improving their success as their experience increases. (Some advocates of machine intelligence go so far as to say that computers are essentially a new life form with nearly limitless possibilities.)

## II. AN OUTLINE OF COMPUTER APPLICATIONS IN CHEMISTRY

There is an information processing revolution in progress. Computers and other information handling devices have changed the world we live in, and they have certainly also changed the way science is done in the latter part of the twentieth century. Information processing of all types—storage, retrieval, copying, transmission, display, transformation—have all been altered by the advent of modern powerful computers. By computer we mean an electronic, digital, stored-program device as has been developed since 1950.

The modern digital computer is a uniquely capable device. It has capabilities to transform information in ways that were previously only possible by humans. The set of properties of digital computers, enumerated in the preceding section, endow them with the full generality for information processing. It has been said that computers constitute the first invention ever that significantly extends the intellectual capability of man.

The computer has influenced the methods used in the pursuit of scientific knowledge so dramatically and in such a ubiquitous way that it almost defies description. The changes are so fundamental that they are often not readily visible. Perhaps an analogy will demonstrate what is meant. The introduction of electricity into the world at the turn of the nineteenth century changed the way science was done in a dramatic and far-reaching way. However, to enumerate these ways is fruitless because of the fundamental way in which all aspects of science were changed. So too with the computer, which has caused changes as fundamental and far-reaching as the introduction of electricity.

The traditional tools of science include modeling, optimization, experimental verification of theories, estimation, parameter studies, and symbolic analysis. All these modes of scientific inquiry have been influenced by computers.

To look more closely at some of the specific influences of computers in chemistry, we break down applications into groups in the following pages. We emphasize that the following discussion is not an attempt to catalog all uses of computers in chemistry, but it is rather an attempt to show the diversity of uses and the fundamental importance of computers in a wide variety of chemical applications. The dividing lines between these groupings are not firm, and so some of the applications discussed could properly fall in more than one group. A few reviews have appeared that deal with the applications of computers in analytical chemistry (10,11,54).

## A. DATA PROCESSING AND NUMBER CRUNCHING

The most widespread use of computers in science is for rapidly and accurately performing calculations that were being done previously by other means that were relatively slow and inaccurate. These functions are not in themselves new, but they are rather new ways of performing old functions. However, the properties of computers generate in them the capability to perform such calculations and data processing tasks with so much speed and accuracy that they affect decision making about what type and how many calculations can be done. That is to say, given modern digital computers, one can routinely perform calculations that were inconceivable in the pre-computer era. In this section we discuss the sheer computational capabilities of computers. Areas of application in science include weather forecasting, nuclear reactor design, economic modeling, environmental monitoring, cryptography, and many others.

The source of this incredible computational facility is the speed of modern digital computers. To compare human potential for calculations with that of a computer, assume that a human can perform one operation per second. Assume that a computer can perform one operation per microsecond. Then, for the human:

$$\frac{1 \text{ operation}}{\text{sec}} \times 3.6 \times 10^3 \frac{\text{sec}}{\text{hr}} \times 8 \frac{\text{hr}}{\text{day}} \times 200 \frac{\text{day}}{\text{yr}} = 5.8 \times 10^6 \frac{\text{operation}}{\text{person-yr}}$$

For the computer:

$$10^6 \frac{\text{operation}}{\text{sec}} \times 3.6 \times 10^3 \frac{\text{sec}}{\text{hr}} = 3.6 \times 10^9 \frac{\text{operation}}{\text{computer-h}}$$

The ratio of these two rates of computation shows that 1 hr of computer computation is equivalent to 620 yr of human computation time. Thus the numerical computations that Gauss spent 20 yr performing could be done today in a time measured in hours. This points out an important consideration that arises from such computational speed: the routine work can be done by computer, freeing the human for the creative and intellectual part of the task.

An excellent example of a field that has been changed drastically by the calculational capabilities of computers is that of X-ray crystallography (24,39). Of course, the automation of X-ray diffractometers and data collection has also been important, but in this section we focus on calculations. In distinction to the past, complete structures can be done in a matter of weeks now, and structures are routinely done in a few months. This represents a time compression of factors of 10–100 over previous practice. By determining a structure is meant locating all the nonhydrogen atoms with an accuracy of 0.01 Å so that the conformation of the molecule is known and so that conclusions can be drawn regarding bond orders, strain, distortions, and so on. The fact that structures of moderate complexity can be determined in weeks to months makes X-ray crystallography a tool that is accessible to synthetic chemists on a routine basis.

Although many advantages of using computers in X-ray crystallography exist, here we are concerned with the calculation phase of this type of work. In the parameter refinement stage of solving a crystal structure, an interactive nonlinear least squares routine must be executed. For a structure of 50 atoms this may involve 450 parameters to be fit to several thousand data points. Large computers make this computation feasible on a routine basis. Additionally, the electron maps that results can be evaluated at tens or hundreds of thousands of locations over the molecule for subsequent output. The display capabilities of computers coupled with graphics display devices allows the convenient output of graphic or pictorial displays of complex structures for examination by the user. Additionally, two (or more) three-dimensional structures can be displayed simultaneously and compared visually or moved relative to one another to test hypotheses regarding shapes, sizes, and interactions, for example, of two structures. All in all, the advent of modern digital computers has remade the field of X-ray crystallography.

An example of a calculation that can be done by hand but that is better done by computer is the estimation of log $P$ (the partition coefficient for a compound between a model lipid phase and an aqueous phase) using the "constructionist approach" implemented in the following equation:

$$\log P = \sum_i a_i f_i + \sum_j c_j$$

where $f_i$ is the fragment constant for the $i$th fragment of the compound, $a_i$ is the number of occurrences of the $i$th fragment, and $c_j$ is the $j$th correction factor (25,42). To apply this method requires the perception of the fragments that make up the compound's molecular structure and many decisions regarding which of the hundreds of available fragment constants are applicable. In view of the fact that log $P$ is an important predictor for a wide variety of biological activities of organic compounds, and in order to facilitate its calculation for large sets of compounds, the procedure for estimation of log $P$ was automated (12). For compounds of moderate size the estimation of log $P$ can be done in 1–2 secs, making feasible studies of large sets of compounds, and so forth. It should be noted that this calculation is not entirely numerical because the molecular structure being used to perform the calculation must be stored and the fragments involved must be perceived in preparation for the calculation of the log $P$ value using the above equation.

Other areas of chemistry where sheer computational power have changed the way things are done include quantum mechanics and statistical mechanics. New research areas have opened up where the quantity of computations provided a fundamental roadblock to progress in the precomputer era. *Ab initio* or even semiempirical quantum mechanical studies are inconceivable without the capability to perform thousands or millions of numerical integrations in an economically feasible manner.

## B. TIME-LIMITED OPERATIONS

Although the great speed of computers makes feasible many operations that were not previously feasible, in the limit this capability leads to the development of wholly new application areas where time limits are rigid. That is, computers make possible some types of experiments and/or computations that are infeasible otherwise. An example would be the necessity to compute midcourse corrections for a space ship traveling to the moon from the earth. Failure to complete the calculation by the time when it must be applied to correct the course would make the result meaningless.

An example of a chemical experimental area where computers have impacted on fundamentally time-limited problems is that of ir spectroscopy. The design, implementation, and use of Fourier transform ir spectrometers (FTIR) has been made possible by the advent of minicomputers which are integral parts of the FTIR instruments (23,23). The minicomputer runs the experiment, that is, controls the interferometer, collects the data possibly with time averaging, and processes them through the Fourier transform to obtain the usual spectrum as a function of wavelength or frequency. For measurements taken on grating spectrometers and FTIR instruments, the $S/N$ ratio of the FTIR spectra is $M^{1/2}$ greater than that taken on the grating instrument. This assumes measurements taken at equal resolution over an equal measurement time with the same detector and on instruments with identical optical throughputs and efficiencies. $M$ is the number of resolution elements. In practice FTIR instruments are not used so often to obtain better spectra in the same experimental time but to obtain spectra of equivalent quality in far less time than would be necessary on a grating spectrometer. The same signal to noise ratio can be obtained $M$ times faster on the FTIR instrument. For experiments where the events being monitored are fast, FTIR measurements can obtain data that could not otherwise be obtained. Examples include the coupling of FTIR to gas chromatographs and kinetic studies where the species being observed are transient. Another case arises for experiments where the sources are weak and measurement times would be unacceptably long with grating instruments. Examples include spectroscopy of astronomical objects, remote sensing, or experiments in the far ir region.

The above discussion refers to the data gathering portion of the FTIR experiment, but the data manipulation part of the experiment is important too. Computing the Fourier transform was once an extremely time-consuming procedure, even with computers, with the time required proportional to $N^2$, where $N$ is the number of data points to be transformed. A new algorithm was discovered in 1965 by Cooley and Tukey (16) that allows the Fourier transform to be calculated in time proportional to $N \log_2 N$. For a 4096 point array, $N^2 = 1.67 \times 10^6$, $N \log_2 N = 4.9 \times 10^4$, and the time savings is a factor of over 300.

This computational efficiency has been a crucial factor in making FTIR an attractive instrumental technique.

### C. EXPERIMENT MANAGEMENT AND CONTROL

Analytical chemical instrumentation has been revolutionized over the past decade by the advent of closed-loop experimental management and control by computers. This involves the direct interfacing of the computer to the instrument so that the results can be monitored as they are produced and control signals can be returned to the instrument. The results often can be manipulated and transformed in complex ways during this on-line operation owing to the difference in rates of the experimental results production and computer operations. Thus complex decision making can be built into the on-line operations. Owing to the flexibility attainable through software alteration, this on-line computer control of experiments offers unique advantages compared to alternative modes of experimentation. Of course, once a set of data is collected, the data can be transformed, correlated, displayed, and so forth in a form that is most meaningful to the experimenter. The improvements in accuracy, speed, and convenience in the performance of complicated experiments can be impressive. Many new experiments, or even experimental areas, have opened up because of these capabilities of computers.

Many areas of analytical instrumentation have participated in the move toward on-line computers being incorporated as part of the instrument. An example is FTIR spectroscopy—an area where the computational capabilities of the incorporated minicomputer are also crucial to the feasibility of the method. The requirements placed upon the minicomputer in FTIR and the advantages that result have been described thoroughly from several points of view in the literature (23). Mass spectrometry also constitutes an excellent example of the impact of minicomputers; a number of manufacturers now offer GC/MS/COMP systems commercially.

### D. OPTIMIZATION

One of the traditional tools of scientific investigation is that of optimization. Given an experiment or a process, an immediate goal is often to improve the efficiency, accuracy, precision, cost, or reproducibility, for example, in such a way as to get the most out of the system under investigation. The advent of computers has dramatically changed the ways in which this can be done in chemistry.

Optimization theory is a field unto itself in applied mathematics, economics, and business. Many of the techniques developed have been applied to chemical systems (44). For example, linear programming constitutes a set of techniques for allocating limited resources among competing activities in an optimal manner. The algebraic function employed as a measure of the desired output of the system is too complex to be optimized directly but must be optimized iteratively. The method can be used even if the functional form describing the interactions of the variables is unknown, as long as the output can be measured. The methods of linear programming have been applied to the optimization of analytical chemical experiments (44,48). In one such application seven experimental variables in a flame spectrometric experiment were varied under computer control using a linear programming search procedure (51). The goal was to maximize the flame spectrometric signal as a function of burner position, fuel flow rate, oxidant flow rate, monochromator settings, and photomultiplier voltage. The simplex method of linear programming has also been used to optimize chromatographic separations (58) and to optimize colorimetric analyses (62).

Other types of optimizations using computer have also been reported For example, in neutron activation analysis the equations describing the signal measured as a function of irradiation time and decay time are well known. However, to obtain the optimum conditions for irradiation and decay times to enhance the selectivity for a particular element of interest in a particular matrix of background elements requires solution of a set of simultaneous equations that cannot be done analytically. The equations must be solved iteratively. Accordingly, a computer program was developed to automate these tedious computations (30), thereby making neutron activation analysis under optimum conditions a much more attractive technique.

## E. INFORMATION STORAGE AND RETRIEVAL

There exists a vast storehouse of chemical information that must be managed. To be useful to chemists it must be accessible. To be accessible it must be stored in an orderly, convenient, and indexed way so that it is economically feasible to access it. Computers are admirably suited to this task, and they have been used for all possible variations of information storage and retrieval of chemical information of all types: bibliographic information, numerical data banks (mostly mass spectra), and chemical structures themselves.

Within the area of storage and retrieval of numerical data, for example, spectra, two advantages accrue due to computer capabilities. First, the data bases to be searched can be very large and therefore have representatives of many structural classes, assuring the user that a wide variety of structural classes have been considered during the matching process. Second, each comparison between an unknown spectrum and a reference spectrum can be a complex calculation. More sophisticated comparisons can be done than would be attempted if the searching were done visually or by some other manual method.

Searches of files of reference mass spectra for the identification of organic compounds has become an important tool in analytical chemistry (8). Mass spectra are of widespread interest in this context because of the number of GC/MS systems that are producing data and the growth of computer-compatible mass spectral reference spectra files. Such search methods can aid the analyst in at least two ways: if the unknown compound is a member of the reference file, the search can reveal this fact and report the identity of the compound; if the unknown is not contained within the file, then a set of spectra that are most similar can be found. The presumption in the second case is that similar compounds have similar spectra and that the matching criterion used shows up this similarity properly.

The NIH-EPA Chemical Information System (CIS) (26,47) is a collection of many data bases and search techniques for provision to users through an interative computer network. The chemical data bases are stored on disc files so that searching need not be in batch mode. Data bases included are mass spectra, $^{13}C$ nmr spectra, X-ray data for crystals and powders, several bibliographic data bases, and others. The system is in a continual state of growth as new data bases are acquired or new types of data begin to be gathered in the chemical community.

One component of CIS is the mass spectral search system (MSSS). Its data base consists of more than 30,000 spectra gathered from several sources. Interactive searching from a computer terminal is done in one of a number of ways: by specification of peaks required to be present or absent in any matching spectrum; by specification of molecular weight or partial or complete molecular formula; or by matching an entire spectrum of an unknown against all members of the file for best fit. This matching is actually done by a method developed by Biemann and co-workers (27). An abbreviated version of the unknown spectrum consisting of the two most intense peaks in each 14 a.m.u. interval is matched against the spectra of the reference file which have previously been similarly abbreviated. A similarity index is computed between the abbreviated unknown spectrum and each reference spectrum; the most similar spectra have the largest similarity indices. Spectra with high indices found are reported along with the *Chemical Abstracts* registry number and ancillary information, including a sketch of the structure or a graphic display of the spectrum if the user's terminal allows it. The MSSS and the remainder of the CIS are accessible through a commercial computer network through the United States, western Europe, and certain other areas of the world.

Two methods for searching very large mass spectral data bases have been developed by McLafferty and co-workers and are called PBM, for probability based matching, and STIRS, for self-training interpretive and retrieval system (41,46,63). These programs compare the spectrum of an unknown compound against the spectra contained in a reference file. However, they also have some capability to make deductions.

The PBM method uses a probability weighing procedure to attach importance to the masses and abundances in the unknown spectrum. It also uses a reverse search procedure wherein each spectrum in the reference file is matched against the unknown spectrum; this allows for identification of spectra for unknowns that are actually mixtures.

The STIRS provides structural information to the user as well as the reference spectra that match most closely. A set of substructural units predicted to be contained in the unknown are found by comparing the unknown spectrum against the "signatures" of these substructures.

Both PBM and STIRS are available to interested users over a time-sharing computer network.

The accumulation of very large data bases of chemical structures within industrial laboratories and in government-gathered files has led to the necessity for developing searching methods to retrieve useful information from the files. A number of computerized searching systems have been reported in the literature. Examples include the structure and nomenclature search system (SANSS) of the NIH/EPA Chemical Information System (26), the substructure searching capabilities of the Upjohn Information System (29), and an information system described by Dyott and co-workers (17). These systems each rely on providing the user with the capability of sketching a substructure and then using it to retrieve structures from the reference file that have the substructure embedded within their structures. Furthermore, these types of systems have links to other types of data so that all the structures thereby retrieved can be checked for activity in a certain biological test or checked for inclusion in a certain list of toxic substances, and so forth. Thus linking of several data bases has been done to provide the user with a flexible, powerful tool for retrieval of information out of massive files in real time.

## F. INTELLIGENT PROBLEM SOLVING

Techniques have been developed over the past several decades that allow computers to perform tasks that are normally considered to require human intelligence to perform. The field of chemistry has supplied some of the problems of this class, which we describe in this section.

The elucidation of chemical structures from spectroscopic data can be broken down into a series of steps as follows: postulate tentative structure(s) that are consistent with the experimental data available; predict the spectra for each of the tentative structures; compare the predicted spectra with the experimentally observed ones and modify the list of tentative structures according to the degree of matching accuracy obtained. This can be restated as the systematic posing, testing, and rejection of hypotheses. A number of computer-assisted approaches to this task have been advanced. One set of methods for approaching the general problem of structure elucidation are those involving library searching, which were described above in a preceding section under information retrieval. In this section we focus on the techniques that have been developed to attack structure elucidation methods by other means.

One example of this area of the application of machine intelligence to chemical structure elucidation is the intensive effort by the Stanford University DENDRAL group. A large number of papers have appeared dealing with the enumeration of isomers, computer-assisted structure elucidation, interpretation of chemical data, and so forth. The underlying thread that runs through these contributions is the application of artificial intelligence techniques to a variety of chemical problems, often with the viewpoint of graph theory imposed on the analysis. The heart of their approach is the CONGEN routine (7), *con*strained and structure *gen*eration. The routine is based on graph theoretical concepts and is capable of generating all possible isomeric structures consistent with a list of constraints input by the user. The constraints are of the nature of numbers of various atom types, presence of substructures, absence of substructures, and so forth. Thus CONGEN is an example of a program that can aid the chemist in generating an exhaustive list of tentative structures consistent with all known constraints, it is available to the chemical community over a time-sharing computer network, and it has been used for a wide variety of research problems. The testing of structural hypotheses by developing spectral information to be compared to the observed data has also been reported (56). Routines were generated to use mass spectral to test the plausibility of the tentative structures against observed mass spectra.

Reports of similar work on computer-assisted structure elucidation have appeared. Clerc and co-workers have focused primarily on the wise use of spectral libraries to aid in structure elucidation (13–15). Sasaki and co-workers (52) have worked on the development of a complete system, CHEMICS-F, to perform the entire task of posing tentative structures, testing them against data, and trimming the list. Munk and co-workers have been developing a system, CASE, for computer-assisted structure elucidation as well (53).

A number of pattern recognition techniques taken from the artificial intelligence literature have been applied to chemical problems. Many classes of chemical problems are well suited to analysis using these methods owing to the nature of the data. These methods can deal with data in which each observation is represented by many more than three variables and therefore cannot be visualized or displayed. These data can be multisource data, with different scales, origins, distributions, and so forth, or discontinuous data. Pattern recognition methods comprise the detection, perception, and recognition of invariant properties among sets of measurements. Application studies of chemical problems using pattern recognition techniques have been reported in a number of areas (30,33,37). Work has been reported in the areas of mass spectrometry (50), ir spectroscopy (66), nmr spectrometry (55), electrochemistry (61), materials science and mixtures analysis (20,40), modeling of chemical experiments (19), and analysis of chromatograms of human biological fluids for pathological conditions (45). Several reports have appeared describing interactive pattern recognition systems for laboratory use (36,64). Pattern recognition methods have also been applied to studies of chemical structure-biological activity relations and drug design (34,60).

## G. GRAPHICS

The effective communication between the user and the computer routine is essential for computations to have great utility. Many forms of data and information, particularly in chemistry, are graphical. The lowest common denominator among chemists is the structural drawing, and thus computers should be able to deal with such structural drawings in order to communicate effectively with chemists.

Interactive graphics devices act as effective interfaces between computers and chemists and do allow the chemists to send information to the computer and receive it from the computer in terms of structural drawings. Many different types of chemical typewriters and other specialized input devices have been described in the chemical literature. Also, many different types of outputs of structural drawings have been reported. These I/O methods have been used as interfaces to many different types of routines including chemical structure elucidation routines, substructure searching routines, and chemical structure information handling routines.

As an example of how widespread computer produced structural drawings have become, we call attention to the fact that crystallography results are now commonly reported in the chemical literature along with ORTEP plots (32). Textbooks (59) now also contain computer-generated ORTEP structure plots. Electron density maps generated by computer are also very common in crystallography.

The production of high-quality stick representations of molecular structures for publication has been reported (4). A routine that generates structural drawings on a teletypewriter from a connection table representation has been reported (6). A routine that generates space-filling representations of molecular structures has been reported (57). Virtually all the chemical structure information handling systems described above, in the section on information storage and retrieval, have graphic input and output for communication with the system user.

## H. CURVE FITTING, SIMULATION, AND MODELING

The existence of conveniently accessible computers makes accessible to experimentalists a variety of curve fitting, simulation, and modeling methodologies that greatly expand their data interpretation powers. By curve fitting is meant the mapping of an algebraic function onto a set of data points. The purpose may be to verify a hypothesis that produced the function or it may be to provide an empirical relation that can be used for interpolation or the taking of derivatives. The standard method involves defining an error function to be the difference between the observed dependent variable values, $y_i$, and the values of the function evaluated for the corresponding independent variable values $f(x_i)$ summing over all observed data, $n$.

$$Q - \sum_{i=1}^{n} [y_i = -f(x_i)]^2$$

The error function is minimized with respect to the parameters of the algebraic function to obtain the fit equation. Simulation involves developing a mathematical model that is implemented in software so as to mimic a real system when the software is executed. The intent is to create a mathematical model whose parameters represent quantities with physical significance. Modeling involves developing a computer-based model that is self-consistent but does not necessarily bear resemblance to reality.

A very large number of curve-fitting papers have appeared in the chemistry literature. Much work has been done with gamma-ray spectra because the data are obtained digitally, conform to well-known algebraic functions, contain random noise imposed on the desired peaks, and are obtained by computerized or computer compatible methods and are therefore easily analyzed. Smoothing of digital data has been widely practiced (67) to improve the visual appearance of digital spectra. An elaborate curve-fitting procedure was reported that developed a continuous, differentiable, analytical function for describing monoenergetic gamma-ray pulse-height distributions from NaI(T1) scintillation counters (38). Using this function a gamma-ray spectrum of 400–4000 points could be represented by 10–20 parameters. An iterative curve-fitting procedure was described for fitting an eight parameter function to chromatographic peak profiles has been reported (9). The resolution of overlapping peaks, that is, deconvolution, has been reported in a number of application areas (1,49). The technique of spline fitting has been applied to chemical data (35).

Simulation is an old and well used method wherein a physical object or event is simulated with a mathematical model in order to study its properties in advance of construction of the device or the running of the experiment. Simulations can be deterministic or stochastic. Deterministic systems are completely defined by mathematical relations containing deterministic variables that are capable of precise measurement and control. Variables that are randomized are stochastic variables. Models or simulations containing variables that are stochastic must utilize mathematical techniques appropriate to probability and statistics. Studying devices through simulation can be economical of time and money because the device can be tested thoroughly before a prototype is constructed. In addition the device can be tested through simulation in situations or regimes that are inaccessible experimentally. For example, a device can be tested through simulation at very high temperatures or very low pressures or under stresses that would not be obtained in real experiments. Many simulations make possible studies that would be totally impossible otherwise. An example of a deterministic simulation of a device that allowed performing simulated experiments that could not be done in the laboratory would be the development of a quadrupole ion trajectory simulation software routine (5). Using this routine the trajectories of individual ions could be observed passing through a quadrupole mass filter. Effects on trajectories of alterations in ion entrance conditions, alternating current phases, and so on could be studied. A different example of simulation is provided by studies of polymers using Monte Carlo simulations (43). Here the molecular level details of polymer growth, for example, can be studied by the generation of large molecules through stochastic simulation and comparison of results with experimental results.

A chemical field where modeling has been widely employed to aid in the interpretation of experiments is kinetics studies. A review has been presented (21) on how computer modeling has been used in gas kinetics studies for experiment design, data interpretation, mechanism development, and derivation of values for elementary reaction rate constants.

The field of modeling also has received attention from chemists. Here the attempt is to define a self-consistent system and then use it to learn something about the real world. A good chemical example is the field of molecular mechanics, force-field studies, or molecular model building. The object is to create routines that develop realistic three-dimensional models of molecules by using classical forces. A strain energy function is defined to be the sum of terms due to bond length deformation, bond angle deformation, torsional angle deformation, and so forth over all the intramolecular interations of the molecule. Then this strain energy function is minimized by moving the atoms about relative to each other searching for a global minimum. The models that result are still abstractions of the real world, but they allow some kinds of inquiries to be made that would be difficult to make otherwise. This field of molecular model building has created a literature that describes the many approaches to the subject (2,3,18,28,65). These methods of developing molecular models are particularly good examples of an entire field of chemistry that could not exist in the absence of large-scale computational facilities.

## III. PROGRAM DEVELOPMENT (PROGRAMMING)

### A. THE NEED FOR COMPUTER PROGRAMS

Computer programming is the rephrasing of a problem in machine terms. The first step is the analysis of the problem. The next is the generation of a procedure (algorithm) for solving that problem that can be implemented on the machine. The final step is coding that procedure into a sequence of machine instructions called a program.

In a sense, the human being acts as an intermediary between experiments which collect data and the computer which processes it. The human's job is to be "boss" by telling the computer what operations to do in what order. The position of "boss" is a demanding one, however.

Normally, one has management control over one of two kinds of devices: (a) simple, unintelligent machines and (b) Intelligent animals. In the former case, most machines have a limited number of operations that can be brought about through the use of a small number of controls. A steam shovel, ir spectrometer, automobile, and other machines respond in a readily understandable fashion to a small set of control variables. In the second case, domestic animals and human beings perform complex tasks with very simple instructions. One tells a horse to "giddap" or slaps the reins on his neck, and a very complicated procedure of running (in one of several gaits) over rough terrain is performed. One does not have to tell the horse how to stop, to go around trees, and so forth. Or one tells a secretary to make an appointment with Mr. Honeycutt and an equally complex series of events occurs which depends totally on the intelligence of the employee.

Computers, however, can perform so many operations so rapidly that a great number of alternative must be considered. However, the "intelligence" for this complex machine must be provided in the computer program. The very strength of the computer—its ability to perform virtually any mathematical or information-processing operation—becomes its disadvantage in that a program must be developed that defines *exactly* the procedure to be followed.

The situation is somewhat like the parlor joke, "How do you sculpt an elephant?" Answer: "Simple—get a big block of granite, and chip off everything that doesn't look like elephant." The basic computer is capable of performing any possible mathematical procedure, (just as the basic block of granite contains all possible sculptures). Now the programmer must extract only that set of operations that will achieve his goal.

## B. WHAT IS A COMPUTER PROGRAM?

A computer program is a set of instructions that will cause a machine to perform a desired mathematical analysis. The most widely used computer language is FORTRAN. For that reason we use a FORTRAN example of a computer program for our discussion.

A routine mathematical analysis is finding the best slope and intercept for fitting a straight line to some data. For example, vapor pressure is related to temperature by the following equation:

$$\ln P = -\frac{\Delta H}{RT} + C \tag{1}$$

where $P$ is vapor pressure, $T$ is absolute temperature, $R$ is the universal gas constant (1.987 cal/mole deg), $\Delta H$ is the change in enthalpy for vaporization, and $C$ is a constant.

If a table of $P$ versus $T$ is measure, then the experimental data can be redefined as

$$Y = \ln P \tag{2}$$

and

$$X = -\frac{1}{RT} \tag{3}$$

to give

$$Y = (\Delta H)X + C \tag{4}$$

Solving for the best slope of $Y$ versus $X$ gives $\Delta H$. By linear least squares analysis the best value is

$$\Delta H = \frac{\sum_i X_i \sum_i Y_i - n \sum_i X_i Y_i}{(\sum X_i)^2 - n \sum X_i^2} \tag{5}$$

where $n$ is the number of points.

The algorithm is fairly simple; one must first compute $\Sigma X_i$, $\Sigma Y_i$, $\Sigma X_i, Y_i$, and $\Sigma X_i^2$ and then use these values to solve equation (5).
A FORTRAN program for implementing this algorithm follows:

| | Statement | Explanation |
|---|---|---|
| C | LEAST-SQUARES PROGRAM FOR DELTA H | A comment (message statement) |
| | DIMENSION X (100), Y (100) | Reserves space for up to 100 values of $X$ and $Y$ |
| | READ (1,9)N | Reads value of $n$ |
| | READ (1,19) (X(I), Y(I), I = 1,N | Reads pairs of $X$ and $Y$ for $i$ from 1 to $n$ |
| | SX = 0.0<br>SY = 0.0<br>SXY = 0.0<br>SX2 = 0.0 | Defines summing variables as zero |
| | DO IO I = 1,N | Causes "loop" through statement "10" $n$ times |
| | SX = SX + X(I) | Adds each successive value of "X" to "SX" |
| | SY = SY + Y(I) | Adds each successive value of "Y" to "SY" |
| | SXY = SXY + X(I)*Y(I) | Adds each successive product of "X" and "Y" to "SXY" |
| 10 | SX2 = SX2 + X(I)**2 | Adds each successive square of "X" to SX2 and terminates loop |
| | DELH = (SX*XY-N*SXY)/((SX)**2-N*SX2) | Calculates $\Delta H$ |
| | WRITE (2,29) DELH | Prints $\Delta H$ |
| | CALL EXIT | Returns computer to system control (terminates program) |
| 9 | FORMAT (15) | Defines forms of "$N$" on data card |
| 19 | FORMAT (2E10.3) | Defines forms of "$X$" and "$Y$" pairs on data card |
| 29 | FORMAT (' DELTA H = ', E10.3) | Defines forms of $\Delta H$ to be printed out |
| | END | Defines end of program |

At first writing, a program seems like a lot of work for solving a problem. It is necessary to punch it into cards (or through a terminal), run it to find the syntax errors, and then test it with data to find the logical errors. Is it worth it? For a very simple calculation, to be performed one time, probably not. However, consider actually doing the above least squares program by hand on 100 points collected on each of 100 compounds—a minimum of 60,700 mathematical operations (subtractions, additions, multiplications, divisions, and squarings) must be performed to process this data. And it can all be accomplished with a single, 20 statement computer program. What's more, the program can be used an unlimited number of times; it does not wear out!

## C. SOURCES OF COMPUTER PROGRAMS

There are three principal ways to acquire a needed computer program. One can borrow (or purchase) a program from someone else, hire a programmer, or write it oneself. Each method has its advantages and disadvantages.

Most major computer centers now have large statistical programming packages such as SPSS (Statistical Package for the Social Sciences); BMDP (Biomedical Programs-UCLA); SSP (Scientific Subroutine Package-IBM); and IMSL (International Mathematics and Statistical Library). Although some familiarity with the computer system is needed to run programs, a naive user can execute very sophisticated programs without being a programmer. Also, organizations such as CONDUIT and the Quantum Chemistry Program Exchange provide documented, tested programs to subscribers. The open literature abounds with references to programs that researchers will provide for little or no cost to anyone who requests a copy. Often one can perform very sophisticated calculations without having to develop the necessary programs. One disadvantage to using these programs, however, is that modification is often difficult and, if the problem to be solved is not perfectly suited for the program, then great difficulties can arise. Further, there is sometimes more than a little effort involved in getting someone else's program to run on one's machine. Finally, one is always left with the possibility that one's data is being erroneously processed by someone else's program. The user of such services is usually best if he or she is a programmer of some competence.

Hiring a programmer also has its advantages and disadvantages. In large laboratories where it is possible to have a customer programming group, the professional programmer can be of great value. This is particularly true when the programmer becomes familiar with particular research areas and research groups. Again, however, the user who understands programming is in a much better position to discuss his or her needs and guarantee the best results.

Programming is becoming sufficiently easy that most scientists would be well served by acquiring at least rudimentary programming skill. Through simplistic languages such as BASIC and use of a simple time-sharing or even microcomputer system, only a few hours are required to start writing programs of some utility. The level of skill developed can then be determined by the needs of the research program. Once some basic skill is acquired, the individual is in a much better position to take advantage of programming services and outside sources of programs.

## D. MAJOR CONSIDERATIONS IN COMPUTER PROGRAMMING

### 1. Languages

There are two major classes of computer languages; assemblers and compilers. Assemblers are constructed to make use of the actual machine functions. That is, assembler statements are usually instructions that can be performed by specific hardware components of the computer. Compilers, on the other hand, are usually more algebraic/English languages designed for ease in programming.

Assembler languages often include an instruction set that corresponds one to one with the hardware instructions of the computer. These may be set of anywhere from 50 to several hundred instructions. They include operations such as "clear register A," "shift left logical," "load byte," and so forth. All assembler languages relate directly to the machine and are the most efficient language in actual operation. However, they are far more difficult to program than compiler languages. Many expert computer users have never learned an assembler language. The assembler is the most powerful language because it uses the actual architecture of the machine. However, it requires a level of expertise far beyond that of compiler programming.

Assemblers require the programmer to specify each individual operation, which means that the very process of printing a number may require a dozen or more instructions. At the crudest level of assembler every operation of the machine is accessible and every function must be performed by the programmer. With more sophisticated assemblers often macro-instruction packages are available that will carry out a series of operations efficiently without requiring the programmer to write lengthly code.

In difficult problems of word manipulation, sorting, and where extreme speed is desired, assembler programming is sometimes mandatory. However, this is much more often the case of the small machines than it is on the large systems. Very little assembler programming is done on the maxi-size machines. In essence, assembler languages require the programmer to "think like the machine."

The original compiler language was FORTRAN. It was developed to give the naive programmer an opportunity to use the machine with minimal effort. The compiler concept is what has made digital computers widely used by the scientific community. Radio would not have been much of a commercial success had all transmissions been restricted to Morse code. Nor would computers be nearly so useful if only assembler languages existed.

The compiler languages (one of which is discussed in some detail in Section III.F) presents pseudo-English/algebraic terms that represent the common steps necessary to implement the solution to a problem. For example, arithmetic expressions are written in the typical algebra of add, subtract, multiply, and divide. Transfers from one part of a program to another are accomplished with "go to (label)" statements. Input and output are accomplished by simple statements including instructions such as "read" and "print."

Because FORTRAN and BASIC are the two most popular compiler languages, some special comments are in order on the very different ways in which they are executed. FORTRAN is a compiler language; meaning that the original FORTRAN program (source program) must be processed (compiled) by a special master program (compiler) and converted into a machine-language program (object program) before it can be run. BASIC, on the other hand, is an interpretive language. Each statement is individually converted as the program executes. This means that BASIC programs can be run immediately, without the sometimes lengthy compiling process. On the other hand, because FORTRAN programs are compiled into efficient machine language code, these programs run much faster. Often FORTRAN programs are 10 times or more faster than BASIC programs. However, BASIC programs are easier to write and modify because the compiling delay is not encountered.

Recently BASIC compilers have been developed that may erase the execution speed advantage of FORTRAN. Under these conditions, one can develop a BASIC program testing execution in the interpretive mode and, when satisfied, compile the program for improved execution speed, thereby having the best of both approaches.

In summary, it is recommended that most scientists become familiar with a compiler language such as FORTRAN or BASIC. Most persons can learn to use a simple compiler language in a matter of a couple of days. Furthermore, they can become experts over a matter of months by programming their own work. However, assembler programming remains an area for the professional and should be approached only if it is necessary.

## 2. Machine Classes

There are three rough designations of the size of computer systems: the large computation center variety such as the IBM 370 Series, CDC 7600 computers, Digital Equipment Corporation PDP-10's, and others; the ubiquitous minicomputer; and the newly arrived microcomputer.

### a. Large-System Computers

Large-system size computers are very complex machines capable of processing many programs simultaneously. They usually offer a variety of languages, several compilers as well as an assembler, and usually have very large storage capability on magnetic tape and disks giving them the capability of dealing with very large problems. A final advantage of the macrocomputer class machine is that many peripherals such as plotters and CRT-type terminals are available.

Along with all this capability comes considerable complexity. It is often fairly difficult to learn the job control language of the large systems. These are the programming steps necessary to set up the machine to run a particular job. Sometimes learning to use this system is more difficult than learning to write the programs necessary to solve the problems. Also, there is a different psychology associated with a large computation center. First, there is the desire to recover costs, so inevitable a biling procedure exists. There are accounting methods to make sure the right people get use of the machine and this inevitably results in an internal machine bureaucracy not unlike the human bureaucracies we construct. Commonly, the large computation center must serve a variety of users and therefore be under the control of no individual. Hence the psychology becomes one of the users going to the machine rather than the machine waiting for the user. More, and larger, problems can be done on macrosystems than with the smaller machines. However, local circumstances often determine whether an individual problem can be solved more quickly on the larger machine.

b. MINICOMPUTERS

For the last dozen years smaller machines, of ever increasing capability, have become widely available. Many of the minicomputers available today are more powerful than computation center machines were 10 years ago. Usually, a minicomputer is under the control of one or a few individuals. On the other hand, the disadvantages are that often the actual computing power is much less than the large systems. Peripherals are often lacking, and the system itself is simpler offering both fewer options for use and less complexity. The minicomputer may be easier to use because of the lack of complexity surrounding it, but capability is often lacking too. For example, it is possible to buy minicomputer systems that have no compiler languages at all, or in many cases the compilers are impoverished or actually have undiscovered and uncorrected errors.

The owner of a minicomputer needs to have some expertise in the computer area. Otherwise, as has frequently been the case, a machine may be bought that ultimately serves no real useful purpose.

c. MICROCOMPUTERS

A new, very exciting development is the microcomputer. These are machines built around a processor that is contained in only one or two integrated circuit chips. Prices are very low, and the capabilities are considerable considering the simplicity. But what was true about the minicomputer is often even more true about the microcomputer. Available software, compilers, and systems are very limited. However, the economics is quickly resulting in a widespread use of microcomputers.

One of the authors has recently had the pleasant experience of buying a hobby microcomputer for less than $1000 and putting it to use on real problems the same day it was purchased. Microcomputers exist that have BASIC and even FORTRAN programming and that author did the computation of the statistical analysis for an entire Doctor of Education dissertation in a period of about 25 hr. The time included about 15 hr of programming, 5 hr of entering data, and 5 hr of computing. The commercially available microcomputer offer an inexpensive route into computer usage for the individual.

### 3. Operating Systems

Operating systems are the control programs that run the computer itself. There are two general classes, "batch" and "time-sharing." Batch programming refers to running a job without intervention of the user. Time-sharing refers to a dialogue of operation between the user and the machine. The former case is the more efficient of machine time. However, in the latter case, the user can change parameters during an operation and often achieve desired results in fewer passes through the machine.

Large machine systems usually have very complex operating systems. These are fully in control of the machine and the user is simply a petitioner to the operating system to run his particular program. The complex operating system means that the large machine can do a variety of things for the user; it also means it can do a variety of things to the user. That is, the user is shielded from getting close to the actual machine architecture, and such things as writing driver programs for magnetic tapes are virtually impossible in many large systems. Because there are many users, it is necessary to defend the system from incursions by individual users.

Minicomputers have operating systems that vary from archaic to very modern. Some support time-sharing or even do background-foreground operations. (This involves holding unimportant jobs in the queue and using time available from major jobs to execute them.) Microcomputers have the same restrictions as minicomputers in that operating systems are often very limited. This is out of necessity, for the microcomputer has only a restricted computational capability, and this can be used either for running the operating system or executing user programs.

### 4. Sociology

Who is in charge of the computer actually varies with the size and cost of the system. On large computer systems there is a necessary bureaucracy and the user is nothing but a customer. On minis and micros it is possible to be the owner of one's system and to use it as one pleases. In many cases large computation centers become so complex that user groups and various public relations personnel are employed to tell people how they can use the system. There is no indication of it happening yet, but it is possible that central systems can grow so large as to be absorbed by their own bureaucracy, like many human organizations. There may, for that reason, be a finite size limit to an effective operating computation centre. A very important point is that the numbers computed on a large machine are no more accurate than those on a small machine. Usually, however, complex answers can be obtained more rapidly on a large system. This should be the criterion for the choice of machine usage, specifically, how cheaply and quickly one can solve one's problem.

### 5. Economics

There is a psychology about computers that says when the machine is not running, it is being wasted. Strangely enough, we rarely worry about a spectrometer or balance in a laboratory that is only being used a few hours out of 24. However, everyone seems very upset when a computer only runs a few hours out of 24.

The computation center, therefore, is usually organized to try to keep the machine running 24 hr/day. Smaller systems may be much more effective if they are simply waiting for their usage. On the major system a charging process is almost inevitable. On the small system it may be a waste of accounting effort to try to recover costs.

Most computation centers have the computer doing the actual accounting and perhaps because the machine is there, usually generate a charging algorithm that is too complex to be explained or even understood. To test this hypothesis, try asking the question, "How much does it cost to run a job?" You will never get a straight answer from a computation center.

Unfortunately, there is no general unit of measurement for job operation. Instead, you will be told that CPU seconds cost so many dollars, disk storage is so much per track, and so forth, and when you get through you will be sure that the retirement program of an insurance company has just been explained to you. The only way to find out is to submit your average job and see how many dollars it costs.

The complex economics of computation centers, combined with the desire to keep the machine running all the time, inevitably generates some scheme of "funny money." This is a way of giving allocations of "dollars" that can be spent at the computation center. The mini- and microcomputers usually can avoid this problem by simply having assigned use priorities.

## E. BASIC PROGRAMMING

As an example of how easily a scientist can learn enough programming to be useful, we present here a brief introduction to BASIC programming. BASIC is a compiler language which is executed one statement at a time by an "interpretive compiler." A small set of eight statements types is described. Each BASIC statement is preceded by a line number and variables are one or two characters long, the first character being alphabetic.

1. 10 DIM X(100), YY(75)

The DIM statement sets aside storage for subscripted variables. The above example reserves 100 locations for the variable *X* and 75 locations for the variable *YY*.

2. 30 INPUT A, B

The INPUT statement requests values of *A* and *B* by printing "?" on the terminal, awaiting a response for *A* and "?," and waiting a response for *B*.

3. 10 LET X = A * B + (C/D) ↑ 2 + 3 * C

The LET statement defines arithmetic operations. Standard operators are: + (addition), = (subtraction), * (multiplication), / (division) and ↑ (exponentiation). Parentheses are used to direct order of operation as necessary. The above statement is the BASIC equivalent of
$X = A \cdot B + (C/D)^2 + 3C$.

4. 70 PRINT X

The PRINT statement causes the values of variables following it to be printed.

5. 100 END

The END statement concludes a program.

6. 50 GOTO 80

The GOTO causes the program to jump to the indicated statement.

7. 70 IF X > Y THEN 270

The IF THEN statement causes a conditional operation. If the argument is true then the following statement is executed; if false, then the following statement is skipped. Another example: 90 IF A = O THEN PRINT X. X is printed only when $A$ is zero.

8. 100 FOR I = TO N

```
100 FOR I = TO N
    :
150 NEXT I
```

The FOR TO statement working with the NEXT statement causes a loop to be executed with $I = 1, 2, 3 \ldots, N$.

Using just these eight simple statements very powerful computer programs can be written and executed in BASIC.

Consider having up to 100 values of $X$ to calculate the mean and standard deviation:

$$\overline{X} = \sum_i \frac{X_i}{n}$$

$$\sigma = \left[ \frac{\Sigma(\overline{X}_i - X)^2}{n - 1} \right]^{1/2}$$

| | |
|---|---|
| 10 DIM X (100) | Reserves space for 100 values of $X$ |
| 20 INPUT N | Reads number of values |
| 30 FOR I = 1 TO N | |
| 40 INPUT X (I) | Reads $n$ values of $X$ |
| 50 NEXT I | |
| 60 LET MX = 0 | Sets summing value to zero |
| 70 FOR I = 1 TO N | |
| 80 LET MX = MX + X (I) | Sums all $X$'s |
| 90 NEXT I | |
| 100 LEXT MX = MX/N | Divides sum by $n$ |
| 110 LET SX = 0 | Sets summing value to zero |
| 120 FOR I = 1 TO N | |
| 130 LET SX = SX + (MX − X(I)↑2 | Sums $(\overline{X} - X)^2$ for all $X$'s |
| 140 NEXT I | |
| 150 LET SX = (SX/N−1)↑(1/2) | Computes standard deviation |
| 160 PRINT "MEAN = ", MX, "STD. Dev. = ", SX | Prints results, statements in quotation marks are printed as labels. |
| 170 END | |

An actual run for five values of $X$ (12, 30, 15, 28, 16) would proceed as follows:

```
? 5    (value of N)
? 12   (value of X1)
? 30   (value of X2)
? 15   (value of X3)
? 28   (value of X4)
? 16   (value of X5)
  MEAN = 20.20 STD. DEV. = 8.20
```

A more extensive example of BASIC programming follows.

This program is a calculation of several stages of fractional distillation. Using Raoult's law, one can calculate the vapor pressure ($P_i$) of any volatile liquid which is acting ideally from its mole fraction ($X_i$):

$$P_i = X_i P_i^0 \tag{6}$$

where $P_i^0$ is the vapor pressure of the pure substance.

Combining the vapor pressures of all components in a mixture gives the total pressure, and from the pressure fraction, the mole fractions of the components in the vapor state can be derived. The composition of the vapor, which is thereby calculated to be in equilibrium with the liquid, gives the results of the first step of distillation. The same calculations can be repeated as often as is desired to see what the effects of many stages of distillation would be.

In this program, the variables $X$ (mole fraction), $V$ (vapor pressure of the pure component), and $P$ (vapor pressure of the individual component) are all dimensioned as 100 so that a mixture of up to 100 components can be used. After spacing down a line, the first PRINT and INPUT pair request and accept the value of the number of components ($N$). Statement 130 tests $N$ to see if the program should be terminated. If $N$ is greater than zero, the number of stages ($N2$) is demanded and entered. Following that, the mole fraction vapor pressures for each of the $N$ pure components are entered in a FOR loop (statement 260). Next, a loop is initiated whose index $J$ *runs from 1 to the total number of stages,* $N2$. For each pass through this loop, the total pressure $PO$ is initialized as $O$, so that the individual pressures can be summed as they are computed in the inner loop (statements 220–250). In this loop, whose index runs over all the components, the individual vapor pressures are calculated by Raoult's law and, at each step, are added to the total pressure. Following the completion of this inner loop, the stage number and total pressure are printed, and a new loop is initiated to calculate and print the mole fractions of the components in the vapor phase from their vapor pressures and the total pressure. The program proceeds in the same manner through each stage of distillation. Finally, it brances back to statements 115–130 to determine if there is another such

```
LIST
100 REM FRACTIONAL DISTILLATION
110 DIM X(100),P(100),V(100)
115 PPRINT
120 PRINT "NUMBER OF COMPONENTS";
125 INPUT N
130 IF N<= 0 THEN 400
135 PRINT "NUMBER OF  STAGES";
140 INPUT N2
145 PRINT
150 PRINT "MOLE FRACTION AND PURE VAPPOR PRESSURE"
160 FOR I= 1 TO N
170 INPUT X(I),V(I)
190 NEXT I
200 FOR J= 1 TO N2
210 LET  P0=0
220 FOR  I=  1 TO N
230 LET P(I)=X(I)*V(I)
240 LET P0=P0+P(I)
250 NEXT I
260 PRINT
270 PRINT "STAGE NUMBER";J;"HAS TOTAL PRESSURE";P
275 PRINT "COMPONENT","FRACTION","PRESSURE"
280 FOR I= 1 TO N
290 LET X(I)*P(I)/P0
300 PRINT  I,X(I),P(I)
310 NEXT I
320 NEXT  J
330 GOTO 115
400 END
1000 LPRINT"LIST"
1010 STOP
```

```
RUN
NUMBER OF COMPONENTS? 3
NUMBER OF STAGES? 10
MOLE FRACTION AND PURE VAPPOR PRESSURE
? .4 , 60
? .5 , 40
? .1 , 70
STAGE NUMBER 1 HAS TOTAL PRESSURE 51
COMPONENT              FRACTION          PRESSURE
       1                .470588           24
       2                .392157           20
       3                .137255           7
STAGE NUMBER 2 HAS TOTAL PRESSURE 53.5294
COMPONENT              FRACTION          PRESSURE
       1                .527472           28.2353
       2                .29304            15.6863
       3                .179487           9.60784
STAGE NUMBER 3 HAS TOTAL PRESSURE 55.9341
COMPONENT              FRACTION          PRESSURE
       1                .565815           31.6484
       2                .209561           11.7216
       3                .224624           12.5641
STAGE NUMBER 4 HAS TOTAL PRESSURE 58.055
COMPONENT              FRACTION          PRESSURE
       1                .584772           33.9489
       2                .144388           8.38245
       3                .27084            15.7236
STAGE NUMBER 5 HAS TOTAL PRESSURE 59.8206
COMPONENT              FRACTION          PRESSURE
       1                .586525           35.0863
       2                .0965473          5.77552
       3                .316928           18.9588
STAGE NUMBER 6 HAS TOTAL PRESSURE 61.2383
COMPONENT              FRACTION          PRESSURE
       1                .574665           35.1915
       2                .0630633          3.86189
       3                .362272           22.185
STAGE NUMBER 7 HAS TOTAL PRESSURE 62.3615
COMPONENT              FRACTION          PRESSURE
       1                .552904           34.4799
       2                .0404502          2.52253
       3                .406646           25.3591
STAGE NUMBER 8 HAS TOTAL PRESSURE 63.2575
COMPONENT              FRACTION          PRESSURE
       1                .524432           33.1742
       2                .0255781          1.61801
       3                .44999            28.4652
STAGE NUMBER 9 HAS TOTAL PRESSURE 63.9883
COMPONENT              FRACTION          PRESSURE
       1                .491744           31.4659
       2                .0159892          1.02312
       3                .492267           31.4993
STAGE NUMBER 10 HAS TOTAL PRESSURE 64.6029
COMPONENT              FRACTION          PRESSURE
       1                .456708           29.5047
       2                9.90002E-03       .63957
       3                .533392           34.4587
```

calculation to be done. The output format provides convenient spacing for labeling the data. An investigation of the sample output shows how the distillation progresses to fractionate in favor of the more volatile components.

Most scientists develop programming skills quite rapidly. The secret is to get over hurdle of writing the first successful program. Then capability quickly increases with experience.

## F. HOW TO PROGRAM

Programming is usually accomplished in four steps; (1) defining the task, (2) developing an algorithm, (3) writing the code, and (4) debugging the program.

1. A complete mathematical analysis of the problem must be accomplished. It is possible to program until the exact problem is defined.
2. Given the mathematical plan, a stepwise procedure must be developed to carry out the calculation. For complex problems, it is often useful to draw flow diagrams in designing the procedure.
3. The program is now coded according to the developed algorithm. Programming languages vary, but a thorough knowledge of any one is usually sufficient to implement most algorithms.
4. After writing the program it must be executed and tested. The compiler usually detects syntactical errors such as uneven numbers of parentheses and other illegal statements. However, logical errors are more difficult to find. For example, a mathematical statement is

$$A = \frac{X + Y}{C + D}$$

and a BASIC statement is 100 LET A = X + Y/C + D. THE BASIC statement is perfectly legal, but will complete $A = X + \frac{Y}{C} + D$. The desired statement would be 100 LET A = (X + Y)/(C + D). Such debugging is accomplished by running known data to produce checkable answers and sometimes adding temporary print statements to check intermediate results.

Computer programming is completely logical. This is very appealing to the scientist who usually finds programming a lot of fun. At the same time it is coldly logical—machines, to date, have not shown intuitive behavior—and can be frustrating because the computer behaves *exactly* as you tell it to.

## IV. SUMMARY

The advent of the high-speed digital computer is bringing dramatic changes to many aspects of analytical chemistry. Problems insoluble a few years ago are now only difficult, and difficult problems have become simple. Programming, however, is a barrier to effective computer usage. Computers are calculating tools so flexible that although they can be instructed to do almost any complicated calculation, they must also be instructed to do the simplest calculation.

Over the past 20 years computers have increased dramatically in capacity and capability and decreased in size and cost, but programming languages have remained very much the same. Although BASIC is certainly simpler to use than FORTRAN and specialty languages have been developed, little progress has been made in making computers easier to program for scientific calculation.

Rather than wait for possibly easier ways to interact with computers, practicing analytical chemists would be well served to acquire some skill of their own in programming, both for their own special applications and to dispel the mysticism surrounding computers and those who use them.

## REFERENCES


1. Allen, G. C., and R. F. McMeeking, "Deconvolution of Spectra by Least-Squares Fitting," *Anal. Chim. Acta*, **103**, 73–108 (1978).
2. Allinger, N. L., "Calculation of Molecular Structure and Energy by Force-Field Methods," in V. Gold, Ed., *Advances in Physical Organic Chemistry*, Vol. 13 Academic, New York, 1976.
3. Altona, C., and D. H. Faber, "Empirical Force Field Calculations. A Tool in Structural Organic Chemistry," *Topics Curr. Chem.*, **45**, 1 (1974).
4. Blake, J. E., N. A. Farmer, R. C. Haines, "An Interactive Computer Graphics System for Processing Chemical Structure Diagrams," *J. Chem. Inf. Comput. Sci.*, **17**, 223 (1977).
5. Campana, J. E., and P. C. Jurs, "Computer Simulation of the Quadrupole Mass Filter," Int. L. Mass. Spectr. Ion Phys., **33**, 119 (1980).
6. Carhart, R. E., "A Model-Based Approach to the Teletype Printing of Chemical Structures," *J. Chem. Inf. Comput. Sci.*, **16**, 82 (1976).
7. Carhart, R. E., D. H. Smith, H. Brown, and C. Djerassi, "Applications of Artificial Intelligence for Chemical Inference. XVII. An Approach to Computer-Assisted Elucidation of Molecular Structure," *J. Am. Chem. Soc.*, **97**, 5755 (1975).
8. Chapman, J. R., *Computers in Mass Spectrometry*, Academic, New York, 1978.
9. Chesler, S. N., and S. P. Cram, "Iterative Curve Fitting of Chromatographic Peaks," *Anal. Chem.*, **45**, 1354–1359 (1973).
10. Childs, C. W., P. S. Hallman, and D. D. Perring, "Applications of Digital Computers in Analytical Chemistry–I," *Talanta*, **16**, 629 (1969).
11. Childs, C. W., P. S. Hallman, and D. D. Perrin, "Applications of Digital Computers in Analytical Chemistry– II," *Talanta*, **16**, 1119 (1969).
12. Chou, J. T., and P. C. Jurs, "Computer-Assisted Computation of Partition Coefficients From Molecular Structures Using Fragment Constants," *J. Chem. Inf. Comput. Sci.*, **19**, 172 (1979).
13. Clerc, J. T., Computer Aided Interpretation of Spectroscopic Data for the Structure Elucidation of Organic Compounds," in E. V. Ludena, N. H. Sabelli, and A. C. Wahl, Eds., *Computers in Chemical Education and Research*, Plenum, New York, 1977.
14. Clerc, J. T., "Computer Aided Spectra Interpretation for Determinging the Structure of Organic Compounds," *Chimia*, **31**, 353 (1977).
15. Clerc, J. T., "Computer Methods for the Spectroscopic Identification of Organic Compounds," *Pure Appl. Chem.*, **50**, 103–106 (1978).
16. Cooley, J. W., and J. W. Tukey, "An Algorithm for the Machine Calculation of Complex Fourier Series," *Math. Comput.*, **19**, 297 (1965).
17. Dyott, T. M., A. M. Edling, C. R. Garton, W. O. Johnson, P. J. McNulty, and G. S. Zander, "An Integrated System for Conducting Chemical and Biological Searches," in W. J. Howe, M. M. Milne, and A. F. Pennell, Eds., *Retrieval of Medicinal Chemical Information*, American Chemical Society, Washington, D.C., 1978.

18. Engler, E. M., J. D. Andose, and P. von R. Schleyer, "Critical Evaluation of Molecular Mechanics," *J. Am. Chem. Soc.*, **95**, 8005 1973).

19. Eskes, A., F. Dupuis, A. Dijkstra, H. DeClercq, and D. L. Massart, "Application of Information Theory and Numerical Taxonomy to the Selection of Gas-Liquid Chromatography Stationary Phases," *Anal. Chem.*, **47**, 2168 (1975).

20. Gaarenstroom, P. D., S. P. Perone, and J. L. Moyers, "Application of Pattern Recognition and Factor Analysis for Characterization of Atmosphereic Particulate Composition in Southwest desert Atmosphere," *Environ. Sci. Technol.*, **11**, 795 (1977).

21. Gardiner, Jr., W. C., "Derivation of Elementary Reaction Rate Constants by Means of Computer Modeling," *J. Phys. Chem.*, **83**, 37 (1979).

22. Griffiths, P. R., *Chemical Infrared Fourier Transform Spectroscopy*, Wiley-Interscience, New York, 1975.

23. Griffiths, P. R., Ed., *Transform Techniques in Chemistry*, Plenum, New York, 1978.

24. Hamilton, W. C., "The Revolution in Crystallography," *Science*, **169**, 133 (1970).

25. Hansch, C., and A. Leo, *Substitutent Constants for Correlation Analysis in Chemistry and Biology*, Wiley, New York, 1979.

26. Heller, S. R., and G. W. A. Milne, "The NIH/EPA Chemical Information System," in W. J. Howe, M. M. Milne, and A. F. Pennell, Eds., *Retrieval of Medicinal Chemical Information*, American Chemical Society, Washington, D.C., 1978.

27. Hertz, H. S., R. A. Hites, and K. Biemann, "Identification of Mass Spectra by Computer-Searching a File of Known Spectra," *Anal. Chem.*, **43**, 681 (1971).

28. Hopfinger, A. J., *Conformational Properties of Macromolecules*, Academic, New York, 1973.

29. Howe, W. J., and T. R. Hagadone, "Progress Toward an On-Line Chemical and Biological Information System at the Upjohn Company," in W. J. Howe, M. M. Milne, and A. F. Pennell, Eds., *Retrieval of Medicinal Chemical Information*, American Chemical Society, Washington, D.C., 1978.

30. Isenhour, T. L., B. R. Kowalski, and P. C. Jurs, "Applications of Pattern Recognition to Chemistry," *CRC Crit. Rev. Anal. Chem.*, **4**, 1 (1974).

31. Isenhour, T. L., and G. H. Morrison, "A Computer Program to Optimize Times of Irradiation and Decay in Activation Analysis," *Anal. Chem.*, **36**, 1089 (1964).

32. Johnson, C. K., *ORTEP: A Fortran Thermal-Ellipsoid Plot Program for Crystal Structure Illustrations*, Oak Ridge National Laboratory, ORNL-3794, Oak Ridge, Tenn., 1965.

33. Jurs, P. C., and T. L. Isenhour, *Chemical Applications of Pattern Recognition*, Wiley-Interscience, New York, 1975.

34. Kirschner, G. L., and B. R. Kowalski, "The Application of Pattern Recognition to Drug Design," in E. J. Ariend, Ed., *Drug Design*, Vol. VIII, Academic, New York, 1978.

35. Klaus, R. L., and H. C. VanNess, "An Extension of the Spline Fit Technique and Applications to Thermodynamic Data," *Am. Inst. Chem. Eng. J.*, **13**, 1132 (1967).

36. Koskinen, J. R. and B. R. Kowalski, "Interactive Pattern Recognition in the Chemical Laboratory," *J. Chem. Inf. Comput. Sci.*, **15**, 119 (1975).

37. Kowalski, B. R., "Measurement Analysis by Pattern Recognition," *Anal. Chem.*, **47**, 1152A (1975).

38. Kowalski, B. R., and T. L. Isenhour, "An Analytical Function for Describing Gamma-Ray Pulse-Height Distributions in NaI(T1) Scintillators," *Anal. Chem.*, **40**, 1196 (1968).

39. Kruger, Carl, "Automated X-Ray Structure Determination as an Analytical Method," *Agnew. Chem.*, **11**, 387 (1972).

40. Kwan, W. O., and B. R. Kowalski, "Classification of Wines by Applying Pattern Recognition to Chemical Composition Data," *J. Food Sci.*, **43**, 1320 (1978).

41. Kwok, K. S., R. Venkataraghavan, and F. W. McLafferty, "Computer-Aided Interpretation of Mass Spectra. III. A Self-Training Interpretative and Retrieval System," *J. Am. Chem. Soc.*, **95**, 4185 (1973).

42. Leo, A., P. Y. C. Jow, C. Silipo, C. Hansch, "Calculation of Hydrophobic Constant (Lop P) from and *f* constants," *J. Med. Chem.*, **18**, 865–868 (1975).

43. Lowry, G. C., *Markov Chains and Monte Carlo Calculations in Polymer Science*, Dekker, New York, 1970.

44. Massart, D. L., and L. Kaufman, "Operations Research in Analytical Chemistry," *Anal. Chem.*, **47**, 1244A (1975).

45. McConnell, M. L., G. Rhodes, U. Watson, and M. Novotny, "Application of Pattern Recognition and Feature Extraction Techniques to Volatile Constituent Metabolic Profiles Obtained by Capillary Gas Chromatography," *J. Chrom.*, **162**, 495 (1979).

46. McLafferty, F. W., and R. Venkataraghavan, "Computer Applications in Mass Spectrometry," in M. L. Gross, Ed., *High Performance Mass Spectrometry: Chemical Applications*, American Chemical Society, Washington, D.C., 1978.

47. Milne, G. W. A., and S. R. Heller, "The NIH-EPA Chemical Information System," in D. H. Smith, Ed., *Computer-Assisted Structure Elucidation*, American Chemical Society, Washington, D.C., 1977.

48. Morgan, S. L., and S. N. Deming, "Simplex Optimization of Analytical Chemical Methods," *Anal. Chem.*, **46**, 1170 (1974).

49. Rechsteiner, C. E., Jr., H. S. Gold, and R. P. Buck, "Analysis of Binary Mixtures by Computer Decomposition of Molecular Fluorescence Spectra," *Anal. Chim. Acta*, **95**, 51–58 (1977).

50. Rotter, H., and K. Varmuza, "Computer-Aided Interpretation of Steroid Mass Spectra by Pattern Recognition Methods. Part III. Computation of Binary Classifiers by Linear Regression," *Anal. Chim. Acta*, **103**, 61 (1978).

51. Routh, M. W., P. A. Seartz, and M. B. Denton, "Performance of the Super Modified Simplex," *Anal. Chem.*, **49**, 1422 (1977).

52. Sasaki, S.-I., H. Abe, Y. Hirota, Y. Ishida, Y. Kudo, S. Ochiai, K. Saito, and T. Yamasaki, "CHEMICS-F: A Computer Program System for Structure Elucidation of Organic Compounds," *J. Chem. Inf. Comput. Sci.*, **18**, 211 (1978).

53. Shelley, C. A., H. B. Woodruff, C. R. Snelling, and M. E. Munk, "Interactive Structure Elucidation," in D. H. Smith, Ed., *Computer-Assisted Structure Elucidation*, American Chemical Society, Washington, D.C. 1977.

54. Shoenfeld, P. S., and J. R. DeVoe, "Statistical and Mathematical Methods in Analytical Chemistry," *Anal. Chem.*, **48**, 403R (1976).

55. Sjostrom, M., and U. Edlund, "Analysis of 13-C NMR Data by Means of Pattern Recognition Methodology," *J. Magn. Res.*, **25**, 285 (1977).

56. Smith, D. H., and R. E. Carhart, "Structure Elucidation Based on Computer Analysis of High and Low Resolution Mass Spectral Data," in M. L. Gross, Ed., *High Performance Mass Spectrometry: Chemical Applications*, American Chemical Society, Washington, D.C., 1978.

57. Smith, G. M., and P. Gund, "Computer-Generated Space-Filling Molecular Models," *J. Chem. Inf. Comput. Sci.*, **18**, 207 (1978).

58. Smits, R., C. Vanroelen, and D. L. Massart, "The Optimization of Information Obtained by Multicomponent Chromatographic Separation Using the Simplex Technique," *Z. Anal. Chem.*, **273**, 1–5 (1975).

59. Stout, G. H., and L. H. Jensen, *X-Ray Structure Determination*, Macmillan, New York, 1968.

60. Stuper, A. J., W. E. Brugger, and P. C. Jurs, *Computer Assisted Studies of Chemical Structure and Biological Function*, Wiley-Interscience, New York, 1979.

61. Thomas, Q. V., R. A. DePalma, and S. P. Perone, "Application of Pattern Recognition Techniques to the Interpretation of Severly Overlapped Voltammetric Data: Studies with Experimental Data," *Anal. Chem.*, **49**, 1376 (1977).

62. Vanroelen, C., R. Smits, P. Van den Winkel, and D. L. Massart, "Application of Factor Analysis and Simplex Technique to the Optimization of a Phosphate Determination via Molybdenum Blue," *A. Anal. Chem.* , **280**, 21–23 (1976).

63. Venkataraghavan, R., H. E. Dayringer, G. M. Pesyna, B. L. Atwater, I. K. Mun, M. M. Cone, F. W. McLafferty "Computer-Assisted Structure Identification of Unknown Mass Spectra," in D. H. Smith, Ed., *Computer-Assisted Structure Elucidation*, American Chemical Society, Washington, D.C., 1977.

64. Wilkins, C. L., "Interactive Pattern Recognition in the Chemical Analysis Laboratory," *J. Chem. Inf. Comput. Sci.*, **17**, 242 (1977).

65. Williams, J. E., P. J. Strang, and P. von R. Schleyer, "Physical Organic Chemistry: Quantitative Conformational Analysis; Calculation Methods," *Ann. Rev. Phys. Chem.*, **19**, 531 (1968).

66. Woodruff, H. B., and M. E. Munk, "A Computerized Infrared Spectral Interpreter as a Tool in Structure Elucidation of Natural Products," *J. Org. Chem.*, **42**, 1976 (1977).

67. Yule, H. P., "Mathematical Smoothing of Gamma-Ray Spectra," *Nucl. Instr. Methods*, **54**, 61 (1967).

## GENERAL REFERENCES

Dickson, T. R., *The Computer and Chemistry*, Freeman, San Francisco, 1968.

Fernbach, S., and A. Taub, Eds., *Computers and Their Role in the Physical Sciences*, Gordon and Breach, New York, 1970.

Isenhour, T. L., and P. C. Jurs, *Introduction to Computer Programming for Chemists*, 2nd ed., Allyn and Bacon, Boston, 1979.

Isenhour, T. L., P. C. Jurs, C. E. Klopfenstein, and C. L. Wilkins, *Introduction to Computer Programming for Chemists: Basic Version*, Allyn and Bacon, Boston, 1975.

Mattson, J. S., H. B. Mark, Jr., and H. C. MacDonald, Eds., *Computer Fundamentals for Chemists*, Dekker, New York, 1973.

Wiberg, K. B., *Computer Programming for Chemists*, Benjamin, New York, 1965.

# SUBJECT INDEX